# ACCURATE RESULTS IN THE CLINICAL LABORATORY

## A Guide to Error Detection and Correction

# ACCURATE RESULTS IN THE CLINICAL LABORATORY

## A Guide to Error Detection and Correction

---

*Edited by*

AMITAVA DASGUPTA, PH.D, DABCC
*Professor of Pathology and Laboratory Medicine*
*University of Texas Health Sciences Center at Houston*
*Houston, TX*

JORGE L. SEPULVEDA, M.D, PH.D
*Associate Professor and Associate Director of Laboratory Medicine*
*Department of Pathology and Cell Biology*
*Columbia University College of Physicians and Surgeons*
*New York, NY*

ELSEVIER

AMSTERDAM • BOSTON • HEIDELBERG • LONDON • NEW YORK • OXFORD
PARIS • SAN DIEGO • SAN FRANCISCO • SINGAPORE • SYDNEY • TOKYO

Elsevier
32 Jamestown Road, London NW1 7BY, UK
225 Wyman Street, Waltham, MA 02451, USA
525 B Street, Suite 1800, San Diego, CA 92101-4495, USA



**British Library Cataloguing-in-Publication Data**
A catalogue record for this book is available from the British Library

**Library of Congress Cataloging-in-Publication Data**
A catalog record for this book is available from the Library of Congress

ISBN: 978-0-12-415783-5

For information on all Elsevier publications
visit our website at www.store.elsevier.com

Typeset by MPS Ltd, Chennai, India
www.adi-mps.com

Printed and bound in United States of America

13 14 15 16 10 9 8 7 6 5 4 3 2 1

# Contents

## 15. Challenges in Confirmation Testing for Drugs of Abuse

LARRY A. BROUSSARD



## 16. Alcohol Determination Using Automated Analyzers: Limitations and Pitfalls

SHEILA DAWLING



## 17. Pre-Analytical Issues and Interferences in Transfusion Medicine Tests

ELENA NEDELCU



## 18. Issues with Immunology and Serology Testing

AMER WAHED AND SEMYON RISIN



## 19. Sources of Errors in Hematology and Coagulation Testing

ANDY NGUYEN AND AMER WAHED



## 20. Challenges in Clinical Microbiology Testing

LAURA CHANDLER



## 21. Sources of Errors in Molecular Testing

LAURA CHANDLER



## 22. Problems in Pharmacogenomics Testing

DINA N. GREENE, CECILY VAUGHN AND ELAINE LYON

# Foreword

Clinicians must make decisions based on information presented to them, both by the patient and by ancillary resources available to the physician. Laboratory data generally provide quantitative information, which may be more helpful to physicians than the subjective information from a patient's history or physical examination. Indeed, with the prevalent pressure for physicians to see more patients in a limited time frame, laboratory testing has become a more essential component of a patient's diagnostic workup, partly as a time-saving measure but also because it does provide information against which prior or subsequent test results, and hence patients' health, may be compared. Tests should be ordered if they could be expected to provide additional information beyond that obtained from a physician's first encounter with a patient and if the results could be expected to influence a patient's care. Typically, clinicians use clinical laboratory testing as an adjunct to their history taking and physical examination to help confirm a preliminary diagnosis, although some testing may establish a diagnosis, such as molecular tests for inborn errors of metabolism. Microbiological cultures of body fluids may not only establish the identity of an infecting organism but also establish the treatment of the associated medical condition. In outpatient practice, clinicians primarily order tests to assist them in their diagnostic practice, whereas for hospitalized patients, in whom a diagnosis has typically been established, laboratory tests are primarily used to monitor a patient's status and response to treatment. Tests of organ function are used to search for drug toxicity, and the measurement of the circulating concentrations of drugs with narrow therapeutic windows is done to ensure that optimal drug dosing is achieved and maintained. The importance of laboratory testing is evident when some physicians rely more on laboratory data than a patient's own assessment as to how he or she feels, opening these physicians to the criticism of treating the laboratory data rather than the patient.

In the modern, tightly regulated, clinical laboratory in a developed country, few errors are likely to be made, with the majority labeled as laboratory errors occurring outside the laboratory. A 1995 study showed that when errors were made, 75% still produced results that fell within the reference interval (when perhaps they should not) [1]. Half of the other errors were associated with results that were so absurd that they were discounted clinically. Such results clearly should not have been released to a physician by the laboratory and could largely be avoided by a simple review by human or computer before being verified. However, the remaining 12.5% of errors produced results that could have impacted patient management. The prevalence of errors may be less now than in the past because the quality of analytical testing has improved, but the ramifications of each error are not likely to be less. The consequences of an error vary depending on the analyte or analytes affected and whether the patient involved is an inpatient or an outpatient. If the patient is an inpatient, a physician, if suspicious about the result, will likely have the opportunity to verify the result by repeating the test or other tests addressing the same physiological functions before taking action. However, if the error occurs with a specimen from an outpatient, causing an abnormal result to appear normal, that patient may be lost to follow-up and present later with advanced disease. Despite the great preponderance of accurate results, clinicians should always be wary of any result that does not seem to fit with the patient's clinical picture. It is, of course, equally important for physicians not to dismiss any result that they do not like as a "laboratory error." The unexpected result should always prompt an appropriate follow-up. The laboratory has a responsibility to ensure that physicians have confidence in its test results while still retaining a healthy skepticism about unexpected results.

Normal laboratory data may provide some assurance to worried patients who believe that they might have a medical problem, an issue seemingly more prevalent now with the ready accessibility of medical information available through computer search engines. However, both patients and physicians tend to become overreliant on laboratory information, either not knowing or ignoring the weakness of laboratory tests in general. A culture has arisen of physicians and patients believing that the published upper and lower limits of the reference range (or interval) of a test define normality. They do not realize that such a range has probably been derived from 95% of a group of presumed healthy individuals, not necessarily selected with respect to all demographic factors or habits that were an appropriate comparative reference for a particular patient. Even if appropriate, 1 in 20 individuals would be expected to have an abnormal result for a single test. In the usual situation in which many tests are ordered together, the probability of abnormal results in a healthy individual increases in proportion to the number of tests ordered. Studies have hypothesized that the likelihood of all of 20 tests ordered at the same time falling within their respective reference intervals is only 36%. The studies performed to derive the reference limits are usually conducted under optimized conditions, such as the time since the volunteer last ate, his or her posture during blood collection, and often the time of day. Such idealized conditions are rarely likely to be attained in an office or hospital practice.

Factors affecting the usefulness of laboratory data may arise in any of the pre-analytical, analytical, or post-analytical phase of the testing cycle. Failures to consider these factors do constitute errors. If these errors occur prior to collection of blood or after results have been produced, while still likely to be labeled as laboratory errors because they involve laboratory tests, the laboratory staff is typically not liable for them. However, the staff does have the responsibility to educate those individuals who may have caused them to ensure that such errors do not recur. If practicing clinicians were able to use the knowledge that experienced laboratorians have about the strengths and weaknesses of tests, it is likely that much more clinically useful information could be extracted from existing tests. Outside the laboratory, physicians rarely are knowledgeable about the intra- and interindividual variation observed when serial studies are performed on the same individuals. For some tests, a significant change for an individual may occur when his or her test values shift from one end of the reference interval toward the other. Thus, a test value does not necessarily have to exceed the reference limits for it to be abnormal for a given patient. If the pre-analytical steps are not standardized when repeated testing is done on the same person, it is more likely that trends in laboratory data may be missed. There is an onus on everyone involved in test ordering and test performance to standardize the processes to facilitate the maximal extraction of information from the laboratory data. The combined goal should be pursuit of information rather than just data. Laboratory information systems provide the potential to integrate all laboratory data that can then be integrated with clinical and other diagnostic information by hospital information systems.

Laboratory actions to highlight values outside the reference interval on their comprehensive reports of test results to physicians with codes such as "H" or "L" for high and low values exceeding the reference interval have tended to obscure the actual numerical result and to cement the concept that the upper and lower reference limits define normality and that the presence of one of these symbols necessitates further testing. The use of the reference limits as published decision limits for national programs for renal function, lipid, or glucose screening has again placed a greater burden on the values than they deserve. Every measurement is subject to analytical error, such that repeated determinations will not always yield the same result, even under optimal testing conditions. Would it then be more appropriate to make multiple measurements and use an average to establish the number to be acted upon by a clinician?

Much of the opportunity to reduce errors (in the broadest sense) rests with the physicians who use test results. Over-ordering leads to the possibility of more errors. Inappropriate ordering—for example, repetitive ordering of tests whose previous results have been normal—or ordering the wrong test or wrong sequence of tests to elucidate a problem should be minimized by careful supervision by attending physicians of their trainees involved in the direct management of their patients. Laboratorians need to be more involved in teaching medical students so that when these students become residents, their test-ordering practices are not learned from senior residents who had learned their habits from the previous generation of residents. Blanket application of clinical guidelines or test order-sets has probably led to much misuse of clinical laboratory tests. Many clinicians and laboratorians have attempted to reduce inappropriate test ordering, but the overall conclusion seems to be that education is the most effective means. Unfortunately, the education needs to be continuously reinforced to have a lasting effect. The education needs to address the clinical sensitivity of diagnostic tests, the context in which they are ordered, and their half-lives. Most important, education needs to address issues of biological variation and pre-analytical factors that may affect test values, possibly masking trends or making the abnormal result appear normal and vice versa.

This book provides a comprehensive review of the factors leading to errors in all the areas of clinical laboratory testing. As such, it will be of great value to all laboratory directors and trainees in laboratory medicine and the technical staff who perform the tests in daily practice. By clearly identifying problem areas, the book lays out the opportunities for improvement. This book should be of equal value to clinicians, as to laboratorians, as they seek the optimal outcome from their care of their patients.

## Reference

[1] Goldschmidt HMJ, Lent RW. Gross errors and workflow analysis in the clinical laboratory. Klin Biochem Metab 1995;3:131–49.

**Donald S. Young, MD, PhD**
*Professor of Pathology & Laboratory Medicine*
*Department of Pathology & Laboratory Medicine*
*University of Pennsylvania Perelman*
*School of Medicine, Philadelphia*

# Preface

Clinical laboratory tests have a significant impact on patient safety and patient management because more than 70% of all medical diagnoses are based on laboratory test results. Physicians rely on hospital laboratories for obtaining accurate results, and a falsely elevated or falsely low value due to interference or pre-analytical errors may have a significant influence on the diagnosis and management of patients. Usually, a clinician questions the validity of a test result if the result does not match the clinical evaluation of the patient and calls laboratory professionals for interpretation. However, clinically significant inaccuracies in laboratory results may go unnoticed and mislead clinicians into employing inappropriate diagnostic and therapeutic approaches, sometimes with very adverse outcomes. This book is intended as a guide to increase the awareness of both clinicians and laboratory professionals about the various sources of errors in clinical laboratory tests and what can be done to minimize or eliminate such errors. This book addresses not only sources of errors in the analytical methods but also various sources of pre-analytical variation because pre-analytical errors account for more than 60% all laboratory errors (Chapter 1). Important pre-analytical variables are addressed in the first three chapters of the book. In Chapter 2, the effects of ethnicity, gender, age, diet, and exercise on laboratory test results are addressed, whereas Chapter 3 discusses the effects of patient preparation and specimen collection. In Chapter 4, specimen misidentification and specimen processing issues are reviewed.

Various endogenous factors, such as bilirubin, lipemia, and hemolysis, can affect laboratory test results, and this important issue is addressed in Chapter 5. Immunoassays are widely used in the clinical laboratory, and more than 100 immunoassays are available commercially for measurement of various analytes. In Chapter 6, various immunoassay formats are discussed with an emphasis on the mechanism of interference of heterophilic antibodies and autoantibodies on immunoassays, especially sandwich immunoassays, and general approaches to eliminate such interference are reviewed.

Many Americans use herbal medicines, and use of these may affect clinical laboratory test results. In addition, certain herbal medicines may cause organ damage, and an unexpected laboratory test result may be the first indication of such organ toxicity. For example, abnormal liver function tests in the absence of a hepatitis infection in an otherwise healthy person may be related to liver toxicity due to use of the herbal sedative kava. These important issues are addressed in detail in Chapter 7.

Clinical chemistry is a vast area of laboratory medicine, responsible for the largest volume of testing in the clinical laboratory and, arguably, affecting a majority of clinical decisions. Sources of errors for measuring common analytes in clinical chemistry are discussed in Chapters 8 and 9, whereas errors in biochemical genetics are discussed in Chapter 10. In Chapter 11, issues concerning measuring various hormones and endocrinology testing are reviewed, whereas Chapter 12 is devoted to challenges in measuring cancer biomarkers.

Therapeutic drug monitoring, drugs of abuse testing, and alcohol determinations are major functions of toxicology laboratories, and there are many interferences in therapeutic drug monitoring, immunoassays used for screening of various drugs of abuse, mass spectrometry methods for drug confirmation, and alcohol determinations using enzymatic assays. These important issues are addressed in Chapters 13–16 with an emphasis on various approaches to eliminate or minimize such interferences.

Sources of errors in hematology and coagulation are addressed in Chapter 17, whereas critical issues in transfusion medicine are addressed in Chapter 18. In Chapter 19, challenges in immunology and serological testings are discussed, whereas sources of errors in microbiology testing and molecular testing are addressed in Chapters 20 and 21, respectively. The particular issues in molecular testing related to pharmacogenomics are addressed in Chapter 22.

The objective of this book is to provide a comprehensive guide for laboratory professionals and clinicians regarding sources of errors in laboratory test results and how to resolve such errors and identify discordant specimens. Error-free laboratory results are essential for patient safety. This book is intended as a practical guide for laboratory professionals and clinicians who deal with erroneous results on a regular basis. We hope this book will help them to be aware of such sources of errors and empower them to eliminate such errors when feasible or to account for known sources of variability when interpreting changes in laboratory results.

We thank all the contributors for taking time from their busy professional demands to write the chapters. Without their dedicated contributions, this project would have never materialized. We also thank our families for putting up with us during the past year while we spent many hours during weekends and evenings writing chapters and editing this book. Finally, our readers will be the judges of the success of this project. If our readers find this book useful, all the hard work of the contributors and editors will be rewarded.

*Amitava Dasgupta*
*Jorge L. Sepulveda*

# List of Contributors

**Amid Abdullah, MD** Department of Pathology and Laboratory Medicine, University of Calgary and Calgary Laboratory Services, Calgary, Alberta, Canada

**Alyaa Al-Ibraheemi, MD** Department of Pathology and Laboratory Medicine, University of Texas Health Sciences Center at Houston, Houston, TX

**Leland Baskin, MD** Department of Pathology and Laboratory Medicine, University of Calgary and Calgary Laboratory Services, Calgary, Alberta, Canada

**Lindsay A.L. Bazydlo, PhD** Department of Pathology, Immunology and Laboratory Medicine, University of Florida College of Medicine, Gainesville, FL

**Michael J. Bennett, PhD** Department of Pathology, University of Pennsylvania Perelman School of Medicine, Evelyn Willing Bromley Endowed Chair in Clinical Laboratories and Pathology, Philadelphia, PA

**Larry A. Broussard, PhD** Department of Clinical Laboratory Sciences, Louisiana State University Health Sciences Center, New Orleans, LA

**Laura Chandler, PhD** Department of Pathology and Laboratory Medicine, Philadelphia VA Medical Center, Philadelphia, PA, and Department of Medicine, Perelman School of Medicine at the University of Pennsylvania, Philadelphia, PA

**Alex Chin, PhD** Department of Pathology and Laboratory Medicine, University of Calgary and Calgary Laboratory Services, Calgary, Alberta, Canada

**Pradip Datta, PhD** Siemens Healthcare Diagnostics, Tarrytown, NY

**Sheila Dawling, PhD** Department of Pathology, Microbiology & Immunology, Vanderbilt University Medical Center, Nashville, TN

**Valerian Dias, PhD** Department of Pathology and Laboratory Medicine, University of Calgary and Calgary Laboratory Services, Calgary, Alberta, Canada

**Dina N. Greene, PhD** Northern California Kaiser Permanente Regional Laboratories, The Permanente Medical Group, Berkeley, CA

**Neil S. Harris, MD** Department of Pathology, Immunology and Laboratory Medicine, University of Florida College of Medicine, Gainesville, FL

**Kamisha L. Johnson-Davis, PhD** Department of Pathology, University of Utah School of Medicine, and ARUP Laboratories, Salt Lake City, UT

**Steven C. Kazmierczak, PhD** Department of Pathology, Oregon Health & Science University, Portland, OR

**Elaine Lyon, PhD** ARUP Institute for Clinical and Experimental Pathology, Salt Lake City, UT, and Department of Pathology, University of Utah, Salt Lake City, UT

**Gwendolyn A. McMillin, PhD** Department of Pathology, University of Utah School of Medicine, and ARUP Laboratories, Salt Lake City, UT

**Christopher Naugler, MD** Department of Pathology and Laboratory Medicine, University of Calgary and Calgary Laboratory Services, Calgary, Alberta, Canada

**Elena Nedelcu, MD** Department of Pathology and Laboratory Medicine, University of Texas Health Sciences Center at Houston, Houston, TX

**Andy Nguyen, MD** Department of Pathology and Laboratory Medicine, University of Texas Health Sciences Center at Houston, Houston, TX

**Octavia M. Peck Palmer, PhD** Department of Pathology and Critical Care Medicine, University of Pittsburgh School of Medicine, Pittsburgh, PA

**Amy L. Pyle, PhD** Nationwide Children's Hospital, Columbus, OH

**Semyon Risin, MD, PhD** Department of Pathology and Laboratory Medicine, University of Texas Health Sciences Center at Houston, Houston, TX

**Cecily Vaughn, MS** ARUP Institute for Clinical and Experimental Pathology, Salt Lake City, UT

**Amer Wahed, MD** Department of Pathology and Laboratory Medicine, University of Texas Health Sciences Center at Houston, Houston, TX

**William E. Winter, MD** Department of Pathology, Immunology and Laboratory Medicine, University of Florida College of Medicine, Gainesville, FL

**Alison Woodworth, PhD** Department of Pathology, Vanderbilt University Medical Center, Nashville, TN

**Donald S. Young, MD, PhD** Department of Pathology and Laboratory Medicine, University of Pennsylvania, Perelman School of Medicine, Philadelphia, PA

CHAPTER

# 1

# Variation, Errors, and Quality in the Clinical Laboratory

*Jorge Sepulveda*

Columbia University Medical Center, New York, New York

## INTRODUCTION

It has been roughly estimated that approximately 70% of all major clinical decisions involve consideration of laboratory results. In addition, approximately 40–94% of all objective health record data are laboratory results [1–3]. Undoubtedly, accurate test results are essential for major clinical decisions involving disease identification, classification, treatment, and monitoring. Factors that constitute an accurate laboratory result involve more than analytical accuracy and can be summarized as follows:

1. The right sample was collected on the right patient, at the correct time, with appropriate patient preparation.
2. The right technique was used collecting the sample to avoid contamination with intravenous fluids, tissue damage, prolonged venous stasis, or hemolysis.
3. The sample was properly transported to the laboratory, stored at the right temperature, processed for analysis, and analyzed in a manner that avoids artifactual changes in the measured analyte levels.
4. The analytical assay measured the concentration of the analyte corresponding to its "true" level (compared to a "gold standard" measurement) within a clinically acceptable margin of error (the total acceptable analytical error (TAAE)).
5. The report reaching the clinician contained the right result, together with interpretative information, such as a reference range and other comments, aiding clinicians in the decision-making process.

Failure at any of these steps can result in an erroneous or misleading laboratory result, sometimes with adverse outcomes. For example, interferences with point-of-care glucose testing due to treatment with maltose-containing fluids have led to failure to recognize significant hypoglycemia and to mortality or severe morbidity [4].

## ERRORS IN THE CLINICAL LABORATORY

Errors can occur in all the steps in the laboratory testing process, and such errors can be classified as follows:

1. Pre-analytical steps, encompassing the decision to test, transmission of the order to the laboratory for analysis, patient preparation and identification, sample collection, and specimen processing.
2. Analytical assay, which produces a laboratory result.
3. Post-analytical steps, involving the transmission of the laboratory data to the clinical provider, who uses the information for decision making.

Although minimization of analytical errors has been the main focus of developments in laboratory medicine, the other steps are more frequent sources of erroneous results. An analysis indicated that in the laboratory, pre-analytical errors accounted for 62% of all errors, with post-analytical representing 23% and analytical 15% of all laboratory errors [5]. The most common pre-analytical errors included incorrect order transmission (at a frequency of approximately 3% of all orders) and hemolysis (approximately 0.3% of all

*Accurate Results in the Clinical Laboratory.*
DOI: http://dx.doi.org/10.1016/B978-0-12-415783-5.00001-3

samples) [6]. Other frequent causes of pre-analytical errors include the following:

- Patient identification error
- Tube-filling error, empty tubes, missing tubes, or wrong sample container
- Sample contamination or collected from infusion route
- Inadequate sample temperature.

Table 1.1 provides a complete list of errors, including pre-analytical, analytical, and post-analytical errors, that may occur in clinical laboratories. Particular attention should be paid to patient identification because errors in this critical step can have severe consequences, including fatal outcomes, for example, due to transfusion reactions. To minimize identification errors, health care systems are using point-of-care identification systems, which typically involve the following:

1. Handheld devices connected to the laboratory information systems (LIS) that can objectively identify the patient by scanning a patient-attached bar code, typically a wrist band.
2. Current laboratory orders can be retrieved from the LIS.
3. Ideally, collection information, such as correct tube types, is displayed in the device.
4. Bar-coded labels are printed at the patient's side, minimizing the possibility of misplacing the labels on the wrong patient samples.

Analytical errors are mostly due to interference or other unrecognized causes of inaccuracy, whereas instrument random errors accounted for only 2% of all laboratory errors in one study [5]. According to that study, most common post-analytical errors were due to communication breakdown between the laboratory and the clinicians, whereas only 1% were due to miscommunication within the laboratory, and 1% of the results had excessive turnaround time for reporting [5]. Post-analytical errors due to incorrect transcription of laboratory data have been greatly reduced because of the availability of automated analyzers and bidirectional interfaces with the LIS [5]. However, transcription errors and calculation errors remain a major area of concern in those testing areas without automated interfaces between the instrument and the LIS. Further developments to reduce reporting errors and minimize the testing turnaround time include autovalidation of test results falling within pre-established rule-based parameters and systems for automatic paging of critical results to providers.

When classifying sources of error, it is important to distinguish between *cognitive errors*, or mistakes, which are due to poor knowledge or judgment, and *noncognitive errors*, commonly known as slips and lapses, due to interruptions in a process that is routine or relatively automatic. Whereas the first type can be prevented by increased training, competency evaluation, and process aids such as checklists or "cheat sheets" summarizing important steps in a procedure, noncognitive errors are best addressed by process improvement and environment re-engineering to minimize distractions and fatigue. Furthermore, it is useful to classify adverse occurrences as *active*—that is, the immediate result of an action by the person performing a task—or as *latent* or *system errors*, which are system deficiencies due to poor design or implementation that enable or amplify active errors. In one study, only approximately 11% of the errors were cognitive, all in the pre-analytical phase, and approximately 33% of the errors were latent [5]. Therefore, the vast majority of errors are noncognitive slips and lapses performed by the personnel directly involved in the process. Importantly, 92% of the pre-analytical, 88% of analytical, and 14% of post-analytical errors were preventable. Undoubtedly, human factors, engineering, and ergonomics—optimization of systems and process redesigning to include increased automation and user-friendly, simple, and rule-based functions, alerts, barriers, and visual feedback—are more effective than education and personnel-specific solutions to consistently increase laboratory quality and minimize errors.

Immediate reporting of errors to a database accessible to all the personnel in the health care system, followed by automatic alerts to quality management personnel, is important for accurate tracking and timely correction of latent errors. In our experience, reporting is improved by using an online form that includes checkboxes for the most common types of errors together with free-text for additional information (Figure 1.1). Reviewers can subsequently classify errors as cognitive/noncognitive, latent/active, and internal to laboratory/internal to institution/external to institution; determine and classify root causes as involving human factors (e.g., communication and training or judgment), software, or physical factors (environment, instrument, hardware, etc.); and perform outcome analysis. Outcomes of errors can be classified as follows:

1. Target of error (patient, staff, visitors, or equipment).
2. Actual outcome on a severity scale (from unnoticed to fatal) and worst outcome likelihood if error was not intercepted, because many errors are corrected before they cause injury. Errors with significant outcomes or likelihoods of adverse outcomes should be discussed by quality management staff to determine appropriate corrective actions and process improvement initiatives.

**TABLE 1.1** Types of Error in the Clinical Laboratory

| **Pre-Analytical** | |
|---|---|
| **TEST ORDERING** | |
| Duplicate order | Order misinterpreted (test ordered ≠ intended test) |
| Ordering provider not identified | Inappropriate/outmoded test ordered |
| Ordered test not performed (include add-ons) | Order not pulled by specimen collector |
| **SAMPLE COLLECTION** | |
| Unsuccessful phlebotomy | Check-in not performed (in the LIS) |
| Traumatic phlebotomy | Wrong patient preparation (e.g., nonfasting) |
| Patient complaint about phlebotomy | Therapeutic drug monitoring test timing error |
| **SPECIMEN TRANSPORT** | |
| Inappropriate sample transport conditions | Specimen damaged during transport |
| Specimen leaked in transit | Specimen damaged during centrifugation/analysis |
| **SPECIMEN IDENTIFICATION** | |
| Specimen unlabeled | Date/time missing |
| Specimen mislabeled: No name or ID on tube | Collector's initials missing |
| Specimen mislabeled: No name on tube | Label illegible |
| Specimen mislabeled: Incomplete ID on tube | Two contradictory labels |
| Wrong specimen label | Overlapping labels |
| Wrong name on tube | Mismatch requisition/label |
| Wrong ID on tube | Specimen information misread by automated reader |
| Wrong blood type | |
| **HIGH PRE-ANALYTICAL TURNAROUND TIME** | |
| Delay in receiving specimen in lab | STAT not processed urgently |
| Delay in performing test | |
| **SPECIMEN QUALITY** | |
| Specimen contaminated with infusion fluid | Hemolyzed |
| Specimen contaminated with microbes | Clotted or platelet clumps |
| Specimen too old for analysis | |
| **SPECIMEN CONTAINERS** | |
| No specimens received/missing tube | Wrong preservative/anticoagulant |
| Specimen lost in laboratory | Insufficient specimen quantity for analysis |
| Wrong specimen type | Tube filling error (too much anticoagulant) |
| Inappropriate container/tube type | Tube filing error (too little anticoagulant) |
| Wrong tube collection instructions | Empty tube |
| **Analytical** | |
| High analytical turnaround time | Test perform by unauthorized personnel |
| Instrument caused random error | Results discrepant with other clinical or laboratory data |
| Instrument malfunction | |
| QC failure | Testing not completed |

(*Continued*)

TABLE 1.1 (Continued)

| | |
|---|---|
| **Pre-Analytical** | |
| QC not completed | Wrong test performed (different from test ordered) |
| **Post-Analytical** | |
| Report not completed | Reported questionable results, detected by laboratory |
| Delay in reporting results | Reported questionable results, detected by clinician |
| Critical results not called | Failure to append proper comment |
| Delay in calling critical results | Read back not done |
| Results reported incorrectly | Results misinterpreted |
| Results reported incorrectly from outside laboratory | Failure to act on results of tests |
| Results reported to wrong provider | |
| **OTHER** | |
| Proficiency test failure | Employee injury |
| Product wastage | Safety failure |
| Product not delivered timely | Environmental failure |
| Product recall | Damage to equipment |

Clearly, efforts to improve accuracy of laboratory results should encompass all of the steps of the testing cycle, a concept expressed as "total testing process" or "brain-to-brain testing loop" [7]. Approaches to achieve error minimization derived from industrial processes include total quality management (TQM); [8] lean dynamics and Toyota production systems; [9] root cause analysis (RCA); [10] health care failure modes and effects analysis (HFMEA); [11,12] failure review analysis and corrective action system (FRACAS) [13]; and Six Sigma [14,15], which aims at minimizing the variability of products such that the statistical frequency of errors is below 3.4 per million. A detailed description of these approaches is beyond the scope of this book, but laboratorians and quality management specialists should be familiar with these principles for efficient, high-quality laboratory operation [8].

## QUALITY IMPROVEMENT IN THE CLINICAL LABORATORY

*Quality* is defined as all the features of a product that meet the requirements of the customers and the health care system. Many approaches are used to improve and ensure the quality of laboratory operations. The concept of TQM involves a philosophy of excellence concerned with all aspects of laboratory operations that impact on the quality of the results. Specifically, TQM approaches apply a system of statistical process control tools to monitor quality and productivity (*quality assurance*) and encourage efforts to continuously improve the quality of the products, a concept known as *continuous quality improvement*. A major component of a quality assurance program is *quality control* (QC), which involves the use of periodic measurements of product quality, thresholds for acceptable performance, and rejection of products that do not meet acceptability criteria. Most notably, QC is applied to all clinical laboratory testing processes and equipment, including testing reagents, analytical instruments, centrifuges, and refrigerators. Typically, for each clinical test, external QC materials with known performance, also known as *controls*, are run two or three times daily in parallel with patient specimens. Controls usually have preassigned analyte concentrations covering important medical decision levels, often at low, medium, and high concentrations. Good laboratory QC practice involves establishment of a laboratory- and instrument-specific mean and standard deviation for each lot of each control and also a set of rules intended to maximize error detection while minimizing false rejections, such as Westgard rules [16]. Another important component of quality assurance for clinical laboratories is participation in proficiency testing (or external quality assessment programs such as proficiency surveys sent by the College of American Pathologists), which involves the sharing of samples with a large number of other laboratories and comparison of the results from each laboratory with its peers, usually involving reporting of the mean and standard deviation (SD) of all the laboratories running the same analyzer/reagent combination. Criteria for QC rules

Quick Occurrence Log

Print Blank Form

PHILADELPHIA VETERANS AFFAIRS MEDICAL CENTER

OCCURRENCE DETECTION PROGRAM - PATHOLOGY AND LABORATORY MEDICINE SERVICE

**Part 1: Occurrence Description**

Provider Name | Lab Area

Event ID 2946 UID (scan specimen barcode):

Erroneous Results Reported | Number of Patients Affected 1 | Test:

Delay in Lab Testing
- TAT > 2h for STAT
- TAT > 4h for outpatient
- TAT > 8h for inpatient

Patient Name

SSN

Event Location:

ED | OR | Inpatient | Outpatient | Other

Event Classification: Pre-Analytic

Occurrence reported by | Person completing report:

Event Description: Wrong preservative / anticoagulant

Free Text Event Description: (Press F7 Key for Spell Check)

An EDTA tube was received for electrolyte analysis. The accession was cancelled and Dr. XXXX was contacted at 12:30.

**Date/Time:**
- Example (Now) 06/26/2012 4:38 PM
- Specimen Collected
- Event Occurred
- Event Discovered
- Event Reported
- Event Corrected

Choose only one
- Wrong name on tube
- Wrong SSN on tube
- 2 contradictory labels
- Mismatch requisition / label
- Hollister Error
- No Name or SSN on tube
- No Name on tube
- No or incomplete SSN on tube
- Date/time missing
- Phlebotomist initials missing
- Label illegible
- Label does not match specimen Type
- No label present
- HANDWRITTEN LABEL
- Hemolysed
- Clotted/ Plt clumps
- Contaminated with IV fluid
- Microbial contamination
- Too old for analysis
- Bad transport conditions (ex: no ice)
- Container leaked in transit
- Specimen damaged in transit
- Specimen damaged in centrifuge
- Order not pulled by specimen collector
- Duplicate Order
- Ordered test <> Intended Test
- No ordering provider
- Unsuccessful phlebotomy
- OTHER
- Traumatic phlebotomy
- Pt complaint about phlebotomy
- Wrong patient preparation (ex.. non-fasting)
- No specimen received or missing tube
- Specimen lost in laboratory
- Wrong specimen type
- Inappropriate container / tube type
- [x] Wrong preservative / anticoagulant
- Insufficient specimen quantity for analysis (QNS)
- Low volume (too much anticoagulant)
- Overfilled (not enough anticoagulant)
- Empty tube
- Incorrect result reported
- Results reported with instrument malfunction
- Results reported with QC failure
- Results reported with QC not completed
- Failure to append proper comment
- Reported questionable results detected by laboratory
- Reported questionable results detected by provider
- Results discrepant with other clinical or laboratory data
- Critical results not called
- Read back not done
- Critical results called too late (>15')

Record: 1 of 1 | Unfiltered | Search

FIGURE 1.1 Example of an error reporting form for the clinical laboratory.

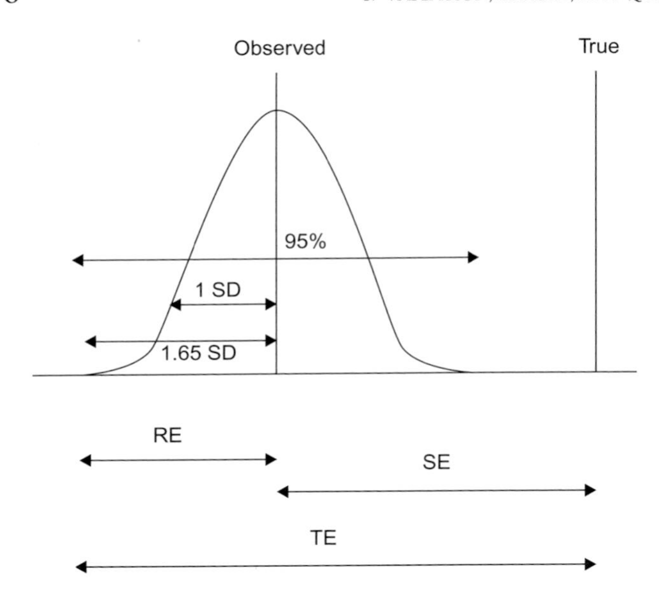


**FIGURE 1.2 Total analytical error (TE) components: random error (RE), or imprecision, and systematic error (SE), or bias, which cause the difference between the true value and the measured value.** Random error can increase or decrease the difference from the true value. Because in a normal distribution, 95% of the observations are contained within the mean ± 1.65 standard deviations (SD), the total error will not exceed bias + 1.65 × SD in 95% of the observations.

and proficiency testing acceptability should take into consideration the concept of total acceptable analytical error because deviations smaller than the total analytical errors are unlikely to be clinically significant and therefore do not need to be detected.

Total analytical error (TAE) is usually considered to combine the following (Figure 1.2): (1) systematic error (SE), or bias, as defined by deviation between the average values obtained from a large series of test results and an accepted reference or gold standard value, and (2) random error (RE), or imprecision, represented by the coefficient of variation of multiple independent test results obtained under stipulated conditions ($CV_a$). At the 95% confidence level, the RE is equal to 1.65 times the $CV_a$ for the method; consequently,

$$TAE = 1.65 \times CV_a + bias$$

Clinical laboratories frequently evaluate imprecision by performing repeated measurements on control materials, preferably using runs performed on different days (between-day precision), whereas bias (or trueness) is assessed by comparison with standard reference materials with assigned values and also by peer comparison, where either the peer mean or median are considered the reference values.

One important concept that some clinicians disregard is that no laboratory measurement is exempt of error; that is, it is impossible to produce a laboratory result with 0% bias and 0% imprecision. The role of technologic developments, good manufacturing practices, proficiency testing, and QC is to minimize and identify the magnitude of the TAE. A practical approach is to consider the clinically acceptable total analytical error or TAAE. Clinical acceptability has been defined by legislation (e.g., the Clinical Laboratory Improvement Act (CLIA)), by clinical expert opinion, and by scientific and statistical principles that take into consideration expected sources of variation. For example, Callum Fraser proposed that clinically acceptable imprecision, or random error, should be less than half of the intraindividual biologic variation for the analyte and less than 25% of the total analytical error [17]. The systematic error, or bias, should be less than 25% of the combined intraindividual ($CV_w$) and interindividual biological ($CV_g$) variation:

$$TAAE_{95\%} < 1.65 \times 0.5 \times CV_w + 0.25 \times \sqrt{CV_w^2 + CV_g^2}$$

Tables of intra- and interindividual biological variation, with corresponding allowable errors, are available and frequently updated [18]. See Table 1.2 for examples. Importantly, the allowable errors may be different at specific medical decision levels because analytical imprecision tends to vary with the analyte concentration, with higher imprecision at lower levels. Also, biological variation may be different in the various clinical conditions, and available databases are starting to incorporate studies of biologic variation in different diseases [18].

A related concept is the *reference change value* (RCV), also called *significant change value* (SCV)—that is, the variability around a measurement that is a consequence of analytical imprecision, within-subject biologic variability, and the number of repeated tests performed [17,19,20]. At the 95% confidence level, RCV can be calculated as follows:

$$RCV_{95\%} = 1.96 \times \sqrt{2} \times \sqrt{CV_a^2 + CV_w^2}$$

Because multiple repeats decrease imprecision errors, if the change is determined from the mean of repeated tests, the formula can be modified to take into consideration the number of repeats in each measurement ($n1$ and $n2$) [20]:

$$RCV_{95\%} = 1.96 \times \sqrt{\frac{2}{n1 \times n2}} \times \sqrt{CV_a^2 + CV_w^2}$$

For example, for a serum creatinine measurement with an analytical imprecision ($CV_a$) of 6.6% and within-subject biologic variation of 5.3%, the RCV at 95% confidence is 23.5% with one measurement for each sample. With two measurements for each sample, the RCV is 11.7%. Therefore, a change between two results that does

TABLE 1.2 Allowable Errors and Reference Change Values for Selected Tests. All Numeric Values are Percentages

| Test | $CV_a$ | $CV_w$ | $CV_g$ | CLIA TAAE | Bio TAAE | Allowable Imprecision | Allowable Bias | $RCV_{95}$ |
|---|---|---|---|---|---|---|---|---|
| Amylase | 5.3 | 8.7 | 28.3 | 30 | 14.6 | 4.4 | 7.4 | 28.2 |
| Alanine aminotransferase | 2.8 | 18.0 | 42.0 | 20 | 26.2 | 9.0 | 11.4 | 50.5 |
| Albumin | 2.6 | 3.1 | 4.2 | 10 | 3.9 | 1.6 | 1.3 | 11.2 |
| Alkaline phosphatase | 4.2 | 6.4 | 24.8 | 30 | 11.7 | 3.2 | 6.4 | 21.2 |
| Aspartate aminotransferase | 2.2 | 11.9 | 17.9 | 20 | 15.3 | 6.0 | 5.4 | 33.5 |
| Bilirubin total | 10.0 | 23.8 | 39.0 | 20 | 31.0 | 11.9 | 11.4 | 71.6 |
| Chloride | 2.4 | 1.2 | 1.5 | 5 | 1.5 | 0.6 | 0.5 | 7.4 |
| Cholesterol | 2.7 | 5.4 | 15.2 | 10 | 8.4 | 2.7 | 4.0 | 16.7 |
| Cortisol | 5.3 | 20.9 | 45.6 | 25 | 29.8 | 10.5 | 12.5 | 59.8 |
| Creatine kinase | 3.6 | 22.8 | 40.0 | 30 | 30.3 | 11.4 | 11.5 | 64.0 |
| Creatinine | 7.6 | 6.0 | 14.7 | 15 | 8.9 | 3.0 | 4.0 | 26.8 |
| Glucose | 3.4 | 6.1 | 6.1 | 10 | 7.2 | 3.0 | 2.2 | 19.4 |
| HDL cholesterol | 3.3 | 7.1 | 19.7 | 30 | 11.1 | 3.6 | 5.2 | 21.7 |
| Iron | 2.5 | 26.5 | 23.2 | 20 | 30.7 | 13.3 | 8.8 | 73.8 |
| Lactate dehydrogenase (LDH) | 2.5 | 8.6 | 14.7 | 20 | 11.4 | 4.3 | 4.3 | 24.8 |
| Magnesium | 2.8 | 3.6 | 6.4 | 25 | 4.8 | 1.8 | 1.8 | 12.6 |
| $pCO_2$ | 1.5 | 4.8 | 5.3 | 8 | 5.7 | 2.4 | 1.8 | 13.9 |
| Protein, total | 2.6 | 2.7 | 4.0 | 10 | 3.5 | 1.4 | 1.2 | 10.4 |
| Thyroxine (T4) | 4.8 | 4.9 | 10.9 | 20 | 7.1 | 2.5 | 3.0 | 19.0 |
| Triglyceride | 3.9 | 20.9 | 37.2 | 25 | 28.0 | 10.5 | 10.7 | 58.9 |
| Urate | 2.9 | 9.0 | 17.6 | 17 | 12.3 | 4.5 | 4.9 | 26.2 |
| Urea nitrogen | 6.2 | 12.3 | 18.3 | 9 | 15.7 | 6.2 | 5.5 | 38.2 |

Source: *Based on data available at http://www.westgard.com/biodatabase1.htm [18].*
$CV_a$, analytical variability in the author's laboratory; $CV_w$, intraindividual variability; $CV_g$, interindividual variability; CLIA TAAE, total allowable analytical error based on Clinical Laboratory Improvement Act (CLIA); Bio TAAE, total allowable analytical error based on interindividual and intraindividual variation. Allowable imprecision = 50% of $CV_w$. Allowable bias $= 0.25 \times \sqrt{CV_w^2 \times CV_g^2}$. $RCV_{95}$, reference change value at 95% confidence based on $CV_w$ and $CV_a$.

not exceed the RCV has a greater than 95% probability that it is due to the combined analytical and intraindividual biological variation; in other words, the difference between the two creatinine results (measured without repeats) should exceed 23.5% to be 95% confident that the change is due to a pathological condition. Conversely, for any change in laboratory values, the RCV formula can be used to calculate the probability that it is due to analytical and biological variation [17,19,20]. See Table 1.2 for examples of RCV at the 95% confidence limit, using published intraindividual variation and the author's laboratory imprecision. Ideally, future LIS should integrate available knowledge and patient-specific information and automatically provide estimates of expected variation based on the previous formulas to facilitate interpretation of changes in laboratory values and guide laboratory staff regarding the meaning of deviations from expected results. In summary, the use of TAAE and RCV brings objectivity to error evaluation, QC and proficiency testing practices, and clinical decision making based on changes in laboratory values.

## CONCLUSIONS

As in other areas of medicine, errors are unavoidable in the laboratory. A good understanding of the sources of error together with a quantitative evaluation of the clinical significance of the magnitude of the error, aided by the establishment of limits of acceptability based on statistical principles of analytical and intraindividual biological variation, are critical to design a quality program to minimize the clinical impact of errors in the clinical laboratory.

## References


[1] Forsman RW. The value of the laboratory professional in the continuum of care. Clin Leadersh Manag Rev 2002;16(6):370–3.

[2] Forsman RW. Why is the laboratory an afterthought for managed care organizations? Clin Chem 1996;42(5):813–16.

[3] Hallworth MJ. The "70% claim": what is the evidence base? Ann Clin Biochem 2011;48(Pt 6):487–8.

[4] Gaines AR, Pierce LR, Bernhardt PA. Fatal Iatrogenic Hypoglycemia: Falsely Elevated Blood Glucose Readings with a Point-of-Care Meter Due to a Maltose-Containing Intravenous Immune Globulin Product. [cited; Available from: <http://www.fda.gov/BiologicsBloodVaccines/SafetyAvailability/ucm155099.htm>; 2009 [06.18.2009]

[5] Carraro P, Plebani M. Errors in a stat laboratory: types and frequencies 10 years later. Clin Chem 2007;53(7):1338–42.

[6] Carraro P, Zago T, Plebani M. Exploring the initial steps of the testing process: frequency and nature of pre-preanalytic errors. Clin Chem 2012;58(3):638–42.

[7] Plebani M, Lippi G. Closing the brain-to-brain loop in laboratory testing. Clin Chem Lab Med 2011;49(7):1131–3.

[8] Valenstein P, editor. Quality management in clinical laboratories. Northfield, IL: College of American Pathologists; 2005.

[9] Rutledge J, Xu M, Simpson J. Application of the Toyota production system improves core laboratory operations. Am J Clin Pathol 2010;133(1):24–31.

[10] Dunn EJ, Moga PJ. Patient misidentification in laboratory medicine: a qualitative analysis of 227 root cause analysis reports in the Veterans Health Administration. Arch Pathol Lab Med 2010;134(2):244–55.

[11] Chiozza ML, Ponzetti C. FMEA: a model for reducing medical errors. Clin Chim Acta 2009;404(1):75–8.

[12] Southard PB, Kumar S, Southard CA. A modified delphi methodology to conduct a failure modes effects analysis: a patient-centric effort in a clinical medical laboratory. Qual Manag Health Care 2011;20(2):131–51.

[13] Krouwer J. Using a learning curve approach to reduce laboratory errors. Accreditation and Quality Assurance: Journal for Quality, Comparability and Reliability in Chemical Measurement 2002;7(11):461–7.

[14] Llopis MA, Trujillo G, Llovet MI, Tarres E, Ibarz M, Biosca C, et al. Quality indicators and specifications for key analytical-extranalytical processes in the clinical laboratory. Five years' experience using the six sigma concept. Clin Chem Lab Med 2011;49(3):463–70.

[15] Gras JM, Philippe M. Application of the six sigma concept in clinical laboratories: a review. Clin Chem Lab Med 2007;45(6):789–96.

[16] Westgard JO, Darcy T. The truth about quality: medical usefulness and analytical reliability of laboratory tests. Clin Chim Acta 2004;346(1):3–11.

[17] Fraser CG. Biological variation: from principles to practice. Washington, DC: AACC Press; 2001.

[18] Westgard J. Desirable Specifications for Total Error, Imprecision, and Bias, derived from intra- and inter-individual biologic variation. [cited 2012; Available from: <http://www.westgard.com/biodatabase1.htm>; 2012.

[19] Kroll MH. Multiple patient samples of an analyte improve detection of changes in clinical status. Arch Pathol Lab Med 2010;134(1):81–9.

[20] Fraser CG. Improved monitoring of differences in serial laboratory results. Clin Chem 2011;57(12):1635–7.

CHAPTER

# 2

# Effect of Age, Gender, Diet, Exercise, and Ethnicity on Laboratory Test Results

*Octavia M. Peck Palmer*

**University of Pittsburgh School of Medicine, Pittsburgh, Pennsylvania**

## INTRODUCTION

Annually, the United States performs approximately 7 billion clinical laboratory tests [1]. Clinical laboratory test results are an indispensable part of the clinician's decision-making process. Accurate laboratory results aid in timely and effective diagnosis, prognosis, treatment, and management of diseases. It is imperative that the *in vitro* diagnostic testing results accurately reflect the *in vivo* physiological processes of the patient. Inaccurate results may lead to unwarranted, invasive testing, postponement of critical therapies, increased patient anxiety, and expensive health care costs. The quality assurance program of each laboratory focuses on providing the highest quality in analytical testing. Pre-analytical (steps prior to analysis), analytical (sample analysis), and post-analytical (steps after analysis) factors can affect the accuracy of serum/plasma analytes measured in the laboratory. The pre-analytical phase refers to the processes that occur prior to blood/body fluid testing. These processes include the phlebotomy collection techniques (sample labeling, tourniquet, and posture), blood/body fluid tube/container types (anticoagulants, gel separators, clot activators, and preservatives), and sample handling (mixing/clotting protocol, temperature, storage, and transport). Nonmodifiable factors such as age, gender, and ethnicity/race (biological factors) must be accounted for in the pre-analysis stage. Patient-related factors such as diet and exercise regimens can be controlled. Standardized patient preparation prior to blood collection can minimize the effects of pre-analytical factors. In some cases, age- and gender-specific reference limits can account for the influence of pre-analytical factors.

This chapter reviews the pre-analytical variables of age, gender, diet, exercise, and ethnicity/race (a surrogate marker for environmental, socioeconomic/demographic, and genetic factors) and their influence on analytes measured in the clinical laboratory. In addition, the chapter discusses other less known effects of fasting, special diets, and nutraceuticals on laboratory tests, with an abbreviated discussion of the influence of genetic factors in response to food and nutraceuticals.

## EFFECTS OF AGE-RELATED CHANGES ON CLINICAL LABORATORY TEST RESULTS

Aging is a complex metabolic process that is not fully understood [2]. Complex physiological changes occur during transitions from the newborn to adult to geriatric stages of life [3]. Understanding the effects of age on laboratory findings can increase diagnostic accuracy. Clinicians must distinguish nonpathologic, age-related changes from pathologic changes. Adult reference ranges for a majority of serum/plasma/urine analytes measured in the laboratory are available [4]. However, complete, standardized, age-specific reference ranges are not available.

In 2000, following the passage of the National Children's Act, the U.S. Congress authorized the National Children's Health Study (NCS). NCS, led by the Eunice Kennedy Shriver National Institute of Child Health and Human Development, is a longitudinal study of 100,000 healthy individuals aged 0–21 years. The American Association of Clinical Chemistry, a collaborator of NCS, funded pilot studies focused on establishing age-specific reference ranges [5].

*Accurate Results in the Clinical Laboratory.*
DOI: http://dx.doi.org/10.1016/B978-0-12-415783-5.00002-5

## Newborn Population

Following birth, arterial blood $pO_2$ rises to approximately 80–90 mmHg. Oxygen consumption is significantly higher in neonates compared to adults. A significant reduction in uric acid concentrations occurs between birth and 6 days of age. Healthy newborns rapidly metabolize glucose as a result of their high red blood cell count, which is not evident in healthy adults [6]. Newborns have increased circulating bilirubin concentrations due to their immature liver. The developing liver is unable to convert bilirubin to bilirubin diglucuronide. Hyperbilirubinemia due to physiologic jaundice is a common condition in newborns and usually resolves within 5–7 days following birth. However, after birth it may be difficult to distinguish this normal physiological phenomenon from hemolytic disease of the newborn [7,8]. Immature kidneys demonstrate vascular resistance, reduced outgoing blood flow from the outer cortex, and reduced glomerular filtration rate (GFR). The kidneys do not efficiently concentrate and dilute urine; regulate acid–base pathways; reabsorb, excrete, or retain sodium; or secrete hydrogen ions [9]. Newborns experience an expanded extracellular fluid volume state. Hypocalcemia usually resolves within the first 2 days of life [10].

## Childhood to Puberty Population

Growth impacts laboratory test results. Two weeks following birth, luteinizing hormone (LH) concentrations increase in both boys and girls, but they decline to prepubertal concentrations by the infants' first birthday. Similarly, follicle-stimulating hormone (FSH) concentrations follow the same trend as LH concentrations after birth but decline to prepubertal concentrations in boys by the first year of life and in girls by the second year of life. Reduced LH and FSH concentrations in the teenage period are not sensitive enough to distinguish between pubertal delay and hypogonadotropic hypogonadism. Gonadal failure indicated by an upward trajectory of LH and FSH concentrations cannot be expected until 10 years of age. Elevated estradiol concentrations are present at birth but rapidly decline during the first week of life to prepubertal concentrations (0.5–5.0 ng/dL for girls and 1.0–3.2 ng/dL for boys). Additional decline to prepubertal concentrations is present by the sixth month in boys and the first year of life in girls [11,12]. Skeletal growth and muscle mass development account, in part, for the increased alkaline phosphatase (ALP), $\gamma$-glutamyl transferase ($\gamma$-GGT), creatinine, and human growth hormone concentrations seen in the childhood to puberty developmental period. The decline in ALP concentrations varies among genders. After the age of 12 years, girls exhibit a decline in ALP, and this decline is apparent in boys after the age of 14 years [13]. Increased circulating ALP concentrations are present during normal growth spurts but also in the setting of bone malignancies (osteoblastic bone cancers, osteomalacia, Paget's disease, and rickets). ALP concentrations are threefold higher in adolescents compared to adults [14]. Increases in creatinine occur between ages 12 and 19 years. Cystatin C concentrations in females decrease during the same age range. Uric acid concentrations continue to decline during the first decade of life [15].

## Adult Population

In both sexes, total cholesterol increases with advancing age (men age 60 years and women age 55 years). In the second decade of life, men have peak uric acid concentrations, which are not detected in women until the fifth decade of life [7].

### *Menopausal (Pre and Post) Period*

Postmenopausal women have increased total cholesterol concentrations, attributed to decreased circulating estrogen. High-density lipoprotein cholesterol (HDL-C) also declines up to 30% [7]. Transition from the peri- to the postmenopausal stage presents dramatic endocrine changes. A strong correlation between age and human chorionic gonadotropin (hCG) is observed [16]. Accurate interpretation of elevated hCG concentrations is critical because appreciable concentrations are present during healthy pregnancy, cancer, or trophoblastic disease [17]. In females, serum hCG concentrations (reference limit hCG $<$ 0.5 mIU/mL) are used to either identify or rule out pregnancy. Knowing the pregnancy status of a patient is essential because invasive medical procedures and medications can have potentially harmful effects on a developing fetus [18]. Slight increases in serum hCG concentrations ($\geq$0.5 mIU/mL) occur in women between the ages of 41 and 55 years. Thus, it is critical to distinguish the origin of the hCG (placental origin vs. pituitary origin). Misinterpretation of elevated hCG concentrations in peri- and postmenopausal women may postpone clinical treatments. In peri- and postmenopausal women (41–55 years old), studies demonstrate that FSH concentrations can help determine the origin of hCG. In peri- and postmenopausal women (41–55 years old) with serum hCG concentrations ranging between 5.0 and 14.0 IU/L, a FSH cutoff of 45.0 IU/L identifies hCG of placental origin with 100% sensitivity and 75% specificity. Importantly, FSH concentrations greater than 45 IU/L are not present in females with hCG of placental origin. FSH reflex testing should only be utilized in pregnancy evaluation of peri- and postmenopausal women (serum hCG concentrations between 5.0 and 14.0 IU/L) [19]. hCG concentrations greater than 14.0 IU/L in this age group indicate pregnancy unless the clinical setting dictates otherwise [16].

### *Geriatric Population*

The aging population is rapidly increasing in the United States. Between the year 2000 (35 million persons) and the year 2010 (40 million persons), the United States experienced a 15% increase in the geriatric population (>65 years or older) [20]. Interpretation of laboratory findings in the geriatric population is challenging due to multiple confounding factors that include (1) physiologic changes that naturally occur with healthy aging, (2) acute and chronic conditions (kidney disease, diabetes, and cardiovascular disease), (3) diets, (4) lifestyles, and (5) medication regimens [21]. After the age of 60 years, albumin concentrations decline each decade, with significant decreases noted in individuals older than 90 years [22]. Low serum calcium concentration in the geriatric population is most commonly caused by low serum albumin concentrations [23]. Protein concentration changes may be entirely due to compromised liver function or poor dietary regimens. Individuals older than 90 years may have decreased total cholesterol concentrations.

Iron perturbations such as decreases in iron storage, serum iron concentrations, and total iron-binding capacity occur during aging. Depletion of iron stores may be followed by increases in serum ferritin and decreases in serum transferrin. Dysregulated liver synthesis during aging may account for the reduced transferrin concentrations [4]. Lack of sufficient dietary iron intake may account for the high prevalence of anemia in the geriatric population. However, iron loss, due to bleeding in the intestinal tract, may also be the culprit for the anemia. Anemia in the geriatric population may, in part, be explained by the age-related decreases in stomach hydrochloric acid (HCl), a key acid responsible for iron absorption in the intestines. Vitamin $B_{12}$ deficiency is also prevalent in geriatrics due to age-related decreases in serum vitamin $B_{12}$ concentrations. The underlying cause of vitamin $B_{12}$ deficiency may be decreased HCI concentrations or chronic atrophic gastritis, which subsequently accounts for limited intrinsic factor and vitamin $B_{12}$ absorption [21].

Age-associated organ function decline correlates with changes in laboratory findings (i.e., reduced creatinine clearance, glucose tolerance, and hypothalamic–pituitary–adrenal axis regulation) that may represent disease or non-disease processes. At least 10% of the healthy geriatric population exhibits physiologic changes that may not be associated with disease. These changes include decreased partial pressure of oxygen in arterial blood (decreases by 25% between the third and eighth decades of life) and magnesium (decreases by 15%) concentrations. Geriatrics may also exhibit elevated serum alkaline phosphatase (increases by 20% between the third and eighth decades of life) and 2-hr postprandial glucose concentrations (after age 40 years, increases 30–40 mg/dL per decade). Increases in cholesterol concentrations (increases by 30–40 mg/dL by age 60 years) and erythrocyte sedimentation rate as high as 40 can be nonpathogenic [7,21].

A 30–40% decline in functioning kidney and the GFR is responsible for reduced creatinine clearance. Creatinine and blood urea nitrogen (BUN) concentrations can overestimate the kidney functioning capacity, as measured by GFR or creatinine clearance, due to reduced muscle mass [24]. Muscle mass degeneration accounts for reduced creatinine production. Serum creatinine concentrations can remain within normal limits despite the underlying diminished renal clearance capacity [21]. Mean creatinine clearance concentrations decrease by 10 mL/min/1.73 $m^2$ per decade and are significantly different between the adult and geriatric populations. The mean creatinine clearance for a 30-year-old individual is approximately 140 mL/min (2.33 mL/sec) per 1.73 $m^2$ of body surface area. In contrast, the mean creatinine clearance for an 80-year-old individual is 97 mL/min (1.62 mL/sec) per 1.73 $m^2$ of body surface area [25]. Small increases in serum aspartate aminotransferase (AST) (18 to 30 U/L) occur between 60 and 90 years of age, whereas peaks in serum alanine aminotransferase (ALT) occur in the fifth decade of life and by the sixth decade gradually decline to concentrations well below those noted in young adults [21]. GGT concentrations rise during aging. A steady increase in serum glucose concentrations and a decrease in glucose tolerance are prevalent in geriatrics. Lower glucose concentrations in geriatrics may be due to poor diet and reduced body mass. Higher serum insulin concentrations are prevalent in elderly adults and may be associated with insulin resistance [21]. In persons older than 75 years, insulin resistance is reportedly responsible for impaired glucose tolerance. The capacity of insulin receptors may be lower in elderly adults. Regarding serum immunoglobulin concentrations, IgA concentration increases slightly in geriatric men, but overall IgG and IgM concentrations gradually decline. Aging compromises the hypothalamic–pituitary–adrenal axis. Aging-related changes include decreases in free thyroxine (T4), triiodothyronine (T3), corticotrophin, and corticosteroid [26]. Specific to men, free testosterone decreases without significant changes in total testosterone concentrations [27,28]. Prostate-specific antigen concentrations increase up to 6.5 ng/mL in men 70 years or older without clinical evidence of prostate cancer [21].

Serum electrolytes, such as potassium and calcium, rise as one ages. Calcium concentration increases in individuals aged 60–90 years in the presence of normal albumin concentrations. However, after the age of 90 years, calcium concentrations gradually decline. Hypocalcemia may be due to a simultaneous drop in serum pH and an increase in parathyroid hormone concentrations. Age significantly impacts lung elastic architecture, alveoli function, and diaphragm strength and significantly alters respiratory function. Thus, the individual has decreased partial pressure of arterial oxygen and increased carbon dioxide pressure and bicarbonate ion concentration [21].

Although age can significantly account for altered clinical laboratory test results, one must consider the overlapping effects caused by disease, such as obesity and hypertension, and/or inadequate dietary intake when interpreting laboratory results that are outside of the reference limits [29]. The abnormal results may highlight age-associated disease processes that require clinical intervention. It is clinically necessary to conduct laboratory studies focused on the systematic effects of aging on serum/plasma/urine analytes. The resulting data will be useful for the development of effective age-specific diagnostic cutoffs.

## EFFECTS OF GENDER-RELATED CHANGES ON CLINICAL LABORATORY TEST RESULTS

Gender encompasses a myriad of complex endocrine and metabolic responses. Gender differences in laboratory analytes can be explained by differential endocrine organ-related functions and skeletal muscle mass [30]. On average, albumin, calcium, magnesium, hemoglobin, ferritin, and iron concentrations are lower in females [7]. A reduction in circulating iron concentrations is, in part, due to blood loss during monthly menses. Mean serum creatinine and cystatin C concentrations are commonly lower in adolescent females compared to adolescent males [15]. Aldolase concentrations are higher in males following the start of puberty. ALP concentrations are higher in girls ages 10–11 years. Boys ages 12–13, 14–15, and 16–17 years have higher ALP concentrations compared to girls in the corresponding age categories. A decline in ALP concentrations begins after age 12 years for girls and 14 years for boys [13]. Menopausal women have higher ALP concentrations compared to males. Serum bilirubin concentrations are lower in women due to decreased hemoglobin concentrations. Females have higher albumin concentrations compared to males of the same age [31]. Lipid profiles are heavily influenced by gender. Total cholesterol concentrations vary not only with age but also with gender. Females younger than age 20 years have higher total cholesterol concentrations compared to males in the corresponding age span. However, between the ages of 20 and 45 years, males commonly have higher total cholesterol concentrations than females. Male peak lipid concentrations occur between the ages of 40 and 60 years, whereas female peak lipid concentrations occur between the ages of 60 and 80 years [32]. Between the ages of 30 and 80 years, mean HDL-C decreases by approximately 30% in females but increases by 30% in males [21,33] These lipid increases may be due to the stimulatory effect of estrogen in women. In contrast, low-density lipoprotein cholesterol (LDL-C) is higher in men. Men also have higher 24-hr urinary excretions of epinephrine, norepinephrine, cortisol, and creatinine compared to women [34]. Women have higher serum GGT and copper and reticulocyte count (due to increased erythrocyte turnover) compared to their male counterparts.

## EFFECTS OF DIET ON CLINICAL LABORATORY TEST RESULTS

Diet may affect test results, whereas starvation also has a profound impact on clinical laboratory test results.

### Food Ingestion-Related Changes on Clinical Laboratory Values

Food ingestion activates *in vivo* metabolic signaling pathways that significantly affect laboratory test results [35]. First, the stomach secretes HCl in response to food consumption, which causes a decrease in plasma chloride concentrations. This mild metabolic alkalotic state (alkaline tide phenomenon) results from exaggerated circulating bicarbonate concentrations in the stomach's venous blood with an accompanying decreased ionized calcium (by 0.05 mmol/L, 0.2 mg/dL) [36]. Second, postprandial-associated impairment in the liver leads to increased bilirubin and enzyme activities. Depending on the content of the meal ingested, the effects on commonly measured analytes may be short- or long-lasting. Thus, an overnight fasting for at least 12 hr is necessary to obtain an accurate representation of *in vivo* glucose, lipids, iron, phosphorus, urate, urea, and ALP concentrations. Interestingly, Lewis *a* secretors (blood groups B and O) experience spikes in ALP concentrations following ingestion of high-fat meals. Lipemia can also interfere with a variety of analytical methods, such as indirect potentiometry. Prior to analysis, lipids can be removed from lipemic samples via ultracentrifugation or by the use of lipid-clearing reagents [37]. Carbohydrate (increases glucose and insulin and decreases phosphorus concentrations) and protein meals (increases cholesterol and growth hormone concentrations within 1 hr of food consumption and also increases glucagon and insulin concentrations) have differential effects on serum analytes. High-protein diets significantly affect various analytes measured in 24-hr urine test. A standard 700-calorie meal markedly increases triglycerides (~50%), AST (~20%), bilirubin and glucose (~15%), and AST concentrations (~10%) [3]. Rapid changes in lipid concentrations are consistent with dietary changes, medications, or disease.

Caffeine intake has significant effects on the human body. Varying concentrations of this stimulant are present in a variety of foods (coffee, tea, chocolate, soft drinks, and energy drinks). The short half-life of caffeine (3–7 hr) also varies among individuals. Caffeine induces catecholamine excretion from the adrenal medulla. In addition, increased gluconeogenesis, which subsequently increases glucose concentrations and impairs glucose tolerance, is evident following caffeine intake. The adrenal cortex is also vulnerable to caffeine's stimulatory effects, as evidenced by increased cortisol, free cortisol, 11-hydroxycorticoids, and 5-hydroxindoleaceatic acid concentrations. Caffeine is responsible for a threefold increase in nonesterified fatty acids, which interfere with the accurate quantification of albumin-bound drugs and hormones. Spuriously high ionized calcium concentrations are present following caffeine ingestion. Caffeine induces elevations in free fatty acids causing a rapid decrease in pH that frees calcium from protein.

Noni juice contains significant amounts of potassium (~56 mEq/L). Ingestion of noni juice leads to hyperkalemia. Specifically, hyperkalemia is apparent in vulnerable populations such as individuals with renal dysfunction and/or populations receiving potassium-increasing regimens such as spironolactone or angiotensin-converting enzyme inhibitors. Bran stimulates bile acid synthesis within 8 hr of ingestion [38]. However, bran inhibits gastrointestinal absorption of vital nutrients, including calcium (decreased by 0.3 mg/dL, 0.08 mmol/L), cholesterol, and triglycerides (decreased by 20 mg/dL, 0.23 mmol/L) [3]. Serotonin (5-hydroxytryptamine) is an ingredient present in a myriad of fruits and vegetables, such as bananas, black walnuts, kiwis, pineapples, and plantains. Bananas markedly increase 24-hr urinary excretion of 5-hydroxyindoleacetic acid in the absence of disease. Avocados suppress insulin secretion, causing impaired glucose tolerance.

## Special Diet-Related Changes on Clinical Laboratory Values

The ketogenic diet is a low-carbohydrate (<40 g/day), moderate-protein, high-fat diet. In the absence of sufficient carbohydrates, the liver converts fat into fatty acids and ketones. Adherence to a ketogenic diet results in elevated blood and urine ketones within several days and diuresis within 2 weeks. Reportedly, a decline in serum triglycerides and an increase in HDL-C occur over several weeks [39]. The nonvegetarian diet has higher plasma ammonia, uric acid, and urea concentrations compared to the vegetarian diet. This diet commonly includes saturated fatty acids. Palmitic acid, a saturated fatty acid, causes a significant rise in plasma cholesterol concentrations. The substitution of saturated fatty acids with polyunsaturated fats and complex carbohydrates can lower LDL-C concentrations. Intake of omega-3 oils may lower triglycerides and very low-density lipoprotein (VLDL) concentrations. Vegetarians have lower LDL-C (approximately 37% lower) and HDL-C (approximately 12%) concentrations compared to nonvegetarians. In contrast, lactovegetarians (vegetarians who consume dairy products) have higher LDL-C (approximately 24% higher) and HDL-C (approximately 7% higher) concentrations compared to vegetarians. Within 20 weeks, a lactovegetarian diet regimen accompanied by low protein and high dietary fiber intake can reduce adrenocortical activity. Lactovegetarians have higher plasma concentrations of dehydroepiandrosterone sulfate (DHEAS) compared to nonvegetarians (individuals who adhere to a moderately protein-rich diet). Moreover, lactovegetarians have reduced urinary 24-hr excretion rates for C-peptide, free cortisol, DHEAS, and total 17-ketosteroid [40]. In middle-aged North American black individuals, reduced urinary 24-hr excretion rates of adrenal and gonadal androgen metabolites occurred following a conversion from the meat-containing Western diet to the vegetarian diet. The fecal fat test, which measures the amount of fat content in stool to diagnose absorption or digestion abnormalities, is susceptible to dietary influences. It is critical that individuals refrain from significant dietary changes before and during sample collection.

The hCG diet consists of hCG sublingual drops or injections paired with a low 500-calorie diet. As previously discussed, hCG can be of placental or nonplacental origin. hCG is evident in placental trophoblastic (hydatidiform mole and choriocarcinoma), gonadal (ovarian, testicular, or extragonadal teratoma), ectopic, or nontrophoblastic tumors. Exogenous hCG may be detectable in the body 10 days post injection/ingestion. Individuals on the hCG diet who received injections of hCG had markedly elevated serum hCG concentrations in the absence of pregnancy or malignancy [41]. It is obvious that individuals on the hCG diet may have unreliable test results. However, the effects of hCG sublingual drops on laboratory tests are unknown. In healthy males, hCG injections (purified urinary and recombinant hCG) stimulate Leydig cells and cause a dose-dependent increase in serum testosterone concentrations [42].

### Fasting/Starvation-Related Changes on Clinical Laboratory Values

Fasting (decreased caloric intake) and starvation (no caloric intake) initiate complex metabolic derangements. Many individuals fast in accordance with culture and religious traditions, so understanding the effects of fasting on laboratory results is paramount. Within 3 days of fasting, glucose concentrations rise by as much as 18 mg/dL despite the body's coordinated efforts to conserve proteins. Subsequently, insulin rapidly declines while glucagon secretion increases in an effort to restore blood glucose to pre-fasting concentrations. The fasting individual undergoes both lipolysis and hepatic ketogenesis. The metabolic acidosis state includes elevated serum acetoacetic acid, β-hydroxybutyrate, and fatty acids and reduced pH, $pCO_2$, and bicarbonate. Focal necrosis of the liver is responsible for reduced hepatic blood flow and impaired glomerular filtration and creatinine clearance; elevated serum ALT, AST, bilirubin, creatinine, and lactate concentrations [3].

The body's reduced energy stores mainly account for significant declines up to 50% in both total and free triiodothyronine concentrations. Fasting differentially affects lipid concentrations. Within 6 days, cholesterol and triglycerides increase while HDL concentrations decrease. Sharp increases up to 15 times the pre-fast plasma in growth hormone concentrations occur early in fasting. Within 3 days of completing a fast, the plasma growth hormone concentration returns to pre-fast levels. Albumin, prealbumin, and complement 3 concentrations decline during an extended fast. However, protein intake following fasting rapidly returns albumin, prealbumin, and complement 3 to pre-fasting concentrations.

Starvation triggers the release of aldosterone and excessive urinary ammonia, calcium, magnesium, and potassium excretion. In contrast, the body's urinary excretion of phosphorus declines. Following a short-term, 14-hr fast, acetoacetate, β-hydroxybutyrate, lactate, and pyruvate blood concentrations begin to rise. Long-term starvation lasting for 40–48 hr causes up to a 30-fold increase in β-hydroxybutyrate. Reportedly, starvation for 4 weeks significantly increased AST, creatinine, and uric acid (20–40%) and decreased GGT, triglycerides, and urea (20–50%).

Upon adequate caloric intake, the body begins to restore blood constituents to pre-fasting concentrations and retains sodium as a result of decreased urinary excretion of both sodium and chloride. Subsequently, aldosterone exceeds fasting concentrations, and urinary excretion potassium slowly returns to normal.

## EFFECTS OF NUTRACEUTICALS ON CLINICAL LABORATORY TEST RESULTS

In 1989, Dr. Stephen DeFelice coined the term *nutraceutical* from the two words "nutrition" and "pharmaceutical." Nutraceuticals, according to the American Nutraceutical Association, include functional foods with health-promoting and disease-preventing benefits. Rigorous safety and efficacy studies are lacking in the field. The pharmacokinetic properties of the commercially available nutraceuticals still need to be elucidated. An estimated 100 million Americans use dietary supplements regularly. Although nutraceuticals exhibit pharmacological effects, patients do not consider them "drugs" and often do not disclose usage to their physicians [43]. How nutraceuticals and conventional drugs interact within the body requires more investigation. Few studies have documented the pharmacokinetics of nutraceuticals and their effects on laboratory results [44]. High-protein supplements cause intermittent abdominal pain. Laboratory studies have reported that high-protein diets can lead to hyperalbuminemia and increased concentrations of AST and ALT. Albumin and liver enzyme activities returned to normal after patients discontinued using the high-protein supplements [45]. Widely used as an antidepressant, St. John's wort (*Hypericum perforatum*) markedly interferes with the metabolism of prescribed drugs. St. John's wort is a potent inducer of P-glycoprotein and cytochrome P450 3A4 (CYP3A4) and, to a lesser extent, CYP1A2 and CYP2C9 [43]. Co-administration of St. John's wort significantly alters concentrations of cyclosporine (transplant rejection) [46], indinavir (HIV inhibitor) [47], and digoxin (P-glycoprotein transporter) [48]. Royal jelly, produced by special glands in the heads of nurse honeybees, is a nutrient-rich food for queen bees. An elderly man undergoing warfarin therapy developed hematuria and an elevated international normalized ratio (7.29) after taking royal jelly supplements for 1 week. The mechanisms by which royal jelly increases the effects of warfarin are not clear. Valerian, prescribed for its antidepressant properties, causes acute hepatotoxicity (elevated ALT, AST, and GGT). Valerian's long-term effects on liver function are unknown. See Chapter 7 for more in-depth discussion on the effects of herbal supplements on clinical laboratory test results.

## EFFECTS OF EXERCISE ON CLINICAL LABORATORY TEST RESULTS

The effect of exercise on laboratory findings varies and highly correlates with the health status of the person [49], temperature, and dietary intake (food or liquid) that occurs during or following exercise [50]. Figure 2.1 shows the frequency distribution of serum creatinine concentrations in athletes and controls (sedentary people) [49]. Alterations in thyroid function occur during high-intensity exercise. Anaerobic exercise elicits increases in T4, free T4, and thyroid-stimulating hormone and decreases in T3 and free T3 [51]. Physical exercise significantly alters plasma volume as a result of fluid volume loss due to sweating and fluid shifts between both intravascular and interstitial bodily compartments [52]. Exercise reduces urinary erythrocyte and leukocyte content and the volume of urine while increasing the urinary protein excretion. Elevated urinary protein will resolve within 24–48 hr. Following exercise, a transient increase in white blood cells, hematocrit, and platelets occurs in parallel with electrolyte abnormalities (serum potassium decreases by 8%), which are present due to the altered hydration state and usually normalize with rehydration. Dehydration causes elevated creatinine and BUN concentrations. In the setting of severe dehydration, a sharp rise in BUN occurs, but creatinine is only mildly elevated. Again, rehydration will gradually decrease these concentrations to normal.

Regular vigorous exercise raises HDL-C and lowers triglycerides, VLDL-C, and LDL-C. AST, ALT, LD, creatinine kinase (CK), and myoglobin significantly increase following weight lifting and can remain elevated for up to 7 days post exercise [53].

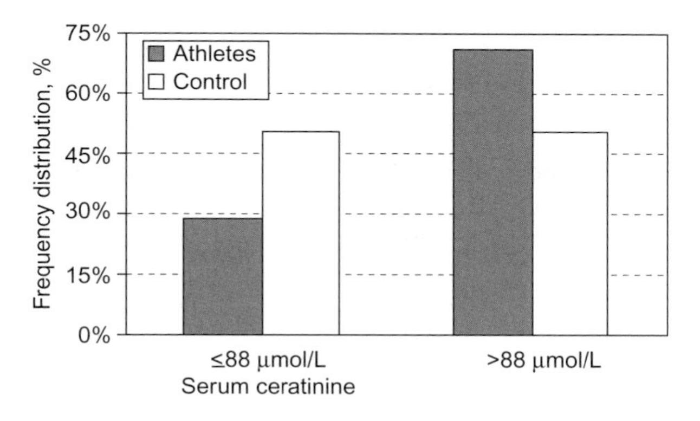


FIGURE 2.1 **Frequency distribution of serum creatinine concentrations in the two groups, athletes and controls.** Data are divided considering as threshold the median value of the control group [88 µmol/L (1.0 mg/dL)]. *Source: Reprinted with permission from the American Association for Clinical Chemistry, publisher of* Clinical Chemistry. *From Banfi, G., Del Fabbro, M. Serum creatinine values in elite athletes competing in 8 different sports: Comparison with sedentary people.* Clinic Chemistry *2006;* ***52****(2), 330–331.*

These findings highlight the importance of refraining from weight lifting prior to clinical laboratory testing. Healthy males who cycled for 30 min (maximal heart rate of 70–75%) and recovered for 30 min had significant increases in hematocrit, red blood cell count, plasma albumin and fibrinogen concentrations, plasma viscosity, and whole blood viscosity [54]. However, the changes were temporary, and concentrations returned to baseline after the 30-min recovery period. In endurance runners, exercise-associated iron deficiency is common. Moderately trained female long-distance runners who underwent long-term endurance exercise (8 weeks) did not have changes in high-sensitivity C-reactive protein, suggesting that inflammation is not a normal process of endurance training. Changes in both serum hepcidin and soluble transferrin receptor may explain the higher prevalence of iron deficiency in this population. Analytes affected by exercise are summarized in Table 2.1.

## EFFECTS OF ETHNICITY/RACE ON CLINICAL LABORATORY TEST RESULTS

Several analytes exhibit race-related changes, and it is important for laboratory professionals to recognize such changes [55]. Total serum protein concentration is usually higher in African Americans than in white individuals, mostly attributable to $\gamma$-globulin concentrations. Serum albumin concentrations, on average, are lower in the African American population compared

**TABLE 2.1** Analytes That Are Affected by Exercise

| Analyte | Effect of exercise |
|---|---|
| Urea | Value may increase after exercise |
| Creatinine | Value may increase after exercise |
| Aspartate aminotransferase | Value may increase after exercise |
| Lactate dehydrogenase | Value may increase after exercise |
| Total creatinine kinase (CK) | Value may increase after exercise |
| CK-MB | Value may increase after exercise |
| Myoglobin | Value may increase after exercise |
| WBC count | Value may increase after exercise |
| Platelet count | Value may increase after exercise |
| Prothrombin time | Value may increase after exercise |
| D-dimer | Value may increase after exercise |
| Packed cell volume | Value may decrease after exercise |
| Activated partial thromboplastin time | Value may decrease after exercise |
| Fibrinogen | Value may decrease after exercise |

to the white population. The activity of CK is usually lower in white individuals compared to African Americans. African American children usually have higher ALP than white children. Serum cystatin C significantly correlated with race/ethnicity in adolescents (ages 12–19 years) [15]. Cystatin C concentrations were higher in non-Hispanic white compared to non-Hispanic black and Mexican Americans. In contrast, creatinine was lower in non-Hispanic white and Mexican Americans compared with non-Hispanic black Americans. In an adult population-based sample (ages 50–67 years), black men had higher 24-hr urinary excretion of creatinine, epinephrine, and norepinephrine compared to white men in the study [34]. In the U.S. Modified Diet of Renal Disease (MDRD) calculation, an ethnicity factor of 1.2 aids in the estimation of GFR in African Americans. However, two large studies conducted in sub-Saharan Africa (Ghana ($N = 944$) and South Africa ($N = 100$)) showed that the MDRD equation performed better in the absence of the ethnicity actor of 1.2 [56]. In a Saudi population ($N = 32$), GFR estimated by MDRD strongly correlated with the measured inulin clearance [57]. However, in a Japanese population ($N = 248$), GFR estimated by the $0.881 \times$ MDRD equation correlated better with measured inulin clearance than with the $1.0 \times$ MDRD equation [58]. Clearly, estimation of GFR by MDRD varies by global region. The variability of GFR may correlate with genetic/environmental factors associated with body muscle mass. Carbohydrate and lipid metabolism also differ between black and white individuals. On average, African Americans exhibit less glucose tolerance than white individuals. However, in a cohort of individuals with normal glucose tolerance, African Americans had higher hemoglobin A1c concentrations compared to white individuals [59]. Significant hematologic differences exist between healthy African Americans and white individuals who are iron sufficient. African Americans have lower hemoglobin concentrations and are more likely to be diagnosed with anemia compared to white individuals [60]. Interpretation of laboratory findings based solely on the presumed effect of ethnicity/race is not appropriate.

# CONCLUSIONS

It is necessary for laboratory professionals to implement age- and gender-specific reference ranges for certain analytes, and such reference range information should be a part of routine reporting of laboratory test results. Nevertheless, diet, exercise, and other factors may alter laboratory test results, and proper investigation must be made to interpret such laboratory test results for proper patient management. Especially for pharmacogenetics testing, ethnic differences are obvious for certain isoenzymes of the cytochrome P450 mixed function oxidase family of enzymes. This important topic is discussed in-depth in Chapter 22.

## References

[1] Wolcott JS, Amanda. Goodman, Clifford. Laboratory Medicine: A National Status Report: The Lewin Group, Falls Church, VA; 2008.
[2] St.-Onge MP, Gallagher D. Body composition changes with aging: the cause or the result of alterations in metabolic rate and macronutrient oxidation? Nutrition 2010;26(2):152–5.
[3] Burtis CAAE, Bruns DE, editors. Tietz textbook of clinical chemistry and molecular diagnostics. 4th ed. St. Louis, MO: Saunders; 2006.
[4] Reference ranges and what they mean. Lab Tests Online. 2001.
[5] Kibak P. The status of laboratory medicine. Clinical laboratory news. August ed. Washington, DC: American Association for Clinical Chemistry; 2008. p. 1–3.
[6] Proytcheva MA. Issues in neonatal cellular analysis. Am J Clin Pathol 2009;131(4):560–73.
[7] Young M, Bermes EW. Specimen collection and other preanalytical variables. In: Burtis CA, Ashwood E, editors. Tietz fundamentals of clinical chemistry. 5th ed. Philadelphia: Saunders; 2001. p. 30–54.
[8] Subcommittee on hyperbilirubinemia. Management of hyperbilirubinemia in the newborn infant 35 or more weeks of gestation. Pediatrics. 2004;114(1):297–316.
[9] Joubert-Huebner E. Transitional Period in the Newborn after Birth. Available at <http://www.ecc-book.com/Transitional_Period.pdf/>; 2010.
[10] Soldin SJWE, Brugnara C, Soldin OP, editors. Pediatric reference intervals. 7th ed. Washington, DC: American Association for Clinical Chemistry; 2011.
[11] Biro FM, Galvez MP, Greenspan LC, Succop PA, Vangeepuram N, Pinney SM, et al. Pubertal assessment method and baseline characteristics in a mixed longitudinal study of girls. Pediatrics 2010;126(3):e583–90.
[12] Sorensen K, Aksglaede L, Petersen JH, Juul A. Recent changes in pubertal timing in healthy Danish boys: associations with body mass index. J Clin Endocrinol Metab 2010;95(1):263–70.
[13] Turan S, Topcu B, Gokce I, Guran T, Atay Z, Omar A, et al. Serum alkaline phosphatase levels in healthy children and evaluation of alkaline phosphatase *z*-scores in different types of rickets. J Clin Res Pediatr Endocrinol 2011;3(1):7–11.
[14] Kutilek S, Cervickova B, Bebova P, Kmonickova M, Nemec V. Normal bone turnover in transient hyperphosphatasemia. J Clin Res Pediatr Endocrinol 2012; [Epub ahead of print].
[15] Groesbeck D, Kottgen A, Parekh R, Selvin E, Schwartz GJ, Coresh J, et al. Age, gender, and race effects on cystatin C levels in U.S. adolescents. Clin J Am Soc Nephrol 2008;3(6):1777–85.
[16] Cole LA, Sasaki Y, Muller CY. Normal production of human chorionic gonadotropin in menopause. N Engl J Med 2007;356 (11):1184–6.
[17] Snyder JA, Haymond S, Parvin CA, Gronowski AM, Grenache DG. Diagnostic considerations in the measurement of human chorionic gonadotropin in aging women. Clin Chem 2005;51 (10):1830–5.
[18] Palmer OM, Grenache DG, Gronowski AM. The NACB laboratory medicine practice guidelines for point of care reproductive testing. J Near-Patient Test Technol 2007;6 (4):265–72.

[19] Gronowski AM, Fantz CR, Parvin CA, Sokoll LJ, Wiley CL, Wener MH, et al. Use of serum FSH to identify perimenopausal women with pituitary hCG. Clin Chem 2008;54 (4):652–6.

[20] Association of aging. Aging Statistics. U.S. Department of Health and Human Services, Washington, DC; 2011.

[21] Brigden ML, Heathcote JC. Problems in interpreting laboratory tests: what do unexpected results mean? Postgrad Med 2000;107(7):145–6 51–2, 55–8 passim.

[22] McLean AJ, Le Couteur DG. Aging biology and geriatric clinical pharmacology. Pharmacol Rev 2004;56(2):163–84.

[23] Cooper MS, Gittoes NJ. Diagnosis and management of hypocalcaemia. BMJ 2008;336(7656):1298–302.

[24] Luggen AS, editor. Laboratory values and implications for the aged. 6th ed. St. Louis: Mosby; 2004.

[25] van der Velde M, Bakker SJ, de Jong PE, Gansevoort RT. Influence of age and measure of eGFR on the association between renal function and cardiovascular events. Clin J Am Soc Nephrol 2010;5(11):2053–9.

[26] Veldhuis JD, Roelfsema F, Iranmanesh A, Carroll BJ, Keenan DM, Pincus SM. Basal, pulsatile, entropic (patterned), and spiky (staccato-like) properties of ACTH secretion: impact of age, gender, and body mass index. J Clin Endocrinol Metab 2009;94 (10):4045–52.

[27] Matsumoto AM. Andropause: clinical implications of the decline in serum testosterone levels with aging in men. J Gerontol Series A, Biol Sci Med Sci 2002;57(2):M76–99.

[28] Harman SM, Metter EJ, Tobin JD, Pearson J, Blackman MR. Longitudinal effects of aging on serum total and free testosterone levels in healthy men: Baltimore Longitudinal Study of Aging. J Clin Endocrinol Metab 2001;86(2):724–31.

[29] Janu MR, Creasey H, Grayson DA, Cullen JS, Whyte S, Brooks WS, et al. Laboratory results in the elderly: the Sydney Older Persons Study. Ann Clin Biochem 2003;40(Pt 3):274–9.

[30] Tipton KD. Gender differences in protein metabolism. Curr Opin Clin Nutr Metab Care 2001;4(6):493–8.

[31] Thalacker-Mercer AE, Johnson CA, Yarasheski KE, Carnell NS, Campbell WW. Nutrient ingestion, protein intake, and sex, but not age, affect the albumin synthesis rate in humans. J Nutr 2007;137(7):1734–40.

[32] Deeg M. Lipid Topics. 2006.

[33] Kelly GS. Seasonal variations of selected cardiovascular risk factors. Altern Med Rev 2005;10(4):307–20.

[34] Masi CM, Rickett EM, Hawkley LC, Cacioppo JT. Gender and ethnic differences in urinary stress hormones: the population-based chicago health, aging, and social relations study. J Appl Physiol 2004;97(3):941–7.

[35] Fraser C. Biological variation: from principles to practice. Washington, DC: AACC Press; 2001.

[36] Niv Y, Fraser GM. The alkaline tide phenomenon. J Clin Gastroenterol 2002;35(1):5–8.

[37] Dimeski G. Interference testing. Clin Biochem Rev 2008;29 (Suppl. 1):S43–8.

[38] Andersson M, Ellegard L, Andersson H. Oat bran stimulates bile acid synthesis within 8 h as measured by 7alpha-hydroxy-4-cholesten-3-one. Am J Clin Nutr 2002;76(5):1111–16.

[39] Yancy Jr WS, Olsen MK, Guyton JR, Bakst RP, Westman EC. A low-carbohydrate, ketogenic diet versus a low-fat diet to treat obesity and hyperlipidemia: a randomized, controlled trial. Ann Intern Med 2004;140(10):769–77.

[40] Remer T, Pietrzik K, Manz F. Short-term impact of a lactovegetarian diet on adrenocortical activity and adrenal androgens. J Clin Endocrinol Metab 1998;83(6):2132–7.

[41] Woolsey CJGJ, Nielsen GL. The effects of the hCG diet on common pregnancy tests. Ogden, UT: Weber State University; 2012.

[42] Handelsman DJ. Clinical review: the rationale for banning human chorionic gonadotropin and estrogen blockers in sport. J Clin Endocrinol Metab 2006;91(5):1646–53.

[43] Hennessy M, Kelleher D, Spiers JP, Barry M, Kavanagh P, Back D, et al. St. Johns wort increases expression of P-glycoprotein: implications for drug interactions. Br J Clin Pharmacol 2002;53 (1):75–82.

[44] Dasgupta A. Herbal supplements: efficacy, toxicity, interactions with Western drugs, and effects on clinical laboratory tests. Hoboken, NJ: Wiley; 2011.

[45] Mutlu EAKA, Mutlu GM. Hyperalbuminemia and elevated transaminases associated with high-protein diet. Scand J Gastroenterol 2006;41:759–60.

[46] Ruschitzka F, Meier PJ, Turina M, Luscher TF, Noll G. Acute heart transplant rejection due to Saint John's wort. Lancet 2000;355(9203):548–9.

[47] Piscitelli SC, Burstein AH, Chaitt D, Alfaro RM, Falloon J. Indinavir concentrations and St John's wort. Lancet 2000;355 (9203):547–8.

[48] Johne A, Brockmoller J, Bauer S, Maurer A, Langheinrich M, Roots I. Pharmacokinetic interaction of digoxin with an herbal extract from St John's wort (*Hypericum perforatum*). Clin Pharmacol Ther 1999;66(4):338–45.

[49] Banfi G, Del Fabbro M. Serum creatinine values in elite athletes competing in 8 different sports: comparison with sedentary people. Clin Chem 2006;52(2):330–1.

[50] Dufour D. Effects of habitual exercise on routine laboratory tests. Clin Chem 1998;44:136.

[51] Ciloglu F, Peker I, Pehlivan A, Karacabey K, Ilhan N, Saygin O, et al. Exercise intensity and its effects on thyroid hormones. Neuro Endocrinol Lett 2005;26(6):830–4.

[52] Ritchie RF, Ledue TB, Craig WY. Patient hydration: a major source of laboratory uncertainty. Clin Chem Lab Med 2007;45 (2):158–66.

[53] Pettersson J, Hindorf U, Persson P, Bengtsson T, Malmqvist U, Werkstrom V, et al. Muscular exercise can cause highly pathological liver function tests in healthy men. Br J Clin Pharmacol 2008;65(2):253–9.

[54] Ahmadizad S, Moradi A, Nikookheslat S, Ebrahimi H, Rahbaran A, Connes P. Effects of age on hemorheological responses to acute endurance exercise. Clin Hemorheol Microcirc 2011;49(1–4):165–74.

[55] Horn PS, Pesce AJ. Effect of ethnicity on reference intervals. Clinical Chemistry 2002;48(10):1802–4.

[56] Eastwood JB, Kerry SM, Plange-Rhule J, Micah FB, Antwi S, Boa FG, et al. Assessment of GFR by four methods in adults in Ashanti, Ghana: the need for an eGFR equation for lean African populations. Nephrol Dial Transplant 2010;25:2178–87.

[57] Al Wakeel JSHD, Al Suwaida A, Tarif N, Chaudhary A, Isnani A, Albedaiwi WA, et al. Validation of predictive equations for glomerular filtration rate in the Saudi population. Saudi J Kidney Dis Transpl 2009;20(6):1030–7.

[58] Imai E, Horio M, Nitta K, Yamagata K, Iseki K, Hara S, et al. Estimation of glomerular filtration rate by the MDRD study equation modified for Japanese patients with chronic kidney disease. Clin Exp Nephrol 2007;11(1):41–50.

[59] Ziemer DC, Kolm P, Weintraub WS, Vaccarino V, Rhee MK, Twombly JG, et al. Glucose-independent, black-white differences in hemoglobin A1c levels: a cross-sectional analysis of 2 studies. Ann Int Med 2010;152(12):770–7.

[60] Beutler E, West C. Hematologic differences between African-Americans and whites: the roles of iron deficiency and alpha-thalassemia on hemoglobin levels and mean corpuscular volume. Blood 2005;106(2):740–5.

CHAPTER

# 3

# Effect of Patient Preparation, Specimen Collection, Anticoagulants, and Preservatives on Laboratory Test Results

*Leland Baskin, Valerian Dias, Alex Chin, Amid Abdullah, Christopher Naugler*

**University of Calgary and Calgary Laboratory Services, Calgary, Alberta, Canada**

## INTRODUCTION

Patient preparation and the specimen type are important pre-analytical factors to consider for laboratory assessment. Although the clinical laboratory has limited capabilities in controlling for the physiological state of the patient, such as biological rhythms and nutritional status, these variables as well as the effect of patient posture, tourniquets, and hemolysis on measurement of analytes must be understood by both the clinical team and laboratory personnel. The most accessible specimen types include blood, urine, and oral fluid. The numerous functions associated with blood make it an ideal specimen to measure biomarkers corresponding to various physiological and pathophysiological processes. Blood can be collected by skin puncture (capillary), which is preferred when blood conservation and minimal invasiveness is stressed, such as in the pediatric population. Other modes of collection include venipuncture and arterial puncture, where issues to consider include the physical state of the site of collection and patient safety. Blood can also be taken from catheters and other intravascular lines, but care must be taken to eliminate contamination and dilution effects associated with heparin and other drugs. Clinical laboratory specimens derived from blood include whole blood, plasma, and serum. However, noticeable differences between these specimen types need to be considered when choosing the optimal specimen type for laboratory analysis. Such important factors include the presence of anticoagulants in plasma and in whole blood, hematocrit variability, and the differences in serum characteristics associated with blood coagulation. Anticoagulants for plasma and/or whole blood collection include ethylenediaminetetraacetic acid (EDTA), heparin, hirudin, oxalate, and citrate, which are available in solid or liquid form. Optimal anticoagulant-to-blood ratios are crucial to prevent clot formation while avoiding interference with analyte measurement, including dilution effects associated with liquid anticoagulants. Given the availability of multiple anticoagulants and additives, blood collection tubes should be filled according to a specified order to minimize contamination and carryover. Other factors to consider regarding blood collection tubes include differences between plastic and glass surfaces, surfactants, tube stopper lubricants, and gel separators, which all affect analyte measurement. The second most popular clinical specimen is urine, which is essentially an ultrafiltrate of blood before elimination from the body and is the preferred specimen to detect metabolic activity as well as urinary tract infections. Proper timing must be ensured for urine collections depending on the need for routine tests, patient convenience, clinical sensitivity, or quantitation. Furthermore, proper technique is required for clean catch samples for subsequent microbiological examination. Certain urine specimens require additives to preserve cellular integrity for cytological analysis and to prevent bacterial overgrowth. It is important to recognize the pre-analytical variables that affect analyte measurement in patient specimens so that properly informed decisions can be made regarding assay

*Accurate Results in the Clinical Laboratory.*
DOI: http://dx.doi.org/10.1016/B978-0-12-415783-5.00003-7

selection and development as well as troubleshooting unexpected outcomes from laboratory analysis.

## BIOLOGICAL RHYTHMS AND LABORATORY TEST RESULTS

Predictable patterns in the temporal variation of certain analytes, reflecting patterns in human needs, constitute biological rhythms. Different analytes have different rhythms, ranging from a few hours to monthly changes. Awareness of such changes can be relevant to proper interpretation of laboratory results. These changes can be divided into circadian, ultradian, and infradian rhythms according to the time interval of their completion.

During a 24-hr period of human metabolic activity, programming of metabolic needs may cause certain laboratory tests to fluctuate between a maximum and a minimum value. The amplitude of change of these circadian rhythms is defined as one-half of the difference between the maximum and the minimum values. Although, in general, these variations occur consistently, alteration in these natural circadian rhythms may be induced by artificial changes in sleep/wake cycles such as those induced by different work shifts. Therefore, in someone working an overnight ("graveyard") shift, an elevated blood iron level taken at midnight would be normal for that individual; however, the norm is for high iron levels to be seen only in early morning.

Patterns of biological variation occurring on cycles less than 24 hr are known as ultradian rhythms. Analytes that are secreted in a pulsatile manner throughout the day show this pattern. Testosterone, which usually peaks between 10:00 a.m. and 5 p.m., is an example of an analyte showing this pattern.

The final pattern of biological variation is infradian. This involves cycles greater than 24 hr. The example most commonly cited is the monthly menstrual cycle, which takes approximately 28–32 days to complete. Constituents such as pituitary gonadotropin, ovarian hormones, and prostaglandins are significantly affected by this cycle.

## ISSUES OF PATIENT PREPARATION

There are certain important issues regarding patient preparation for obtaining meaningful clinical laboratory test results. For example, glucose testing must be done after the patient has fasted overnight. These issues are discussed in this section.

### Fasting

The effects of meals on blood test results have been known for some time. Increases in serum glucose, triglycerides, bilirubin, and aspartate aminotransferase are commonly observed after meal consumption. On the other hand, fasting will increase fat metabolism and increase the formation of acetone, β-hydroxybutyric acid, and acetoacetate both in serum and in urine. Longer periods of fasting (more than 48 hr) may result in up to a 30-fold increase in these ketone bodies. Glucose is primarily affected by fasting because insulin keeps the serum concentration in a tight range (70–110 mg/dL). Diabetes mellitus, which results from either a deficiency of insulin or an increase in tissue resistance to its effects, manifests as an increase in blood glucose levels. In normal individuals, after an average of 2 hr of fasting, the blood glucose level should be below 7.8 mmol/L (140 mg/dL). However, in diabetic individuals, fasting serum levels are elevated and thus constitute one criterion for making the diagnosis of diabetes. Other well-known examples of analytes showing variation with fasting interval include serum bilirubin, lipids, and serum iron.

### Body Position

Physiologically, blood distribution differs significantly in relation to body posture. Gravity pulls the blood into various parts of the body when recumbent, and the blood moves back into the circulation, away from tissues, when standing or ambulatory. These shifts directly affect certain analytes due to dilution effects. This process is deferential, meaning that only constituents of the blood that are nondiffusible will rise because there is a reduction in plasma volume upon standing from a supine position. This includes, but is not limited to, cells, proteins, enzymes, and protein-bound analytes (thyroid-stimulating hormone, cholesterol, T4, and medications such as warfarin). The reverse will take place when shifting from erect to supine because there will be a hemodilution effect involving the same previously mentioned analytes. Postural changes affect some groups of analytes in a much more profound way—at times up to a twofold increase or decrease depending on whether the sample was obtained from a supine or an erect patient. Most affected are factors directly influencing hemostasis, including renin, aldosterone, and catecholamines. It is vital for laboratory requisitions to specify the need for supine samples when these analytes are requested.

## DIFFERENCES BETWEEN WHOLE BLOOD, PLASMA, AND SERUM SPECIMENS FOR CLINICAL LABORATORY ANALYSIS

Approximately 8% of total human body weight is represented by blood, with an average volume in

**TABLE 3.1** Principal Components of Plasma

| Component | Quantity | Units |
|---|---|---|
| Sodium | 144 | mmol/L |
| Potassium | 4 | mmol/L |
| Bicarbonate | 25 | mmol/L |
| Chloride | 105 | mmol/L |
| Hydrogen ions | 40 | mmol/L |
| Calcium | 2.5 | mmol/L |
| Magnesium | 0.8 | mmol/L |
| Inorganic phosphate | 1.1 | mmol/L |
| Glucose | 4.5 | mmol/L |
| Cholesterol | 2.0 | g/L |
| Fatty acids | 3.0 | g/L |
| Total protein | 70–85 | g/L |
| Albumin | 45 | g/L |
| α-Globulins | 7 | g/L |
| β-Globulins | 8.5 | g/L |
| γ-Globulins | 10.6 | g/L |
| Fibrinogen | 3 | g/L |
| Prothrombin | 1 | g/L |
| Transferrin | 2 | g/L |

females and males of 5 and 5.5 L, respectively [1]. Blood consists of a cellular fraction, which includes erythrocytes, leukocytes, and thrombocytes, and a liquid fraction, which transports these elements throughout the body. Blood vessels interconnect all the organ systems in the body and play a vital role in communication and transportation between tissue compartments. Blood serves numerous functions, including delivery of nutrients to tissues; gas exchange; transport of waste products such as metabolic by-products for disposal; communication through hormones, proteins, and other mediators to target tissues; and cellular protection against invading organisms and foreign material. Given these myriad roles, blood is an ideal specimen for measuring biomarkers associated with various physiological conditions, whether it is direct measurement of cellular material and surface markers or measurement of soluble factors associated with certain physiological conditions.

Plasma consists of approximately 93% water, with the remaining 7% composed of electrolytes, small organic molecules, and proteins. Various constituents of plasma are summarized in Table 3.1. These analytes are in transit between cells in the body and are present in varying concentrations depending on the physiological state of the various organs. Therefore, accurate analysis of the plasma is crucial for obtaining information regarding diagnosis and treatment of diseases. In clinical laboratory analysis, plasma can be obtained from whole blood through the use of anticoagulants followed by centrifugation. Consequently, plasma specimens for the clinical laboratory contain anticoagulants such as heparin, citrate, EDTA, oxalate, and fluoride. The relative roles of these anticoagulants in affecting analyte measurements are discussed later in this chapter. In contrast to anticoagulated plasma specimens, serum is the clear liquid that separates from blood when it is allowed to clot. Further separation of the clear serum from the clotted blood can be achieved through centrifugation. Given that fibrinogen is converted to fibrin in clot formation during the coagulation cascade, serum contains no fibrinogen and no anticoagulants.

In the clinical laboratory, suitable blood specimens include whole blood, plasma, and serum. Key differences in these sample matrices influence their suitability for certain laboratory tests (Table 3.2).

## Whole Blood

In addition to the obvious advantage of whole blood for the analysis of cellular elements, these specimens are also preferred for analytes that are concentrated within the cellular compartment. Erythrocytes can be considered to be a readily accessible tissue with minimal invasive procedures and may more accurately reflect tissue distribution of certain analytes. Examples of such analytes include vitamins, trace elements, and certain drugs (Table 3.3). Erythrocytes are the most abundant cell type in the blood. In adults, 1 μL of blood contains approximately 5 million erythrocytes compared to 4000–11,000 leukocytes and 150,000–450,000 platelets [4]. The volumetric fraction of erythrocytes is clinically referred to as the hematocrit, expressed as a percentage of packed erythrocytes in a blood sample after centrifugation. The normal range for adult males is 41–51%, and that for adult females is 36–45% [4]. Clearly, alterations in hematocrit will directly alter the available plasma water concentration, which in turn affects the measurement of water-soluble factors in whole blood.

A major use for whole blood specimens is for point-of-care analysis. Although point-of-care meters can be located in the clinical laboratory, the primary advantage of this technology is near-patient testing, which offers rapid and convenient analysis and the use of small sample volumes while the clinician is examining the patient. The most common point-of-care specimens are taken by skin puncture. These blood samples are

TABLE 3.2 Components of Clinical Whole Blood, Plasma, and Serum Matrices

| Whole Blood | Plasma | Serum |
|---|---|---|
| Cellular elements | | |
| Erythrocytes | | |
| Leukocytes | | |
| Thrombocytes | | |
| Proteins | Proteins | Proteins (excluding fibrinogen) |
| Electrolytes | Electrolytes | Electrolytes |
| Nutrients | Nutrients | Nutrients |
| Waste (metabolites) | Waste (metabolites) | Waste (metabolites) |
| Hormones | Hormones | Hormones |
| Gases | Gases | Gases |
| *May* contain anticoagulants | *Contains* anticoagulants | Contains *no* anticoagulants |
| Patient on therapeutics | | |
| Specimen additive | | |

composed of a mixture of blood from the arterioles, venules, and capillaries and may be diluted with interstitial fluid and intracellular fluid. Furthermore, the extent of dilution with interstitial and intracellular fluid is also affected by the hematocrit. Given these physiological differences, analytes measured in whole blood do not exactly match results obtained from analysis of plasma or serum samples. Indeed, there is less water inside erythrocytes compared to the plasma; therefore, levels of hydrophilic analytes such as glucose, electrolytes, and water-soluble drugs will be lower in the capillary whole blood [5].

As mentioned previously, it is apparent that changes in both hematocrit level and plasma water content contribute to the discrepancy in analyte measurements between whole blood and plasma methods. However, in the case of point-of-care glucose meters, it was proposed and later adopted by the International Federation of Clinical Chemistry that a general conversion factor of 1.11 be applied to obtain plasma-equivalent glucose molarity [6,7]. Although this was an attempt to produce more harmonized results regarding glucose measurement and reduce clinical misinterpretations, the application of a general conversion factor does not take into account the wide variations in both hematocrit and plasma water levels exhibited by some patient subpopulations. Indeed, the proportion of total errors exceeding 10 and 15% in glucose measurements has been found to increase with patient acuity [8]. For this reason, interpretation of analyte measurements in whole blood should be sensitive to the hemodynamic status of the patient.

## Plasma Versus Serum Specimens

Although it is clear that there are certain advantages to whole blood specimens in point-of-care and hematological testing, plasma and serum are the preferred blood specimens for measuring soluble factors in the clinical laboratory. In addition to their previously mentioned discrepancies in composition, plasma and serum exhibit variations in the concentration of certain analytes. Certainly, the coagulation cascade contributes to consumption of some substances (e.g., fibrinogen, platelets, and glucose) and to the release of others (e.g., potassium, lactate, lactate dehydrogenase, phosphate, and ammonia). For example, the presence of fibrinogen in plasma contributes significantly to the higher levels (±5%) of total protein compared to serum [9]. Conversely, the release of elements or cell lysis associated with the coagulation cascade is responsible for the increase in potassium (±6%), inorganic phosphate (±11%), ammonia (±38%), and lactate (±22%) in serum compared to plasma [9]. Furthermore, anticoagulants, preservatives, and other additives that aid or inhibit coagulation may interfere with the assay, as discussed later. Also, the presence of fibrinogen may interfere with chromatic detection or binding in immunoassays or the appearance of a peak that may simulate a false monoclonal protein in the gamma region during protein electrophoresis [9,10].

## Serum Versus Plasma for Clinical Laboratory Tests

There are many advantages to using plasma over serum for clinical laboratory analysis. However, for

TABLE 3.3 Examples of Analytes Measured in Blood Cell Lysates [2,3]

| Hematology | Vitamins | Trace Elements | Drugs | Toxic Elements |
|---|---|---|---|---|
| Hemoglobin | **Direct measurement** | Chromium | Cyclosporine | Cyanide |
| Red cell indices | Vitamin E | Selenium | Sirolimus (rapamycin) | Lead |
| Porphyrias | Vitamin $B_1$ (thiamine) | Zinc | Tacrolimus (FK506, Prograf) | Mercury |
| Cytoplasmic porphyrin metabolic enzyme activity | Vitamin $B_2$ (riboflavin) | | | |
| | Also FMN, FAD | | | |
| | Vitamin $B_6$ (pyridoxine, pyridoxamine, pyridoxal) | | | |
| | PLP | | | |
| | Biotin | | | |
| | Folic acid (folate) | | | |
| | Pantothenic acid | | | |
| | **Functional activity** | | | |
| | Vitamin $B_1$ (thiamine) | | | |
| | Transketolase | | | |
| | Vitamin $B_2$ (riboflavin) | | | |
| | FAD-dependent glutathione reductase | | | |
| | Vitamin $B_6$ (pyridoxine, pyridoxamine, pyridoxal) | | | |
| | AST, ALT activity | | | |
| | Vitamin $B_{12}$ (cyanocobalamin) | | | |
| | Deoxyuridine suppression test | | | |
| | Niacin | | | |
| | NAD/NADP ratio | | | |

ALT, alanine aminotransferase; AST, aspartate aminotransferase; FAD, flavin adenine dinucleotide; FMN, riboflavin-5′-phosphate; NAD, nicotinamide adenine dinucleotide; NADP, nicotinamide adenine dinucleotide phosphate; PLP, pyridoxal-5′-phosphate.

some analytes, serum is preferred over plasma. These issues are addressed in this section.

### *Larger Sample Volume*

If blood is allowed to clot and is then centrifuged, approximately 30–50% of the original specimen volume is collected as serum. Conversely, plasma constitutes approximately 55% of the volume of uncoagulated blood after centrifugation. Therefore, the higher yield associated with plasma samples is generally preferred, especially when sample volume may be critical as in the case of the pediatric population, smaller patients, or in special cases in which blood volume needs to be conserved.

### *Less Pre-Analytical Delay*

The process of clotting requires at least 30 min under normal conditions without coagulation accelerators. Furthermore, coagulation may still occur postcentrifugation in serum samples. Therefore, another advantage of plasma is that analyte determinations can be achieved in whole blood prior to plasma separation provided that a suitable anticoagulant has been used. For example, an anticoagulated whole blood specimen may be used for point-of-care measurements followed by plasma separation, which would avoid the delay associated with obtaining an additional specimen for laboratory analysis.

### *Reduction of in Vitro Hemolysis*

In addition to the time delay associated with blood clotting, there is an increased risk of lysis and consequent false increases in many intracellular analytes such as potassium, iron, and hemoglobin released from erythrocytes in serum specimens. Therefore, it is advised to separate the serum as quickly as possible. Conversely, plasma separation can be achieved at higher centrifugal speeds without risking the initiation of hemolysis and thrombocytolysis [9].

### *Advantages of Serum Over Plasma*

Anticoagulants and additives in plasma specimens can directly interfere with the analytical characteristics of the assay, protein binding with the analyte of interest, and sample stability. Furthermore, liquid anticoagulants may lead to improper dilution of the sample. For example, blood drawn in tubes with sodium citrate is diluted by 10%, but this may increase depending on whether the draw is complete. Moreover, incomplete mixing with anticoagulants can lead to the risk of clot formation. Also, the choice of anticoagulant will depend on their respective influences on the various assays offered by the clinical laboratory, and tubes with anticoagulants and additives are often more expensive.

## DIFFERENT ANTICOAGULANTS AND PRESERVATIVES, ORDER OF DRAW, PROBLEMS WITH GEL INTERFERENCE IN SERUM SEPARATOR TUBES, AND SHORT DRAWS

Selection and use of evacuated blood collection tubes (BCT) for specimen collection is a major pre-analytical process impacting patient results in the clinical laboratory. Appropriate quality processes must be maintained to ensure accurate and reliable laboratory results. Kricka *et al.* [11] demonstrated how laboratories can establish their own processes using control materials to verify interferences in the BCT they use. Bowen *et al.* [12] provided a comprehensive review of the different BCT components that can contribute to analytical errors in the clinical laboratory. Table 3.4 summarizes the components of BCT and indications for use. The inclusion of additives is necessary for reducing pre-analytical variability and for faster turnaround times in the laboratory. BCT are available with a variety of labeling options and stopper colors as well as a range of draw volumes. Additives are designated to preserve or stabilize analytes by inhibiting metabolic enzymes. Additives such as clot activators are added to BCT to facilitate clotting primarily in plastic BCT. Stoppers and stopper lubricants are generally manufactured to facilitate capping, de-capping, and blood flow during collection and to minimize adsorption. Separator gels are used to separate packed blood cells from serum or plasma. Anticoagulants are used to prevent coagulation of blood or blood proteins. For a summary of collection additives, refer to Table 3.5.

**TABLE 3.4** Components in Evacuated Blood Collection Tubes and Indications for Use

| Components | Indications |
|---|---|
| Tube wall | Plastic or glass: Plastic preferable for safety reasons |
| Stopper | Inert plasticizers |
| Surfactants | Silicone minimizes adsorption of analytes, cells |
| Stopper lubricants | Ease of capping, de-capping |
| Clot activator | Promote clotting to obtain serum in plastic collection tubes |

### Plastic and Glass Tubes

Most clinical laboratories have routinely moved away from glass to plastic BCT to comply with occupational health and safety standards because plastic BCT are safer to use and reduce potential exposure to blood [13]. Compared to glass, the polyethylene tetraphthalate materials used in BCT are generally unbreakable, can withstand high centrifugation speeds, are inert to adsorption of analytes, are lighter, and can be easily incinerated for disposal. In contrast to plastic surfaces, the relatively hydrophilic surface of glass is less resistant to adherence of platelets, fibrin, and clotted blood components and ideal for blood flow. Also, the silica surface of glass greatly facilitates the clotting of blood, allowing for a cleaner separation of the clot from serum during centrifugation. For these reasons, the interior surface of plastic collection tubes and stoppers is now routinely spray coated with surfactants and silicate polymers to make the surface properties similar to those of glass. Minor clinically significant differences between glass and plastic exist for a variety of different tests ranging from general chemistry to special chemistry, molecular testing, and hematology.

### Surfactants

Commercially available BCT may contain different types of surfactants that are often not listed in the manufacturer's package insert but are commonly silicone-based polymers. Although these are considered to be inert, there have been reports of interferences in clinical assays. Generally, surfactants can bind nonspecifically, displace from solid matrix, and complex with or mask detection of signal antibodies in immunoassay reagents, contributing to increases in absorbance and turbidity to cause interferences [14]. Bowen *et al.* [15,16] showed that the surfactant Silwet L-720 used in Becton Dickinson SST collection tubes gave falsely elevated total triiodothyronine (TT3) by the Immulite 2000/2500 immunoassay. One mechanism

TABLE 3.5 Blood Collection Tubes, Additives, and General Applications

| Tube Type | Additive | Stopper Color | Application |
|---|---|---|---|
| *SERUM* | | | |
| Glass | None | Red | Tests that cannot be collected into SST tubes |
| Plastic | | Royal blue (red band label) | Trace elements serum (copper, zinc, aluminum, chromium, nickel) |
| Plastic | Clot activator | Red/black | Tests that cannot be collected with gel; some therapeutic drugs (antidepressants) |
| Plastic<br>Also called SST | Clot activator + gel separator | Gold | Many chemistry and immunochemistry tests; hepatitis tests |
| Plastic | Thrombin + gel separator | Orange | Rapid clotting (5 min); general chemistry tests for acute care requiring urgent turnaround time |
| *PLASMA OR WHOLE BLOOD* | | | |
| Plastic<br>Also called PST | Lithium-heparin + gel separator | Light mint green | Most chemistry tests for acute care requiring fast turnaround time, ammonia |
| Heavy metal free | $Na_2EDTA$ or $K_2EDTA$ | Royal blue (with lavender or blue band on label) | Trace elements blood (lead, arsenic, cadmium, cobalt, manganese, mercury, molybdenum, thallium) |
| Plastic | Acid citrate dextrose solution "A" (ACDA) | Pale yellow | Flow cytometry (CD3, CD4, CD8); HLA typing |
| Plastic | Sodium fluoride, potassium oxalate, iodoacetate | Gray | Preserves glucose up to 5 days; lactate, glucose (tolerance) |
| Plastic | Sodium, lithium, or ammonium heparin (no gel) | Dark green | Amino acids, blood gases, glucose phosphate dehydrogenase (G6PD) |
| Plastic | Sodium citrate | Light blue | PT (INR), PTT, other coagulation tests |
| Glass | Sodium citrate | Black | Erythrocyte sediment rate |
| Glass | Sodium heparin | Dark green | Toxicology and nutritional tests |
| | $K_2EDTA$ or $K_3EDTA$ | Lavender | Routine hematology, pretransfusion (blood bank), hemoglobin A1c, antirejection drugs, parathyroid hormone |

shown was that increasing surfactant concentrations dose dependently desorbed the capture antibody from the solid phase among other nonspecific effects. However, this was manufacturer method dependent because TT3 levels were unaffected by the AxSYM immunoassay, in which antibodies are adsorbed onto the solid phase with more robust binding [15,16]. This manufacturer confirmed similar interferences for a variety of other immunoassays (folate, vitamin $B_{12}$, follicle-stimulating hormone, hepatitis B surface antigen, cancer antigen 27, and cortisol) on a variety of different instrument platforms and has since decreased the surfactant content to reduce this interference [17,18].

## Stoppers and Stopper Lubricants

Stopper lubricants containing glycerol or silicone make capping and de-capping of tubes easier, as well as minimize adherence of cells and clots to stoppers. Standard red-topped evacuated clot tubes are contaminated with zinc, aluminum, and magnesium, and all contain varying amounts of heavy metals [19]. Components such as tris(2-butoxyethyl) phosphate from rubber stopper tubes leaching into the blood during collection have been shown to displace drugs from binding proteins in blood [20]. Manufacturers have reformulated stopper plasticizer content to minimize this effect. Triglyceride assays that measure glycerol can be falsely elevated by such effect. Stopper components and stopper lubricants can also be a potential source of interference with mass spectrometry-based assays [21].

## Serum Separator Gel Tubes

Separator gels are thixotropic materials that form a physicochemical barrier after centrifugation in BCT to separate packed cells from plasma or serum. Delay in separating clot results in intracellular leakage of potassium, phosphate, magnesium, and lactate dehydrogenase into serum, plasma, or whole blood [22]. Ease of use of a single centrifugation step, improved specimen analyte stability, reduced need for aliquoting, convenient storage, and transport in a single primary tube are reasons for preferential usage of SST in the clinical laboratory. It is very important to follow manufacturer protocols for laboratory conditions such as proper tube mixing after collection, centrifugation acceleration/deceleration speeds, temperature, and storage conditions. Noncompliance may result in unexpected degradation of separator gel or release of gel components. Gel and oil droplets interfere with accuracy and liquid-level sensing of instrument pipettes, coat cuvettes, or bind to solid phase in heterogeneous immunoassays. Re-centrifugation, inadequate blood-draw volume, storage time, temperature, and drug adsorption may affect laboratory results. Hydrophobicity of the drug, length of time on separator gel, storage temperature, and methodology sensitivity are important considerations with regard to the stability of therapeutic drugs in BCT. Dasgupta *et al.* [23] demonstrated that hydrophobic drugs such as phenytoin, phenobarbital, carbamazepine, quinidine, and lidocaine are adsorbed onto the gel with significant decreases (ranging from 5.9 to 64.5%) in serum Vacutainer SST tubes. Reformulation of the separator gel in SST II tubes significantly reduced absorption and improved performance [24]. Schouwers *et al.* [25] demonstrated minimal effect of separator gel in Starstedt S-Monovette serum tubes for the collection of four therapeutic drugs (amikacin, vancomycin, valproic acid, and acetaminophen) and eight hormones and proteins.

## Anticoagulants

Anticoagulants are used to prevent coagulation of blood or blood proteins to obtain plasma or whole blood specimens. The most routinely used anticoagulants are EDTA, heparin (sodium, ammonium, or lithium salts), and citrates (trisodium and acid citrate dextrose). Anticoagulants can be powdered, crystallized, solids, or lyophilized liquids. The optimal anticoagulant:blood ratio is essential to preserve analytes and prevent clot or fibrin formation via various differing mechanisms.

In most clinical laboratories, potassium EDTA is the anticoagulant of choice for the complete blood count, as recommended by the International Council of Standardization in Hematology [26] and the Clinical and Laboratory Standards Institute [27]. Dipotassium, tripotassium, or disodium salts of EDTA are used as dry or liquid additives in final concentrations ranging from 1.5 to 2.2 mg/mL blood when the evacuated BCT is filled correctly to its stated draw volume. EDTA acts as a chelating agent to bind cofactor divalent cations (mainly calcium) to inhibit enzyme reactions involved in the clotting cascade—specifically the conversion of prothrombin to thrombin and subsequent inhibition of the thrombolytic action on fibrinogen to fibrin necessary for clot formation [28]. For this reason, EDTA plasma is not recommended for coagulation tests such as prothrombin time (PT) and activated partial thromboplastin time (aPTT) [29]. EDTA is an excellent preservative of blood cells and morphology parameters. Stability is 48 hr for hemoglobin and 24 hr for erythrocytes. Because the hypertonic activity by EDTA can alter erythrocytic indices and hematocrit, smears should be made within 2 or 3 hr of the blood draw. The white blood cell count remains stable for at least 3 days in EDTA anticoagulated blood stored at room temperature. EDTA is adequate for platelet preservation; however, morphological changes occur over time [30]. Clotting can result if there is insufficient EDTA relative to blood. This is usually caused by overfilling the vacuum tube or poor solubility of EDTA (most commonly with disodium salts) [31]. EDTA draws water from cells to artifactually dilute plasma and is generally not recommended for general chemistry tests. Specifically, EDTA chelates other metallic ions such as copper, zinc, or magnesium and alters cofactor-dependent activity of many enzymes, such as alkaline phosphatase and creatine kinase, and hence is not used for these chemistry assays. EDTA is also used for blood bank pretransfusion testing; flow cytometry; hemoglobin A1c; and most common immunosuppressive anti-rejection drugs, such as cyclosporine, tacrolimus, sirolimus, and everolimus. Whole blood EDTA in BCT transported on ice is preferable for the collection of unstable hormones susceptible to proteolysis *in vitro*, such as corticotropin, parathyroid hormone, C-peptide, vasoactive peptide (VIP), antidiuretic hormone, carboxy-terminal collagen cross-links, calcitonin, renin, procalcitonin, and unstable peptides such as cytokines. Spray-dried potassium EDTA is the preferred anticoagulant for quantitative proteomic and molecular assay protocols such as viral nucleic acid extraction, gene amplification, or sequencing [28].

Heparin, a heterogeneous mixture of anionic glycosaminoglycans, inactivates serine proteases involved in the coagulation cascade—primarily thrombin and factors II (prothrombin) and Xa—through an antithrombin-dependent mechanism. For this reason, heparinized plasma is not used for coagulation tests. Typically, lyophilized or solid lithium, sodium, or ammonium salts of heparin are added to BCT at varying final concentrations of 10–30 USP units/mL of blood [27]. Hygroscopic heparin formulations are used instead of solutions to avoid dilution effects. Heparin is the recommended anticoagulant for many chemistry tests requiring whole blood or plasma because chelating properties and effects on water shifts in cells are minimal. Heparinized plasma is useful for tests requiring faster turnaround times because it does not require clotting, minimizing the risk of sample pipetting interference due to fibrin microclots [32]. Heparin is the only anticoagulant recommended for the determination of pH blood gases, electrolytes, and ionized calcium [33]. Lithium heparin is commonly used instead of sodium heparin for general chemistry tests [34]. Obviously, additives may directly affect the measurement of certain analytes. For example, lithium heparin plasma can have lithium levels in the toxic range, greater than 1.0 mmol/L; when filled correctly, sodium heparin tubes have sodium levels 1 or 2 mmol/L higher; and ammonium heparin increases measured ammonia levels [29]. Heparin should not be used for coagulation tests and is not recommended for protein electrophoresis and cryoglobulin testing because of the presence of fibrinogen, which co-migrates with $\beta_2$ monoclonal proteins. Information on tube type performance on many analytical tests as well as specimen stability characteristics is available from manufacturers upon request. The Becton Dickinson Diagnostic Preanalytical Division publishes clinical "white papers" for most of their various BCT products [35].

The BCT recommended by the Clinical and Laboratory Standards Institute (CLSI; H21-A5) for coagulation testing is trisodium citrate buffered (to maintain the pH) or unbuffered, available as 3.2% (or 3.8%) concentrations. The combination of sodium citrate and citric acid is called buffered sodium citrate. The recommended preservative ratio is 9:1 (blood:citrate). Citric acid and dextrose should not be used because this combination will dilute the plasma and cause hemolysis. Different citrate concentrations can have significant effects on PTT and PT assays, resulting in variable reagent responsiveness. It is necessary for labs using the international normalization ratio (INR) to ensure the same citrate concentration is used for the determination of the International Sensitivity Index. Sodium citrate acts primarily to chelate calcium, and coagulation factors are unaffected. The binding effect of citrate can be reversed by recalcifying the blood or derived plasma to its normal state. This reversible action makes it highly desirable for clotting and factor assay studies. Citrate also has minimal effects on cells and platelets and is used for platelet aggregation studies.

Hirudin is a single-chain, carbohydrate-free polypeptide derived from the leech (*Hirudo medicinalis*). Hirudin binds irreversibly to the fibrinogen recognition site of thrombin without the involvements of cofactors to prevent transformation of fibrinogen to fibrin. The use of hirudin as a more potent anticoagulant is promising for general chemistry, hematology, and molecular testing, although it is not readily commercially available [36].

Thrombin initiates clotting in the presence of calcium. Rapid 5-min Clot Serum Tubes (RST) containing thrombin as an additive are now commercially available from Becton Dickinson [37]. Laboratories find these ideal for obtaining serum for assays requiring a fast turnaround time. Strathmann *et al.* [38] showed that RST tubes have fewer false-positive results and better reproducibility compared to lithium heparin PST tubes for 28 general chemistry tests and immunoassays.

Potassium oxalate is used in combination with sodium fluoride and sodium iodoacetate to inhibit enzymes involved in the glycolytic pathway. Potassium oxalate chelates calcium and calcium-dependent enzymes and reactions to act as an anticoagulant. Sodium fluoride inhibits enolase and iodoacetate inhibits glyceraldehyde-3-phosphate enzyme activity to prevent metabolism of glucose and ethanol. Oxalate BCT is useful specifically for glucose, lactate, and ethanol tests. Glycolysis enzyme inhibition is not immediate and may be delayed up to 4 hr after collection, allowing glucose levels to fall by 5–7% per hour at room temperature. For this reason, the use of fluoride anticoagulants is undesirable for the collection of neonatal glucose specimens in capillary whole blood unless the specimens are transported on ice.

## ORDER OF DRAW OF VARIOUS BLOOD COLLECTION TUBES

To avoid erroneous results, BCT must be filled or used during phlebotomy in a specified order. A standardized order of draw (OFD) minimizes carryover contamination of additives between tubes. Table 3.6 shows an example of the OFD for blood collection as used at Calgary Laboratory Services. Many laboratories have established their own protocols for the OFD for multiple tube collections, with slight variations based on CLSI recommendations. The general order of draw is as follows:

1. Microbiological blood culture tubes
2. Trace element tubes (nonadditive)
3. Citrated coagulation tubes
4. Non-anticoagulant tubes for serum (clot activator, gel or no gel)
5. Anticoagulant: heparin tubes (with or without gel)
6. Anticoagulant: EDTA tubes

**TABLE 3.6** Blood Collection Tubes by Order of Draw

| Order of Draw | Color of Stopper | Invert | Additive | Comments/Common Tests |
|---|---|---|---|---|
| 1 | Clear | Not required | No additive | Tube used *only* as a discard tube |
| 2 | Blood culture bottle | Invert gently to mix | Bacterial growth medium and activated charcoal | When a culture is ordered along with any other blood work, the blood cultures *must* be drawn first |
| 3 | Yellow | 8–10 times | Sodium polyanethol sulfonate (SPS) | Tube used for mycobacteria (AFB) blood culture |
| 4 | Royal blue (with red band on label) | Not required | No additive | Tube used for copper and zinc |
| 5 | Red glass | Not required | No additive | Tube used for serum tests that *cannot* be collected in serum separator tubes (SST), such as tests performed by tissue typing. *Note*: Red plastic tubes are preferable for lab tests |
| 6 | Light blue | 3–4 times | Sodium citrate anticoagulant | Tube used mainly for PT (INR), PTT, and other coagulation studies |
| 7 | Black glass | 3–4 times | Sodium citrate anticoagulant | Tube used for ESR *only* |
| 8 | Red | 5 times | Clot activator, and no anticoagulant | Tube used for serum tests that *cannot* be collected in SST tubes, such as tests performed by tissue typing |
| 9 | Gold | 5 times | Gel separator and clot activator | Usually referred to as "SST" (serum separator tube). After centrifugation, the gel forms a barrier between the clot and the serum |
| 10 | Dark green Glass (with rubber stopper) | 8–10 times | Sodium heparin anticoagulant | Tube used for antimony |
| 11 | Dark green | 8–10 times | Sodium heparin anticoagulant | Tube used mainly for amino acids and cytogenetics tests |
| 12 | Light green (mint) | 8–10 times | Lithium heparin anticoagulant and gel separator | Usually referred to as "PST" (plasma separator tube). After centrifugation, the gel forms a barrier between the blood cells and the plasma. Tube used mainly for chemistry tests on acute care patients |
| 13 | Royal blue (with blue band on label) | 8–10 times | $K_2EDTA$ anticoagulant | Tube used for trace elements |
| 14 | Royal blue (with lavender band on label) | 8–10 times | $Na_2EDTA$ anticoagulant | Tube used for lead |
| 15 | Lavender | 8–10 times | EDTA anticoagulant | Tube used mainly for complete blood count, pretransfusion testing, hemoglobin A1c, and antirejection drugs |
| 16 | Pale yellow | 8–10 times | Acid citrate dextrose solution "A" (ACDA) | Tube used for flow cytometry testing |
| 17 | Gray | 8–10 times | Sodium fluoride and potassium oxalate anticoagulant | Tube used for lactate |

Blood collection tubes must be filled in a specific sequence to minimize contamination of sterile specimens, avoid possible test result error caused by carryover of additives between tubes, and reduce the effect of microclot formation in tubes. When collecting blood samples, allow the tube to fill completely to ensure the blood:additive ratio necessary for accurate results. Gently invert each tube the required number of times immediately after collection to adequately mix the blood and additive. Never pour blood from one tube into another tube. See www.Calgarylabservices.com for the most current version of this document [39].

7. Acid citrate dextrose tubes
8. Glycolytic inhibitor tubes.

Tubes with additives must be thoroughly mixed by gentle inversion as per manufacturer-recommended protocols. Erroneous test results may be obtained when the blood is not thoroughly mixed with the additive. A discard tube (plastic/no additive) is sometimes used to remove air and prime the tubing when a winged blood collection kit is used. Tubes for microbiological blood cultures are filled first to avoid bacterial contamination from epidermal flora. When trace metal testing on serum is ordered, it is advisable to use trace element tubes. Royal-blue Monoject trace element BCT are available for this purpose. These tubes are free from trace and heavy metals; however, it is advisable to consult the manufacturer's package insert to determine upper tolerable limits of trace and heavy metal contamination to determine acceptability for clinical diagnostic use. Citrated tubes are used next in the OFD because plastic tubes contain a silica clot activator that may cause interference with coagulation clotting factors. Serum BCT without and with gel is drawn next in the OFD before plasma anticoagulant non-gel or gel tubes to reduce anticoagulant contamination. The OFD for tubes with anticoagulants is heparin, EDTA, and, lastly, glycolytic inhibitors. In potassium EDTA, contamination is postulated to occur by direct transfer of blood to other tubes, backflow for potassium EDTA-containing tubes to other tubes (incorrect OFD), or syringe needle contamination. Calam and Cooper [40] originally reported that incorrect OFD results in hyperkalemia and hypocalcemia. Cornes *et al.* [41] also showed that *in vitro* contamination by potassium EDTA is a relatively common but often unrecognized cause of spurious hyperkalemia, hypomagnesaemia, hypocalcemia, hypophosphatemia, and hyperferritinemia. However, Sulaiman *et al.* [42] reported that when using the Sarsted S-Monovette system, there are no effects of EDTA contamination from an incorrect OFD, and they suggested that the ease of use of the phlebotomy venesection system or the experience of the phlebotomists are more important considerations. Likewise, a study concluded that collection for coagulation tests after serum collection using clot activators in collection tubes (silica particles and thrombin) showed minimal effects [43]. The recommended CLSI order of the draw for microcollection tubes to prevent microclot formation and platelet clumping is blood gases, EDTA tubes, and other additive tubes for plasma or whole blood and serum [44]. The combined effects of fluoride and oxalate may inhibit enzyme activity in some immunoassays or interfere with sodium, potassium, chloride, lactic acid, and alkaline dehydrogenase enzyme measurements. EDTA plasma yields potassium levels greater than 14 mmol/L and total calcium levels less than 0.1 mmol/L [29]. Manufacturers use different tube colors for easy visual identification. The universal colors are lavender for EDTA, blue for citrate, red for serum, green for heparin, and gray for fluoride.

Inadequate blood volume is a common cause of sample rejection in the laboratory. CLSI recommends that the draw volume be no more than 10% below the stated draw volume. The excess amount of additive to blood volume has the potential to adversely affect the accuracy of test results. Some studies have shown that underfilled coagulation tubes have excess citrate that can neutralize calcium in the reagent to give falsely prolonged PT and hence inaccurate INR [31]. However, for automated hematology, underfilled tubes containing powdered potassium EDTA had no effect on blood counts [45]. Inappropriately high heparin: blood ratios can cause prolonged clotting times and increased fibrin microclots. Gerhardt *et al.* [46] suggest that binding of heparin to troponins decreases immunoreactivity and hence lowers cardiac troponin levels by 15% compared to serum. In contrast, Daves *et al.* [47], using tubes from a different manufacturer, demonstrated no interference from heparin but, rather, from the separator gel in Terumo Venosafe tubes.

## COLLECTION SITES; DIFFERENCES BETWEEN ARTERIAL, CAPILLARY, AND VENOUS BLOOD SAMPLES; AND PROBLEMS WITH COLLECTIONS FROM CATHETERS AND INTRAVENOUS LINES

Although a wide range of specimens are occasionally analyzed in the clinically laboratory, the primary specimen is blood in one form or another. Blood is composed of two broad types of components—liquid (~55%) and cellular (~45%). The cellular components consist of erythrocytes, leukocytes, and platelets. While *in vitro*, the liquid component is plasma and consists primarily of water (~93%) and proteins (~7%) with dissolved water-soluble molecules and a smaller lipid-based fraction [48,49]. Exposure to various substances including glass and plastic initiates the coagulation cascade. Consumption of the clotting factors converts plasma to serum. In addition, many of the cellular components are bound by the clot [50]. The primary purpose of blood is transportation of nutrients, oxygen, and metabolic waste products through the body. Arteries are blood vessels that carry blood away from the heart so arterial blood, with the exception of that in the pulmonary artery, has uniform high oxygen content [49]. Similarly, because veins carry blood toward the heart, venous blood, except that in the pulmonary vein, has decreased oxygen and glucose as well as increased carbon dioxide ($CO_2$), lactic acid, ammonia, and acidity (lower pH). The exact composition of venous blood varies throughout the body and depends on the metabolic activity of the tissues that a specific vein drains [49].

Serum is generally used for most clinical laboratory analyses. To obtain serum, blood is allowed to clot for approximately 20 min, after which serum is separated from the clot by centrifugation. For some systems, plasma or whole blood is required or preferred. Blood collected by skin puncture, so-called capillary blood, is primarily a mixture of arterial blood and interstitial and intracellular fluids. Because arteriolar pressure is greater than that of capillaries and venules, arterial blood will predominate in these samples. Its content will vary with the relative proportion of these components [49,51]. Because the source of blood has no consequence to most analyses, venous blood is usually preferred because of its ease of collection [50]. Arterial blood is necessary only for the measurement of arterial blood gases ($pO_2$, $pCO_2$, and pH) in order to assess oxygenation status in critically ill patients and those with pulmonary disorders. Although venous blood may yield an adequate assessment of pH status, it does not accurately reflect the arterial $pO_2$ and alveolar $pCO_2$ status but, rather, that of the extremity from which it is drawn [49].

Laboratory analysis of blood specimens may be affected by the technique used to collect the sample. Multiple sites may be used for the collection of venous blood, but the sites chosen usually depend on age and condition of the individual, amount of blood needed, and the analyses to be performed.

## Skin Puncture

Skin puncture is the usual method for collection of blood from infants as well as some older pediatric patients (<2 years of age). In some specific situations, such as obesity, severe burns, thrombotic tendencies, and point-of-care testing, skin puncture may also be used on adults [49,51]. In infants, the heel is the primary site of collection, with the earlobe or finger often being used in older patients [49]. As mentioned previously, it is essentially mildly diluted arterial blood, but it will suffice for most analyses. Due to its similarity to arterial blood, in most patients it may be used to assess pH and $pCO_2$ regardless of site of collection, but for measurement of $pO_2$, only blood collected from the earlobe is recommended [49]. If used for blood gas measurement, care must be taken to minimize exposure to ambient air while collecting the specimen [48,51]. A few assays cannot be performed on these specimens, including erythrocyte sedimentation rate and coagulation studies and blood cultures [48].

## Venipuncture

The median cubital vein in the antecubital fossa is the most commonly used site due to its accessibility and size, followed by the neighboring cephalic and basilic veins [13,49,51,52]. Veins on the dorsal surface of the hand and wrist, radial aspect of the wrist, followed by dorsal and lateral aspects of the ankle are also used, but these should only be used if one can demonstrate good circulation [51,52]. Sites to be avoided include arms ipsilateral to a mastectomy; scarred skin and veins; fistulas; sites distal to intravenous (IV) lines; and edematous, obese, and bruised areas [13].

## Arterial Puncture

In order, the radial, brachial, and femoral arteries are the preferred sites for arterial puncture [13,49,51]. Sites to be avoided include those that are irritated, edematous, and inflamed or infected. Although skin puncture provides a similar specimen to arterial blood, for neonates, specimens for blood gas analysis are best collected from an umbilical artery catheter [52].

## Indwelling Catheters and Intravenous Lines

For single phlebotomy, it is generally better to avoid an area near an IV line [13]. A site in a different extremity or distal to the IV line is preferred. However, for patients periodically requiring numerous specimens, collecting blood through IV lines and indwelling catheters, including central venous lines or arterial catheters, offers the advantage of facilitating this without phlebotomy [49]. This allows staff without extensive phlebotomy skills to collect blood, thereby freeing experienced phlebotomists to concentrate on other patients. Unfortunately, this creates an inherent risk for improper specimen collection. In order to avoid contamination and dilution with IV fluid; it is recommended that the valve be closed for at least 3 min prior to specimen collection [52]. In order to clear the IV fluid from the line, approximately 6–10 mL should be withdrawn and discarded [49,52]. Because heparin is often in a line to maintain patency, larger volumes may need to be discarded for coagulation studies [49]. Blood drawn from these lines should not be cultured due to the high risk of contamination from bacteria growing in the line [49].

## Contamination

IV fluid is typically composed of water containing various electrolytes, glucose, and occasionally other substances. Therefore, contamination of a specimen with this fluid falsely elevates concentrations of these analytes, but at the same time, contamination causes dilution of the specimen. Thus, values of analytes that are not present in the IV fluid should be decreased [48]. Skin antisepsis is typically accomplished with isopropanol or iodine compounds [51,52]. Isopropanol is generally recommended. The site should be allowed to dry for 30–60 sec to minimize the risk of interference with alcohol assays. Iodine compounds have been noted to affect some assays and probably should be avoided for chemistry studies [51]. In particular, povidone iodine can falsely elevate potassium, phosphorous, and uric acid in specimens collected by skin puncture [51].

### Anticoagulants

For coagulation assays, the proper ratio of blood to citrate is critical. For this reason, the tube must be filled to the required volume. Excess air allowed to enter the tube will limit the quantity of blood, thereby perturbing this ratio [51].

### Tourniquet Effect

Application of a tourniquet to approximately 60 mmHg pressure causes anaerobic metabolism and thus may elevate the lactate and ammonia and lower the pH [52]. Tissue destruction may cause the release of intracellular components such as potassium and enzymes. Venous stasis due to prolonged tourniquet application (>3 min] may cause significant hemoconcentration with an 8–10% increase in several enzymes, proteins, protein-bound substances, and cellular components [51]. Prolongation of venous occlusion from 1 to 3 min was documented to increase total protein by 4.9%, iron by 6.7%, lipids by 4.7%, cholesterol by 5.1%, AST by 9.3%, and bilirubin by 8.4% and to decrease potassium by 6.2% [53]. In addition, stress, hyperventilation, and muscle contractions (e.g., repeated fist clenching) may elevate analytes such as glucose, cortisol, muscle enzymes, potassium, and free fatty acids. For these reasons, it is important to limit venous occlusion to less than 1 min if possible [52].

### Hemolysis

Hemolysis will elevate the concentration of any constituent of erythrocytes and may slightly dilute constituents present in low levels in erythrocytes. It becomes significant when the serum concentration of hemoglobin surpasses 20 mg/dL. Typically, hemolysis elevates aldolase, acid phosphatase, isocitrate dehydrogenase, lactate dehydrogenase, potassium, magnesium, ALT, hemoglobin, and phosphate [49,52]. AST is elevated slightly [52]. Colorimetric assays are also affected by the increased absorbance of light by Hgb in the 500- to 600-nm range [49]. The effects of hemolysis are further discussed in Chapter 5.

## URINE COLLECTION, TIMING, AND TECHNIQUES

The examination of urine for diagnostic purposes dates back at least to Roman times [54], when it formed one of the cornerstones of the ancient "diagnostic laboratory" (in essence, the practitioners' senses of sight, smell, and taste). Urinalysis remains one of the key diagnostic tests in the modern clinical laboratory, and as such, proper timing and collection techniques are important.

Urine is essentially an ultrafiltrate of blood, which is supplied to the kidneys at a rate of approximately 120–125 mL/min. As a result, approximately 1 or 2 L of urine is produced daily. The kidneys are responsible for many homeostatic functions, including acid–base balance, electrolyte balance, fluid balance, and the elimination of nitrogenous waste products. In addition to these regulatory functions, the kidneys produce the hormones erythropoietin and calcitriol [1,25(OH)$_2$ vitamin D$_3$] as well as the enzyme renin. Not surprisingly, the complexity of the kidneys makes them susceptible to a number of toxic, infectious, hypoxic, and autoimmune insults. Fortunately, there is a considerable amount of redundancy built into the renal apparatus such that individuals can function normally with only a single kidney.

Examination of urine may take several forms: microscopic, chemical (including immunochemical), and electrophoresis. Each of these may yield important diagnostic information about the health of the genitourinary system or about substances (e.g., drugs of abuse) that are found in the blood.

### Timing of Collection

Three different timings of collection are commonly encountered. The most common is the random or "spot" urine collection. This collection is performed at the patient's convenience and is generally acceptable for most routine urinalysis and culture. However, if it would not unduly delay diagnosis, the first voided urine in the morning is generally the best sample. This is because the first voided urine is generally the most concentrated and contains the highest concentration of sediment [49,55]. The third timing of collection is the 12- or 24-hr collection. This is the preferred technique for quantitative measurements such as creatinine, electrolytes, steroids, and total protein. The usefulness of these collections is limited, however, by poor patient compliance.

### Specimen Labeling

As with all laboratory specimens, proper labeling is essential. This should include labeling both the requisition and the specimen container with the patient's name and additional unique identifiers as required by the receiving laboratory, the ordering practitioner's name, and the date and time of collection.

### Clean Catch Specimen

For most urine testing, a clean catch specimen is optimal. The technique as described for males is to retract the foreskin (if present) and clean the glans with a mild antiseptic solution. With the foreskin still retracted, allow the bladder to partially empty. This will serve to flush bacteria and other material from the urethra. Finally, collect a resulting "midstream" sample for testing. The technique for females is similar: Separate the labia minora and clean the urethral meatus with mild soap followed by rinsing. Sterile cotton balls are commonly used for this purpose. Continue holding the labia minora apart and partially empty the bladder. Finally, collect a midstream sample for testing. Practically, the cleansing step is often poorly performed or skipped altogether. Indeed, even without prior cleansing, contamination rates in one study were not increased relative to a clean catch control group (23 and 29%, respectively) [56]. For reasons other than bacteriologic examination (e.g., pregnancy testing), the cleansing step may be omitted.

### Catheterization

In situations in which the patient cannot provide a clean catch specimen, catheterization represents another option. Due to the attendant risks (infection and mechanical damage), this technique is performed only when necessary and only by trained personnel. Attention must be paid to sterile technique and the selection of a properly sized catheter.

### Suprapubic Aspiration

More invasive still is the technique of suprapubic puncture of the bladder and needle aspiration of the urine sample. This technique has been advocated for infants and young children to confirm a positive test from an adhesive bag or container sample [57]. In the experience of the author, this is rarely done. A suprapubic aspiration sample may also be obtained in conjunction with the placement of a suprapubic catheter for the relief of urethral obstruction.

### Adhesive Bags

Urine collection from infants and young children prior to toilet training can be facilitated through the use of disposable plastic bags with adhesive surrounding the opening. When used in an outpatient setting, these are checked regularly by the parents, and the urine is transferred to a sterile container as soon as it is noticed in the bag.

### Specimen Handling, Containers, and Preservatives

For point-of-care urinalysis (e.g., urine dipstick and pregnancy testing), any clean and dry container is acceptable. Disposable sterile plastic cups and even clean waxed paper cups are often employed. If the sample is to be sent for culture, the specimen should be collected in a sterile container. For routine urinalysis and culture, the containers should not contain preservative. For specific analyses, some preservatives are acceptable. A list of these is given in Schumann and Friedman [57]. The exception to this is for timed collections in which hydrochloric acid, boric acid, or glacial acetic acid is used as a preservative.

Storage of the sample at room temperature is generally acceptable for up to 2 hr. After this time, the degradation of cellular and some chemical elements becomes a concern. Likewise, bacterial overgrowth of both pathologic and contaminating bacteria may occur with prolonged storage at room temperature. Therefore, if more than 2 hr will elapse between collections and testing of the urine specimen, it must be refrigerated. Storage in refrigeration for up to 12 hr is acceptable for urine samples destined for bacterial culture. Cellular degradation in cytologic urine specimens may be inhibited by refrigeration for up to 48 hr [57]. However, the optimal technique is timely mixture of the cytologic urine sample with equal parts of a preservative: 50–70% ethanol has traditionally been used for this purpose. With the advent of liquid-based cytology platforms, ethanol has been largely replaced by proprietary preservative solutions and collection containers designed for the specific liquid-based cytology platform used. Consultation with the laboratory processing the sample is necessary to ensure the proper preservative is used. In all cases, formalin is an unsuitable preservative for cytology specimens.

## CONCLUSIONS

Many pre-analytical variables affect clinical laboratory test results. Therefore, it is very important to be aware of such factors in order to investigate potentially erroneous test results. Because blood and urine are the two major specimens analyzed in clinical laboratories, this chapter focused on pre-analytical issues that affect laboratory test results. Oral fluid and hair specimens are used for testing for drugs of abuse, but 90% of testing for abused drugs is still conducted using urine specimens. There are commercially available devices for collecting oral fluids that are straightforward to use. Collection of hair specimen is also relatively simple. Cerebrospinal fluids are also tested for certain analytes, but usually clinicians collect such specimens.

## References


[1] Sherwood L. The blood. In: Sherwood L, editor. Human physiology: from cells to systems. Belmont, CA: Thomson Brooks/Cole; 2004.

[2] Moyer TP, Shaw LM. Therapeutic drugs and their management. In: Burtis CA, Ashwood ER, Bruns DE, editors. Tietz textbook of clinical chemistry and molecular diagnostics. 4th ed. St. Louis: Saunders; 2006.

[3] Shenkin A, Baines M, Fell GS, Lyon TDG. Vitamins and trace elements. In: Burtis CA, Ashwood ER, Bruns DE, editors. Tietz textbook of clinical chemistry and molecular diagnostics. 4th ed. St. Louis: Saunders; 2006.

[4] Vajpayee N, Graham S, Bem S. Basic examination of blood and bone marrow. In: McPherson R, Pincus M, editors. Henry's clinical diagnosis and management by laboratory methods. 21st ed. Philadelphia: Saunders; 2007.

[5] Nichols JH. Point-of-care testing. In: Clarke W, Dufour DR, editors. Contemporary practice in clinical chemistry. 2nd ed. Washington, DC: AACC Press; 2006.

[6] D'orazio P, Burnett RW, Fogh-Andersen N, Jacobs E, Kuwa K, Kulpmann WR, et al. Approved IFCC recommendation on reporting results for blood glucose: International Federation of Clinical Chemistry and Laboratory Medicine Scientific Division, Working Group on Selective Electrodes and Point-of-Care Testing (IFCC-SD-WG-SEPOCT). Clin Chem Lab Med 2006;44:1486–90.

[7] Fogh-Andersen N, D'orazio P. Proposal for standardizing direct-reading biosensors for blood glucose. Clin Chem 1998;44:655–9.

[8] Lyon ME, Lyon AW. Patient acuity exacerbates discrepancy between whole blood and plasma methods through error in molality to molarity conversion: "Mind the gap!". Clin Biochem 2011;44:412–17.

[9] Guder WG, Narayanan S, Wisser H, Zawta B. Plasma or serum? Differences to be considered. Diagnostic samples: from the patient to the laboratory. 4th ed. Weinheim: Wiley-VCH; 2009.

[10] Hallbach J, Hoffmann GE, Guder WG. Overestimation of albumin in heparinized plasma. Clin Chem 1991;37:566–8.

[11] Kricka LJ, Park JY, Senior MB, Fontanilla R. Processing controls in blood collection tubes reveals interference. Clin Chem 2005;51:2422–3.

[12] Bowen RA, Hortin GL, Csako G, Otañez OH, et al. Impact of blood collection devices on clinical chemistry assays. Clin Biochem 2010;43:24–5.

[13] Ernst DJ. Applied phlebotomy. Baltimore: Lippincott Williams & Wilkins; 2005.

[14] Selby C. Interference in immunoassay. Ann Clin Biochem 1999;36:704–21.

[15] Bowen RA, Chan Y, Cohen J, Rehak NN, et al. Effect of blood collection tubes on total triiodothyronine and other laboratory assays. Clin Chem 2005;51:424–33.

[16] Bowen RA, Chan Y, Ruddel ME, Hortin GL, et al. Immunoassay interference by a commonly used blood collection tube additive, the organosilicon surfactant Silwet L-720. Clin Chem 2005;51:1874–82.

[17] VS7336 BD White Paper: A Comparison of Adjusted BD Vacutainer SST Glass Tubes and Adjusted BD Vacutainer SST Plus Tubes in Various Configurations with BD Vacutainer Serum Glass Tubes for Cortisol, Total T3, Total T4, and TSH on the DPC Immulite Analyzer. <http://www.bd.com/vacutainer/pdfs/techbulletins/Assay_Interference_Global_23september2004_VS7313.pdf>; (accessed March 2012).

[18] Stankovic AK, Parmar G. Assay interferences from blood collection tubes: a cautionary note. Clin Chem 2006;52:1627–8.

[19] Rodushkin I, Odman F. Assessment of the contamination from devices used for sampling and storage of whole blood and serum for element analysis. J Trace Elem Med Biol 2001;14:40–5.

[20] Shah VP, Knapp G, Skelly JP, Cabana BE. Interference with measurements of certain drugs in plasma by a plasticizer in Vacutainer tubes. Clin Chem 1982;28:2327–8.

[21] Drake SK, Bowen RA, Remaley AT, Hortin GL. Potential interferences from blood collection tubes in mass spectrometric analyses of serum polypeptides. Clin Chem 2004;50:2398–401.

[22] Boyanton Jr BL, Blick KE. Stability studies of twenty-four analytes in human plasma and serum. Clin Chem 2002;48:2242–7.

[23] Dasgupta A, Dean R, Saldana S, Kinnaman G, et al. Absorption of therapeutic drugs by barrier gels in serum separator blood collection tubes. Volume–and time-dependent reduction in total and free drug concentrations. Am J Clin Pathol 1994;101:456–61.

[24] Bush VJ, Janu MR, Bathur F, Wells A, et al. Comparison of BD vacutainer SST Plus tubes with BD SST II Plus tubes for common analytes. Clin Chim Acta 2001;306:139–43.

[25] Schouwers S, Brandt I, Willemse J, Van Regenmortel N, et al. Influence of separator gel in Sarstedt S-Monovette® serum tubes on various therapeutic drugs, hormones, and proteins. Clin Chim Acta 2012;413:100–4.

[26] International Council for Standardization in Haematology (ICSH). Recommendations for single-use evacuated containers for blood specimen collection for hematological analyses. Lab Hematol 2002;8:1–4.

[27] Clinical and Laboratory Standards Institute. Approved Standard,-Fifth Edition H1-A5 Tubes and additives for venous blood specimen collection. Wayne, PA: Clinical and Laboratory Standards Institute; 2003.

[28] Banfi G, Salvagno GL, Lippi G. The role of ethylenediamine tetraacetic acid (EDTA) as in vitro anticoagulant for diagnostic purposes. Clin Chem Lab Med 2007;45:565–76.

[29] Lippi G, Banfi G, Buttarello M, Ceriotti F, et al. Recommendations for detection and management of unsuitable samples in clinical laboratories. Clin Chem Lab Med 2007;45:728–36.

[30] Green D, McMahon B, Foiles N, Tian L. Measurement of hemostatic factors in EDTA plasma. Am J Clin Pathol 2008;130:811–15.

[31] Chuang J, Sadler MA, Witt DM. Impact of evacuated collection tube fill volume and mixing on routine coagulation testing using 2.5-ml (pediatric) tubes. Chest 2004;126:1262–6.

[32] O'Keane MP, Cunningham SK. Evaluation of three different specimen types (serum, plasma lithium heparin and serum gel separator) for analysis of certain analytes: clinical significance of differences in results and efficiency in use. Clin Chem Lab Med 2006;44:662–8.

[33] NCCLS: procedures for the collection of Arterial Blood Specimens; approved standard, Fourth Edition H11-A4. NCCLS, Wayne, PA: 2004.

[34] Brandhorst G, Engelmayer J, Götze S, Oellerich M, et al. Pre-analytical effects of different lithium heparin plasma separation

tubes in the routine clinical chemistry laboratory. Clin Chem Lab Med 2011;49:1473–7.

[35] Preanalytical Systems Library of Clinical White Papers. <http://paswhitepapers.bd.com>; (accessed March 2012).

[36] Menssen HD, Brandt N, Leben R, Müller F, et al. Measurement of hematological, clinical chemistry, and infection parameters from hirudinized blood collected in universal blood sampling tubes. Semin Thromb Hemost 2001;27:349–56.

[37] Dimeski G, Masci PP, Trabi M, Lavin MF, et al. Evaluation of the Becton-Dickinson rapid serum tube: does it provide a suitable alternative to lithium heparin plasma tubes?. Clin Chem Lab Med 2010;48:651–7.

[38] Strathmann FG, Ka MM, Rainey PM, Baird GS. Use of the BD Vacutainer rapid serum tube reduces false-positive results for selected Beckman coulter Unicel DxI immunoassays. Am J Clin Pathol 2011;136:325–9.

[39] <http://www.calgarylabservices.com/files/HealthcareProfessionals/Specimen_Collection/BloodCollectionTubes.pdf>; (November 2012).

[40] Calam RR, Cooper MH. Recommended "order of draw" for collecting blood specimens into additive-containing tubes. Clin Chem 1982;28:1399.

[41] Cornes MP, Ford C, Gama R. Spurious hyperkalemia due to EDTA contamination: common and not always easy to identify. Ann Clin Biochem 2008;45:601–3.

[42] Sulaiman RA, Cornes MP, Whitehead SJ, Othonos N, et al. Effect of order of draw of blood samples during phlebotomy on routine biochemistry results. J Clin Pathol 2011;64:1019–20.

[43] Fukugawa Y, Ohnishi H, Ishii T, Tanouchi A, et al. Effect of carryover of clot activators on coagulation tests during phlebotomy. Am J Clin Pathol 2012;137:900–3.

[44] Clinical laboratory standards institute: procedures for the collection of diagnostic blood specimens by venipuncture; approved standard, Sixth Edition H3-A6; Wayne, PA: Clinical Laboratory Standards Institute; 2007.

[45] Xu M, Robbe VA, Jack RM, Rutledge JC. Under-filled blood collection tubes containing K2EDTA as anticoagulant are acceptable for automated complete blood counts, white blood cell differential, and reticulocyte count. Int J Lab Hematol 2010;32:491–7.

[46] Gerhardt W, Nordin G, Herbert AK, Burzell BL, et al. Troponin T and I assays show decreased concentrations in heparin plasma compared with serum: lower recoveries in early than in late phases of myocardial injury. Clin Chem 2000;46:817–21.

[47] Daves M, Trevisan D, Cemin R. Different collection tubes in cardiac biomarkers detection. J Clin Lab Anal 2008;22:391–4.

[48] McCall RE, Tankersley CM. Phlebotomy essentials. 3rd ed. Baltimore: Lippincott Williams & Wilkins; 2005.

[49] Henry JB. Clinical diagnosis and management by laboratory methods. 20th ed. Philadelphia: Saunders; 2001.

[50] Pickard NA. Chapter 2. Collection and handling of patient specimens. In: Kaplan LA, Pesce AJ, editors. Clinical chemistry: theory, analysis and correlation. 2nd ed. St. Louis, MO: Mosby; 1989.

[51] Garza D, Becan-McBride K. Phlebotomy handbook: blood collection essentials. 7th ed. Upper Saddle River, NJ: Pearson; 2005.

[52] Young DS, Bernes EW, Haverstick DM. Chapter 2. Specimen collection and processing. In: Burtis CA, Ashwood ER, Bruns DE, editors. Tietz textbook of clinical chemistry and molecular diagnostics. 4th ed. St. Louis, MO: Saunders; 2006.

[53] Statland BE, Bokelund H, Winkel P. Factors contributing to intraindividual variation of serum constituents: 4. Effects of posture and tourniquet application on variation of serum constituents in healthy subjects. Clin Chem 1974;20:1513–19.

[54] Iorio L, Avagliano F. Observations on the liber medicine orinalibus by hermogenes. Am J Nephrol 1999;19:185–8.

[55] Kark RM, Lawrence JR, Pollack VE, et al. A primer of urinalysis. 2nd ed. New York: Harper & Row; 1963.

[56] Lifshitz E, Kramer L. Outpatient urine culture: does collection technique matter? Arch Intern Med 2000;160:2537–40.

[57] Schumann GB, Friedman SK. Wet urinalysis: interpretations, correlations and implications. Chicago: ASCP Press; 2003.

CHAPTER

# 4

# Sample Processing and Specimen Misidentification Issues

*Alison Woodworth*[*], *Amy L. Pyle*[†]

[*]Vanderbilt University Medical Center, Nashville, Tennessee

[†]Nationwide Children's Hospital, Columbus, Ohio

## INTRODUCTION

Every step of handling and processing laboratory specimens should be closely monitored to ensure ideal conditions for testing and producing meaningful test results. Like all areas of health care, the laboratory is subject to error, especially in the pre-analytical phase of testing, which includes specimen collection, labeling, transportation, storage, and processing [1]. The 2000 Institute of Medicine report titled "To Err Is Human: Building a Safer Health System" estimated that an alarming 44,000–98,000 deaths occur in the United States each year as the result of medical errors [2]. Although most of those deaths were due to medication errors, laboratory errors do account for some mortality, either directly or indirectly [3,4]. Advances in information systems, automation, and instrumentation have vastly reduced the analytical error rate; however, there is still much room for improvement, particularly in pre-analytical processes, which are the source of the majority of laboratory errors [2]. Laboratory errors can be difficult to quantify due to poor reporting and recognition of events. Studies of laboratory errors show error rates from one error in every 33–50 events in one laboratory to as low as one error in every 283 laboratory results [5]. Mistakes occurring prior to specimen analysis accounted for the majority of laboratory errors. Estimates of pre-analytical error rates among all laboratory errors range from 31.6% to as high as 84.5%, with more than 90% of those due to mistakes originating in patient care units (e.g., phlebotomy; see Chapter 3) [5,6]. This chapter reviews errors in sample processing and identification, from immediately after sample collection to just before testing, and examines strategies to prevent such errors.

## TRANSPORTATION

All specimens must travel from the collection site to processing and/or testing sites. This ranges from short trips down the hall to long cross-country trips to reference laboratories. No matter the nature of transportation, laboratory staff must be cognizant of time, temperature, and turbulence, which can all influence specimen integrity.

### Transportation Time

Many obstacles prevent timely transportation from collection site to laboratory. Frequently, samples are collected at remote off-site locations and transported to a main laboratory for processing. Specimens can get lost or misplaced for hours. Tubes can get stuck or delayed in pneumatic tube systems. Couriers may get stuck in traffic while samples sit in the car. A nurse or phlebotomist may slip a tube into his or her pocket. No matter the reason, sample processing is often delayed, and this can be a source of pre-analytical error. Many compounds are subject to variation with time, especially prior to separation of serum/plasma from cells. Every reasonable effort should be made to reduce transport time between drawing and processing the sample.

Certain analytes are particularly susceptible to change with time, particularly if a sample is exposed to cellular components (Table 4.1). Blood cells retain *in vitro* metabolic activity, especially when maintained at room temperature (RT) or higher, which can alter the chemical constituents of a sample. In uncentrifuged,

*Accurate Results in the Clinical Laboratory.*
DOI: http://dx.doi.org/10.1016/B978-0-12-415783-5.00004-9

whole blood samples kept at RT, glucose concentrations fall at 5–7% per hour due to ongoing glycolysis by the blood cells [7–9]. Even in samples that have been centrifuged, but not separated, glucose in serum or plasma will continue to drop, falling outside a clinically significant change range in less than 4 hr [10]. Likewise, lactate increases concurrently with the fall in glucose, as pyruvate is reduced to lactate during glycolysis [11]. Samples with bacterial growth or leukocytosis undergo more glycolysis, rapidly decreasing glucose and elevating lactate [11–13]. Care should be taken to avoid prolonged time to processing, especially in patients with bacteremia and/or leukocytosis, because artifactual hypoglycemia may otherwise be misinterpreted and prompt an unnecessary workup for hypoglycemia [14,15]. The same phenomenon is well-known in another setting: Low cerebrospinal fluid (CSF) glucose (hypoglycorrhachia) is a hallmark of meningitis with bacteria and white cells in the CSF [16].

Elevations in intracellular erythrocyte components can occur via transmembrane diffusion of cellular constituents when serum or plasma is maintained on the clot or cells. Phosphate is approximately seven times more concentrated in red cells than plasma, and it leaks from red cells into plasma with extended storage time [13]. The same is true for potassium, although the intracellular gradient may be maintained if the cells are kept at RT and can consume glucose to generate the adenosine triphosphate (ATP) required to feed the $Na/K^{+}$-ATPase [13]. Chloride and carbon dioxide concentrations fall over time, likely due to the chloride bicarbonate shift into red cells (see Chapter 8) [10]. Other components that are concentrated inside cells, such as lactate dehydrogenase (LDH) and aspartate aminotransferase (AST), may leak out; however, as long as the red cells remain intact, a significant rise in enzyme concentration is not observed [10,13]. Although infrequently caused during handling and transportation, hemolysis can also significantly affect laboratory tests, such as potassium, AST, and LDH [17]. See Chapter 5 for more information on hemolysis.

In addition to depletion via metabolism, many analytes are simply unstable *in vivo* and *in vitro*, and they remain intact for a relatively short time after specimen collection. Often, these are short peptide hormones, such as adrenocorticotropic hormone (ACTH) and brain natriuretic peptide, which are rapidly degraded [18–20]. Other hormones, such as insulin and parathyroid hormone, are subject to degradation, although at a slower rate [13,21]. Finally, insignificant increases may be observed in several analytes, such as sodium, due to hemoconcentration, as water moves into the cells as the samples stand [10].

The Clinical and Laboratory Standards Institute (CLSI) guidelines for specimen handling and processing recommend separating plasma or serum from the cells within 2 hr of sample collection [9,22]. Most analytes are stable for longer than 2 hr (Table 4.1), so rejection of all specimens received 2 hr or more after draw is unnecessary. If laboratories were to strictly follow guidelines, specimen processing stations, including centrifuges, should be placed at every location where blood is drawn so samples could be immediately processed and serum or plasma mechanically separated from the cells. This is not feasible in most large medical centers and clinical practices. Therefore, laboratories need sound strategies for identifying specimens of poor integrity and policies for accepting and rejecting those samples. Specimen integrity must be maintained from draw to analysis. Tactics to optimize specimen integrity include implementation of strategies to reduce transport time, use of appropriate tubes, and development of strict guidelines for specimen storage conditions during transport. Reducing transit time preserves sample stability and shortens turnaround time. For inpatient specimens, robotic couriers and pneumatic tube systems can cut down on staffing and decrease transport time to the laboratory. These solutions may provide an alternative to opening a satellite lab in far sites from a medical center. One study estimated that robotic couriers could decrease turnaround time by 34% in a 591-bed hospital [23]. Pneumatic tube systems can send carriers containing laboratory specimens, paperwork, pharmaceuticals, and more throughout a hospital at high speeds. Thus, pneumatic tube systems are in wide use in medical centers throughout the world [24–26].

Gel separator tubes were introduced to provide a single, closed system to draw, process, test, and store samples (see Chapter 3 for a discussion about sample containers). The tubes contain a thixotropic gel, which has a density intermediate between plasma or serum and cells. Upon centrifugation, the gel moves to the interface between the liquid and cellular components of the sample [27]. Use of serum and plasma separator tubes extends the stability of most chemistry analytes [22,28–30]. However, many drugs, such as phenytoin, phenobarbital, carbamazepine, lidocaine, and tricyclic antidepressants, absorb into the gel, so these tubes should not be used for therapeutic drug monitoring [22,31,32]. Gel separator tubes offer a practical option for rapidly separating plasma or serum from cells, and they slow many of the time-dependent processes that reduce specimen integrity [33]. These tubes have the additional advantage of helping to prevent labeling errors. Because samples can be drawn, tested, and stored all in the same tube, there is less need to aliquot and re-label, reducing labeling and misidentification errors.

When highly accurate glucose results are necessary, such as for glucose tolerance tests, samples should always be drawn into tubes containing sodium

**TABLE 4.1** Stability of Common Chemistry and Immunochemistry Analytes with Varying Time, Temperature, and Tube Types[a]

| Analyte | Serum | Heparin Plasma | EDTA Plasma | Urine | <24 hr | 24 hr | 48–72 hr | ≥14 days | References |
|---|---|---|---|---|---|---|---|---|---|
| ACTH | | | X | | 4°C, RT[b] | | | | [119] |
| α-Fetoprotein | X | | | | | | | 4°C, RT | [120] |
| Albumin | X | X | X | | | | 4°C, RT | | [10,13,121–123] |
| Aldosterone | | | X | | | 4°C, RT | | | [119] |
| Alkaline phosphatase | X | X | | | | | 4°C, RT | | [10,121,122] |
| ALT | X | X | X | | | | 4°C, RT | | [10,122–124] |
| Amylase | X | | | | | | 4°C, RT | | [13,121,122] |
| Apo A1 and B | | | X | | | | RT | | [123] |
| AST | X | X | | | | | 4°C, RT | | [10,121,122,124] |
| Bilirubin, total | X | | | | | | 4°C, RT | | [122] |
| BNP | | | X | | | RT | 4°C | −20°C | [30,125,126] |
| BUN | X | X | | | RT[c] | | 4°C, RT | | [122,124] |
| Calcium | X | X | | | | | 4°C, RT | | [10,122] |
| Catecholamines | | | | X | | | | 4°C, −20°C | [127] |
| Cholesterol, total | X | X | X | | | | RT | | [10,30,121,123] |
| hCG | X | | | | | | | 4°C, RT | [120] |
| Creatine kinase | X | | X | | | | 4°C, RT | | [121,123] |
| Carbon dioxide | X | X | | | | RT | | | [10,30] |
| Creatinine | X | X | X | | | RT[c] | 4°C, RT | | [30,122,123] |
| | | | | X | | | | 4°C, −20°C | [128] |
| Cortisol | X | X | | | | RT[c] | RT | | [33,121] |
| C-reactive protein | | | X | | | | 4°C, RT | | [129] |
| Estradiol | | | X | | | 4°C, RT | | | [119] |
| Estriol, unconjugated | X | | | | | | | 4°C, RT | [120] |
| Ferritin | X | X | | | | RT | | | [33] |
| GGT | X | X | | | | | 4°C, RT | | [10,121,122] |
| Growth hormone | | | X | | | 4°C, RT | | | [119] |
| Glucagon | | | X | | | 4°C, RT | | | [119] |
| Glucose | X | X | | | 4°C, RT[c] | 4°C | | | [10,122,130] |
| Hemoglobin A1c | | | X | | | | RT | | [131] |
| HDL | X | X | | | | RT | | | [10,30] |
| Homocysteine | | X | | | | | 4°C | | [132] |

(*Continued*)

TABLE 4.1 (Continued)

| Analyte | Serum | Heparin Plasma | EDTA Plasma | Urine | <24 hr | 24 hr | 48–72 hr | ≥14 days | References |
|---|---|---|---|---|---|---|---|---|---|
| Lactate | X | X | | | RT | | | | [10] |
| LDH | X | X | | | RT | 4°C | | | [10,122] |
| Microalbumin | | | | X | | | | 4°C, −20°C | [133] |
| Metanephrines | | | | X | | | RT | | [127] |
| Phosphorus | X | X | | | | RT | 4°C | | [10,122] |
| Potassium | X | X | | | 4°C | RT[c] | RT | | [30] |
| Protein, total | X | X | X | | | | 4°C, RT | | [10,122,123] |
| Sodium | X | X | | | 4°C | | RT | | [10,30,122] |
| Triglycerides | X | X | | | | | RT | | [10,30,121,122] |
| TSH | X | X | | | | RT | | | [33] |
| Free T4 | X | X | | | | RT | | | [33] |
| Uric acid | X | X | | | | | 4°C, RT | | [10,122] |
| Vitamin $B_{12}$ | X | X | | | | RT | | | [33] |

[a]*This list is not exhaustive, but it shows the results from several studies using different times, temperatures, and specimen types. Most data shown represent the ending time point for the respective studies and not necessarily the time at which the analyte stability fails.*
[b]*Room temperature (21–25°C).*
[c]*Denotes different stability in unseparated samples.*
ACTH, adrenocorticotropic hormone; ALT, alanine aminotransferase; AST, aspartate aminotransferase; BNP, brain natriuretic peptide; BUN, blood urea nitrogen; GGT, γ-glutamyl transferase; hCG, human chorionic gonadotropin; HDL, high-density lipoprotein; LDH, lactate dehydrogenase; TSH, thyroid-stimulating hormone.

fluoride and potassium oxalate ("gray-top tubes") [34]. Fluoride and oxalate inhibit glycolysis, preventing artifactually low glucose results [35]. Prevention of glycolysis ensures that lactate concentrations also remain stable, making the gray-top tube the preferred tube type for lactate samples. It takes up to 2 hr for fluoride and oxalate to completely inhibit glycolysis, so some glucose loss can still occur [36]. After 2 hr in fluoride- and oxalate-containing blood collection tubes, glucose is stable for 72 hr [35]. Because the antiglycolytics come as salts with sodium and potassium, these tubes are not suitable for electrolyte determinations; they may also interfere with certain enzymatic assays, such as the urease method for blood urea nitrogen (BUN) [22]. Regardless of the benefit of glycolysis inhibition, immediate separation of plasma or serum from the cells remains the best means for stabilizing glucose [34]. It is important to ensure proper procedure to maintain specimen integrity, especially during long periods in transit, for preventing pre-analytical errors and providing optimal test results.

## Effects of Temperature

Maintenance of a well-controlled transport environment is essential to reduce pre-analytical error. Manipulation of transport temperature may increase analyte stability. Lower temperatures generally enhance analyte stability, but care must be taken because one temperature does not fit all analytes. Extreme heat denatures proteins and can diminish enzyme activity [37]. Lower temperatures inhibit metabolic processes such as glycolysis. Thus, most analytes are more stable at 4°C than RT (Table 4.1). Some analytes must be chilled because they rapidly degrade at RT, including ammonia, lactate, pyruvate, parathyroid hormone-related protein, and gastrin. These specimens should be chilled immediately after collection, and this temperature should be maintained throughout the pre-analytical phase [22,38].

Generally, most analytes are more stable at lower temperatures; however, there are several notable exceptions. Cold inhibits glycolysis, which provides the ATP required for cell surface $Na^{+}/K^{+}$ pumps. Without ATP, intracellular potassium accumulates and begins to leak out of cells over time. Thus, extracellular potassium may be artificially elevated in specimens stored in the cold for longer than 6 hr [10,22]. Likewise, if a sample is maintained at RT, glucose is consumed to sustain glycolysis. This process maintains the appropriate potassium concentration but falsely decreases glucose [13]. This makes transporting specimens collected for a basic metabolic panel difficult, but the problem is solved by separating the plasma from the cells before transporting.

Testing is often intentionally delayed—for example, for batch analyses or to send specimens to reference laboratories—and specimens may be frozen to preserve sample integrity until they are tested. Some enzymes are sensitive to freezing temperatures. When stored in liquid nitrogen, AST, alanine aminotransferase, alkaline phosphates, γ-glutamyl transferase, and LDH remain stable, whereas amylase increases and creatinine kinase (CK) activity decreases [39]. CK activity decreases significantly when frozen to −20°C for even short periods [40]. Certain analytes, such as cryoglobulins, must be maintained at body temperature and require transit in a warm water bath, kept at approximately 37°C [41].

CLSI guidelines recommend using an ice bath to rapidly and effectively cool a specimen. Using a mixture of ice and water will ensure good contact between the tube and the cooling medium [22]. Care should be taken to prevent direct contact of the sample with ice, especially dry ice, because hemolysis can occur with temperature extremes [22]. When transporting samples outside, special considerations should be made for the weather. For example, on very hot days, samples should be sealed in a plastic bag and then immersed in ice slurry to provide better cooling than ice packs alone. On the other hand, in very cold climates, coolers may not require additional ice packs or may need extra insulation to keep the sample from freezing [42,43].

Humidity is also a consideration, particularly for samples that will be exposed to outside air, such as dried blood spots. High humidity speeds the degradation of some analytes in dried blood spots, including biotinidase, galactose-1-phosphate uridyltransferase, glucose-6-phosphate dehydrogenase and thyroxine [44–46]. High humidity may also cause inaccurate results on self-monitoring blood glucose meters [22]. Humidity can be prevented by transporting samples in plastic bags with desiccant packets, and it can be monitored with humidity indicator cards [47,48].

## Effects of Handling and Turbulence

Moving a specimen from the draw site to the laboratory for processing may involve one or more forms of transportation. Outreach specimens are often brought to the main laboratory by courier; although not perfect, courier services are the preferred method for transporting samples from remote locations [49]. Compared to mailing specimens, couriers are faster and can provide better control of the sample environment, both temperature and jostling [49]. Inpatient samples, however, are frequently delivered to the laboratory via pneumatic tube system or robots. Robotic couriers replace full-time employees and transport samples efficiently and safely [23]. Pneumatic tube systems are widely used because they swiftly transport samples within hospitals, thereby reducing turnaround time [50,51]. Pneumatic tube systems can move at speeds upwards of 7.5 m/sec (that is more than 17 mph) and often go through rapid accelerations, sharp corners, and abrupt decelerations. The biomechanical forces experienced by the samples in pneumatic tubes can affect the quality of transported specimens. There is an increased risk of container breakage or leakage, particularly when they are glass and/or difficult to close (e.g., urine collection cups). The mechanical forces imposed by a pneumatic tube system can also directly damage blood cells, leading to hemolysis. This is particularly true of specimens from patients with hematologic disorders, on chemotherapy, and those with other conditions that cause red and white cell fragility [24,51].

LDH is a good marker for evaluating exposure to turbulence because it is quickly released from traumatized red cells. LDH activity may rise in a sample, even without significant hemolysis [26]. Along with elevating LDH, pneumatic tube systems induce elevations in plasma hemoglobin, AST, and potassium (Table 4.2) [10,24–26]. Full tubes are less subject to damage by agitation compared to partially filled tubes [25]. In particular, pneumatic tube-induced changes in LDH ranged from −1.0 to 13.9% for full tubes compared to 8.6–30.7% for half-filled tubes [13].

Considerations must be made for the proper use of pneumatic tube systems to prevent spills and specimen damage. Samples should be placed in sealed plastic bags prior to transport. Chilled specimens must be double sealed to prevent ice or water leakage. Tubes should not be placed directly into the ice bag because the water may cause labels to fall off and may leak into the tubes. Finally, container seals should be evaluated prior to transport. Poorly sealed containers should not be sent through pneumatic tube systems. Because a risk of spilling remains, some institutions choose to restrict types of specimens sent in pneumatic tubes. For example, specimens soaking in formalin may not be permitted in the carriers.

**TABLE 4.2** Analytes Affected by Agitation with Transportation

| Analyte | Effect of Transportation | References |
|---|---|---|
| Lactate dehydrogenase | ++ | [24,25] |
| Potassium | + | [24–26] |
| AST | + | [24] |
| Phosphorus | + | [24] |
| Plasma hemoglobin | + | [25] |

Specimen damage due to turbulence in a pneumatic tube system can be prevented in numerous ways. First, when new systems are designed, care should be taken to avoid rapid acceleration and deceleration [24,51]. Second, tubes should be completely filled [13]. Third, samples should be well packed into pneumatic tubes prior to transporting. Towels, foam pads, bubble wrap, and other means have been used to wrap specimens to prevent jostling. In a small quality improvement study, five conditions were examined: control (remained in laboratory with no transport), no padding, or wrapped in bubble wrap, a towel, or a sheet of foam ($n = 5$, each condition). Also, LDH was measured for all the samples. LDH increased significantly for samples transported without any padding as well as for those with either bubble wrap or foam in the carrier. Only when the specimens were tightly wrapped in a towel was there only a minimal increase in LDH [52].

*CASE REPORT* A glomerular filtration rate study was performed in which the patient was administered technetium 99 m (6-hr half-life), and blood was sent via pneumatic tube system at 30-min intervals to the laboratory to measure the circulating radioactivity using a gamma counter. The laboratory was notified in advance that the procedure would be done and to prepare for the testing. When no samples arrived 1 hr after expected, the technologist called the floor to inquire about the samples. He was told the samples were sent via pneumatic tube to the laboratory; however, they were never received. An investigation was launched to find the missing pneumatic tube carrier, including searching for carriers trapped in the system. A runner was sent to each tube station in the hospital, and eventually the missing carrier was found in a clinic that was closed at the time the carrier arrived. Unfortunately, the samples were found too late to be used, and the procedure had to be repeated. This case is an example of a pitfall of pneumatic tube systems and human error: Sometimes carriers get sent to the wrong tube stations. Risks such as this may be avoided by restricting which stations carriers can be sent between as well as clearly posting station numbers at each tube station.

## Shipping to Reference Labs

Shipping samples is inevitable, especially for specialized tests that are not performed in most clinical laboratories. Therefore, systems should be in place for sending specimens to distant reference labs. As with transporting samples from nearby locations, time, temperature, and handling must be considered when shipping samples over greater distances. In most cases, plasma or serum samples should be separated from cells and aliquot into a separate tube rather than shipping the primary tube. Certain tests may have special specimen requirements. Always consult the reference laboratory's test directory for appropriate specimen type and storage conditions prior to collection of send-out samples. Specimens may also be placed into a secondary container or packing to reduce turbulence. Generally, samples are sent either frozen or refrigerated in a Styrofoam container, with walls at least 1-in. thick, to ensure adequate insulation. Refrigerated specimens should be sent with sufficient frozen packs to keep the interior of the container between 0° and 10°C [53]. Frozen specimens should be sent with dry ice. One solid chunk of dry ice ($1 \times 3 \times 4$ in.) should be enough to keep a sample frozen for 48 hr [38]. Staff must be properly trained for the shipping of biological specimens and dry ice. Biological specimens should be treated as infectious agents and therefore are subject to specific laws and regulations; dry ice is considered a hazardous material to ship and thus requires special considerations. Overnight or next-day shipping reduces transit time and preserves specimen integrity.

## Special Case: Blood Gases and Ionized Calcium

Specimens collected for blood gas determinations require special care because the analytes are very sensitive to time, temperature, and handling. In standing whole blood samples, pH falls at a rate of 0.04–0.08/hr at 37°C, 0.02–0.03/hr at 22°C, and <0.01/hr at 4°C. This drop in pH is concordant with decreased glucose and increased lactate. In addition, $pCO_2$ increases approximately 5.0 mmHg/hr at 37°C, 1.0 mmHg/hr at 22°C, and 0.5 mmHg/hr at 4°C. At 37°C, $pO_2$ decreases by 5–10 mmHg/hr, but it decreases by only 2 mmHg/hr at 22°C. Metabolic activity in the specimen is affected by temperature as well as the number of metabolically active cells present: Specimens with leukocytosis and/or thrombocytosis may demonstrate a spurious hypoxemia because leukocytes and platelets consume oxygen *in vitro* [54,55]. $pO_2$ loss is best prevented by immediate analysis because even rapid cooling may not sufficiently reduce metabolism [55,56].

Ideally, all blood gas specimens should be measured immediately and never stored [56]. A plastic syringe, transported at RT, is recommended if analysis will occur within 30 min of collection. If testing is delayed for more than 30 min, specimens should be collected in a glass syringe and immediately immersed and kept in a mixture of water and crushed ice to chill the specimens [57]. Plastic contracts with cooling, making pores large enough for atmospheric oxygen to cross into the tube, but not carbon dioxide [58]. Glass syringes are recommended for delayed analysis because glass does not allow the diffusion of oxygen or carbon dioxide [58–61].

Blood gas analyzers reheat samples to 37°C for analysis to recapitulate physiological temperature. However, for patients with abnormal body temperature, either hyperthermia due to fever or induced hypothermia in patients undergoing cardiopulmonary bypass, a temperature correction should be made to determine accurate pH, $pO_2$, and $pCO_2$ results [56,62]. This adjustment is particularly important in hypothermia, in which the temperature change has a marked impact on pH, $pO_2$, and $pCO_2$ [56]. Blood gas instrument manufacturers use various but similar equations. Equations 4.1–4.3 are recommended by CLSI for temperature corrections [56,62]:

$$\mathrm{pH} = \mathrm{pH_m} - [0.0147 + 0.0065 \times (\mathrm{pH_m} - 7.40)](T - 37^\circ) \tag{4.1}$$

where pH is patient's adjusted pH, $T$ is patient's temperature, and $\mathrm{pH_m}$ is measured pH.

$$p\mathrm{CO_2} = p\mathrm{CO_{2m}} \times 10^{0.019(T-37^\circ)} \tag{4.2}$$

where $pCO_2$ is patient's adjusted $pCO_2$, $T$ is patient's temperature, and $pCO_{2m}$ is measured $pCO_2$.

$$pO_2 = pO_{2m} \times 10^{\left[\frac{(5.49\times10^{-11}pO_2^{3.88})+0.071}{(9.72\times10^{-9}pO_2^{3.88})+2.30}\right](T-37^\circ)} \tag{4.3}$$

where $pO_2$ is patient's adjusted $pO_2$, $T$ is patient's temperature, and $pO_{2m}$ is measured $pO_2$.

Blood gas specimen exposure to air should be minimized because gases in the sample will rapidly equilibrate with atmospheric air. Anaerobic technique should be used to draw all blood gas samples. However, even with the most careful collection, air bubbles can arise from the syringe hub dead space. Bubbles must be completely expelled from the specimen prior to transport because the $pO_2$ will be significantly increased and $pCO_2$ decreased within 2 min [63].

Rapid turnaround of blood gas results is important for patient care and requires swift specimen transport to the laboratory. However, blood gas specimens are particularly sensitive to handling and transport. Transit through a pneumatic tube system has been shown to significantly change $pO_2$, especially when there are air bubbles in the sample, although only very small changes have been noted for $pCO_2$ and pH [64]. Therefore, it is recommended that if specimens must be transported by pneumatic tube, all air bubbles must be completely removed prior to transport [62].

Ionized calcium is often measured by ion-sensitive electrodes in blood gas analyzers. Ionized calcium is inversely related to pH: Decreasing pH decreases albumin binding to calcium, thereby increasing free, ionized calcium. Therefore, specimens sent to the lab for ionized calcium determinations should be handled with the same caution as other blood gas samples because preanalytical errors in pH will impact ionized calcium results [65–67]. Additional information on blood gases and ionized calcium is discussed in Chapter 8.

## THE EFFECTS OF CENTRIFUGATION ON LABORATORY RESULTS

Due to instability of certain analytes in unprocessed serum or plasma, CLSI recommends that plasma or serum be separated from cells as soon as possible but definitely within 2 hr of collection [22]. Centrifugation is an integral part of specimen processing. Improper centrifugation techniques may lead to erroneous results for several laboratory tests.

Appropriate preparation of specimens prior to centrifugation is required to ensure accurate laboratory results. Serum specimens must be allowed ample time to clot prior to centrifugation. Tubes with clot activators require sufficient mixing and at least 30 min of clotting time, whereas serum tubes may require up to 60 min. Patients taking anticoagulants may require longer clotting times [22,68–70]. To ensure efficient release of additive/anticoagulant, plasma specimens must be mixed gently according to manufacturers' instructions [69,70]. Insufficient mixing leads to inefficient clotting in serum tubes and platelet clumping and/or clotting in plasma tubes, leading to inadequate separation of plasma and/or serum from cellular material during the centrifugation phase [69,70]. Specimens should not be opened prior to or during centrifugation in order to avoid evaporation [68]. Serum specimens requiring anaerobic conditions, such as ionized calcium and pH, should not be exposed to air prior to centrifugation. Exposure to air causes loss of $CO_2$, leading to increased pH and decreased ionized calcium [67,68,71].

***CASE REPORT*** A 40-year-old male was admitted to the hospital with a serum potassium concentration of 8.0 mmol/L (mEq/L) obtained on a specimen with no detectable hemolysis. The patient was given potassium-lowering drugs and another serum specimen was collected for potassium analysis. After treatment, the serum potassium concentration was 7.5 mmol/L, with no hemolysis. The patient became confused and developed muscle cramps and vomiting. The doctor requested a point-of-care whole blood potassium; the result of this assay was 2.7 mmol/L. The potassium-lowering therapy was promptly discontinued. On examination of the complete blood count, the white blood count was $20 \times 10^9$/L (normal, 4.5–11.0 $\times 10^9$/L), and the platelet count was

$480 \times 10^9$/L (normal, 150–350 $\times 10^9$/L). The elevated serum potassium results were due to an error in the centrifugation process. The laboratory scientists determined that the g-force used for centrifugation was too high, causing platelet lysis, particularly in a patient with elevated counts. The normal potassium results from whole blood were accurate [72].

Laboratories should select appropriate centrifugation speed, time, and temperature for each analyte. Inadequate centrifugation speed may lead to interferences with numerous assays in the clinical laboratory. Specimens spun at speeds too slow for the tube or for not enough time may not have adequate separation of serum from the clot or plasma from cellular components. Centrifugation is required to remove all microclots and fibrin strands from serum. Erroneous results may occur for numerous chemistry and immunochemical assays in the presence of such materials. Invisible microfibers or other particulate matter in incompletely centrifuged serum or plasma interferes with assays such as troponin [73,74]. Platelet contamination of the plasma in inadequately separated specimens leads to inaccurate results in coagulation studies [75]. Furthermore, centrifugation at slow speeds leads to increased "trapped plasma" in the red cell fraction, leading to abnormal results for analytes found in red cells (i.e., potassium and glucose-6-phosphate dehydrogenase). Specimens spun too fast are subject to *in vitro* hemolysis and/or cell lysis and release of intracellular constituents such as potassium [76,77].

In order to avoid abnormal results due to improper centrifugation, all laboratories should establish appropriate centrifugation time and speed for tube type, centrifuge, and rotor [22]. Laboratories should calculate relative centrifugal force (RCF), not revolutions per minute (RPM), for each centrifuge model, rotor, head, and the radius of the rotor in order to determine appropriate speed [22]. The following is the equation for RCF, expressed as multiples of gravitational force (g):

$$\text{RCF}\,(g) = (1.118 \times 10^{-5}) \times \text{radius of the rotor (cm)} \times (\text{RPM})^2 \quad (4.4)$$

Manufacturers provide recommendations for appropriate RCF and spin times for individual tube types [69,70].

Specimens should be centrifuged at appropriate temperatures to ensure specimen integrity prior to analysis. The internal temperature of a centrifuge may become hot with activity, leading to degradation of temperature-sensitive compounds such as ACTH. High temperature during centrifugation may also induce hemolysis. Centrifugation at refrigerated temperatures is not appropriate for all specimens. Laboratories should only perform potassium measurements on specimens stored and/or centrifuged in the cold for less than 2 hr because potassium leaks from cells with prolonged cold exposure [22,78]. Gel polymers in plasma and serum separator tubes are often temperature dependent; thus, centrifugation of these tubes in a chilled centrifuge may result in inefficient barrier formation. Laboratories should consult manufacturers' guidelines for spin temperatures for any barrier tubes [69]. Ionized calcium and pH in serum are affected by centrifugation temperatures, where ionized calcium results change by 0.006 mmol/L per 1°C [79]. Tight control of internal temperature is critical during the centrifugation phase. Centrifuge temperature should not deviate more than ±2.5°C for ionized calcium or pH [67]. Unless specimens are heat labile (in which case they should be maintained at even cooler temperatures), CLSI recommends fixing centrifuge temperatures at 20–22°C [22,68]. Heat-sensitive specimens arriving to the laboratory chilled should be spun in the cold [22].

Laboratories should consult tube manufacturers' literature when deciding which centrifuge is appropriate for their specimen. No matter the type of centrifuge, they should always be balanced prior to processing specimens. Improper balancing can lead to tube cracking, breakage, and/or inadequate separation of serum or plasma from cells as well as centrifuge damage. The use of fixed-angle rotors, particularly with separator tubes, will cause the gel to form at an angle. Angled gel formation may indicate an inadequate barrier seal, leading to mixing of serum or plasma with cells; numerous abnormalities are associated with storage on the clot or cells. Furthermore, centrifugation in fixed-angled rotors for prolonged times may induce hemolysis [78]. Swinging bucket rotors allow for a more reliable barrier seal and will not cause hemolysis at appropriate speed and temperatures, and these are recommended by most tube manufacturers [22,69,70].

***CASE REPORT*** Physicians from a community hospital alerted the central laboratory about numerous cases of hyperkalemia in an otherwise healthy patient population. None of the specimens had significant hemolysis. At the time, phlebotomy and processing practices were to collect potassium requests into serum separator tubes, allow them to clot for 20–60 min, and then centrifuge the tubes according to manufacturer's instructions. The processed serum separator tubes (SSTs) were then delivered to the central laboratory the following day, re-centrifuged, and analyzed. The laboratory decided to discontinue re-centrifugation and the hyperkalemia phenomenon was no longer observed. Furthermore, in a large timed study, the investigators demonstrated that re-centrifugation after prolonged storage in processed SSTs resulted in falsely elevated potassium [72].

Re-centrifugation of tubes is not recommended by CLSI and tube manufacturers [22,69,70]. As in the case report, several other studies demonstrated falsely elevated potassium after re-centrifugation [72,80,81]. According to manufacturers, re-centrifugation of gel separator tubes disrupts the barrier and may allow mixing of cell components with separated plasma. Release of potassium may occur as the result of cell lysis. Hira *et al.* proposed that a small portion of plasma remains with the cell fraction after centrifugation of gel separator tubes [72]. Extended exposure to cells allows leakage of potassium into that plasma fraction. Re-centrifugation causes this potassium-rich fraction to mix with the plasma layer. Longer initial centrifugation times reduce this effect, presumably by reducing the amount of plasma in the cell fraction [80].

Until recently, no study had examined the effect of re-centrifugation on other common chemical and immunochemical analytes. In-house studies demonstrated that re-centrifugation of plasma separator tubes (PSTs) not only caused spurious changes in potassium but also caused changes in other analytes, including creatinine, glucose, bilirubin, and AST (Table 4.3) [81]. Interestingly, in this study potassium was not increased in respun tubes stored at RT. Concentrations increased with increasing time at 4°C, supporting the claim that a small amount of plasma remains in the cell fraction of PSTs and becomes $K^+$ rich with increasing refrigeration time. Glucose is significantly lower in re-centrifuged tubes, but only in those stored at RT. Glycolysis occurs in the plasma fraction exposed to the cells, and when the cell-exposed plasma is mixed with previously separated plasma, glucose concentrations are significantly lowered. Total bilirubin and AST increase in fractions stored at RT and 4°C, and this is likely due to hemolysis with centrifugation. Creatinine increases in specimens stored at RT after re-centrifugation are likely due to an analytical interference with the Jaffe reaction [76,77].

To avoid erroneous results, re-centrifugation should be avoided. This may be difficult for laboratories with large outreach programs and automation. Such laboratories might consider using a code in their laboratory information system (LIS) directing these tubes to bypass the centrifuge or dedicate a lane in the input/output buffer for prespun tubes. Should laboratories want to clarify prespun specimens, serum or plasma can be aliquoted into a separate tube and centrifuged.

## EFFECT OF STORAGE CONDITIONS ON LABORATORY RESULTS

Delays in specimen analysis and physician requests for additional testing of a specimen make it important for laboratories to establish specific in-laboratory storage conditions. Numerous types of interferences result from inappropriate storage of specimens in the laboratory, including inappropriate storage temperatures/times, interfering substances, and light sensitivity. Inappropriate storage temperatures or times most often result in specimen degradation in whole blood and unseparated serum/plasma specimens. CLSI recommends the following as "general" guidelines for in-lab specimen storage. Serum/plasma should be separated from cellular components immediately after centrifugation, either by transferring to a new tube or by use of physical separators such as gel [22]. Separated specimens can be stored, tightly capped to avoid evaporation and concentration, up to 8 hr at RT (preferably 20–25°C) and up to 48 hr at 4°C; after 48 hr, specimens should be frozen at −20°C [22,82]. Samples should be snap frozen on dry ice or liquid nitrogen to avoid gradient formation [83]. Prior to analysis, specimens should be thawed at RT because heating may denature analytes. Gentle inversion can remove gradients formed with freezing. Centrifugation will sediment cellular material and/or fibrin strands that form upon freezing [84]. Although repeat freeze–thaw cycles are not recommended by CLSI [22], very few analytes are affected by this process [85].

Several analytes cannot be stored according to the CLSI "general" recommendations. Whole blood specimens should not be centrifuged. Storage of whole blood depends on the analyte. Freezing of whole blood induces hemolysis and is not recommended for hematological or blood gas specimens. Lamellar body counts are significantly decreased with freezing; thus, amniotic fluid specimens should be stored at 4°C [86]. In unseparated plasma specimens, prolonged storage at 4°C induces leakage from the cells and falsely elevates potassium [78]. Catecholamines are released from lysed red blood cells and reuptake is slowed, falsely elevating results with prolonged storage at 4°C [87]. Cryoactivation of pro-renin with long-term storage in the cold (refrigerated or on ice) falsely elevates plasma renin activity assays [88]. Specimens collected for LDH isoenzyme testing should be stored at RT prior to analysis because freezing and long-term storage at 4°C results in loss of LDH-5 activities [68]. Prothrombin time is significantly increased with prolonged frozen storage, and these specimens should only be stored refrigerated or at RT [89]. Because of the number of exceptions to the general in-lab specimen handling procedures, laboratories should consult manufacturers' package inserts for appropriate specimen storage conditions prior to analysis. Laboratories should conduct in-house stability studies prior to changing approved in-lab storage conditions [84].

Some studies suggest that utilization of gel separator tubes eliminates the need to physically separate plasma/serum from cells for short-term in

**TABLE 4.3** Stability of Common Chemical and Immunochemical Analytes in Serum or Heparin Plasma Separator Tubes after Re-Centrifugation[a]

| Analyte | Specimen | | Time Before Recentrifugation | | | References |
|---|---|---|---|---|---|---|
| | Serum | Heparin Plasma | <4 hr | <12 hr | Up to 24 hr | |
| Potassium | | X | | | RT[b] | [81] |
| | X | | | 4°C | | [72,80] |
| Sodium | | X | | | RT, 4°C | [81] |
| Chloride | | X | | | RT, 4°C | [81] |
| $CO_2$ | | X | | | RT, 4°C | [81] |
| BUN | | X | | | RT, 4°C | [81] |
| Creatinine | | X | RT | | 4°C | [81] |
| Glucose | | X | RT | | 4°C | [81] |
| Calcium | | X | | | RT, 4°C | [81] |
| tProt | | X | | | RT, 4°C | [81] |
| Albumin | | X | | 4°C | RT | [81] |
| tBili | | X | RT, 4°C | | | [81] |
| AlkP | | X | | | RT, 4°C | [81] |
| AST | | X | RT, 4°C | | | [81] |
| ALT | | X | RT | | 4°C | [81] |
| HDL-C | | X | | | RT, 4°C | [81] |
| Chol | | X | | | RT, 4°C | [81] |
| Trig | | X | | | RT, 4°C | [81] |
| LDL-C | | X | | | RT, 4°C | [81] |
| TSH | | X | | | RT, 4°C | [81] |
| fT4 | | X | | 4°C | RT | [81] |
| Ferritin | | X | | | RT, 4°C | [81] |
| VitB12 | | X | | | RT, 4°C | [81] |

[a]*Stability was assessed based on values outside the significant change limits after recentrifugation compared to values immediately before recentrifugation [81].*
[b]*Room temperature (21–25°C).*
ALT, alanine aminotransferase; AlkP, alkaline phosphatase; AST, aspartate aminotransferase; BUN, blood urea nitrogen; Chol, cholesterol; fT4, free thyroxine; HDL-C, high-density lipoprotein cholesterol; LDL-C, low-density lipoprotein cholesterol; tBili, total bilirubin; tProt, total protein; Trig, triglycerides; TSH, thyroid-stimulating hormone; VitB12, vitamin $B_{12}$.

lab storage [22]. Prolonged contact with gel does interfere with some analytes, especially drugs and estradiol, for which decreased concentrations may be observed due to absorption into the gel polymer [22,31]. Gel separator tubes should not be used for these analytes. Furthermore, plasma separator tubes may not be appropriate for storage of common chemistry analytes at 4°C for more than 48 hr [90]. After long-term storage, barrier seals should be inspected on all separator tubes prior to analysis. Laboratories should consult manufacturers' instructions for a list of analytes that are stable in these tubes for long-term storage.

Some analytes require preservatives in order to maintain in-laboratory stability. Catecholamines in urine degrade significantly at 4°C after 48 hr at pH 6.0, but they are stable frozen or at 4°C when preserved with acid to achieve pH 2.0–3.0 [91]. Studies suggest that glucose is more stable in citrated blood compared to fluoride [92]. Addition of protease inhibitors such as aprotinin stabilizes peptide hormones such as parathyroid-related protein [75].

Exposure to fluorescent light rapidly degrades bilirubin and other heme products [68]. In addition, vitamins such as carotene and vitamin A and also red blood cell folate are degraded by light. In plasma, some drugs, such as nifedipine and chloramphenicol [93], are light sensitive. Specimens collected for analysis of light-sensitive analytes should be protected from light by collecting them in brown containers or wrapping them in aluminum foil during transport and storage in the laboratory.

## EFFECT OF CROSS-CONTAMINATION OF SPECIMENS ON LABORATORY RESULTS

Cross-contamination of specimens occurs when a foreign pathogen, chemical, or other material is introduced into a patient specimen. Cross-contamination can lead to inaccurate results in most clinical laboratories, but it is most often a concern in the microbiology and molecular diagnostics laboratories. Specimens processed in laboratories may be contaminated with bacteria and/or DNA [94]. Prevention measures include specimen preparation in dedicated hoods, dedicated pipettes and the use of barrier pipette tips, and frequent decontamination procedures [95]. For complete discussions of interference with microbiology and molecular testing, see Chapters 20 and 21, respectively.

Cross-contamination of specimens can occur at the time of collection, processing, or analysis. Collection errors may involve intravenous fluid or syringe contamination with incorrect specimen draw order [69,70,96]. See Chapter 3 for more details on interferences at specimen collection.

Cross-contamination of specimens during aliquoting may also cause errors in laboratory tests. These types of contaminants can be reduced by changing disposable pipette tips with each aliquot and by aliquoting patient specimens one at a time. Cross-contamination may also occur on automated analyzers just prior to analysis if instrument sample probes do not use disposable pipette tips. Carryover is most common for analytes with large measuring ranges such as human chorionic gonadotropin and in ultrasensitive assays such as hepatitis antigens [97]. Documented carryover checks are required by the College of American Pathologists (CAP) when validating a new method and at intervals outlined by individual laboratories. Instrument manufacturers have reduced carryover cross-contaminations by introduction of wash steps and/or disposable pipette tips for sample probes [97].

*CASE REPORT* A urine sample was sent to the laboratory for a calcium:creatinine ratio. A 1-mL aliquot of the urine was taken to acidify prior to testing. The technologist adjusted the urine pH to approximately 4.0 with acid, and the sample was submitted for testing. The chemistry analyzer produced undetectable results for both calcium and creatinine, preventing calculation of the ratio. This was significantly different from the previous result so the technologist retraced her sample preparation steps and discovered that she had used the wrong bottle of acid to pH the specimen. Rather than using 0.1 N HCl, she had used a less concentrated acetic acid solution that required a much higher volume to adjust the pH and thus diluted the sample. Another aliquot was taken from the original sample cup, and the tests were repeated following proper preparation procedures. The repeat analysis produced results similar to the previous determination.

Specimen cross-contamination may occur with incorrect preservation of urine specimens. Prior to analysis, urine may be preserved with a variety of additives. The type of additive depends on the laboratory and assay [98]. For many analytes, the preservatives can be added upon arrival to the clinical laboratory. Addition of the incorrect additive may contaminate the urine specimen and interfere with results. In the case described previously, acetic acid was added to urine to adjust the pH to the optimal point for analysis, when in fact hydrochloric acid is recommended. Because a higher volume of acetic acid was required, the specimen was significantly diluted. It is equally likely that a sample might be alkalinized when it should be acidified. Laboratories can identify incorrect preservatives by measuring pH of all specimens prior to analysis. This type of urine contamination can be prevented by making up 24-hr urine jugs containing preservatives for patients. Laboratories should clearly outline procedures for processing and storage of 24-hr urine specimens, including adding appropriate preservatives.

## CONSEQUENCES OF SPECIMEN MISIDENTIFICATION ON LABORATORY TESTING

Accurate patient and specimen identification is required for quality patient care. Regulatory agencies such as The Joint Commission (TJC) have made it a top priority in order to ensure patient safety. Patient and specimen misidentification occurs in numerous phases of the testing process. During the pre-analytical phase, specimen identification begins when the patient presents to the hospital, doctor's office, or phlebotomist. Accurate identification requires collection of at least two unique identifiers from the patient and ensuring that those match the patient's prior records. If a patient is unable to provide identifiers (i.e., neonate or patient in a coma), a family member or nurse should verify the identity of the patient. Information on laboratory requisitions or electronic orders must also match patient information in the patient's chart or electronic medical record. Specimens should not be collected unless all identification discrepancies have been resolved [99]. At specimen collection, phlebotomists should ensure that the area is cleared of identification information from other patients. The sample(s) should be collected and labeled in front of the patient. The specimen should be sent to the laboratory with the test request. Upon acceptance into the laboratory, the identifiers on the specimen should match the requisition and/or electronic order. Non-bar-coded specimens should be accessioned, labeled with a bar code (or relabeled, if necessary), processed (either manually or on an automated line), and sent for analysis. Identification of the specimen should be carefully maintained during centrifugation, aliquoting, and analysis. Most laboratories use bar-coded labeling systems to preserve sample identification.

Patient and specimen ID errors can also occur during the analytical phase. Automated analyzers use bar codes to identify specimens during analysis and results reporting. For manual assays, laboratories should carefully match identifiers on specimens with work lists. Laboratories should carefully monitor repeats, dilutions, add-ons, and reflex testing, particularly if these are manual processes. In addition, procedures should be in place to ensure that bar codes are accurately printed because poorly or misprinted bar codes may be read incorrectly by laboratory instruments [100]. Such procedures may include regular cleaning and servicing of label printers.

Misidentification also occurs in the post-analytical phase of testing. Results from automated analyzers are electronically transferred to middleware or the LIS, where rules may dictate whether results are autoverified or require attention from a technologist or pathologist. With manual assays, results are manually entered and technologists must match patient identifiers on specimens, work lists, or result printouts with information in the LIS. Most LISs are interfaced with hospital information systems to report results in individual patient's charts. In the absence of electronic medical records, laboratory representatives must print laboratory results and fax or mail them to treating physicians.

Misidentification errors can occur at any point in this complex process, leading to outcomes ranging from harmless to severe. Outcomes are far less likely to be severe if the error is identified and fixed prior to reporting the results. For this reason, groups such as TJC, CAP, and the Institute of Medicine have made misidentification errors a priority [101,102]. Since the implementation of many related initiatives, error rates have been reduced, but problems persist [102].

*CASE REPORT* An endocrinologist contacted the clinical laboratory regarding discrepant laboratory results. Her patient was a 30-year-old female with a past medical history of acromegaly and resected pituitary tumor. Although stable after surgery, the patient had occasional headaches but no visual field disturbances. Pertinent laboratory values included the following: insulin-like growth factor-1 (IGF-1), 1265 ng/mL (normal range, 114–492 ng/mL); and growth hormone (GH), 0.1 ng/mL (normal, <8.0 ng/mL). Upon investigation, the IGF-1 testing was performed on plasma, whereas the GH assay was performed on serum. Both specimens were manual aliquots made by specimen processing staff in the laboratory. Testing of the primary tubes retrieved from refrigerated storage revealed a GH result of 40 ng/mL and IGF-1 of 1195 ng/mL. This case is an example of specimen misidentification that occurred during manual aliquoting in the laboratory. When possible, laboratories should use automated specimen processing equipment, including an aliquoting device. If unavailable or not appropriate for a particular analyte, laboratories should employ a strict procedure for manual aliquoting that ensures careful attention to patient identification. This case prompted the clinical laboratory to implement a system in which specimens were aliquoted one at a time and not in a batch. In addition, the processing technologist is asked to initial all aliquots.

In this case, the mislabeled specimen was identified because of a discrepancy between laboratory values and the clinical picture. No treatment decision was made based on these results. However, it is best to identify mislabels prior to reporting the results. Specimens that arrive with multiple labels should be carefully checked to ensure that all identifiers match. Furthermore, paper requisitions should be matched to specimen tubes. Suspicious specimens should be rejected or investigated by blood typing, DNA testing, or delta checking [99]. In many cases, there are no obvious signs of misidentification on the tube or requisition; therefore, laboratories must utilize other tools to identify these specimens.

*CASE REPORT* A patient reported to a busy clinic for preoperative laboratory work. Specimens were collected for complete blood count/differential, electrolytes, and a coagulation panel. The samples were received by and processed in the laboratory. Prior to release of results, an instrument flag revealed that a delta check rule had failed. The patient's hemoglobin value changed from 12.3 to 8.7 g/dL within 48 hr. This prompted an investigation by the technologist, who first confirmed that the patient had no recent transfusion history. She sent an aliquot of the sample to the blood bank, where it was confirmed that the blood type in the aliquot did not match that in the patient's record. The error was a result of misidentification of specimens. All tests ordered for the patient were canceled, and specimens were redrawn and tests repeated. Repeat results revealed a hemoglobin of 12.2 g/dL. In this case, the combination of delta checks and blood typing confirmed a misidentification error and potentially averted an unnecessary transfusion prior to surgery. Furthermore, this case emphasizes the need for a clear hospital-wide policy on correction of misidentified specimens outlining times when it is appropriate to change results, recollect, or confirm original results.

Delta checks are a simple way to detect mislabels. A delta check is a process of comparing a patient's result to his or her previous result for any one analyte over a specified period of time [103,104]. The difference or "delta," if outside pre-established rules, may indicate a specimen mislabel or other pre-analytical error. Laboratories should determine the difference in concentration (or relative change) as well as the time interval that is most appropriate for each analyte's delta check. With the increasing use of automation and middleware, delta checks have become a common practice for core laboratories [104,105]. Delta checks are most often set up for assays with little intraindividual variation that are tightly regulated within patients, including mean corpuscular volume (MCV), creatinine, BUN, bilirubin, and total protein [103]. Simulation studies demonstrated that delta checks for MCV, hematocrit (HCT), BUN, and creatinine are the most sensitive for detecting mislabeled specimens [105]. Furthermore, medical centers should establish their own delta checks based on their individual patient population. For example, delta checks for creatinine and BUN may not be appropriate for a large dialysis clinic, whereas HCT delta checks may be ineffective in a large hematology/oncology practice. Furthermore, with the previously identified mislabel rate of approximately 1/1000 in 2006 [106], only MCV has a high enough positive predictive value to detect all mislabels [105]. Advancements in middleware may allow us to design multianalyte delta checks with better predictive ability.

Delta checks are not appropriate for all analytes because of high intraindividual variabilities. For example, GH from a previous case report shows diurnal variation: Concentrations at night are significantly higher than in the morning. Thus, laboratories should employ measures to prevent mislabels at their source. Laboratories should adopt a strict specimen rejection policy to reduce entry of questionable specimens into the analytical process [99]. For example, laboratories may decide to reject all specimens that arrive unlabeled or that show disagreement between the requisition and label on the specimen unless they are irreplaceable (i.e., CSF specimens collected from infants or neonates, surgical specimens, etc.). Furthermore, laboratories are required by CAP to conduct periodic audits of patient records from requisition to result into the patient chart to search for errors. Intermittent and continual audits of patient ID bands are also helpful for reducing misidentification [99]. Data show that when laboratories monitor their identification errors on a regular basis, fewer errors occur [107]. Finally, utilization of new technology significantly reduces error rates. Reduction of manual steps removes opportunity for error. Bar coding linked with electronic order reduces manual steps at specimen collection and accessioning [102]. Errors do still exist with bar coding, prompting some laboratories and automation platforms to adopt radio frequency identification (RFID) [99,100,108]. Laboratory automation reduces manual steps, such as aliquoting and manual order and results entry, thereby reducing misidentification errors [102].

*CASE REPORT* A 68-year-old male presented to the hospital with sharp abdominal pain. The patient underwent an appendectomy and received 1 unit of type A blood. The patient developed disseminated intravascular coagulation and died 24 hr after receiving the transfusion. Postmortem analysis of the patient's blood revealed that he was actually type O. The patient had been sharing a room with another patient whose blood was type A. The specimen sent to the blood bank had been inappropriately labeled [109].

This is a rare case in which the wrong patient's blood was in the tube sent to the blood bank. Deaths related to ABO incompatible acute transfusion reactions occur in approximately 1/600,000 patients [110]. The most common error leading to a transfusion reaction is misidentification of a specimen at collection [111], and it is responsible for up to 22% of ABO incompatible transfusions [110]. The International Society of Blood Transfusion Biomedical Excellence for Safer Transfusion Committee performed a study examining the rate of ABO and RhD (rhesus blood group D antigen) testing that did not match previous result(s); this phenomenon was termed "wrong blood in tube" (WBIT) and occurred in approximately 1/1786 specimens [112,113]. An analysis of 122 clinical laboratories by CAP found that WBIT occurred in approximately 1/2500 samples [114]. These and other studies prompted TJC to place correct patient identification and prevention of misidentification-related transfusion errors as the first goal among their National Patient Safety Goals [101].

To reduce the rate of WBIT and misidentification errors in the blood bank, TJC recommends the use of two unique identifiers for each patient when collecting and administering blood components [101]. The American Association for Blood Banks requires two separate determinations of blood type for patients receiving blood in facilities using computer crossmatch. CAP checklists require phlebotomists to positively identify patients by verifying two unique identifiers before specimen collection. Furthermore, they require that all specimens be labeled at the time of collection [115]. Since international focus on misidentification prevention began in the early 2000s, numerous institutions have published methods to successfully reduce WBIT, including (1) matching requisitions to patient information on the accompanying specimen and in the electronic order system [107], (2) implementation of an electronic error monitoring system [106], (3) introduction of patient identification technologies such as bar coding and RFID [102,115–117], and (4) ABO/Rh testing analysis of two separate specimens for all patients with no blood type on file [118]. For more details on misidentification errors in transfusion medicine, see Chapter 17.

Despite great successes in reducing WBIT and other specimen misidentification errors, they still persist [102]. Laboratorians and hospital personnel should implement and continue vigilant error review and process improvement programs to prevent morbidity and mortality associated with specimen misidentification.

## CONCLUSIONS

Because most errors occur in the pre-analytical phase of laboratory testing, it is important to have robust procedures in place in the laboratory to eliminate various errors that may occur in this phase. Proper blood collection procedure, transport of specimens, and timely centrifugation ensure not only good specimen integrity but also accurate test results. Sample misidentification may cause serious errors in laboratory test results and may cause significant morbidity or even mortality, especially if a blood typing specimen is mislabeled. Good laboratory practice involves careful attention not only in the analytic phase but also in the pre- and post-analytical phases.

## References

[1] Lippi G, Banfi G, Buttarello M, Ceriotti F, Daves M, Dolci A, et al. Recommendations for detection and management of unsuitable samples in clinical laboratories. Clin Chem Lab Med 2007;45(6):728–36.

[2] Lippi G, Guidi GC, Mattiuzzi C, Plebani M. Preanalytical variability: the dark side of the moon in laboratory testing. Clin Chem Lab Med 2006;44(4):358–65.

[3] Food and Drug Administration. Fatalities reported to FDA following blood collection and transfusion: annual summary for fiscal year 2011. Rockville, Maryland: Food and Drug Administration; 2011.

[4] Laposata M, Dighe A. "Pre-pre" and "post-post" analytical error: high-incidence patient safety hazards involving the clinical laboratory. Clin Chem Lab Med 2007;45(6):712–9.

[5] Bonini P, Plebani M, Ceriotti F, Rubboli F. Errors in laboratory medicine. Clin Chem 2002;48(5):691–8.

[6] Wiwanitkit V. Types and frequency of preanalytical mistakes in the first Thai ISO 9002:1994 certified clinical laboratory, a 6-month monitoring. BMC Clin Pathol 2001;1(1):5.

[7] Kumar A, Roberts D, Wood KE, Light B, Parrillo JE, Sharma S, et al. Duration of hypotension before initiation of effective antimicrobial therapy is the critical determinant of survival in human septic shock. Crit Care Med 2006;34(6):1589–96.

[8] Chan AY, Swaminathan R, Cockram CS. Effectiveness of sodium fluoride as a preservative of glucose in blood. Clin Chem 1989;35(2):315–17.

[9] Laessig RH, Indriksons AA, Hassemer DJ, Paskey TA, Schwartz TH. Changes in serum chemical values as a result of prolonged contact with the clot. Am J Clin Pathol 1976;66(3):598–604.

[10] Boyanton Jr. BL, Blick KE. Stability studies of twenty-four analytes in human plasma and serum. Clin Chem 2002;48(12):2242–7.

[11] Astles R, Williams CP, Sedor F. Stability of plasma lactate *in vitro* in the presence of antiglycolytic agents. Clin Chem 1994;40(7):1327–30.

[12] Sazama K, Robertson EA, Chesler RA. Is antiglycolysis required for routine glucose analysis? Clin Chem 1979;25(12):2038–9.

[13] Zhang DJ, Elswick RK, Miller WG, Bailey JL. Effect of serum-clot contact time on clinical chemistry laboratory results. Clin Chem 1998;44(6 Pt 1):1325–33.

[14] Goodenow TJ, Malarkey WB. Leukocytosis and artifactual hypoglycemia. JAMA 1977;237(18):1961–2.

[15] Arem R, Jeang MK, Blevens TC, Waddell CC, Field JB. Polycythemia rubra vera and artifactual hypoglycemia. Arch Intern Med 1982;142(12):2199–201.

[16] Viola GM. Extreme hypoglycorrhachia: not always bacterial meningitis. Nat Rev Neurol 2010;6(11):637–41.

[17] Carraro P, Servidio G, Plebani M. Hemolyzed specimens: a reason for rejection or a clinical challenge? Clin Chem 2000;46 (2):306–7.

[18] Davidson NC, Coutie WJ, Struthers AD. N-terminal proatrial natriuretic peptide and brain natriuretic peptide are stable for up to 6 hours in whole blood *in vitro*. Circulation 1995;91 (4):1276–7.

[19] Tsuji T, Imagawa K, Masuda H, Haraikawa M, Shibata K, Kono M, et al. Stabilization of human brain natriuretic peptide in blood samples. Clin Chem 1994;40(4):672–3.

[20] Evans MJ, Livesey JH, Ellis MJ, Yandle TG. Effect of anticoagulants and storage temperatures on stability of plasma and serum hormones. Clin Biochem 2001;34(2):107–12.

[21] Oddoze C, Lombard E, Portugal H. Stability study of 81 analytes in human whole blood, in serum and in plasma. Clin Biochem 2012;45(6):464–9.

[22] Kiechle FL, Betsou F, Blakeney J, Calam RR, Catalasan IM, Raj P, et al. Procedures for the handling and processing of blood specimens for common laboratory tests; approved guideline. Wayne, PA: Clinical and Laboratory Standards Institute; 2010.

[23] Rossetti MD, Felder RA, Kumar A. Simulation of robotic courier deliveries in hospital distribution services. Health Care Manag Sci 2000;3(3):201–13.

[24] Streichert T, Otto B, Schnabel C, Nordholt G, Haddad M, Maric M, et al. Determination of hemolysis thresholds by the use of data loggers in pneumatic tube systems. Clin Chem 2011;57 (10):1390–7.

[25] Steige H, Jones JD. Evaluation of pneumatic-tube system for delivery of blood specimens. Clin Chem 1971;17(12):1160–4.
[26] Ellis G. An episode of increased hemolysis due to a defective pneumatic air tube delivery system. Clin Biochem 2009;42 (12):1265–9.
[27] BD Vacutainer (R) Evacuated Blood Collection System. Package Insert; 2002.
[28] Laessig RH, Hassemer DJ, Westgard JO, Carey RN, Feldbruegge DH, Schwartz TH. Assessment of the serum separator tube as an intermediate storage device within the laboratory. Am J Clin Pathol 1976;66(4):653–7.
[29] Li Z, Yan C, Yan R, Zheng X, Feng Z. Evaluation of BD Vacutainer SST™ II plus tubes for special proteins testing. J Clin Lab Anal 2011;25(3):203–6.
[30] O'Keane MP, Cunningham SK. Evaluation of three different specimen types (serum, plasma lithium heparin and serum gel separator) for analysis of certain analytes: clinical significance of differences in results and efficiency in use. Clin Chem Lab Med 2006;44(5):662–8.
[31] Dasgupta A, Yared MA, Wells A. Time-dependent absorption of therapeutic drugs by the gel of the Greiner Vacuette blood collection tube. Ther Drug Monit 2000;22(4):427–31.
[32] Bush V, Blennerhasset J, Wells A, Dasgupta A. Stability of therapeutic drugs in serum collected in vacutainer serum separator tubes containing a new gel (SST II). Ther Drug Monit 2001;23 (3):259–62.
[33] Chance J, Berube J, Vandersmissen M, Blanckaert N. Evaluation of the BD Vacutainer((R)) PS II Blood collection tube for special chemistry analytes. Clin Chem Lab Med 2009.
[34] Shi RZ, Seeley ES, Bowen R, Faix JD. Rapid blood separation is superior to fluoride for preventing *in vitro* reductions in measured blood glucose concentration. J Clin Pathol 2009;62 (8):752–3.
[35] Chan AY, Swaminathan R, Cockram CS. Effectiveness of sodium fluoride as a preservative of glucose in blood. Clin Chem 1989;35(2):315–17.
[36] Mikesh LM, Bruns DE. Stabilization of glucose in blood specimens: mechanism of delay in fluoride inhibition of glycolysis. Clin Chem 2008;54(5):930–2.
[37] Bais R, Panteghini M. Principles of clinical enzymology. In: Burtis CA, Ashwood ER, Bruns DE, editors. Tietz textbook of clinical chemistry and molecular diagnostics. 4th ed. St. Louis, MO: Saunders; 2006.
[38] Young DS, Bermes EW, Haverstick DM. Specimen collection and processing. In: Burtis CA, Ashwood ER, Bruns DE, editors. Tietz textbook of clinical chemistry and molecular diagnostics. 4th ed. St. Louis, MO: Saunders; 2006. p. 41–58.
[39] Jung K, Bader K, Grutzmann KD. Long-term stability of enzymes in human serum stored in liquid nitrogen. Enzyme 1984;31(4):209–16.
[40] Lev EI, Hendler I, Siebner R, Tashma Z, Wiener M, Tur-Kaspa I. Creatine kinase activity decrease with short-term freezing. Enzyme Protein 1994;48(4):238–42.
[41] Motyckova G, Murali M. Laboratory testing for cryoglobulins. Am J Hematol 2011;86(6):500–2.
[42] Nichols JH. Preanalytical variation. In: Clarke W, Dufour DR, editors. Contemporary practice in clinical chemistry. Washington, DC: AACC Press; 2006.
[43] Jensen EA, Stahl M, Brandslund I, Grinsted P. Stability of heparin blood samples during transport based on defined preanalytical quality goals. Clin Chem Lab Med 2008;46(2):225–34.
[44] Freer DE. Observations on heat/humidity denaturation of enzymes in filter-paper blood spots from newborns. Clin Chem 2005;51(6):1060–2.
[45] Fujimoto A, Okano Y, Miyagi T, Isshiki G, Oura T. Quantitative Beutler test for newborn mass screening of galactosemia using a fluorometric microplate reader. Clin Chem 2000;46 (6 Pt 1):806–10.
[46] Waite KV, Maberly GF, Eastman CJ. Storage conditions and stability of thyrotropin and thyroid hormones on filter paper. Clin Chem 1987;33(6):853–5.
[47] Adam BW, Lim TH, Hall EM, Hannon WH. Preliminary proficiency testing results for succinylacetone in dried blood spots for newborn screening for tyrosinemia type I. Clin Chem 2009;55(12):2207–13.
[48] Chace DH, Adam BW, Smith SJ, Alexander JR, Hillman SL, Hannon WH. Validation of accuracy-based amino acid reference materials in dried-blood spots by tandem mass spectrometry for newborn screening assays. Clin Chem 1999;45(8):1269–77.
[49] Babic N, Zibrat S, Gordon IO, Lee CC, Yeo KT. Effect of blood collection tubes on the incidence of artifactual hyperkalemia on patient samples from an outreach clinic. Clin Chim Acta 2012;413:1454–8.
[50] Guss DA, Chan TC, Killeen JP. The impact of a pneumatic tube and computerized physician order management on laboratory turnaround time. Ann Emerg Med 2008;51(2):181–5.
[51] Felder RA. Preanalytical errors introduced by sample-transportation systems: a means to assess them. Clin Chem 2011;57(10):1349–50.
[52] Justice T. Personal Communication. 2011.
[53] Lippi G, Lima-Oliveira G, Nazer SC, Moreira ML, Souza RF, Salvagno GL, et al. Suitability of a transport box for blood sample shipment over a long period. Clin Biochem 2011;44 (12):1028–9.
[54] Mehta A, Lichtin AE, Vigg A, Parambil JG. Platelet larceny: spurious hypoxaemia due to extreme thrombocytosis. Eur Respir J 2008;31(2):469–72.
[55] Fox MJ, Brody JS, Weintraub LR. Leukocyte larceny: a cause of spurious hypoxemia. Am J Med 1979;67(5):742–6.
[56] Scott MG, LeGrys VA, Klutts JS. Electrolytes and blood gases. In: Burtis CA, Ashwood ER, Bruns DE, editors. Tietz textbook of clinical chemistry and molecular diagnostics. 4th ed. Philadelphia, PA: Saunders; 2006.
[57] Blonshine S, Fallon KD, Lehman CM, Sittig S. Procedures for the collection of arterial blood specimens; approved standard–fourth edition. 4th ed. Wayne, PA: CLSI; 2004.
[58] Knowles TP, Mullin RA, Hunter JA, Douce FH. Effects of syringe material, sample storage time, and temperature on blood gases and oxygen saturation in arterialized human blood samples. Respir Care 2006;51(7):732–6.
[59] Mahoney JJ, Harvey JA, Wong RJ, Van Kessel AL. Changes in oxygen measurements when whole blood is stored in iced plastic or glass syringes. Clin Chem 1991;37(7):1244–8.
[60] Toffaletti JG. Blood gasses and electrolytes. 2nd ed. Washington, DC: AACC Press; 2002.
[61] Pretto JJ, Rochford PD. Effects of sample storage time, temperature and syringe type on blood gas tensions in samples with high oxygen partial pressures. Thorax 1994;49(6):610–12.
[62] (CLSI). CaLSI. Blood Gas and pH Analysis and Related Measurements; Approved Guideline–Second Edition. CLSI document C46-A2. 2009.
[63] Biswas CK, Ramos JM, Agroyannis B, Kerr DN. Blood gas analysis: effect of air bubbles in syringe and delay in estimation. Br Med J (Clin Res Ed) 1982;284(6320):923–7.
[64] Victor Peter J, Patole S, Fleming JJ, Selvakumar R, Graham PL. Agreement between paired blood gas values in samples transported either by a pneumatic system or by human courier. Clin Chem Lab Med 2011;49(8):1303–9.
[65] Toffaletti J, Blosser N, Kirvan K. Effects of storage temperature and time before centrifugation on ionized calcium in blood collected in plain vacutainer tubes and silicone-separator (SST) tubes. Clin Chem 1984;30(4):553–6.

[66] Endres DB, Rude RK. Mineral and bone metabolism. In: Burtis CA, Ashwood ER, Buns DE, editors. Tietz textbook of clinical chemistry and molecular diagnostics. 4th ed. Philadelphia, PA: Saunders; 2006.
[67] D'Orazio P, Toffaletti JG, Wandrup J. Ionized calcium determinations: precollection variables, specimen choice, collection, and handling; approved guideline–second edition. 2nd ed. Wayne, PA: Clinical and Laboratory Standards Institute.
[68] Haverstick DM, Groszbach AR. Specimen collection and processing. In: Burtis CA, Ashwood ER, Bruns DE, editors. Tietz textbook of clinical chemistry and molecular diagnostics. 5th ed. St. Louis, MO: Saunders; 2012. p. 145–62.
[69] BD Vacutainer (R) Evacuated blood collection system: for *in vitro* diagnostic use. Franklin Lakes, NJ: Becton Dickinson; 2010. p. 1–9.
[70] Evacuated blood collection system: for *in vitro* diagnostic use. Austria: Greiner Bio-One GmbH; 2011. p. 1–6.
[71] Madiedo G, Sciacca R, Hause L. Air bubbles and temperature effect on blood gas analysis. J Clin Pathol 1980;33(9):864–7.
[72] Hira K, Shimbo T, Fukui T. High serum potassium concentrations after recentrifugation of stored blood specimens. N Engl J Med 2000;343(2):153–4.
[73] Kazmierczak SC, Sekhon H, Richards C. False-positive troponin I measured with the abbott AxSYM attributed to fibrin interference. Int J Cardiol 2005;101(1):27–31.
[74] Strathmann FG, Ka MM, Rainey PM, Baird GS. Use of the BD vacutainer rapid serum tube reduces false-positive results for selected Beckman Coulter Unicel DxI immunoassays. Am J Clin Pathol 2011;136(2):325–9.
[75] Narayanan S. The preanalytic phase: an important component of laboratory medicine. Am J Clin Pathol 2000;113(3):429–52.
[76] Lippi G, Blanckaert N, Bonini P, Green S, Kitchen S, Palicka V, et al. Haemolysis: an overview of the leading cause of unsuitable specimens in clinical laboratories. Clin Chem Lab Med 2008;46(6):764–72.
[77] Lippi G, Plebani M, Di Somma S, Cervellin G. Hemolyzed specimens: a major challenge for emergency departments and clinical laboratories. Crit Rev Clin Lab Sci 2011;48(3):143–53.
[78] Stankovic AK, Smith S. Elevated serum potassium values: the role of preanalytic variables. Am J Clin Pathol 2004;121 (Suppl):S105–12.
[79] D'Orazio P, Toffaletti JG, Wandrup J. Ionized calcium determinations: precollection variables, specimen choice, collection, and handling; approved guideline–second edition. Wayne, PA: Clinical and Laboratory Standards Institute.
[80] Hue DP, Culank LS, Toase PD, Maguire GA. Observed changes in serum potassium concentration following repeat centrifugation of Sarstedt Serum Gel Safety Monovettes after storage. Ann Clin Biochem 1991;28(Pt 3):309–10.
[81] Pyle AL, Dawling S, Woodworth A. Effect of repeated centrifugation of PST tubes on common chemistry analytes. Clin Chem 2010;56:A24.
[82] Burtis CA, Begovich JM, Watson JS. Factors influencing evaporation from sample cups and assessment of their effect on analytical error. Clin Chem 1975;21(13):1907–17.
[83] Myers GL, Eckfeldt JH, Greenberg N, Levine JB, Miller WG, Wiebe DA. Preparation and validation of commutable frozen human serum pools as secondary reference materials for cholesterol measurement procedures; approved guideline. Wayne, PA: NCCLS; 1999.
[84] Livesey JH, Ellis MJ, Evans MJ. Pre-analytical requirements. Clin Biochem Rev 2008;29(Suppl. 1):S11–15.
[85] Comstock GW, Burke AE, Norkus EP, Gordon GB, Hoffman SC, Helzlsouer KJ. Effects of repeated freeze-thaw cycles on concentrations of cholesterol, micronutrients, and hormones in human plasma and serum. Clin Chem 2001;47(1):139–42.
[86] Lockwood CM, Crompton JC, Riley JK, Landeros K, Dietzen DJ, Grenache DG, et al. Validation of lamellar body counts using three hematology analyzers. Am J Clin Pathol 2010;134 (3):420–8.
[87] Boomsma F, Alberts G, van Eijk L, Man in 't Veld AJ, Schalekamp MA. Optimal collection and storage conditions for catecholamine measurements in human plasma and urine. Clin Chem 1993;39(12):2503–8.
[88] Sealey JE. Plasma renin activity and plasma prorenin assays. Clin Chem 1991;37(10 Pt 2):1811–19.
[89] Lippi G, Franchini M, Montagnana M, Salvagno GL, Poli G, Guidi GC. Quality and reliability of routine coagulation testing: can we trust that sample? Blood Coagul Fibrinolysis 2006;17(7):513–9.
[90] Brandhorst G, Engelmayer J, Gotze S, Oellerich M, von Ahsen N. Pre-analytical effects of different lithium heparin plasma separation tubes in the routine clinical chemistry laboratory. Clin Chem Lab Med 2011;49(9):1473–7.
[91] Roberts NB, Higgins G, Sargazi M. A study on the stability of urinary free catecholamines and free methyl-derivatives at different pH, temperature and time of storage. Clin Chem Lab Med 2010;48(1):81–7.
[92] Gambino R, Piscitelli J, Ackattupathil TA, Theriault JL, Andrin RD, Sanfilippo ML, et al. Acidification of blood is superior to sodium fluoride alone as an inhibitor of glycolysis. Clin Chem 2009;55(5):1019–21.
[93] Hammett-Stabler CA, Johns T. Laboratory guidelines for monitoring of antimicrobial drugs. National academy of clinical biochemistry. Clin Chem 1998;44(5):1129–40.
[94] Forbes BA, Banaiee N, Beavis KG, Brown-Elliott BA, Latta PD, Elliott LB, et al. Laboratory detection and identification of mycobacteria; approved guideline. Wayne, PA: CLSI.
[95] Sabath DE, Algar E, Bhattacharyya PK, Bijwaard KE, Hong T, Lindeman N, et al. Nucleic acid amplification assays for molecular hematopathology; approved guideline–second edition. Wayne, PA: Clinical and Laboratory Standards Institute; 2012.
[96] Sulaiman RA, Cornes MP, Whitehead SJ, Othonos N, Ford C, Gama R. Effect of order of draw of blood samples during phlebotomy on routine biochemistry results. J Clin Pathol 2011;64(11):1019–20.
[97] Boyd JC, Hawker CD. Automation in the clinical laboratory. In: Burtis CA, Ashwood ER, Bruns DE, editors. Tietz textbook of clinical chemistry and molecular diagnostics. 5th ed. St. Louis, MO: Saunders; 2012. p. 469–86.
[98] Nichols JH. Preanalytical variation. In: Clark W, editor. Contemporary practice in clinical chemistry. 2nd ed. Washington, DC: AACC Press; 2011. p. 1–12.
[99] Woodcock SM, Bettinelli R, Fedraw LA, Fischer C, Henricks WH, Mann P, et al. Accuracy in patient and sample identification; approved guideline. Wayne, PA: Clinical and Laboratory Standards Institute.
[100] Snyder ML, Carter A, Jenkins K, Fantz CR. Patient misidentifications caused by errors in standard bar code technology. Clin Chem 2010;56(10):1554–60.
[101] The Joint Commission. National Patient Safety Goals Effective January 1, 2012, Hospital Accreditation Program. Effective January 1, 2012 ed. Oak Terrace, IL: The Joint Commission; 2012. p. 1–13.
[102] Dunn EJ, Moga PJ. Patient misidentification in laboratory medicine: a qualitative analysis of 227 root cause analysis reports in the Veterans Health Administration. Arch Pathol Lab Med 2010;134(2):244–55.
[103] Klee GG, Westgard JO. Quality management. In: Burtis CA, Ashwood ER, Bruns DE, editors. Tietz textbook of clinical chemistry and molecular diagnostics. 5th ed. St. Louis, MO: Saunders; 2012. p. 163–204.

[104] Ladenson JH. Patients as their own controls: use of the computer to identify "laboratory error". Clin Chem 1975;21(11):1648–53.

[105] Strathmann FG, Baird GS, Hoffman NG. Simulations of delta check rule performance to detect specimen mislabeling using historical laboratory data. Clin Chim Acta 2011;412(21–22):1973–7.

[106] Wagar EA, Tamashiro L, Yasin B, Hilborne L, Bruckner DA. Patient safety in the clinical laboratory: a longitudinal analysis of specimen identification errors. Arch Pathol Lab Med 2006;130(11):1662–8.

[107] Valenstein PN, Raab SS, Walsh MK. Identification errors involving clinical laboratories: a College of American Pathologists Q-Probes study of patient and specimen identification errors at 120 institutions. Arch Pathol Lab Med 2006;130(8):1106–13.

[108] Francis DL, Prabhakar S, Sanderson SO. A quality initiative to decrease pathology specimen-labeling errors using radiofrequency identification in a high-volume endoscopy center. Am J Gastroenterol 2009;104(4):972–5.

[109] Aleccia J. Patients Still Stuck with Bill for Medical Errors. 2008 2/29/2008 8:26:51 AM ET [cited 2012 06/28/2012]; Available from: <http://www.msnbc.msn.com/id/23341360/ns/health-health_care/t/patients-still-stuck-bill-medical-errors/#.T-yk5vVibJs>.

[110] Linden JV, Paul B, Dressler KP. A report of 104 transfusion errors in New York State. Transfusion 1992;32(7):601–6.

[111] O'Neill E, Richardson-Weber L, McCormack G, Uhl L, Haspel RL. Strict adherence to a blood bank specimen labeling policy by all clinical laboratories significantly reduces the incidence of "wrong blood in tube". Am J Clin Pathol 2009;132(2):164–8 [quiz 306].

[112] Dzik WH, Murphy MF, Andreu G, Heddle N, Hogman C, Kekomaki R, et al. An international study of the performance of sample collection from patients. Vox Sang 2003;85(1):40–7.

[113] Murphy MF, Stearn BE, Dzik WH. Current performance of patient sample collection in the UK. Transfus Med 2004;14(2):113–21.

[114] Grimm E, Friedberg RC, Wilkinson DS, AuBuchon JP, Souers RJ, Lehman CM. Blood bank safety practices: mislabeled samples and wrong blood in tube—a Q-Probes analysis of 122 clinical laboratories. Arch Pathol Lab Med 2010;134(8):1108–15.

[115] Valenstein PN, Sirota RL. Identification errors in pathology and laboratory medicine. Clin Lab Med 2004;24(4):979–96 [vii].

[116] Wald H, Shojania K. Making health care safety: a critical analysis of patient safety practices. In: AfHRa, editor. Quality. Washington DC: U.S. Department of Health and Human Services; 2001. p. 487–500.

[117] Porcella A, Walker K. Patient safety with blood products administration using wireless and bar-code technology. AMIA Annu Symp Proc 2005:614–18.

[118] Figueroa PI, Ziman A, Wheeler C, Gornbein J, Monson M, Calhoun L. Nearly two decades using the check-type to prevent ABO incompatible transfusions: one institution's experience. Am J Clin Pathol 2006;126(3):422–6.

[119] Ellis JM, Livesey JH, Evans MJ. Hormone stability in human whole blood. Clin Biochem 2003;36(2):109–12.

[120] Bujišić N. Effects of serum-clot contact time on second-trimester prenatal screening markers and their stability in serum. J Med Biochem 2010;29(2):84–8.

[121] Chu SY, MacLeod J. Effect of three-day clot contact on results of common biochemical tests with serum. Clin Chem 1986;32(11):2100.

[122] Ono T, Kitaguchi K, Takehara M, Shiiba M, Hayami K. Serum-constituents analyses: effect of duration and temperature of storage of clotted blood. Clin Chem 1981;27(1):35–8.

[123] Clark S, Youngman LD, Palmer A, Parish S, Peto R, Collins R. Stability of plasma analytes after delayed separation of whole blood: implications for epidemiological studies. Int J Epidemiol 2003;32(1):125–30.

[124] Tanner M, Kent N, Smith B, Fletcher S, Lewer M. Stability of common biochemical analytes in serum gel tubes subjected to various storage temperatures and times pre-centrifugation. Ann Clin Biochem 2008;45(Pt 4):375–9.

[125] Pereira M, Azevedo A, Severo M, Barros H. Long-term stability of endogenous B-type natriuretic peptide during storage at −20°C for later measurement with Biosite Triage assay. Clin Biochem 2007;40(15):1104–7.

[126] Wu AHB, Packer M, Smith A, Bijou R, Fink D, Mair J, et al. Analytical and clinical evaluation of the bayer advia centaur automated b-type natriuretic peptide assay in patients with heart failure: a multisite study. Clin Chem 2004;50(5):867–73.

[127] Willemsen JJ, Ross HA, Lenders JWM, Sweep FCGJ. Stability of urinary fractionated metanephrines and catecholamines during collection, shipment, and storage of samples. Clin Chem 2007;53(2):268–72.

[128] d'Eril GM, Valenti G, Pastore R, Pankopf S. More on stability of albumin, N-acetylglucosaminidase, and creatinine in urine samples. Clin Chem 1994;40(2):339–40.

[129] Hartweg J, Gunter M, Perera R, Farmer A, Cull C, Schalkwijk C, et al. Stability of soluble adhesion molecules, selectins, and c-reactive protein at various temperatures: implications for epidemiological and large-scale clinical studies. Clin Chem 2007;53(10):1858–60.

[130] Gambino R, Piscitelli J, Ackattupathil TA, Theriault JL, Andrin RD, Sanfilippo ML, et al. Acidification of blood is superior to sodium fluoride alone as an inhibitor of glycolysis. Clin Chem 2009;55(5):1019–21.

[131] Rohlfing CL, Hanson S, Tennill AL, Little RR. Effects of whole blood storage on hemoglobin a1c measurements with five current assay methods. Diabetes Technol Ther 2012;14(3):271–5.

[132] Calam RR, Mansoor I, Blaga J. Homocysteine stability in heparinized plasma stored in a gel separator tube. Clin Chem 2005;51(8):1554–5.

[133] Innanen VT, Groom BM, de Campos FM. Microalbumin and freezing. Clin Chem 1997;43(6):1093–4.

CHAPTER

# 5

# Hemolysis, Lipemia, and High Bilirubin

## Effect on Laboratory Tests

*Steven C. Kazmierczak*

**Oregon Health & Science University, Portland, Oregon**

## INTRODUCTION

A wide variety of pre-analytical factors can adversely impact the integrity of specimens that are submitted for analysis, including improper or incorrect use of collection containers, excessive time delay from specimen collection to analysis, failure to store specimens at an appropriate temperature prior to analysis, failure to shield the specimen from direct light, and collection of the specimen at the wrong time of day or at an inappropriate time following the administration of certain medications [1]. Prevention of medical errors in health care has received a great deal of attention since the publication in 2000 of the report from the Institute of Medicine that estimated that medical errors result in approximately 44,000–98,000 preventable deaths and 1 million excess injuries each year in U.S. hospitals [2]. Therefore, accurate laboratory test results are important in order to prevent medical errors because diagnoses of many diseases today are based on laboratory test results. Although medication errors are most often cited as the main cause of medical errors, inappropriate treatment of patients due to incorrect test results caused by interfering substances has been noted to be a factor contributing to medical errors [3–5]. Interference in clinical assays is often underestimated and often undetected in clinical laboratories [6]. Pre-analytical errors due to endogenous interfering substances are perhaps one of the most common causes of errors that occur in laboratory testing.

In principle, interferences that affect the spectrophotometric measurement of a sample can be reduced by use of an adequately blanked analytical method. However, this is often not practical or easy to implement. In addition, endogenous interfering substances can cause not only spectral interference but also chemical interferences in some assays. Also, endogenous interference due to hemolysis can increase the concentrations of those analytes that are present in the erythrocytes. Although common endogenous interferences such as hemolysis, lipemia, and icterus are known to interfere with photometric assays, interference with turbidimetric methods and immunoassays has also been reported.

The prevalence of endogenous interfering substances seen in patient samples submitted for analysis can be significant, but the actual frequency at which interferences occur can be difficult to estimate. One study that investigated the prevalence of endogenous interferences seen in outpatients found that 9.7% of samples submitted for analysis contained at least one endogenous interfering substance [7]. Of the samples considered to have some type of endogenous interfering substances, 76% were considered to be lipemic, 16.5% were hemolyzed, and 5.5% were icteric. However, significant differences in the incidence of endogenous interferences have been noted with respect to where patients are located within a hospital setting.

Observations by both physicians and clinical laboratory staff suggest that the rate of hemolysis in samples collected in emergency departments significantly exceeds that in samples drawn in other hospital locations. Another study that evaluated a total of 4021 samples found that hemolyzed samples were more frequently seen in samples collected in the emergency department compared to samples obtained from the medical unit [8]. Of the 2992 specimens collected in the emergency department, 372 (12.4%) were hemolyzed, whereas of the 1029 samples from the medical unit, 16

*Accurate Results in the Clinical Laboratory.*
DOI: http://dx.doi.org/10.1016/B978-0-12-415783-5.00005-0

(1.6%) were hemolyzed. The use of trained phlebotomists to collect blood from patients in the medical unit, whereas the emergency department utilized nurses not formally trained in phlebotomy practices, was suggested to play a significant role in the differences seen in the rates of hemolysis.

The incidence of endogenous interfering substances seen in specimens submitted to the clinical laboratory is dependent on a number of factors, including the patient population being served (i.e., neonates, diabetics, elderly, patients on total parenteral nutrition, inpatients vs. outpatients, etc.), use of skilled phlebotomists versus minimally trained health care providers, and elapsed time from collection of sample to processing and analysis. Also important is the mechanism that is used to identify the presence of endogenous interfering substances. Visual inspection of samples for identification of interfering substances is still in use by laboratories, although this practice is rapidly being supplanted by the use of instruments with the capability to detect and quantify the amount of interference present. Manual visual detection of endogenous interferences is noted to suffer from significant lack of agreement between individuals and also vastly underestimates the actual number of samples that have levels of endogenous interfering substances that can cause assay interference. Also, increased concentrations of bilirubin can result in underestimation by visual means of the amount of plasma hemoglobin that is present in hemolyzed samples. This situation is frequently seen in samples collected from newborns, who often show increased bilirubin concentrations and whose samples are often hemolyzed. Thus, the use of automated systems for detection of endogenous interfering substances is imperative if proper evaluation of these types of samples is to be accomplished. One study that utilized an algorithm for the detection and processing of clinically or analytically relevant amounts of hemolysis found that automated detection of relevant hemolysis was approximately 70-fold higher compared to when sample hemolysis was assessed by manual detection [8].

Despite the vast number of publications that have addressed the problems of endogenous interfering substances, there is still a significant lack of understanding concerning sources of endogenous interferences. For example, artificial substitutes such as Intralipid used to mimic lipemia often do not behave the same as samples with native lipemia [9]. Despite this, virtually all studies designed to assess the effect of lipemia utilize Intralipid to evaluate this interference. Other aspects of interference testing often ignored or overlooked include whether the interfering substance produces similar interference effects at different analyte concentrations. Evaluation of endogenous interfering substances should be performed at several different concentrations of analyte. For example, a 10% bias in the measured concentration of 100 mg/dL of glucose due to 300 mg/dL of plasma hemoglobin may be insignificant when the glucose concentration in the sample is 200 mg/dL.

## EFFECT OF HEMOLYSIS ON LABORATORY TESTS

Hemolysis is the disruption of the erythrocyte membrane with release of hemoglobin and other intracellular components into the surrounding serum or plasma. Intracellular components released in high concentrations from erythrocytes include enzymes such as lactate dehydrogenase (LDH) and aspartate aminotransferase (AST), as well as electrolytes such as potassium and magnesium. In cases of massive hemolysis, the release of intracellular fluid from erythrocytes can even result in the dilution of serum or plasma analytes, such as sodium, that are usually present in low concentrations within erythrocytes.

Normally, serum and plasma contain very low concentrations of free hemoglobin, with plasma containing less than 2 mg/dL and serum less than 5 mg/dL. Visual detection of hemolysis does not occur until the free hemoglobin concentration is greater than 20–70 mg/dL. The ability to visually detect hemoglobin in serum or plasma is affected by the concentration of other compounds present in the sample. For example, serum or plasma from patients who are jaundiced can mask hemolysis when hemolysis is estimated by visual means. Thus, samples with moderate hemolysis and free hemoglobin concentrations of 100–150 mg/dL may go undetected when visual assessment of samples is performed [10].

Hemolysis can be caused by a variety of mechanisms, including physical disruption of cells and by immunological and chemical means. Hemolysis can be divided into *in vivo* hemolysis and *in vitro* hemolysis. When red cell lysis occurs within the body with subsequent release of intracellular components into the plasma, this is termed *in vivo* hemolysis. In contrast, when lysis of cells following the collection of blood occurs, it is termed *in vitro* hemolysis. Common causes of *in vivo* and *in vitro* hemolysis are shown in Table 5.1.

### *In Vivo* Hemolysis

*In vivo* hemolysis can be categorized on the basis of where red cell destruction occurs. Intravascular hemolysis occurs when erythrocytes are destroyed while the cells are still within the vascular system,

**TABLE 5.1** Potential Causes of *in Vivo* and *in Vitro* Hemolysis

| |
|---|
| ***IN VIVO HEMOLYSIS*** |
| Extravascular hemolysis |
| Enzyme deficiencies (e.g., glucose-6-phosphate dehydrogenase deficiency) |
| Hemoglobinopathies (e.g., sickle cell, thalassemia) |
| Erythrocyte membrane defects (e.g., hereditary spherocytosis) |
| Infection (e.g., *Bartonella*, *Babesia*, malaria) |
| Autoimmune hemolytic anemia |
| Other (e.g., hypersplenism, liver disease) |
| Intravascular hemolysis |
| Microangiopathy (e.g., prosthetic heart valve, thrombotic thrombocytopenic purpura) |
| Transfusion reaction |
| Infection (e.g., sepsis, severe malaria) |
| Paroxysmal cold hemoglobinuria |
| Paroxysmal nocturnal hemoglobinuria |
| ***IN VITRO HEMOLYSIS*** |
| Excessive aspiration force (blood drawn too vigorously, especially through small or superficial veins) |
| Catheter partially obstructed |
| Blood forced into the tube from syringe |
| Specimen frozen |
| Mechanical damage (e.g., shaking, excessive force in pneumatic tubes) |
| Delay in analysis |

*Source: Data from Garby and Noyes [11].*

whereas extravascular hemolysis is due to destruction of red cells by the phagocytic system in the liver, spleen, or bone marrow. It is important to recognize and differentiate between *in vivo* and *in vitro* hemolysis because analytes such as potassium or LDH that might not be reported in samples showing *in vitro* hemolysis should be reported in samples with *in vivo* hemolysis. For example, increased potassium in a sample with *in vitro* hemolysis can be assumed to be artifactual and not representative of the patient's true potassium concentration. However, increased potassium measured in a patient with *in vivo* hemolysis represents the true intravascular potassium in the patient.

*In vivo* intravascular hemolysis is typically characterized by increased LDH, decreased haptoglobin, and increased urine hemoglobin concentration. The increased LDH is the result of release of the enzyme from erythrocytes, whereas the decreased haptoglobin is the result of binding of haptoglobin to free hemoglobin. Hemoglobin, which is a tetramer, is rapidly broken down in plasma to dimers, with the resultant hemoglobin–haptoglobin complex being rapidly cleared. The half-life of the hemoglobin–haptoglobin complex is approximately 10–30 min due to rapid elimination by the monocyte–macrophage system, whereas the half-life of free haptoglobin is 5 days [11]. Haptoglobin becomes saturated when free hemoglobin concentrations exceed approximately 150 mg/dL. Plasma hemoglobin not bound to haptoglobin is readily filtered through the glomerulus and will be excreted in the urine, resulting in hemoglobinuria. Note that low haptoglobin concentrations without *in vivo* hemolysis may be seen in newborns and young children and in those individuals with haptoglobin deficiency [12].

The free hemoglobin will produce a positive reaction for heme protein when measured using a urine dipstick. The free hemoglobin dimers that remain in circulation are oxidized to form methemoglobin, which dissociates to produce free heme and globin chains. The oxidized free heme binds to hemopexin and is removed from circulation by the liver, spleen, and bone marrow. Heme is further metabolized to eventually form unconjugated bilirubin. Measurement of hemopexin, methemoglobin, and unconjugated bilirubin can be helpful in the differentiation of *in vivo* intravascular or extravascular hemolysis from *in vitro* hemolysis. *In vivo* hemolysis due to immunohemolytic causes such as ABO transfusion reactions, paroxysmal cold hemoglobinuria, and idiopathic autoimmune hemolytic anemia or due to use of certain drugs can result in further hemolysis of sensitized erythrocytes during the process of blood collection, clot formation, and centrifugation. Thus, the collection of a sample anticoagulated with heparin can help eliminate further hemolysis that might occur *in vitro* during clot formation. Recommended criteria have been published for helping differentiate *in vivo* from *in vitro* hemolysis [13]:

- Collect both serum and plasma samples.
- Because anticoagulation of blood helps minimize *in vitro* hemolysis, perform all tests on plasma whenever *in vivo* hemolysis is suspected.
- Measure hemoglobin and potassium concentrations and LDH activity in the serum and plasma specimens.
- Specimens with increased LDH activity and hemoglobin concentrations but normal potassium concentrations indicate *in vivo* hemolysis.
- Measure haptoglobin, hemopexin, and the reticulocyte count to confirm *in vivo* hemolysis.

*CASE REPORT*[1] A 58-year-old female was admitted with a diagnosis of paroxysmal nocturnal hemoglobinuria (PNH). Results of her physical examination were normal. Laboratory results obtained on days 1, 4, 5, and 7 of her admission are as follows:

| Test (Reference Interval) | Specimen | Day 1 | Day 4 | Day 5 | Day 7 |
|---|---|---|---|---|---|
| Potassium (3.2–4.6 mmol/L) | P | | | 4.6 | 3.9 |
| | S | 4.9 | 5.5 | 5.7 | 4.6 |
| Lactate dehydrogenase (133–248 U/L) | P | | | 2830 | 3020 |
| | S | 2630 | 3170 | 3000 | 3305 |
| Hemoglobin (0–50 mg/L) | P | 52 | | 162 | 132 |
| | S | 297 | 301 | 221 | 210 |
| Haptoglobin (410–2100 mg/L) | S | <150 | <150 | <150 | <150 |
| Reticulocytes (0.5–1.5%) | | 10.6 | 9.5 | | |

P, plasma; S, serum.

Results for the plasma samples collected from this patient demonstrated increased plasma LDH and hemoglobin and a normal potassium concentration. These findings, along with low serum haptoglobin and increased reticulocyte count, strongly suggest *in vivo* hemolysis. Of interest was the finding of higher serum potassium and hemoglobin concentrations and LDH activity compared with plasma. These findings suggest that further lysis of erythrocytes occurred during the process of clotting of serum samples, with subsequent release of potassium, LDH, and hemoglobin into the serum. Thus, this patient had *in vivo* hemolysis due to PNH. In addition, there is concomitant *in vitro* hemolysis following the collection of blood into tubes not containing anticoagulant, with further lysis of erythrocytes occurring during the process of clot formation. It may be advantageous to collect blood from patients with immunohemolytic anemia into tubes containing an anticoagulant such as heparin in order to prevent further hemolysis of cells that might occur during clotting.

## *In Vitro* Hemolysis

Poor specimen quality is recognized as the most frequent source of errors in the pre-analytical phase of testing. Among samples submitted to the clinical laboratory for testing and that are deemed unsuitable for analysis, *in vitro* hemolysis accounts for approximately 40–70% of the cases [1,14,15]. *In vitro* hemolysis can occur at a variety of stages, including during phlebotomy, sample handling and processing, and storage. In addition, red cell fragility may be more pronounced in some patients, resulting in a higher likelihood of *in vitro* hemolysis. Table 5.2 lists the various causes of *in vitro* hemolysis.

*In vitro* hemolysis can cause a positive or negative bias in an analyte. The mechanisms causing bias with *in vitro* hemolysis include proteolysis of analytes due to release of intracellular compounds, release of thromboplastic substances, dilution of the analyte due to release of cytoplasmic contents, release of the analyte from erythrocytes, and analytical interference due to hemoglobin and other intracellular substances [15]. Analytes such as potassium, LDH, and AST that are

**TABLE 5.2** Common Causes of *in Vitro* Hemolysis and the Various Stages in which They Occur

| |
|---|
| Phlebotomy |
| Collection from catheter |
| Collection of capillary blood (finger stick or heel stick) |
| Needle gauge |
| Length of time that tourniquet is used |
| Fist clenching by patient |
| Tube underfilled |
| Vigorous mixing of sample following collection |
| Lack of mixing following collection |
| Transfer from syringe into Vacutainer tube |
| Specimen transport |
| Transport via pneumatic tube |
| Significant time delay in specimen transport to laboratory |
| Transport of specimen at inappropriate temperature |
| Specimen processing |
| Significant time delay between receipt of specimen and centrifugation |
| Centrifuge temperature extremes |
| Speed of centrifugation |
| Re-centrifugation of previous centrifuged specimen |
| Poor barrier separation if gel barrier tubes used |
| Specimen storage |
| Improper storage temperature following analysis |
| Length of time that specimen is stored following analysis |

[1] Adapted from Blank *et al.* [13].

present within erythrocytes at concentrations greater than approximately 10 times those seen in extracellular fluids will cause an increase following hemolysis. Also, the greater concentration for some analytes, such as potassium, neuron-specific enolase, and acid phosphatase, observed in serum compared with plasma is due to release of these compounds from platelets during fibrin clot formation [16]. Interference from hemoglobin may be due to the spectral properties of this compound, which has an absorbance peak at 420 nm and shows significant absorbance between 340 and 440 nm and between 540 and 580 nm. In addition to the spectral interference effects of hemoglobin, this compound may interfere due to the reactivity of its iron atoms, which can participate in oxidation-reduction reactions or in reactions that involve hydrogen peroxide (see Chapter 8 for examples of tests using hydrogen peroxide and peroxidase assays). Although many analytes are subject to interference effects from *in vitro* hemolysis, the influence of *in vitro* hemolysis on measured potassium concentrations is probably the most widely recognized. Based on the frequency at which potassium is measured in the clinical laboratory, along with the serious consequences of misdiagnosis of hypokalemia or hyperkalemia, this analyte is widely recognized as being affected by *in vitro* hemolysis.

The prevalence of *in vitro* hemolysis can vary widely depending on the patient population being tested; whether trained phlebotomists or inexperienced individuals are collecting the sample; and whether the sample is processed on-site or is sent to a remote site for processing, with significant time delays between collections and processing. With respect to the type of patient population that is being tested, the prevalence of *in vitro* hemolysis in outpatients has been found to be approximately 90 times less than that of samples collected from patients in the emergency department, in which studies often show *in vitro* hemolysis to be present in approximately 10% of samples [17]. Other hospital locations associated with a high prevalence of *in vitro* hemolysis include pediatric and neonatal wards, in which finger stick and heel stick samples are often collected and which are associated with high rates of *in vitro* hemolysis.

Poor blood collection practices contribute to *in vitro* hemolysis. A study that evaluated the root causes of *in vitro* hemolysis found that approximately 80% of the samples with *in vitro* hemolysis were attributed to aspirating blood too vigorously through a needle into a syringe (31%), collecting blood from a butterfly needle into a syringe (20%), collecting from an intravenous catheter into a syringe (17%), or collecting from an infusion port into a syringe (12%) [14]. Of interest, errors in handling, including freezing of specimens, accounted for 1% of the causes of *in vitro* hemolysis.

The use of automated analyzers for measuring plasma hemoglobin in patient blood samples has proven to be a reliable means of identifying *in vitro* hemolysis. Some instruments report *in vitro* hemolysis using a semiquantitative scale, whereas others report the actual plasma hemoglobin concentration. However, once identified, laboratories utilize a variety of different mechanisms for dealing with specimens that are hemolyzed, including outright rejection of hemolyzed samples, analysis of hemolyzed samples and reporting of results with a disclaimer stating that those analytes affected by *in vitro* hemolysis may be incorrect, or correction or adjustment of the measured analyte concentration using a correction factor based on the magnitude of hemolysis. The vast majority of laboratories either reject samples with *in vitro* hemolysis or analyze the samples and report the results with a comment stating that the results may not be accurate.

Reporting a result that has been corrected or adjusted based on the magnitude of *in vitro* hemolysis is not a recommended approach [18]. The wide range of correction factors that have been proposed for correcting measured potassium for *in vitro* hemolysis demonstrates how problematic the use of correction factors can be. For example, correction factors that have been proposed for adjusting potassium in samples with *in vitro* hemolysis range from an increase of 0.20 mmol/L of potassium per 100 mg/dL of plasma hemoglobin to an increase of 0.51 mmol/L per 100 mg/dL of plasma hemoglobin [18,19]. The wide range of correction factors that have been proposed highlights the fact that factors influencing hemolysis are not as simplistic as they seem. Factors such as interindividual variability in erythrocyte hemoglobin concentrations or in the concentration of erythrocytes, the effect of erythrocyte age on intracellular concentrations of various analytes, and differences in red cell membrane fragility likely contribute to the differences that are seen in recommended correction factors [19–21]. For example, decreased erythrocyte concentrations may lead to faster flow of cells through the needle during phlebotomy, resulting in increased shear forces and cell membrane rupture [22]. Also, variability in erythrocyte membrane permeability as a result of disease can impact the effects of *in vitro* hemolysis. Older erythrocytes have greater permeability to cations compared to younger cells, and they contain approximately half the potassium content of younger cells [23,24]. Hemolysis that occurs as a result of mechanical trauma to cells during blood collection is likely to cause lysis of older cells containing different concentrations of certain analytes, whereas hemolysis induced by lysis of all cells—due, for example, to freezing and thawing of blood—will induce lysis of all cells. In addition, conditions such as chronic lymphocytic leukemia have been associated with increased fragility of leukocyte membranes, with release of intracellular contents from these cells during the collection of blood [25].

The mechanisms causing *in vitro* hemolysis during collection of blood from a patient may be very different from the mechanisms that have been employed in studies designed to investigate the effect of *in vitro* hemolysis. Mechanisms used to simulate *in vitro* hemolysis include osmotic lysis of cells, freeze–thaw cycles, physical disruption of cells by forcing through a small-bore needle, physical disruption of clotted blood with a wooden applicator stick, and homogenization of whole blood in a blender [19,22,24,26–29]. In addition, some studies remove platelets and leukocytes prior to inducing hemolysis, whereas other studies do not. The use of these different methods to simulate *in vitro* hemolysis probably accounts for the wide variability in results that is sometimes reported regarding the effects of *in vitro* hemolysis on measured analyte concentrations.

Evaluation of the effects of hemolysis on laboratory test results using paired blood samples collected during the same phlebotomy draw, in which one sample is hemolyzed and the other sample is free of hemolysis, shows that the effects of *in vitro* hemolysis can be very different from those in studies in which hemolysis is introduced by artificial means [18]. Methods used to mimic *in vitro* hemolysis, such as the addition of a whole blood lysate prepared by the freezing and thawing of whole blood or osmotic lysis of cells following the addition of distilled water to packed cells, typically result in the lysis of all cells, including reticulocytes, mature erythrocytes, platelets, and leukocytes. However, *in vitro* hemolysis due to mechanical disruption of cells during blood collection typically does not result in lysis of all cells. Older erythrocytes are more prone to shear stresses that occur during sample collection compared with younger cells [30]. Shear stress that occurs during phlebotomy can result in the formation of erythrocyte membrane pores that allow the leak of small ions such as potassium but block passage of larger molecules such as LDH, AST, and hemoglobin [31]. Thus, the use of plasma hemoglobin as a marker of *in vitro* hemolysis may not show a direct relationship with the loss of small ions, such as potassium, from cells due to shear stress forces.

***CASE REPORT*** A 74-year-old male was seen by his physician for a routine physical. A general chemistry screen consisting of electrolytes and liver and renal function tests was ordered, and a blood sample was collected into a plain evacuated tube. The phlebotomist mentioned to the physician that she had a difficult time collecting blood from the patient due to the lack of "good veins," and she needed to use a 23-gauge needle to collect the sample because she was out of 21-gauge needles. After collecting the sample, the phlebotomist placed the tube of blood in a rack and took her lunch break. She returned 1 hr later, and she centrifuged the sample and removed the serum from the cells. The technologist who analyzed the sample observed that the serum had a pink to red color but failed to note this observation on the report that was sent to the physician. The following are laboratory results obtained on serum:

| Test (Reference Interval) | Result |
|---|---|
| Glucose (≤99 mg/dL) | 55 |
| Potassium (3.2–4.6 mmol/L) | 5.8 |
| Lactate dehydrogenase (133–248 U/L) | 322 |
| Aspartate transaminase (≤48 U/L) | 62 |
| Alanine aminotransferase (≤55 U/L) | 18 |
| Creatinine kinase (≤200 U/L) | 93 |

The results obtained from this patient were consistent for a sample that had been delayed in processing to remove the serum from cells and that was also hemolyzed. The relatively low glucose suggested some delay in processing, with metabolism of glucose by cells. The rate of disappearance of glucose in the presence of blood cells has been reported to be approximately 10 mg/dL per hour, but the rate increases with glucose concentration, temperature, and white blood cell count. The increased AST and LDH could be indicative of liver or skeletal muscle injury, but the normal alanine aminotransferase and normal creatinine kinase values suggested that liver or skeletal muscle injury was not present in this patient and that AST and LDH were likely increased due to *in vitro* hemolysis. The increased potassium was also consistent with *in vitro* hemolysis. In addition, the lengthy delay in processing of the sample could have led to leakage of potassium out of cells into the serum. The use of small-bore needles like the one used for collection of blood from this patient is discouraged because needles larger than 21 gauge have been found to be associated with increased rates of *in vitro* hemolysis. Small-bore needles can cause a larger vacuum force being applied to the blood, causing increased shear stress on cells and thus causing them to rupture.

## LIPEMIA

Unlike *in vitro* hemolysis, lipemia is something not easy to avoid. The following are frequently associated with lipemia: ethanol use; diabetes mellitus; hypothyroidism; chronic renal failure; pancreatitis; primary biliary cirrhosis; total parenteral nutrition; and the use of medications such as steroids, estrogen, and protease inhibitors. Interference from lipemia is due to the ability of lipoprotein particles to scatter light and displace plasma water in the sample. The scattering of light due to the turbidity caused by triglycerides can increase the absorbance values of solutions, thereby decreasing the operating scale for colorimetric methods. The interference effects due to lipemia can cause either increased or decreased values depending on whether a sample blank is used. If the turbidity is too great, no measurement may be possible due to limits on the linear range of the assay [32]. Methods based on nephelometry can be adversely affected as a result of light scatter.

The displacement of plasma water by high triglyceride concentrations can interfere with methods that are based on measurement in the aqueous portion of the sample and assume that all samples contain approximately 93% plasma water. Samples containing very high concentrations of triglycerides can contain much less plasma water per unit volume of sample due to the volume displacement effects of triglycerides. Pseudohyponatremia, caused by increased triglyceride concentrations in samples in which sodium is measured using an indirect method involving dilution of the sample prior to analysis, is one common example of interference encountered with lipemic samples (see Chapter 8 for more details on the water exclusion effect). Following centrifugation of a lipemic sample, or if the sample is allowed to stand for a period of time prior to analysis, the lipids in the sample are not evenly distributed but tend to concentrate within the upper portion of the sample. Thus, analytes dissolved in the aqueous phase of the sample will be at much lower concentrations in the upper layer, which is predominantly occupied by lipid. The converse is true for lipids and lipid-soluble components such as drugs that are present in higher concentrations in the upper layer consisting primarily of lipid.

In addition to the interference effects of lipemia caused by interference in spectrophotometric analysis and by displacement of plasma water, lipids can also cause interference by physical–chemical mechanisms. An analyte that is located primarily in the lipid layer may not be as accessible to reagents used in the measurement of the analyte or may not be accessible to an antibody used in immunoassay methods. Also, lipids may affect electrophoretic and chromatographic separation of analytes [33].

Lipemia causing test interference can be the result of recent food intake, altered or deranged lipid metabolism, or infusion of lipid-containing solutions. Plasma triglyceride concentrations, primarily in the form of chylomicrons, increase substantially between 1 and 4 hr after eating, and the chylomicrons and chylomicron remnants remain elevated for 6–12 hr afterwards. Patients should be fasted for at least 12 hr before blood samples are collected. For patients who are receiving parenteral infusion of lipids, the treatment should be stopped for at least 8 hr prior to collection of blood.

In whole blood, it is difficult to visually detect lipemia unless the triglyceride concentration is greater than 1000 mg/dL. In serum or plasma, visual detection is usually observed when triglyceride concentrations are greater than 300 mg/dL, although the ability of triglycerides to cause turbidity is dependent on whether the triglycerides are present primarily in the form of chylomicrons, very low-density lipoprotein (VLDL), or low-density lipoprotein (LDL) particles. Most analyzers are able to evaluate samples for lipemia by measuring at wavelengths above 600 nm.

A variety of methods have been proposed for removing lipids from serum or plasma. High-speed centrifugation is often performed because this procedure produces a clear infranatant. Other methods used include the extraction of lipids with organic solvents and the precipitation of lipids using various compounds. Some manufacturers add detergents or enzymes to their reagent in order to remove turbidity. Use of centrifugation to separate lipids depends on the type of lipids contributing to the interference. If lipemia is due to chylomicrons, centrifugation at 12,000 *g* can help separate chylomicrons. However, if the lipids present consist primarily of VLDL or LDL particles, much higher centrifugation is required. Lipemia can be removed from EDTA samples used for hematology by centrifugation. However, the cell-free supernatant that is removed must be replaced by an equal volume of saline.

The investigation of interference due to lipemia is difficult because the light scattering properties of chylomicrons and VLDL particles are dependent on the size of the particles present. VLDL is broken into three size categories referred to as small VLDL (27–35 nm), intermediate VLDL (35–60 nm), and large VLDL (60–200 nm). Only the intermediate and large VLDL particles are effective at scattering light. Chylomicrons are a heterogeneous mixture of particles that range in size from 70 to 1000 nm. Because VLDL particles and chylomicrons vary greatly in size and triglyceride content, measured triglyceride concentrations correlate poorly with sample turbidity [34].

The use of artificial fat emulsions such as Intralipid to mimic lipemia is often employed to establish the effect of lipemia on assay interference. Particles in Intralipid range in size from 200 to 600 nm. Thus, Intralipid misses the range of values for large VLDL particles and misses the lower and upper ranges seen with chylomicrons [35]. Therefore, extreme care should be exercised when interpreting the results of interference studies that use samples with Intralipid or other synthetic emulsions added to mimic lipemia [9]. Such artificially derived samples may not behave in the same manner as true lipemic samples obtained from patients.

*CASE REPORT* A 20-year-old female with a history of diabetes presented to the emergency department 7 months following discontinuation of her insulin therapy. She was diagnosed with diabetic ketoacidosis, acute pancreatitis, and severe hypertriglyceridemia. Values obtained upon admission and 24 hr later are as follows:

| Test (Reference Interval) | Result, Admission | Result, 24 hr |
|---|---|---|
| Glucose (≤ 99 mg/dL) | 312 | 158 |
| Triglycerides (<150 mg/dL) | 15,300 | 13,500 |
| Sodium (134–143 mmol/L) | 125 | 122 |
| Lipase (< 85 U/L) | 458 | 616 |
| Amylase (< 200) | 154 | 955 |

The sodium in this patient was measured using indirect potentiometry, in which the sample is diluted prior to analysis by the ion-selective electrode. This method of sodium analysis is known to be adversely affected by the presence of compounds such as lipids, which displace the aqueous portion of the sample. Use of an alternate method for measuring sodium would be appropriate in this case. This patient presented with severe abdominal pain and was diagnosed with pancreatitis. The elevated lipase measured upon admission is consistent with pancreatitis. However, the normal amylase measured in this same sample was an unexpected finding. Centrifugation of the sample to remove the lipemia and re-analysis of the sample showed the amylase in the sample collected 24 hr following admission to be 955 U/L. The laboratory director confirmed that the method used for analysis of amylase is affected by lipemia, which causes falsely decreased values [36].

## ICTERUS

Increased concentrations of bilirubin are another source of endogenous interference. Bilirubin has high absorbance between wavelengths of 340 and 500 nm. Thus, methods that rely on measurement at these wavelengths are prone to interference effects from bilirubin. In addition to the spectral properties of bilirubin, this compound can also react chemically with reagents. For example, bilirubin can react in oxidase/peroxidase-based assays such as those used for measurement of glucose, cholesterol, triglycerides, creatinine, and uric acid [37]. Bilirubin reacts with hydrogen peroxide formed in the test system, resulting in lower than expected test results. Also, bilirubin can interfere with dyes that bind to albumin. The reduction in absorption of bilirubin due to oxidation in an alkaline environment is the main cause for bilirubin interference in some versions of the Jaffe method for creatinine. In a strongly acidic environment, the absorption of conjugated bilirubin shifts to the ultraviolet wavelengths that can result in interference with phosphate when the phosphomolybdate method is used [38].

Visual inspection of samples for the detection of hyperbilirubinemia is not very sensitive. Samples containing hemolysis make detection even more difficult. Automated detection of bilirubin by analyzers that assess the presence of common interfering substances is the recommended method for identifying samples that are icteric. Spectral interference from bilirubin can be eliminated by the use of blanking procedures. However, chemical interference caused by bilirubin cannot be eliminated by blanking. The addition of potassium ferrocyanide or bilirubin oxidase to the reagent can eliminate interference from bilirubin in methods based on formation of hydrogen peroxide (see Chapter 8).

## METHODS FOR EVALUATING THE EFFECT OF ENDOGENOUS INTERFERING SUBSTANCES ON LABORATORY TESTS

Guidelines have been established for the assessment of interference effects, and various experimental designs have been advocated by others [4,39–41]. One common method used to evaluate the effect of an interfering substance is to measure the analyte of interest by use of an alternate method known to not be affected by the interfering substance. The most common approach used to assess interference effects is to add serially higher concentrations of the interfering substance to aliquots of the same matrix and then measure the substance of interest in each aliquot. The effect of the interfering substances can be assessed by use of regression analysis by plotting measured analyte concentration versus the concentration of interfering substances. If the slope obtained from the regression analysis differs by some predetermined value (i.e., ± 10%), then interference is determined to be present. Unfortunately, most of the experimental models used to assess the effect of interfering substances make the assumption that interference is not related to the concentration of the analyte being measured [39–41]. For example, a 0.5-mg/dL positive bias caused by 150 mg/dL of plasma hemoglobin observed in a sample with a baseline total bilirubin concentration of 0.5 mg/dL would represent a 100% increase in total bilirubin. This finding would be extrapolated to infer that a sample with a baseline total bilirubin concentration of 10.0 mg/dL would show a measured concentration of 20 mg/dL if *in vitro* hemolysis of 150 mg/dL of plasma hemoglobin was present. Unfortunately, this type of analysis may be inappropriate and misleading when applied to analytes where the effect of interfering substances is dependent on analyte concentration.

At least three types of interference have been demonstrated to contribute to the effect of interfering substances [42]: analyte-dependent, where the magnitude of the interference effect is dependent on the concentration of the analyte of interest; analyte-independent, where the magnitude of the interference effect is constant regardless of the analyte concentration; and a combination of the previous two, where the effect of an interfering substance is dependent on both the concentration of the analyte and the concentration of the interfering substance. A model for assessing the relative contribution of each of these three types of interference effects has been described [42]. Briefly, this model involves the creation of a series of aliquots containing the analyte of interest at various concentrations. Each aliquot is then subdivided, and various concentrations of suspected interfering substance are added. This results in a matrix containing the analyte of interest at several different concentrations, with each concentration containing the suspected interfering substance at various concentrations.

## CONCLUSIONS

The effects of endogenous interfering substances on laboratory test results have been widely studied, often with conflicting or inconclusive results. Endogenous interfering substances can cause erroneous test results and lead to inappropriate patient care if acted upon. Collection of another sample not containing the interfering substance, or providing an alert that the test result may be inaccurate due to the presence of an interfering substance, is recommended. Many instruments now evaluate and report the level of endogenous interfering substances that may be present. Utilization of these automated methods for assessing endogenous interfering substances should help identify samples that may be inappropriate for analysis. Manufacturers' information on the effects of endogenous interfering substances on specific test methods is often a useful guide as to which analytes are subject to these effects. In addition to the information provided by manufacturers, a number of guidelines are available that can help laboratorians evaluate the effects of interfering substances.

## References


[1] Jones BA, Calam RR, Howanitz PJ. Chemistry specimen acceptability: a college of american pathologists Q-probes study of 453 laboratories. Arch Pathol Lab Med 1997;121:19–26.

[2] Kohn Linda T, Corrigan Janet M, Donaldson Molla S. To err is human: building a safer health system. Washington, DC: Committee on Quality of Health Care in America. Institute of Medicine. National Academy Press; 2000.

[3] Plebani M, Carraro P. Mistakes in a stat laboratory: types and frequency. Clin Chem 1997;43:1348–51.

[4] Kroll MH, Elin RJ. Interference with clinical laboratory analyses. Clin Chem 1994;40:1996–2005.

[5] Witte DL, VanNess SA, Angstadt DS, Pennell BJ. Errors, mistakes, blunders, outliers, or unacceptable results: how many? Clin Chem 1994;43:1352–6.

[6] Simundic AM, Topic E, Nikolac N, Lippi G. Hemolysis detection and management of hemolyzed specimens. Biochem Med 2010;20:154–9.

[7] Ryder K, Glick M, Glick S. Incidence and amount of turbidity, hemolysis, and icterus in serum from outpatients. Lab Med 1991;22:415–18.

[8] Vermeer HJ, Thomassen E, deJonge N. Automated processing of serum indices used for interference detection by the laboratory information system. Clin Chem 2005;51:244–7.

[9] Bornhorst JA, Roberts RF, Roberts WF. Assay-specific differences in lipemic interference in native and intralipid-supplemented samples. Clin Chem 2004;50: 2197–2001.

[10] Kazmierczak SC, Robertson AF, Briley KP. Comparison of hemolysis in blood samples collected using an automatic incision device and a manual lance. Arch Pediatr Adolesc Med 2002;156:1072–4.

[11] Garby L, Noyes WD. Studies on hemoglobin metabolism: I. The kinetic properties of the plasma hemoglobin pool in normal man. J Clin Invest 1959;38:1479–83.

[12] Van Lente F, Marchand A, Galen RS. Evaluation of a nephelometric assay for haptoglobin and its clinical usefulness. Clin Chem 1979;25:2007–10.

[13] Blank DW, Kroll MH, Ruddel ME, Elin RJ. Hemoglobin interference from *in vivo* hemolysis. Clin Chem 1995;31:1566–9.

[14] Carraro P, Servidio G, Plebani M. Hemolyzed specimens: a reason for rejection or a clinical challenge? Clin Chem 2000;46:306–7.

[15] Lippi G, Blanckert N, Bonini P, et al. Haemolysis: an overview of the leading cause of unsuitable specimens in clinical laboratories [review]. Clin Chem Lab Med 2008;46:764–72.

[16] Luddington R, Peters J, Baker P, Baglin R. The effect of delayed analysis or freeze-thawing on the measurement of natural anticoagulants, resistance to activated protein C and markers of activation of the haemostatic system. Thromb Res 1997;87:577–81.

[17] Lippi G, Salvagno GL, Favaloro EJ, Guidi GS. Survey on the prevalence of hemolytic specimens in an academic hospital according to collection facility: Opportunities for quality improvement [letter]. Clin Chem Lab Med 2009;47:616–18.

[18] Mansour MMH, Azzazy HME, Kazmierczak SC. Correction factors for estimating potassium concentrations in samples with in vitro hemolysis: a detriment to patient safety. Arch Pathol Lab Med 2009;133:960–6.

[19] Hawkins R. Variability in potassium/hemoglobin ratios for hemolysis correction [letter]. Clin Chem 2002;48:796.

[20] Romero PJ, Romero EA, Winkler MD. Ionic calcium content of light dense human red cells separated by Percoll density gradients. Biochim Biophys Acta 1997;1323:23–8.

[21] Kazmierczak SC, Castellani WJ, Van Lente F, Hodges ED, Udis B. Effect of reticulocytosis on lactate dehydrogenase isoenzyme distribution in serum: *in vivo* and *in vitro* studies. Clin Chem 1990;36:1638–41.

[22] Dimeski G, Clague AE, Hickman PE. Correction and reporting of potassium results in hemolyzed samples. Ann Clin Biochem 2005;42:119–23.
[23] Hentschel WM, Wu LL, Tobin GO, et al. Erythrocyte cation transport activities as a function of cell age. Clin Chim Acta 1986;157:33–43.
[24] Brydon WB, Roberts LB. The effect of hemolysis on the determination of plasma constituents. Clin Chim Acta 1972;41:435–8.
[25] Colussi G, Ciprini D. Pseudohyperkalemia in extreme leukocytosis. Am J Nephrol 1995;15:450–2.
[26] Jay DW, Provasek D. Characterization and mathematical correction of hemolysis interference in selected Hitachi 717 assays. Clin Chem 1993;39:1804–10.
[27] Owens H, Siparsky G, Baja L, Hampers LC. Correction of factitious hyperkalemia in hemolyzed specimens. Am J Emerg Med 2005;23:872–5.
[28] Lippi G, Salvagno GL, Montagnana M, Brocco G, Guidi GC. Influence of hemolysis on routine clinical chemistry testing. Clin Chem Lab Med 2006;44:311–16.
[29] Pai SH, Cyr-Manthey M. Effects of hemolysis on chemistry tests. Lab Med 1991;22:408–10.
[30] Rifkind JM, Araki K, Hadley EC. The relationship between the osmotic fragility of human erythrocytes and cell age. Arch Biochem Biophys 1983;222:582–9.
[31] Kinosita Jr K, Tsong TY. Hemolysis of human erythrocytes by a transient electric field. Proc Natl Acad Sci 1977;74:1923–9.
[32] Artiss JD, Zak B. Problems with measurement caused by high concentrations of serum lipids. CRC Crit Rev Clin Lab Sci 1987;25:19–41.
[33] Creer MH, Ladenson J. Analytical error due to lipemia. Lab Med 1983;14:351–5.
[34] Sonntag O, Glick MR. Serum-Index und interferogram—Ein neuer weg zur prufung und darsetllung von interferenzen durch serumchromogene. Lab Med 1989;13:77–82.
[35] Kroll MH. Evaluating interference caused by lipemia [editorial]. Clin Chem 2004;50:1969–70.
[36] Hahn SJ, Park JH, Lee JH, et al. Severe hypertriglyceridemia in diabetic ketoacidosis accompanied by acute pancreatitis: case report. J Korean Med Sci 2010;25:1375–8.
[37] Perlstein MT, Thibert RJ, Watkins R, Zak B. Spectrophotometric study of bilirubin and hemoglobin interactions in several hydrogen peroxide generating procedures [abstract]. Clin Chem 1977;23:1133.
[38] Duncanson GO, Worth HGI. Pseudohypophosphatemia as a result of bilirubin interference. Ann Clin Biochem 1990;27: 263–7.
[39] Clinical and Laboratory Standards Institute. Interference testing in clinical chemistry: approved guideline – second edition. Wayne, PA: Clinical and Laboratory Standards Institute; 2005 [CLSI document EP7-A2].
[40] Galteau MM, Siest G. Drug effects in clinical chemistry: 2. Guidelines for evaluation of analytical interferences. J Clin Chem Clin Biochem 1984;22:275–9.
[41] Letellier G, Desjarlais F. Analytical interference of drugs in clinical chemistry: 1. Study of twenty drugs on seven different instruments. Clin Biochem 1985;18:345–51.
[42] Kroll MH, Ruddel M, Blank DW, Elin RJ. A model for assessing interference. Clin Chem 1987;33:1121–3.

CHAPTER

# 6

# Immunoassay Design and Mechanisms of Interferences

*Pradip Datta*

Siemens Healthcare Diagnostics, Tarrytown, New York

## INTRODUCTION

Immunoassays are used in clinical laboratories for analysis of a variety of analytes, including hormones, serum proteins, antibodies to infectious or allergic agents, therapeutic drug monitoring, and drugs of abuse testing. These immunoassays exhibit high sensitivity and a broad analytical range. In the early days of immunoassays, the methods were laborious, requiring skilled labor. However, revolutionary developments in immunoassay automation during the past 20 years resulted in fast and effective ways of analyzing many analytes, and currently more than 100 immunoassays are commercially available. Most immunoassay methods use specimens without any pretreatment; are easy to use; and are run on fully automated, continuous, high-throughput, random access systems. These assays use very small amounts of sample volumes (10-100 μL); reagents may be stored in the analyzer; most have stored calibration curves on the automated analyzer system, often stable for 1 or 2 months; and results can be reported in 10–30 min. Immunoassays offer fast throughput, automated rerun, autoflagging (to alert for poor specimen quality such as hemolysis), and high sensitivity and specificity. Results can be reported directly into laboratory information systems. Immunoassays measure the analyte concentration in a specimen by forming a complex with a specific binding molecule, which in most cases is an analyte-specific antibody (or a pair of specific antibodies). The complex generates signal (e.g., sandwich immunoassay), which is then converted into the analyte concentration via a "calibration curve." The immunoreaction is further utilized in various formats and labels, giving a whole series of immunoassay technologies, systems, and options. However, like every analytical method, immunoassays suffer from interferences that are described later in this chapter.

## IMMUNOASSAY METHODS AND ASSAY PRINCIPLES

Immunoassays are homogeneous or heterogeneous in design, and they have different assay formats and different types of signal generation (Table 6.1). Immunoassays are classified by assay format as either competition or immunometric (commonly referred to as "sandwich") assays. Competition immunoassays work best for analysis of small molecules, requiring a limited amount of a single analyte-specific antibody and labeled analyte. In competition immunoassays, the analyte in the specimen competes with the labeled analyte for limited antibody binding sites, and the signal is measured either without separation of bound labeled antigen–antibody complexes from free labeled antigen–antibody complexes (homogenous format) or after separation of bound from free labeled antigen–antibody complexes (heterogeneous format). On the other hand, sandwich immunoassays are used mostly for large molecules, such as proteins or peptides, and utilize two different specific antibodies in the assay design. In sandwich immunoassays, the analyte in the specimen binds to two different antibodies that recognize two separate binding sites in the antigen molecule (different epitopes); one antibody may be conjugated to a solid phase and the other to a label. The bound complex is separated from other assay components by a proper washing protocol, and the relevant amount of label produces the signal. The signal

*Accurate Results in the Clinical Laboratory.*
DOI: http://dx.doi.org/10.1016/B978-0-12-415783-5.00006-2

TABLE 6.1 Examples of Various Types of Commercially Available Immunoassays

| Immunoassay Types | Example | Assay Signal |
|---|---|---|
| Competition Immunoassays (for small molecules: | FPIA* | Fluorescence polarization |
| Molecular weight <1000 Dalton) | (Abbott) Therapeutic drugs Abused drugs | |
| | EMIT®* (Siemens) Therapeutic drugs Abused drugs | Absorbance at 340 nm (Enzyme Modulation) |
| | CEDIA®* (Thermo Fisher) Therapeutic drugs Abused drugs | Colorimetry (Enzyme Modulation) |
| | KIMS®* (Roche) Abused drugs | Optical detection in the visible wavelength region |
| | LOCI* (Siemens) Various analytes | Chemiluminescence |
| Sandwich TIA* (Siemens, Roche) | Turbidimetry, latex | |
| (Analytes, MW > 1000 D). Serum Proteins | CLIA (Siemens) Hormones, proteins | Chemiluminescence |
| | CLIA (Abbott) Hormones, proteins | Chemiluminescence |
| | CLIA (Beckman) Hormones, proteins | Chemiluminescence |
| | CLIA (Roche) Hormones, proteins | Electro-chemiluminescence |
| | CLIA- ELISA (Siemens) Hormones, proteins | Chemiluminescent |

*Homogeneous assays.*

in sandwich assays is directly proportional to the analyte concentration; with low background noise, this type of immunoassay can be highly sensitive, capable of detecting very low concentrations of the analyte. The signals generated are mostly optical—absorbance, fluorescence, or chemiluminescence.

Depending on the need of separation between the bound labels (labeled antigen–antibody complex) versus free labels, the immunoassays may be homogeneous or heterogeneous. In the former, the bound label has different properties than the free label, and no physical separation between the two is needed. For example, in fluorescent polarization immunoassay (FPIA), the free label (which is a relatively small molecule—a few hundred daltons) has different Brownian motion than when label is complexed to a large antibody (≥140 kDa). This difference in the fluorescence polarization properties of the label is utilized to quantify the analyte from the signal generated [1]. In another type of homogeneous immunoassay, an enzyme is the label, and the activity of the enzyme may be lost if the labeled antigen is complexed with an antibody. This format is the basis of the enzyme multiplied immunoassay technique (EMIT) and cloned enzyme donor immunoassay (CEDIA) methods [2,3]. In the EMIT method, the label enzyme, glucose-6-phosphate dehydrogenase, is active unless the labeled antigen is bound to the antibody. The active enzyme reduces nicotinamide adenine dinucleotide (NAD) to NADH, and the absorbance is monitored at 340 nm. Similarly, in the CEDIA method, two genetically engineered inactive fragments of the enzyme β-galactosidase are coupled to the antigen and the antibody reagents. When they combine, the active enzyme is produced, and the substrate—a chromogenic galactoside derivative—produces the assay signal. In a third commonly used format of homogeneous immunoassay (turbidimetric immunoassay (TIA)), analytes (antigen) or their analogs are coupled to colloidal particles made of latex, for example [4]. Because antibodies are bivalent, the latex particles agglutinate in the presence of the antibody. However, in the presence of free analyte in the specimen, there is less agglutination and the resulting turbidity can be monitored as end point or as rate. Another example of homogeneous chemiluminescent immunoassay technology is the luminescent oxygen channeling immunoassay (LOCI), in which the immunoassay reaction is irradiated with light generating singlet oxygen molecules in microbeads (Sensibeads) coupled to the analyte. When bound to the respective antibody molecule, also coupled to another type of bead (Chemibead), Chemibeads react with singlet oxygen and chemiluminescence signals are generated, proportional to the concentration of the analyte–antibody complex. This technology is used in the Siemens Dimension Vista automated assay system [5]. In the kinetic interaction of microparticle in solution (KIMS) assay, in the absence of antigen molecules, free antibodies bind to drug microparticle conjugates forming particle aggregates that result in an increase in absorption, which is optically measured at various visible wavelengths (500–650 nm). When antigen molecules are present in the specimen, they bind with free antibody molecules and prevent formation of particle aggregates, resulting in diminished absorbance in proportion to the drug concentration. The On-Line Drugs of Abuse Testings immunoassays marketed by Roche Diagnostics are based on the KIMS format.

In heterogeneous immunoassays, on the other hand, the bound label is physically separated from the unbound label, and the generated signal is measured.

The separation is often done magnetically, where the reagent analyte (or its analog) is provided coupled to paramagnetic particles (PMPs), and the antibody is labeled. Conversely, the antibody may also be provided conjugated to the PMPs, and the reagent analyte may carry the label. After separation and washing, the bound label is reacted with other reagents to generate the signal. This is the mechanism in many chemiluminescent immunoassays (CLIA), in which the label may be a small molecule that generates a chemiluminescent signal. Examples of immunoassay systems in which the chemiluminescent labels generate signals by chemical reaction are the ADVIA Centaur from Siemens and the Architect from Abbott [6]. An example in which the small label is activated electrochemically is the ELECSYS automated immunoassay system from Roche Diagnostics [7].

The label may also be an enzyme (enzyme-linked immunosorbent assay (ELISA)) that generates chemiluminescent, fluorometric, or colorimetric signal depending on the enzyme substrates used. Examples of commercial automated assay systems using ELISA technology and chemiluminescent labels are Immulite from Siemens and ACCESS from Beckman-Coulter [8,9]. Another type of heterogeneous immunoassay uses polystyrene particles. If these particles are micro-sizes, the type of assay is called microparticle enhanced immunoassay [10]. In older immunoassay formats, the labels were radioactive: radioimmunoassay or RIA. Today, RIA is rarely used due to safety and waste disposal issues involving radioactive materials.

The main reagent in the immunoassay is the binding molecule, which is most commonly an analyte-specific antibody or its fragment. Several types of antibodies or their fragments are used in immunoassays, including polyclonal antibodies, which are raised in animals after the analyte (as antigen) along with an adjuvant are injected into the animals. Small-molecular-weight analytes are most commonly injected as conjugates to a large protein. The animal's sera is monitored for the appearance of analyte-specific antibodies, and when a sufficient concentration of the antibody is reached, the animal is bled. The serum can be used as the analyte-specific binder in an immunoassay; however, in most cases, antibodies are purified from serum and used in clinically available assays. Because there are many clones of the antibodies specific for the analyte, these antibodies are called polyclonal. In newer technologies, a plasma cell of the animal can be selected to produce the optimum antibody and then can be fused to an immortal cell. The resulting tumor cell grows uncontrollably, producing only the single clone of the desired antibody. Such antibodies, called monoclonal antibodies, may be grown in live animals or cell culture. There are several benefits of monoclonal antibodies over polyclonal ones. First, the characteristics of polyclonal antibodies are dependent on the animal producing the antibodies; if the source individual animal must be changed, the resultant antibody may be quite different. Second, because polyclonal antibodies constitute many antibody clones, polyclonal antibodies may have less specificity than monoclonal antibodies [11–13]. Sometimes, instead of using the whole antibody, fragments of the antibody, generated by digestion of the antibody by peptidases (e.g., Fab, Fab′, or their dimeric complexes), are also used as reagents.

The other main reagent component of the immunoassay is the label. There are many different kinds of labels, generating different kinds of signals. For example, use of acridinium ester labels, when treated with peroxide, produces chemiluminescent signals. As described previously, an enzyme may be used as the label, which can generate different types of signals depending on the substrate used for the enzyme.

Although the immunoassay methods are now widely used, there are limitations and drawbacks to these methods. One of them is the limitation of the binding molecule (i.e., the antibody) in terms of its specificity. Many of the endogenous metabolites of the analyte, especially if it is a drug molecule, may have structural recognition motifs that are very similar to that of the analyte. There are also other molecules different from the analyte but that produce comparable recognition motifs as the analyte. These molecules are generally called cross-reactants. When present in the sample, these molecules produce falsely elevated (or, in some instances, falsely lower) results in the relevant immunoassay [11–13]. Because monoclonal antibodies are more specific, immunoassays employing such antibodies in reagents suffer less from cross-reactivity than do polyclonal antibody-based assays, although this may not be the case for all analytes. Other components in a specimen (e.g., bilirubin, hemoglobin, protein, or lipid) may interfere in the immunoassay by interfering with the assay signal and thus produce incorrect assay results. A third type of immunoassay interference involves endogenous human antibodies in the specimen, which may interfere with the assay reagent components, assay antibodies, or the antigen labels. Such interference may be due to the presence of heterophilic antibodies or various human anti-animal antibodies in the specimen.

## SPECIMEN TYPES USED IN IMMUNOASSAYS

Serum and plasma are the most common types of specimens used in immunoassays. Whole blood specimens must be used for some analytes, such as the immunosuppressant drugs (cyclosporine, tacrolimus, sirolimus, and everolimus), although the immunosuppressant drug mycophenolic acid can be monitored in serum or plasma. Urine is the most commonly used specimen in drugs of abuse testing. Urine samples are less frequently affected by hemoglobin or icterus. Turbidity interference is possible in urine, but the cause is most likely bacterial growth or urate precipitation. Preservatives in urine, such as acetic acid, boric acid, or alkali, may interfere in some urine assays. Cerebrospinal fluid (CSF) specimens are used for monitoring the integrity of the blood–brain barrier (by analyzing plasma proteins) or infection in CSF. The most common CSF interfering substance is blood contamination with hemolysis. Interference from turbidity is also possible in such specimens. Other types of specimens used for immunoassays are saliva, sweat, tears, ascitic and stomach fluids, and bronchial secretions. Hair and nail specimens have been used to provide evidence of a longer term history of drug abuse [14,15]. Amniotic fluid, cord blood, meconium, and breast milk have been used to determine fetal and perinatal exposure to drugs [16].

## EXAMPLES OF IMMUNOASSAYS

Immunoassays are commercially available for a variety of analytes: proteins, hormones, tumor markers, rheumatoid factor, troponin, small peptides, steroids, and drugs, including the following:

- Anemia markers: ferritin, folate, vitamin $B_{12}$
- Autoantibodies: to diagnose and monitor autoimmune diseases
- Cardiovascular disease markers: troponin, B-natriuretic peptide, myoglobin, etc.
- Diabetes markers: insulin, C-peptide, HbA1c
- Drugs of abuse
- Hormones: reproductive, gastrointestinal, metabolic, etc.
- Infectious diseases: antibodies to and antigens from infectious micro-organisms
- Liver disease markers
- Therapeutic drug monitoring: antibiotics, anticoagulation, anticonvulsant, antidepression, anti-inflammation, cardioactive, or neoplastic drugs
- Thyroid markers: T4, T3, thyroid-stimulating hormone (TSH), etc.
- Tumor markers: cancers of breast, colon, prostate, lung, ovary, etc.

## PITFALLS IN IMMUNOASSAYS

Although immunoassays are widely used in the clinical laboratory, they suffer from the following types of interferences, rendering false-positive or false-negative results:

1. Endogenous interfering components that interfere nonspecifically via "matrix effects" or in signal generation; for examples, bilirubin, hemoglobin, lipids, and paraproteins may interfere with immunoassays using serum or plasma specimens.
2. Interferences from the endogenous and exogenous components, which cross-react with the antibodies used to detect the analyte.
3. System or method-related errors, such as pipetting probe contamination and carryover.
4. Prozone (or "hook") effect: Depending on the concentrations of reagent antibodies used in the assay, very high levels of antigen may reduce the concentrations of "sandwich" (antibody 1–antigen–antibody 2) complexes (which generate the assay signal), instead forming mostly single antibody–antigen complexes. The hook effect causes significant false-negative results.
5. Heterophilic interference is caused by endogenous human antibodies in the sample. These antibodies bind to assay components, generating false results. These interferences may also be caused by autoantibodies, macro-analytes (endogenous conjugates of analyte and antibody), macro-enzymes, and rheumatoid factors.

It is important to recognize the sources of such discordant results and conduct follow-up studies to provide clinically meaningful results. In Chapter 5, interference of endogenous bilirubin and lipemia was discussed along with the effect of hemolysis on clinical laboratory test results. Many drug metabolites cross-react with the antibody against the parent drug—for example, the cross-reactivity of carbamazepine-10,11-epoxide, an active metabolite of carbamazepine, with various carbamazepine immunoassays. Digoxin immunoassays suffer from interferences from endogenous digoxin-like immunoreactive substances as well as from a variety of drugs (spironolactone, potassium canrenoate, and their common metabolite canrenone) and certain herbal supplements. See Chapter 13 for an in-depth discussion of interferences in therapeutic drug monitoring assays. In this chapter, the interference of heterophilic antibodies with various immunoassays is discussed in detail because this topic is not addressed in-depth in other chapters.

### Interference from Heterophilic Antibodies

Heterophilic antibodies are human antibodies that interact with assay antibodies, causing false-positive or false-negative results. The heterophilic antibodies are polyclonal and heterogeneous in nature, and they comprise the following types: (1) heterophilic antibodies, which interact poorly and nonspecifically with the assay antibodies; (2) anti-animal antibodies, which interact strongly and specifically with the assay antibodies; (3) autoantibodies, which are endogenous human antibodies that interfere with an assay; (4) therapeutic antibodies, which are antibodies given therapeutically that interfere with an assay; (5) macro-analytes or macro-enzymes, which are oligomeric or polymeric conjugates of an analyte and/or conjugated with endogenous antibody; and (6) rheumatoid factors. Heterophilic antibodies may arise in a patient in response to exposure to certain animals or animal products, due to infection by bacterial or viral agents, or nonspecifically.

Although many of the immunoglobulin (Ig) clones in normal human serum may display anti-animal antibody properties, only those antibodies with sufficient titer and affinity toward the reagent antibody used in assay may cause clinically significant interference. Among the anti-animal antibodies, the most common are human anti-mouse antibodies (HAMA) because of the wide use of murine monoclonal antibody products in therapy or imaging. Heterophilic antibody and anti-animal antibody interferences are often classified together as heterophilic antibody interferences. Such interferences have been mostly found with immunometric sandwich assays and less often with competition assays. Sample dilution and depletion or removal of interfering antibodies has been recommended to remove heterophilic antibody interference. A patient history of exposure to animals or animal products or autoimmune diseases alerts laboratory professionals to the possibility of encountering heterophilic antibody interference in an assay.

Because heterophilic antibodies are found mainly in serum, plasma, or whole blood, but not in urine, such interference is absent in analysis of urine specimen for the same analyte. This provides an excellent way to detect the interference for analytes that may be present in both matrices. For example, many case studies with false-positive human chorionic gonadotropin (hCG) in serum/plasma have been described in the literature. In every case, if β-hCG could have been measured in parallel urine samples, the false results and the resulting dire consequences could have been easily avoided [17,18].

Heterophilic antibody interference may cause critical impact and clinical misjudgment, resulting in unnecessary follow-up testing and unneeded but potentially dangerous therapy, leading to significant patient morbidity, especially when due to a false-positive hCG (also a cancer marker) measurement in serum without investigating a parallel urine specimen. The fact that such interferences may not be suspected from the patient history, or that such an effect may be transient in nature, complicates the responsibility of the clinical laboratory to report accurate patient results.

### *Mechanism of Heterophilic Antibody Interference*

The interfering antibodies exert their effect by interacting with the assay antibody/antibodies or the analyte. In many cases, especially with heterophilic antibodies, the interactions among interfering antibodies and assay antibodies are nonspecific and with lower avidity. In such cases, if the assay employs equilibrium conditions, no interference is found. However, most immunoassays are performed on automated analyzers. Because clinicians demand increasingly faster laboratory results, most autoanalyzers use immunoassays under nonequilibrium conditions and thus are more prone to assay interferences.

In the sandwich-type immunoassays, heterophilic antibodies can form the "sandwich complex" even in the absence of the target antigen, generating mostly false-positive results. However, if the interfering antibody binds with only one of the assay reagent antibodies (capture or label), false-negative results are observed, although less frequently than false-positive results. In general, competition immunoassays are less affected by antibody interference. Two-site, immunometric sandwich assays (e.g., cardiac troponin I and human chorionic gonadotropin) are more affected by heterophilic antibody interference.

### *Interference from Heterophilic Antibody*

Heterophilic antibodies are poorly defined human antibodies that cause interference by noncompetitive binding mostly to the Fc region of assay antibodies; however, instances of heterophilic antibody binding to other parts of the assay antibody (e.g., idiotope or the "hinge" region) have also been reported. Heterophilic antibodies are found more often in sick and hospitalized patients, with reported prevalences of 0.2–15%. However, during approximately the past decade, most commercial assays have started incorporating blocking reagents against heterophilic antibodies in their assay reagent formulation, reducing interference from heterophilic antibodies. Although a 2002 study based on a literature review concluded that the incidence of false results arising from interference of heterophilic antibodies in various immunoassays is only 0.05% [19], a 2005 study reported a prevalence of 0.2–3.7% of heterophilic antibody interference (measured by assay responses with and without an interference-blocking reagent) in eight automated tumor marker immunoassays [20]. Heterophilic antibodies may be present in serum in response to microbial infections in patients [21]. However, interferences from specific anti-animal antibodies and complement proteins may often be reported together as heterophilic antibody interference. Because heterophilic antibodies are very heterogeneous in nature, and their concentrations may differ among individuals, no blocking-reagent can guarantee 100% protection against such interference. Thus, it is important to note that although the frequency of heterophilic antibody interference may have been reduced, recently reported cases have a large magnitude of interference, affecting the clinical interpretation of test results. There are many examples of incorrect results caused by heterophilic antibodies, including calcitonin [22], thyroid [23], and hormones and tumor markers [24]. Examples of false-negative α-fetoprotein (AFP) concentrations in serum, thus causing incorrect diagnoses in second-trimester Down syndrome screening, have been documented by Mannings et al. [25]. The authors used a different AFP assay to identify the incorrect results. Because the false-negative results were corrected after storage of the samples at 4ºC for 1 week, the authors concluded that complement proteins acted as interfering factors.

### *Interference from Human Anti-Animal Antibodies*

Human anti-animal antibodies (HAAA) arise most often when the patient is exposed to a specific animal antigen. In the majority of cases, the exposure is from diagnostic (e.g., tumor targeted imaging agents) or therapeutic applications of tumor-specific monoclonal antibodies. Because these antibodies are mostly murine in source, the most prevalent examples of HAAA are HAMA, which interfere with mouse antibody-based immunoassays [26,27]. Digibind, used in treating life-threatening digitalis toxicity, is the Fab fragment of sheep anti-digoxin antibodies. Therapeutic insulin made from pigs may cause generation of anti-pig antibodies in patients. However, currently, most insulin preparations used in drug therapy are bioengineered in order to eliminate this problem. Factor VIII, which is used in therapy, is also prepared from pigs. Many vaccines are generated in rabbits or chickens (eggs). Anti-animal antibodies may also arise from contact with animals (e.g., animal husbandry or keeping of animals as pets) [28] and the transfer of dietary antigens across the gut wall in conditions such as celiac disease [29].

HAAA can belong to the IgG, IgA, IgM, or, rarely, IgE class. When HAAA are elucidated by animal immunoglobulins, HAAA can have anti-idiotype or anti-isotype specificity. Anti-idiotype antibodies are directed against the hypervariable region of the immunoglobulin molecule, which binds the antigen, and anti-isotype antibodies are directed against the constant regions. The anti-idiotype antibodies may again generate endogenous anti-anti-idiotype antibodies. Assuming the "mirror-image" principle of idiotypic antibodies, the anti-anti-idiotype antibody could recognize the original analyte and may bind to it. However, in general, most HAAA are anti-isotype antibodies.

The magnitude and duration of HAAA vary widely. Serum concentrations of HAAA range from micrograms to grams per liter. The HAAA may be transient lasting a few days to months and years [30]. The prevalence estimates of HAAA, especially HAMA, vary widely from <1% to 80% among different hospitalized patients or outpatients. Several commercial assay kits are available for estimation of HAMA in human serum or plasma [31]. However, due to the heterogeneous nature of HAMA, a negative HAMA test result does not confirm the absence of all types of HAMA in a sample.

HAMA interferes mostly with immunoassays that use murine antibodies in the assay design. As increasingly more monoclonal antibodies (most common source is mouse) have been used in commercial immunoassays, the impact of HAMA interference has become a serious clinical issue. Although HAMA concentration is usually <10 μg/mL, HAMA concentrations as high as 1000 μg/mL have been reported. Because HAMA arise from exposure of patients to mouse antibodies, cancer patients who may have been exposed to these antibodies as part of imaging or therapeutic agents have higher prevalences of HAMA (40–70%) than other patients. HAMA can be IgG (most common), IgM, IgA, or IgE and can be directed to any part of the monoclonal antibody used in the assay (Fc, Fab, idiotope, etc.). HAMA incidences and concentrations are increasing with the increased use of diagnostic or therapeutic use of monoclonal murine antibodies. The current tendency to use therapeutic humanized antibodies is expected to reverse this trend.

In addition to mouse, rabbit and goat are also used to generate assay antibodies. Therefore, like HAMA, immunoassay interference caused by human anti-rabbit (HARA) and anti-goat antibodies has also been described. HARA interference was shown in transthyretin, haptoglobin, and C-reactive protein assays [32].

### *Interference from Rheumatoid Factors*

Rheumatoid factors (RFs) are IgM-type antibodies that interact with assay antibodies at the Fc area. RFs are present in serum from more than 70% of patients with rheumatoid arthritis. RFs are also found in patients with other autoimmune diseases. RF concentration increases in infection or inflammation. RF interference follows the same mechanism as interference from other types of antibodies. Therefore, in two-antibody immunometric assays, RFs bridge the capture and label antibodies without involving the antigen and generate a false-positive signal, thus spuriously elevating the value of the analyte. In single antibody competition-type immunoassays, RFs bind to the assay antibody, preventing its reaction to the label reagent through steric hindrance, thus generating false-positive results. If RFs are suspected to cause interference, the patient's history needs to be examined to determine if the RF concentration is expected to be elevated in the patient's serum. RF concentration in the serum or plasma can also be measured using commercially available immunoassays. RFs can be removed from the sample by the many separation steps described later in this chapter. In a study in which RF interference produced false-positive cardiac troponin I results in an immunometric assay, the authors inactivated RFs by incubating the sample with anti-RF antibody and the interference was eliminated [33].

### *Interference from Autoantibodies and Therapeutic Antibodies*

Autoantibodies are endogenous patient antibodies that bind to either the analyte or one of the reagents used in the assay. Such antibodies are more commonly observed in patients suffering from autoimmune diseases. Autoantibodies are often identified by gel filtration separation or polyethylene glycol precipitation of immunoreactive components from affected sera.

#### AUTOANTIBODIES TO THE ANALYTE

The autoantibodies may bind to the analyte-label conjugate in a competition-type immunoassay, producing false-positive results. Alternatively, they may bind to the analyte in a sandwich assay or competition assay that uses a label containing an analog of the analyte, giving false-negative results. An example has been described in a cardiac troponin I assay (a marker for cardiovascular diseases; see Chapter 9) in which an autoantibody directed against troponin caused negative interference [34]. On the other hand, Verhoye *et al.* [35] reported three patients with false-positive thyrotropin results that were caused by interference from an autoantibody against thyrotropin. The interfering substance in the affected specimens was identified as autoantibody by gel filtration chromatography and polyethylene glycol precipitation, followed by immunoreaction.

#### MACRO-ANALYTES

Often, the analyte may conjugate with autoantibodies to create macro-analytes, which may generate incorrect immunoassay results. For example, macroamylasemia and macroprolactinemia may produce incorrect results in amylase and prolactin assays, respectively. In macroprolactinemia, the hormone prolactin conjugates with itself and/or with its autoantibody to create macroprolactin in the patient's circulation. The macro-analyte is physiologically inactive but often interferes with many prolactin immunoassays, generating false-positive prolactin results [36]. Such interference may be removed by polyethylene glycol precipitation. Another example of macro-analyte is a false-positive AxSYM troponin I result in an asymptomatic patient caused by an autoantibody–troponin complex. The interference was not observed in four other troponin I immunoassays. Serial dilution and treatment with mouse serum failed to resolve the discordance, indicating the absence of HAMA interference. Gel filtration chromatography and polyethylene glycol precipitation studies identified the immunoreactive component as an IgG–troponin complex [37]. On the other hand, the measurement of a macro-analyte of squamous cell carcinoma antigen with IgM was used to assess the risk of hepatocellular carcinoma in patients with cirrhosis by using an immunoassay targeted to detect the macro-analytes as tumor markers [38].

#### AUTOANTIBODY TO A COMPONENT IN THE REAGENT

In one example, an endogenous anti-avidin antibody interfered with a theophylline assay that used the avidin–biotin system [39]. In this competition-type immunoassay, the autoantibody interacted with avidin in the reagent, interfering in complex formation and causing false-positive results. Of course, if the assay were sandwich type, the reduced signal would have caused false-negative results. Autoantibodies to various labeled enzymes (e.g., alkaline phosphatase and peroxidase) have been reported, and these antibodies may interact with the label enzyme in ELISA, thus falsely elevating the results in a competition-type immunoassay.

Therapeutic antibodies are used to bind and inactivate toxic components from circulation or directed against tumor antigens or immunologic receptors. For example, Digibind (Glaxo/Burroughs Wellcome), composed of Fab fragments from ovine anti-digoxin antibodies, is used to treat life-threatening digoxin overdoses, and it interferes with most digoxin assays. This interference is removable by ultrafiltration or protein precipitation [40] (see Chapter 13).

## HOW TO DETECT AND CORRECT HETEROPHILIC ANTIBODY INTERFERENCES

If a test result does not correlate with the clinical picture, false-positive or false-negative test results can be suspected, but when false results are subtle and/or plausible, the results could be misleading. For example, a "normal" result may truly be "abnormal," and a disease state of a patient could be missed due to interference in an immunoassay by heterophilic antibodies. Use of Bayesian logic based on the prevalence of a disease may help identify false results in the diagnosis of a disease [41]. Various ways of detecting heterophilic antibody and correcting interferences due to the presence of heterophilic antibody in the specimen are summarized in Table 6.2.

Assay development scientists incorporate steps such as sample blanking and robust assay design to minimize interferences, including matrix effects arising from protein and other nonspecific constituents in the specimen. Thus, during the development of multiplexed cytokine assays, researchers screened for and added appropriate blockers in the reagents to reduce heterophilic interference [42]. When suspected, the interfering substance may be removed from the specimen by specific agents, ultrafiltration, gel filtration chromatography, precipitation, or centrifugation prior to reanalysis. Alternatively, the specimen may be analyzed by a different method for the same analyte, which is known to be free from such interference.

If a discordant result is suspected to be caused by interference from some endogenous antibody, the best practices to confirm such interference include (1) dilution linearity study with the specimen; (2) examination of the patient history (exposure to immunogenic animals or animal products and history of hyperactive immune system); (3) assaying the sample, if possible, using a different immunoassay utilizing different antibodies/reagents; and (4) treating the sample to block the interference or remove the interfering antibody, and repeating the assay. This strategy is exemplified in a false-positive TSH result leading to thyroxin overdose. The incorrect result was traced to RF interference because the sample showed nonlinear dilution. The interference was removed by treating the sample with a heterophile blocking reagent. In addition, a correct result was also obtained using a different immunoassay [43]. An example of nonlinear dilution for specimens with heterophilic antibody interference is shown in Figure 6.1, which shows the effect of successive dilutions of a HAMA-containing sample (spiked with 32 μg/mL of theophylline) versus those

**TABLE 6.2** Different Sources of Heterophilic Interference, their Detection, and Reduction

| Antibody | Detection | Reduction |
|---|---|---|
| Heterophilic antibody (Weak Interference) | Serial dilution producing non-linear results | **1.** Non-specific animal serum 'Cocktail' of animal sera<br>**2.** Blocking agent changes result Serial dilution or blocking agent |
| Heterophilic antibody* (Strong interference) | Serial dilution producing non-linear results | **1.** Serial dilution but preferably by using blocking agent. |
| Complement Proteins (Specific, strong interference) | Complement assay | **1.** Heat inactivation<br>**2.** Use Fab or F(ab') antibodies in assay design |
| Rheumatoid Factor (RF)* | Test for RF | Treat with anti-RF antibody |

**For small non-protein bound analytes, use ultrafiltration or solid phase conjugated to Protein A or Protein G to remove interfering antibodies.*

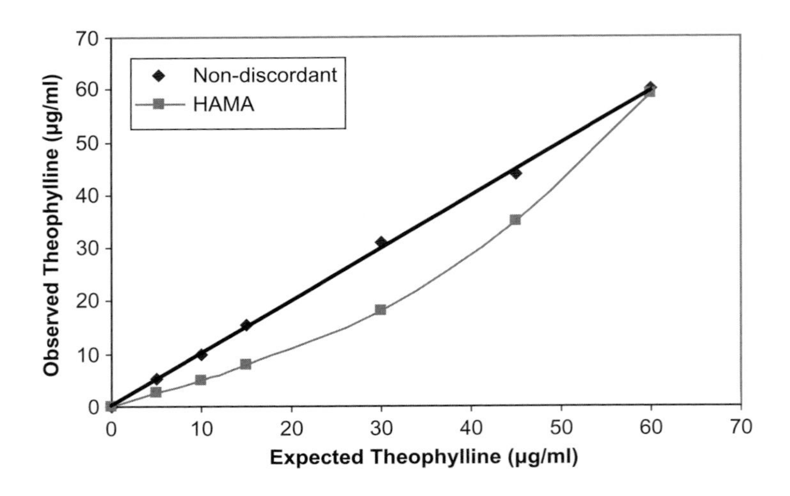


**FIGURE 6.1** HAMA interference detected by sample serial dilution.

of a serum-based calibrator for the assay (60 μg/mL) in a theophylline immunoassay using a mouse anti-theophylline antibody. The HAMA sample interferes with the assay and reads 59 μg/mL when assayed undiluted (false-positive value). After successive dilutions (1.3-, 2-, 4-, 6-, and 12-fold with the assay diluent), the interfering antibody is diluted enough so as not to cause any interference in the assay. The slope of a line fitted through the lowest three dilutions indicates a theophylline concentration of 31.1 μg/mL, close to the original spike value (Datta *et al.*, unpublished data). However, dilutions do not always provide the correct analyte value in the sample because of increased imprecision in the low end of the assay and because of the "matrix effect" between the calibrator matrix and a patient sample.

As described previously, a patient history of any exposure to animal antibodies, illness, or exposure to animals should also alert for heterophilic antibodies or HAAA as possible sources for inaccurate results. At that time, the assay insert should be examined for the types of antibodies and heterophilic antibody blockers used in the assay.

There are various types of commercial or home-brew blockers for heterophilic antibody or HAAA [44,45]. The blocker can be nonimmune animal serum, polyclonal antibody, polymerized IgG, nonimmune mouse monoclonals, or a mixture of monoclonal antibodies or fragments of IgG [Fc, Fab, or F(antibody')$_2$] preferably from the same species used to raise the reagent antibodies. As expected, although the nonspecific heterophilic interference can be mitigated by addition of nonimmune serum to the reagents, purified IgG, preferably of the same subtype as used in assay, is better than serum in reducing the more specific and stronger binding HAAA.

Several blocking agents are commercially available: Immunoglobulin Inhibiting Reagent (IIR; Biorecalamation), Heterophilic Blocking Reagent (HBR; Scantibodies), Heteroblock (Omega Biologicals), and MAB 33 (monoclonal mouse IgG1) and Poly MAB 33 (polymeric monoclonal IgG1/Fab; Boehringer Mannheim). IIR is a proprietary formulation of high-affinity anti-animal antibody, and HBR is monoclonal mouse anti-human IgM. A suspected discordant sample (e.g., a sample giving false-positive hCG results) may be separately incubated with the blocker and then re-assayed [44]. Reinsberg [45] studied the efficacy of various blocking reagents in eliminating HAMA interference. In another example, a clinically discordant false-positive serum myoglobin result (where another cardiac marker concentration, such as troponin I, was negative) was attributed to HAMA interference. The interference could be removed by the use of HBR [46]. Most commercial assay reagents include such blockers, but due to the heterogeneous nature of the interfering antibodies, no blocker can guarantee success in all samples.

Nonspecific and weaker heterophilic antibody interferences can even be mitigated by incubating the sample with any nonimmune animal serum (even one different from the source of the assay antibody). Use of a "cocktail of animal sera" to reduce heterophilic interference has been suggested [31]. Thus, heterophilic interference in a lutropin immunoassay was reduced equally by adding mouse, sheep, or goat sera to the sample [47]. If the interference is complement mediated involving Fc parts of the interfering antibodies, heat inactivation of the sample (56°C for 30 min) may remove the interference [48]. A limitation of this method is that not all analytes can survive such an antibody denaturing process.

As previously mentioned, a general concept of prudent assay design is to use Fab or F(ab')$_2$ fragments of the analyte-specific antibody, thereby reducing interference from human anti-isotype antibodies [48]. Kuroki *et al.* [49] used human/mouse chimeric antibody in a carcinoembryonic antigen assay to reduce HAMA interferences. Another interesting concept is to use chicken antibodies in the assay because to date no heterophilic antibody interference against assays using chicken antibodies has been reported in the literature [50]. By changing the reaction temperature and allowing a longer time to achieve equilibrium, it may be possible to reduce heterophilic antibody interference but not HAAA interference [51]. However, with modern autoanalyzers, changing reaction temperature or reaction time is mostly not possible.

If the previously described methods to correct the interference do not work or cannot be applied, the interference can be resolved by removing the interfering substance and re-assaying the clean specimen. A simple solution to remove antibody interference is selective removal of the antibodies from a sample. This can be achieved by selective adsorption of human IgG by a solid phase containing protein A or protein G [52]. However, this does not work if the majority of the interfering antibodies are of the IgM type. Alternately, the antibody fraction in the sample may be precipitated out with a polyethylene glycol reagent (preferably PEG 6000) [35]. Low-molecular-weight analytes, if they are not highly protein bound, may be extracted away from the interfering immunoglobulins by preparation of protein-free filtrates via ultrafiltration or protein precipitation (using trichloroacetic acid, sulfosalicylic acid, or ammonium sulfate). The centrifugal ultrafiltration is a fast and relatively easy method that uses 10- or 30-kDa cutoff filter membrane ultracentrifugal cartridges (e.g., Amicon's Microcon or Centricon). For digoxin assays, this step may not only remove antibody interference but also remove interference from endogenous digoxin-like immunoreactive substances (~95% protein bound) or Digibind [53]. On the other hand, the harsh conditions of acid precipitation may damage many analyte molecules.

## CONCLUSIONS

Immunoassays on automated systems are widely used in today's clinical laboratories. Various types of immunoassays have been developed to analyze all types and sizes of antigens. The assays use photometric, luminometric, or fluorometric signals and homogeneous or heterogeneous reaction types. Serum and plasma are the main specimen types used for immunoassays. Other types of specimens have also been used. Immunoassays are used not only on the central laboratory analyzers but also on the patient bedside point-of-care systems. Despite the excellent sensitivity and specificity of the immunoassays, they suffer from interferences: serum constituents, cross-reactants, or endogenous antibodies. There are various ways to detect such interferences and obtain accurate results.

### References

[1] Jolley ME, Stroupe SD, Schwenzer KS, et al. Fluorescence polarization immunoassay: III. An automated system for therapeutic drug determination. Clin Chem 1981;27:1575–9.

[2] Hawks RL, Chian CN, eds. Urine testing for drugs of abuse. Rockville, MD: National Institute of Drug Abuse (NIDA). Department of Health and Human Services; NIDA research monograph 1986; 73.

[3] Jeon SI, Yang X, Andrade JD. Modeling of homogeneous cloned enzyme donor immunoassay. Anal Biochem 2004;333:136–47.

[4] Datta P, Dasgupta A. A New turbidimetric digoxin immunoassay on the ADVIA 1650 analyzer is free from interference by spironolactone, potassium canrenoate, and their common metabolite canrenone. Ther Drug Monit 2003;25:478–82.

[5] Snyder JT, Benson CM, Briggs C, et al. Development of NT-proBNP, troponin, TSH, and FT4 LOCI(R) assays on the new dimension (R) EXL with LM clinical chemistry system [Abstract B135]. Clin Chem 2008;54:A92.

[6] Dai JL, Sokoll LJ, Chan DW. Automated chemiluminescent immunoassay analyzers. J Clin Ligand Assay 1998;21:377–85.

[7] Forest J-C, Masse J, Lane A. Evaluation of the analytical performance of the boehringer mannheim elecsys 2010 immunoanalyzer. Clin Biochem 1998;31:81–8.

[8] Babson AL, Olsen DR, Palmieri T, et al. The Immulite assay tube: a new approach to heterogeneous ligand assay. Clin Chem 1991;37:1521–2.

[9] Christenson RH, Apple FS, Morgan DL. Cardiac troponin I measurement with the ACCESS immunoassay system: analytical and clinical performance characteristics. Clin Chem 1998;44:52–60.

[10] Montagne P, Varcin P, Cuilliere ML, Duheille J. Microparticle-enhanced nephelometric immunoassay with microsphere-antigen conjugate. Bioconjug Chem 1992;3:187–93.

[11] Datta P, Larsen F. Specificity of digoxin immunoassays toward digoxin metabolites. Clin Chem 1994;40:1348–9.

[12] Datta P. Oxaprozin and 5-(*p*-hydroxyphenyl)-5-phenylhydantoin interference in phenytoin immunoassays. Clin Chem 1997;43:1468–9.

[13] Datta P, Dasgupta P. Bidirectional (positive/negative) interference in a digoxin immunoassay: importance of antibody specificity. Ther Drug Monit 1998;20:352–7.

[14] Pichini S, Pacifici R, Altieri I, Pellegrini M, Zuccaro P. Determination of opiates and cocaine in hair as trimethylsilyl derivatives using gas chromatography-tandem mass spectrometry. J Anal Toxicol 1999;5:343–8.

[15] Palmeri A, Pichini S, Pacifici R, Zuccaro P, Lopez A. Drugs in nails: physiology, pharmacokinetics and forensic toxicology. Clin Pharmacokinet 2000;38:95–110.

[16] Dickson PH, Lind A, Studts P, Nipper HC, Makoid M, et al. The routine analysis of breast milk for drugs of abuse in a clinical toxicology laboratory. J Forensic Sci 1994;39:2341–5.

[17] Albersen A, Kemper-Proper E, Thelen MHM, et al. A case of consistent discrepancies between urine and blood human chorionic gonadotropin measurements. Clin Chem Lab Med 2011;49:1029–32.

[18] Braunstein G. False positive serum human chorionic gonadotropin results: causes, characteristics, and recognition. Am J Obstet Gynecol 2002;187:217–24.

[19] Levinson S, Miller J. Towards a better understanding of heterophile (and the like) antibody interference with modern immunoassays. Clin Chim Acta 2002;325:1–15.

[20] Preissner CM, Dodge LA, O'Kane DJ, et al. Prevalence of heterophilic antibody interference in eight automated tumor marker immunoassays. Clin Chem 2005;51:208–10.

[21] Covinsky M, Laterza O, Pfeifer JD, Farkas-Szallasi T, Scott MG. An IgM antibody to *Escherichia coli* produces false-positive results in multiple immunometric assays. Clin. Chem 2000;46:1157–61.

[22] Papapetron PD, Polymeris A, Karga H, et al. Heterophilic antibody causing falsely high serum calcitonin values. J Endocrinol Invest 2006;29:919–23.

[23] Chin KP, Pin YC. Heterophilic antibody interference in thyroid assay. Intern Med 2008;47:2033–7.

[24] Tate J, Ward G. Interferences in immunoassay. Clin Biochem Rev 2004;25:105–20.

[25] Mannings L, Trow S, Newman J, et al. Interferences in the autoDELFIA hAFP immunoassay and effect on second-trimester down's syndrome screening. Ann Clin Biochem 2011;48:438–40.

[26] Miller RA, Maloney DG, McKillop J, Levy R. *In vivo* effects of murine hybridoma monoclonal antibody in a patient with T-cell leukemia. Blood 1981;58:78–86.

[27] Grossman H. Clinical applications of monoclonal antibody technology. Urol Clin N Am 1986;13:465–74.

[28] Berglund L, Holmberg NG. Heterophilic antibodies against rabbit serum causing falsely elevated gonadotropin levels. Acta Obstet Gynecol Scand 1989;68:377–8.

[29] Falchuck KR, Iselbacher KJ. Circulating antibodies to bovine albumin in ulcerative colitis and crohn's disease: characterization of the antibody response. Gastroenterol 1976;70:5–8.

[30] Kazmierczak SC, Catrou PG, Briley KP. Transient nature of interference effects from heterophilic antibodies: examples of interference with cardiac marker measurements. Clin Chem Lab Med 2000;38:33–9.

[31] Kricka LJl. Human anti-animal antibody interferences in immunological assays. Clin Chem 1999;45:942–56.

[32] Benoist JF, Orbach D, Biou D. False increase in C-reactive protein attributable to heterophilic antibodies in two renal transplant patients treated with rabbit antilymphocyte globulin. Clin Chem 1999;45(5):616–18.

[33] Dasgupta A, Banerjee SK, Datta P. False positive troponin I in the MEIA due to the presence of rheumatoid factors in serum. Am J Clin Pathol 1999;112:753–6.

[34] Eriksson S, Halenius H, Pulkki K, et al. Negative interference in cardiac troponin I immunoassays by circulating troponin autoantibodies. Clin. Chem 2005;51:839–47.

[35] Verhoye E, Avd Bruel, Delanghe JR, et al. Spuriously high thyrotropin values due to anti-thyrotropin antibody in adult patients. Clin Chem Lab Med 2009;47:604–6.

[36] Kavanagh L, McKenna TJ, Fahie-Wilson MN, et al. Specificity and clinical utility of methods for determination of macroprolactin. Clin Chem 2006;52:1366–72.

[37] Michielsen ECHJ, Bisschops PGT, Janssen MJW. False positive troponin result caused by a true macro-troponin. Clin Chem Lab Med 2011;49:923–5.

[38] Zulin J, Veggiani GL, Pengo P, et al. Experimental values of specificity of squamous cell carcinoma antigen-IgM assay in patients with cirrhosis. Clin Chem Lab Med 2010;48:217–33.

[39] Banfi G, Pontillo M, Sidoli A, et al. Interference from antiavidin antibodies in thyroid testing in a woman with multiendocrine neoplasia syndrome type 2B. J Clin Ligand Assay 1995;18:248–51.

[40] Dasgupta A, Wells A, Datta P. Effect of digoxin fab-antibody on the measurement of total and free digitoxin by fluorescence polarization and a new chemiluminescent immunoassay. Ther Drug Monit 1999;21:251–5.

[41] Ismail AAA, Ismail AA, Ismail Y. Probabilistic bayesian reasoning can help identifying potentially wrong immunoassays results in clinical practice: even when they appear "not-unreasonable". Ann Clin Biochem 2011;48:65–71.

[42] Martins T, Pasi B, Litwin C, Hill H. Heterophile antibody interference in a multiplexed fluorescent microsphere immunoassay for quantitation of cytokines in human serum. Clin Vaccine Immunol 2004;11:325–9.

[43] Georges A, Charrie A, Raynaud S, et al. Thyroxine overdose due to rheumatoid factor interference in thyroid stimulating hormone assay. Clin Chem Lab Med 2011;49:873–5.

[44] Butler SA, Cole LA. Use of heterophilic antibody blocking agent (HBT) in reducing false-positive hCG results. Clin Chem 2001;47:1332–3.

[45] Reinsberg. Different efficacy of various blocking reagents to eliminate interferences by HAMA in a 2-site immunoassay. Clin Biochem 1996;29:145–8.

[46] Bonetti A, Monica C, Bonaguri C, et al. Interference by heterophilic antibodies in immunoassay: wrong increase of myoglobin values. Acta Biomed 2008;79:140–3.

[47] Sampson M, Ruddel M, Zwig M, Elin RJ. Falsely high concentration of serum lutropin measured with the abbott Mx. Clin Chem 1994;40:1976.

[48] Vaidya HC, Beatty BJ. Eliminate interference from heterophilic antibodies in a two-site immunoassay for CKMB by using F $(ab')_2$ conjugate and polyclonal mouse IgG. Clin Chem 1992;38:1737.

[49] Kuroki M, Matsumoto Y, Arakawa F, et al. Reducing interference from heterophilic antibodies in a two-site immunoassay for carcinoembryonic antigen (CEA) by using a human/mouse chimeric antibody to CEA as the tracer. J Immunol Methods 1995;180:81–91.

[50] Span PN, Grebenchtchikov N, Geurts-Moespot J, Sweep CGJ. Screening for interference in immunoassays. Clin Chem 2003;49:1708–9.

[51] Miller JJ, Valdez R. Approaches to minimizing interference by cross-reacting molecules in immunoassays. Clin Chem 1991;37:144–53.

[52] Liendo C, Ghali JK, Graves SW. A new interference in some digoxin assays: anti-murine heterophilic antibodies. Clin Pharmacol Ther 1996;60:593–8.

[53] Dasgupta A, Biddle DA, Wells A, Datta P. Positive and negative interference of the Chinese medicine Chan Su in serum digoxin measurement—Elimination of interference by using a monoclonal chemiluminescent digoxin assay or monitoring free digoxin concentration. Am J Clin Pathol 2000;114: 174–9.

CHAPTER

# 7

# Effect of Herbal Remedies on Clinical Laboratory Tests

*Amitava Dasgupta*

University of Texas Health Sciences Center at Houston, Houston, Texas

## INTRODUCTION

According to the Dietary Supplement Health and Education Act of 1994, herbal remedies sold in the United States are classified as food supplements. Manufacturers of herbal remedies are not allowed by law to claim any medical benefit from these products, but at the same time they are not under surveillance of the U.S. Food and Drug Administration (FDA). In Germany, however, German Commission E has some control over marketing of herbal supplements because the commission publishes monographs prepared by an interdisciplinary committee using historical information; chemical, pharmacological, clinical, and toxicological study findings; case reports; epidemiological data; and unpublished manufacturers' data. If an herbal supplement has an approved monograph, it can be marketed. European Directive 2004/24/EC, released in 2004 by the European Parliament and also by the Council of Europe, provides the basis for regulation of herbal supplements in the European market. This directive requires that authorization be obtained from the national regulatory authorities of each European country in which herbal medicines are to be released in the market and that these products must be safe. The safety of a supplement is established based on published scientific literature, and when the data on safety are not sufficient, this is communicated to consumers. In Europe, there will be two kinds of herbal supplements in the future: (1) herbal supplements with well-established safety and efficacy and (2) traditional herbal supplements that do not have a recognized level of efficacy but are relatively safe [1]. The Australian government also created a Complementary Medicine Evaluation Committee in 1997 to address regulatory issues regarding herbal remedies. In Canada, the federal government implemented a policy in 2004 to regulate natural health products and naturopaths; many traditional Chinese medicine practitioners, homeopaths, and Western herbalists are concerned that this policy will eventually affect their access to the products they need to practice effectively [2].

In the United States, the sale of herbal remedies significantly increased from $200 million in 1988 to more than $3.3 billion in 1997. Within the European community, sales of herbal remedies are also widespread, with estimated annual sales of $7 billion in 2001 [3]. Sales of herbal supplements were estimated to be $15.7 billion in 2000, and in 2003 sales increased to an estimated $18.8 billion [4]. Currently, the global annual sales of herbal remedies are estimated to be $60 billion, representing almost 20% of the overall pharmaceutical market [5]. The popularity of herbal supplements is steadily increasing among the general population in the United States. It is estimated that approximately 20,000 herbal products are available in the United States, and in one survey, approximately one out of five adults reported using an herbal supplement within the past year. The 10 most commonly used herbal supplements are echinacea, ginseng, ginkgo biloba, garlic, St. John's wort, peppermint, ginger, soy, chamomile, and kava [6]. In general, higher use of herbal supplements is found among more educated, higher-income, white, and older females [7]. Use of herbal supplements can affect laboratory test results by various mechanisms:

- Physiological effects: When an herbal remedy alters normal physiological functions of the body or causes organ damage, the unexpected laboratory test may provide the first indication of toxicity of

*Accurate Results in the Clinical Laboratory.*
DOI: http://dx.doi.org/10.1016/B978-0-12-415783-5.00007-4

the herb. An example is elevated liver enzymes due to use of kava.
- Drug–herb interactions may lead to unexpected levels of a therapeutic drug causing either treatment failure or drug toxicity. Therapeutic drug monitoring is very useful in identifying such clinically important drug–herb interactions. An example is the finding of unexpectedly low levels of cyclosporine due to interaction of the herbal antidepressant St. John's wort with cyclosporine in a patient who previously showed therapeutic levels.
- A component of an herbal remedy may be mistaken as a drug by the antibody used in the immunoassay for that particular drug.
- If an herbal remedy is contaminated with a heavy metal (lead, mercury, or arsenic) or a Western drug, the unexpected presence of that drug in the blood of the patient may confuse the doctor. In addition, heavy metal toxicity may result from taking herbal supplements.

## ELEVATED LIVER FUNCTION TEST DUE TO USE OF HERBAL SUPPLEMENTS

The liver is the largest organ in the body and is responsible for many important physiological functions, including metabolism of drugs and toxin for elimination from the body. Many potentially toxic compounds enter the body through the gastrointestinal tract and are immediately transported via the portal vein to the liver. Therefore, the liver sustains the highest exposure to some toxins and is susceptible to damage. Liver function tests are used for diagnosis of adequate liver function as well as to identify any potential injury. Liver function tests include enzymes, bilirubin, proteins, and coagulation factors. These analytes are chosen because they are readily released in large quantities following cellular injury. Measurement of the serum or plasma activities of the enzymes aspartate aminotransferase (AST), alanine aminotransferase (ALT), $\gamma$-glutamyl transferase (GGT), and alkaline phosphatase (ALP) are routinely performed to detect liver injury. These intracellular enzymes are released when hepatocytes are damaged, but none of these are specific to liver. GGT, along with ALP, also reflects injury to biliary cells. Total bilirubin and its conjugated and unconjugated forms are useful in differentiating cases of jaundice. Various herbal products, such as kava, chaparral, comfrey, pennyroyal, black cohosh, mistletoe, and green tea extract, can cause liver damage. Despite known toxicity and warnings from authorities, these herbal supplements are still available in health food stores.

### Kava

Kava is the most commonly cited herb related to liver toxicity. Kava, an herbal sedative with antianxiety or calming effects, is prepared by extracting the rhizomes of *Piper methysticum*, a South Pacific plant. The active ingredients of the plant, known as kavalactones, are found mostly in the rhizomes [8]. Early scientific research indicated that kava was as efficacious as antidepressant drugs and tranquilizers in treating anxiety disorders [9]. In fact, kava extracts were considered safe alternatives to drug therapy for treating anxiety disorders before 1998. However, by 2003, 11 cases of hepatic failure had been reported related to the use of kava, including 7 patients requiring liver transplantation and 4 deaths. In 2003, kava was banned in the European Union and Canada, and the FDA issued another warning. By 2009, more than 100 cases of hepatotoxicity had been linked to kava exposure. Many have followed coingestion with alcohol, which appears to potentiate the hepatoxicity [10]. It has been proposed that kava hepatotoxicity is not observed after consumption of traditional aqueous extract of kava as consumed by native people of the Pacific Islands during ceremonies, but hepatotoxicity is observed after consumption of commercially prepared kava extracts that may employ ethanol or acetone during the extraction process. Teschke *et al.* [11] reported that subsequent cases analyzed by the World Health Organization and published reports revealed that traditional aqueous extracts used in New Caledonia, Australia, the United States, and Germany may also cause hepatotoxicity. The authors further commented that the primary cause of hepatotoxicity is possibly attributed to poor quality of raw material as well as mold hepatotoxins present in kava roots [11]. Pipermethysticine, a toxic alkaloid present in kava leaves and stems, may contaminate kava products during production if quality control is poor and only kava roots are used to prepare extract. Flavokavin, another toxic alkaloid present in kava roots, may also cause toxicity. Hepatocellular toxicity results from mitogen-activated protein kinase signaling leading to oxidative stress and apoptosis. The chemical content of kava products and therefore their potential for toxicity vary according to plant age, parts used, cultivating conditions, geographic location, and growth conditions of the plant [12].

Hepatotoxicity associated with kava consumption is usually reflected by increases in the serum activities of ALT, AST, and GGT. A 50-year-old male who exceeded the maximum recommended dose of three capsules daily for 2 months was found to have AST and ALT levels 60- to 70-fold higher than the reference ranges. Hepatitis serology for this patient was negative, and kava use was determined to be the cause of liver damage. This patient eventually received a liver transplant [13]. Ingestion of kava may cause temporary yellowing of the skin, hair, and nails. A 70-year-old male who took kava products for anxiety for 2 or 3 weeks experienced itching several hours after sun exposure and developed plaques on his face, chest, and back. A similar case involved a 52-year-old female who presented with papules and plaques on her face, arms, back, and chest after taking kava products for 3 weeks. In both cases, a skin biopsy revealed lymphocytic infiltration of the dermis with destruction of the sebaceous glands [14]. Although liver damage is the most widely documented toxicity of kava, several cases involving central nervous system depression have been reported when the herbal is combined with other sedatives and hypnotics. One such case report described a 54-year-old male who became comatose after 3 days of kava ingestion while taking alprazolam, cimetidine, and terazosin. It was thought that the adverse reaction was related to a pharmacodynamic interaction between kava and alprazolam [15].

*CASE REPORT* A 42-year-old healthy white male presented to the hospital with weakness, loss of appetite, and jaundice. He admitted alcohol consumption of no more than one drink per day but denied use of any medication or abuse of any drug. However, he reported that he returned from a 20-day honeymoon vacation in the Samoan Islands 3 weeks before going to the hospital. The patient showed signs of scleral and skin jaundice and also had abnormal liver function tests, including AST of 1602 U/L, ALT of 2841 U/L, GGT of 121 U/L, ALP of 285 U/L, and lactate dehydrogenase of 460 U/L. His total bilirubin was 9.3 mg/dL. Serological tests for hepatitis, cytomegalovirus, and Epstein–Barr virus were negative. Abdominal ultrasound showed a hyperechoic liver structure with normal biliary ducts. Because the available information did not provide any satisfactory cause of his disease, the patient was interviewed again and he admitted that during his trip to the Samoa Islands he had repeatedly participated in a kava ceremony and consumed a total of 2 or 3 L of traditional kava extract. The patient was discharged 19 days after admission, and after 36 days he completely recovered based on repeated ultrasound scan and laboratory test results [16].

*CASE REPORT* A 52-year-old female presented to her local doctor with a 2-week history of fatigue, nausea, and increasing jaundice. For the past 3 months, she had taken an herbal supplement (Kava 1800 Plus) prescribed by a naturopath for anxiety. She took one tablet three times daily labeled as containing 60 mg kavalactones, 50 mg *Passiflora incarnata*, and 100 mg *Scutellaria laterifloria*. She was then referred to the author's hospital, and on admission her serum ALP was 190 U/L, ALT 4539 U/L, serum albumin 3.4 g/dL, and bilirubin 12.2 mg/dL. All tests for hepatitis A–C, Epstein–Barr virus, and cytomegalovirus were negative. Over subsequent weeks, her condition deteriorated, and on day 17 of admission to the hospital she underwent liver transplant. Unfortunately, the procedure was complicated by excessive bleeding and the patient died. It was determined that her liver failure was due to use of kava [17].

## Comfrey and Coltsfoot

Comfrey is a perennial plant whose leaves and roots are traditionally used for wound healing; repairing broken bones; and in the treatment of arthritis, gout, and psoriasis. However, there is no scientific evidence to support these claims. Comfrey contains pyrrolizidine alkaloids, which are well-known hepatotoxins, and Russian comfrey is even more toxic than European and Asian comfrey due to a higher content of toxic alkaloids. Yeong *et al.* [18] described a case involving a 23-year-old male presenting with severe veno-occlusive disease and hypertension, who subsequently died from liver failure. The patient was a vegetarian who had used comfrey leaves as a dietary supplement.

Pyrrolizidine alkaloids and their N-oxides are found in comfrey as well as in a variety of related herbs. Coltsfoot (*Tussilago farfara*) is another herb closely related to comfrey that also contains a pyrrolizidine alkaloid (senecionine). Veno-occlusive disease has been reported in a newborn infant of a woman who ingested coltsfoot-containing tea.

*CASE REPORT* A female infant was referred to a neonatal intensive care unit 5 days after birth due to jaundice, massive hepatomegaly, and ascites. The infant was delivered by cesarean section at 36 weeks of gestation due to an unconfirmed suspicion of premature separation of placenta. Initial biochemical tests showed AST was 3725 U/L, ALT 760 U/L, and bilirubin 9.6 mg/dL; however, all tests for hepatitis, Epstein–Barr virus, and cytomegalovirus were negative. When no cause of liver failure in the infant was found, the mother admitted that she had used an herbal tea daily throughout her entire pregnancy. Toxicological analysis of the herbal tea using thin-layer chromatography revealed the presence of senecionine, a pyrrolizidine alkaloid. The herbal tea was found to contain 10 different plants, of which coltsfoot was 9% of the preparation by weight. Although the mother did not experience any adverse effects, the immature liver of the fetus was unable to detoxify the alkaloids, leading to the development of the veno-occlusive disease. Unfortunately, the infant died 38 days after birth [19].

## Germander

Germander (*Teucrium chamaedrys*) is an aromatic plant in the "mint family" belonging to the genus *Teucrium*. Its blossoms are used as a folk medicine to treat dyspepsia, diabetes, and gout. However, chronic use of germander may cause hepatotoxicity. In general, the toxic effects of germander are first seen within 9 weeks of use and are manifested by jaundice and elevated liver enzymes (ALT and AST). After discontinuation of the herbal, recovery may take 6 weeks to 6 months. The mechanism of toxicity is thought to be related to diterpenoid compounds that are metabolized to more potent toxins within the liver [20]. The major hepatotoxic diterpene found in germander is teucrin A, which is bioactivated by the cytochrome P450 enzymes in the liver [21]. Another hepatotoxic ingredient of germander is teuchmaedryn A. More than 45 cases of hepatotoxicity following germander use have been reported in France, including one fatality [22].

***TWO CASE REPORTS*** In the first case, a 55-year-old female became ill after taking 1600 mg/day of germander for 6 months in an effort to reduce her cholesterol. Her laboratory findings included total bilirubin 13.9 mg/dL, direct bilirubin 8.4 mg/dL, AST 1180 U/L, ALT 1500 U/L, and ALP 164 U/L; however, all serological tests for hepatitis (A–C) were negative. Her hepatitis resolved within 2 months of discontinuing the germander. In the second case, a 45-year-old female took germander (260 mg/day) for weight loss. Her laboratory studies also suggested hepatotoxicity: total bilirubin 3.5 mg/dL, direct bilirubin 2.2 mg/dL, AST 417 U/L, ALT 451 U/L, and ALP 79 U/L. She discontinued the herbal and her health condition improved. After 4 months, she felt much better and resumed the germander. Within the week, she again developed indications of liver toxicity: total bilirubin 17.0 mg/dL, direct bilirubin 13.2 mg/dL, AST 1245 U/L, ALT 784 U/L, and ALP 89 U/L. Hepatitis tests were again negative. The patient stopped using germander again, and her liver function tests improved over 3 months [23].

## Chaparral and Pennyroyal

Chaparral is a plant that grows in the southwestern United States and northern Mexico, and its leaves are used as an herbal therapy in the treatment of a wide variety of symptoms from cold sores to muscle pain.

Unfortunately, chaparral has been associated with severe hepatotoxicity as documented by several reports of chaparral-associated hepatitis. Sheikh *et al.* [24] reviewed 18 cases of suspected chaparral toxicity reported to the FDA and confirmed 13 cases of hepatotoxicity related to the herbal. Clinical presentation included significant elevations in liver enzyme activities and other biochemical markers of hepatic injury 3–52 weeks after ingestion of chaparral. In most cases, liver function tests return to the reference range after cessation of use, but at least 2 cases in which fulminate hepatitis resulted suggest irreversible injury can occur.

Pennyroyal (*Mentha pulegium*) is a plant in the mint genus whose leaves release a spearmint-like fragrance when crushed. Because of the strong fragrance, it is often found as an additive to bath products and in aromatherapy. Traditionally, pennyroyal has been brewed as a tea to be ingested in small amounts as an abortifacient and emmenagogue. The plant and oil contain several components including pulegone, which is metabolized by the liver to a more toxic compound, methofuran, and depletes the hepatic glutathione stores (similar to that induced by overdose of acetaminophen). A review of 18 cases of toxicity found that ingestion of as little as 10 mL of pennyroyal oil can cause severe toxicity [25]. Interestingly, the antidote used in acetaminophen overdose, *N*-acetylcysteine, has been used successfully in treating pennyroyal toxicity [26].

## Other Supplements That Cause Liver Damage

The most dangerous Oriental weight loss product is ma huang, which contains ephedra alkaloids. Ma huang may cause severe cardiotoxicity, hepatotoxicity, and even death. Ephedrine found in ma huang and related weight loss products may be responsible for the toxic effects observed. Oriental weight loss products Chaso and Onshido contain *N*-nitrosofenfluramine and should be avoided. Other Oriental herbal supplements containing multiple herbs and several kampo medicines (notably Dai-saiko-to and Sho-saiko-to) are known to cause liver injuries. Jin-bu-huan, a sedative analgesic, is also known to cause liver damage. Shou-wu-pian, a proprietary Chinese medicine used for treating dizziness, back pain, and constipation, may also cause hepatitis. Shen-min, another Chinese medicine, has also been associated with drug-induced hepatitis [20].

LipoKinetix is a weight loss aid that contains phenylpropanolamine, caffeine, yohimbine, diiodothyronine, and sodium usniate. In 2002, seven patients who were using LipoKinetix developed acute hepatitis [27]. Sodium usniate found in LipoKinetix is derived from usnic acid, which has known liver toxicity. Usnic acid is also present in kombucha tea (also known as

Manchurian mushroom or Manchurian fungus tea), which is prepared by brewing kombucha mushrooms in sweet black tea. Although kombucha tea is considered a healthy elixir, the limited evidence currently available raises safety concerns, especially regarding potential hepatotoxicity and the possibility of life-threatening lactic acidosis. A 22-year-old male newly diagnosed with HIV experienced shortness of breath and was febrile to 103°F within 15 hr of drinking kombucha tea and was admitted to the hospital. His serum creatinine was 2.1 mg/dL, and he experienced life-threatening lactic acidosis (lactate 1.9 mmol/L (116.2 mg/dL); upper limit of normal: 19.8 mg/dL) [28].

Many other herbs may cause some liver damage. For example, skullcap is used as a sedative, calming agent, and also recommended for nervous tension, epilepsy, and hysteria. There are several cases of hepatotoxicity involving skullcap in combination with another herbal. One case involved skullcap in combination with valerian, whereas another involved skullcap in combination with gingko biloba [29]. Gotu kola (Sanskrit: Mandukaparni; *Centella asiatica*) has been used in Indian Ayurvedic medicine for a long time for the treatment of hypertension and wound healing. This preparation contains the pentacyclic triterpenic saponosides, asiaticoside and madecassoside, which may cause liver damage. In one study, the authors presented three cases of women ages 61, 52, and 49 years who presented with high ALT (1193, 1694, and 324 U/L, respectively), ALP (503, 472, and 484 U/L, respectively), and bilirubin (4.23, 19.89, and 3.9 mg/dL, respectively). All patients improved after discontinuation of *C. asiatica* [30].

## HERBAL SUPPLEMENTS CAUSING KIDNEY DAMAGE

In 1993, rapidly progressing kidney damage was reported in a group of young women who were taking pills containing Chinese herbs while attending a weight loss clinic in Belgium. It was discovered that one prescription Chinese herb had been replaced by another Chinese herb containing aristolochic acid, a known toxin to kidney [31]. This was the first comprehensive document of nephropathy associated with the use of aristolochic acid, which is now referred to as "Chinese herbal nephropathy" characterized by progressive fibrosing interstitial nephritis leading to renal failure and severe anemia. Later, there were many reports of kidney damage due to the use of herbal supplements containing aristolochic acid. In most of the reported cases, when a patient discontinued the use of herbal supplement containing aristolochic acid, renal disease may proceed to end-stage renal disease [32]. Aristolochic acid is also a known carcinogen, and despite FDA warning, it is still sold in the United States [33]. Kong *et al.* [34] reported a case of a married couple in which the husband developed Fanconi's syndrome and his wife developed end-stage renal failure and anemia after using Chinese herbs containing aristolochic acid. However, none of their five children, who never used the herbal product, showed renal insufficiency, indicating that the renal failure was not related to any genetic factor but was due to use of the herbal supplements containing aristolochic acid. Today, aristolochic-acid-containing herbal supplements are the major cause of herb-induced renal failure, although several other herbs may also have nephrotoxicity.

Use of other herbal supplements has also been associated with nephrotoxicity. Although licorice is relatively safe if used as a flavoring agent in candies, its chronic use as a supplement may cause hypokalemia, which can lead to rhabdomyolysis and eventually acute kidney injury. Yohimbine may cause lupus-like syndrome and proteinuria. Willow bark contains salicin, which is metabolized to salicylate. Chronic use of willow bark product may cause nephrotoxicity. Bladderwrack is a large brown alga that is a common food in Japan. Long-term use of bladderwrack tablets may cause chronic kidney disease [5]. The National Kidney Foundation stated that wormwood plant, sassafras, Chinese herbal medicine tung shueh, and horse chestnut may be toxic to kidney. In addition, people suffering from chronic renal disease should not take alfalfa, aloe, bayberry, blue cohosh, broom, buckthorn, capsicum, cascara, coltsfoot, dandelion, ginger, ginseng, horsetail, licorice, mate, nettle, noni juice, rhubarb, senna, and vervain. According to the National Kidney Foundation, herbals such as chaparral, comfrey, ephedra (ma huang), lobelia, mandrake, pennyroyal, pokeroot, sassafras, senna, and yohimbe are unsafe for all people, and these herbal products should be avoided (http://www.kidney.org/atoz/atozItem.cfm?id=123). Several herbal supplements are known to cause hematuria and loss of protein in urine (proteinuria). Examples of these herbs are aloe juice from leaf, kava, saffron, etc. Many other herbs, such as calamus, chaparral, horse chestnut seed, and wormwood oil, may cause nephrotoxicity [35,36]. In addition, several herbs also have diuretic effects and may cause irritation to cells of kidney (renal epithelial cells). These herbs—including asparagus roots, lovage root, parsley herb and root, watercress, and white sandalwood—should be avoided in people with any kidney problems [37]. Herbs that may cause kidney damage are shown in Table 7.1.

***CASE REPORT*** A 75-year-old male presented with a 2-month history of lethargy, poor appetite, and nausea. He used Chinese herbal supplements ("long dan

**TABLE 7.1** Herbal Supplements Causing Unexpected Laboratory Test Results

| Laboratory Test Results | Herbal Supplement |
|---|---|
| Elevated liver enzymes/bilirubin and other tests showing liver damage | Kava, chaparral, comfrey, germander, pennyroyal, lipokinetic, skullcap, valerian, ma huang |
| Elevated creatinine, BUN, and other tests indicating kidney damage | Aristolochic acid containing Chinese herbs, wormwood oil, sassafras, horse chestnut, calamus, bladderwrack, white sandalwood oil, juniper berry, yohimbine, willow bark, licorice |
| Hypoglycemia | Ginsengs (Asian, Korean, American), *Coccinia indica*, bitter melon (*Momordica charantia*), fenugreek seeds, *Gymnema sylvestre*, aloe vera, prickly pear cactus, fig leaf, Milk Thistle, Holy basil (*Ocimum sanctum*), nopal (*Opunita streptacantha*), chromium, Asian herbal supplements contaminated with oral hypoglycemic agents |
| Abnormal thyroid profile | Kelp |

xie gan wan," "lei gong teng," and "ke yin wan") to treat his psoriasis for 3 years and observed significant improvement. Laboratory investigation revealed renal failure with a serum creatinine level of 965 μmol/L (10.9 mg/dL; upper limit of normal: 1.2 mg/dL) and urea nitrogen of 43.1 mmol/L (259.6 mg/dL; upper limit of normal: 18 mg/dL). Renal ultrasound demonstrated unobstructed, small kidneys with cortical thinning bilaterally. A renal biopsy showed severe tubulointerstitial fibrosis and atrophy. These findings were consistent with chronic exposure to a nephrotoxin. The patient's Chinese remedies were sent for further analysis by high-performance liquid chromatography, which revealed the presence of aristolochic acid in the Chinese medicine "long de xie gan wan." Hemodialysis was initiated and later converted into ambulatory peritoneal dialysis. The patient died 4 years after initial presentation following withdrawal from dialysis [38].

*CASE REPORT* A 41-year-old Chinese man with a history of urinary tract infection purchased four boxes of Chinese herbal preparations "fen qing wu lin wan," with each box containing 40 small packets. The preparations were taken twice daily (one small packet each time), and after 2 days he had loose stools four or five times a day daily and abdominal pain. On day 23, his abdominal pain was so severe that he went to the hospital. His serum creatinine was 484 μmol/L (5.5 mg/dL) and urea nitrogen was 16.8 mmol/L (101.2 mg/dL). He also had oliguria (200 mL urine per day) and died from multiple organ failure 1½ months after initiation of taking herbal supplement. Pathological examination during autopsy revealed acute tubular necrosis with significant protein cast, acute hemorrhagic necrotic enteritis with hematoceles in the ileum, severe bilateral pulmonary hemorrhage with blood accumulation in the bronchi, mild cardiac hypertrophy, mild edema and congestion in the brain, mild fatty degeneration of the liver, and anemic spleen. The product "fen qing wu lin wan" contains the Chinese herb "guan mu tong," which is the dried stem of *Aristolochia manshuriensis* containing aristolochic acid. If such herb is consumed in excessive dosage or for a long period, aristolochic acid may cause acute necrosis of the renal tubules as well as chronic interstitial nephritis [39].

## HERBAL REMEDIES AND HYPOGLYCEMIA

In humans, glucose is the primary fuel for the brain and muscles; the normal range of glucose is 70–99 mg/dL after overnight fasting, whereas fasting glucose greater than 126 mg/dL measured on two different occasions may indicate diabetes mellitus. Yeh *et al.* [40] reviewed data from 108 clinical trials examining 36 herbs (single or combination) and nine vitamin/mineral supplements. They determined that there are still insufficient data to draw definitive conclusions regarding the efficacy of herbal supplements in controlling diabetes, but the best evidence of efficacy for controlling blood glucose is for *Coccinia indica* and American ginseng. Other supplements that show promise are *Gymnema sylvestre*, *Aloe vera*, vanadium, *Momordica charantia*, and nopal. Chromium is the most widely studied supplement for controlling blood glucose [40]. *Coccinia indica* (ivy ground) is a creeping plant that grows wildly in many areas of India, and it is used in Ayurvedic medicine for treating diabetes. Many formulations containing *Coccinia indica* are available in India and in the United States. Cinnamon-6 herbal formulation (Herbal Pride) contains this herb as one of its many components. However, the FDA has not evaluated the efficacy of this product in controlling type 2 diabetes.

Several different plant species are referred to as ginseng, including Asian ginseng, Korean ginseng, Siberian ginseng, Japanese ginseng, and American ginseng. Two long-term trials administering American ginseng for 8 weeks reported decreases in fasting blood glucose in subjects. Glycosylated hemoglobin, another marker of long-term glucose control, was also decreased in subjects taking American ginseng [40]. Various studies also indicate reduction in blood glucose after taking Korean or Asian ginseng. The use of garlic in lowering blood sugar is controversial, although a few studies have shown positive effects. Fenugreek may also be effective in reducing blood sugars, but studies reported in the literature were not always well designed [40]. Fig leaves (*Ficus carica*) ingested during breakfast may be effective in lowering blood sugar [41]. Although not an herbal supplement, chromium, a trace element, is required for the maintenance of normal metabolism of glucose. Chromium deficiency may cause glucose intolerance, although most diabetic patients are not chromium deficient. Chromium picolinate may be effective in lowering blood glucose, although some studies have shown no benefit of chromium in reducing blood sugar.

*CASE REPORT* A 46 year-old female was diagnosed with non-insulin-dependent diabetes 10 months ago, and her serum glucose was stabilized with diet, exercise, and glipizide. She reported to the clinic with a 3-day episode of low glucose levels in serum ranging from 60 to 85 mg/dL using a home glucose monitoring device. In the clinic, her nonfasting glucose using finger stick was only 55 mg/dL. She was not taking any other medication, and there was no change in her lifestyle. The doctor treated her with glucagon to increase her serum glucose level, and when her serum glucose concentration reached 110 mg/dL, she felt much better. At that point, she admitted taking several herbal supplements: ginseng for lowering blood glucose and cholesterol, garlic to prevent heart disease, astragalus for lowering blood glucose, and juniper berry tea for aiding in digestion. Such herbal supplements were responsible for her abnormally low serum glucose and her symptoms [42].

*CASE REPORT* A 58-year-old male with type 2 diabetes mellitus was treated with metformin 1000 mg twice daily and extended-release glipizide 10 mg daily and showed good glycemic control because his fasting glucose ranged from 113 to 132 mg/dL and his two hemoglobin A (1c) values were 6.8 and 6.7%, respectively, obtained during a 1-year period. He denied using any herbal supplement. One month later, he reported four hypoglycemic events, with glucose readings between 49 and 68 mg/dL. At that point, glipizide was discontinued. One month later, he reported no hypoglycemic events, but during reconciliation he admitted consuming crude prickly pear cactus pads daily for 2 months for glucose control. Patients with type 2 diabetes mellitus receiving hypoglycemic agents should be routinely counseled about the use of herbal supplements in order to avoid hypoglycemic events [43].

Adulteration of herbal antidiabetic products with pharmaceuticals is a significant problem. This is particularly associated with herbal supplements manufactured in some Asian countries. In one report from Hong Kong, the authors found undisclosed pharmaceuticals in several herbal supplements, including glibenclamide in 22 products, phenformin in 18 products, metformin in 6 products, rosiglitazone in 6 products, gliclazide in 2 products, glimepiride in 2 products, nateglinide in 1 product, and repaglinide in 1 product [44]. Other pharmaceuticals (nonhypoglycemic agents) were also detected in some products. Some products also contained up to four pharmaceuticals. The authors concluded that patients taking such adulterated herbal supplements could be at risk for serious adverse events and even fatality [44].

*CASE REPORT* A 56-year-old Indonesian tourist from Jakarta presented to the emergency department of Royal Perth Hospital, Australia, 3 days after his arrival to Australia with flu-like symptoms. On arrival of the ambulance, his capillary blood glucose level was 2.1 mmol/L (37.8 mg/dL), and he received 1 mg of glucagon intramuscularly. During transfer to the hospital, he was also given aspirin (300 mg) and isosorbide mononitrate (10 mg, sublingually). On arrival at the hospital, his capillary blood glucose level was 4.3 mmol/L (77.4 mg/dL). On admission to the hospital, he reported having diet-controlled type 2 diabetes mellitus, ischemic heart disease (with coronary artery bypass surgery 12 years previously), hypertension, and hypercholesterolemia. He was taking amlodipine, aspirin, and atorvastatin, but he denied taking any medications for his diabetes. He was given a meal containing a sandwich but soon after his venous glucose concentration was reported by the laboratory to be 2.9 mmol/L (52.2 mg/dL), indicating persistent hypoglycemia despite being given sugary drinks and then increasing dextrose infusions and intermittent boluses of dextrose. Finally, a 50 mL per hour infusion of 50% dextrose was required to prevent his hypoglycemia. At that point, the patient admitted that although he was not taking any hypoglycemic agents for diabetes, he was taken a traditional Chinese medicine for diabetes called "ZhenQi" that he had purchased in Malaysia. He had been taking this medication for the past 5 years, initially taking five capsules per day but then increasing the dose to three capsules three times daily for the past 2 years. The label on the bottle of this preparation listed the ingredients as ginseng, pearl, ram's horn, bark, and "frog extract." He had started a new bottle of the preparation recently and felt lethargic, cold and tremor. Serum taken during the period of hypoglycemia (when his glucose level was

52.2 mg/dL) had elevated levels of C-peptide (3.80 nmol/L; normal range, 0.20–0.90 nmol/L) and insulin (50 mU/L; normal range, 3–26 mU/L). Analysis of the herbal medication capsules by gas chromatography and mass spectrometry revealed the presence of glibenclamide, a known hypoglycemic agent. Fortunately, the patient responded to the treatment and was discharged from the hospital on day 3. This case report indicates an episode of severe hypoglycemia due to adulteration of a Chinese herbal formula with a pharmaceutical [45].

## LICORICE AND HYPOKALEMIA

Licorice prepared from the root of *Glycyrrhiza glabra* has a sweet taste and it is used in various candies and sometimes also in cough suppressant syrup. It is believed that licorice is effective in treating ulcer, but licorice has glycyrrhizinic acid, which affects the endocrine system and may also cause hypertension. Mechanistically, glycyrrhizinic acid inhibits the enzyme (11β-hydroxysteroid dehydrogenase) that converts cortisol to cortisone in the kidney, allowing cortisol to act as a mineralocorticoid promoting salt retention and potassium wastage. Individuals taking blood pressure medications should not use licorice. Even in normotensive people, excessive intake of licorice may cause hypertension, hypokalemia, and related electrolyte imbalance requiring medical attention. Licorice should not be used by anyone taking digoxin because licorice-induced electrolyte imbalance may cause severe digoxin toxicity, especially in the elderly. An 84-year-old man who was taking digoxin experienced severe toxicity from using licorice requiring hospitalization [46]. Long-term licorice ingestion is a well-known cause of secondary hypertension and hypokalemia. Lin *et al.* [47] reported a case of an Asian man who ingested tea flavored with 100 g of natural licorice root containing 2.3% glycyrrhizic acid daily for 3 years and presented to the hospital with paralysis, metabolic alkalosis, and severe hypokalemia (serum potassium, 1.8 mmol/L). His renal potassium wasting continued for 2 weeks after discontinuation of licorice and therapy with spironolactone and potassium chloride supplement. Meltem *et al.* [48] reported a case of a 21-year-old man who ingested the powderized over-the-counter product "Tekumut" for 2 weeks to quit smoking and presented to the hospital with muscle weakness. The product contained licorice. Although licorice usually induces hypokalemia that is mild in nature, licorice use may rarely cause paralysis, rhabdomyolysis, or ventricular fibrillation leading to death if untreated [48].

*CASE REPORT* A 31-year-old, previously healthy Caucasian male was admitted to the hospital due to progressive weakness of the legs and muscle cramping that developed over a few hours. His clinical examination revealed paraparesis without sensory loss, but the laboratory report revealed severe hypokalemia (1.5 mmol/L; normal range, 3.6–4.5 mmol/L), hypophosphatemia (0.42 mmol/L; normal range, 0.87–1.45 mmol/L), and mild hypomagnesemia (0.65 mmol/L; normal range, 0.7–1.1 mmol/L). His creatine kinase level was elevated (3425 U/L; normal, <190 U/L), indicating rhabdomyolysis. His creatinine was 1.2 mg/dL. The patient admitted consuming approximately 1½ packets of original Fisherman's Friend menthol eucalyptus lozenges daily for the past 3 months. These lozenges contained licorice, and his consumption of glycyrrhizinic acid, the active ingredient in licorice, was approximately 60–90 mg/day. Despite discontinuation of these lozenges, his potassium level remained low. The patient was initially treated with parenteral and enteral potassium as well as spironolactone. Within 24 hr, his potassium level began to rise, and his neurological symptoms disappeared. After 3 days, the patient was asymptomatic; at this point, his potassium level had increased to 2.9 mmol/L and the diagnosis was hypokalemia and secondary hyperaldosteronism suggestive of Gitelman syndrome. Although he was discharged from the hospital at that point, he was later readmitted to the hospital for recurring symptoms despite treatment with aldosterone antagonist and supplement of potassium and magnesium. The trigger for the exacerbations could be his consumption of alcohol, lemon juice, and iced tea in excessive amounts. Finally, vomiting and failure to replace salt led to a serum potassium level of 1.0 mmol/L, and the patient was temporarily placed on mechanical ventilation. This case demonstrates the serious consequences of ingesting excessive amount of licorice and of dietary choices [49].

## KELP AND ABNORMAL THYROID FUNCTION TESTS

Kelp is prepared from seaweed and contains a very high amount of iodine. Seaweed or kelp is a part of the regular diet in some Asian countries, especially Japan. In addition, kelp is used as an ingredient in many Oriental medicines. Teas *et al.* [50] studied the iodine content in commonly available seaweeds from commercial sources in the United States and observed that the iodine content varied widely from 16 to more than 8165 micrograms in one uncooked serving. Iodine is water soluble and cooking makes the iodine content of cooked seaweed difficult to estimate, but the authors commented that it is possible that in some Asian seaweed dishes the iodine content may exceed 1100 micrograms/day, the upper tolerable iodine level in humans.

Eating kelp supplements, especially kelp containing Oriental medicines, may be problematic because excess iodine may cause thyroid malfunction, especially in people taking thyroxin supplements. In healthy individuals, the thyroid gland has an intrinsic autoregulatory mechanism to handle excess iodine intake. The oxidation of iodine, which is essential for synthesis of thyroid hormone, is inhibited in the presence of high iodine concentration in blood. This acute inhibitory effect, known as the Wolff–Chaikoff effect, lasts approximately 48 hr, followed by normalization of normal thyroid hormone synthesis. However, patients with underlying thyroid disorders may not be able to adopt excessive iodine in circulation. These patients may develop either hypothyroidism or hyperthyroidism. Individuals with elevated sensitivity, such as patients with Grave's disease or chronic autoimmune thyroiditis, may develop hypothyroidism after intake of excess iodine. In contrast, patients with low sensitivity, such as those with endemic goiter or iodine deficiency, may develop hyperthyroidism [51]. A 72-year-old woman with normal thyroid gland developed hyperthyroidism after ingesting kelp tablets. The thyroid functions returned to normal 6 months after discontinuation of kelp [52]. In another case report, a 32-year-old woman with goiter presented to the hospital with symptoms of hyperthyroidism. Her problem developed 4 weeks after taking kelp-containing herbal tea following the recommendation of a Chinese herbal practitioner [51].

## DRUG–HERB INTERACTION AND UNEXPECTED DRUG LEVELS IN ROUTINE DRUG MONITORING

For optimal therapeutic benefit of a drug, a certain blood level of the drug known as the therapeutic range must be maintained in an individual. If the drug concentration is less than the therapeutic level, treatment failure may result, whereas drug concentrations higher than the therapeutic range may cause unwanted drug toxicity. This is particularly problematic for drugs with a narrow therapeutic range that require routine monitoring for optimal patient management. Several herbal supplements are known to cause significant interactions with many drugs, causing either treatment failure or drug toxicity. The most documented drug–herb interaction is that of St. John's wort with many drugs causing treatment failure. Most drug–herb interactions are pharmacokinetic in nature, where the herbal supplement affects the metabolism of a drug. For example, St. John's wort increases the hepatic metabolism of many drugs, resulting in subtherapeutic drug levels. Drug–herb interactions may also be pharmacodynamic in nature, where one herbal product may stimulate the pharmacological response to a drug. Fortunately, only a fraction of over-the-counter medications and prescription medications account for most of the interactions with herbal supplements. Sood *et al.* [53] surveyed 1818 patients and identified 107 drug–herb interactions that had clinical significance. The five most common herbal supplements (St. John's wort, ginkgo, garlic, valerian, and kava) accounted for 68% of such interactions, and four different classes of prescription drugs (antidepressants, antidiabetic, sedatives, and anticoagulation medications) accounted for 94% of all clinically significant interactions. In one report, the authors identified 32 drugs that interact with herbal supplements [54]. These drugs, including drugs that are routinely monitored, are listed in Table 7.2. Therapeutic drug monitoring is useful for identifying certain clinically significant drug–herb interactions because a patient may not disclose the use of herbal supplements to the physician. Shi and Klotz [55] commented that drug–herb interaction certainly increases the risk of therapy with drugs that have narrow therapeutic ranges, such as warfarin, cyclosporine, and digoxin. Fortunately, therapy with warfarin is monitored routinely using the international normalization ratio (INR), whereas digoxin and cyclosporine are subjected to routine therapeutic drug monitoring.

### Interactions of St. John's Wort with Various Drugs

St. John's wort, an herbal antidepressant, is sold in the United States as an alcoholic or dried extract of hypericum, a perennial aromatic shrub with bright yellow flowers that bloom from June to September. The flowers are believed to be most abundant and brightest on approximately June 24, the day traditionally believed to be the birthday of St. John the Baptist, hence the name St. John's wort. Hypericin, hyperforin, and quercetin are major constituents in St. John's wort that mediate reduction of intestinal absorption as well as bioavailability of many drugs by induction of intestinal P-glycoprotein drug efflux pump (MDR1: multidrug resistant 1 expression) and induction of activities of both intestinal and liver cytochrome P450 mixed function oxidase (CYP) enzymes responsible for the metabolism of many drugs. Hyperforin appears to play a major role in activating pregnane X receptor, leading to the transcriptional activation of genes that regulate the activities of CYP3A4 and other cytochrome subtype enzymes responsible for drug metabolism. Therefore, St. John's wort preparations with low

**TABLE 7.2** Drugs (Including Drugs that are Subjected to Routine Monitoring) that are Known to Interact with Herbal Supplements

| Class of Medication | Drug | Routine Monitoring? |
|---|---|---|
| Analgesic including NSAIDs | Acetaminophen, salicylate, ibuprofen, naproxen | No, except salicylate and acetaminophen in overdose |
| | Tolmetin, indometacin | No |
| Anticancer | Imatinib, irinotecan | No |
| Anticoagulant | Warfarin, phenprocoumon | INR monitoring |
| Antifungal | Voriconazole, erythromycin | No |
| Antiasthmatic | Theophylline | Yes |
| Antihistamine | Fexofenadine | No |
| Antiepileptic | Phenytoin, phenobarbital | Yes (both drugs) |
| Antidepressants | Amitriptyline, imipramine, lithium | Yes (all three drugs) |
| | Mirtazapine, venlafaxine fluoxetine[a] | No |
| | Sertraline,[a] paroxetine,[a] alprazolam | No |
| | Midazolam, haloperidol, phenelzine | No |
| | Nefazodone | No |
| Antiretroviral | Indinavir,[a] ritonavir,[a] saquinavir[a] | No |
| | Atazanavir, lamivudine, nevirapine[a] | No |
| Anti-parkinson | Levo-dopa | No |
| Cardioactive drugs | Digoxin, propranolol | Yes (both drugs) |
| | Verapamil,[a] nifedipine | No |
| Cough suppressant | Dextromethorphan | No |
| Cholesterol-lowering drugs | Pravastatin, atorvastatin, simvastatin | No |
| Hypoglycemic agents | Tolbutamide, gliclazide, metformin, tolbutamide | No |
| Diuretics | Hydrochlorothiazide | No |
| Immunosuppressants | Cyclosporine, tacrolimus | Yes (both drugs) |
| Oral contraceptives | Ethinyl estradiol/desogestrel | No |
| Synthetic opioid | Methadone | Yes |

[a]*Not monitored routinely but may be monitored for selected patient populations.*
INR, international normalization ratio; NSAIDs, nonsteroidal anti-inflammatory drugs.

hyperforin content (<1%) may not cause any clinically significant interactions with any drugs [56]. In general, it is considered that St. John's wort induces CYP3A4, CYP2E1, and CYP2C19, with no effect on CYP1A2, CYP2D6, and CYP2C9. However, St. John's wort also induces clearance of drugs that are not metabolized via liver enzymes, such as digoxin, fexofenadine, and talinolol, which are well-known P-glycoprotein substrates [57]. Although most drug–herb interactions involving St. John's wort are pharmacokinetic in nature, pharmacodynamic interactions of certain antidepressants with St. John's wort have also been reported. Important interactions of St. John's wort with various drugs are summarized in Table 7.3.

Most important clinically pharmacokinetic drug interactions with St. John's wort involve immunosuppressants, warfarin, and antiretrovirals because treatment failure may have serious consequences for these patients. Transplant recipients taking cyclosporine or tacrolimus may face acute organ rejection due to self-medication with St. John's wort because St. John's wort induces the metabolism of both drugs, thus reducing whole blood concentrations of these drugs, sometimes by more than 50%. Ernst [58] reported that St. John's wort can endanger the success of organ transplantation due to its interaction with cyclosporine, which has led to several cases of organ rejection. Alscher and Klotz [59] reported a case study in which a 57-year-old kidney

**TABLE 7.3** Important Pharmacokinetic and Pharmacodynamic Drug Interaction with St. John's Wort

| Class of Drug | Name of Drug | Comment |
|---|---|---|
| *PHARMACOKINETIC INTERACTIONS* | | |
| Immunosuppressants | Cyclosporine, tacrolimus | Reduced level due to induction of CYP3A4 but both drugs are routinely monitored by TDM |
| Protease inhibitors | Atazanavir, lopinavir, indinavir | Reduced level due to induction of CYP3A4 |
| NNRTI | Nevirapine | Reduced level due to induction of CYP3A4 |
| Anticoagulants | Warfarin, phenprocoumon | Increased metabolism and reduced efficacy |
| Cardioactive | Digoxin, verapamil | Reduced level for both drugs but only digoxin is subjected to TDM |
| Calcium channel blocker | nifedipine | Reduced level due to induction of CYP3A4 |
| β-Blocker | talinolol | Reduced level due to induction of P-glycoprotein |
| Management of angina | ivabradine | Reduced level due to induction of CYP3A4 |
| Anticonvulsant | Mephenytoin | Increased urinary excretion of metabolites due to induction of CYP2C19 |
| Antihistamine fexofenadine | | Decreased level due to induction of P-glycoprotein |
| Benzodiazepines | Alprazolam, midazolam, quazepam | Reduced blood level due to induction of CYP3A4 |
| Anticancer | Irinotecan, imatinib | Reduced level due to induction of CYP3A4 |
| Tricyclic antidepressant | Amitriptyline | Reduced level due to CYP3A4 induction by this drug is routinely monitored by TDM |
| Cholesterol-lowering drugs | Atorvastatin, simvastatin | Reduced efficacy but lipid profile determination can identify such drug–herb interaction |
| Synthetic opioid | Methadone | Reduced level due to induction of CYP3A4 |
| Oral contraceptives | Norethindrone, ethinyl estradiol | Reduced level may lead to failure of contraception |
| Proton pump inhibitor | Omeprazole | Reduced level due to induction of CYP2C19 |
| Antidiabetic | Gliclazide | Reduced efficacy but routine glucose monitoring can identify the interaction |
| *PHARMACODYNAMIC INTERACTION* | | |
| SSRIs | Fluoxetine, sertraline, paroxetine | Possibility of serotonin syndrome |
| SNRI | Venlafaxine | Possibility of serotonin syndrome |
| Antidepressants | Bupropion, buspirone, nefazodone | Possibility of serotonin syndrome |
| Anti-migraine agent | Eletriptan | Possibility of serotonin syndrome |

NNRTI, non-nucleoside reverse-transcriptase inhibitor; SNRI, serotonin-norepinephrine reuptake inhibitor; SSRIs, selective serotonin reuptake inhibitors; TDM, therapeutic drug monitoring.

transplant patient taking cyclosporine (125–150 mg/day) and prednisolone (5 mg/day) showed therapeutic cyclosporine trough levels (100–130 ng/mL) in the past 2 years. This patient suddenly demonstrated a subtherapeutic cyclosporine concentration of 70 ng/mL despite increasing the daily cyclosporine dose to 250 mg. The patient admitted to taking an herbal tea mixture for depression that contained St. John's wort. Five days after discontinuing the herbal tea, his cyclosporine level had increased to 170 ng/mL. Then the dose was reduced to 175 mg/day to readjust his cyclosporine whole blood concentration, and his trough cyclosporine level was 130 ng/mL. Mai *et al.* [60] reported that the hyperforin content of St. John's wort determines the magnitude of interaction between St. John's wort and cyclosporine. Patients who received low hyperforin-containing St. John's wort showed minimal changes in pharmacokinetic parameters and needed no dose adjustment. In contrast, patients who received high amounts of hyperforin-containing St. John's wort needed dose increases within 3 days in order to maintain the trough therapeutic concentration of cyclosporine. Significant reduction in area under the curve (AUC) for tacrolimus was also observed in 10 stable renal transplant patients receiving St. John's wort. The maximum concentration of tacrolimus was also reduced from a mean of 29.0 ng/mL to 22.4 ng/mL following co-administration of St. John's wort [61]. Interestingly, the pharmacokinetic parameters of mycophenolic acid, another immunosuppressant, were not affected by co-administration of St. John's wort [62].

Warfarin (Coumadin) is an anticoagulant drug used in the treatment of a variety of conditions including heart disease such as atrial fibrillation, preventing blood clots in lower extremities (deep vein thrombosis), preventing stroke, and pulmonary embolism. Warfarin therapy should be carefully controlled by measuring the clotting capacity of blood (using INR). A patient attending a Coumadin clinic must avoid St. John's wort because of potential failure of warfarin therapy due to increased clearance of warfarin. Jiang *et al.* [63] studied the interaction of warfarin and St. John's wort using 12 healthy subjects and concluded that St. John's wort significantly induced clearance of both S- and R-warfarin, which in turn resulted in a significant reduction in the pharmacological effect of racemic warfarin.

***CASE REPORT*** A 85-year-old male with a history of hypertension, previous anterior wall myocardial infarction, and atrial fibrillation had been receiving warfarin (5 mg daily) for 1 year. After watching a television program on herbal remedies, he decided to start taking St. John's wort. One month later, he presented to the emergency room with upper gastrointestinal bleeding. His hemoglobin was 7.9 g/dL, hematocrit 23%, and INR 6.2. He was treated with fresh frozen plasma and blood transfusion, and his bleeding was controlled. The patient was instructed not to use any herbal supplement. Although St. John's wort reduces the efficacy of warfarin by increasing its hepatic metabolism, this rare case indicates the opposite effect in which the patient developed a bleeding episode due to interaction of St. John's wort with warfarin. The ingredients in St. John's wort may potentiate the anticoagulant effect of warfarin in sensitive individuals instead of reducing its efficacy [64].

Patients suffering from AIDS and receiving highly active antiretroviral therapy (HAART) must not consume St. John's wort or other herbal supplements due to the possibility of treatment failure from drug–herb interactions. Clinically significant interactions of antiretroviral agents with St. John's wort have been documented. Therefore, patients with AIDS taking amprenavir, atazanavir, zidovudine, efavirenz, indinavir, lopinavir, nelfinavir, nevirapine, ritonavir, and saquinavir must avoid concomitant use of St. John's wort [65]. St. John's wort was shown to reduce the AUC of the HIV-1 protease inhibitor indinavir by a mean of 57% [66]. Reduced concentration of nevirapine due to administration of St. John's wort has also been reported [67].

The interaction between St. John's wort and digoxin is of clinical significance. Johne *et al.* [68] reported that use of St. John's wort for 10 days resulted in a 33% decrease in peak and 26% decrease in trough serum digoxin concentrations. The mean peak digoxin concentration was 1.9 ng/mL in the placebo group and 1.4 ng/mL in the group taking St. John's wort. Clearance of imatinib mesylate, an anticancer drug, is also increased due to administration of St. John's wort, resulting in reduced clinical efficacy of the drug. In a study of 10 healthy volunteers, 2 weeks of treatment with St. John's wort significantly reduced maximum plasma concentration by 29% and the AUC by 32%. The half-life of the drug was reduced by 21% [69]. St. John's wort also showed significant interaction with another anticancer drug, irinotecan. In one study of 5 patients, ingestion of St. John's wort (900 mg/day) for 18 days resulted in an average 42% reduction in the concentration of SN-38, the active metabolite of irinotecan. This reduction also caused decreased myelosuppression [70]. Many benzodiazepines are metabolized by CYP3A4 and, as expected, St. John's wort reduces plasma concentrations of alprazolam, midazolam, and quazepam. St. John's wort also reduces the plasma concentration of the antiepileptic drug mephenytoin, but it does not interact with the antiepileptic drug carbamazepine.

Theophylline is metabolized by CYP1A2, CYP2E1, and CYP3A4. Reduced plasma concentrations of theophylline due to intake of St. John's wort have also been reported [57]. Fortunately, theophylline is subjected to routine therapeutic drug monitoring, and such interaction can be suspected from observing reduced theophylline levels in a patient who previously showed therapeutic theophylline levels. Reduced plasma levels of methadone were also observed in the presence of St. John's wort. Long-term treatment with St. John's wort (900 mg/day) for a median period of 31 days (range, 14–47 days) decreased the trough concentrations of methadone by an average of 47% in four patients. Two patients experienced withdrawal symptoms due to reduced plasma levels of methadone [71].

St. John's wort has significant interaction with oral contraceptives. Murphy *et al.* [72] studied the interaction between St. John's wort and oral contraceptives by investigating pharmacokinetics of norethindrone and ethinyl estradiol using 16 healthy women. Treatment with St. John's wort (300 mg three times a day for 28 days) resulted in a 13–15% reduction in dose exposure from oral contraceptives. Breakthrough bleeding increased in the treatment cycle, as did evidence of follicle growth and probable ovulation. The authors concluded that St. John's wort increased metabolism of norethindrone and ethinyl estradiol and thus interfered with contraceptive effectiveness. St. John's wort also induces both CYP3A4-catalyzed sulfoxidation and 2C19-dependent hydroxylation of omeprazole [73]. Tannergren *et al.* [74] reported that repeated administration of St. John's wort significantly decreased the bioavailability of R- and S-verapamil. This effect is caused by induction of first-pass metabolism by CYP3A4 most likely in the gut. Xu *et al.* [75] reported that treatment with St. John's wort significantly increased the apparent clearance of gliclazide, a drug used in the treatment of patients with type 2 diabetes mellitus. Sugimoto *et al.* [76] reported interactions of St. John's wort with cholesterol-lowering drugs simvastatin and pravastatin. In a double-blind, cross-over study using 16 healthy male volunteers, the authors demonstrated that use of St. John's wort (900 mg/day) for 14 days decreased peak serum concentrations of simvastatin hydroxyl acid, the active metabolite of simvastatin, from an average of 2.3 ng/mL in the placebo group to 1.1 ng/mL in the group taking St. John's wort. On the other hand, St. John's wort did not influence plasma pravastatin concentration.

*CASE REPORT* A 59-year-old black male presenting to the physician with hyperlipidemia was initially treated with pravastatin but was later changed to rosuvastatin 10 mg/day dosage and marked improvement was observed in his lipid profile. However, a lipid profile 6 months later showed markedly increased cholesterol (165 mg/dL in April 2007 but 237 mg/dL in October 2007) and low-density lipoprotein (LDL) cholesterol (99 mg/dL in April 2007 but 162 mg/dL in October 2007). The patient admitted to taking St. John's wort capsules (two capsules per day containing 300 mg St. John's wort, 80 mg rosemary, and 40 mg spirulina) daily for insomnia starting in June 2007. He was asked to stop taking St. John's wort immediately, and 4 months later he showed significantly reduced cholesterol (163 mg/dL) and LDL cholesterol (95 mg/dL). Although simvastatin and atorvastatin are metabolized by CYP3A4 and St. John's wort reduces efficacy of both drugs, rosuvastatin is primarily eliminated through biliary excretion. However, 10% of the drug is metabolized by CYP2C9 and CYP2C19, and most likely St. John's wort induces CYP2C9- and CYP2C19-mediated clearance of rosuvastatin. The authors speculated that it was unlikely that rosemary or spirulina affected metabolism of rosuvastatin [77].

Although pharmacokinetic interactions of drugs with St. John's wort are more common, there are also clinically significant pharmacodynamic interactions of certain antidepressants with St. John's wort, which exerts its antidepressant effect by inhibiting uptake of neurotransmitter serotonin, norepinephrine, and dopamine. Hyperforin, an active component of St. John's wort, is responsible for its effect. Many antidepressant drugs also exert their pharmacological effects by a similar mechanism. Therefore, if a patient takes antidepressant medication such as fluoxetine, sertraline, paroxetine, or venlafaxine along with St. John's wort, the serotonin syndrome may occur [78]. Taking St. John's wort along with buspirone may also cause serotonin syndrome [79]. Important drug–herb interactions involving St. John's wort are listed in Table 7.3.

*CASE REPORT* A 50-year-old female with asthma and chronic depression was taking paroxetine 40 mg/day. She was a nondrinker and did not use any tranquilizers. She stopped taking paroxetine 10 days prior to her clinic visit and was switched to St. John's wort 600 mg powder each day. The night after her clinic visit, she slept poorly and then took 20 mg paroxetine along with St. John's wort in order to sleep better. Approximately noon the next day, her sister visited her and found her incoherent, groggy, slow-moving, and almost unable to get out of her bed. When the author examined her at 2p.m., her physical examination and laboratory reports were unremarkable but she was groggy and lethargic. The patient presented with a clinical syndrome resembling sedative/hypnotic intoxication after taking both St. John's wort and paroxetine [80].

## Herbal Supplements That Interact with Warfarin

Warfarin therapy must be critically monitored by measuring INR because warfarin is known to interact with many Western drugs and herbal supplements. Although potential interaction between a Western drug and warfarin can be easily avoided because clinicians are well aware of such interactions, interaction of warfarin with an herbal supplement can be dangerous because patients may often take herbal supplements without realizing that that combination of the supplement with warfarin may cause harm. Important interactions of warfarin with various herbal supplements are summarized in Table 7.4.

Warfarin acts by antagonizing the co-factor function of vitamin K. Although the clinical efficacy of warfarin varies with intake of vitamin K and genetic polymorphisms that modulate expression of CYP2C9, the isoform responsible for clearance of S-warfarin and several herbal supplements also have significant effects

**TABLE 7.4** Herbal Products and Food that Interact with Warfarin Therapy

| Effect on Warfarin | Herbal Supplement |
|---|---|
| Potentiates effect of warfarin and increases risk of bleeding with warfarin therapy. Increased INR may be an indication of such warfarin–herb interaction | Angelica root, arnica flower, anise, bogbean, borage seed oil, bromelain, boldo, borage, chamomile, coenzyme Q10, danshen, dong quai, devil's claw, fenugreek, feverfew, garlic, ginger, grape seed, ginkgo biloba, horse chestnut, licorice, lovage root, meadowsweet, passionflower herb, papaya, fish oil supplements, evening primrose oil, royal jelly, saw palmetto, willow bark |
| Decreases effect of warfarin | St. John's wort, Green tea |
| Reduced INR may be an indication of such warfarin–herb interaction | Milk thistle, goldenseal |

on the metabolism of warfarin. As mentioned previously, St. John's wort has a clinically significant interaction with warfarin. Herbal supplements that may potentiate the effect of warfarin and thus increase the risk of bleeding include angelica root, arnica flower, anise, borage seed oil, bromelain, chamomile, fenugreek, feverfew, garlic, ginger, horse chestnut, licorice root, lovage root, meadowsweet, passionflower herb, poplar, and willow bark [81]. The anticoagulant effect of warfarin also increases if combined with antiplatelet herbs such as danshen and ginkgo biloba. Conversely, vitamin-K-containing supplements such as green tea extract may antagonize the anticoagulant effect of warfarin. The INR was increased in a patient treated with warfarin for atrial fibrillation when he started taking the coumarin-containing herbal products boldo and fenugreek. His INR returned to normal 1 week after discontinuation of herbal supplements [82,83]. Increased anticoagulation due to interaction between warfarin and danshen has been reported [84]. There is a case report of an 87-year-old African American man receiving warfarin therapy who was admitted to the hospital with an INR of 6.88. The INR was increased to 7.29 during the hospital stay, whereas his previous INR values ranged from 1.9 to 2.4 (therapeutic range: 2–3). The patient admitted taking an herbal supplement called royal jelly 1 week prior to his hospital admission, and the elevated INR in this patient was probably related to the use of royal jelly [85]. Consumption of coenzyme Q10 and ginger also appears to increase the risk of bleeding with warfarin therapy [86]. Chan *et al.* [87] observed that patients with nonvalvular atrial fibrillation who received warfarin had little knowledge about the potential interaction of herbal supplements and food with warfarin, and patients who consumed herbal supplements at least four times per week had suboptimal anticoagulation control with warfarin.

*CASE REPORT* A 46-year-old female receiving a total weekly dosage of 56 mg of warfarin had an INR between 1.6 and 2.2 during the past 4 months. While taking the same dosage of warfarin, her INR increased to 4.6 after she drank approximately 1.5 quarts (1420 mL) of cranberry juice for 2 days. Two weeks after she discontinued drinking cranberry juice, her INR was reduced to 2.3, and it varied between 1.4 and 2.5 in the following 3 months while she took 56 mg of warfarin per week. At a subsequent visit after drinking 2 quarts of cranberry juice (1893 mL) for 3 days, her INT was again elevated to 6.5. At that time, warfarin was discontinued. Her INR was reduced to 1.86 after 3 days. Seven days later, her INR was again elevated to 3.2 with 56 mg of warfarin per week. Her two episodes of highly elevated INR were most likely due to interaction of cranberry juice with warfarin [88].

## Other Clinically Important Drug–Herb Interactions

Although the literature regarding drug–herb interactions is growing at a rapid pace, many reported interactions are of limited clinical significance and many herbal supplements, such as black cohosh, saw palmetto, echinacea, hawthorn, and valerian, seem to cause minor risk to patients receiving pharmacotherapy. However, a few herbs, notably St. John's wort, may cause adverse events sufficiently serious to endanger patients' health [89]. There are also many significant drug–herb interactions involving warfarin. Fortunately, INR determination for warfarin can identify such clinically significant drug–herb interactions. Ginkgo biloba may cause bleeding episodes in patients taking nonsteroidal anti-inflammatory drugs such as aspirin and ibuprofen, and it may reduce the effects of the antiepileptic drugs valproic acid and phenytoin. Garlic may reduce the efficacy of the antiretroviral drug saquinavir. Kava should not be taken with benzodiazepines due to the possibility of serious adverse effects, including coma [57]. Significant interactions of various herbs with other drugs are summarized in Table 7.5.

**TABLE 7.5** Other Clinically Significant Drug–Herb Interactions

| Herb | Interacting Drugs | Effect |
|---|---|---|
| Ginkgo biloba | Aspirin, ibuprofen | Bleeding |
| | Omeprazole | Reduced plasma level |
| | Ritonavir | Reduced plasma level |
| | Trazodone | Coma |
| | Phenytoin, valproic acid | Reduced concentration |
| Garlic | Saquinavir | Reduced effect |
| | Chlorpropamide | Hypoglycemia |
| | Ibuprofen | Bleeding |
| Ginseng | Phenelzine | Insomnia, headache, irritability |
| Kava | Alprazolam, paroxetine | Lethargic state |
| | Levodopa | Reduced effect |

## ADULTERATION OF HERBAL SUPPLEMENTS WITH WESTERN DRUGS

Although uncommon for herbal supplements prepared in the United States, herbal supplements from Asia and Indian Ayurvedic medicine may be contaminated with Western drugs. In this case, drug overdose may occur in an unsuspecting person taking such herbal supplements because the presence of such Western drugs is not disclosed in package inserts. Some Chinese herbal supplements intended for diabetic patients may be contaminated with hypoglycemic agents, and taking such products may cause severe hypoglycemia. This topic was discussed previously. However, Asian herbal supplements intended to treat other symptoms may also be contaminated with Western drugs. In one report, the authors analyzed 2069 samples of traditional Chinese medicines collected from eight hospitals in Taiwan, and in 618 samples (23.7%) they found undeclared Western pharmaceuticals, most commonly caffeine, acetaminophen, indometacin, hydrochlorothiazide, and prednisolone [90,91]. Most of these herbal supplements were used to alleviate pain, inflammation, or symptoms of arthritis. Pharmaceuticals such as acetaminophen, indometacin, and prednisolone can achieve these therapeutic effects. However, contaminated herbal supplements can also cause severe drug overdoses due to adulteration of the supplement with a drug.

*CASE REPORT* A 33-year-old female with an 8-year history of epilepsy was managed with valproate, carbamazepine, and phenobarbital but was never prescribed phenytoin. One month before admission, she started consuming three proprietary Chinese medicines in addition to her prescription medicines. She followed instructions for taking Chinese medicines for almost 1 month and then became comatose and was admitted to the hospital. Serum drug level assays on the second day of admission surprisingly showed a toxic phenytoin level of 48.5 μg/mL (therapeutic range, 10–20 μg/mL). She was treated conservatively, and after 10 days her clinical signs of phenytoin toxicity disappeared and she did not suffer any neurological damage. Analyses of three Chinese proprietary medicines showed the presence of 41 mg of phenytoin in jue dian shen ying wan (orange capsule), whereas the other two Chinese medicines were adulterated with carbamazepine and valproate. The patient consumed six orange capsules for almost 1 month, causing her severe phenytoin toxicity. Unfortunately, the manufacturer's information stated that these preparations only contained Chinese medicines for controlling epilepsy [92].

Savaliya *et al.* [93] reported that Indian Ayurvedic medicines may be contaminated with both steroidal and nonsteroidal anti-inflammatory drugs. Dexamethasone and diclofenac were detected in 10 Ayurvedic products out of 58 preparations analyzed. In addition, piroxicam was detected in 1 product, and dexamethasone alone was detected in 1 product. Many Indian Ayurvedic medicines are also contaminated with heavy metals such as lead, arsenic, or mercury either due to the manufacturing process or because the heavy metal (known as bhasma in Sanskrit) is a component of Ayurvedic medicine.

*CASE REPORT* A 58-year-old female from India who was residing in the United States presented to the emergency department with a 10-day history of progressively worsening postprandial lower abdominal pain and nausea accompanied by vomiting. She was healthy but suffered from well-controlled non-insulin-dependent diabetes mellitus and hypertension. On admission, her physical exam was unremarkable except for abdominal tenderness in the lower quadrants. Laboratory tests indicated a normochromic normocytic anemia with hemoglobin of 7.7 g/dL, hematocrit of 22.6%, and mean corpuscular volume (MCV) of 87 fL with normal iron status. A computed tomography scan of the abdomen and pelvis showed no specific abnormalities and the patient was discharged; however, she returned to the hospital 5 days later with worsening abdominal pain, nausea, and bilious vomiting. Physical exam was remarkable for diffuse abdominal tenderness and pale conjunctivae. The laboratory evaluation was notable for anemia with hemoglobin of 8.8 g/dL, hematocrit of 23.5%, MCV of 87 fL, and corrected reticulocyte count of 7%. The patient was admitted, and review of her peripheral blood smear demonstrated normochromic, normocytic anemia with extensive coarse basophilic stippling of the erythrocytes. Her heavy metal screening tests showed an elevated blood lead level of 102 μg/dL (normal, <10 μg/dL). Zinc protoporphyrin was subsequently found to be elevated at 912 μg/dL (normal, <35 μg/dL). Her diagnosis was severe lead poisoning. At that point, the patient disclosed that she had been taking an Indian Ayurvedic medicine called Jambrulin obtained from Unjha pharmacy through a family member in India. She had been taking two pills daily over a period of 5 or 6 weeks in an effort to enhance control of her diabetes. She stopped taking the medication approximately 2 weeks prior to admission because of the abdominal pain. The patient was instructed not to take Jambrulin and received dimercaptosuccinic acid, an oral lead chelator, at a dose of 10 mg/kg three times a day for 5 days followed by 10 mg/kg twice a day for 2 weeks. At the end of chelation therapy, her blood lead level was significantly decreased to 46 μg/dL and her symptoms were resolved. When Ayurvedic medicine pills were tested, they showed approximately 21.5 mg of lead per pill. The pills were also sent to the Connecticut Department of Public Health Adult Blood Lead Epidemiology and Surveillance Program and Public Health Laboratory and were found to contain approximately 3.5% lead by weight or 35,000 μg/g [94].

Heavy metals and pesticides are also frequently present as contaminants in commonly prescribed raw Chinese herbal medicines. Harris *et al.* [95] reported that out of 334 samples representing 126 different Chinese herbal medicines analyzed, all 334 samples contained at least one heavy metal (lead, arsenic, chromium, mercury, or cadmium), whereas 115 samples (34%) had detectable levels of all five heavy metals tested. In addition, 42 different pesticides were detected in 108 samples (36.7%). It is also possible that poisoning after consuming raw herbal supplement may occur due to mistaken identity of the plant. Lin *et al.* [96] reported an outbreak of foxglove leaf poisoning when nine people mistakenly drank tea prepared from foxglove leaves instead of drinking tea made from comfrey leaves because comfrey leaves resemble foxglove leaves. Significant cardiac toxicity developed in three individuals, and digoxin concentrations varied from 4.4 to 135.9 ng/mL in these nine individuals. Patients were also treated with Digibind, and all patients recovered.

## CONCLUSIONS

The popularity of herbal remedies among the general population is on the rise, and such practice also increases the risk of herbal supplement-induced liver damage, other organ damage, as well as drug–herb interactions. Because of the perception that herbal supplements are safe, the majority of people do not disclose their use of herbal supplements to their health care professionals. Mehta *et al.* [97] reported that overall, only 33% of herbal and dietary supplement users reported disclosing their use of herbal supplements to their conventional health care providers. Therefore, the clinical laboratory may play an important role in helping clinicians to identify a potential drug–herb interaction. For example, abnormal liver function tests in a healthy individual during a routine physical examination may indicate use of kava or other herbal supplements known to cause liver damage. In addition, an elevated cholesterol level in a patient taking statin, which controlled his or her cholesterol level in the past, may be indicative of lower efficacy of the statin drug due to its lower serum levels secondary to a drug–herb interaction. Similarly, hypoglycemia in a patient receiving a hypoglycemic agent may also be related to a drug–herb interaction. If during routine drug monitoring, the observed drug level is significantly lower than the previous measurements and if noncompliance can be ruled out, it may be an indication of a potential drug–herb interaction. The most probable cause is use of St. John's wort, and on discontinuation of St. John's wort, the drug level usually returns to pre-herbal supplement use levels within 2 weeks. Similarly, observing an unusual INR during routine monitoring of a patient taking warfarin is also indicative of a potential interaction between warfarin and an herbal supplement [98].

In addition, many drugs that are not routinely monitored also interact with herbal supplements, and these herb–drug interactions are more difficult to detect by laboratory test. Because of the serious consequences of treatment failure from drug–herb interactions, transplant recipients, patients receiving HAART for AIDS treatment, as well as patients receiving warfarin or any related anticoagulants must refrain from using any herbal supplements.

## References

[1] Calapai G. European legislation on herbal medicines: a look into the future. Drug Saf 2008;31:428–31.
[2] Moss K, Boon H, Ballantyne P, Kachan N. New Canadian natural health product regulations: a qualitative study on how CAM practitioners perceive they will be impacted. BMC Complement Altern Med 2006;10(6):18.
[3] Mahady GB. Global harmonization of herbal health claims. J Nutr 2001;131:1120S–3S.
[4] Kelly JP, Kaufman DW, Kelley K, Rosenberg L, et al. Recent trends in use of herbal and other natural products. Arch Intern Med 2005;165:281–6.
[5] Jha V. Herbal medicines and chronic kidney disease. Nephrology 2010;15:10–17.
[6] Bent S. Herbal medicine in the United States: review of efficacy, safety and regulation. J Gen Intern Med 2008;23:854–9.

[7] Egan B, Hodgkins C, Shepherd R, Timotijevic L, et al. An overview of consumer's attitudes and beliefs about food supplements. Food Funct 2011;2:747–52.
[8] Jamieson DD, Duffield PH, Cheng D, Duffield AM. Composition of central nervous system activity of the aqueous and lipid extract of kava (*Piper methysticum*). Arch Int Pharmacodyn 1989;301:66–80.
[9] Scherer J. Kava-Kava extract in anxiety disorders: an outpatient observational study. Adv Ther 1998;15:261–9.
[10] Li XZ, Ramzan I. Role of Ethanol in Kava Hepatotoxicity Phytother Res 2010;24:475–80.
[11] Teschke R, Sarris J, Schwetzer I. Kava hepatotoxicity in traditional and modern use: the presumed pacific kava paradox hypothesis revisited. Br J Clin Pharmacol 2012;73:170–4.
[12] Rowe A, Zhang LY, Ramzan I. Toxicokinetics of kava. Adv Pharmacol Sci 2011;326724 (open access journal).
[13] Escher M, Desmeules J. Hepatitis associated with kava, a herbal remedy. Br Med J 2001;322:139.
[14] Jappe U, Frankle I, Reinhold D, Gollnick HP. Sebotrophic drug reaction resulting from kava-kava extract therapy; A new entity?. J Am Acad Dermatol 1998;38:104–6.
[15] Almedi JC, Grimsley EW. Coma from the health food store: interaction between kava and alprazolam. Ann Intern Med 1996;125:940–1.
[16] Christl SU, Seifert A, Seeler D. Toxic hepatitis after consumption of traditional kava preparation. J Travel Med 2009;16:55–6.
[17] Cow PJ, Connelly NJ, Hill RL, Crowley P, et al. Fatal fulminant hepatic liver failure induced by a natural therapy containing kava. Med J Aust 2003;178:442–3.
[18] Yeong ML, Swinburn B, Kennedy M, Nicholson G. Hepatic veno-occlusive disease associated with comfrey ingestion. J Gastroenterol Hepatol 1990;5:211–14.
[19] Roulet M, Laurini R, Rivier L, Calame A. Hepatic veno-occlusive disease in newborn infant of a woman drinking herbal tea. J Pediatr 1988;112:433–6.
[20] Seeff LB. Herbal hepatotoxicity. Clin Liver Dis 2007;11:577–96.
[21] Kouzi SA, McMurtry RJ, Nelson SD. Hepatotoxicity of germander (*Teucrium chamaedrys* L) and one of its constituent neoclerodane diterpenes teucrin a in the mouse. Chem Res Toxicol 1994;7:850–6.
[22] Zhou SF, Xue CC, Yu XQ, Wang G. Metabolic activation of herbal and dietary constituents and its clinical and toxicological implications: an update. Curr Drug Metab 2007;8:526–53.
[23] Laliberte L, Villeneuve JP. Hepatitis after use of germander, a herbal remedy. Can Med Assoc J 1996;154:1689–92.
[24] Sheikh NM, Philen RM, Love LA. Chaparral associated hepatotoxicity. Arch Intern Med 1997;157:913–19.
[25] Nisbet BC, O'Connor RE. Black cohosh induced hepatitis. Del Med J 2007;79:441–4.
[26] Anderson IB, Mullen WH, Meeker JE, Khojasteh-Bakht SC, et al. Pennyroyal toxicity: measurement of toxic metabolites levels in two cases and review of literature. Ann Intern Med 1996;124:726–34.
[27] Favreau JT, Ryu ML, Braunstein G, Orshansky G, et al. Severe hepatotoxicity associated with use of dietary supplement. Ann Intern Med 2002;136:590–5.
[28] SungHee Kole A, Jones HD, Christensen R, Gladstein J. A case of Kombucha tea toxicity. J Intensive care med 2009;24:205–7.
[29] Whiting PW, Clouston A, Kerlin P. Black cohosh and other herbal remedies associated with acute hepatitis. Med J Aust 2002;177:440–3.
[30] Jorge OA, Jorge AD. Hepatotoxicity associated with the ingestion of Centella Asiatica. Rev Esp Enferm Dig 2005;97:115–24.
[31] Vanhaelen M, Vanhaelen-Fastre R, Nut P, et al. Rapidly progressive interstitial renal fibrosis in young women: association with slimming regimen including Chinese herb. Lancer 1993;341:387–91.
[32] Chang CH, wand YM, Yang AH, Chiang SS. Rapidly progressive intestinal renal fibrosis associated with Chinese herbal medicines. Am J Nephrol 2001;21:441–8.
[33] Gold LS, Slone TH. Aristolochic acid, an herbal carcinogen sold on the web after FDA alert. N Engl J Med 2003;349:1576–7.
[34] Kong PI, Chiu YW, Kuo MC, Chen Sc, et al. Aristolochic acid nephropathy due to herbal drug intake manifested differently as Fanconi's syndrome and end stage renal failure in a 7-year follow up. Clin Nephrol 2008;70:537–41.
[35] Jellin J, Gregory P, Batz F , et al. Pharmacist's letter/prescriber's letter. Natural medicines comprehensive database. 4th edition Stockton, CA: Therapeutic Research Faculty; 2002.
[36] Blowey DL. Nephrotoxicity of over the counter analgesics, natural medicines, and illicit drugs. Adolesc Med 2005;16:31–43.
[37] Myhre MJ. Herbal remedies, nephropathies and renal disease. Nephron Nurs J 2000;27:473–8.
[38] Chau W, Ross R, Li JY, Yong TY, et al. Nephropathy associated with use of a Chinese herbal product containing aristolochic acid. Med J Aust 2011;194:367–8.
[39] Shaohua Z, Ananda S, Ruxia Y, Liang R, et al. Fatal renal failure due to Chinese herb "GuanMu Tong" (*Aristolochia manshuriensis*): autopsy finding and review of literature. Forensic Sci Int 2010;199:e5–7.
[40] Yeh GY, Eisenberg DM, Kaptchuk TJ, Phillips RS. Systematic review of herbs and dietary supplements for glycemic control in diabetes. Diabetes Care 2003;26:1277–94.
[41] Serraclara A, Hawkins F, Perez C, Dominguez E, et al. Hypoglycemic action of an oral fig leaf decoction in type 1 diabetic patients. Diabetic Res Clin Pract 1998;39:19–22.
[42] Karch AM, Jarch FE. The herb garden: remember to ask your patients all about preparations. Am J Nurs 1999;99:12.
[43] Sobieraj DM, Freyer CW. Probable hypoglycemic adverse drug reaction associated with prickly pear cactus, glipizide, and metformin in a patient with type 2 diabetes. Ann Pharmacother 2010;44:1334–7.
[44] Ching CK, Lam YH, Chan AY, Mak TW. Adulteration of herbal antidiabetic products with undeclared pharmaceuticals: a case series in Hong Kong. Br J Clin Pharmacol 2011;73:795–800.
[45] Goudie AM, Kaye JM. Contaminated medication precipitating hypoglycemia. Med J Aust 2001;175:256–7.
[46] Harada T, Ohtaki E, Misu K, Sumiyoshi T, et al. Congestive heart failure cause by digitalis toxicity in an elderly man taking licorice containing Chinese herbal medicine. Cardiology 2002;98:218.
[47] Lin SH, Yang SS, Chau T, Halperin ML. An usual cause of hypokalemia paralysis: chronic licorice ingestion. Am J Med Sci 2003;325:153–6.
[48] Meltem AC, Figen C, Nalan MA, Mahir K, et al. A hypokalemic muscular weakness after licorice ingestion: a case report. Cases J 2009;17(2):8053.
[49] Knobel U, Modarres G, Schneemann M, Schmid C. Gitelman's syndrome with persistent hypokalemia—Don't forget licorice, alcohol, lemon juice, iced tea and salt depletion: a case report. J Med Case Reports 2011;14:312.
[50] Teas J, Pino S, Critchley A, Braverman LE. Variability of iodine content in common commercially available edible seaweed. Thyroid 2004;14:836–41.
[51] Mussig K, Thamer C, Bares R, Lipp HP, et al. Iodine induced thyrotoxicosis after ingestion of kelp containing tea. J Gen Intern Med 2006;21:C11–14.
[52] Shilo S, Hirsch HJ. Iodine induced hyperthyroidism in a patient with a normal thyroid gland. Postgrad Med J 1996;62:661–2.
[53] Sood A, Sood R, Brinker FJ, Mann R, et al. Potential for interaction between dietary supplements and prescription medications. Am J Med 2008;121:207–11.

[54] Yang XX, Hu ZP, Duan W, Zhu YZ, et al. Drug–herb interactions: eliminating toxicity with hard drug design. Curr Pharmaceutical Design 2006;12:4649–64.
[55] Shi S, Klotz U. Drug interactions with herbal medicines. Clin Pharmacokinetic 2012;51:77–104.
[56] Viachojannis J, Cameron M, Churbasik WS. Drug interactions with St. John's wort products. Pharmacol Res 2011;63:254–6.
[57] Izzo AA, Ernst E. Interaction between herbal medicines and prescribed drugs. Drugs 2009;69:1777–98.
[58] Ernst E. St.John's wort supplements endanger the success of organ transplantation. Arch Surg 2002;137:316–19.
[59] Alscher DM, Klotz U. Drug interaction of herbal tea containing St. John's wort with cyclosporine [letter]. Transpl Int 2003;16:543–4.
[60] Mai I, Bauer S, Perloff ES, Johne A, et al. Hyperforin content determines the magnitude of the St. John's wort–cyclosporine drug interaction. Clin Pharmacol 2004;76:330–40.
[61] Hebert MF, Park JM, Chen YL, Akhtar S, Larson AM. Effects of St John's wort (*Hypericum perforatum*) on tacrolimus pharmacokinetics in healthy volunteers. Clin Pharmacol 2004;44:89–94.
[62] Mai I, Stormer E, Bauer S, Kruger H , et al. Impact of St. John's wort treatment on the pharmacokinetics of tacrolimus and mycophenolic acid in renal transplant patients. Nephrol Dial Transplant 2003;18:819–22.
[63] Jiang X, Williams KM, Liauw WS, Ammit AJ, et al. Effect of St. John's wort and ginseng on the pharmacokinetics and pharmacodynamics of warfarin in healthy subjects. Br J Clin Pharmacol 2004;57:592–9.
[64] Uygur Bayramicli O, Kalkay MN, Oskay Bozkaya E, Dogan Kose E, et al. St.John's wort (*Hypericum perforatum*) and warfarin: dangerous liaisons. Turk J Gastroenterol 2011;22:115 [Letter to the editor].
[65] van den Bout-van den Beukel CJ, Koopmans PP, van der Ven AJ, De Smet PA, Burtger DM. Possible drug metabolism interactions of medicinal herbs with antiretroviral agents. Drug Metab Rev 2006;38:477–514.
[66] Piscitelli SC, Burstein AH, Chaitt D, Alfaro RM, Fallon J. Indinavir concentrations and St. John's wort. Lancet 2000;355:547–8.
[67] de Maat MM, Hoetelmans RM, Math t RA, Van Gorp EC, et al. Drug interaction between St. John's wort and nevirapine. AIDS 2001;15:420–1.
[68] Johne A, Brockmoller J, Bauer S, Maurer A, et al. Pharmacokinetic interaction of digoxin with an herbal extract from St John's wort (*Hypericum perforatum*). Clin Pharmacol Ther 1999;66:338–45.
[69] Smith P. The influence of St. John's wort on the pharmacokinetics and protein binding of imatinib mesylate. Pharmacotherapy 2004;24:1508–14.
[70] Mathijssen RH, Verweij J, de Bruijn P, Loos WJ, Sparreboom A. Effects of St. John's wort on irinotecan metabolism. J Natl Cancer Inst 2002;94:1247–9.
[71] Eich-Hochli D, Oppliger R, Golay KP, Baumann P, Eap CB. Methadone maintenance treatment and St. John's wort: a case study. Pharmacopsychiatry 2003;36:35–7.
[72] Murphy PA, Kern SE, Stanczyk FZ, Westhoff CL. Interaction of St. John's wort with oral contraceptives: effects on the pharmacokinetics of norethindrone and ethinyl estradiol, ovarian activity and breakthrough bleeding. Contraception 2005;71:4102–408.
[73] Wang LS, Zhou G, Zhu B, Wu J, et al. St.John's wort induces both cytochrome P450 3A4 catalyzed sulfoxidation and 2 C19 dependent hydroxylation of omeprazole. Clin Pharmacol Ther 2004;75:191–7.
[74] Tannergren C, Engman H, Knutson L, Hedeland M, et al. St. John's wort decreases the bioavailability of R and S-verapamil through induction of the first pass metabolism. Clin Pharmacol Ther 2004;5:298–309.
[75] Xu H, Williams KM, Liauw WS, Murray M, et al. Effects of St. John's wort and CYP2C9 genotype on the pharmacokinetics and pharmacodynamics of gliclazide. Br J Pharmacol 2008;153:1579–86.
[76] Sugimoto K, Ohmori M, Tsuruoka S, Nishiki K, et al. Different effect of St. John's wort on the pharmacokinetics of simvastatin and pravastatin. Clin Pharmacol Ther 2001;70:518–24.
[77] Gordon RY, Becker DJ, Rader DJ. Reduced efficacy of rosuvastatin by St. John's wort [letter to the editor]. Am J Med 2009;122:e1–2.
[78] Singh YN. Potential for interaction of kava and St. John's wort with drugs. J Ethnopharmacol 2005;100:108–13.
[79] Dannawi M. Possible serotonin syndrome after combination of buspirone and St. John's wort. J Psychopharmacol 2002;16:401.
[80] Gordon JB. SSRIs and St. John's wort: possible toxicity. Am Fam Physician 1998;57:950–3.
[81] Heck AM, DeWitt BA, Lukes AL. Potential interaction between alternative therapies. Am J Health Syst Pharm 2000;57:1221–7.
[82] Lambert JP, Cormier A. Potential interaction between warfarin and boldo-fenugreek. Pharmacotherapy 2002;21:509–12.
[83] Tam LS, Chan Tym Leung WK, Critchley JA. Warfarin interaction with Chinese traditional medicines: danshen and methyl salicylate medicated oil. Aust NZ J Med 1995;25:238.
[84] Yu CM, Chan JC, Sanderson JE. Chinese herbs and warfarin potentiation by danshen. J Intern Med 1997;25:337–9.
[85] Lee NJ, Fermo JD. Warfarin and royal jelly interaction. Pharmacotherapy 2006;26:583–6.
[86] Shalansky S, Lynd L, Richardson K, Ingaszewski A, et al. Risk of warfarin related bleeding events and supra therapeutic international normalized ratios associated with complementary and alternative medicines: a longitudinal study. Pharmacotherapy 2007;27:1237–47.
[87] Chan HT, So LT, Li SW, Siu CW. Effect of herbal consumption on time in therapeutic range of warfarin therapy in patients with atrial fibrillation. J Cardiovasc Pharmacol 2011;58:87–90.
[88] Hamann GI, Campbell JD, George CM. Warfarin–cranberry interaction. Ann Pharmacother 2011;45:e17.
[89] Izzo AA. Interactions between herbs and conventional drugs: overview of clinical data. Med Princ Pract 2012;21:404–28.
[90] Huang WF, Wen KC, Hsiao ML. Adulteration by synthetic therapeutic substances of traditional Chinese medicine in Taiwan. J Clin Pharmacol 1997;37:344–50.
[91] Vanderstricht BI, Parvasis OE, Vanhaelen-Fastre RJ. Remedies may contain cocktail of active drugs. Br Med J 1994;308:1162.
[92] Lau KK, Lai CK, Chan AYW. Phenytoin poisoning after using Chinese proprietary medicines. Hum Exp Toxicol 2000;19:385–6.
[93] Savaliyaa AA, Prasad B, Raijada DK, Singh S. Detection and characterization of synthetic steroidal and non-steroidal antiinflammatory drugs in Indian Ayurvedic/herbal products using LC-MS/TOF. Drug Test Anal 2009;1:372–81.
[94] Gunturu KS, Nagarajan P, McPhedran P, Goodman TR, et al. Ayurvedic herbal medicine and lead poisoning. J Hematol Oncol 2011;20(4):51.
[95] Harris ES, Cao S, Littlefield BA, Craycroft JA, et al. Heavy metal and pesticides content in commonly prescribed individual raw Chinese herbal medicines. Sci Total Environ 2011;409:4297–305.
[96] Lin CC, Yang CC, Phua DH, Deng JF, et al. An outbreak of foxglove leaf poisoning. J Chin Med Assoc 2010;73:97–100.
[97] Mehta DH, Gardiner PM, Phillips RS, McCarthy EP. Herbal and dietary supplement disclosure to health care providers by individual with chronic conditions. J Altern Complement Med 2008;14:1263–9.
[98] Dasgupta A, Bernard DW. Complementary and alternative medicines: effects on clinical laboratory tests. Arch Pathol Lab Med 2006;130:521–8.

CHAPTER

# 8

# Challenges in Routine Clinical Chemistry Testing

## Analysis of Small Molecules

*Jorge Sepulveda*

**Columbia University Medical Center, New York, New York**

## INTRODUCTION

General or routine chemistry is the area of laboratory medicine dealing with measuring commonly ordered analytes, usually in serum or plasma, and is responsible for the highest volume of testing in clinical laboratories. In the author's medical center (Philadelphia Veterans Affairs Medical Center), routine clinical chemistry tests account for 85% of the test volume in the clinical laboratory, with the top nine tests (creatinine, glucose, urea, calcium, potassium, chloride, sodium, carbon dioxide, and thyroid-stimulating hormone) adding to 34% of the total test volume. The ability to handle large volumes of testing is due to the availability of highly automated, high-throughput analyzers, often linked to a robotic laboratory automation system that identifies, processes, and distributes specimens to the analyzers with high efficiency.

The analytical methods used in routine chemistry include the following assay types:

1. Chemical methods use chemical reactions to produce a measurable product, most commonly yielding changes in color or optical absorbance proportional to the analyte levels. An example is the Jaffe reaction to measure creatinine, in which creatinine reacts with a chromogen to form a colored product.
2. Electrochemical reactions result in changes in redox potential or in exchange of electrons at electrodes with consequent current generation proportional to the analyte concentration. An example is an oxygen-sensitive electrode used to measure $O_2$ consumed in a glucose oxidase reaction, which is proportional to the glucose concentration.
3. Enzymatic assays use enzymes to measure analyte concentrations or substrates to measure enzyme levels. The action of the enzyme on the substrate generates a product that can be measured by a chemical method. Sometimes a cascade of enzymes is necessary to generate a measurable product. An example of an enzymatically measured analyte is creatinine, assayed by a cascade of creatininase, creatinase, sarcosine oxidase, and peroxidase enzymes. Examples of clinically useful enzyme analytes are alanine and aspartate aminotransferases, alkaline phosphatase, creatinine kinase, lactate dehydrogenase (LDH), lipase, and amylase.
4. Ligand assays use proteins with high affinity for the analyte to measure its concentration. Most often, these assays are immunoassays, in which antibodies or antibody fragments are used to bind the ligand, although in a few cases, other specific-binding proteins are used. Examples of immunoassays include drug testing and monitoring assays, measurements of cardiac biomarkers such as cardiac troponins and B-type natriuretic peptides, and assays for hormones and tumor biomarkers. Examples of ligand assays using non-antibody high-affinity binding proteins include assays for folates and vitamin $B_{12}$. Whereas many of the low-volume ligand assays are part of "specialized chemistry" often handled in separate areas of the clinical laboratory, with the advent of multifunctional high-throughput analyzers able to handle a multitude of assay methodologies, many of the high-volume immunoassays have migrated to the highly automated routine chemistry area.

*Accurate Results in the Clinical Laboratory.*
DOI: http://dx.doi.org/10.1016/B978-0-12-415783-5.00008-6

Some methods use combinations of these categories. For example, a glucose assay may use glucose oxidase to generate $O_2$, which is then measured by an oxygen-specific electrode in an electrochemical reaction. In this chapter, frequent sources of inaccurate results in commonly ordered small molecule analytes, focusing on common pre-analytical variables (Table 8.1) and analytical interferences (Table 8.2), are addressed. In Chapter 9, common errors in enzymes and other protein assays are reviewed and in Chapter 10, sources of inaccuracy in biochemical genetics are discussed. Chapters 11 and 12 focus on challenges in endocrinology testing and tumor markers testing, respectively. In many laboratories, common therapeutic drugs, as well as initial toxicology screens, are done in routine chemistry by high-throughput analyzers and immunoassays, and these are reviewed in Chapters 13–15. Comprehensive listings of pre-analytical variables and interferences in laboratory testing are available in reference books [1–3].

The most commonly ordered tests in clinical chemistry are often ordered as panels. The Centers for Medicare and Medicaid Services, formerly called Health Care Financing Administration (HCFA), recognizes six panels of tests delineated by the American Medical Association CPT editorial board (Table 8.3), of which the Basic Metabolic Panel (BMP) comprising five electrolytes (sodium, potassium, chloride, carbon dioxide, and calcium) and three small organic analytes (glucose, urea, and creatinine) is the most commonly ordered panel of tests. Note that all the test components of a panel need to be medically justified for the panel to be reimbursed.

## CREATININE ANALYSIS

The small molecule creatinine is a product of the catabolism of creatine phosphate, a major source of rapidly available reserve energy, particularly in striated muscle and brain. Creatine phosphate can donate a phosphate group to adenosine diphosphate (ADP) to form adenosine triphosphate (ATP), a reaction catalyzed by creatine phosphokinase (CPK). Both creatine phosphate and creatine are converted nonenzymatically to creatinine at an almost steady rate of 2% per day [4].

Striated muscle and neural tissues have the highest requirements for quickly available energy and, therefore, the highest concentrations of creatine. Given the much larger volume of skeletal muscle, this tissue is the source of up to 94% of creatinine [4], and in healthy individuals the blood levels of both creatinine and CPK are proportional to skeletal muscle mass.

Creatinine is widely used as a marker of renal function—specifically as a surrogate for the glomerular filtration rate (GFR). The GFR is the volume of urine filtered by the glomeruli per unit of time. Clearance of a substance is defined as the amount of plasma cleared of that substance per minute, and it can be calculated by the following formula:

$$\text{Clearance} = \frac{U \times V}{P}$$

where $U$ is the urinary concentration, $P$ is the plasma concentration, and $V$ is the flow of urine in milliliters per minute. Clearance equals the GFR if the following assumptions are correct for the substance:

1. Production and plasma concentrations are constant.
2. There is no metabolism of the substance.
3. Excretion is exclusively by renal filtration, without secretion or reabsorption.

Whereas these assumptions are mostly correct for creatinine in healthy individuals, they may become less valid in patients with impaired renal function, leading to lower than expected plasma creatinine levels and overestimation of the GFR. It has been estimated that 16–66% of the creatinine formed in patients with chronic renal failure is subject to metabolism, mostly by reconversion to creatine [4]. Tubular secretion accounts for only 7–10% of the creatinine excretion in healthy individuals, but it increases proportionally to the degree of creatinine levels. Likewise, extrarenal excretion, particularly by degradation mediated by fecal bacteria, is increased in patients with chronic renal failure. Conversely, creatinine production decreases with higher blood levels of creatinine, possibly due to feedback inhibition of creatine synthesis. Despite these limitations, creatinine measurement in the plasma is one of the most widely used tests in clinical chemistry and a cost-effective approach to screening and evaluating kidney disease.

Creatinine clearance can be calculated by measuring both plasma and urine creatinine concentrations together with an assessment of urinary flow, usually by measuring the volume of urine excreted per 24 hr. Although directly measuring creatinine clearance should be a more accurate estimation of the GFR, in practice it suffers from inaccuracy resulting from compounding the errors of three measurements and from difficulties in properly timing urine collection.

Interestingly, estimation of the GFR based on measurement of plasma creatinine alone has generally shown better reproducibility and more accurate estimation of GFR than creatinine clearance. A widely used formula recommended by the National Kidney Foundation is the Modification of Diet in Kidney Disease (MDRD) equation that takes into consideration, in addition to plasma creatinine, age, sex, and race (Caucasian vs. African American) [5]. The use of age, sex, and race accounts for the influence of these factors in creatinine production, mainly through varying

TABLE 8.1 Common Pre-Analytical Sources of Variation in Small Molecule Analytes[a]

| | Na | K | Mg | iCa | Ca | Cl | $CO_2$ | Phos | Glu | Bili | Lac | $NH_4$ | Trig | Chol | LDL-C | HDL-C | UA | BUN | CR |
|---|---|---|---|---|---|---|---|---|---|---|---|---|---|---|---|---|---|---|---|
| Elderly | = | = | = | ↓/= | = | = | = | =/↓ | =/↑ | = | | | = | = | =/↑ | =/↓ | =/↑ | =/↑ | ↑ |
| Female/male | | | = | =/↓ | = | | | =/↑ | ↓ | ↑ | ↓ | | ↓ | ↓ | ↓ | ↑ | =/↓ | ↓ | ↓ |
| Meals | = | 0–5 | = | = | = | = | 4 | 15 | 50 | 16 | | | 78 | = | = | = | 0–3 | = | 0–50 |
| Starvation/malnutrition | = | =/↓ | | | | = | ↓ | | ↓ | 240 | | 17 | ↓35 | ↓10 | | | 5–20 | ↓20 | 0–20 |
| Obesity | = | = | | = | = | | ↓ | =/↓ | =/↑ | | ↑ | | ↑ | =/↑ | = | ↓ | ↑ | | |
| Exercise | 0–3 | Var | Var | =/↑ | Var | =/↑ | =/↓ | Var | Var | ↑ | ↑ | 300 | Var | Var | Var | ↑ | ↑ | Var | =/↑ |
| Caffeine | | ↓ | | ↑ | | | ↓ | | ↑ | | | | | = | = | = | | | |
| Smoking | = | | | | | | | | = | ↑ | | ↑ | =/↑ | Var | Var | ↓ | = | ↓ | = |
| Alcoholism | =/↑ | ↓ | Var | ↓ | | | ↓ | ↓ | Var | ↑ | ↑ | | 20 | 10 | ↓40 | ↑ | =/↑ | | ↓ |
| Pregnancy | = | = | ↓10 | ↓ | ↓10 | = | ↓ | Var | =/↓ | =/↓ | | = | ↑ | ↑ | ↑ | ↑ | Var | ↓ | =/↓ |
| Menstruation | ↓ | | 5 | | | ↑ | ↓ | ↓ | | | | | = | ↑ | | = | | | |
| Anxiety | | | | | | | | | ↑ | | ↑ | | ↑ | ↑ | | = | | ↑ | = |
| Supine to upright | = | = | 4 | = | 5–10 | = | = | = | = | | | = | 12 | 10 | 10 | 10 | = | = | = |
| Tourniquet | 3 | ↑ | ↑ | =/↑ | 3 | | ↓4 | ↓3 | ↓4 | 8 | ↑ | ↑ | 7 | 5–20 | | | | | |
| Serum/plasma | = | 5 | = | ↓ | = | = | = | 2–8 | = | = | 5–20 | 30 | = | = | = | = | = | = | = |
| Storage of whole blood at room temperature | = | ↑ | 10 | Var | ↓3 | ↓ | ↓ | =/↑ | ↓ | = | ↑ | ↑ | =/↓ | = | = | = | = | = | = |

[a]*Numbers represent typical percentage increase (↑) or decrease (↓) due to the pre-analytical variable. Var, variable effect, depending on the study. For details, see references [2,3].*

**TABLE 8.2** Common Interferents in Small Molecule Chemistry Assays[a]

**Strong redox agents**
Acetaminophen
Acetoacetate
Ascorbic acid
Azide
Bilirubin
Bromide
Catecholamines
Cysteine
Diatrizoate X-ray contrast agents
Dopamine
Gentisic acid
Glutathione
Guanidine
Hydralazine
Hydroxyurea
Iodide
Isoniazid
Lactate
Methylene blue
*N*-acetylcysteine
Nitrates
Procainamide
Sulfasalazine
Tetracycline
Thiocyanate
Uric acid

**Glucose assay**
Bilirubin
Fructose
Galactose
Hemoglobin
Hemolysis
Icodextrine
Lipemia
Maltose
Mannitol
Mannose
Oxygen levels
pH levels
Strong redox agents
Xylose

**L-Lactate assay**
Citrate
D-Lactate
Glycolate
Glyoxylate
Oxalate

**Urea assay**
Ammonium
Fluoride
Hemolysis
Levodopa
Methylbenzethonium

**Uric acid assay**
Bilirubin
Borate
Hemolysis
Strong redox agents
Theophylline metabolites
Xanthine

**Ammonia assay**
Alanine aminotransferase
Glucose $>600$ mg/dL
Hemolysis

**Bilirubin assay**
Amphotericin B
Levodopa
Methotrexate
Nitrofurantoin
Naproxen
Sulfasalazine
Hemoglobin
Light
Strong redox agents
Paraproteins
Lipemia

**Cholesterol assay**
Strong redox agents

**HDL cholesterol assay**
Strong redox agents
Paraproteins
Triglycerides

**Triglycerides assay**
Strong redox agents
Heparin
Glycerol

**Sodium, potassium assay**
Strong redox agents

**Chloride assay**
β-Hydroxybutyrate
Bicarbonate
Bromide
Heparin
Iodide
Lactate
Salicylates
Strong redox agents
Thiocyanate

**Total $CO_2$ assay**
Hemoglobin $>800$ mg/dL
Bilirubin $>30$ mg/dL
Paraproteins

**Total calcium assay**
Bilirubin
Calcium contaminants
Citrate
EDTA
Hemolysis
Lipemia
Oxalate

**Total magnesium assay**
Calcium
Heparin

**Phosphate assay**
Alteplase
Bilirubin
Fluoride
Hemolysis
Heparin
Lipemia
Liposomal amphotericin B
Mannitol
Paraproteins
Tissue plasminogen activator

[a] *Creatinine interferents are shown in Table 8.4. For details, see text.*

average muscle mass. The estimated GFR (eGFR) is calculated by the following formula, where serum creatinine (SCr) is reported in milligrams per deciliter:

$$\text{eGFR (mL/min/1.73 m}^2) = 175 \times (\text{sCr})^{-1.154} \times (\text{Age})^{-0.203} \times (0.742 \text{ if female}) \times (1.212 \text{ if African American})$$

This formula is valid for individuals 18 years or older. For children, an alternative formula should be used to calculate eGFR—most commonly the revised Schwartz equation [6]:

$$\text{eGFR (mL/min/1.73 m}^2) = (0.41 \times \text{height in cm})/\text{sCr}$$

**TABLE 8.3** Automated Multichannel Chemistry Panels Recognized by HCFA/CMS

| Test | Test Abbreviation | Basic Metabolic Panel | Comprehensive Metabolic Panel | Electrolyte Panel | Renal Function Panel | Hepatic (Liver) Function Panel | Lipid Panel |
|---|---|---|---|---|---|---|---|
| Panel abbreviation | | BMP | CMP | Lytes | RFP | LFP, LFT | |
| Creatinine | Cr | X | X | | X | | |
| Urea nitrogen | BUN | X | X | | X | | |
| Glucose | GLU | X | X | | X | | |
| Sodium | Na | X | X | X | X | | |
| Potassium | K | X | X | X | X | | |
| Chloride | Cl | X | X | X | X | | |
| Carbon dioxide | $CO_2$ | X | X | X | X | | |
| Anion gap | AG | [a] | [a] | [a] | [a] | | |
| Calcium | Ca | X | X | | X | | |
| Phosphate | P | | | | X | | |
| Albumin | Alb | | X | | X | X | |
| Total protein | TP | | X | | | X | |
| Alkaline phosphatase | ALP | | X | | | X | |
| Alanine aminotransferase | ALT | | X | | | X | |
| Aspartate aminotransferase | AST | | X | | | X | |
| Total bilirubin | TBili | | X | | | X | |
| Direct bilirubin | DBili | | | | | X | |
| Triglycerides | TG | | | | | | X |
| Total cholesterol | TC | | | | | | X |
| High-density lipoprotein cholesterol | HDL-C | | | | | | X |
| Low-density lipoprotein cholesterol | LDL-C | | | | | | [a] |
| Non-HDL cholesterol (calculated) | Non-HDL-C | | | | | | [a] |
| CPT code | | 80048 | 80053 | 80051 | 80069 | 80076 | 80061 |

[a]Calculated parameters. LDL-C may also be determined by a direct measurement.

To avoid rounding errors, creatinine results should be reported to two decimal places (e.g., 1.25 mg/dL). Note that these formulas use only one laboratory measurement, sCr, and are normalized to the average adult body surface area (BSA) of 1.73 $m^2$. Although increased BSA is typically associated with larger muscle mass, and therefore higher creatinine production, it also correlates with larger kidney size and thus increased GFR. These two factors tend to cancel each other out, resulting in a relatively linear inverse correlation of serum creatinine levels and BSA-normalized GFR.

This MDRD formula is applicable to isotope dilution mass spectrometry (IDMS)-traceable creatinine assays. IDMS methods are considered the gold standard for creatinine measurement and suffer from less interference than other methods, but they are available only in a few laboratories. However, the use of IDMS assayed calibrators based on an international standard (SRM 967: Creatinine in Frozen Human Serum), created by the National Institute for Standards and Technology, allows different methods to be "traceable" to the gold standard method, thus improving accuracy and standardization of the various creatinine assays. In general, compared to non-IDMS-traceable methods, IDMS-traceable methods show lower results, as expected from higher specificity of mass spectrometry for creatinine. The coefficient of the MDRD formula was changed from 186, which was based on non-IDMS-traceable creatine assays, to 175 to account for the lower creatinine levels measured with IDMS-traceable methods. Because the GFR is inversely proportional to sCr, the revised formula results in comparable eGFR. Currently, most, if not all, laboratories in the United States use IDMS-traceable calibrators and the revised MDRD formula.

## Limitations of the MDRD Equation

There are certain limitations of the MDRD equation:

1. The formula underestimates the true GFR in the normal and near normal range. If the eGFR is greater than 60 mL/min/1.73 $m^2$, laboratories should not report the numeric calculated measurement; instead, they should report eGFR as "$>$60 mL/min/1.73 $m^2$." Note that all creatinine-based screening methods miss patients with mild degrees of renal damage (chronic kidney disease category 2; GFR = 60–89 mL/min/1.73 $m^2$). In this range, large changes in GFR result in small changes in creatinine, usually within the reference range. In contrast, with severe renal failure, small changes in GFR result in large changes in creatinine levels.
2. As previously stated, this equation applies only to individuals 18 years or older. Although the original study did not include patients older than 70 years, the formula is considered useful in older individuals as a screen for renal disease.
3. In general, the MDRD equation is a more accurate estimation of GFR than directly measured creatinine clearance. However, carefully measured creatinine clearance or an alternative method of GFR estimation, such as inulin clearance, is preferable in patients with the following:
   a. Unusual dietary intake (vegetarian, low-meat diet, Atkins diet, and creatine supplements). Meat creatine is converted to creatinine by cooking.
   b. Unusual creatinine production due to advanced liver insufficiency or in individuals with extreme body size or muscle mass, including body builders, severe obesity, and muscle wasting or loss (e.g., severe malnutrition, cachexia, frailty, amputation, paraplegia, rhabdomyolysis, muscle dystrophy, and other neuromuscular disorders).
   c. Treatment with drugs that inhibit creatinine secretion, such as cimetidine, trimethoprim, pyrimethamine, dapsone, and possibly phenacemide, salicylates, and vitamin D derivatives [7]. Normally, tubular secretion accounts for approximately 10% of creatinine excretion, and the use of drugs that inhibit secretion can cause small increases in serum creatinine without an actual decrease in GFR. Note that the MDRD equation was derived in chronic renal failure patients with presumably higher degrees of tubular secretion of creatinine; therefore, the underestimation of GFR may be larger than 10%. Tubular secretion of creatinine is particularly elevated in recipients of kidney transplantation, leading to poor performance of the MDRD equation in estimating GFR in these patients [8].
   d. The MDRD equation, like all creatinine-based estimates of GFR, is applicable only to individuals with stable creatinine levels. It does not apply to patients with rapidly changing creatinine levels, including pregnant women and acutely ill patients, particularly those with acute renal failure.
4. Most currently available drug dosage guidelines are based on GFR estimations using older approaches, such as creatinine clearance or the Cockcroft–Gault formula, which is based on non-IDMS-traceable methods. However, a large study concluded that there is little difference between drug dosages calculated with the various methods of GFR estimation [9], leading the National Kidney Disease Education Program to recommend that for most patients, the eGFR based on the MDRD (or based on the Cockcroft–Gault equation) should be used. If using the MDRD-based eGFR in very large or very small patients, the eGFR (reported in mL/min/1.73 $m^2$) should be multiplied by the patient's BSA (in $m^2$) to obtain the patient's individual estimated GFR (in mL/min).

## Creatinine Assay Methods

Creatinine assays can be classified as chemical, based on the Jaffe reaction, or enzymatic methods.

### *Jaffe-Based Methods*

In the Jaffe reaction, first described in 1886, picrate ions react with creatinine under alkaline conditions to yield an orange-red product absorbing in the 490- to 520-nm wavelength range. This reaction is not specific to creatinine, and more than 50 compounds have been described that can produce a similar chromogen with

**TABLE 8.4** Substances Interfering with Modern Creatinine Assays[a]

| Interferent | Jaffe | Creatininase |
|---|---|---|
| 5-Flurocytosine | | Yes |
| Acetaminophen | | Yes |
| Acetoacetate | Yes | |
| Acetone | Yes | |
| Acetylcysteine | | Yes |
| Albumin (>4 g/dL) | Yes | |
| Ascorbate | Yes | Minimal |
| Bilirubin (>20 mg/dL) | Yes | Yes |
| Bromide (from halothane) | | Yes |
| Catecholamines | | Yes |
| Cefaclor | Yes | |
| Cefoxitin | Yes | |
| Cefpirome | Yes | |
| Cephalothin | Yes | |
| Cephazolin | Yes | |
| Creatine | Yes | Yes |
| Dipyrone (methimazole) | | Yes |
| Dobesilate | | Yes |
| Dobutamine | | Yes |
| Dopamine | | Yes |
| Fluorescein | Yes | |
| Glucose (>300 mg/dL) | Yes | |
| Glutamine | Yes | |
| Glutathione | Yes | |
| Hemoglobin (>500 mg/dL) | Yes | Yes (Hgb >900 mg/dL) |
| Hemoglobin F | Yes | |
| Hydroxyurea | | Yes |
| Intralipid (1000 mg/dL) | <10% | <10% |
| L-Dopa | Yes | Yes |
| Lidocaine | | Yes |
| Methyldopa | | Yes |
| Paraprotein | Yes | Yes |
| $p\text{CO}_2$ | | Yes[b] |
| Proline | Yes | Yes |
| Protein (>12 g/dL or <3 g/dL) | <20% | |
| Pyruvate | Yes | |
| Rifampicin | | Yes |
| Uric acid | Yes | |

(*Continued*)

**TABLE 8.4** (Continued)

[a]*The various methods are grouped based on the Jaffe reaction or creatininase, and interferences are variable, depending on the specific method used. For complete information, consult the manufacturer's package insert, as well as reference books [1–3].*

[b]*iSTAT only; approximately 5% for each 10-mmHg deviation from 40 mmHg.*

alkaline picrate ions. These substances, called non-creatinine chromogens, include proteins, ketoacids such as pyruvate, acetoacetate, ascorbic acid, glucose, cephalosporins, and other reducing substances, and they can account for up to 20% of the apparent creatinine concentration in the plasma of healthy individuals. Some of these compounds are increased in the blood of patients with chronic renal and liver disease, adding to the difficulties of accurately measuring creatinine and eGFR in these patients. Possibly the most clinically significant interferences with the Jaffe reaction occur in patients with diabetic or starvation-associated ketoacidosis, but the exact amount of interference varies with the particular creatinine assay used.

To reduce interference from non-creatinine chromogens, most modern assays in automated analyzers use a kinetic approach. Some non-creatinine chromogens, such as proteins, react quickly (within 20 sec) with the picrate ions, whereas others, such as acetoacetate, react slowly and do not form significant amounts of chromogen until 60–100 sec after the reaction is initiated. Chromogen formation measured between 20 and 60 sec is mostly derived from creatinine; therefore, the rate of chromogen formation in this time period is used for measuring creatinine levels in current Jaffe-based assays. Further minimization of non-creatine chromogen interferences is achieved by careful optimization of reagent concentration, temperature, and measurement wavelengths. Another approach, the "compensated" Jaffe assay, automatically subtracts a fixed number (e.g., 0.20 mg/dL) from the final result to account for the average contribution of non-creatinine chromogens. However, this approach assumes that this contribution is constant and does not account for changes in the concentration of non-creatinine chromogen or sample matrix, such as in elderly individuals with lower protein concentrations and lower creatinine production. Because non-creatinine chromogens do not appear to interfere significantly when creatinine is assayed in urine, the compensation approaches should not be used when assaying creatinine in urine, at the risk of causing significant negative biases in urine creatinine and creatinine clearance calculations.

An often overlooked interference with Jaffe-type creatinine assays is the presence of hemoglobin F in the plasma/serum [10]. Most methods are insensitive to even significant degrees of hemolysis (e.g., 500 mg/dL of plasma-free hemoglobin) because hemoglobin A quickly converts to a brown pigment under alkaline conditions, which is blanked out of the reaction rate measurement. However, hemoglobin F is alkali-resistant and slowly changes color in a Jaffe reaction. Individuals with high amounts of hemoglobin F, such as newborns and patients with thalassemias or hereditary persistence of hemoglobin F, can have significant interference in Jaffe-based creatinine assays when a hemolyzed sample is assayed.

Another major source of interference in the Jaffe reaction is bilirubin, in both the conjugated and the unconjugated form, which inhibits the reaction between picrate and creatinine and results in falsely low creatinine measurements. Attempts to minimize the effect of this interference include deproteinization, which also removes bilirubin; blanked kinetic measurements; and the addition of buffering ions, surfactants, and ferrocyanide or bilirubin oxidase to convert bilirubin to biliverdin, which does not significantly interfere. With these modifications, modern Jaffe-based creatinine assays do not show significant interference from bilirubin levels less than 20–25 mg/dL.

Table 8.4 lists many of the substances reported to interfere with modern Jaffe-based creatinine assays. A complete medication list and consultation with the particular laboratory performing the assay is important to determine the occurrence and magnitude of possible interferences. Note that many of the interfering drugs are accumulated in the urine to levels much higher than in plasma, and therefore urine creatinine determinations may be more susceptible to error from a variety of interferents.

*CASE REPORTS* A 40-year-old woman with insulin-dependent diabetes mellitus was admitted for ketoacidosis [11]. On admission, her blood pressure was low (100/60 mmHg), heart rate was elevated (124 beats per minute), and respirations were fast (20/min) and deep, suggesting metabolic acidosis with respiratory compensation. Laboratory values at admission included hyperglycemia (592 mg/dL), hyponatremia (134 mM), hypochloridria (90 mM), low serum bicarbonate (6 mM), and an elevated anion gap of 36 mM, consistent with ketoacidosis. Blood urea nitrogen (BUN) was 18 mg/dL and creatinine was 3.2 mg/dL, resulting in an unusual BUN:creatinine ratio of 5.6. The normal BUN:creatinine ratio is between 10 and 20 and may be increased in cases of prerenal causes of acute renal injury, including gastrointestinal bleeding and dehydration, whereas intrarenal damage, such as acute tubular necrosis, may reduce urea reabsorption, thus lowering the BUN:creatinine ratio to less than 10. This patient had signs of dehydration, with decreased turgor and dry mucous membranes, and the BUN:creatinine ratio did not appear consistent with the clinical presentation. Absorption of creatinine with an exchange resin to purify it from interfering chromogens yielded a true creatinine concentration of 1.0 mg/dL at admission, in contrast with 3.2 mg/dL by an automated Jaffe method. Addition of acetoacetate, but not β-hydroxybutyrate, to normal serum in concentrations exceeding 2 mM resulted in a linear increase in creatinine levels with the automated Jaffe method, implicating acetoacetate in the falsely high creatinine results. In addition, it is possible that high glucose and acetone levels associated with ketoacidosis further contributed to spuriously high creatinine levels.

In addition to diabetes, ketoacidosis can be associated with fasting. A study of five individuals who fasted for 96 hr showed an increase in mean serum creatinine levels from 1.0 to 1.7 mg/dL, corresponding to an average increase in acetoacetate levels of 0.03 to 1.39 mM [12]. The correlation between acetoacetate and creatinine measured by a Jaffe method was very high, with a coefficient of 0.95. No such correlation was seen with an enzymatic creatinine method, which was unaffected by acetoacetate.

Since these cases were reported in the 1980s, improvements have been made in the Jaffe reaction to minimize interference by acetoacetate, mainly by restricting the kinetic measurement to an optimal time range. Current Jaffe-based methods do not show significant interference with acetoacetate levels below 20 mM, although some rapid creatinine methods designed for STAT use, such as the Beckman modular creatinine assay (CREAm), can show a positive bias of approximately 0.07 mg/dL for each 1 mM of acetoacetate elevation [13]. Notably, the Jaffe-based creatinine assay performed in the cuvette side of the Beckman DxC analyzer shows no interference with acetoacetate levels below 20 mM. Because acetoacetate levels in diabetic ketoacidosis range from 1.2 to 9 mM, averaging $3.5 \pm 1.4$ mM [11,14], this interference remains a problem with some creatinine assays.

A 6-year-old female had bone marrow transplant for acute myelogenous leukemia and developed severe graft-versus-host disease and severe hyperbilirubinemia, with plasma bilirubin reaching 55.3 mg/dL [15]. She was treated with hemodialysis for renal insufficiency secondary to a large bladder hematoma with obstructive uropathy. As part of assessing the efficiency of hemodialysis, the sieving coefficient, calculated from creatinine concentration in dialysate of 0.7 mg/dL divided by plasma creatinine of 0.3 mg/dL, equaled 2.33, which is clearly abnormal because 100%

efficiency will result in a coefficient of 1.0. When the same calculation was applied to plasma and dialysate urea, the sieving coefficient was 1.0. The creatinine was initially measured with a Jaffe-based method, and a repeat measurement using an enzymatic method resulted in both plasma and dialysate creatinine levels of 0.7 mg/dL and therefore a sieving coefficient of 1.0. This case highlights the susceptibility of the Jaffe-based assays to negative interference by high levels of bilirubin, and the potential for overestimation of hemodialysis efficiency, because the measurement in the plasma, but not the dialysate, is susceptible to this interference. It is not clear why bilirubin does not cross the dialysis membrane, although binding of bilirubin to albumin and electrostatic repellence between negative charges in both conjugated bilirubin and membrane are possible explanations.

### *Enzymatic Creatinine Assays*

To obviate the problems with the Jaffe-based methods and increase specificity for creatinine, several assays use enzymatic reactions to obtain a measurable product. Two enzymes can act on creatinine: creatininase, which converts creatinine to creatine, and creatinine deiminase, which generates *N*-methylhydantoin and $NH_3$. To convert one of these products to a photometrically measurable substance, a cascade of other enzymes must be used. For example, in the most commonly used method, creatininase generates creatine and creatinase converts creatine to sarcosine, which in turn generates hydrogen peroxide and formaldehyde after treatment with sarcosine oxidase. Subsequently, peroxidase catalyzes the oxidation of a chromogen to a colored product by $H_2O_2$.

Whole blood analyzers, such as the Abbott iSTAT, Radiometer ABL800, and Nova StatSensor, use a similar creatininase/creatinase/sarcosine oxidase cascade and electrodes that generate current upon oxidation by $H_2O_2$ to measure creatinine in whole blood. In the case of the Radiometer, two electrodes are used: One electrode is associated with the full cascade of enzymes and measures $H_2O_2$ generation from creatinine, whereas a second electrode omits the creatininase enzyme and measures background $H_2O_2$ generation from creatine and other substances, thus significantly minimizing interferences [16]. Rapid creatinine measurements in whole blood analyzers have become increasingly important in the expedite screening of patients undergoing emergency radiologic procedures using contrast dyes, as a mechanism to prevent contrast-induced nephropathy (CIN), because patients with eGFR less than 60 mL/min/1.73 $m^2$ are at higher risk of CIN.

Another approach used in automated analyzers to minimize interference by reaction intermediates is to measure reaction rates with all enzymes but without creatininase and then add creatininase to measure creatinine-dependent chromogen generation. As with other hydrogen peroxide ($H_2O_2$)-based methods, the results of enzymatic creatinine assays that generate $H_2O_2$ are subject to interference by strong reducers, such as ascorbic acid and bilirubin, which compete with the chromogen for $H_2O_2$. Interference by ascorbic acid can be overcome by ascorbate oxidase, whereas bilirubin oxidase or ferrocyanide can minimize interference by bilirubin.

Alternatively to using peroxidase and measuring $H_2O_2$ generation, another method uses formaldehyde dehydrogenase to oxidize formaldehyde to acetic acid while converting nicotinamide adenine dinucleotide ($NAD^+$) to NADH, which can be measured by absorption at 340 nm. In another $H_2O_2$-free approach, creatinine deiminase converts creatinine to *N*-methylhydantoin and ammonia, and either of these compounds can be converted by subsequent steps into a chromogen. The simpler methods use glutamate dehydrogenase to convert ammonia and α-ketoglutarate to glutamate while oxidizing NADH to $NAD^+$. The use of recombinant creatinine deaminase from the bacterium *Tissierella creatinini* has improved the specificity of the assay because this enzyme is highly specific for creatinine, in contrast to creatinine deiminase from other sources that can react with cytosine and related compounds. However, none of these methods have found widespread routine use in clinical laboratories.

Most interfering substances reported with the Jaffe methods do not significantly interfere with enzymatic creatinine assays at physiologic or therapeutic levels, with the exception of catecholamines, levodopa, α-methyldopa, rifampicin, and calcium dobesilate (Doxium and Dobest), a vasoprotective drug used in the treatment of diabetic retinopathy and chronic venous insufficiency (see Table 8.4).

*CASE REPORT* A 36-year-old male with severe heart failure and progressive renal failure had initial creatinine levels of 4.6, 4.4, and 4.3 mg/dL in the first 2 days of admission to the hospital. Given the cardiovascular situation, the patient was transferred to the intensive care unit (ICU) and treated with dopamine to improve blood flow and kidney function. The creatinine values in the next 2 days in the ICU fluctuated more than 25% with no correlation to the patient's unchanged clinical status. When the samples were analyzed by both the Roche enzymatic assay and a Jaffe-based method, spurious decreases in creatinine levels were seen with fresh samples tested with the Roche method but not with samples stored for more than 24 hr. When dopamine was added at a molar ration of

0.25:1 to creatinine, the negative bias was −67% with the Roche enzymatic assay, −13% with the Vitros enzymatic assay, and between −3 and −5% with three Jaffe-based methods [17]. It is well recognized that strong reducing substances, such as dopamine, dobutamine, and other catecholamines, can interfere with enzymatic reactions that use $H_2O_2$ and peroxidase as intermediate steps to color generation. This interference is due to either complex formation with the chromogen, in the case of dopamine and methods using 4-aminophenazone, or reactivity between the catecholamine and $H_2O_2$, with consequent depletion of the peroxide and falsely low results [18]. Instability of the catecholamines explains why samples older than 24 hr showed no negative interference by these substances [17].

## UREA ANALYSIS

Urea is another substance excreted primarily (90%) by the kidney, and because it was initially measured as urea nitrogen, it is reported in the United States as BUN, even though all modern assays measure urea concentration and not nitrogen. To convert BUN to urea, in milligrams per deciliter, multiply by 2.14. Urea is formed by the urea cycle, the end pathway for disposal of ammonia mostly derived from protein catabolism. The enzymes of the urea cycle are most abundant in the liver, and they are present in the kidney and other tissues in much smaller amounts.

In contrast to creatinine, passive tubular reabsorption of urea is significant, particularly in situations of impaired urinary flow; such as with dehydration and reduced blood volume, and therefore urea clearance is lower than the GFR. In situations of high urine flow, such as pregnancy, diuretic treatment, and hyperosmolar loads, urea reabsorption is minimal and urea clearance approaches the GFR. A similar situation is seen in advanced renal failure, with high osmotic diuresis. Despite the superiority of creatinine for estimation of GFR, the measurement of urea has a role in the following clinical situations:

1. When creatinine measurements are misleading due to interferences or physiologic situations, such as extremes of muscle mass, that affect the relationship of creatinine and GFR.
2. To help determine the mechanism of renal insufficiency by comparing the elevation in urea and creatinine, as reflected by the BUN:creatinine ratio.
   - **a.** Increased BUN:creatinine ratio
     - **i.** Prerenal disease, with hypovolemia and/or impaired kidney perfusion, leads to increased urea reabsorption and elevation of BUN disproportionally to the decrease in GFR and increase in creatinine, which often is in the normal range. Therefore, the BUN:creatinine ratio is elevated, typically greater than 20. Examples of systemic conditions associated with prerenal azotemia include bleeding, burns, shock, cirrhosis, congestive heart failure, anaphylaxis, dehydration, prolonged diarrhea, vomiting, and sweating. Examples of localized impairment of renal blood flow include aortic coarctation, aneurysm, dissection, and renal artery occlusion by thromboembolism or stenosis.
     - **ii.** Conditions that increase urea production (high-protein diet, high catabolic states such as fever, infection, burns, hyperthyroidism, acute myocardial infarction, postsurgery (especially gastrointestinal surgery), anti-anabolic drugs such as tetracycline and corticosteroids, and gastrointestinal bleeding with absorption and catabolism of blood proteins) or decrease creatinine production (e.g., low muscle mass) can also increase the BUN:creatinine ratio. Due to these multifactorial effects, the BUN:creatinine ratio is of low value in discriminating the cause of renal insufficiency in acutely ill patients, although a high BUN:creatinine ratio is associated with worse outcomes in these patients [19,20].
     - **iii.** Postrenal disease, such as obstruction, stenosis, or reflux in the urinary outflow tract (ureters, bladder, and urethra), will also lead to reduced urinary flow and increases in both BUN and creatinine, with the BUN:creatinine ratio usually, but not always, greater than 15.
   - **b.** Decreased BUN:creatinine ratio (< 10)
     - **i.** Conditions that decrease urea reabsorption: high urinary flow (e.g. pregnancy and osmotic diuresis), acute tubular necrosis, interstitial nephritis, and sickle cell disease.
     - **ii.** Conditions with decreased urea synthesis (negative nitrogen balance): severe liver disease, malnutrition, low protein intake, protein malabsorption (e.g., celiac disease), protein loss (nephrotic syndrome), increased protein synthesis (late pregnancy, childhood, convalescence, acromegaly, anabolic drugs such as androgens, growth hormone).
     - **iii.** Conditions with increased creatinine production (bodybuilders, rhabdomyolysis, etc.).
   - **c.** Normal BUN:creatinine ratio (10–20)
     - **i.** Most cases of renal failure.
3. Urinary urea measurements, in conjunction with an assessment of protein intake, can be used as a crude

estimation of proper nitrogen balance for nutritional purposes. For example, net nitrogen balance can be calculated by the following formula:

$$\text{Nitrogen balance} = \text{protein intake } (g/\text{day})/6.25 - \text{urea nitrogen } (g/\text{day}) + 4$$

## Urea Assay Methods

Virtually all current routine methods for urea measurement use urease to convert urea to ammonium and bicarbonate ions. The detection of ammonium formation, which is proportional to the concentration of urea, is achieved by one of the following methods:

1. Change in conductivity: The Beckman modular assay uses the rate of change in conductivity of the diluted sample when ammonium ($NH_4^+$), bicarbonate ($HCO_3^-$), and hydroxyl ($OH^-$) are generated by urease from non-ionic urea.
2. Ammonium-selective potentiometry: In some whole blood analyzers, urease is immobilized on a membrane, and the generation of ammonium is measured by current generated at the ammonium-selective electrode.
3. Reaction with quinolinium dye: The Ortho Vitros dry chemistry method uses an ammonium-permeable membrane to filter the ammonium generated by the urease reaction into the color-forming layer, where it reacts with a quinolinium chromogen, generating color measured by reflectance spectrometry at 670 nm.
4. Glutamate dehydrogenase: This enzyme converts ammonium and α-ketoglutarate to glutamate and in the process oxidizes NADH to $NAD^+$ with the consequent decrease in absorption at 340 nm.

In all these methods, an obvious interference to control for is endogenous or exogenous (anticoagulant) ammonium, which will cause an increase in urea corresponding to the ammonium concentration. Normally, the concentration of ammonia in plasma (10–200 μM) is much lower than the concentration of urea (1–10 mM), but it may become a problem in certain metabolic disorders, including those associated with liver dysfunction, and in acidic or bacterially contaminated urine. The problem can be solved by taking a sample blank reading with all reagents except urease. High amounts of fluoride inhibit urease and may cause falsely low BUN results, but the amount present in gray-top tubes (~2.5 mg/mL) does not interfere significantly. Likewise, ammonium heparin anticoagulant at 14 units/mL shows no significant interference with BUN.

Other interferences reported include levodopa (−3 mg/dL at 40 μg/mL) and methylbenzethonium chloride (−5 mg/dL at 0.5 mg/dL) for the modular Beckman assay. No significant interference is seen with hemolysis (at 500–800 mg/dL hemoglobin), bilirubin (30–100 mg/dL), and lipemia (500 mg/dL Intralipid) in various BUN assays. The Vitros assay shows a bias of 1.1 mg/dL with 500 mg/dL of plasma hemoglobin.

# AMMONIA ASSAY

Ammonia is generated in the body predominantly from the action of intestinal bacteria, which deaminate urea and amino acids. In patients with gastric *Helicobacter pylori* infections, the urease activity of this organism can be an important source of circulating ammonia, particularly in patients with cirrhosis [21]. The ammonia generated in the intestine flows into the liver through the portal vein and is metabolized in the urea cycle. In excess, ammonia can be a potent neurotoxin.

Plasma ammonia levels are measured most commonly in two clinical situations: to monitor metabolic disorders in infants, such as deficiencies in urea cycle enzyme and related pathways, and in advanced liver disease to monitor the risk of hepatic encephalopathy. The latter indication is controversial because the correlation between ammonia levels and encephalopathy is poor, and routine measurement of plasma ammonia is not recommended in patients with known liver disease, although it may be useful in patients with encephalopathy of unclear etiology [22].

## Pre-Analytical Factors

Diets with high protein content may result in increased ammonia generation, mainly reflected by increased urinary excretion, whereas starvation will decrease ammonia by 17% [23]. Exercise increases ammonia generation approximately threefold [22]. Subjects should abstain from smoking because it increases ammonia blood levels by approximately 10 μmol/L/hr [22]. Certain drugs, such as ammonium salts, barbiturates, opioids, valproic acid, and asparaginase, can increase ammonia generation. Patients treated with fluids containing glutamine, such as for irrigation post-transurethral prostatectomy, can have significant increases in ammonemia [22].

Clenching fist, prolonged tourniquet application, and improper cleaning of the skin during blood collection can result in spuriously high ammonia. Of particular concern are capillary samples collected from infants, in which sweat contamination and higher frequency of hemolysis and tissue damage can artificially increase ammonia in the sample [22,24]. Arterial ammonia levels are higher than venous and correlate better with liver function; therefore, arterial specimens are preferable in this condition [22].

Blood samples for ammonia analysis should be collected in tubes containing ethylenediaminetetraacetic acid (EDTA) or heparin as anticoagulant, and obviously ammonium heparin or ammonium oxalate should not be used. Serum is not recommended because ammonia is produced by platelet lysis during the clotting process. The tubes containing the blood samples should be placed immediately on ice slurry and centrifuged as soon as possible to prevent generation of ammonia *in vitro*, 90% of which originates from red cells, platelets, and leukocytes [22,25]. In healthy individuals, ammonia accumulates in whole blood at a rate of approximately 4 μmol/L/hr, compared to 5 and 25 μmol/L/hr at 20° and 37°C, respectively [25]. Analysis should proceed expeditiously because ammonia levels increase in plasma even at 0°C. After cellular separation, ammonia accumulates in plasma at a rate of approximately 0.4 μmol/L/hr [26]. Conditions associated with elevated levels of γ-glutamyl transferase (GGT) in the plasma, such as alcoholism and hepatobiliary diseases, can have up to a 35-fold increase in the rate of ammonia generated *in vitro* from glutamine and peptide deamidation in the plasma, even at 0°C [26]. The addition of 2 mM borate and 5 mM serine to inhibit GGT can prevent ammonia generation in patients with high levels of GGT [26]. The Urea Cycle Disorders Conference Group issued a consensus statement suggesting that blood should be collected in a prechilled ammonia-free tube, kept on ice, and centrifuged within 15 min of collection [24]. Plasma should be frozen at −70°C if not assayed immediately. Collection of capillary samples and delay in processing the samples were found to be the most common causes of falsely elevated ammonia results in children, which occurred in approximately 28% of the patients with values higher than the reference range [24].

### Assay Methodology for Ammonia

The most commonly used methods for ammonia determination use glutamate dehydrogenase to convert ammonia and α-ketoglutarate to glutamate while oxidizing NADPH (or NADH) to $NADP^+$ (or $NAD^+$) with a resulting decrease in absorption at 340 nm. Alternatively, in the Vitros dry chemistry method, ammonia is isolated from the rest of the sample solution with a semipermeable membrane that subsequently reacts with bromophenol blue, causing a change in reflection density measured at 600 nm.

Atmospheric ammonia may cause an artificial increase in ammonia levels. Hemolyzed samples are not recommended because of higher rates of spontaneous ammonia generation *in vitro*. Glucose at 600 mg/dL (33.3 mM) can cause a decrease of 8–40 μmol/L in ammonia concentration using the Vitros method. High levels of alanine aminotransferase have been reported to cause a positive interference in enzymatic ammonia assays in patients with liver failure, possibly by catalyzing the conversion of the alanine to pyruvate, which is then converted by LDH into lactate, consuming NADH [27]. Sample blanking is recommended to correct for interference caused by dehydrogenases and other substances in the plasma able to oxidize NAD(P)H [28]. Interestingly, methods using NADPH are much less susceptible to nonspecific interferences than those using NADH [29].

## URIC ACID ANALYSIS

Uric acid is the product of nucleic acid catabolism, as a major breakdown product of purine nucleosides. Accumulation of uric acid in joints leads to inflammatory arthritis known as gout, whereas excessive uric acid excretion can lead to urate nephropathy and kidney stones. The final step in uric acid formation is the conversion of xanthine to uric acid catalyzed by xanthine oxidase, and inhibition of this enzyme by allopurinol can be effective in reducing circulating uric acid levels. Uric acid is excreted by the kidneys by distal tubular secretion, a process that can be enhanced by uricosuric drugs such as probenecid, sulfinpyrazone, azapropazone, tiaprofenate, and several others and inhibited by diuretics and organic acids.

Causes of hyperuricemia include several congenital metabolic disorders, such as Lesch–Nyhan syndrome resulting from deficiency of hypoxanthine-guanine-phosphoribosyltransferase, excess purine intake (e.g., red meat, liver, kidney, and sardines), excess nucleic acid turnover (leukemias, myeloma, tumor lysis syndrome, trauma, and excessive exercise), or decreased excretion caused by renal disease or inhibitors of urate secretion (organic acids including lactate and acetoacetate, salicylates, thiazides, and lead poisoning). Interestingly, prolonged starvation and other ketoacidosis states can lead to more than a 20% increase in uricemia, due to a combination of decreased intake and increased catabolism of nucleic acids, and decreased excretion due to inhibition by ketoacids [30]. In the case of acute alcohol ingestion, the associated increased purine catabolism, dehydration, and lactic acidosis further contribute to increased uricemia [31].

Causes of hypouricemia include decreased synthesis (deficiency or inhibition of xanthine oxidase or its molybdenum co-factor, severe liver disease, purine synthesis inhibitors such as 6-mercaptopurine, and azathioprine) and decreased renal reabsorption (Fanconi syndrome, interstitial nephritis, radiologic contrast nephropathy, and overtreatment with uricosuric drugs).

### Analytical Considerations for Urea Analysis

Virtually all routine assays for uric acid use uricase to convert uric acid to allantoin with generation of $H_2O_2$ and $CO_2$ followed by peroxidase-catalyzed oxidation of a chromogen by the $H_2O_2$. As with other peroxidase-based methods, interference by ascorbic acid and bilirubin is common. For example, the Beckman assay shows a negative bias of 10% with 10 mg/dL of bilirubin and of 0.3 mg/dL with 1.5 mg/dL of ascorbate. The Vitros dry chemistry method obviates bilirubin interference (up to 27 mg/dL) by removal of proteins from the color-forming layer, but it is still affected by high ascorbate levels (−17% bias at 100 mg/dL). In the Roche method, ascorbate oxidase is used to reduce interference by ascorbic acid. Other interferants reported with uricase/peroxidase methods include hemolysis, theophylline metabolites, catecholamines, methylene blue, sulfasalazine, gentisic acid, hydralazine, and other reducing substances (see also the section on creatinine). Borate preservatives, sodium azide, and ascorbate in urine interfere with the urease reaction and are frequent causes of misleading urinary urate measurements [32]. Urine pH less than 8.0 can result in urate precipitation and inaccurate urinary urate measurement.

*CASE REPORT* A 30-year-old patient with Burkitt's lymphoma was hospitalized for dyspnea, night sweats, fatigue, and early satiety [33]. Tumor staging indicated the patient was in stage D. The patient's admission serum uric acid levels were 14.1 mg/dL, and the patient was treated for 3 days with allopurinol, resulting in a decline of uricemia to 5.1 mg/dL. After a single dose of cyclophosphamide treatment, resulting in rapid tumor lysis and disappearance of his chest masses in 2 days, uric acid levels increased to approximately 10 mg/dL as measured by a chemical method but only 6 mg/dL as measured by a uricase method. The discrepancy correlated with high serum levels of xanthine, an inhibitor of the uricase reaction, which increased from a prechemotherapy level of 10 mg/dL to a peak of 140 mg/dL. Normally, xanthine is rapidly cleared by the kidney, but in cases of rapid tumor lysis in which conversion to uric acid is inhibited by allopurinol, it can reach higher levels, especially if associated with renal insufficiency, and cause falsely low uric acid measurements with uricase methods.

## GLUCOSE ANALYSIS

When point-of-care (POC) and home testing assays are included, glucose represents the most frequently measured analyte in clinical settings. As the most important source of energy for most tissues, including the brain, levels of glucose in the blood are highly regulated and deviations from normality have severe consequences. Hypoglycemia is defined as blood glucose levels less than 45 mg/dL or by the Whipple's triad of low glycemia (<70 mg/dL) with symptoms of hypoglycemia (palpitations, sweating, tremors, weakness, headache, fainting, etc.) and disappearance of the symptoms when glucose is normalized. Hypoglycemia can have many causes:

1. Hyperinsulinic hypoglycemia, characterized by low ketones and free fatty acids
   a. High-insulin secretion (associated with high levels of C-peptide), which includes drugs (sulfonylurea, meglitides, quinine, and pentamidine), insulinoma, islet cell dysplasia, and very rarely β cell-stimulating autoantibodies.
   b. Low-insulin secretion (low C-peptide): excessive exogenous insulin administration, anti-insulin antibodies, insulinomimetic antibodies, and ectopic insulin-like growth factor production by tumors.
2. Non-hyperinsulinic hypoglycemia
   a. With elevated free fatty acids and ketones: Failure to produce or obtain glucose or excess consumption of glucose leads to increased fatty acid oxidation and ketonemia—drugs (ethanol, salicylates, and propanolol), severe liver disease, hormone deficiency (growth hormone, cortisol, thyroid, and pituitary), extrapancreatic malignancy, and malnutrition, inborn errors of metabolism (galactosemia, fructosemia, glycogen storage disorders, gluconeogenesis defects, various organic acidurias, etc.).
   b. Inborn diseases of fatty acid oxidation: Compensatory increase in glucose consumption results in hypoglycemia, low ketone levels, normal to elevated free fatty acids, and low acylcarnitine levels.

Hyperglycemia is usually a chronic disease caused by insulin deficiency with pancreatic β cell destruction (diabetes mellitus type 1) or peripheral insulin resistance, often associated with partial insulin deficiency (diabetes type 2). Rare inherited forms of diabetes include defects in pancreatic genesis, insulin secretion, and glucose sensing in the β cells, termed maturity-onset diabetes of the young and often involving transcription factors such as HNF1, HNF4, KLF1, PAX4, and IPF1; defects in insulin or its receptor genes; and as part of complex genetic disorders such as Down, Klinefelter, Turner, and Prader–Willi syndromes. Chronic hyperglycemia is also often associated with diseases of the exocrine pancreas (hemochromatosis, chronic pancreatitis, and cystic fibrosis), endocrinopathies (Cushing's syndrome, acromegaly, glucagonoma, pheochromocytoma, and hyperthyroidism), infections (congenital rubella or cytomegalovirus, coxsackievirus B, mumps, sepsis, and meningitis), drugs (thiazides, steroids, dilantin, statins, etc.), toxins (streptozotocin and other fungal toxins), and, rarely, anti-insulin receptor antibodies or the stiff-person syndrome with anti-glutamate-decarboxylase autoantibodies [34].

The American Diabetic Association (ADA) has defined criteria for diagnosis of diabetes mellitus, to include any of the following [34,35]:

1. Hemoglobin A1c (HbA1c) ≥ 6.5%.
2. Fasting plasma glucose ≥ 126 mg/dL (fasting defined as no caloric intake for ≥ 8 hr).
3. Oral glucose tolerance test (OGTT), with 75-g load of glucose, plasma glucose ≥ 200 mg/dL at 2 hr.
4. Random plasma glucose ≥ 200 mg/dL with classic symptoms of hyperglycemia (polyuria, polydipsia, or unexplained weight loss).

It is important to note that criteria 1–3 should be confirmed by repeated testing on a different day, unless the patient has two or more of the criteria or in cases of obvious diabetes, defined by plasma glucose >500 mg/dL, HbA1c >7.0%, ketoacidosis, nonketotic coma, and classic symptoms of hyperglycemia. In most cases, the OGTT is unnecessary to diagnose or exclude diabetes, as the combination of fasting glucose, random glucose, and HbA1c offers sufficient sensitivity and specificity.

Patients with intermediate values between the upper limit of the reference range and the ADA criteria are at risk for developing diabetes and its complications:

1. Hemoglobin A1c (HbA1c): 6.0–6.5%. Some experts advocate an HbA1c of 5.7–6.5% to define prediabetic risk [34].
2. Impaired fasting glucose: 100–125 mg/dL.
3. Impaired glucose tolerance—random plasma glucose or 2-hr OGTT: 140–199 mg/dL.

For diagnosing gestational diabetes, more stringent criteria have been defined [36]:

1. Fasting plasma glucose ≥ 92 mg/dL.
2. 1-hr plasma glucose (75-g load) ≥ 180 mg/dL.
3. 2-hr plasma glucose (75-g load) ≥ 153 mg/dL.

## Pre-Analytical Considerations for Glucose Measurement

The ADA guidelines recommend fasting glucose as one of the screening tests for diabetes. Patients should abstain from solid food for at least 8 hr but should maintain normal liquid intake, otherwise dehydration and hemoconcentration could artificially increase the measured glucose level. It is recommended that patients abstain from alcohol, smoking, caffeine, and any herbs or medications known to affect glucose metabolism, and that collection of blood is performed in the morning to avoid the effects of diurnal variation. For the OGTT, the patient should consume an unrestricted diet containing at least 150 g/day of carbohydrates for 3 days and fast for 10–16 hr before the test is performed [37]. Anxiety, recumbence, inactivity, acute illness, incomplete ingestion of the glucose solution, and performance of the test in the afternoon can cause misleading results with the OGTT.

Perhaps the most common source of error in glucose measurements in inpatients is contamination by intravenous fluids containing a high concentration of glucose [38]. For example, 50% dextrose in water (DW) contains 50 g/dL of glucose, which is 500-fold higher than the normal glycemia. Therefore, even a 0.1% contamination with 50% DW could increase apparent glucose levels by 50 mg/dL. Sampling should never be done proximal to the infusion site and preferably should be done in the contralateral site. The infusion should be stopped for 1 or 2 min prior to collection. If it must be collected from the catheter, it should be flushed with an isotonic saline volume a minimum of 10 times the intraluminal catheter volume, and the first 5 mL of blood should be discarded or diverted.

Prolonged tourniquet application and fist clenching and unclenching can result in glucose consumption by the underperfused tissues, leading to lower glucose levels. For example, tourniquet application for 6 min resulted in a drop of plasma glucose by nearly 4% [3].

Serum or plasma should be separated from cells within 30 min to minimize consumption of glucose by blood cells, which is a source of variability in glucose results much higher than analytical imprecision. If separation cannot be done within 30 min, it is recommended that the sample be placed on ice or an appropriate glycolysis inhibitor be present [35]. In the past, sodium fluoride was recommended, usually in combination with an anticoagulant such as potassium oxalate (gray-top tubes), to prevent glycolysis because fluoride is a potent inhibitor of the enolase enzyme in the glycolytic pathway. However, inhibition does not occur during the first 1 or 2 hr due to poor permeability of the red cell membrane to fluoride, and consumption of 2–10 mg/dL per hour of glucose may be seen for up to 3 or 4 hr. Consensus guidelines for diagnosis of diabetes mellitus recommended the use of tubes containing citrate, EDTA, and fluoride for effective inhibition of glucose consumption at 24–37°C [35]. However, these tubes are not currently available in the United States, and one alternative is to keep gray-top tubes on ice if centrifugation is delayed more than 30 min, particularly in situations associated with increased *in vitro* glucose consumption, including polycythemia (e.g., in newborns), leukocytosis ($>$60,000/μL), thrombocytosis, sepsis, and the use of nonsterile blood collection tubes. Other body fluids, such as cerebrospinal fluid, pleural fluid, and peritoneal fluid, should be immediately analyzed for glucose because of glucose consumption by cells. Urine should be treated with acetic acid, borate, thymol, or sodium azide and refrigerated to avoid bacterial glucose consumption.

Fasting glucose levels in arterial blood are 2–5 mg/dL higher than those in capillary blood and 5–10 mg/dL higher than those in venous blood. However, after a meal, increased glucose utilization by peripheral tissues stimulated by the postprandial insulin surge can result in capillary or arterial blood glucose levels 20–70 mg/dL higher than those in venous blood [37]. Low perfusion associated with shock, hypotension, and vascular disorders can also cause similar discrepancies between arterial, capillary, and venous blood as slow blood flow leads to increased glucose extraction by the tissues. These discrepancies are frequent sources of confusion for clinical providers because glucose can be measured nearly simultaneously in hospitalized patients as part of an arterial blood gas profile, with POC glucose meters sampling capillary blood, and in the central laboratory high-throughput analyzers using venous blood serum or plasma.

Often, the discrepancy between whole blood analyzers, such as the blood gas instruments or the glucose meters, and plasma or serum glucose is attributed to the difference between glucose levels in whole blood and in the plasma fraction. This is true if glucose is measured in lysed blood. Whereas glucose is freely permeable and equilibrates at similar concentrations in plasma and the intracellular water fraction, the water content in red blood cells is only 71% of the total volume compared to 93% in the plasma, resulting in a 10–15% decrease in glucose concentration compared to plasma, depending on hematocrit. However, most meters measure glucose in unlysed whole blood (i.e., in the water fraction of plasma); therefore, the glucose concentrations are equivalent to those measured in plasma. In true whole blood analyzers using lysed blood, a correction factor is typically applied to report glycemia in terms of milligrams per deciliter of plasma [39].

## Methodology for Glucose Testing

Virtually all glucose assays use enzymatic approaches to generate a measurable product proportional to the glucose concentration. The following enzymes can act on glucose and have been utilized to measure glucose levels:

1. Glucose oxidase: β-D-glucose + $O_2 \rightarrow$ gluconate + $H_2O_2$
2. Hexokinase: D-glucose (and other hexoses) + ATP → glucose (hexose)-6-phosphate; usually coupled with glucose-6-phosphate dehydrogenase:
   i. Glucose-6-phosphate + $NAD(P)^+$ → 6-phosphogluconolactone + NAD(P)H
3. Glucose dehydrogenase: glucose + $NAD(P)^+$ → gluconolactone + NAD(P)H

As described previously, the change in concentration of $NAD(P)^+$ versus NAD(P)H can be assayed spectrophotometrically at 340 nm and can form the basis for most methods using hexokinase or glucose dehydrogenase. Alternatively, NAD(P)H can be used to reduce a tetrazolium chromogen to a colored product or to reduce an electron donor, such as flavin adenine dinucleotide or pyrroloquinoline quinone (PQQ), resulting in measurable current. Careful calibration and quality control are required because of instability of co-factors, particularly ATP, $NAD^+$ and $NADP^+$, in solution.

Glucose oxidase methods rely on one of the following approaches to produce a measurable signal:

1. Measuring oxygen consumption with an $O_2$-specific electrode: This approach is used in the Beckman DxC/LX20 modular assay. To minimize spontaneous $O_2$ generation from $H_2O_2$, molybdate, iodide, and ethanol are included to enhance catalase-mediated inactivation of the peroxide.
2. Measuring $H_2O_2$ formation by using peroxidase to convert a chromogen to a colored product: Many chromogens have been used, including *o*-dianisidine and 4-aminoantipyrine.
3. Measuring $H_2O_2$ formation by oxidation of a platinum electrode, generating current proportional to the glucose levels: This method can be used in whole blood analyzers because it avoids optical readings as well as issues with oxygen consumption by blood cells.

### *Interferences in Glucose Assays*

#### GLUCOSE OXIDASE

Interference by strong redox agents is expected in glucose oxidase methods using peroxidase, similarly to other methods (e.g., uric acid methods). These include endogenous substances such as uric acid, bilirubin, hemoglobin, glutathione, and acetoacetate and

TABLE 8.5 Pharmaceuticals Containing Sugars That Cause False-Positive Elevations in Glucose Tests with GDH-PQQ Assays

| Manufacturer | Trade Name | Product | Interfering Substance | Interferent Concentration (g/dL) or Maximum Dose |
|---|---|---|---|---|
| Talecris Biotherapeutics | Gamimune N 5%[a] | Immune globulin intravenous (human) | Maltose | 9–11 |
| Octapharma Pharmazeutika Produktionsges m.b.H. | Octagam | Immune globulin intravenous (human) | Maltose | 10 |
| Cangene Corporation | WinRho SDF Liquid | Rh (D) immune globulin intravenous (human) | Maltose | 10 |
| Cangene Corporation | None | Vaccinia immune globulin intravenous (Human) | Maltose | 10 |
| Cangene Corporation | HepaGam B | Hepatitis immune globulin (human) | Maltose | 10 |
| GlaxoSmithKline | Bexxar | Tositumomab and 131I-tositumomab | Maltose | 10 |
| Baxter | Extraneal | Peritoneal dialysis solution | Icodextrine | 7.5 |
| Baxter | Adept | Adhesion reduction solution | Icodextrine | 4 |
| Bristol-Myers Squibb | Orencia | Abatacept | Maltose | 2-g dose |
| NERL Diagnostics | D-Xylose USP | D-Xylose | Xylose | 25-g dose |
| | | Galactose tolerance test | Galactose | 40-g dose |

[a]*Discontinued.*

exogenous chemicals such as ascorbic acid, acetaminophen, tetracycline, and catecholamines.

With glucose oxidase methods relying on oxygen consumption, high oxygen levels in the blood can cause falsely low glucose results; conversely, low oxygen levels associated with high altitude can cause falsely high results. This method should not be used in whole blood due to highly variable oxygen levels and consumption by red cells. The reaction is also dependent on pH levels, with acid pH causing increased glucose levels and high pH causing low glucose results.

## HEXOKINASE

These methods are more susceptible to interferences that affect absorption at 340 nm, including turbidity due to lipemia, hyperbilirubinemia, and hemolysis. In addition to spectral interference by hemoglobin, hemolysis is associated with the release of several interfering substances from blood cells, including phosphates that inhibit hexokinase, glucose-6-phosphate, glucose-1-phosphate, fructose-6-phosphate, and enzymes competing for glucose-6-phosphate and NADP. High levels of fructose (e.g., in some OGTT preparations) also interfere with hexokinase methods: Ingestion of 2 g/kg body weight of fructose raises glycemia by 10 mg/dL for 1 or 2 hr [37].

## GLUCOSE DEHYDROGENASE

This enzyme is not entirely specific for glucose, and other sugars, such as maltose, galactose, xylose, and mannose, can interfere in a manner related to the particular formulation of the assay.

Tang *et al.* [40] investigated several POC glucose meters for possible interference by a list of 30 commonly used drugs, and they established that ascorbic acid (3–9 mg/dL), acetaminophen (2–10 mg/dL), dopamine (2–4 mg/dL), and mannitol (500 mg/dL) caused interference with various glucose meters. With the exception of acetaminophen and mannitol, these levels are much higher than typical therapeutic concentrations. However, in certain situations associated with rapid drug infusion or decreased excretion, such as in critically ill patients with renal failure, interferences by these drugs may become significant.

*CASE REPORT* An 86-year-old male with a diagnosis of diabetes was admitted for rapidly progressive cellulitis of the foot with necrotizing fasciitis and sepsis [41]. He was treated with above-the-knee amputation, dialysis for oliguria and progressive renal insufficiency, insulin, antibiotics, and intravenous immunoglobulin (Octagam) for off-label sepsis treatment. The Octagam was administered over a 7-hr period, during which a POC device used to measure glucose levels showed increasing glycemia peaking at 535 mg/dL 4 hr into the infusion. The patient was then treated with 12 units/hr of intravenous insulin, which was titrated toward 24 units/hr by the next day. The patient then became increasingly unresponsive, and a POC glucose level reading was 115 mg/dL, whereas a central laboratory assay resulted in a glucose of 12 mg/dL. The patient remained unresponsive despite 50% dextrose and insulin administration, and the patient eventually died due to irreversible neurological brain damage.

Several other similar cases reported to the U.S. Food and Drug Administration (FDA), including 7 fatalities, led this agency to issue a safety warning in October 2005 alerting patients and health care providers to the possibility of falsely elevated glucose results with several blood testing devices due to interference from intravenous products containing maltose, galactose, or xylose. This problem is unique to devices using glucose dehydrogenase (GDH) coupled with oxidation of PQQ for production of a colored product. Subsequently, the FDA reported 13 deaths associated with GDH–PQQ glucose testing, 10 of which occurred in patients who received Extraneal (icodextrin) peritoneal dialysis solution and 3 that occurred in patients who received maltose infusions, and issued a public health notification in August 2009 reiterating the possibility of fatal errors with GDH–PQQ glucose meters [42]. The problem is particularly severe when large amounts of interfering sugars are being infused in the patient; for example, a total dose of 25 g maltose can result in peak blood maltose levels of approximately 100–150 mg/dL, which decay with a half-life of approximately 90 min [43]. Table 8.5 shows some pharmaceuticals containing substances interfering with the GDH–PQQ glucose assays.

## ANALYSIS OF ELECTROLYTES

Electrolytes are substances that form ions in aqueous solutions, therefore conducting current. In clinical biochemistry, this term usually refers to clinically significant small ions that account for most of the conductivity of body fluids, and it excludes other charged substances such as amino acids, peptides, proteins, and nucleic acids. The most frequently measured electrolytes have critical physiologic roles and tightly regulated concentrations. These include the cationic metals sodium ($Na^+$), potassium ($K^+$), magnesium ($Mg^{2+}$), and calcium ($Ca^{2+}$) and the anions chloride ($Cl^-$), phosphate ($HPO_4{}^{2-}$), and bicarbonate ($HCO_3{}^-$), of which $Na^+$, $K^+$, $Cl^-$, and $HCO_3{}^-$ are the most commonly assayed, usually as an "electrolyte panel" (see Table 8.3). It is beyond the scope of this chapter to discuss the clinical implications of electrolyte measurements, and the reader is referred to excellent textbooks on the topic [37,44]; However, a few key concepts are important to understand physiopathologic and pre-analytical considerations of relevance to avoid misinterpreting electrolyte levels.

With the exception of calcium and magnesium, which are albumin-bound, the major electrolytes are present almost exclusively as free ionized species in body fluids. Calcium and magnesium can be measured as the total concentration in the fluid or as the free ionized fraction. The other electrolytes are measured as free ions.

Cellular membranes, which are predominantly hydrophobic bilayers, do not allow for free movement of electrolytes between the intracellular and extracellular compartments. As a result, movement of electrolytes occurs mostly through protein complexes called pores, which can have open or closed states and can be subject to regulation. In addition, ions can be actively moved across pores by active transporters called pumps to maintain physiologic important gradients that are associated with electric charges critical for many cellular and organ functions, such as muscle contraction and nerve conduction. For example, sodium is actively pumped out of cells, whereas potassium is actively transported into the intracellular compartment mostly by $Na^+/K^+$-ATPases, which exchange three $Na^+$ for two $K^+$ and therefore create the membrane resting potential with net negative charges intracellularly. This process alone is responsible for approximately one-third of energy consumption in most cells and even higher in neurons. Interference with this active mechanism can lead to perturbations in nerve conduction or muscle contraction, such as when digoxin inhibits the $Na^+/K^+$-ATPase, resulting in decreased membrane potential and release of calcium from intracellular stores.

Active transporters also play a critical role in the kidney tubules, where $K^+$ and $H^+$ are actively transported into the tubular lumen in exchange for $Na^+$, thus regulating blood levels of sodium and potassium and acid–base balance. In addition, $Cl^-$ ions move freely with $Na^+$ to maintain electrical balance. As sodium and chloride ions move across cell membranes, they carry water molecules and thus regulate cellular water content and volume. As the major ions in the extracellular compartment, $Na^+$ and $Cl^-$ are responsible for most of the osmotic pressure across cell membranes, which can be calculated by the following formula:

$$\text{Plasma osmolality (in mOsm/kg)} = 2 \times Na^+ + \text{glucose}/18 + \text{BUN}/2.8$$

where urea and glucose (expressed in mg/dL) are the major non-ionic molecules contributing to osmotic pressure, and $2 \times Na^+$ represents the combined effect of sodium and its partner anions. This formula is useful to calculate the osmolal gap, which is the difference between measured osmolality (usually assayed by freezing point depression osmometers) and calculated osmolality. Differences greater than 10 mOsm/L indicate the presence of unmeasured osmotically active substances, such as alcohols, ethylene glycol, acetone, mannitol, hypergammaglobulinemia, and hypertriglyceridemia, or point to a measurement error in sodium, glucose, BUN, or osmolality.

As with regulation of osmotic pressure in cells, sodium transport in the kidney is a major factor in water retention and blood volume. Conversely, the balance between water and sodium content in the extracellular compartment determines sodium concentration. Therefore, the kidney transporters are major targets of hormonal systems that affect plasma sodium concentration and water content, as well as blood volume and blood pressure, including antidiuretic hormone (ADH), natriuretic peptides, and the renin–angiotensin–aldosterone system.

An important factor in interpreting plasma electrolyte concentrations is the effect of nonelectrolyte substances, such as glucose, that do not freely cross the plasma membrane. Elevated glucose levels in the plasma can cause increased osmotic pressure in the extracellular fluids because glucose does not quickly equilibrate with the intracellular fluid in the absence of insulin. Therefore, high glycemia can cause a shift of water from the intracellular to the extracellular fluid, resulting in dilution of all extracellular solutes. Particularly in the case of sodium, which has a narrow range in plasma, it is advisable to correct the plasma sodium for dilution due to hyperglycemic osmotic expansion in order to avoid misinterpreting the measured hyponatremia as either an excess of total body water or a salt deficiency. The following formula can be used to correct dilutional hyponatremia due to hyperglycemia ($Na^+$ in mM and glucose in mg/dL):

$$Na^+(\text{corrected}) = Na^+(\text{measured}) + 0.016 \times \text{glucose} - 1.6$$

In contrast to tight regulation across cellular membranes, electrolyte ions move freely between the intravascular and extravascular compartments. Intercellular junctions in capillaries can regulate the distribution of macromolecules, such as proteins, between the extravascular and the intravascular compartments. Usually, the concentration of proteins in the plasma is higher than that in the extravascular compartment and contributes to pulling water into the intravascular compartment, forming a pressure gradient called oncotic or colloid osmotic pressure. If there is a reduction of plasma protein content or an increase in vascular permeability, fluid moves out of the intravascular compartment and causes edema.

In general, electrolyte concentrations are not significantly affected by changes of plasma proteins or intravascular volume, with the exception of total calcium and magnesium. For example, a shift from the supine to the upright position with increased intravascular hydrostatic pressure causes flow of intravascular fluid to the interstitial space, resulting in a decrease in plasma volume of approximately 12%. Under normal endothelial permeability conditions, molecular complexes greater than 4 nm in diameter (e.g., most proteins, including albumin, with a 7-nm diameter) will be retained in the intravascular compartment; therefore, their concentration will be increased proportionally. Most electrolytes are not affected, but calcium, being approximately 40% protein bound, will increase by 5–10% (0.2–0.8 mg/dL) when moving from the supine to the upright position, whereas magnesium, being less protein bound, will increase by approximately 4% [3]. A similar situation with shift to interstitial space and hemoconcentration occurs when the tourniquet is kept for more than 1 min, especially with repeated fist clenching, during blood collection from the arm. In this case, in addition to total calcium and magnesium, potassium will increase as much as 2 mM due to muscle activity [45], whereas glucose and phosphate will decrease due to intracellular consumption.

Free ionized calcium comprises 48–50% of the total calcium, whereas 40% is protein bound (32% bound to albumin and 8% to globulins) and 10–12% complexed with other anions, such as phosphate, bicarbonate, and lactate. Approximately 25–30% of plasma magnesium is complexed with albumin, whereas 15–20% is complexed with anions and 50–55% exists in the free form. Only the free ions are physiologically active. Therefore, allowances should be made for the albumin concentration to correctly evaluate measurements of total calcium and magnesium. Because total levels of divalent cations linearly correlate with albumin concentrations below the reference range (<4 mg/dL), several formulas have been used to estimate the levels of total calcium and magnesium with albumin in the normal range, such as follows (electrolytes in mg/dL, albumin in g/dL)[46]:

$$\text{Calcium (corrected)} = \text{total calcium (measured)} - 0.8 \times \text{albumin} + 3.2$$

$$\text{Magnesium (corrected)} = \text{total magnesium (measured)} - 0.123 \times \text{albumin} + 0.492$$

However, the reliability of these formulas has been questioned, particularly in renal failure and critically ill patients with several factors affecting electrolyte levels, including altered pH and anion levels. Excess $H^+$ causes a decrease in negative charges in calcium-binding proteins, leading to dissociation of protein-bound calcium and increased free $Ca^{2+}$. For each decrease of 0.1 pH units, free $Ca^{2+}$ increases by 0.2 mg/dL. The reverse changes occur in alkalemia. Other factors affecting the accuracy of these formulas include the effect of substances that complex with calcium or compete for binding to albumin, such as free fatty acids, bilirubin, heparin, drugs, and several organic anions including lactate and ketoacids. In the case of critically ill patients, calcium correction formulas showed very poor sensitivity to detect hypocalcemia, and direct free $Ca^{2+}$ measurements should be done [47]. In the case of renal failure patients, phosphate levels showed a significant effect in the proportion of free calcium and should be taken into consideration [48]. Direct measurement of free $Ca^{2+}$ is also preferable in cases of liver transplantation, malignancy, and myeloma and in neonates [37].

Some studies indicate a strong correlation of total and free $Mg^{2+}$, but there is poor correlation between plasma total magnesium or free $Mg^{2+}$ and total body magnesium because only 0.3% of total body magnesium is present in plasma. Whole body magnesium may be depleted up to 20% with maintained normal magnesemia. Because total Mg assays have better specificity for magnesium and are cheaper, more widely available, and better standardized, they may be preferable to free $Mg^{2+}$ assays for assessment of plasma magnesium. The most accurate way to diagnose magnesium deficiency is to perform an Mg tolerance test, in which 24-hr urine Mg excretion is measured before and after the patient is injected with 1.2 mmol of Mg/kg body weight. A recovery of less than 80% of the injected magnesium in the urine is indicative of magnesium deficiency in the absence of diuretic treatment, renal disease, or drugs, such as cyclosporine and tacrolimus, that cause urinary Mg wasting [49].

A similar situation occurs with calcium because plasma calcium correlates poorly with total body calcium—99% of the body calcium is trapped as calcium phosphate (hydroxyapatite) in the bone. In contrast with magnesium, intracellular concentrations of calcium are much lower than plasma concentrations; therefore, hemolysis does not lead to an increase in calcium levels, although hemoglobin can cause spectral interference with colorimetric methods.

## Physiologic Pre-Analytical Issues

Circadian variation in hormones such as ADH, corticotrophin (ACTH), cortisol, aldosterone, renin, and epinephrine along with individual rhythms associated with meals, exercise, and sleep affect electrolyte excretion, the movement of ions and water across cell membranes, and the distribution of fluid between the intra- and extravascular compartments. For example, potassium levels in the plasma can vary by more than 0.6 mM, with peaks between 8 a.m. and 4 p.m. and nadirs between 8 p.m. and 1 a.m. [50]. This variation may have more to do with shifts to the intracellular compartment than rates of excretion because urinary levels tend to parallel serum $K^+$ concentration [50]. Circadian variation in the plasma levels of phosphate is also observed, with peaks in the afternoon and evening, immediately after major meals, and the nadir in the fasting morning specimens.

As major intracellular electrolytes, the levels of both potassium and phosphate increase after organic food ingestion, with phosphate levels in plasma increasing by approximately 15% and potassium by approximately 5% 2 hr after a standard meal [3]. Standard meals do not have a significant effect on sodium, chloride, calcium, or magnesium, although magnesium levels tend to be higher in vegetarians and vegans.

Exercise stimulates release of potassium mainly from skeletal muscles, leading to rapid increases in venous potassium averaging 0.8–1.2 mM and peaking as high as 8 mM in arterial blood during intense exercise [51]. Well-trained athletes have increased sodium pump density in their muscles and show lower increases in plasma $K^+$ during exercise, whereas patients with heart failure and muscular dystrophies have lower muscle $Na^+/K^+$ pump density [51]. Higher catecholamine levels can induce intracellular potassium influx by stimulating the $Na^+/K^+$-ATPase via $\beta_2$-adrenergic receptors, causing a quick drop in serum potassium levels postexercise and during stress (e.g., in patients with acute myocardial infarction) [51]. The hypokalemic effect of $\beta$-adrenergic receptors can be attenuated by caffeine and stimulation of $\alpha$-adrenergic receptors in the liver [51]. Similarly, peak insulin levels after meals or glucose infusion tend to reduce potassium levels because insulin leads to increased activity of the $Na^+/K^+$ pump and increased cellular potassium influx. Due to their synergistic effects in activating the $Na^+/K^+$ pump, insulin and $\beta_2$ agonists are used in the treatment of hyperkalemia.

The levels of plasma phosphate also reflect the net movement of phosphate ions across cell membranes as well as dietary intake. For example, phosphate levels increase immediately after meals and start declining after 2 or 3 hr as insulin stimulates glucose and phosphate intracellular influx. Metabolic or respiratory acidosis leads to hydrolysis of intracellular organic phosphates with release into the extracellular fluid. On the other hand, high intracellular phosphate consumption with associated hypophosphatemia is common in states of enhanced cell growth, such as during childhood, acromegaly, and tumor proliferation. Hypophosphatemia is frequent in the postoperative phase due to multiple factors, such as respiratory alkalosis, insulin, and diuretics. Catecholamines also cause decreases in phosphatemia, which tends to parallel serum $K^+$ during exercise [52]. Magnesium is released from muscles leading to mild increases during intense exercise, whereas well-trained athletes tend to have decreased magnesemia with endurance exercise. Although total calcium does not change with exercise, the lactic acidosis associated with intense exercise can result in increases in free $Ca^{2+}$ as the lower pH dissociates calcium from proteins.

Several drugs can affect electrolyte levels. For a comprehensive listing of drug effects on clinical testing, see Young [1]. For example, angiotensin converting enzyme (ACE) inhibitors, angiotensin receptor blockers, and potassium-sparing diuretics are associated with higher risk of hyperkalemia, whereas $\beta_2$ agonists, theophylline, chloroquine, thiopental, thiazides, and other loop diuretics can lead to hypokalemia. Interestingly, consumption of caffeine-containing sugary sodas has been associated with hypokalemia due to inhibition of the $Na^+/K^+$ pump by the caffeine metabolite theophylline, insulin secretion, and osmotic diarrhea [51].

Acute ingestion of ethanol can also have significant effects on electrolytes. First, ethanol inhibits secretion of hypothalamic ADH, resulting in increased free water diuresis, which can lead to dehydration and hypernatremia; it also significantly increases aldosterone, leading to sodium retention and loss of potassium. Local effects at the kidney level may also lead to increased reabsorption of sodium and chloride. Second, alcohol is metabolized to acetate, with resulting metabolic acidosis with low plasma bicarbonate, compounded by ketoacidosis related to inhibition of glucose utilization. In chronic alcoholics, phosphate and magnesium deficiencies are common.

Bicarbonate in plasma is in equilibrium with dissolved carbon dioxide and hydrogen ion (pH):

$$CO_2 \leftrightarrow HCO_3^+ + H^+$$

Situations with acidosis generate an excess of hydrogen ions and shift the previous equation to the left with decrease in plasma bicarbonate, whereas alkalosis has the opposite effect. Similarly, an increase in carbon dioxide concentrations (e.g., due to respiratory insufficiency) leads to a shift to the right, with an increase in bicarbonate levels and compensatory decrease in pH. The lungs can contribute to acid–base regulation by controlling the amount of $CO_2$ excretion. In addition to the short-term buffering and respiratory mechanisms, acid–base balance is highly regulated by the kidney, where the excretion of hydrogen ions and bicarbonate can be modified by several mechanisms, including alteration of the rate of absorption of $Na^+$ in exchange for $H^+$ and $K^+$ and production of ammonia from glutamine to trap $H^+$ for excretion.

In order to maintain ionic balance, the loss of bicarbonate in the body leads to an increase in chloride anions, whereas bicarbonate retention results in loss of chloride. The exception to this rule is when the loss of bicarbonate is caused by metabolic acidosis resulting from accumulation of organic acids such as lactate or ketoacids. In this case, the organic acids balance the cations, and the chloride concentration is minimally affected. Conversely, changes in plasma chloride levels result in inverse changes in bicarbonate concentration. For example, loss of chloride by vomiting of gastric fluid will result in an increase in plasma bicarbonate concentration and metabolic alkalosis, whereas ingestion of chloride will result in metabolic acidosis with low bicarbonate. Measurement of chloride and bicarbonate (or total $CO_2$) even in venous plasma is useful to identify and classify possible acid–base disturbances.

In addition to renal regulation, chloride responds to changes in plasma $CO_2$ in the following manner: As $CO_2$ accumulates in plasma, it moves freely into the red cells, where it is rapidly converted into $HCO_3^-$ by carbonic anhydrase. Because there is no such enzyme in plasma, the rate of conversion to $HCO_3^-$ is much lower in plasma, creating a huge gradient that forces bicarbonate out of the cells. Because the plasma membrane is impermeable to $H^+$ and hydrogen ions are tightly captured by red cell proteins, particularly deoxyhemoglobin, chloride, which can move freely across the membrane, must move intracellularly to maintain ionic balance (chloride shift). This phenomenon is relevant to measurement of chloride in blood samples: With improperly capped blood samples stored for prolonged periods before cell separation, the loss of $CO_2$ is followed by conversion of intracellular bicarbonate to $CO_2$, which diffuses out of the cells; as intracellular bicarbonate concentration decreases, chloride shifts intracellularly to maintain ionic balance, causing artificially low plasma chloride measurements.

The balance between the four major electrolytes is best illustrated by the anion gap formula:

$$Na^+ + K^+ = Cl^- + HCO_3^- + \text{anion gap}$$

The anion gap is caused by unmeasured anions such as proteins and organic acids. In a healthy individual, unmeasured anions account for approximately 12–15 mEq/L. The anion gap will increase if unmeasured anions are in excess, such as with lactic acidosis, and will decrease if unmeasured cations, such as immunoglobulins, are significantly increased. It is good laboratory practice to monitor the anion gap for possible analytical errors in the measurement of components of the electrolyte panel.

In addition to the changes in $HCO_3^-$ and $Cl^-$, acidemia causes intracellular shifts of $K^+$, as $H^+$ and $K^+$ compete for intracellular anions. As $H^+$ moves intracellularly, $K^+$ moves out, and on average, plasma $K^+$ will rise approximately 0.6 mM (range, 0.2–0.7 mM) for each 0.1 unit of fall in plasma pH [53]. This increase in $K^+$ is much lower with organic acidemia, such as in lactic- or ketoacidosis, perhaps because these anions will buffer $H^+$ intracellularly. The reverse phenomenon is true because hyperkalemia will cause a mild metabolic acidosis. On the other hand, metabolic alkalosis and hypokalemia are similarly interconnected, with a drop of approximately 0.4 mmol $K^+$ per 0.1 unit of pH increase. Also, it is important to note that in situations associated with peripheral ischemia, such as in hypothermia, vascular disease, thrombosis, and shock, a combination of acidosis and hypoxia with insufficiency of the $Na^+/K^+$ pump will result in higher potassium levels in peripheral vein samples, and a more physiologic assessment of kalemia should be done from central vein sampling if available. Phosphate levels also tend to respond to pH in the same manner as $K^+$, with acidosis causing hydrolysis of intracellular phosphates and increased phosphatemia, whereas alkalosis results in activation of glycolysis and consumption of phosphate by the cell.

As mentioned previously, pH also has an important effect on the fraction of free $Ca^{2+}$, with every change in 0.1 unit of pH resulting in the opposite change in free $Ca^{2+}$ of approximately 0.2 mg/dL. It is important to concurrently report the pH of the sample for proper interpretation of $Ca^{2+}$. The clinician must distinguish *in vivo* changes in pH causing true shifts in free $Ca^{2+}$ from spurious changes in free $Ca^{2+}$ resulting from sample pH adulteration. Most commonly, samples lose $CO_2$ upon air exposure, with a corresponding increase in pH and falsely decreased $Ca^{2+}$. Prolonged storage of uncentrifuged whole blood can also lead to glucose consumption by blood cells, with corresponding lactic acid production resulting in pH drop and spurious $Ca^{2+}$ increase. Every effort should be made to avoid these sources of misleading results in free $Ca^{2+}$ determination. Ideally, specimens should entirely fill a minimally heparinized and tightly capped plastic blood gas syringe and be analyzed within 15–30 min.

## Specimen Issues

Specimens for electrolyte analysis include whole blood, plasma, serum, and urine. Sometimes, assessment of electrolytes in other body fluids is important. For example, chloride levels in sweat are used to diagnose cystic fibrosis. Because the levels are lower than in plasma, the methods have to be validated for detecting levels as low as 10 mM in sweat. Likewise, other body fluids, such as gastrointestinal fluids, may be analyzed to identify electrolyte disturbances specific to the organ of interest, but the results are not as well standardized or comparable, and matrix effects need to be taken into consideration when applying different methods to the analysis. For urinary and fecal fluids, timed collections are indicated to assess rates of electrolyte excretion.

Arterial and capillary samples differ significantly from venous blood in calcium (−2.4 to 4.6%), sodium (0 to −2.3%), and chloride (−1.8 to 2.0%) measurements [37]. It is important to avoid contamination by infusion fluids, one of the most common causes of misleading laboratory results in hospitalized patients. In particular, contamination with sodium citrate infusions has a significant effect on free $Ca^{2+}$ and pH measurements, even after discarding the first 9 mL of blood from the catheter [54]. In addition to binding $Ca^{2+}$, citrate contamination can also lower $Na^+$ by forming weak complexes with $Na^+$ [55]. Other less obvious sources of contamination that interfere with sodium measurements include sodium or lithium heparin (which weakly chelates $Na^+$ and reduces $Na^+$ levels), benzalkonium heparin-coated catheters, benzalkonium antiseptic cleaning agents, and ticarcillin, which contains a high concentration of sodium [55].

The clotting process leads to release of intracellular fluid, mostly from platelets, and therefore the serum contains higher levels of intracellular electrolytes than plasma, such as potassium (+0.3 to 0.4 mM or 5 to 15%) and phosphate (+2 to 8%), whereas other electrolytes are only minimally affected (<1%) [37]. There is a linear relationship between the serum–plasma difference in potassium levels and the platelet count (~0.15 mM per 100,000/μL) [56]. Similarly, patients with high leukocytosis, particularly with fragile cells such as in chronic lymphocytic and chronic myelomonocytic leukemias, commonly have pseudohyperkalemia. In patients with high platelet or leukocyte counts, collection of samples in heparinized tubes with avoidance of hemolysis is indicated to avoid spurious hyperkalemia due to the clotting process. However, in certain cases, heparin-induced lysis of leukemic cells can result in higher $K^+$ levels in plasma than in serum ("reverse pseudohyperkalemia").

Centrifugation at very high speeds or in fixed-angle rotors can induce cell rupture and cause pseudohyperkalemia. Re-centrifugation of blood samples after prolonged storage is discouraged because the small amount of serum remaining below the gel separator is rich in potassium and will move to the original serum layer, falsely increasing $K^+$ [57].

Contamination from intracellular electrolytes is much magnified if there is hemolysis, leading to a significant increase in potassium and phosphate, with very minor increases in magnesium. Because the potassium concentration in red cells is 105 mM, even a small amount of hemolysis results in significant increases in plasma potassium. On average, the amount of potassium released by *in vitro* hemolysis correlates with free hemoglobin, increasing approximately 0.2–0.5 mM for each 0.1 g/dL of plasma hemoglobin [58]. However, formulas for correcting potassium based on estimation of free hemoglobin are highly variable and should not be used. Instead, laboratories should establish a hemolysis threshold, by visual or spectrophotometric inspection, above which the artifactual increase in potassium is unacceptable, leading to specimen rejection. Hemoglobin levels of 0.1 g/dL or greater (corresponding to lysis of 0.7% of the red cells in an individual with a hemoglobin of 15 g/dL) are usually associated with clinically significant spurious increases in potassium. More problematic are whole blood analyzers, such as blood gas and POC devices, where hemolysis cannot be easily recognized. Whole blood potassium levels should be regarded with suspicion if inconsistent with the clinical situation, and in these cases centrifugation of the specimen to assess hemolysis is indicated. In particular, in the absence of renal insufficiency, hyperkalemia is rare, and elevated $K^+$ should prompt a search for causes of spurious hyperkalemia [45]. Table 8.6 summarizes the various causes of pseudohyperkalemia.

In contrast to potassium, the intracellular magnesium concentration is only two- or threefold the plasma levels; therefore, common amounts of hemolysis do not significantly increase magnesium levels in plasma. Calcium levels are also not significantly

**TABLE 8.6** Causes of Pseudohyperkalemia[a]

| *RELEASE FROM CELLS* IN VITRO |
|---|
| Prolonged tourniquet time (up to 3 mM) |
| Forearm exercise and fist clenching (1–2 mM) |
| Delayed transportation and centrifugation |
| Cold storage/transportation of uncentrifuged samples |
| Familial pseudohyperkalemia: Autosomal dominant hereditary stomatocytosis with leakage of $K^+$ at $<37°C$; immediate separation is required |
| Leukocytosis (can also temporarily cause low $K^+$ at 25–30°C) |
| Thrombocytosis (serum $K^+$ > plasma $K^+$ by ~0.15 mmol/100,000 platelets/μL) |
| Hemodialysis: Up to 50% have serum $K^+$ > plasma $K^+$, even with normal cell counts. |
| Reverse pseudohyperkalemia: Plasma $K^+$ > serum $K^+$ in leukocyte malignancies fragilized by heparin |
| *IN VITRO HEMOLYSIS (AST AND LDH SHOULD BE ELEVATED)* |
| Ethanol contamination |
| Difficult venipuncture |
| Inappropriate needle diameter |
| Syringe and needle collection |
| Excessive storage/transport temperature ($<30°C$) |
| Excessive centrifugation speed |
| Prolonged contact with wooden applicator sticks used to remove fibrin (up to 0.5 mM) |
| *CONTAMINATION* |
| Collection through catheter or above potassium-infusion site |
| $K_2$-EDTA contamination, wrong tube type or order of tube collection (also low calcium) |
| *DILUTION* |
| Blood sample from arm with infusion lacking $K^+$ (e.g., DW5) causes dilution. |

[a]*Suspect if patient has no predisposing factors, signs, or symptoms of hyperkalemia, such as numbness, chronic renal failure, abnormal electrocardiogram, treatment with loop diuretics, spironolactone, and ACE inhibitors.*

affected by hemolysis, although free $Ca^{2+}$ can be affected by binding from intracellular proteins and changes in specimen pH. For colorimetric calcium, phosphate, and magnesium assays, spectral interferences with the analytical measurement are possible with significant hemolysis, hyperbilirubinemia, or lipemia.

Anticoagulants may significantly affect electrolyte measurements. Obviously, anticoagulants such as sodium or potassium heparin, citrate, or oxalate should not be used for electrolyte measurements. Likewise, divalent cation chelators such as EDTA, citrate, and oxalate should not be used for calcium or magnesium determinations. Collection of blood for electrolyte determination should precede the use of anticoagulants containing sodium, potassium, or chelators due to the possibility of contamination of the needle. Less obviously, heparin may bind calcium and magnesium unless all binding sites are saturated, and therefore balanced heparin should be used for measurement of free calcium or magnesium in plasma or whole blood. While unexpected hypocalcemia is a clue to contamination by potassium EDTA in cases of suspected pseudohyperkalemia, the best approach is to directly assay for serum EDTA in practices where high rates of EDTA contamination are observed [59,60].

*CASE REPORT* A primary care physician was called in the evening because one of her patients was contacted to go urgently to the hospital due to a potassium level of 7.2 mM [45]. This 72-year-old male was being treated with lisinopril for hypertension, and his previous potassium levels ranged from 4.5 to 5.0 mM. A repeat specimen in the hospital showed a level of 4.8 mM and normal BUN and creatinine, an electrocardiogram was normal, and the patient was discharged home. Subsequent investigation indicated that the primary care practice was sending the uncentrifuged samples in a metal container to the central laboratory and exposing the specimens to near freezing temperatures for hours during transportation.

Storage of blood samples before serum or plasma separation leads to progressive leakage of potassium from cells. In contrast with the attenuating effect on glucose consumption, storage at 4°C inhibits the $Na^+/K^+$-ATPase responsible for keeping potassium inside the cell and magnifies the artificial increase in blood $K^+$ levels. Plasma kept in contact with blood cells for 8 hr at 4°C shows an average increase in $K^+$ of approximately 1 mM, minimal change at 23°C, and a decrease of approximately 1 mM at approximately 30°C [3]. Exposure to temperatures higher than 30°C for several hours can lead to consumption of glucose and cell disruption, with consequent leakage of potassium into the plasma. Samples with high leukocyte counts can exaggerate these effects, with increased potassium leakage at low or high temperatures and increased intracellular shifts at 25–30°C. Because of the possibility of either pseudohyperkalemia or pseudohypokalemia with changes in ambient temperature, centrifugation of whole blood samples with a gel separator is indicated within 1 hr of collection if prolonged storage or transportation is expected before analysis.

In addition to potassium, phosphate levels tend to increase upon unrefrigerated storage of whole blood due to the action of cellular phosphatases on organic phosphates. In contrast, levels of $HCO_3^-$ will decrease as the tubes lose $CO_2$ after being exposed to air (partial pressure of carbon dioxide ($pCO_2$) decreases ~15–20% per hour at room temperature in de-capped tubes) [61]. Because $CO_2$ accounts for only 3–10% of the total $CO_2$ in plasma, this loss of $CO_2$ results in only moderate decreases in serum $HCO_3^-$ or total $CO_2$ (~0.6–1 mM per hour) [61]. However, in large-volume laboratories, the loss of $HCO_3^-$ may become significant because the tubes may sit open for more than 1 hr. This loss of $CO_2$ is accentuated at temperatures higher than room temperature and attenuated in refrigerated serum specimens. The loss of $CO_2$ occurs even in capped tubes if the filling is not complete. A study of 10-mL Vacutainer tubes showed that for each extra milliliter of air above the sample in an underfilled tube, the $HCO_3^-$ will decline by 0.5 or 0.6 mM and the anion gap will increase by 0.2 or 0.3 mM [62].

## Analytical Issues

There are many analytical issues concerning analysis of electrolytes. In this section, such issues are discussed.

### *Ion-Specific Electrodes*

Virtually all major high-throughput and POC analyzers utilize ion-specific electrodes (ISEs) to measure $Na^+$, $K^+$, and $Cl^-$. Furthermore, free $Ca^{2+}$ and $Mg^{2+}$ are measured with ISEs. A special type of ISE is a pH meter, which determines the concentration of $H^+$ ions. In general, ISEs contain an electrolyte-selective membrane that allows only the ion of interest to move across the membrane, thus creating an electromotive force proportional to the ion concentration. There are two main types of ISEs: Indirect ISEs dilute the sample 1:20 to 1:34 before measurement, whereas direct ISEs use undiluted samples. Whereas direct ISEs measure the electrolyte concentration in the plasma water fraction, indirect ISEs measure electrolyte concentrations in the diluted sample and multiply the results by the dilution factor. In 2011, more than 80% of the laboratories in the United States used indirect ISEs for measuring sodium, potassium, and chloride. In samples with significantly altered plasma water fraction, results with indirect ISEs can be misleading. POC and whole blood gas analyzers, as well as the thin-film electrode technology (Ortho/Vitros) and the Roche Integra systems, use direct ISEs and can be an alternative for rare samples with significantly altered plasma water fraction.

## ADJUSTMENT FOR PLASMA WATER WITH INDIRECT METHODS

In healthy individuals, plasma water occupies approximately 93% of the plasma volume. When solute concentrations are measured in plasma water, an adjustment factor of 93% is applied to report measurements in units of plasma. For example, a measured sodium concentration in plasma water of 156 mM is corrected to 145 mM, reported as plasma sodium concentration. Methods that measure solute concentration only in plasma water without dilution (direct methods) yield the same results independently of the true plasma water fraction and true plasma concentration of the solutes. Because the physiological activity of electrolytes and other solutes is proportional to their concentration in plasma water, this is an accepted convention and an appropriate approach to measuring the solutes. The problem occurs with methods that measure analytes after significant dilution of the plasma sample because that solute is now diluted in a much larger volume of water (indirect methods). For example, a change in plasma water fraction from 93% to 80% will result in a reduction in sodium levels measured with indirect methods of 12% relative to the measurement with direct methods (Figure 8.1). To calculate the amount of change in plasma water fraction, and consequently in sodium and other solutes, several formulas are useful.

Plasma water fraction (PWF) can be estimated as follows [63]:

$$\begin{aligned} \text{PWF\%} &= 99.1 - 0.001 \\ &\quad \times (\text{triglycerides} + \text{cholesterol in mg/dL}) \\ &\quad - 0.73 \times \text{total protein } (g/\text{dL}) \end{aligned}$$

If both direct and indirect sodium measurements are available and the diluent volume (in ml/ml of plasma) is known, PWF can be better estimated by the following formula [64]:

$$\text{PWF \%} = 93.3 \times \text{diluent} \times \frac{\text{indirect Na}^+}{\text{direct Na}^+ \times (0.933 + \text{diluent}) - 0.933 \times \text{indirect Na}^+}$$

Assuming a normal PWF of 93%, any plasma solute concentration can be calculated as follows:

$$\text{True concentration} = \frac{\text{measured concentration} \times 93\%}{\text{calculated PWF\%}}$$

Table 8.7 shows different values of calculated PWF depending on the plasma levels of triglycerides, cholesterol, and total protein. It can be seen that only extremes of protein concentrations have a significant water exclusion effect and that triglycerides and cholesterol even at the highest pathological levels have a minimal effect on plasma water and true solute concentrations. The error of less than 3% caused by plausible plasma lipid concentrations is below the total allowable error for sodium (±4 mM) and therefore only a minor source of inaccuracy. A rare exception is hypercholesterolemia caused by high levels of lipoprotein X in patients with severe cholestasis [65]. In one case report, a sodium level of 145 mM measured by direct potentiometry was reported as 124 mM by an indirect method [66]. The patient's total protein was 5.1 g/dL, triglycerides 208 mg/dL, and cholesterol 1836 mg/dL, resulting in calculated plasma water of 93%. However, high levels of lipoprotein X appear to occupy a larger volume than other lipoproteins and result in disproportional water exclusion. Using the direct–indirect measurement difference and a dilution factor of 1/33, plasma water was calculated as 80%. More common causes of pseudohyponatremia (or pseudonormonatremia in patients with true low plasma $Na^+$) are elevated protein levels, whereas pseudohypernatremia occurs in patients with hypoproteinemia. A good approach to investigate hyponatremia is to directly measure serum osmolality: In the presence of pseudohyponatremia, the measured plasma osmolality will be normal, resulting in a gap to calculated osmolality; true hyponatremia will have a corresponding decrease in plasma osmolality.

Direct ISEs are more susceptible to protein buildup and need frequent washing and calibration. In a patient with extremely high glucose levels (2916 mg/dL), a positive interference (+9 mM) with sodium measurement was seen with a direct sodium ISE but not with indirect ISE methods [67].

The $Cl^-$ electrodes use an organic polymeric salt-exchanger membrane to achieve selectivity for $Cl^-$. However, other lipophylic anions, including the halides bromide and iodide, thiocyanate (increased in smokers, consumers of milk and vegetables such as broccoli, and after nitroprusside treatment), salicylates, heparin, lactate, bicarbonate, and β-hydroxybutyrate can also interact with the membrane and cause falsely elevated $Cl^-$ results [55]. These high-polarity membranes are particularly susceptible to protein precipitation and need careful maintenance.

High levels of ascorbic acid (~12 mM) can interfere with $Na^+$ (+43%), $K^+$ (+58%), $Ca^{2+}$ (+103%), and $Cl^-$ (−33%) Beckman Synchron LX20 ISEs, possibly by redox reactivity at the ISE membrane [68]. Other

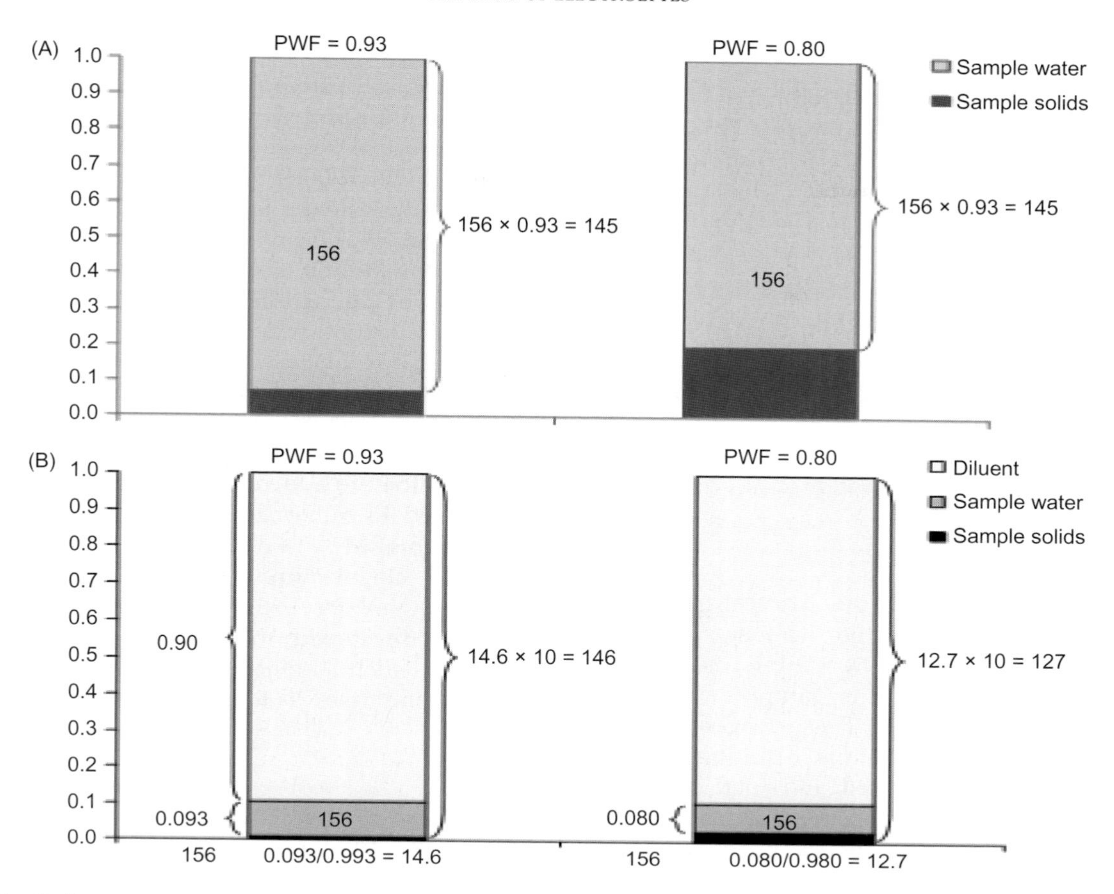


FIGURE 8.1 **Sodium concentrations in the plasma water fraction (PWF) of normal individuals (left) and in patients with high amounts of plasma solids (right) as measured by direct ISEs (A) or indirect ISEs after 1/10 dilution (B).** Note that a PWF of 80% is shown for illustration purposes because it would need highly unusual lipid concentrations >14,000 mg/dL, total protein >25 mg/dL, or a combination of these (e.g., triglycerides = 3000 mg/dL with cholesterol = 500 mg/dL and total protein = 12 mg/dL).

TABLE 8.7 Calculated Plasma Water Fraction (PWF) and True Sodium Concentration for a Measured Sodium Concentration of 140.0 mM Assayed with an Indirect Method in Plasma Containing Different Levels of Protein, Triglycerides, or Cholesterol

| Protein Level (g/dL) | Triglycerides (mg/dL) | Cholesterol (mg/dL) | Calculated PWF (%) | Estimated True Sodium (mM) | % Error |
|---|---|---|---|---|---|
| 7 | 60 | 200 | 93.7 | 138.9 | 0.8 |
| 20 | 60 | 200 | 84.2 | 154.6 | −10.4 |
| 10 | 60 | 200 | 91.5 | 142.2 | −1.6 |
| 4 | 60 | 200 | 95.9 | 135.7 | +3.1 |
| 7 | 3000 | 200 | 90.8 | 143.4 | −2.5 |
| 7 | 60 | 1000 | 92.9 | 140.1 | −0.1 |

strong redox agents, such as *N*-acetylcysteine, cysteine, glutathione, guanidine, procainamide, and sodium azide, show similar interferences.

### *Total $CO_2$*

Two main methods are in use:

*Potentiometry*: The plasma is diluted, acidified to convert all $HCO_3^-$ to $CO_2$, and the $pCO_2$ is measured with a $pCO_2$ electrode. Most instruments use indirect potentiometry, whereas Vitros analyzers use direct potentiometry in thin-film electrodes. These ISE methods are subject to interferences by strong redox substances (see Table 8.2). Note that whole blood gas analyzers do not measure $HCO_3^-$ directly; rather, $pCO_2$ and pH are measured by potentiometry, whereas $HCO_3^-$ is calculated from the equilibrium equation of Henderson–Hesselbach:

$$CO_2 + H_2O \leftrightarrow HCO_3^- + H^+$$

*Enzymatic methods*: The plasma is alkalinized to convert all $CO_2$ to $HCO_3^-$, and the $HCO_3^-$ is used by phosphoenolpyruvate (PEP) carboxylase to convert PEP to oxaloacetate, which is then reduced to malate by malate dehydrogenase, while generating $NAD^+$ from NADH. The decrease in NADH concentration, measured by absorbance at 340 nm, is proportional to the amount of $HCO_3^-$. Interferences are minimal and only seen at very high levels of hemolysis (free hemoglobin >800 mg/dL), hyperbilirubinemia (>30 mg/dL), or with specimen turbidity caused by paraprotein precipitation [69].

### *Calcium Assays*

Total calcium is measured by chemical methods in most analyzers, except in the Beckman Synchron instruments, which use an indirect ISE to measure ionized calcium after diluting the sample with strong calcium complexing reagents. These reagents dissociate $Ca^{2+}$ from proteins and anions, resulting in a standardized molar ratio of ionized and total calcium in the solution at equilibrium. The chemical methods use a colorimetric approach after releasing protein-bound calcium by acidification followed by complexing calcium with a chromogen. Some methods use an arsenazo III dye under alkaline conditions, whereas others use *o*-cresolphthalein in acidic pH to produce a colored complex with calcium. Reaction conditions and buffers have been optimized to enhance selectivity for calcium over magnesium and improve linearity.

Interferences with calcium methods include calcium chelators in the sample, either as inappropriate anticoagulants such as EDTA, citrate, and oxalate, or *in vivo* as a result of transfusion with citrate-containing blood products or apheresis with citrate-anticoagulant infusion. Heparin reduces free ionized calcium (~0.01 mM per unit/mL of heparin) but has no significant effect on total calcium. The use of balanced or calcium-titrated lyophilized heparin reduces but does not eliminate this effect, whereas lithium–zinc heparin does not show significant interference with free $Ca^{2+}$ determinations [37,55]. Spectral interferences by high levels of hemolysis, bilirubin, and lipemia are common with colorimetric methods and minimal with ISE methods. In case of significant spectral interference, it is possible to treat the samples with ethyleneglycoltetraacetic acid (EGTA) to chelate all the calcium and subtract the results from the initial measurement. Contamination by calcium compounds can occur in glass tubes or cork-capped tubes, but these are rarely used in clinical practice.

### *Magnesium Assays*

As with calcium, total magnesium is commonly measured by spectrophotometry after forming complexes with dyes, such as arsenazol, calmagite, magon, or methylthymol blue. Specificity for magnesium is not absolute, and calcium interference is reduced by chelator such as EGTA. Free $Mg^{2+}$ can be measured by magnesium-specific electrodes, although these are not as reliable as calcium ISEs and suffer from interference from free calcium, heparin, certain tube additives and silicone lubricants, and thiocyanate. To correct for calcium interference, $Ca^{2+}$ is simultaneously measured by a parallel calcium-specific electrode, and the results are used to adjust the $Mg^{2+}$ measurement. As with calcium, albeit to a lower extent, changes in pH should be avoided to minimize disturbing the proportion of free and total $Mg^{2+}$. For most clinical purposes, assays for total magnesium are more reliable and cost-effective than free $Mg^{2+}$ measurements. The new heparins developed to minimize interferences with free $Ca^{2+}$, such as zinc or lithium heparin, can cause spurious increases in total magnesium and should be avoided.

### *Phosphate Assays*

Inorganic phosphate is commonly assayed by reaction of phosphate ions with ammonium molybdate to form a phosphomolybdate complex that is then measured spectrophotometrically, either by direct UV absorption at 340–363 nm or after reaction with a reducing agent to increase stability and color formation, measured at 600–700 nm [37]. Methods using UV absorption are more susceptible to interference by lipemia, hemolysis, and hyperbilirubinemia. Analyzers that utilize dry-film slides to remove bilirubin, lipids, hemoglobin, and other proteins from the analytical layer are less susceptible to these interferences.

A common cause of interference in phosphate assays is the presence of monoclonal immunoglobulins and paraproteins associated with multiple myeloma and Waldenström macroglobulinemia, which interfere with the colorimetric reaction and cause spurious hyperphosphatemia and, rarely, false hypophosphatemia [70,71]. Hyperphosphatemia in the presence of normal calcemia and renal function and in the absence of other causes of hyperphosphatemia, such as hypoparathyroidism and acromegaly, should prompt suspicion for a monoclonal gammopathy [72]. Other less common interferences include Alteplase, a catheter flushing solution containing high phosphorus excipient, liposomal amphotericin B, heparin, mannitol, fluoride, and tissue plasminogen activator.

# BLOOD GASES ANALYSIS

Whole blood gas analyzers provide rapid assessment of the patient's acid–base and ventilation status, which are essential in the treatment and monitoring of critically ill patients, as well as some patients with chronic metabolic and respiratory diseases. Typically, these analyzers offer direct measurements of the partial pressures of oxygen ($pO_2$) and carbon dioxide ($pCO_2$) in blood, as well as pH and electrolytes ($Na^+$, $K^+$, $Cl^-$, and $Ca^{2+}$), hemoglobin (with or without co-oximetry for oxy-, deoxy-, carboxy-, met-, and fetal-hemoglobin measurement), and a variety of calculated parameters, including $HCO_3^-$ concentration and hematocrit. Many blood gas analyzers also offer measurements of other substances of interest in critical care medicine, including glucose, lactate, creatinine, urea, and bilirubin. In this section, the methodologies and sources of error in pH, blood gas, and co-oxymetry measurements are discussed; electrolytes and metabolites are reviewed in other sections.

## Pre-Analytical Issues

Arterial blood samples are necessary for evaluation of oxygen status because the $pO_2$ and derived calculated parameters are significantly different between arterial and venous blood. In most patients, arterial $pO_2$ is 60–70 mmHg higher than venous $pO_2$. However, in many patients, assessing the oxygen status is not as critical and can be estimated from pulse oxymetry or transcutaneous monitoring of $SaO_2$, and venous blood can effectively be used to assess the acid–base status, avoiding the difficulties and complications of arterial puncture. Several studies examined the differences between venous and arterial blood gases and concluded that in mildly to moderately ill patients without severe hemodynamic or cardiorespiratory failure, pH, $pCO_2$, and calculated $HCO_3^-$ were sufficiently accurate to estimate acid–base status [73–76]. Arterial–venous differences for pH averaged +0.032 (range, 0.009–0.050), whereas $pCO_2$ averaged −4.9 (range, −1.6 to −7.0 mmHg) and calculated $HCO_3^-$ averaged +1.40 (range, −0.15 to +1.88 mM).

Samples for blood gas analysis should be collected in gas-impermeable syringes, coated with lyophilized heparin, and sealed after collection to prevent gas exchange. Syringes should be held tip up, lightly tapped, and a drop of blood ejected to remove all air bubbles before placing a tight cap. Drawing blood through a butterfly infusion set can increase $pO_2$ by diffusion through the plastic lines [77]. Liquid heparin, which equilibrates with ambient air, can have a significant effect on blood gas analysis, leading to decreases in $pCO_2$ and increases in $pO_2$. Likewise, large bubbles or entrapped air will cause $pO_2$ to equilibrate with ambient air (with $pO_2 = 160$ mmHg at sea level) and $pCO_2$ to decrease ($pCO_2$ of ambient air = 0.25 mmHg). The change in $pCO_2$ is usually mild because of the capability of the plasma bicarbonate to buffer any changes in $pCO_2$. However, even a small bubble comprising 0.5% of the blood volume will cause an increase of 8 mmHg in $pO_2$ [78]. Due to rapid changes in air pressure and agitation, this problem is exacerbated if the samples are transported in pneumatic tube systems, even when no bubbles are visible [79]. Therefore, samples should be hand-delivered to the laboratory when accuracy of $pO_2$ is essential.

Samples should be analyzed as soon as possible. If the analysis is delayed more than 30 min, the samples should be kept on ice to minimize cellular respiratory activity. Samples kept at room temperature in gas-impermeable containers show an average decrease in $pO_2$ of 2–4 mmHg/hr, whereas $pCO_2$ increases approximately 1 mmHg/hr and pH decreases 0.02 to 0.03 units [37,80]. These changes are accentuated at higher temperatures and in patients with markedly elevated leukocyte, platelet, or erythrocyte counts; for example, patients with greater than 100,000 leukocytes/μL can have a drop in $pO_2$ of 40 mmHg in 5 min. When the specimen is placed on ice, these changes are reduced by more than 50%, except in samples collected in plastic syringes, which have different degrees of permeability to oxygen from ambient air and ice water. Plastic syringes that allow some gas exchange on ice result in changes in $pO_2$ of 2–15% (either increase or decrease depending on whether the patient's $pO_2$ is lower or higher than ambient) [80,81]. Plastic syringes should be kept at room temperature and analyzed in less than 30 min. If delays are unavoidable, gas-impermeable glass or plastic syringes kept on ice should be used.

Prolonged application (more than 1 min) of a tourniquet for collection of venous blood, especially when coupled with fist clenching, can lead to oxygen consumption and lactic acidosis with a drop in pH. Application of tourniquet for 30 sec followed by release before collection is recommended [82].

## Analytical Issues

Methods for blood gas analysis rely on electrochemical measurements, either by potentiometry, in the case of the $pCO_2$ and pH electrodes, or by amperometry, in the case of the $pO_2$ electrode. The pH electrode has a glass membrane that exchanges metal ions with protons on both sides of the membrane. A change in $H^+$ concentration on the sample side of the membrane generates a potential difference across the membrane that is proportional to the pH. The $pCO_2$ electrode has a membrane permeable only to uncharged molecules such as $CO_2$, $O_2$, and $N_2$. When $CO_2$ diffuses across the membrane into the electrolyte solution, it dissociates into $HCO_3^-$ and $H^+$, which generate a potential difference across the $H^+$-impermeable membrane that is proportional to the $pCO_2$. In the $pO_2$ electrode, oxygen diffuses across an oxygen-permeable membrane and is reduced to water on the cathode (usually a platinum rod), consuming electrons and $H^+$ ions. To complete the current, in the anode silver ions are oxidized to $Ag^+$ with electron release and subsequently complexed with $Cl^-$ to form a deposit of AgCl on the anode. The size of the current generated from the consumption of electrons during oxygen reduction is proportional to the amount of oxygen in the sample ($pO_2$).

Bicarbonate concentration is calculated from the measurement of pH and $pCO_2$ using the following Henderson–Hesselbach equation ($pCO_2$ in mmHg and $HCO_3^-$ in mM):

$$\text{pH} = pK' + \log \frac{HCO_3^-}{\alpha \times pCO_2}$$

In most patients, $pK' = 6.103$ and $\alpha = 0.0306$, but these factors depend on temperature and the ionic strength of the solution. Marked changes in ionic strength of the sample, as well as extremes of protein and lipid concentration, can affect the $CO_2$ solubility coefficient [37]. Therefore, in some patients, particularly acutely ill children, calculated $HCO_3^-$ based on $pCO_2$ and pH measurements in blood gas analyzers is less reliable than total $CO_2$ measured by the routine chemistry analyzers.

Oxygen parameters are calculated from directly measured $pO_2$ and hemoglobin parameters; the following are examples (Hgb in g/dL and $pO_2$ in mmHg):

- Oxygen saturation
  $$(SaO_2) = \frac{\text{oxyhemoglobin}}{\text{deoxyhemoglobin} + \text{oxyhemoglobin}}$$
- Fraction of oxygenated hemoglobin ($FO_2Hb$) = oxyhemoglobin/total hemoglobin
- Blood oxygen content ($ctO_2$) = $1.36 \times FO_2Hb \times Hgb + 0.0031 \times pO_2$

It is important to understand the difference between $FO_2Hb$ and $SaO_2$ because they differ significantly in patients with significant amounts of abnormal hemoglobins, such as carboxy- or methemoglobin, which cannot carry oxygen. Oxygen saturation is commonly measured by pulse oxymetry, which simply measures oxyhemoglobin ($O_2Hb$) and deoxyhemoglobin (HHb) by transdermal spectrophotometry at two wavelengths. $SaO_2$ can also be calculated from empirical formulas using pH, $pO_2$, and hemoglobin, but these formulas assume normal oxygen binding by hemoglobin and are unreliable in patients with changes in the $O_2$–hemoglobin dissociation curve. In contrast, the more sophisticated co-oxymeter analyzers scan absorbance of a whole blood hemolysate at wavelengths between 535 and 670 nm and calculate the fraction of each hemoglobin fraction by their characteristic absorption profile.

In most patients, $FO_2Hb$ and $SaO_2$ are roughly interchangeable. For example, in a patient with 94% $O_2Hb$ and 3% HHb, $SaO_2 = 94/97 = 96.9\%$. However, in a patient with 85% $O_2Hb$, 3% HHb, and 12% carboxyhemoglobin (COHb), $SaO_2 = 85/88 = 96.6\%$, which is significantly different from $FO_2Hb$ of 85%. With pulse oxymetry alone, this patient may be erroneously assumed to have normal blood oxygen content, whereas co-oxymetry would demonstrate the impairment in oxygen content and therefore the decreased oxygen delivery to the tissues.

Measurement of blood gases is highly dependent on atmospheric pressure and ambient temperature, which can vary during the day. Therefore, frequent calibration, usually every 30 min, is performed to account for these fluctuations. In some electrodes, exposure of the patient to other gases, such as NO and halothane, can spuriously increase $pO_2$ measurements.

Co-oxymetry depends on photometric scan and therefore is affected by substances that absorb in the range of 535–670 nm, such as methylene blue, which is used in the treatment of methemoglobinemia and absorbs strongly in the 550- to 700-nm region. A concentration of 25 mg/L of methylene blue will spuriously increase $FO_2Hb$ and reduce methemoglobin results by approximately 4–8 percent points, which may lead to misleading interpretation of the efficacy of methylene blue treatment [83]. Other dyes, such as Patent Blue V, Evans Blue, and Cardio Green, show variable amounts of interference with co-oxymetry, usually less than 1 percentage point. Highly lipemic samples can also be affected; for example, a 4% concentration of Intralipid will increase carboxy- and methemoglobin by 0.5–0.9%. Some co-oxymeters measure hemoglobin F fraction, which is more susceptible to interference by abnormal pH, blue dyes, and lipemia, and at high levels seen in neonates and hereditary persistence of fetal hemoglobin, it can cause decreases in HHb and $SaO_2$ results. Other colored substances causing significant interference with co-oxymetry are vitamin $B_{12}$ preparations. For example, a concentration of 200 μg/mL of cyanocobalamin (more than 10-fold maximum attainable serum levels) can cause a 2 percentage point decrease in $FO_2Hb$ measured by the Radiometer ABL800.

# LACTATE ANALYSIS

Evaluation of lactic acidosis is important in the assessment of many critically ill patients. Lactic acid production increases when tissues are subject to anaerobic conditions (tissue hypoxia), such as with hemodynamic impairment or respiratory failure. Lactic acidosis due to tissue hypoxia is classified as type A, whereas the rarer type B is due to metabolic diseases interfering with mitochondrial function (e.g., diabetes, drugs such as ethanol, methanol, and salicylates, and several inborn errors of metabolism). Although L-lactate is the predominant isomer generated by human cells, intestinal bacteria can produce large amounts of D-lactate, which is slowly metabolized by human tissues. In certain situations, such as with jejunoileal bypass or other causes of short bowel syndrome, impaired carbohydrate absorption leads to increased production of D-lactate by colonic Gram-positive anaerobic bacteria and accumulation in the systemic circulation with metabolic acidosis. D-Lactate is not measured by assays using enzymes specific for L-lactate.

Several studies associate lactic acid levels with mortality. For example, in patients admitted to the ICU, elevated lactate at admission was associated with a 24% mortality, and patients with persistent lactic acidosis after 24 hr of treatment had 82% mortality [84]. Routine monitoring of lactate is indicated in acutely ill patients at risk for tissue hypoxia due to either decreases in oxygen delivery or increases in oxygen consumption.

Even minor exercise can rapidly increase lactate levels, and a minimum resting period of 30 min is indicated before sample collection. Hyperventilation, catecholamine stimulation, and infusion of intravenous glucose or bicarbonate also lead to increased lactate production.

Lactate measurements are typically available in whole blood gas analyzers and reported with arterial blood gas profile. However, if assessment of oxygen status is not essential, venous blood is preferable for assessment of lactic acidosis to avoid the complications of an arterial puncture. Several studies show that venous lactate levels are slightly higher and correlate strongly with arterial lactate levels. For example, a study in emergency department patients showed a regression $R^2$ for the equation arterial lactate = 0.996 × venous lactate − 0.259 [85]. This equation can be used to predict arterial lactate, or clinical cutoff values can be slightly adjusted for venous lactate. In cases of impaired hemodynamics (e.g., in shock), the arterial–venous lactate difference will increase and can actually be used as a measure of tissue perfusion.

A study in healthy volunteers showed that application of a tourniquet for up to 5 min before blood collection had no effect on lactate levels [86]. The same study also determined that storage on ice is not necessary if the sample is analyzed within 15 min. However, prolonged tourniquet application together with repeated fist clenching will increase lactate and should be avoided. Samples should be stored on ice if separation from cells cannot be achieved in less than 15 min. Alternatively, samples may be collected in gray-top tubes containing sodium fluoride and potassium oxalate, which show a minimal increase (averaging 0.3 mM) after 8 hr at room temperature [87].

## Analytical Issues

Lactate measurements are typically performed either in whole blood by amperometry or in serum by colorimetric methods. The whole blood analyzers and some high-throughput instruments use an electrode separated from the sample by a multilayer membrane. Lactate crosses the outer membrane and is oxidized to pyruvate by lactate oxidase immobilized between the outer and inner membranes. The reaction consumes $O_2$ and produces $H_2O_2$, which crosses the inner membrane and is oxidized at the platinum anode, whereas the silver cathode is reduced to produce a current proportional to the amount of lactate. Potent redox agents such as cysteine, fluoride, bromide, iodide, thiocyanate, isoniazid, acetaminophen, and hydroxyurea may show interference with this method.

The colorimetric methods use either L-lactate oxidase or L-lactate dehydrogenase. The lactate oxidase generates $H_2O_2$, which then oxidizes a chromogen in a reaction catalyzed by peroxidase to generate a product measurable by spectrophotometry. Lactate dehydrogenase methods rely on the conversion of lactate to pyruvate at high pH, with generation of NADH and increased absorption at 340 nm. D-Lactate is not measured by these assays and requires a specific method using D-lactate dehydrogenase.

Because the lactate oxidase is not entirely specific for lactate, some methods based on this enzyme (i.e., the Radiometer ABL) are subject to positive interference by citrate, oxalate, glycolate, and glyoxylate [88]. The latter three substances are ethylene glycol metabolites and may result in spuriously high lactate levels in patients who ingested ethylene glycol. For a list of household products containing ethylene glycol, see the National Institutes of Health reference [89]. Other lactate oxidase methods (e.g., iSTAT, Abbott Architect, Ortho Vitros, Siemens Advia, Roche Modular, and Beckman Synchron) show less interference, whereas LDH-based methods (e.g., Siemens

Dimension) are not subject to interference by these metabolites. Based on these differences, a "lactate gap" between the Radiometer measurement and methods unaffected by glycolate and glyoxylate may be indicative of ethylene glycol poisoning [88,90]. The magnitude of the interference may be variably dependent on the method and the age of the lactate electrode [91]; therefore, it should not be used as a quantitative approach to estimating amounts of ethylene glycol.

Patients with elevated LDH may show spurious decreases in lactate concentrations due to its conversion to pyruvate *in vitro*. To avoid this interference, plasma or serum samples should be frozen or analyzed rapidly after collection. Not all assays show this interference; for example, up to 4000 IU/L of LDH does not interfere with the Beckman Synchron lactate assay.

*CASE REPORT* A 72-year-old male was admitted to the ICU for cardiac failure and accidental ingestion of a large amount of propylene glycol. Laboratory results included an anion gap of 27 mEq/L, metabolic acidosis with pH = 7.16, and low amounts of propylene glycol in the blood. An arterial blood gas measurement using the Radiometer ABL 700 analyzer revealed 39 mM lactate, whereas a serum sample measured with the Vitros analyzer showed normal lactate levels. Measurement with a D-lactate-specific kit resulted in D-lactate levels greater than 100 mM [92]. This case highlights that propylene glycol can be metabolized to D-lactate in the absence of intestinal malabsorption, and that the Radiometer method is not entirely specific for L-lactate and shows positive interference by D-lactate.

# BILIRUBIN ANALYSIS

Bilirubin is an end product of heme metabolism. The major source for heme production is hemoglobin; therefore, bilirubin generation depends on the rates of hemoglobin synthesis and degradation. Bilirubin originating from heme metabolism in the various tissues, predominantly from the phagocytic system and ineffective erythropoiesis, binds noncovalently to albumin and is carried by the circulation to the liver, where it is further metabolized by conjugation to glucuronic acid, a reaction catalyzed by the uridine diphosphate glucuronyltransferase enzyme coded by the UGT1A1 gene. Excretion of conjugated bilirubin then occurs via the biliary system into the intestines, where 95% of it is reabsorbed into the circulation while the remainder is further metabolized by intestinal bacteria into urobilinogen, which can also be reabsorbed (~20%). Circulating urobilinogen and conjugated bilirubin are soluble in the plasma and can be excreted by the kidneys.

Main causes of hyperbilirubinemia include the following:

1. Defects in conjugation, including severe liver insufficiency, genetic defects in UGT1A1 (Crigler–Najjar and Gilbert syndromes), and inhibition of UGT1A1, leading to increased blood unconjugated bilirubin levels
2. Increased heme catabolism—for example, associated with severe hemolysis—leading to elevated blood levels of unconjugated bilirubin
3. Defects in biliary excretion ("cholestasis"), caused by diseases interfering with liver or biliary architecture, which cause increases predominantly in conjugated bilirubinemia.

Accurate measurement of unconjugated bilirubin is of particular importance in neonates because this substance penetrates the blood–brain barrier and can cause severe encephalopathy, known as kernicterus. Determination of conjugated bilirubin is also critical in the evaluation of neonatal jaundice because severe diseases such as sepsis, biliary atresia, and hepatitis are associated with elevated conjugated bilirubin. In adults, measurement of total bilirubin is a relatively sensitive test for liver and biliary dysfunction. In hepatobiliary diseases, discrimination between conjugated and unconjugated hyperbilirubinemia can help distinguish pure cholestastic conditions, which elevate conjugated bilirubin, from those involving hepatocytes, which typically lead to increases in both conjugated and unconjugated bilirubin. However, it is often unnecessary because other markers of hepatic or biliary disease will be abnormal.

## Pre-Analytical Issues

Starvation and caloric malnutrition can be associated with large increases in unconjugated bilirubin, especially in patients with Gilbert syndrome. Several substances affect bilirubin glucuronidation, and unconjugated bilirubin can rise as a result of elevated thyroid hormones, estrogens, oral contraceptives, fatty acids in breast milk, antibiotics such as novobiocin or gentamicin, or anti-retroviral agents such as indinavir and atazanavir. Other drugs, such as sulfonamides and ibuprofen, raise unconjugated bilirubin by displacing it from albumin binding sites. Recent alcohol ingestion, smoking, meals, and exercise have all been associated with mild increases in bilirubin [2,3].

Prolonged tourniquet application or standing can increase unconjugated bilirubin by approximately 10% due to its binding to albumin [3]. Because bilirubin is

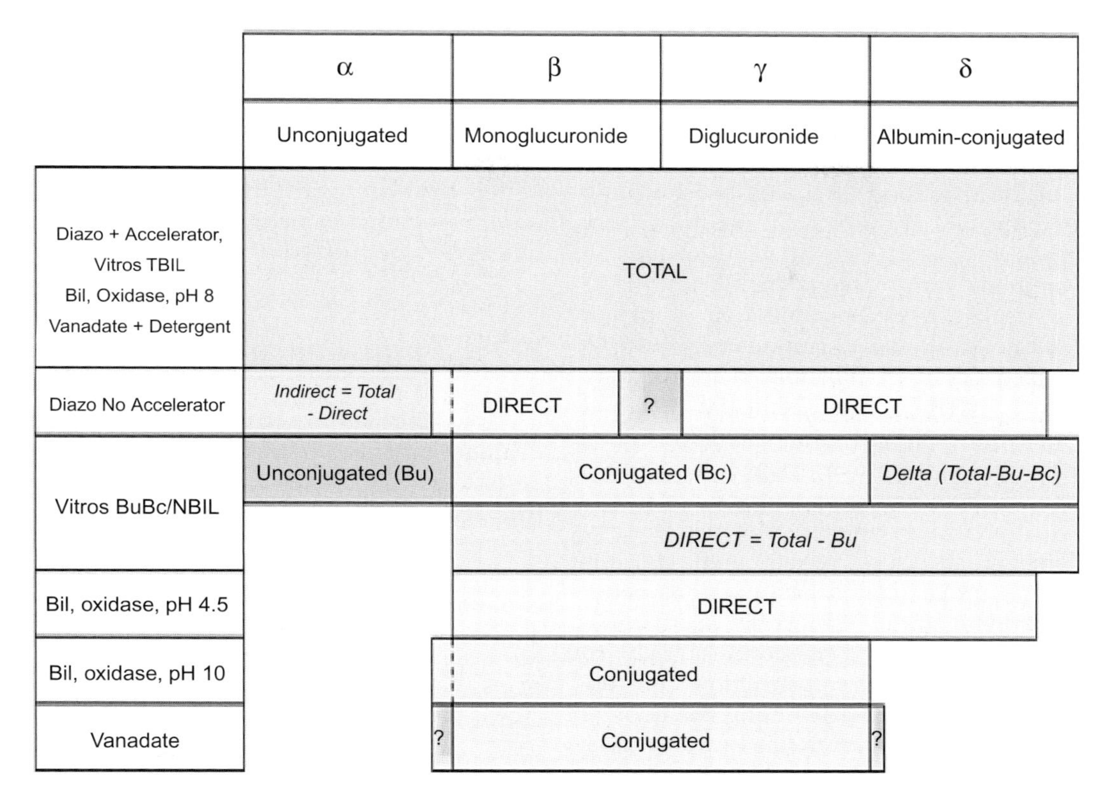


FIGURE 8.2 **Distribution of bilirubin fractions measured with diazo assays derived from the Jendrassik and Grof method, from the Vitros BuBc dry chemistry method, or from the vanadate and bilirubin oxidase methods.** Calculated fractions are depicted in italics. The graph shows that 10–25% of δ-bilirubin may not react with direct methods [109], whereas 1–10% of unconjugated bilirubin may react with diazo or bilirubin oxidase methods. "Var" refers to a variable portion of conjugated bilirubin that is not measured by diazo direct methods. The question mark reflects the unknown reactivity of vanadate oxidation methods with α and δ fractions.

sensitive to degradation by light, samples should be shielded from exposure to light during pre-analytical processing if analysis is delayed. A decrease of as much as 50% can occur with 1 or 2 hr of exposure to light [2].

## Analytical Issues

Fractionation by chromatography shows that bilirubin circulates in four main forms (Figure 8.2) [93]:

1. Unconjugated bilirubin, loosely bound to albumin (α-bilirubin)
2. Bilirubin monoglucuronide (β-bilirubin)
3. Bilirubin diglucuronide (γ-bilirubin)
4. Bilirubin conjugated to albumin (δ-bilirubin).

In normal individuals, only unconjugated bilirubin (α) is present in the serum. The delta fraction occurs by substitution of albumin for the glucuronide moiety of conjugated bilirubin and is increased in neonates and in patients with prolonged hyperbilirubinemia. Whereas the half-life of fractions α–γ is short (minutes in normal individuals), δ-bilirubin has the half-life of albumin (2 or 3 weeks) and complicates assessment of recovery of jaundice. Bilirubin is determined most commonly by a colorimetric reaction between diazotized sulfanilic acid and bilirubin—the so-called "diazo reaction." Initially described by Erlich in 1883, the reaction has been subject to several modifications to increase reactivity with unconjugated bilirubin. Most current methods are derived from the diazo method of Jendrassik and Grof (J/G), in which caffeine benzoate is added to the reaction to displace unconjugated bilirubin from albumin and accelerate its reaction with the diazo chromogen. Other less commonly used accelerators include methanol, dyphylline, and various surfactants. When the accelerator is present, all of the bilirubin fractions (α–δ) react quickly with the diazo dye, and the result is labeled as "total bilirubin." When the accelerator is omitted, unconjugated bilirubin reacts very slowly, the conjugated fractions (β, γ) react quickly but incompletely, and δ-bilirubin reacts to a variable extent; the result is labeled "direct bilirubin" (Figure 8.2). Direct methods can variably measure between 1 and 10% of unconjugated bilirubin, so their specificity for conjugated fractions is more important than their accuracy [94] because unconjugated bilirubin is estimated from "indirect bilirubin," calculated by subtracting "direct" from "total" bilirubin.

More recent methods include spectrophotometric determination of the oxidation of conjugated bilirubin by vanadate or bilirubin oxidase to biliverdin, with a resulting decrease in absorbance at 425–460 nm. In both of these methods, only conjugated bilirubin fractions (β and γ) are oxidized. Manipulation of pH or addition of detergents allows unconjugated bilirubin to also be oxidized for total bilirubin determination (Figure 8.2). In the Ortho "dry chemistry" method (BuBc slide), the absorbance of conjugated (β and γ) bilirubin is differentiated from that of unconjugated (α) bilirubin by spectrophotometric scanning in the presence of a cationic mordant (Figure 8.2). Because δ-bilirubin is not detected in these recent methods, the conjugated (Bc) and unconjugated (Bc) fractions better represent the functional status of bilirubin production, conjugation, and excretion and respond quickly to changes such as removal of biliary obstruction. One disadvantage of spectrophotometric scanning methods is the susceptibility to spectral interference by substances such as amphotericin B, levodopa, methotrexate, nitrofurantoin, sulfasalazine, hemoglobin, and bilirubin isomers originating from light exposure.

The classic diazo method is particularly susceptible to interferences due to the presence of serum substances such as hemoglobin, immunoglobulins (paraproteins), lipemia, acetoacetate, and ascorbic acid [2,71]. More recent methods using different diazo reagents with serum blanking, bilirubin oxidase, or vanadate are less susceptible to such interferences.

*CASE REPORT* A 37-year-old male was admitted for a suicide attempt by ingestion of naproxen in combination with ethanol. A liver profile showed only mild elevation in aspartate aminotransferases (66 U/L) and γ-glutamyl-transpeptidase (53 U/L), with normal alanine aminotransferase and alkaline phosphatase, and a total bilirubin concentration of 20.5 mg/dL [95]. There was no past medical history or physical evidence of liver disease. Naproxen serum levels were 138 μg/mL—well above the therapeutic level. Measurement of bilirubin with the Vitros BuBc method shows levels of Bu and Bc below 0.1 mg/dL. Another similar case showed a naproxen level of 167 μg/mL, and total bilirubin determinations with two classic J/G methods were 13.9 and 10.7 mg/dL, whereas an alternative diazo method using 2,4-dichlorophenyl diazonium tetrafluoroborate showed only 0.5 μg/mL of total bilirubin. In both these patients, direct bilirubin was 0.1 mg/dL or less. The interference appears to be mostly due to the naproxen metabolite *O*-desmethylnaproxen reacting with the classic J/G diazotized sulfanilic reagent in the presence of caffeine [96].

Methods have been developed to measure free, non-protein-bound unconjugated bilirubin levels, which better correlate with the severity of acute bilirubin encephalopathy in neonates [97]. However, these are not well standardized or widely used.

## LIPID PROFILES ANALYSIS

Lipid profiles are commonly used in the routine evaluation of cardiovascular risk, given the high correlations of hypercholesterolemia and hypertriglyceridemia and cardiovascular risk. A standard lipid profile includes determination of serum or plasma total cholesterol (TC), high-density lipoprotein-associated cholesterol (HDL-C), low-density lipoprotein-associated cholesterol (LDL-C), and total triglycerides (TG). In many laboratories, LDL-C is calculated from the Friedewald equation (LDL-C = TC − HDL-C − TG/5), which correlates well with gold standard determinations of LDL-C by ultracentrifugation when TG are less than 400 mg/dL. This calculation is based on the following assumptions: (1) The vast majority of cholesterol is transported by LDL, HDL, and very low-density lipoproteins (VLDL) in fasting plasma; (2) most TG are associated with VLDL in fasting specimens; and (3) the normal ratio of TG to cholesterol in VLDL is 5. However, these assumptions are not always true.

### Fasting Versus Nonfasting Lipid Profiles

Cardiovascular risk assessment and monitoring of lipid-lowering therapy based on LDL-C and TG should be performed in specimens collected after a period of 12–14 hr of overnight fasting without any dietary intake except for water and medication because TG (predominantly in the form of chylomicrons) remain elevated after meals for several hours. In nonfasting specimens and other conditions with elevated TG, non-VLDL particles carrying TG change the ratio of TG to cholesterol and invalidate the Friedewald formula, usually leading to overestimation of LDL-C. In these cases, a direct assay for LDL-C can be used, although direct LDL-C results in nonfasting specimens may not be as predictive of cardiovascular events in women [98]. Recently, emphasis has been placed on non-HDL-C (TC − HDL-C) as a secondary target of therapy because it includes not only LDL-C but also other atherogenic lipoproteins, such as intermediate-density lipoproteins [99]. Importantly, TC, HDL-C, and non-HDL-C can be accurately determined in nonfasting specimens. Some studies also indicate that nonfasting TG, which peak approximately 4 or 5 hr postprandially, may be better predictors of cardiovascular events [100], but the exact cutoffs and postprandial intervals have not been well-defined. Together with determination of apolipoproteins A1 and B100, it is possible that in the future, nonfasting lipid profiles will be acceptable for cardiovascular risk assessment.

## Other Pre-Analytical Considerations

Although cholesterol and triglycerides are considered small molecules, they circulate as components of large lipoprotein complexes; therefore, plasma levels are subject to dilutional variation in protein concentration induced by a change to the upright position (~7–16% increase) and prolonged tourniquet application (5–20% higher). Pregnancy is associated with a decrease of cholesterol and triglycerides in the first trimester but a gradual increase afterwards [2]. The nephrotic syndrome and hypothyroidism are associated with elevated TC, predominantly due to LDL-C. Stress, inflammation, infections, and other acute disorders are associated with increases in triglycerides and cholesterol, whereas acute coronary syndromes result in decreases of TC and LDL-C and 20–30% increases in TG within 24–48 hr, leading to the recommendation that lipid profiles should be avoided after 24 hr of hospitalization or initiation of an acute event [101]. A study using fasting samples indicated that the magnitude of change after an acute myocardial infarction is small, with a mean decrease in TC and LDL-C of 2% at 24 hr of admission and a subsequent increase of 6% in the next 2 days, whereas TG and HDL-C showed even smaller changes [102]. However, other studies showed larger reductions in TC, non-HDL-C, direct LDL-C, TG, and apolipoprotein A1 in the first 18–42 hr post-admission, whereas apolipoprotein B100 and HDL-C did not change significantly [101,103]. The best recommendation is to perform lipid profiles at admission of patients with suspected acute coronary syndromes.

## Analytical Issues

Virtually all currently available methods for measuring cholesterol rely on the production of hydrogen peroxide from cholesterol in an enzymatic reaction catalyzed by cholesterol oxidase. Interestingly, this enzyme is not specific for cholesterol, and plant sterols can also be reacting. However, even in individuals ingesting large amounts of plant sterols, the plasma concentration (μg/dL) is minuscule compared to cholesterol (mg/dL) [104].

In order to measure both free cholesterol and esterified cholesterol, samples are treated with cholesteryl ester hydrolase to release cholesterol from fatty acid esters. The $H_2O_2$ produced by the oxidase reaction is then typically measured by a peroxidase catalyzed reaction. Therefore, all previously described interferences with peroxidase-based methods (e.g., hemoglobin, bilirubin, ascorbic acid, and uric acid) also affect cholesterol measurements, although these interferences are minimized by current approaches, such as the use of multiple wavelengths, kinetic measurements, and bilirubin oxidase.

For the determination of HDL-C, either HDL is physically separated from other lipoproteins with precipitation methods or in homogeneous assays the cholesterol associated with HDL is distinguished from the cholesterol associated with other lipoproteins by a variety of different approaches. Separation methods include precipitation of non-HDL cholesterol with dextran sulfate or a heparin–manganese complex, followed by centrifugation and analysis of the supernatant. These methods are labor-intensive and have been largely replaced with homogeneous methods. Approaches used in homogeneous methods, often used in combinations, include (1) detergent solubilization of the cholesterol in HDL while inhibiting release from other lipoproteins with polyanions; (2) steric inhibition of the reaction of non-HDL cholesterol with the enzymes by the use of antibodies, for example, to apoB100 contained in LDL, VLDL, and chylomicrons; (3) modification of cholesterol esterase and oxidase by linking polyethylene glycol moieties to the enzymes, which excludes access to cholesterol in non-HDL lipoproteins; and (4) selective oxidation of cholesterol from non-HDL with subsequent destruction of $H_2O_2$ by catalase, followed by addition of a catalase inhibitor and a detergent releasing cholesterol from HDL. Similar approaches are used to measure LDL-C by "direct" homogeneous assays. The main issue with homogeneous methods is the specificity for HDL-C or LDL-C and the degree of reactivity with other lipoproteins, particularly in patients with abnormal amounts of unusual lipoproteins (e.g., in patients with type III hyperlipidemia with excess remnant lipoproteins, cholestasis, and chronic liver or kidney disease). These methods are also subject to varying degrees of interference in samples with high levels of TG [105,106]. A comparison of various homogeneous methods for LDL-C and HDL-C determination showed that most methods met the acceptable total error goals in healthy individuals, but all methods failed to attain these goals in patients with cardiovascular disease, dyslipidemias, and hypertriglyceridemia [106].

*CASE REPORT* A 64-year-old male presented with anemia (hemoglobin, 11.7–12.0 g/dL) first noted 4 months earlier [107]. Other abnormal laboratory results included serum HDL of 10 mg/dL (46 mg/dL 3 years earlier) and a total protein of 9.5 g/dL with albumin of 3.8 g/dL. Total cholesterol was 98 mg/dL, with 75 mg/dL in the LDL-C fraction. Measurement of HDL-C in a slide chemistry system showed levels of 44 mg/dL. The patient was diagnosed with Waldenström's macroglobulinemia with biclonal IgM-κ and IgM-λ paraproteins. Interference from paraproteins in HDL-C has been noted with precipitation, liquid homogeneous, and also solid-phase homogeneous assays [71].

Like cholesterol analysis, triglyceride assays use a mix of enzymes to produce a measurable product. The first reaction involves hydrolysis of the triglycerides with lipase, releasing free fatty acids and glycerol. The next reaction involves glycerol kinase, which uses ATP to phosphorylate glycerol to glycerophosphate. The measureable product can then be produced by several approaches: (1) measuring ADP produced from ATP with pyruvate kinase and LDH, with production of $NAD^+$; (2) using glycerophosphate oxidase to produce dihydroxyacetone and $H_2O_2$, followed by a typical peroxidase reaction; and (3) using glycerophosphate dehydrogenase to produce dihydroxyacetone phosphate and $NADH^+$. The major problem with these methods is the presence of endogenous glycerol in the sample. In most patients, particularly outpatients, the amount of free glycerol in the plasma is negligible, but it can be significantly increased in patients with severe hypertriglyceridemia, diabetes, chronic kidney disease, total parenteric nutrition, heparin treatment, and congenital deficiency of glycerol kinase, which can be asymptomatic in adults [108]. Contamination of the sample with glycerol from the collection tubes or hand creams and soaps can also lead to falsely high TG [2]. This problem can be obviated by use of serum blanks, in which glycerol is measured by omitting the lipase enzyme, although at the cost of reduced sample throughput. Note that lipolysis can occur in samples stored at room temperature or contaminated with heparin (which induces lipase); in these samples, falsely low TG results would occur with glycerol blanking assays.

## CONCLUSIONS

Pre-analytical sources of variation are the most frequent causes of inaccurate results in clinical chemistry. For many analytes, awareness of physiologic variation, proper patient preparation, and consistent collection practices—including time of the day, fasting, posture, tourniquet application, and collection techniques—is essential to obtain reliable results and correctly interpret any changes observed. Although good manufacturing practices, availability of reference standard materials, quality control, and proficiency testing have minimized errors and inaccuracies in the analytical phase, many methods remain subject to interferences, especially in unusual clinical situations. It is important to question all results that do not fit the clinical picture; consult laboratory specialists, manufacturer's instructions, and reference literature about possible sources of interference; and attempt to eliminate interferents by appropriate sample treatment or by retesting with another methodology that may not be subject to the same interferences.

## References


[1] Young DS. Effects of drugs on clinical laboratory tests. 3rd ed. Washington, DC: AACC Press; 2000.

[2] Young DS. Effects of preanalytical variables on clinical laboratory tests. 3rd ed. Washington, DC: AACC Press; 2007.

[3] Guder WG, Narayanan S, Wisser H, Zawta B. Diagnostic samples: from the patient to the laboratory. The impact of preanalytical variables on the quality of laboratory results. 4th ed. Weinheim, Germany: Wiley-Blackwell; 2009.

[4] Wyss M, Kaddurah-Daouk R. Creatine and creatinine metabolism. Physiol Rev 2000;80(3):1107–213.

[5] Levey AS, Coresh J, Greene T, Marsh J, Stevens LA, Kusek JW, et al. Expressing the modification of diet in renal disease study equation for estimating glomerular filtration rate with standardized serum creatinine values. Clin Chem 2007;53(4):766–72.

[6] Schwartz GJ, Munoz A, Schneider MF, Mak RH, Kaskel F, Warady BA, et al. New equations to estimate GFR in children with CKD. J Am Soc Nephrol 2009;20(3):629–37.

[7] Andreev E, Koopman M, Arisz L. A rise in plasma creatinine that is not a sign of renal failure: which drugs can be responsible? J Intern Med 1999;246(3):247–52.

[8] Maillard N, Mehdi M, Thibaudin L, Berthoux F, Alamartine E, Mariat C. Creatinine-based GFR predicting equations in renal transplantation: reassessing the tubular secretion effect. Nephrol Dial Transplant 2010;25(9):3076–82.

[9] Stevens LA, Nolin TD, Richardson MM, Feldman HI, Lewis JB, Rodby R, et al. Comparison of drug dosing recommendations based on measured GFR and kidney function estimating equations. Am J Kidney Dis 2009;54(1):33–42.

[10] Peake M, Whiting M. Measurement of serum creatinine—Current status and future goals. Clin Biochem Rev 2006;27 (4):173–84.

[11] Molitch ME, Rodman E, Hirsch CA, Dubinsky E. Spurious serum creatinine elevations in ketoacidosis. Ann Intern Med 1980;93(2):280–1.

[12] Mascioli SR, Bantle JP, Freier EF, Hoogwerf BJ. Artifactual elevation of serum creatinine level due to fasting. Arch Intern Med 1984;144(8):1575–6.

[13] Kemperman FA, Weber JA, Gorgels J, van Zanten AP, Krediet RT, Arisz L. The influence of ketoacids on plasma creatinine assays in diabetic ketoacidosis. J Intern Med 2000;248(6):511–17.

[14] Sheikh-Ali M, Karon BS, Basu A, Kudva YC, Muller LA, Xu J, et al. Can serum beta-hydroxybutyrate be used to diagnose diabetic ketoacidosis?. Diabetes Care 2008;31(4):643–7.

[15] Chadha V, Garg U, Warady BA, Alon US. Sieving coefficient inaccuracies during hemodiafiltration in patients with hyperbilirubinemia. Pediatr Nephrol 2000;15(1-2):33–5.

[16] Andersson A-C, Strandberg K, Becker C, Hägglöf-Persson A, Lundström G, Thämlitz R, et al. Interference testing of the creatinine sensor in the ABL837 FLEX analyzer. Point of Care 2007;6(2):139–43.

[17] Saenger AK, Lockwood C, Snozek CL, Milz TC, Karon BS, Scott MG, et al. Catecholamine interference in enzymatic creatinine assays. Clin Chem 2009;55(9):1732–6.

[18] Karon BS, Daly TM, Scott MG. Mechanisms of dopamine and dobutamine interference in biochemical tests that use peroxide and peroxidase to generate chromophore. Clin Chem 1998;44 (1):155–60.

[19] Rachoin JS, Daher R, Moussallem C, Milcarek B, Hunter K, Schorr C, et al. The fallacy of the BUN:creatinine ratio in critically ill patients. Nephrol Dial Transplant 2011;27:2248–55.

[20] Beier K, Eppanapally S, Bazick HS, Chang D, Mahadevappa K, Gibbons FK, et al. Elevation of blood urea nitrogen is predictive of long-term mortality in critically ill patients independent of "normal" creatinine. Crit Care Med 2011;39(2):305–13.

[21] Miyaji H, Ito S, Azuma T, Ito Y, Yamazaki Y, Ohtaki Y, et al. Effects of *Helicobacter pylori* eradication therapy on hyperammonaemia in patients with liver cirrhosis. Gut 1997;40(6):726–30.

[22] Dufour DR, Lott JA, Nolte FS, Gretch DR, Koff RS, Seeff LB. Diagnosis and monitoring of hepatic injury: I. Performance characteristics of laboratory tests. Clin Chem 2000;46 (12):2027–49.

[23] Degoutte F, Jouanel P, Begue RJ, Colombier M, Lac G, Pequignot JM, et al. Food restriction, performance, biochemical, psychological, and endocrine changes in judo athletes. Int J Sports Med 2006;27(1):9–18.

[24] Urea Cycle Disorders Conference Group. Consensus statement from a conference for the management of patients with urea cycle disorders. J Pediatr 2001;138(1 Suppl):S1–5.

[25] da Fonseca-Wollheim F. Preanalytical increase of ammonia in blood specimens from healthy subjects. Clin Chem 1990;36 (8 Pt 1):1483–7.

[26] da Fonseca-Wollheim F. Deamidation of glutamine by increased plasma gamma-glutamyltransferase is a source of rapid ammonia formation in blood and plasma specimens. Clin Chem 1990;36(8 Pt 1):1479–82.

[27] Herrera DJ, Moore S, Heap S, Preece MA, Griffiths P. Can plasma ammonia be measured in patients with acute liver disease? Ann Clin Biochem 2008;45(4):426–8.

[28] Herrera DJ, Hutchin T, Fullerton D, Gray G. Non-specific interference in the measurement of plasma ammonia: importance of using a sample blank. Ann Clin Biochem 2010;47(Pt 1):81–3.

[29] da Fonseca-Wollheim F, Heinze KG. Which is the appropriate coenzyme for the measurement of ammonia with glutamate dehydrogenase? Eur J Clin Chem Clin Biochem 1992;30 (9):537–40.

[30] Lloyd-Mostyn RH, Lord PS, Glover R, West C, Gilliland IC. Uric acid metabolism in starvation. Ann Rheum Dis 1970;29 (5):553–5.

[31] Yamamoto T, Moriwaki Y, Takahashi S. Effect of ethanol on metabolism of purine bases (hypoxanthine, xanthine, and uric acid). Clin Chim Acta 2005;356(1-2):35–57.

[32] Wood WG. The determination of uric acid in urine—Forgotten problems rediscovered in an external quality assessment scheme. Clin Lab 2009;55(9-10):341–52.

[33] Hande KR, Perini F, Putterman G, Elin R. Hyperxanthinemia interferes with serum uric acid determinations by the uricase method. Clin Chem 1979;25(8):1492–4.

[34] American Diabetes Association. Diagnosis and classification of diabetes mellitus. Diabetes Care 2010;33(Suppl. 1):S62–9.

[35] Sacks DB, Arnold M, Bakris GL, Bruns DE, Horvath AR, Kirkman MS, et al. Executive summary: guidelines and recommendations for laboratory analysis in the diagnosis and management of diabetes mellitus. Clin Chem 2011;57(6):793–8.

[36] Metzger BE, Gabbe SG, Persson B, Buchanan TA, Catalano PA, Damm P, et al. International association of diabetes and pregnancy study groups recommendations on the diagnosis and classification of hyperglycemia in pregnancy. Diabetes Care 2010;33 (3):676–82.

[37] Burtis CA, Ashwood ER, Bruns DE. Tietz textbook of clinical chemistry and molecular diagnostics. 5th ed. St. Louis, MO: Elsevier; 2012.

[38] Watson KR, O'Kell RT, Joyce JT. Data regarding blood drawing sites in patients receiving intravenous fluids. Am J Clin Pathol 1983;79(1):119–21.

[39] D'Orazio P, Burnett RW, Fogh-Andersen N, Jacobs E, Kuwa K, Kulpmann WR, et al. Approved IFCC recommendation on reporting results for blood glucose: international federation of clinical chemistry and laboratory medicine scientific division, working group on selective electrodes and point-of-care testing (IFCC-SD-WG-SEPOCT). Clin Chem Lab Med 2006;44 (12):1486–90.

[40] Tang Z, Du X, Louie RF, Kost GJ. Effects of drugs on glucose measurements with handheld glucose meters and a portable glucose analyzer. Am J Clin Pathol 2000;113(1):75–86.

[41] Gaines AR, Pierce LR, Bernhardt PA. Fatal Iatrogenic Hypoglycemia: Falsely Elevated Blood Glucose Readings with a Point-of-Care Meter Due to a Maltose-Containing Intravenous Immune Globulin Product. 2009 06/18/2009 [cited; Available from: <http://www.fda.gov/BiologicsBloodVaccines/SafetyAvailability/ucm155099.htm>.

[42] Food and Drug Administration. FDA Public Health Notification: Potentially Fatal Errors with GDH-PQQ* Glucose Monitoring Technology. 2009 03/09/2012 [cited; Available from: <http://www.fda.gov/MedicalDevices/Safety/AlertsandNotices/PublicHealthNotifications/ucm176992.htm>.

[43] Schleis TG. Interference of maltose, icodextrin, galactose, or xylose with some blood glucose monitoring systems. Pharmacotherapy 2007;27(9):1313–21.

[44] Schrier R. Renal and electrolyte disorders. Philadelphia: Lippincott; 2010.

[45] Smellie WS. Spurious hyperkalaemia. BMJ 2007;334 (7595):693–5.

[46] Kroll MH, Elin RJ. Relationships between magnesium and protein concentrations in serum. Clin Chem 1985;31 (2):244–6.

[47] Dickerson RN, Alexander KH, Minard G, Croce MA, Brown RO. Accuracy of methods to estimate ionized and "corrected" serum calcium concentrations in critically ill multiple trauma patients receiving specialized nutrition support. J Parenter Enteral Nutr 2004;28(3):133–41.

[48] Ferrari P, Singer R, Agarwal A, Hurn A, Townsend MA, Chubb P. Serum phosphate is an important determinant of corrected serum calcium in end-stage kidney disease. Nephrology 2009;14(4):383–8.

[49] Tong GM, Rude RK. Magnesium deficiency in critical illness. J Intensive Care Med 2005;20(1):3–17.

[50] Solomon R, Weinberg MS, Dubey A. The diurnal rhythm of plasma potassium: relationship to diuretic therapy. J Cardiovasc Pharmacol 1991;17(5):854–9.

[51] Clausen T. Hormonal and pharmacological modification of plasma potassium homeostasis. Fundam Clin Pharmacol 2010;24(5):595–605.

[52] Ljunghall S, Joborn H, Rastad J, Akerstrom G. Plasma potassium and phosphate concentrations—Influence by adrenaline infusion, beta-blockade and physical exercise. Acta Med Scand 1987;221(1):83–93.

[53] Adrogue HJ, Madias NE. Changes in plasma potassium concentration during acute acid–base disturbances. Am J Med 1981;71(3):456–67.
[54] Cardinal P, Allan J, Pham B, Hindmarsh T, Jones G, Delisle S. The effect of sodium citrate in arterial catheters on acid-base and electrolyte measurements. Crit Care Med 2000;28(5):1388–92.
[55] Dimeski G, Badrick T, John AS. Ion selective electrodes (ISEs) and interferences—A review. Clin Chim Acta 2010;411(5-6):309–17.
[56] Wulkan RW, Michiels JJ. Pseudohyperkalaemia in thrombocythaemia. J Clin Chem Clin Biochem 1990;28(7):489–91.
[57] Hira K, Shimbo T, Fukui T. High serum potassium concentrations after recentrifugation of stored blood specimens. N Engl J Med 2000;343(2):153–4.
[58] Mansour MMH, Azzazy HME, Kazmierczak SC. Correction factors for estimating potassium concentrations in samples with *in vitro* hemolysis: a detriment to patient safety. Arch Pathol Lab Med 2009;133(6):960–6.
[59] Hawkins RC. EDTA contamination of serum samples: common but not necessarily significant. Ann Clin Biochem 2011;48(Pt 5):478.
[60] Davidson DF. EDTA analysis on the roche MODULAR analyser. Ann Clin Biochem 2007;44(Pt 3):294–6.
[61] Kirschbaum B. Loss of carbon dioxide from serum samples exposed to air: effect on blood gas parameters and strong ions. Clin Chim Acta 2003;334(1-2):241–4.
[62] Herr RD, Swanson T. Pseudometabolic acidosis caused by underfill of Vacutainer tubes. Ann Emerg Med 1992;21(2):177–80.
[63] Waugh WH. Utility of expressing serum sodium per unit of water in assessing hyponatremia. Metabolism 1969;18(8):706–12.
[64] Nguyen MK, Ornekian V, Butch AW, Kurtz I. A new method for determining plasma water content: application in pseudohyponatremia. Am J Physiol Renal Physiol 2007;292(5):F1652–6.
[65] Sivakumar T, Chaidarun S, Lee HK, Cervinski M, Comi R. Multiple lipoprotein and electrolyte laboratory artifacts caused by lipoprotein X in obstructive biliary cholestasis secondary to pancreatic cancer. J Clin Lipidol 2011;5(4):324–8.
[66] Turchin A, Seifter JL, Seely EW. Mind the gap. N Engl J Med 2003;349(15):1465–9.
[67] Al-Musheifri A, Jones GR. Glucose interference in direct ion-sensitive electrode sodium measurements. Ann Clin Biochem 2008;45(Pt 5):530–2.
[68] Meng QH, Irwin WC, Fesser J, Massey KL. Interference of ascorbic acid with chemical analytes. Ann Clin Biochem 2005;42(Pt 6):475–7.
[69] Goldwasser P, Manjappa NG, Luhrs CA, Barth RH. Pseudohypobicarbonatemia caused by an endogenous assay interferent: a new entity. Am J Kidney Dis 2011;58(4):617–20.
[70] Lovekar S, Chen JL. A 90-year-old man with hyperphosphatemia. Am J Kidney Dis 2011;57(2):342–6.
[71] Dalal BI, Brigden ML. Factitious biochemical measurements resulting from hematologic conditions. Am J Clin Pathol 2009;131(2):195–204.
[72] Loh TP, Saw S, Sethi SK. Hyperphosphatemia in a 56-year-old man with hypochondrial pain. Clin Chem 2010;56(6):892–5.
[73] Toftegaard M, Rees SE, Andreassen S. Evaluation of a method for converting venous values of acid–base and oxygenation status to arterial values. Emerg Med J. 2009;26(4):268–72.
[74] Herrington WG, Nye HJ, Hammersley MS, Watkinson PJ. Are arterial and venous samples clinically equivalent for the estimation of pH, serum bicarbonate and potassium concentration in critically ill patients? Diabet Med 2012;29(1):32–5.
[75] Yildizdas D, Yapicioglu H, Yilmaz HL, Sertdemir Y. Correlation of simultaneously obtained capillary, venous, and arterial blood gases of patients in a paediatric intensive care unit. Arch Dis Child 2004;89(2):176–80.
[76] Kelly AM, Kerr D, Middleton P. Validation of venous $pCO_2$ to screen for arterial hypercarbia in patients with chronic obstructive airways disease. J Emerg Med 2005;28(4):377–9.
[77] Thelin OP, Karanth S, Pourcyrous M, Cooke RJ. Overestimation of neonatal $PO_2$ by collection of arterial blood gas values with the butterfly infusion set. J Perinatol 1993;13(1):65–7.
[78] Lu JY, Kao JT, Chien TI, Lee TF, Tsai KS. Effects of air bubbles and tube transportation on blood oxygen tension in arterial blood gas analysis. J Formos Med Assoc 2003;102(4):246–9.
[79] Collinson PO, John CM, Gaze DC, Ferrigan LF, Cramp DG. Changes in blood gas samples produced by a pneumatic tube system. J Clin Pathol 2002;55(2):105–7.
[80] Beaulieu M, Lapointe Y, Vinet B. Stability of $PO_2$, $PCO_2$, and pH in fresh blood samples stored in a plastic syringe with low heparin in relation to various blood-gas and hematological parameters. Clin Biochem 1999;32(2):101–7.
[81] Knowles TP, Mullin RA, Hunter JA, Douce FH. Effects of syringe material, sample storage time, and temperature on blood gases and oxygen saturation in arterialized human blood samples. Respir Care 2006;51(7):732–6.
[82] Cengiz M, Ulker P, Meiselman HJ, Baskurt OK. Influence of tourniquet application on venous blood sampling for serum chemistry, hematological parameters, leukocyte activation and erythrocyte mechanical properties. Clin Chem Lab Med 2009;47(6):769–76.
[83] Gourlain H, Buneaux F, Borron SW, Gouget B, Levillain P. Interference of methylene blue with CO-oximetry of hemoglobin derivatives. Clin Chem 1997;43(6 Pt 1):1078–80.
[84] Smith I, Kumar P, Molloy S, Rhodes A, Newman PJ, Grounds RM, et al. Base excess and lactate as prognostic indicators for patients admitted to intensive care. Intensive Care Med 2001;27(1):74–83.
[85] Mikami A, Ohde S, Deshpande G, Mochizuki T, Otani N, Ishimatsu S. Can we predict arterial lactate from venous lactate in the emergency department? Critical Care 2012;16(Suppl 1):P259.
[86] Jones AE, Leonard MM, Hernandez-Nino J, Kline JA. Determination of the effect of *in vitro* time, temperature, and tourniquet use on whole blood venous point-of-care lactate concentrations. Acad Emerg Med 2007;14(7):587–91.
[87] Astles R, Williams CP, Sedor F. Stability of plasma lactate *in vitro* in the presence of antiglycolytic agents. Clin Chem 1994;40(7 Pt 1):1327–30.
[88] Brindley PG, Butler MS, Cembrowski G, Brindley DN. Falsely elevated point-of-care lactate measurement after ingestion of ethylene glycol. CMAJ 2007;176(8):1097–9.
[89] National Institutes of Health. Household Products Database: Ethylene Glycol. 2011 October, 2011 [cited 2012; Available from: <http://hpd.nlm.nih.gov/cgi-bin/household/search?queryx = 107-21-1&tbl = TblChemicals&prodcat = all>.
[90] Porter WH, Crellin M, Rutter PW, Oeltgen P. Interference by glycolic acid in the Beckman synchron method for lactate: a useful clue for unsuspected ethylene glycol intoxication. Clin Chem 2000;46(6 Pt 1):874–5.
[91] Chiu WW. Glycolate interference on radiometer ABL analyser is influenced by age of lactate electrode. CMAJ 2007;176: 1097–9.
[92] Jorens PG. Accuracy of point-of-care measurements. CMAJ. 2007;177(9):1070.
[93] Ostrea Jr. EM, Ongtengco EA, Tolia VA, Apostol E. The occurrence and significance of the bilirubin species, including delta bilirubin, in jaundiced infants. J Pediatr Gastroenterol Nutr 1988;7(4):511–16.

[94] Lo SF, Doumas BT. The status of bilirubin measurements in U. S. laboratories: why is accuracy elusive? Semin Perinatol 2011;35(3):141–7.

[95] Al Riyami N, Zimmerman AC, Rosenberg FM, Holmes DT. Spurious hyperbilirubinemia caused by naproxen. Clin Biochem 2009;42(1-2):129–31.

[96] Dasgupta A, Langman LJ, Johnson M, Chow L. Naproxen metabolites interfere with certain bilirubin assays. Am J Clin Pathol 2010;133(6):878–83.

[97] Ahlfors CE, Wennberg RP, Ostrow JD, Tiribelli C. Unbound (free) bilirubin: Improving the paradigm for evaluating neonatal jaundice. Clin Chem 2009;55(7):1288–99.

[98] Mora S, Rifai N, Buring JE, Ridker PM. Comparison of LDL cholesterol concentrations by Friedewald calculation and direct measurement in relation to cardiovascular events in 27,331 women. Clin Chem 2009;55(5):888–94.

[99] Grundy SM, Cleeman JI, Merz CN, Brewer Jr. HB, Clark LT, Hunninghake DB, et al. Implications of recent clinical trials for the National Cholesterol Education Program Adult Treatment Panel III guidelines. Circulation 2004;110(2):227–39.

[100] Ridker PM. Fasting versus nonfasting triglycerides and the prediction of cardiovascular risk: do we need to revisit the oral triglyceride tolerance test? Clin Chem 2008;54 (1):11–13.

[101] Fresco C, Maggioni AP, Signorini S, Merlini PA, Mocarelli P, Fabbri G, et al. Variations in lipoprotein levels after myocardial infarction and unstable angina: the LATIN trial. Ital Heart J. 2002;3(10):587–92.

[102] Pitt B, Loscalzo J, Ycas J, Raichlen JS. Lipid levels after acute coronary syndromes. J Am Coll Cardiol 2008;51(15):1440–5.

[103] Henkin Y, Crystal E, Goldberg Y, Friger M, Lorber J, Zuili I, et al. Usefulness of lipoprotein changes during acute coronary syndromes for predicting postdischarge lipoprotein levels. Am J Cardiol 2002;89(1):7–11.

[104] Gylling H, Hallikainen M, Nissinen MJ, Simonen P, Miettinen TA. Very high plant stanol intake and serum plant stanols and non-cholesterol sterols. Eur J Nutr 2010;49(2):111–17.

[105] Perera NJ, Burns JC, Perera RS, Lewis B, Sullivan DR. Adjustment of direct high-density lipoprotein cholesterol measurements according to intercurrent triglyceride corrects for interference by triglyceride-rich lipoproteins. J Clin Lipidol 2010;4(4):305–9.

[106] Miller WG, Myers GL, Sakurabayashi I, Bachmann LM, Caudill SP, Dziekonski A, et al. Seven direct methods for measuring HDL and LDL cholesterol compared with ultracentrifugation reference measurement procedures. Clin Chem 2010;56(6):977–86.

[107] Murali MR, Kratz A, Finberg KE. Case records of the Massachusetts General Hospital. Case 40-2006: a 64-year-old man with anemia and a low level of HDL cholesterol. N Engl J Med 2006;355(26):2772–9.

[108] Walmsley TA, Potter HC, George PM, Florkowski CM. Pseudohypertriglyceridaemia: a measurement artefact due to glycerol kinase deficiency. Postgrad Med J. 2008;84(996):552–4.

[109] Cascavilla N, Falcone A, Sanpaolo G, D'Arena G. Increased serum bilirubin level without jaundice in patients with monoclonal gammopathy. Leuk Lymphoma 2009;50(8):1392–4.

CHAPTER

# 9

# Challenges in Routine Clinical Chemistry Analysis

## Proteins and Enzymes

*Jorge Sepulveda*

**Columbia University Medical Center, New York, New York**

## INTRODUCTION

In this chapter, the source of inaccurate results in commonly ordered enzymes and other protein biomarker assays, excluding hormones and tumor markers, is addressed. See Chapters 11 and 12 for sources of errors in endocrine tests and measurement of tumor markers, respectively. In this chapter, issues with albumin and total protein measurements, commonly ordered as part of the liver panel or the comprehensive metabolic panel, are addressed. In addition, two protein biomarkers used in the evaluation of cardiovascular disease—cardiac troponin and B-type atrial natriuretic peptide—are also discussed.

## ALBUMIN AND TOTAL PROTEIN

Measurement of total protein and albumin is useful in several clinical situations, including the following:

1. Diagnosing hypoalbuminemia resulting from acute or chronic inflammation, liver insufficiency, renal or gastrointestinal losses, protein malnutrition, etc.
2. Albumin levels can be of value in calculating free electrolyte (calcium), drug (phenytoin), and hormone (testosterone and cortisol) concentrations.
3. Screening for monoclonal gammopathies by detecting an increase in globulins as measured by the difference between total protein and albumin. For more sensitive detection of monoclonal gammopathies, direct measurement of various immunoglobulins, serum and urine protein electrophoresis, and immunofixation techniques are indicated (see Chapter 18).
4. High albumin, total protein, and hematocrit levels can suggest acute dehydration, whereas a proportional decrease of albumin, total protein, and hematocrit can suggest hemodilution.

### Pre-Analytical Issues

As mentioned previously (Chapter 8), recumbency leads to fluid shifts from the extravascular to the intravascular compartment, resulting in mild hemodilution and a decrease in protein concentrations up to 10%. This decrease can be more pronounced in patients predisposed to edema (e.g., cardiac and liver insufficiency). Similarly, prolonged application of a tourniquet can lead to hemoconcentration with an up to 8% increase in albumin and total protein. Similar effects are seen with all the protein biomarkers described in this chapter. Pregnancy is also associated with hemodilution and a decrease in total protein by as much as 1.3 g/dL in the third trimester.

### Analytical Issues and Interferences

Total protein is commonly measured in plasma or serum by the biuret method, which takes advantage of the reaction of peptide bonds in proteins with cupric ions at alkaline pH to form a colored product that absorbs at 540–560 nm. Therefore, this method assumes a constant ratio of peptide bonds to mass of proteins, which is not true for all proteins present in

*Accurate Results in the Clinical Laboratory.*
DOI: http://dx.doi.org/10.1016/B978-0-12-415783-5.00009-8

serum but is sufficiently accurate for clinical use. The range of measurement is typically from 1 to 10 g/dL; therefore, it cannot be used to measure low levels of protein in body fluids such as urine and cerebrospinal fluid. For these situations, turbidimetric methods using protein precipitation with trichloroacetic acid, sulfosalicylic acid, or benzethonium chloride or dye-binding methods using Coomassie blue or pyrogallol red-molybdate are employed.

Dextran, carbenicillin, hemoglobin, and bilirubin can have mild positive interferences with some total protein methods, whereas lipemia at high levels (>3+), methylbenzethonium, aminophenazone, sulfasalazine, and fluorescein may cause negative interference. Total protein and albumin measurements are particularly susceptible to error due to poor mixing of thawed samples.

Albumin is usually measured in automated analyzers by its capacity to bind a dye with resulting change in absorption. Two dyes are commonly used—bromcresol green (BCG) and bromcresol purple (BCP). Although generally comparable, albumin values measured with BCG tend to be falsely elevated (as much as 1 g/dL) in hypoalbuminemic patients undergoing hemodialysis, compared to an albumin immunoassay [1], and falsely low above the normal range due to nonspecific binding to globulins [2]. On the other hand, for most patients, BCP results are more comparable to immunoassay. In a few individuals with advanced renal failure, a uremic toxin (CMPF; 3-carboxy-4-methyl-5-propyl-2-furanpropionic acid) can bind to albumin and compete with BCP [3], thus causing artificially low results. This interference appears to be present particularly in patients undergoing hemodialysis and is not observed in patients undergoing peritoneal dialysis or patients with renal failure who are not subjected to dialysis [4]. Similarly, albumin-bound bilirubin interferes with the BCP method [5]. Despite these problems, the BCP method is recommended for general use because it is more specific for albumin and better correlates with immunoassay or gold standard nephelometry, even in hemodialysis patients. The following formula for converting BCP to BCG results works well in both renal insufficiency and nonrenal patients:

$$\mathrm{Alb_{BCG}} = 0.55 + \mathrm{Alb_{BCP}}(\text{in g/dL})$$

This formula is applicable even when serum albumin is less than 3 g/dL [6]. Formulas for calcium correction using BCP-measured albumin are different from the standard formula described in Chapter 7; for example (calcium in mg/dL, albumin in g/dL) [7,8],

$$\text{Calcium (corrected)} = \text{total calcium (measured)} - 0.4 \times \text{albumin} + 1.2$$

The BCG method, but not BCP, is negatively affected by the presence of high levels (280 U/mL) of heparin [9], clofibrate, and phenylbutazone in the sample. High-dose penicillin-G causes negative interference in the BCP method [10]. Some BCP methods show up to 25% positive interference by 400 mg/dL of free hemoglobin, whereas others do not show this interference.

## ALANINE AND ASPARTATE AMINOTRANSFERASES ANALYSIS

Aspartate aminotransferase (AST), also known as serum glutamate-oxaloacetate transaminase (SGOT), and alanine aminotransferase (AST), also known as serum glutamate-pyruvate transaminase (SGPT), are important enzymes commonly measured to identify and follow-up liver damage. Aminotransferases are key enzymes in amino acid metabolism, and they work by transferring amino groups between the various amino acids and 2-oxoglutarate ($\alpha$-ketoglutarate), which functions as an acceptor of the amino group to yield glutamate in a reaction involving pyridoxal-5′-phosphate (P5P), derived from vitamin $B_6$, as a coenzyme. The reaction is reversible, and glutamate can function as an amine donor to various ketoacids to form amino acids—for example, alanine from pyruvate with AST and aspartate from oxaloacetate with ALT.

The AST enzyme is coded by two different genes, GOT1 and GOT2. The GOT1 gene codes for c-AST, which is present in the cytosol and most abundant in cardiac and skeletal muscle and red blood cells, whereas the GOT2 gene codes for m-AST, a mitochondrial isoform of AST most abundant in hepatocytes. The catalytic activities of the two AST isoforms are similar, but they can be differentiated by immunoassays or by modified enzymatic assays because m-AST is highly susceptible to proteolysis by proteinases K. The m-AST isoform has a longer half-life in serum (~87 hr) than does c-AST. Whereas c-AST predominates in the plasma of healthy individuals (~88–95% of the AST activity) and patients with mild liver disease, m-AST is predominant in liver parenchyma (~80% of total AST) and increases more than c-AST in acute hepatitis and more severe liver disease [11]. It is likely that the cytoplasmic location allows c-AST to leak into the systemic circulation with mild, reversible cellular damage, whereas the more abundant m-AST requires injury to the mitochondria (e.g., associated with liver ischemia or severe necrosis). In alcoholic hepatitis, the ratio of m-AST to c-AST is also typically elevated, perhaps indicating a particular susceptibility of mitochondria to ethanol-induced damage, whereas in uncomplicated viral hepatitis the predominant lesion is in the cell membrane, and therefore c-AST predominates.

The GPT gene codes for ALT1, the cytosolic isoform of ALT (c-ALT), whereas the related gene GPT2 codes for the mitochondrial isoform, ALT2 (m-ALT). ALT1 is expressed mainly in liver and kidney, and it is weakly expressed in intestine, heart, skeletal muscle, and salivary glands. It is the predominant isoform in serum from healthy individuals, whereas ALT2 is the predominant ALT isoform expressed in skeletal muscle and heart [12,13]. In rodents, ALT2 appears to be present in liver and seems to undergo a more significant increase than ALT1 after liver damage or steatosis [14,15], but it does not appear to be expressed in human liver [12,13,16]. ALT2 is also expressed in adrenal gland, adipose tissue, neurons, endocrine pancreas, and prostate, where it is induced by androgens [13]. ALT1, but not ALT2, is inducible in the liver by peroxisome proliferator-activated receptor (PPAR) $\alpha$ agonists, such as fenofibrate [17]. A similar induction of ALT activity is seen in conditions associated with increased neoglucogenesis, such as fasting, insulin resistance, type 2 diabetes, and the metabolic syndrome [18,19]. Obesity can increase levels of AST and ALT by approximately 40–50%. It is unclear whether the available enzymatic assays show any differential reactivity with ALT1 versus ALT2, although recombinant purified ALT1 had 15-fold higher activity than ALT2 in one study [20]. The different cellular localization, tissue specificity, and inducibility of the two ALT isoforms raise the possibility that isoform-specific immunoassays may have better clinical usability than the currently used enzymatic assay. Despite the presence in other tissues, isolated elevations of ALT are rarely seen in the absence of liver disease.

Both AST and ALT are sensitive markers of liver damage, with ALT being more specific due to its restricted tissue expression and AST being more sensitive because it is approximately two or three times more abundant in liver than ALT. The ratio of AST/ALT (de Ritis ratio) is commonly used to help differentiate various causes of liver damage (Table 9.1). Various parameters affect the ratio:

1. Extent of extra-liver disease: Because AST is widely expressed and ALT is predominantly from liver, any conditions involving systemic injury, especially involving skeletal muscles, heart, kidney, lung, pancreas, intestine, or erythrocytes, will increase the ratio. Hemolysis occurring *in vitro* will also increase AST with little effect on ALT.
2. Severity of damage: In normal individuals and patients with reversible injury involving the cellular membranes, thought to occur through cytoplasmic blebbing, c-AST and c-ALT predominate in roughly equal amounts. Severe injury to mitochondria, such as in liver necrosis or degeneration due to drug toxicity or ischemia, leads to a massive release of the much more abundant m-AST, which increases the de Ritis ratio.
3. Elimination of ALT compared to AST: In general, the half-life of ALT is much higher (~2 days) than that of AST (< 24 hr); therefore, although AST may predominate during an acute episode of toxic or

**TABLE 9.1** Conditions Affecting AST and ALT Levels

| Condition | Level of AST | AST/ALT Ratio | Comment |
|---|---|---|---|
| Healthy | <40 U/L | 0.7–1.4 | |
| Acetaminophen toxicity | >10,000 U/L | >1[a] | Slower increase in 24–48 hr |
| Ischemic injury | >10,000 U/L | >1[a] | Rapid rise in <24 hr |
| Alcoholic hepatitis | <300 U/L | >2 | |
| Autoimmune hepatitis | <10 × ULN | <2 | |
| Biliary obstruction | <5 × ULN | <2 | Return to normal in 72 hr |
| Cholangitis | <10 × ULN | <2 | |
| Cirrhosis | <10 × ULN | >1 | Levels may be normal or decreased |
| Hemochromatosis | <10 × ULN | <2 | |
| Hepatitis, viral, acute | >300 U/L, 10–100 × ULN | <1 | AST remains elevated for 7–14 days; ALT decays ~10%/day, remains elevated 27 ± 16 days |
| Hypo- and hyperthyroidism | <3 × ULN | <2 | |
| Infectious mononucleosis | 50–800 U/L | <2 | |
| Liver metastases | <5 × ULN | >1 | |
| Nonalcoholic steatosis | <10 × ULN | <2 | |
| Primary biliary cirrhosis | <10 × ULN | >2 | |
| Reye's syndrome | | >2 | |
| Wilson's disease | <10 × ULN | >2 | |

[a] *AST/ALT > 1 initially, then reverts to <1.*
ULN, upper limit of reference range.

ischemic liver injury, during the recovery phase the ratio may become less than 1. Diseases leading to m-AST release also prolong the elevation of AST activity because m-AST has a longer half-life than c-AST.

4. The extent of liver synthetic function affects ALT more than AST because ALT is more liver-specific. Situations with reduced liver function, such as in chronic cirrhosis, tend to have elevated de Ritis ratios due to lower ALT production. In some cases of chronic liver disease, as well as in fulminant liver failure, ALT concentrations may decrease below the reference range.
5. Any induction of ALT expression—for example, treatment with PPARα agonists or insulin resistance: Obese patients and those with the metabolic syndrome tend to have higher levels of ALT (40–50%) compared to AST [21].
6. The amount of P5P in circulation: Using assays that do not supplement the reaction with exogenous P5P, this becomes a rate-limiting factor, with the ALT enzyme being more susceptible to low levels of P5P seen, for example, with chronic alcoholism, malnutrition, and renal failure, the latter due to P5P binders in the plasma. Also, m-AST is more susceptible to loss of P5P than is c-AST, and assays using P5P supplementation tend to have higher m-AST/c-AST ratios.

Typically, patients with alcoholic hepatitis or severe alcohol-induced liver disease such as cirrhosis have a de Ritis ratio greater than 2 [22], whereas in uncomplicated acute or chronic liver disease the ratio is usually less than 2. Possible mechanisms involved in elevation of the ratio in severe alcoholic liver disease include mitochondrial oxidative damage, decreased liver ALT production, decreased P5P, and extrahepatic damage such as alcoholic myopathy. The ratio is also helpful in distinguishing alcoholic from non-alcoholic liver steatosis, especially when coupled with other determinations such as red cell mean corpuscular volume and γ-glutamyl transferase [23], although the ratio tends to be high in advanced cases of alcoholic and non-alcoholic cirrhosis [24]. In contrast, acute viral hepatitis tends to be associated with de Ritis ratios less than 1 and aminotransferase levels greater than 300 IU/L. Patients with a de Ritis ratio less than 0.6 in acute viral hepatitis had better prognoses than those with ratios greater than 1.2 [25].

Many drugs affect AST and ALT levels, and monitoring of transaminase elevation is routine when evaluating or following therapy with potentially hepatotoxic drugs. Some drugs also increase ALT and AST levels by inducing its expression. Commonly used drugs that may affect ALT and AST levels include acetaminophen, aminoglycosides, angiotensin-converting enzyme (ACE) inhibitors, anticonvulsants, cephalosporins, clofibrate, fenofibrate, fluconazole, ganciclovir, griseofulvin, isoniazid, isoniazid, macrolides, nicotinic acid, nonsteroidal anti-inflammatory drugs, omeprazole, opiates, statins, sulfonamides, and sulfonylureas.

The following are additional patient-related factors to take into consideration when evaluating ALT or AST measurements:

1. ALT and, more important, AST levels in normal individuals tend to parallel muscle mass and therefore are 10–20% higher in males, African Americans, and athletes.
2. Intense exercise can increase AST and ALT levels, usually less than threefold, subsiding within 1–7 days [26].
3. Both hypothyroidism and hyperthyroidism tend to increase AST levels.
4. Levels of AST and ALT increase with age until approximately 40 years and then decline approximately 10–20% to the age of 60 years. Both AST and ALT tend to be higher in elderly individuals—approximately 10% after age 70 years [27].
5. Circadian variation affects ALT by approximately 45%, peaking in the afternoon and a nadir at night, whereas AST varies by less than 10%, although some studies show no significant variation [27].
6. Both AST and ALT are elevated in obese patients (40–50% higher) and those with hypertriglyceridemia or high-sugar or high-fat diets [27,28].
7. Hospitalization by itself and even placebo treatment may be associated with mild elevations in transaminases, particularly ALT [29,30].

*CASE REPORT* A 32-year-old female was referred to a hepatologist for possible liver biopsy [31]. She had a history of AST levels between 120 and 500 U/L persisting for more than 7 years without any symptoms, family history, or significant past medical history of liver disease, except for a limited acute Epstein–Barr virus infection at age 14 years, which caused a reversible 10-fold elevation of ALT and AST. Medical examination was normal, an abdominal ultrasound was unremarkable, and serum levels of other liver enzymes including ALT, hepatitis and autoimmune serology, ferritin, ceruloplasmin, and thyrotropin were normal. The patient denied taking any medications or herbal supplements, except for oral contraceptives. Discontinuation of the contraceptive resulted in no change in AST levels. Before performing the liver biopsy, the patient's serum sample was treated with protein A Sepharose, which binds immune complexes, resulting in a change in AST activity from 459 to 57 U/L, whereas another healthy control sample showed no change. The elevation in AST was attributed to macro-AST, and the patient was subject to no further studies.

Although rarely seen, macro-AST is more common in patients with liver disease, particularly those with autoimmune, viral, or neoplastic liver disease, but it can be found in healthy individuals and persist for several years [31,32]. Isolated increases in AST without ALT or creatine kinase elevations suggest macro-AST, which can be eliminated by polyethylene glycol precipitation, ultrafiltration, dialysis, and exclusion or affinity chromatography.

### Specimen Processing

Serum or plasma should be separated from cells to avoid leakage of enzymes from red blood cells, which increases markedly after 24 hr at room temperature. The ALT and AST activities tend to decay when specimens are stored at room temperature, particularly in specimens with low P5P levels, because apoenzymes are more susceptible to degradation than are holoenzymes [33]. Because the enzymes are irreversibly degraded, lower activities may be seen even with methods using P5P supplementation. However, several studies show only minimal decay of the enzymes at room temperature, when refrigerated, or when frozen for at least 3 days [27,34–36]. Repeated freezing and thawing may decrease ALT activity.

### Methodology

Transaminase levels are measured by supplying 2-oxoglutarate and the specific amino acid (L-aspartate for AST and L-alanine for ALT), which generate glutamate and the respective ketoacids. In some methods (e.g., Vitros), P5P is supplied to enhance the activity and remove circulating P5P levels as an analytical variable. A signal is generated by dehydrogenases specific to the ketoacid generated in the transaminase reaction. For AST, the oxaloacetate is converted to malate by malate dehydrogenase, with generation of $NAD^+$, whereas the pyruvate resulting from the ALT reaction is converted to lactate by lactate dehydrogenase (LDH). Because the method uses serum blanking and all the components are in vast excess to usual plasma levels of the corresponding endogenous substances, interference by pyruvate, lactate, and LDH is uncommon. In rare cases with very high levels of LDH (>12,000 U/L), the preincubation step during serum blanking may consume all the NADH substrate and cause a very low result or an analyzer flag. A 20-fold dilution of the sample will mitigate this problem. In the Beckman SYNCHRON assay, pyruvate levels of less than 6 mg/dL (normal levels are < 1 mg/dL) show minimal interference. However, pyruvate can be generated from lactate and cause interference with ALT measurements in samples in which plasma was in contact with cells for more than 48 hr, especially in samples with high LDH levels, which converts lactate to pyruvate [36]. Other interferences include heparin greater than 50 U/L with the Vitros method [37] and also significant lipemia, hyperbilirubinemia, and hemolysis (particularly AST, which is 10–30 times more abundant in red cells than in plasma) [27].

## γ-GLUTAMYL TRANSFERASE AND ALKALINE PHOSPHATASE ANALYSIS

γ-Glutamyl transferase (GGT) and alkaline phosphatase (ALP) are often measured to determine if a patient has intra- or extrahepatic cholestasis because these enzymes are abundant in the plasma membrane of hepatocytes, particularly in the biliary canaliculi and the basolateral surface of biliary epithelia. ALP is the product of one of three genes—ALPI, expressed in the intestine; ALPP, expressed in placenta; and ALPL, expressed in most tissues, particularly in liver, kidney, and bone, in which it can be differentially glycosylated, resulting in changes in protein conformation, stability, and clearance. For example, ALPP is heat stable at 65°C, the bone form of ALPL, derived from osteoblasts, is extensively glycosylated and heat labile, and liver ALP is heat stable at 56°C. Heterogeneity of glycosylation affects clearance of the various forms of ALP by the liver galactose receptor [38].

Most plasma ALP activity derives from liver, with up to 50% from bone and a small amount of intestinal form. The release of ALP into the circulation during cholestatic disease may be mediated by increased phospholipase activity induced by bile acids because the enzyme is anchored to the membrane by a glycosyl phosphatidyl inositol (GPI) moiety [39]. The various isoforms of ALP can be separated by electrophoresis, isoelectric focusing, wheat germ lectin precipitation, high-performance liquid chromatography, or specific immunoassays; however, it is simpler to determine GGT levels because this enzyme is increased in liver but not bone disease. The liver form of ALPL is elevated in cholestatic disease (up to 12 times normal), particularly in extrahepatic obstruction, whereas parenchymal liver disease, such as viral, alcoholic, or autoimmune hepatitis, usually increases ALP less than 3-fold. Liver metastases can also result in high levels of ALP, depending on the extent of the infiltration. The bone ALPL is elevated in growing children and in disorders with osteoblastic activity and bone remodeling, such as Paget's disease (up to 25-fold), rickets and osteomalacia (2- to 4-fold), and in osteogenic bone cancer (very high levels). Elevated ALPP is seen in pregnancy (2- or 3-fold) and some tumors (Regan isoenzymes), and ALPI is increased in liver cirrhosis and blood group O or B secretors.

GGT is abundantly expressed in kidney, liver, pancreas, and intestine, but most of the circulating activity in healthy individuals is of hepatic origin. Like ALP, GGT is cleared by the liver, and various glycosylation isoforms are cleared at different rates by the liver asialoglycoprotein receptor [40]. Interestingly, GGT is inducible in the liver by a variety of xenobiotic agents, most notably microsomal inducers such as the anticonvulsants phenytoin and phenobarbital, oral contraceptives, cimetidine, furosemide, and methotrexate, and in patients with chronic ethanol exposure, who can have hepatic GGT levels approximately 3-fold higher than normal [41]. Chronic alcoholism may also be associated with increased release of GGT into the circulation because serum levels can be as high as 5–30 times the upper limit of normal. In general, it is thought that GGT levels reflect oxidative stress and glutathione consumption in the liver, which makes sense given the role of GGT in the metabolism of glutathione, a key antioxidative agent in the liver. The same liver and biliary diseases that cause increased ALP will also cause GGT elevations to a similar extent, although GGT tends to be more sensitive than ALP to cholestasis, with elevations averaging 12-fold in greater than 93% of cases compared to 3-fold for ALP [42]. Elevations of GGT originating from other organs can be seen in acute and chronic pancreatitis or pancreatic cancer and mildly in chronic kidney disease, diabetes, acute myocardial infarction, hyperthyroidism, rheumatoid arthritis, and lung disease [42,43].

The following additional pre-analytical factors should be taken into consideration when interpreting ALP and GGT measurements [27,34,42]:

- Normal levels of ALP increase in growing children with a peak approximately fivefold the adult levels at approximately age 13 years, mostly due to the bone isoenzyme. The levels then decay until age 25 years when they reach relatively stable levels, similar in men and women. GGT levels increase linearly with age until approximately 40–50 years and are approximately 40–50% higher in men than in women.
- In African Americans, GGT activity is approximately twice that of Caucasians, whereas ALP is 10–15% higher.
- Obesity increases ALP and GGT by approximately 20–30%, and in individuals with a body mass index greater than 30 $m^2$, levels of GGT may be 50% higher.
- Diurnal variation of less than 10% in ALP and GGT levels has been reported with activities highest in the morning and lowest in the afternoon.
- Ingestion of fatty meals can increase intestinal ALP as much as 30 U/L for approximately 12 hr, especially in Lewis-positive secretors of blood types O or B.
- Starvation can decrease GGT, whereas meals cause an initial decrease followed by a small increase.
- Smoking increases GGT by 10–50% and ALP by approximately 10%.
- Oral contraceptives increase GGT and decrease ALP by approximately 20%, whereas pregnancy increases ALP 200–300% and decreases GGT by approximately 20–30%.
- ALP and GGT activities are generally stable in separated serum or plasma at room temperature, 4C or frozen for at least 7 days. Frozen serum or plasma should be thawed for at least 18 hours at room temperature to achieve full ALP enzyme reactivation.

## Analytical Issues

Most commercially available methods for ALP analysis use the chromogenic substrate 4-nitrophenyl phosphate, which upon loss of the phosphate group catalyzed by ALP under alkaline conditions forms a yellow product that absorbs at 405–410 nm. Massive blood transfusion, plasmapheresis, or citrate anticoagulant inhibits ALP activity because citrate chelates the required zinc and magnesium ion co-factors. Citrate, oxalate, and fluoride also inhibit GGT activity up to 15% and should be avoided. The GGT assay also uses a chromogenic substrate such as $\gamma$-glutamyl-*p*-nitroaniline, which is cleaved by GGT into *p*-nitroaniline, measured at 405–120 nm, and a glutamyl group that is transferred to an acceptor such as glycylglycine.

Hemolysis and heparin may interfere with some ALP and GGT methods, whereas others show no such interferences. In patients with hepatobiliary diseases, macro-GGT may occur as complexes with IgA autoantibodies and apolipoproteins A and B [44]. Binding to lipoproteins may increase GGT half-life from approximately 9 to 20 hr [40]. IgM paraproteins in heparinized plasma may also interfere with some GGT assays [45]. Macro-ALP has also been rarely observed, without clear clinical significance, although increased prevalence in patients with inflammatory bowel disease and chronic peritoneal dialysis has been suggested [46,47].

## AMYLASE AND LIPASE ANALYSIS

These enzymes are measured to assess pancreatic disease, and there is considerable debate regarding the utility of amylase versus lipase for this purpose. In general, lipase appears more sensitive and specific for pancreatic disease. Amylase is highly expressed in salivary glands from the AMY1 gene (S-amylase), whereas pancreatic amylase (P-amylase) originates from the AMY2A and AMY2B genes. There is high interindividual variation in the levels of amylase in part due to copy number variations of the AMY1 gene (from 2 to 15 diploid copies), which may be related to the ability to adapt to high-starch diets and minimize the risk of insulin resistance and diabetes [48]. The amylase enzyme hydrolyzes the α-1,4-glucosidic linkage between two glucose residues in polysaccharides such as starch and amylose.

Pancreatic lipase is expressed from the PNLIP gene, specifically in exocrine pancreas. Other lipase genes code for enzymes, such as intestinal or hepatic lipase and lipoprotein lipase, which are not measured appreciably with current pancreatic lipase assays. Pancreatic lipase hydrolyzes triglycerides, releasing the fatty acids at positions 1 and 3 and resulting in a 2-monoacylglycerol, and it requires activation by bile acids and the presence of a small protein co-factor secreted by the pancreas, colipase.

Amylase is also weakly expressed in a variety of other tissues, and elevations can be seen in a variety of conditions, including the following:

1. Abdominal lesions, leading to increased absorption of P-amylase from the intestinal lumen, including various gastric and intestinal lesions, appendicitis, peritonitis, abdominal trauma, and abdominal aortic aneurysm.
2. Ovary and fallopian tubes express S-amylase, and elevations can be seen in serous and mixed serous and mucinous ovary cysts and tumors, pelvic inflammatory disease, and ruptured ectopic pregnancy.
3. Miscellaneous other conditions can lead to hyperamylasemia, including pregnancy, head injury, septic shock, burns, bulimia, anorexia nervosa, acute alcohol or organophosphate intoxication, diabetic ketoacidosis, cardiac surgery, pheochromocytoma, thymoma, and some lung and colon tumors. Elevations can also be seen in approximately 20% of patients subjected to a variety of abdominal or extra-abdominal surgical interventions.

In healthy individuals, approximately 40–50% of serum amylase activity is of pancreatic origin, whereas 55–60% derives from salivary glands. Amylase is a small protein (54–62 kDa) and is partially eliminated by glomerular filtration; therefore, it is present in normal urine and elevations can be seen in renal failure.

Peter Wilding and colleagues [49] first described persistent elevations of amylase due to the formation of a high-molecular-weight complex with immunoglobulins, thus avoiding glomerular filtration and significantly increasing the half-life of plasma amylase activity. Macroamylase is probably the most common of the macro-enzymes causing abnormal test results, often without any subjacent clinical abnormalities, and it is present in 2–5% of patients with elevated amylase, usually of S-type. One helpful indication for the presence of macroamylase is calculation of the amylase-to-creatinine clearance ratio (ACR):

$$\text{ACR} = \frac{\text{urine amylase} \times \text{serum creatinine}}{\text{serum amylase} \times \text{urine creatinine}}$$

An ACR ratio less than 1% suggests macroamylasemia. Whereas ratios greater than 5% suggest acute pancreatitis, renal disease, or diabetic ketoacidosis, measurement of urinary amylase and calculation of the ACR ratio is too nonspecific for clinical use and should be restricted to cases of suspected macroamylasemia. Although much rarer than macroamylasemia, cases of macrolipase have also been reported, including complexes with IgG, IgA, and $\alpha_2$-macroglobulin [50].

In contrast, pancreatic lipase measurements offer increased specificity and sensitivity for pancreatic disease and in general should replace amylase determinations for the diagnosis and monitoring of pancreatic disease. At least one prominent institution has discontinued offering amylase for pancreatic disease evaluation [51]. In acute pancreatitis, both amylase and lipase typically rise 4–8 hr after the onset of the symptoms and peak at 24–48 hr. Due to its shorter half-life, amylase returns to normal range within 3–5 days, whereas lipase tends to persist for 1 or 2 weeks. In most studies, the sensitivity of lipase for pancreatitis varies from 85 to 100%, whereas amylase averages 85–95%. In cases of chronic pancreatic disease with pancreatic insufficiency, production of pancreatic enzymes may be impaired. Because levels of lipase in pancreatic tissue are approximately four times higher than amylase, normal or decreased levels of amylase are seen more often than lipase in chronic pancreatic insufficiency.

Although much less common than with amylase, nonpancreatic sources of lipase also exist, mainly related to conditions involving the gastrointestinal and biliary tracts, such as cholecystitis and gastroenteritis. Lipase, a 48-kDa protein, is also filtered by the glomeruli and may be increased in renal insufficiency, but in contrast to amylase, it is fully reabsorbed by the renal tubules and not normally found in urine.

In general, extensive investigation of elevations of amylase or lipase less than threefold the upper limit of normal in patients with nonspecific abdominal pain is of low diagnostic yield and probably not cost-effective [52]. Some individuals show levels of amylase and lipase that range from 1 to 16 times the upper limit of normal, with marked day-to-day fluctuations, without any symptoms or evidence of pancreatic disease. This condition has been termed Gullo's syndrome and in some case seems to have a hereditary component [53]. In approximately 50% of cases of asymptomatic hyperamylasemia, a significant pathological pancreatic or extrapancreatic disease can be present; therefore, a thorough diagnostic workup may be indicated [54].

### Analytical Issues

Amylase is usually measured by a variety of methods. One of the simplest methods, the Ortho Vitros amylase slide, uses the natural substrate starch covalently linked to a dye, dried in the spreading layer, which upon addition of the sample is hydrolyzed by amylase to release the dye into the reagent layers, where it is measured by reflectance spectrophotometry. Some liquid chemistry methods use a series of enzymatic reactions to convert the product of an amylase reaction (usually one or more maltose residues) to a measureable product (usually NADH from a glucose-6-phosphate dehydrogenase reaction). One problem with these assays is that the stoichiometry of the reaction may shift depending on the different maltose oligomers present. Other assays use a variety of chromogenic maltose oligomers to release a measurable dye (typically 4-nitrophenyl) upon hydrolysis by amylase with or without further hydrolysis by $\alpha$-glycosidases. The International Federation of Clinical Chemistry (IFCC) reference method uses a heptamaltose substrate covalently linked to a protecting ethylidene residue to avoid slow hydrolysis of the intact substrate by $\alpha$-glycosidase in the reagent mix.

A variety of substrates and reaction conditions have also been used for pancreatic lipase measurement. For example, one Beckman SYNCHRON method uses the following sequence of reactions leading to formation of a quinone dye from the oxidative coupling of *N*-ethyl-*N*-(2-hydroxy-3-sulfopropyl)-*m*-toluidine (TOOS) with 4-aminoantipyrine (4-AAP):

1. Pancreatic lipase: 1,2-Diglyceride $+ H_2O \rightarrow$ 2-monoglyceride + fatty acid
2. Monoglyceride lipase: 2-Monoglyceride $+ H_2O \rightarrow$ glycerol + fatty acid
3. Glycerol kinase: Glycerol + ATP $\rightarrow$ glycerol-3-phosphate + ADP
4. Glycerophosphate oxidase: Glycerol-3-phosphate $+ O_2 \rightarrow$ dihyroxyacetone-phosphate $+ H_2O_2$
5. Horseradish peroxidase: $2\ H_2O_2$ + 4-AAP + TOOS $\rightarrow$ quinone diimine dye $+ 4\ H_2O$

The Ortho Vitros method uses a similar sequence of enzymes, but the substrate is 1-oleoyl-2,3-diacetylglycerol and the final acceptor of $H_2O_2$ oxidation is a leuco dye. The Vitros substrate may be more specific for intestinal than pancreatic lipase and may be interfered by lipoprotein lipase in heparinized samples [55], but the addition of colipase and bile salts in the current assay makes it very specific for pancreatic lipase. The Panteghini method uses a completely different approach resulting in direct formation of a colored dye (methylresorufin) from a synthetic substrate (1,2-*O*-dilauryl-rac-glycero-3-glutaic acid-(6′-methylresorufin)-ester) [56].

Few analytical interferences have been reported for amylase and lipase assays. Icodextrin, a polysaccharide used in peritoneal dialysis, may inhibit amylase and interfere with some amylase assays, particularly those using 4-nitrophenyl substrates. Notably, other assays unaffected by icodextrin showed true amylase levels to be reduced in patients undergoing peritoneal dialysis [57]. An inhibitor present with pancreatitis associated with severe hypertriglyceridemia interfered with starch-iodine amylase assays, but most modern assays, including the starch-based Vitros method, are unaffected by triglyceride levels of at least 800 mg/dL. Very high levels of glycerol, greater than 160 mg/dL (normal levels $<0.6$ mg/dL), can cause interferences in some lipase methods, particularly those that use diglyceride substrates [27,58].

## LACTATE DEHYDROGENASE ANALYSIS

LDH is a ubiquitously expressed enzyme with limited clinical sensitivity and specificity for a variety of diseases involving tissue damage. The main clinical indications for measuring LDH include monitoring of disease activity in cases of hemolytic and megaloblastic anemias, thrombotic thrombocytopenic purpura, lung disease, and some tumors, particularly lymphomas and germinal cell cancers. Some clinicians use LDH elevations as an early indicator of tissue damage. However, due to its nearly ubiquitous expression, LDH is elevated in a large number of conditions and shows poor specificity for any particular clinical situations. For example, intense exercise can result in elevations in LDH of 30–50% [27]. LDH fractionation can be used to somewhat improve specificity because there are five isoforms with some tissue specificity: LDH1 predominates in heart, kidneys, and red cells; LDH5 predominates in liver and skeletal muscle; and LDH2, −3, and −5 are present in a variety of different tissues. Fractionation can occasionally be helpful to identify the source of an LDH elevation, but due to considerable overlap of tissue expression, it does not provide sufficient diagnostic accuracy for significant clinical use. Measuring LDH in other body fluids may help point to exudates due to infectious, inflammatory, necrotic, or malignant processes. Because body fluids typically contain much lower amounts of LDH than serum, a ratio of body fluid/plasma LDH greater than 0.6 or a fluid LDH level greater than 200 IU/L is more consistent with severe hemorrhage or an exudate [59]. For example, increased LDH levels in the cerebrospinal fluid may indicate bacterial or viral meningitis if hemorrhage or loss of blood−brain barrier function can be excluded.

LDH catalyzes the conversion of lactate to pyruvate, whereas $NAD^+$ is reduced to NADH or vice versa. Reaction rates are monitored by measuring absorbance at 340 nm. Either the forward (lactate to pyruvate) or the reverse reaction (pyruvate to lactate) can be used, although the IFCC reference assay uses the forward reaction because it is less susceptible to substrate exhaustion and loss of linearity.

The most important pre-analytical consideration is avoidance of hemolysis because red cells contain LDH levels approximately 50- to 150-fold higher than those of plasma, and even minor amounts of hemolysis can significantly increase measured plasma LDH levels. For this reason, LDH is a sensitive marker of *in vivo* hemolysis. Similarly, serum contains up to 40% higher levels of LDH than plasma due to release from platelets during clotting. However, serum may be a preferred specimen, especially if platelet-poor plasma cannot be consistently achieved. Plasma or serum should be separated from cells in less than 1 hr to avoid leakage of LDH from cellular elements. The LDH activity is unstable in refrigerated or −20°C frozen serum; therefore, it is preferable to store specimens at room temperature, which maintains LDH activity for up to 3 days. Storage of specimens on heparinized tubes with gel-based plasma separator results in an approximately 5% loss of activity [60].

## CREATINE KINASE ANALYSIS

Creatine kinase (CK) catalyzes the transfer of phosphate between ATP and creatine to generate creatine phosphate and ADP or vice versa. Creatine phosphate serves as an important reservoir of high-energy phosphates in striated muscle, brain, retina, inner ear, spermatozoa, and, to a lesser degree, smooth muscle, which are tissues that can consume ATP rapidly. Logically, CK is highly abundant in these tissues, particularly heart and skeletal muscle, and is used as the most sensitive marker for skeletal muscle damage, or rhabdomyolysis. In cases of acute rhabdomyolysis, CK levels as high as 1000-fold normal can be seen. Table 9.2 lists causes of rhabdomyolysis and other conditions associated with elevations of CK in plasma, including pre-analytical variables.

Various forms of CK exist, including the mitochondrial enzyme, which predominantly catalyzes the transfer of phosphate from ATP to creatine, and the "cytosolic" form, which catalyzes the reverse reaction, regenerating ATP from phosphocreatine. Three isoenzymes of cytosolic CK occur, with different tissue expression, each composed of a dimer of two subunits—either B (brain type), coded by the CKB gene, or M (muscle type), coded by CKM. The CK-BB homodimer is abundant in brain and at lower levels in a variety of tissues; the CK-MM homodimer is abundant (98% of CK activity) in skeletal muscle; and CK-MB heterodimers are most abundant in heart, in which they comprise approximately 25−30% of CK activity. In the past, elevations of CK-MB were used in the diagnosis of myocardial infarction, but it has been essentially superseded with the more sensitive and specific cardiac troponin measurements.

Pre-analytical variables affecting CK measurements are listed in Table 9.2. Note that most of these cause relatively minor changes in CK and that acute rhabdomyolysis is associated with large increases in plasma CK. However, they may become relevant when following the activity of chronic muscle disease such as polymyositis or Duchenne dystrophy.

Methods for CK analysis are based on either the forward reaction (creatine → phosphocreatine) or the reverse reaction (phosphocreatine → creatine). The cytosolic enzyme favors the reverse reaction and therefore it proceeds approximately sixfold faster. The ATP generated can be measured by coupling to a pair of reactions, first using hexokinase to phosphorylate glucose to glucose-6-phosphate and then using glucose-6-phosphate dehydrogenase to generate NADPH and 6-phosphoglutonate. Alternatively, in the Vitros dry chemistry method, ATP produced by CK can be measured by a series of reactions involving glycerol kinase catalyzed production of L-α-glycerophosphate, which is then converted to dihydroxyacetone phosphate by glycerophosphate oxidase, with production of $H_2O_2$. Subsequently, the $H_2O_2$ is measured by oxidation of a leuco dye. Because adenylate kinase, which is present in large amounts in erythrocytes, can convert ADP to ATP in the absence of CK, modern assays include

TABLE 9.2 Causes of CK Elevation

| |
|---|
| **Rhabdomyolysis** |
| Crush syndrome, blunt trauma, extensive surgery, intramuscular injections, bee stings |
| Seizures, tetanus, exercise, childbirth, status asthmaticus, intense coughing |
| Malignant hyperthermia, prolonged hypothermia |
| Ischemic injury (compression as a result of a coma, sickle cell disease, intoxication) |
| Metabolic, inflammatory, and dystrophic myopathies |
| Toxins: quails, snake and insect venoms, buffalo fish, tetanus, toluene, gasoline, carbon monoxide, halothane |
| Drugs: alcohol, cocaine, heroin, amphetamine, phencyclidine, colchicines, corticosteroids, fibrates, isoniazide, statins, neuroleptics, opioids, zidovudine, etc. |
| **Cardiac damage** |
| Myocardial infarction |
| Other causes (see Table 9.3) |
| **Brain damage** |
| Extensive brain infarction |
| Acute psychotic reaction |
| Head injury, subarachnoid hemorrhage |
| **Other tissue damage** |
| Hypoxic shock |
| Cancer of prostate, bladder, gastrointestinal tract, lung, breast, uterus, testis |
| Pulmonary embolism, gastrointestinal ischemic necrosis |
| **Other causes of CK elevation** |
| Hypothyroidism |
| Reye's syndrome |
| Sepsis, typhoid fever |
| Smoking |
| CK levels correlate with muscle mass: higher in males and African Americans, lower in females and elderly individuals |
| Last trimester of pregnancy |
| Neonates and infants during first year of life |
| **Pre-analytical factors** (for details, see refs. [27,34,96]) |
| Prolonged tourniquet application (~10% increase after 6 min) |
| Hemolysis |
| Repeated freeze and thaw |
| Exposure to 56°C |

AMP and adenosine pentaphosphate as adenylate kinase inhibitors. Because CK is reversibly inactivated by oxidation of cysteinyl residues, *N*-acetylcysteine is added to reverse any oxidized sulfhydryl groups. In addition to the reversible oxidation of sulfhydryl groups, progressive irreversible inactivation of CK can slowly occur, and samples stored at room temperature for more than 2 days, or refrigerated for more than 7 days, show significant decreases [27,61]. Although small degrees of hemolysis (<500 mg/dL) can be tolerated, because the red cells contain no CK and adenylate kinase is inactivated, severe hemolysis releases high enough concentrations of ADP, phosphates, glucose-6-phosphate, and other enzymes that act on the reagent components to cause significant interferences.

## CARDIAC TROPONIN ANALYSIS

Troponins (I and T) are key components of striated muscle sarcomeres, where they regulate muscle contraction in response to calcium levels. Troponin exists as a trimeric complex: Troponin C senses calcium levels and induces a conformation change in troponin I, which in turn induces a change in tropomyosin to allow actin to interact with myosin for muscle contraction; troponin T anchors the complex to tropomyosin. Three genes code for the three isoforms of troponin I: slow twitch skeletal muscle TNNI1, fast twitch skeletal muscle TNNI2, and cardiac TNNI3. Similarly, three genes code for troponin T: the skeletal muscle TNNT1 and TNNT2 and the cardiac TNNT3. Sufficient differences exist in primary amino acid sequence at the N-terminus of cardiac troponin I and troponin T to be able to separate the cardiac isoforms from skeletal troponins with immunoassays. Cardiac troponin assays were first used in clinical practice in 1992, when monoclonal antibody sandwich immunoassays for cardiac troponin T (cTnT) and cardiac troponin I (cTnI) applicable to routine clinical detection of myocardial injury were first described [62,63]. Since then, cardiac troponin assays (cTnT and cTnI) have become the standard for detection of acute myocardial infarction, and in 2007 the European Society of Cardiology, American College of Cardiology Foundation, American Heart Association, and World Heart Federation Task Force established elevation of cardiac troponins as one key defining element in the diagnosis of acute myocardial infarction (AMI) [64].

Myocardial infarction is defined as loss of cardiac myocytes (necrosis) caused by prolonged ischemia, and AMI is a type of myocardial infarction occurring between 6 hr and 7 days after the ischemic event. An AMI can be diagnosed by a rise and/or fall in markers of myocardial damage in the blood, including cardiac troponins, with at least one measurement above the 99th percentile of normal population and one of the following criteria:

1. Clinical: Ischemic symptoms
2. Pathology: Myocardial cell death
3. Electrocardiography
   a. Evidence of myocardial ischemia (new ST-segment changes or new left bundle branch block (LBBB))
   b. Evidence of loss of electrically functioning tissue (Q waves)
4. Imaging
   a. Reduction or loss of tissue perfusion
   b. Cardiac wall motion abnormalities.

Acute myocardial infarctions can be classified into five clinical categories [64]. The criteria listed previously pertain to the first two clinical types of AMI, most commonly seen in emergency departments:

Type 1: Spontaneous AMI resulting from ischemia due to atherosclerotic coronary artery disease
Type 2: AMI secondary to ischemia due to increased oxygen demand (e.g., arrhythmias and hypertension) or decreased oxygen supply (e.g., anemia, coronary spasm, embolism, and hypotension)

In addition, AMI diagnoses can be made in the following clinical situations:

Type 3: Sudden, unexpected cardiac death, often with history of symptoms of ischemia, accompanied by electrocardiographic evidence of ischemia or evidence of fresh thrombus by angiography or autopsy
Type 4: Post percutaneous coronary intervention increase in biomarker greater than threefold the 99th percentile of the reference population
Type 5: Coronary artery bypass grafting with an increase in biomarker greater than threefold the 99th percentile of the reference population and new Q waves or new LBBB, angiographic evidence of coronary occlusion, or imaging evidence of new loss of viable myocardium.

The more broad category of acute coronary syndrome (ACS; Figure 9.1) is defined as any constellation of clinical symptoms that are compatible with AMI, and it includes unstable angina, without elevation of cardiac biomarkers, and AMI, characterized by elevation of biomarkers, further classified by the electrocardiographic presentation as follows:

1. ST-elevation AMI (STEMI)
2. Non-ST-elevation AMI (NSTEMI)
3. Q-wave AMI, which can be diagnosed only after 24 hr.

In STEMI, rapid electrocardiogram-based diagnosis is critical for timely percutaneous intervention, and troponin measurements are not essential. In NSTEMI, the diagnosis relies on troponin measurements. In cases in which rapid intervention may be considered, fast turnaround time for reporting troponin results is essential. In addition to cardiac ischemia and AMI, any condition resulting in myocardial injury can potentially elevate cardiac troponin levels. A comprehensive list of causes of myocardial damage and troponin elevation is presented in Table 9.3.

A common misconception is that renal failure causes troponin elevation in the absence of myocardial injury, for example, by decreasing elimination of troponin or troponin fragments. This does not appear to be the case, and it is well demonstrated that patients with renal insufficiency and troponin elevation do have myocardial damage and increased risk for

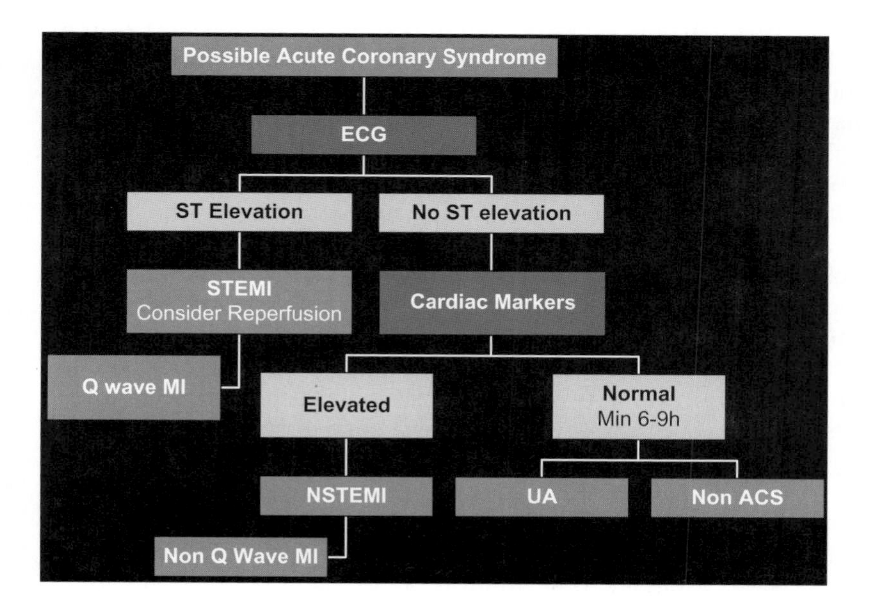


FIGURE 9.1 Classification of acute coronary syndromes (ACS).
MI, myocardial infarction; NSTEMI, non-ST elevation MI; STEMI, ST elevation MI; UA, unstable angina.

**TABLE 9.3** Causes of Cardiac Troponin Elevation

| |
|---|
| **Acute myocardial infarction** |
| **Direct trauma to the heart** |
| Ablation, angioplasty, cardiac surgery, implantable device discharges, pacemakers, cardioversion |
| **Myocardial toxins** |
| Adriamycin, 5-fluorouracil, cocaine, amphetamines, Herceptin, snake venoms |
| **Wall stress and subendocardial hypoperfusion** |
| Pulmonary embolism, severe systemic or pulmonary hypertension, aortic valve disease, aortic dissection, hypertrophic cardiomyopathy, acute or chronic heart failure, tachy- or bradyarrhythmias, strenuous exercise, shock and severe hypotension |
| **Systemic disease with myocardial injury** |
| Severe renal failure, respiratory failure, sepsis, acute central nervous system damage (stroke, hemorrhage), rhabdomyolysis, extensive burns (total body surface area > 30%); other critically ill patients, especially with diabetes, hypothyroidism, transplant vasculopathy |
| **Other causes of cardiac injury** |
| Cardiac inflammation (pericarditis, myocarditis) |
| Infiltrative cardiomyopathy (amyloidosis, hemochromatosis, sarcoidosis, scleroderma) |
| Coronary vasospasm, including Takotsubo ballooning stress cardiomyopathy |

adverse cardiovascular events [65–67]. However, decreased elimination of cTnT in heart failure patients with renal insufficiency may play a role in the degree of troponin elevation originating from myocardial damage [68].

Another misconception is that re-expressed fetal skeletal troponin T will cause elevations in cardiac troponin T measurements in skeletal muscle disease. Although this may have been the case due to nonspecificity of the antibodies against cTnT, assays have been optimized to exclude cross-reactivity against all skeletal troponins, and expression of cTnT has been excluded in a variety of muscle diseases [69]. It is possible that rarely in certain muscle diseases, true re-expression of cTnT may occur and cause elevation even with high-sensitivity fourth-generation cTnT assays [70]. However, high-sensitivity cTnI levels remain normal in these patients.

In general, troponin elevations originating from myocardial damage not associated with acute ischemia do not follow a typical rise and fall observed during an AMI: With high-sensitivity assays, troponin elevations can be seen as soon as 2 or 3 hr after initiation of the ischemic event, peak at approximately 24 hr, and remain elevated for 7–14 days. The troponin levels at 24–72 hr correlate with infarct size, although this is better evaluated with imaging studies. Re-infarction can be detected even in the presence of troponin elevation by noting an increase in serial troponin levels.

A key concept in measuring troponins is functional sensitivity, or the minimum concentration of troponin that can be detected by an assay with imprecision of 10% or less. The consensus definition of AMI requires that troponin assays have a functional sensitivity below the upper limit of the reference range, which is defined as the 99th percentile of a healthy population [64]. It is important to note that the latest generation cardiac troponin immunoassays are able to reliably detect plasma levels of troponins in a majority of normal individuals—that is, the functional sensitivity is significantly below the 99th percentile ("high-sensitivity" assays). It is also important to note that excluding apparently healthy individuals with cardiovascular risk factors or increased B-natriuretic peptide in order to determine the 99th percentile will improve the sensitivity for cardiovascular risk detection in patients with ACS [71].

It is well-established that patients with troponin elevations above the reference range are at short- and long-term risk for cardiovascular events and may benefit from therapeutic intervention [64]. It is also possible that even in apparently healthy individuals, detectable levels of troponin may indicate minor cardiac damage and increased risk of adverse cardiovascular events in the future [72]. This use of cardiac troponins for risk assessment should be distinguished from its diagnostic value for detection of AMI.

The following are helpful criteria to interpret troponin results and distinguish evolving AMI from non-ischemic causes of troponin elevation:

1. The pretest probability of ACS and coronary artery disease (CAD) should be taken into consideration: In a patient with nonspecific symptoms and low pretest probability of cardiovascular disease, a cardiac troponin level in the reference range is reassuring and may allow the patient to be discharged from the emergency department.
2. In a patient with a high probability of ACS (typical chest pain, history of CAD, etc.), if an initial cardiac troponin is negative or very mildly elevated, an increase greater than 30% in cTnI levels measured 6 hr later is consistent with AMI [73]. Other criteria based on delta changes have been proposed; for example, an absolute change of at least 9.2 ng/L (0.0092 ng/mL) with a high-sensitivity cTnI assay had the highest area under the receiver operator curve for diagnosis of NSTEMI [74]. Importantly, these delta criteria must be optimized for each assay, time window, and patient population [75].

3. Most non-ischemic causes of myocardial damage result in relatively stable, mild elevations of troponins, whereas rapidly increasing, significant elevations of cardiac troponin (e.g., cTnI > 0.5 ng/mL) are virtually diagnostic of AMI.
4. Increments in troponins may not be seen if patients present with AMI more than 24 hr after development of symptoms. In some cases, a decrease in serial troponin results may indicate that the patient suffered an AMI one or more days ago.

Reversible ischemia induced by exercise does not appear to result in significant troponin release, although there is some controversy regarding whether minor amounts of intact troponins or degradation products can be released with reversible ischemia or even tachycardia-induced cardiac myocyte stretching [76]. After heavy exercise, such as in marathon runners, troponin levels are typically elevated, in some cases above 0.5 ng/mL, but it is unclear whether they indicate some degree of cardiac damage [77] and may instead reflect cardiomyocyte membrane leakage associated with mild cardiac inflammation [78]. Rapid normalization of troponin levels 24 hr after a race is a good indicator that the troponin elevation represents a benign, reversible process [79].

## Analytical Issues

Cardiac troponins are released in the circulation in various forms; including holo-complex of cTnI/cTnT/cTnC, free intact molecules, and a variety of degradation products and post-translationally modified forms, including phosphorylated molecules [80,81]. This heterogeneity complicates the development of standardized reference materials and comparability of the troponin assays because there are differences in the ability of the various antibodies to recognize the multiple forms [80]. When the classification of patients according to elevations above the 99th percentile was compared between four different assays, 3 or 4% were differently classified between cTnI assays and 15–17% between cTnI and cTnT assays [82].

Potential pitfalls and interferences in troponin measurements are common with other immunoassays. See Chapter 6 for an in-depth discussion of immunoassay design and interferences observed in immunoassays. For example, heterophilic antibodies, rheumatoid factors, and anti-animal antibodies in the serum of certain individuals can cause persistent falsely elevated or depressed troponin results, which can vary depending on the specific immunoassay used [83]. Similarly, autoantibodies directed against cardiac troponins can cause falsely decreased (up to 20-fold) measurements [84]. Cardiac troponin I macrocomplexes with immunoglobulins causing persistent elevation in the absence of cardiac pathology have been described [85].

Severe hemolysis typically interferes with immunoassays, particularly those using fluorometric or electrochemiluminescence detection. Some assays are more susceptible to particles such as fibrin strands or bubbles in the sample. Because measurements in serum and plasma are not always comparable, the same sample type should be used consistently in serial measurements. Samples should be promptly analyzed or refrigerated because troponin is relatively unstable at room temperature and susceptible to *in vitro* degradation by proteases. Because the proteolytic fragments may or may not be detected by a particular assay, the manufacturer's instructions for specimen processing should be followed.

# B-TYPE NATRIURETIC PEPTIDE ANALYSIS

The B-type natriuretic peptide (BNP) is an important regulator of renal sodium excretion, as well as cardiac function, through the action of BNP-receptors (NPRA and NPRB) coupled to GMP cyclase. Injection of recombinant BNP (Nasiritide) can be used clinically to improve cardiac function in patients with heart failure (HF). In normal hearts, BNP is produced in atrial but not ventricular myocytes as pro-$BNP_{1-108}$ (intact pro-BNP), which is stored in atrial secretory granules. In situations of atrial dilation or increased wall tension, intact pro-BNP is cleaved by proteases into an inactive N-terminal fragment containing amino acid residues 1–76 (NT-pro-$BNP_{1-76}$) and the bioactive peptide $BNP_{32}$, which contains the C-terminal 32 amino acids. Once secreted in circulation, BNP has a very short half-life of 13–20 min due to cellular uptake, renal excretion, and inactivation by endothelial-bound neutral endopeptidases. In contrast, NT-pro-BNP has a half-life of 25–70 min, mostly determined by renal excretion; therefore, it is measured at higher levels in plasma. In the case of renal insufficiency, the levels of BNP and more extensively NT-pro-BNP are elevated.

BNP can be produced by the ventricular myocytes in situations of chronic myocardial stress (stretch and wall tension) and myocardial remodeling associated with AMI and cardiac failure. Therefore, measurements of BNP or NT-pro-BNP have been used in the evaluation of patients with AMI and cardiac insufficiency. A good correlation between the levels of natriuretic peptides and the severity of HF (American Heart Association classes I–IV) can be observed, and natriuretic peptides can be in the diagnosis and follow-up of acute and chronic HF. However, these assays should be complementary and not replace other clinical diagnostic tools for HF diagnosis and therapy. In general, BNP and NT-pro-BNP are best used when the cause of acute dyspnea is clinically unclear because in patients with obvious HF or very low probability of HF, the predictive values of natriuretic peptide measurements are poor [86]. When clinical uncertainty exists, natriuretic peptide values are most helpful if levels are markedly elevated—for example, BNP > 500 pg/mL (suggesting HF) or < 100 pg/mL (unlikely to be acute HF). Corresponding diagnostic values for NT-pro-BNP are >450 (in those younger than 50 years), >900 (in those 50–75 years old), or >1800 pg/mL (in those older than 75 years) for ruling in HF versus <300 pg/mL for exclusion [87]. Intermediate levels have the lowest predictive values, and the patient may be a candidate for further evaluation with specialist assessment and echocardiography depending on the clinical assessment.

Because BNP responds to ventricular wall stress, it can be increased in many other conditions other than cardiac failure, including the following:

1. Other cardiac diseases
   a. Acute coronary syndrome
   b. Valvular heart disease
   c. Left ventricular hypertrophy with or without arterial hypertension
   d. Atrial fibrillation
   e. Inflammatory cardiac disease
2. High-output stress (such as tachycardia, hypoxemia, anemia, sepsis, hyperthyroidism, hyperaldosteronism, Cushing's syndrome, and advanced liver cirrhosis with ascites)
3. Pulmonary disease and pulmonary vascular conditions (such as embolism, severe pulmonary hypertension, chronic obstructive pulmonary disease, and cor pulmonale)
4. Acute or chronic renal failure (eGFR < 60 mL/min/1.73 $m^2$)
5. Severe neurological disease, such as subarachnoid hemorrhage, stroke, and trauma.

Other useful guidelines for the use of BNP or NT-pro-BNP include the following [88]:

1. Given high physiologic intraindividual variation (~30–50%) and analytical imprecision (up to 15%), only changes greater than 40–100% compared to the previous result are clinically significant [89–91].
2. The test should not be ordered more than once per week because real changes related to pathophysiology or therapy require time to occur.
3. Patients with chronic HF may have elevated natriuretic peptide levels and acute dyspnea of noncardiac origin.
4. The levels of natriuretic peptide elevations do not differentiate between HF due to left ventricular systolic dysfunction and HF with preserved ejection fraction, although values tend to be lower in patients with diastolic HF.
5. Cutoffs for the use of NT-pro-BNP have been published by the International Collaborative for NT-proBNP Study (ICON) [87]:
   a. For monitoring therapy and risk stratification in HF, it is reasonable to collect a baseline sample and a second prior to discharge for risk assessment and monitor therapeutic effectiveness. Patients are at lower risk with reductions greater than 50%. These measurements identify those in need of more aggressive management. For example, the threshold with the best balance of sensitivity and specificity for 1-year mortality is approximately 1000 pg/mL for NT-pro-BNP. An admission NT-pro-BNP concentration greater than 5180 pg/mL is strongly predictive of death by 76 days.
   b. If estimated glomerular filtration rate is less than 60 mL/min/1.73 $m^2$, a higher threshold (e.g., NT-pro-BNP < 1200 pg/mL) is best for exclusion of HF.
   c. For risk stratification in pulmonary embolism, values below 1000 pg/mL indicate good prognosis. Persistent elevations of NT-pro-BNP greater than 7500 pg/mL after 24 hr or less than 50% decrease indicate right ventricular dysfunction and a poor prognosis.
   d. For risk stratification in ACS, values less than 1115 pg/mL indicate a high probability for recovery of left ventricular function. Values greater than 1170 pg/mL for men and greater than 2150 pg/mL for women identify high-risk patients.

False-negative results—that is, normal natriuretic peptide levels despite the presence of HF—can be observed in the following situations:

1. Obesity, diuretics, ACE inhibitors, β-blockers, aldosterone receptor blockers, and aldosterone antagonists can reduce levels and decrease sensitivity for HF.
2. Patients with end-stage HF can have low levels of circulating natriuretic peptides.
3. Lack of elevation despite acute hemodynamic compromise and/or severe acute pulmonary congestion can occur in patients who present within the first hour of the onset of pulmonary edema caused by acute mitral regurgitation, in patients with preserved ejection fractions, and in those with constrictive pericarditis without intrinsic heart disease.

### Pre-Analytical Considerations

Due to poor stability in serum, BNP assays should be performed in EDTA anticoagulated plasma and samples should not be stored for more than 4 hr at room temperature. NT-pro-BNP measurements can be performed in either serum or plasma, and the sample is stable for at least 72 hr at room temperature. Serum samples show approximately 10% higher levels of NT-pro-BNP compared to EDTA-anticoagulated samples [92].

### Analytical Issues

Like troponin, natriuretic peptide measurements are immunoassays and suffer from the same limitations discussed in Chapter 6. There is considerable heterogeneity in the exact peptides that are measured by the various BNP and NT-pro-BNP assays [93]. "BNP" assays typically measure intact pro-$BNP_{1-108}$ in addition to $BNP_{32}$, although it is unclear if this affects the diagnostic ability of BNP assays. In plasma, $BNP_{32}$ quickly loses the N-terminal three amino acids to yield $BNP_{3-32}$, which may not be fully measured with some BNP assays. No other proteolytic fragments appear to circulate in patients with HF [93,94].

All "NT-pro-BNP" assays measure NT-pro-$BNP_{1-76}$ and pro-$BNP_{1-108}$. However, intact NT-pro-$BNP_{1-76}$ has not been detected in circulation; instead, several other N-proteolytic fragments originating from cleavage at both the N-terminus and the C-terminus of NT-pro-$BNP_{1-108}$ can be present in plasma and more extensively in serum. For this reason, the best assays for pro-BNP should be directed at the central amino acids. Current U.S. Food and Drug Administration-licensed assays use a sandwich immunoassay approach, with the capture antibody directed against amino acids 1–21 and the detection antibody directed against amino acids 39–50.

Glycosylation of the central portion of pro-BNP has been described, which may affect interaction with antibodies directed against amino acids 39–50 and significantly increase renal excretion. Oligomerization of pro-BNP peptides can occur through leucine zipper motifs at the N-terminus of pro-BNP [95], potentially hiding or exposing epitopes and changing the half-life and therefore the levels of measured BNP or NT-pro-BNP.

## CONCLUSIONS

Proteins and enzymes are useful biomarkers, and only the most commonly assayed proteins in clinical laboratory medicine were discussed in this chapter. Enzyme assays are subject to a variety of interferences depending on the particular approach used for product detection, whereas immunoassays for biomarkers suffer from common pitfalls, including interference by heterophilic antibodies, rheumatoid factors, autoantibodies, hemolysis, lipemia, and fibrin particles and bubbles. Occasionally, falsely elevated levels of protein biomarkers or enzymes occur due to complexes with immunoglobulins, the most common examples being macroamylase and macro-CK. As with all other tests, particular attention to each manufacturer's specifications for pre-analytical and analytical test characteristics is essential to avoid errors and explain discrepancies with other clinical parameters.

### References


[1] Carfray A, Patel K, Whitaker P, Garrick P, Griffiths GJ, Warwick GL. Albumin as an outcome measure in haemodialysis in patients: the effect of variation in assay method. Nephrol Dial Transplant 2000;15(11):1819–22.

[2] Xu Y, Wang L, Wang J, Liang H, Jiang X. Serum globulins contribute to the discrepancies observed between the bromocresol green and bromocresol purple assays of serum albumin concentration. Br J Biomed Sci 2011;68(3):120–5.

[3] Mabuchi H, Nakahashi H. Underestimation of serum albumin by the bromcresol purple method and a major endogenous ligand in uremia. Clin Chim Acta 1987;167(1):89–96.

[4] Beyer C, Boekhout M, van Iperen H. Bromcresol purple dye-binding and immunoturbidimetry for albumin measurement in plasma or serum of patients with renal failure. Clin Chem 1994;40(5):844–5.

[5] Ihara H, Nakamura H, Aoki Y, Aoki T, Yoshida M. Effects of serum-isolated vs synthetic bilirubin-albumin complexes on dye-binding methods for estimating serum albumin. Clin Chem 1991;37(7):1269–72.

[6] Clase CM, St Pierre MW, Churchill DN. Conversion between bromcresol green- and bromcresol- purple measured albumin in renal disease. Nephrol Dial Transplant 2001;16(9):1925–9.

[7] Labriola L, Wallemacq P, Gulbis B, Jadoul M. The impact of the assay for measuring albumin on corrected ("adjusted") calcium concentrations. Nephrol Dial Transplant 2009;24(6):1834–8.

[8] Jain A, Bhayana S, Vlasschaert M, House A. A formula to predict corrected calcium in haemodialysis patients. Nephrol Dial Transplant 2008;23(9):2884–8.

[9] Meng QH, Krahn J. Lithium heparinised blood-collection tubes give falsely low albumin results with an automated bromcresol green method in haemodialysis patients. Clin Chem Lab Med 2008;46(3):396–400.

[10] Ono M, Aoki Y, Masumoto M, Hotta T, Uchida Y, Kayamori Y, et al. High-dose penicillin G-treatment causes underestimation of serum albumin measured by a modified BCP method. Clin Chim Acta 2009;407(1–2):75–6.

[11] Wada H, Kamiike W. Aspartate aminotransferase isozymes and their clinical significance. Prog Clin Biol Res 1990;344:853–75.

[12] Yang RZ, Blaileanu G, Hansen BC, Shuldiner AR, Gong DW. cDNA cloning, genomic structure, chromosomal mapping, and functional expression of a novel human alanine aminotransferase. Genomics 2002;79(3):445–50.

[13] Lindblom P, Rafter I, Copley C, Andersson U, Hedberg JJ, Berg A-L, et al. Isoforms of alanine aminotransferases in human tissues and serum—Differential tissue expression using novel antibodies. Arch Biochem Biophys 2007;466(1):66–77.

[14] Yang R-Z, Park S, Reagan WJ, Goldstein R, Zhong S, Lawton M, et al. Alanine aminotransferase isoenzymes: molecular cloning and quantitative analysis of tissue expression in rats and serum elevation in liver toxicity. Hepatology 2009;49(2): 598–607.

[15] Liu R, Pan X, Whitington PF. Increased hepatic expression is a major determinant of serum alanine aminotransferase elevation in mice with nonalcoholic steatohepatitis. Liver Int 2009;29 (3):337–43.

[16] Glinghammar B, Rafter I, Lindstrom AK, Hedberg JJ, Andersson HB, Lindblom P, et al. Detection of the mitochondrial and catalytically active alanine aminotransferase in human tissues and plasma. Int J Mol Med 2009;23(5):621–31.

[17] Thulin P, Rafter I, Stockling K, Tomkiewicz C, Norjavaara E, Aggerbeck M, et al. PPARα regulates the hepatotoxic biomarker alanine aminotransferase (ALT1) gene expression in human hepatocytes. Toxicol Appl Pharmacol 2008;231(1):1–9.

[18] Koike T, Miyamoto M, Oshida Y. Alanine aminotransferase and γ-glutamyltransferase as markers for elevated insulin resistance-associated metabolic abnormalities in obese Japanese men younger than 30 years of age. Obes Res Clin Pract 2010;4 (1):e73–9.

[19] Iacobellis G, Moschetta A, Buzzetti R, Ribaudo MC, Baroni MG, Leonetti F. Aminotransferase activity in morbid and uncomplicated obesity: predictive role of fasting insulin. Nutr Metab Cardiovasc Dis 2007;17(6):442–7.

[20] Liu L, Zhong S, Yang R, Hu H, Yu D, Zhu D, et al. Expression, purification, and initial characterization of human alanine aminotransferase (ALT) isoenzyme 1 and 2 in High-five insect cells. Protein Expr Purif 2008;60(2):225–31.

[21] Oh SY, Cho YK, Kang MS, Yoo TW, Park JH, Kim HJ, et al. The association between increased alanine aminotransferase activity and metabolic factors in nonalcoholic fatty liver disease. Metabolism 2006;55(12):1604–9.

[22] Nyblom H, Berggren U, Balldin J, Olsson R. High AST/ALT ratio may indicate advanced alcoholic liver disease rather than heavy drinking. Alcohol Alcohol 2004;39(4): 336–9.

[23] Dunn W, Angulo P, Sanderson S, Jamil LH, Stadheim L, Rosen C, et al. Utility of a new model to diagnose an alcohol basis for steatohepatitis. Gastroenterology 2006;131(4):1057–63.

[24] Nyblom H, Björnsson E, Simrén M, Aldenborg F, Almer S, Olsson R. The AST/ALT ratio as an indicator of cirrhosis in patients with PBC. Liver Int 2006;26(7):840–5.

[25] Gitlin N. The serum glutamic oxaloacetic transaminase/serum glutamic pyruvic transaminase ratio as a prognostic index in severe acute viral hepatitis. Am J Gastroenterol 1982;77(1): 2–4.

[26] Pettersson J, Hindorf U, Persson P, Bengtsson T, Malmqvist U, Werkström V, et al. Muscular exercise can cause highly pathological liver function tests in healthy men. Br J Clin Pharmacol 2008;65(2):253–9.

[27] Young DS. Effects of preanalytical variables on clinical laboratory tests. 3rd ed. Washington, DC: AACC Press; 2007.

[28] Purkins L, Love ER, Eve MD, Wooldridge CL, Cowan C, Smart TS, et al. The influence of diet upon liver function tests and serum lipids in healthy male volunteers resident in a Phase I unit. Br J Clin Pharmacol 2004;57(2):199–208.

[29] Narjes H, Nehmiz G. Effect of hospitalisation on liver enzymes in healthy subjects. Eur J Clin Pharmacol 2000;56(4):329–33.

[30] Cai Z, Christianson A, Ståhle L, Keisu M. Reexamining transaminase elevation in Phase I clinical trials: the importance of baseline and change from baseline. Eur J Clin Pharmacol 2009;65(10):1025–35.

[31] Chtioui H, Mauerhofer O, Gunther B, Dufour JF. Macro-AST in an asymptomatic young patient. Ann Hepatol 2010;9(1):93–5.

[32] Orlando R, Carbone A, Lirussi F. Macro-aspartate aminotransferase (macro-AST): a 12-year follow-up study in a young female. Eur J Gastroenterol Hepatol 2003;15(12):1371–3.

[33] Rej R. Apparent stability of aminotransferases: influence of the analytical method. Clin Chem 1991;37(1):131–2.

[34] Guder WG, Narayanan S, Wisser H, Zawta B. Diagnostic samples: from the patient to the laboratory. The impact of preanalytical variables on the quality of laboratory results. 4th ed. Weinheim, Germany: Wiley-Blackwell; 2009.

[35] Heins M, Heil W, Withold W. Storage of serum or whole blood samples? Effects of time and temperature on 22 serum analytes. Eur J Clin Chem Clin Biochem 1995;33(4):231–8.

[36] Boyanton BL, Blick KE. Stability studies of twenty-four analytes in human plasma and serum. Clinical Chemistry 2002;48(12): 2242–7.

[37] Donnelly JG, Soldin SJ, Nealon DA, Hicks JM. Is heparinized plasma suitable for use in routine biochemistry? Pediatr Pathol Lab Med 1995;15(4):555–9.

[38] Blom E, Ali MM, Mortensen B, Huseby NE. Elimination of alkaline phosphatases from circulation by the galactose receptor. Different isoforms are cleared at various rates. Clin Chim Acta 1998;270(2):125–37.

[39] Deng JT, Hoylaerts MF, De Broe ME, van Hoof VO. Hydrolysis of membrane-bound liver alkaline phosphatase by GPI-PLD requires bile salts. Am J Physiol 1996;271(4 Pt 1):G655–63.

[40] Grostad M, Huseby NE. Clearance of different multiple forms of human gamma-glutamyltransferase. Clin Chem 1990;36(9): 1654–6.

[41] Ishii H, Ebihara Y, Okuno F, Munakata Y, Takagi T, Arai M, et al. Gamma-glutamyl transpeptidase activity in liver of alcoholics and its localization. Alcohol Clin Exp Res 1986;10(1): 81–5.

[42] Dufour DR, Lott JA, Nolte FS, Gretch DR, Koff RS, Seeff LB. Diagnosis and monitoring of hepatic injury: I. Performance characteristics of laboratory tests. Clin Chem 2000;46(12): 2027–49.

[43] Ryu S, Chang Y, Kim DI, Kim WS, Suh BS. Gamma-glutamyltransferase as a predictor of chronic kidney disease in nonhypertensive and nondiabetic Korean men. Clin Chem 2007;53(1):71–7.

[44] Artur Y, Wellman-Bednawska M, Jacquier A, Siest G. Complexes of serum gamma-glutamyltransferase with apolipoproteins and immunoglobulin A. Clin Chem 1984;30(5):631–3.

[45] Dimeski G, Carter A. Rare IgM interference with Roche/Hitachi Modular glucose and gamma-glutamyltransferase methods in heparin samples. Clin Chem 2005;51(11):2202–4.

[46] McTaggart MP, Rawson C, Lawrence D, Raney BS, Jaundrill L, Miller LA, et al. Identification of a macro-alkaline phosphatase complex in a patient with inflammatory bowel disease. Ann Clin Biochem 2012;49:405–7.

[47] Zetterberg H. Increased serum concentrations of intestinal alkaline phosphatase in peritoneal dialysis. Clin Chem 2005;51(3): 675–6.

[48] Mandel AL, Breslin PA. High endogenous salivary amylase activity is associated with improved glycemic homeostasis following starch ingestion in adults. J Nutr 2012;142(5):853–8.

[49] Wilding P, Cooke W, Nicholson G. Globulin-bound amylase: a cause of persistently elevated levels in serum. Ann Intern Med 1964;60:1053.

[50] Taes YE, Louagie H, Yvergneaux J-P, De Buyzere ML, De Puydt H, Delanghe JR, et al. Prolonged hyperlipasemia attributable to a novel type of macrolipase. Clin Chem 2000;46 (12):2008–13.

[51] Wu AH, Gross S. Further insight on amylase and lipase. Med Lab Obs 2006;38(2):6 [author reply].

[52] Byrne MF, Mitchell RM, Stiffler H, Jowell PS, Branch MS, Pappas TN, et al. Extensive investigation of patients with mild elevations of serum amylase and/or lipase is "low yield.". Can J Gastroenterol 2002;16(12):849–54.

[53] Mariani A. Chronic asymptomatic pancreatic hyperenzymemia: Is it a benign anomaly or a disease? JOP 2010;11(2):95–8.

[54] Pezzilli R, Morselli-Labate AM, Casadei R, Campana D, Rega D, Santini D, et al. Chronic asymptomatic pancreatic hyperenzymemia is a benign condition in only half of the cases: a prospective study. Scand J Gastroenterol 2009;44(7):888–93.

[55] Tetrault GA. Lipase activity in serum measured with Ektachem is often increased in nonpancreatic disorders. Clin Chem 1991;37(3):447–51.

[56] Panteghini M, Bonora R, Pagani F. Measurement of pancreatic lipase activity in serum by a kinetic colorimetric assay using a new chromogenic substrate. Ann Clin Biochem 2001;38(Pt 4): 365–70.

[57] Anderstam B, Garcia-Lopez E, Heimburger O, Lindholm B. Determination of alpha-amylase activity in serum and dialysate from patients using icodextrin-based peritoneal dialysis fluid. Perit Dial Int 2003;23(2):146–50.

[58] Steinhauer JR, Hardy RW, Robinson CA, Daly TM, Chaffin C, Konrad RJ. Comparison of non-diglyceride- and diglyceride-based assays for pancreatic lipase activity. J Clin Lab Anal 2002;16(1):52–5.

[59] Joseph J, Badrinath P, Basran G, Sahn S. Is albumin gradient or fluid to serum albumin ratio better than the pleural fluid lactate dehydroginase in the diagnostic of separation of pleural effusion? BMC Pulm Med 2002;2(1):1.

[60] Bailey IR. Effect on 16 analytes of overnight storage of specimens collected into heparinised evacuated tubes with plasma separator. Ann Clin Biochem 1990;27(Pt 1):56–8.

[61] Wilding P, Zilva JF, Wilde CE. Transport of specimens for clinical chemistry analysis. Ann Clin Biochem 1977;14(6):301–6.

[62] Bodor GS, Porter S, Landt Y, Ladenson JH. Development of monoclonal antibodies for an assay of cardiac troponin-I and preliminary results in suspected cases of myocardial infarction. Clin Chem 1992;38(11):2203–14.

[63] Katus HA, Looser S, Hallermayer K, Remppis A, Scheffold T, Borgya A, et al. Development and *in vitro* characterization of a new immunoassay of cardiac troponin T. Clin Chem 1992;38 (3):386–93.

[64] Thygesen K, Alpert JS, White HD, on behalf of the Joint ESC/ACCF/AHA/WHF Task Force for the Redefinition of Myocardial Infarction. Universal definition of myocardial infarction. Circulation 2007;116(22):2634–53.

[65] Hayashi T, Obi Y, Kimura T, Iio K, Sumitsuji S, Takeda Y, et al. Cardiac troponin T predicts occult coronary artery stenosis in patients with chronic kidney disease at the start of renal replacement therapy. Nephrol Dial Transplant 2008;23(9): 2936–42.

[66] Babuin L, Jaffe AS. Troponin: the biomarker of choice for the detection of cardiac injury. CMAJ 2005;173(10):1191–202.

[67] Jacobs LH, van de Kerkhof J, Mingels AM, Kleijnen VW, van der Sande FM, Wodzig WK, et al. Haemodialysis patients longitudinally assessed by highly sensitive cardiac troponin T and commercial cardiac troponin T and cardiac troponin I assays. Ann Clin Biochem 2009;46(Pt 4):283–90.

[68] Tsutamoto T, Kawahara C, Yamaji M, Nishiyama K, Fujii M, Yamamoto T, et al. Relationship between renal function and serum cardiac troponin T in patients with chronic heart failure. Eur J Heart Fail 2009;11(7):653–8.

[69] Ricchiuti V, Apple FS. RNA expression of cardiac troponin T isoforms in diseased human skeletal muscle. Clin Chem 1999;45 (12):2129–35.

[70] Jaffe AS, Vasile VC, Milone M, Saenger AK, Olson KN, Apple FS. Diseased skeletal muscle: a noncardiac source of increased circulating concentrations of cardiac troponin T. J Am Coll Cardiol 2011;58(17):1819–24.

[71] Keller T, Ojeda F, Zeller T, Wild PS, Tzikas S, Sinning CR, et al. Defining a reference population to determine the 99th percentile of a contemporary sensitive cardiac troponin I assay. Int J Cardiol 2012;May:4 [Epub ahead of print].

[72] Eggers KM, Jaffe AS, Lind L, Venge P, Lindahl B. Value of cardiac troponin I cutoff concentrations below the 99th percentile for clinical decision-making. Clin Chem 2009;55(1):85–92.

[73] Apple FS, Pearce LA, Smith SW, Kaczmarek JM, Murakami MM. Role of monitoring changes in sensitive cardiac troponin I assay results for early diagnosis of myocardial infarction and prediction of risk of adverse events. Clin Chem 2009;55(5): 930–7.

[74] Mueller M, Biener M, Vafaie M, Doerr S, Keller T, Blankenberg S, et al. Absolute and relative kinetic changes of high-sensitivity cardiac troponin T in acute coronary syndrome and in patients with increased troponin in the absence of acute coronary syndrome. Clin Chem 2012;58(1):209–18.

[75] Apple FS, Morrow DA. Delta cardiac troponin values in practice: are we ready to move absolutely forward to clinical routine? Clin Chem 2012;58(1):8–10.

[76] White HD. Pathobiology of troponin elevations: do elevations occur with myocardial ischemia as well as necrosis? J Am Coll Cardiol 2011;57(24):2406–8.

[77] Regwan S, Hulten EA, Martinho S, Slim J, Villines TC, Mitchell J, et al. Marathon running as a cause of troponin elevation: a systematic review and meta-analysis. J Interv Cardiol 2010;23 (5):443–50.

[78] Saravia SG, Knebel F, Schroeckh S, Ziebig R, Lun A, Weimann A, et al. Cardiac troponin T release and inflammation demonstrated in marathon runners. Clin Lab 2010;56(1-2):51–8.

[79] Traiperm N, Gatterer H, Wille M, Burtscher M. Cardiac troponins in young marathon runners. Am J Cardiol 2012;110:594–8.

[80] Labugger R, Organ L, Collier C, Atar D, Van Eyk JE. Extensive troponin I and T modification detected in serum from patients with acute myocardial infarction. Circulation 2000;102(11): 1221–6.

[81] Peronnet E, Becquart L, Poirier F, Cubizolles M, Choquet-Kastylevsky G, Jolivet-Reynaud C. SELDI-TOF MS analysis of the cardiac troponin I forms present in plasma from patients with myocardial infarction. Proteomics 2006;6(23):6288–99.

[82] Ungerer JP, Marquart L, O'Rourke PK, Wilgen U, Pretorius CJ. Concordance, variance, and outliers in 4 contemporary cardiac troponin assays: implications for harmonization. Clin Chem 2012;58(1):274–83.

[83] Eriksson S, Ilva T, Becker C, Lund J, Porela P, Pulkki K, et al. Comparison of cardiac troponin I immunoassays variably affected by circulating autoantibodies. Clin Chem 2005;51(5): 848–55.

[84] Savukoski T, Engstrom E, Engblom J, Ristiniemi N, Wittfooth S, Lindahl B, et al. Troponin-specific autoantibody interference in different cardiac troponin I assay configurations. Clin Chem 2012;58(6):1040–8.

[85] Legendre-Bazydlo LA, Haverstick DM, Kennedy JL, Dent JM, Bruns DE. Persistent increase of cardiac troponin I in plasma without evidence of cardiac injury. Clin Chem 2010;56(5): 702–5.

[86] Packer M. Should B-type natriuretic peptide be measured routinely to guide the diagnosis and management of chronic heart failure?. Circulation 2003;108(24):2950–3.

[87] Januzzi JL, van Kimmenade R, Lainchbury J, Bayes-Genis A, Ordonez-Llanos J, Santalo-Bel M, et al. NT-proBNP testing for

diagnosis and short-term prognosis in acute destabilized heart failure: an international pooled analysis of 1256 patients. European Heart Journal 2006;27(3):330–7.

[88] Thygesen K, Mair J, Mueller C, Huber K, Weber M, Plebani M, et al. Recommendations for the use of natriuretic peptides in acute cardiac care. European Heart Journal 2011;33:2001–6.

[89] Mair J, Falkensammer G, Poelzl G, Hammerer-Lercher A, Griesmacher A, Pachinger O. B-type natriuretic peptide (BNP) is more sensitive to rapid hemodynamic changes in acute heart failure than N-terminal proBNP. Clin Chim Acta 2007;379(1-2): 163–6.

[90] Schou M, Gustafsson F, Nielsen PH, Madsen LH, Kjaer A, Hildebrandt PR. Unexplained week-to-week variation in BNP and NT-proBNP is low in chronic heart failure patients during steady state. Eur J Heart Fail 2007;9(1):68–74.

[91] Bruins S, Fokkema MR, Romer JW, Dejongste MJ, van der Dijs FP, van den Ouweland JM, et al. High intraindividual variation of B-type natriuretic peptide (BNP) and amino-terminal proBNP in patients with stable chronic heart failure. Clin Chem 2004;50(11):2052–8.

[92] Sokoll LJ, Baum H, Collinson PO, Gurr E, Haass M, Luthe H, et al. Multicenter analytical performance evaluation of the Elecsys proBNP assay. Clin Chem Lab Med 2004;42(8): 965–72.

[93] Apple FS, Panteghini M, Ravkilde J, Mair J, Wu AH, Tate J, et al. Quality specifications for B-type natriuretic peptide assays. Clin Chem 2005;51(3):486–93.

[94] Shimizu H, Masuta K, Aono K, Asada H, Sasakura K, Tamaki M, et al. Molecular forms of human brain natriuretic peptide in plasma. Clin Chim Acta 2002;316(1-2):129–35.

[95] Seidler T, Pemberton C, Yandle T, Espiner E, Nicholls G, Richards M. The amino terminal regions of proBNP and proANP oligomerise through leucine zipper-like coiled-coil motifs. Biochem Biophys Res Commun 1999;255(2):495–501.

[96] Young DS. Effects of drugs on clinical laboratory tests. 3rd ed. Washington, DC: AACC Press; 2000.

CHAPTER

# 10

# Sources of Inaccuracy in Biochemical Genetics Testing

*Michael J. Bennett*

University of Pennsylvania Perelman School of Medicine, Philadelphia, Pennsylvania

## INTRODUCTION

The mainstay of biochemical genetics analysis includes measurement of body fluid amino acids, urine organic acids, and plasma acylcarnitine. These tests are used initially to diagnose a number of metabolic diseases, some in the newborn period, and subsequently are used to monitor patient response to treatment following diagnosis. Typical methods include an initial chromatographic separation step followed by specific biomarker analysis that may involve spectrophotometric, fluorometric, and sometimes mass spectrometric identification and subsequent quantification. Interferences may come from a variety of exogenous or endogenous sources, including dietary components, drugs, components of the metabolism of the gastrointestinal biome, and xenobiotics. Frequently, the use of separation technology and specific detection such as mass spectrometry can identify these components so that they do not cause interference with targeted biomarkers; occasionally, however, and in a method-dependent manner, interferences do occur that create interpretive concerns. This is particularly true in assays of urinary organic acids that typically are used in an untargeted manner when investigating patients for a large number of potential disorders. Inappropriate sample collection or lack of knowledge of the clinical condition such as prandial status and knowledge of concurrent therapy despite having positive identification methodology can also result in erroneous measurement of some of these biomarkers. These potential interferences can impact interpretation of results by indicating a wrong diagnosis or a positive diagnosis in a normal individual. The issues of interference are diminished in patients with a known diagnosis when specific targeted biomarkers are being used to monitor therapy.

## ANALYSIS OF AMINO ACIDS

The traditional method for the measurement of body fluid amino acids for more than four decades has been ion-exchange chromatography with postcolumn derivatization with ninhydrin for detection followed by spectrophotometric quantification [1,2]. In a College of American Pathologists (CAP) Biochemical Genetics proficiency program, 66% of respondents reported use of this methodology. Other methodologies included high-pressure liquid chromatography in 22% of the laboratories in the CAP survey, [3] and increasing numbers of laboratories are using ultraperformance liquid chromatography (UPLC) [4] or tandem mass spectrometric analysis [5]. Plasma or serum represents the major body fluid that is used for amino acid analysis. For a few specific metabolic disorders or clinical indications, urine or cerebrospinal fluid may be the preferred sample type. Optimizing sample type is very important for amino acid analysis because interferences differ for the different body fluids and for the different methodologies. Urine tends to have the most interference, particularly when urine analysis is performed using ninhydrin-based ion-exchange or high-performance liquid chromatography (HPLC) systems. Ninhydrin reacts with primary amines, such as amino acids and ammonium, and also some secondary amines, such as the imino acids proline and hydroxyproline, which form a different color with ninhydrin but can be detected when dual-wavelength detection

*Accurate Results in the Clinical Laboratory.*
DOI: http://dx.doi.org/10.1016/B978-0-12-415783-5.00010-4

is applied. A number of ninhydrin-positive interferences have been identified, [6] including penicillins, cephalosporins, α-methyldopa, and levodopa. The degree of interference and the amino acid with which ninhydrin-positive compounds interfere are very much dependent on the running conditions used by the amino acid analyzer and on the make and model of the analyzer; therefore, they should be evaluated on an individual laboratory basis. The most commonly encountered ninhydrin-positive interference comes from metabolites of penicillin-based antibiotics such as ampicillin, which generates a number of positively staining metabolites that create interpretive difficulty particularly in the measurement of urine amino acids [7]. Figure 10.1 shows a chromatogram of an HPLC ninhydrin detection system of urine from a patient taking ampicillin. The ampicillin peak obliterates tyrosine and overlaps significantly with phenylalanine in this system. Depending on the analytical system being used, this interference could be elsewhere in the chromatogram. Glutathione, which is released from red blood cells when hydrolyzed, is a major source of potential interference with sample hydrolysis [8]. Laboratories should also be aware that new drugs are continually being developed and that some of them may cause unique interferences. For example, the novel anticonvulsant gabapentin is increasingly being prescribed for patients with seizure disorders that are refractive to other anticonvulsants. Gabapentin interferes with the measurement of histidine in many ninhydrin-based systems [9]. Table 10.1 lists a number of significant commonly seen ninhydrin-positive interferences. Data for interferences with HPLC-based detection systems other than those using ninhydrin are not readily available because there are few users and no historical database, but it is likely that there will be similar degrees of interference from the same classes of compounds. HPLC– or UPLC–mass spectrometric methodologies can overcome the effects of many interferences because the positive identification of true peaks using spectral analysis can determine the purity of a compound and the degree of interference when there is co-elution of interfering peaks [5]. Many mass spectrometric processes for amino acids based on flow injection technology, which does not have a chromatographic component, are unable to distinguish isobaric forms of the amino acids because the only determinate of identification is based on a common molecular transition and mass of the particular amino acids. Thus, the isobaric amino acids leucine, isoleucine, alloisoleucine, and hydroxyproline give a single signal, and in order to confirm a diagnosis of maple syrup urine disease (branched-chain 2-ketoacid dehydrogenase deficiency) through detection and measurement of the pathognomonic biomarker alloisoleucine, a separate analysis is required using a form of separation chromatography [10].

## ANALYSIS OF ORGANIC ACIDS

Organic acid analysis, which is performed on urine samples, is undertaken using gas chromatography–mass spectrometry (GC-MS). In the most recent CAP Biochemical Genetics proficiency report, 99% of reporting laboratories used GC-MS technology, whereas only one laboratory used an LC-MS approach. The advantage of having a mass spectrometric identification system coupled to a separation step is that there is a very low probability of true analytical interference. Many compounds in urine are not related to endogenous metabolism, including dietary and drug metabolites, products of the metabolism of the human biome (bacterial metabolites), and a variety of xenobiotic compounds [11]. The use of full-scan data analysis coupled to the many mass spectral library sources allows us to readily distinguish the most important diagnostic metabolites from metabolites that are derived from other sources, such as bacterial metabolism. In the current climate of human biome and metabolome discovery, the technology that has been used for more than 30 years to identify organic acids is now being adapted to metabolomic discovery and the identification of new disease biomarkers [12]. The area in the field of urinary organic acid analysis in which interference is likely to be an issue is not in the analytical component but, rather, in the interpretation of the data. Thus, the presence of C6–C10 medium-chain dicarboxylic acids (adipic, suberic, and sebacic acids) may indicate a possible defect in the oxidation of fatty acids [13] but may also be an artifact of fasting ketosis, medium-chain triglyceride feeding, or collection of samples into a plastic container [14]. Valproic acid, which is frequently the first-line drug for pediatric seizure disorders, is a C8 branched-chain fatty acid that is metabolized to generate up to six different organic acid metabolites and their corresponding glucuronides that can be readily identified by their mass spectral properties, thus minimizing the effects of analytical interference (Figure 10.2). However, valproate also inhibits some mitochondrial metabolic pathways and may also generate biomarkers that in the absence of the drug would indicate possible mitochondrial metabolic diseases and thus confound sample interpretation and patient diagnosis. Glycerol is frequently identified in urine organic acid profiles and is mostly derived artifactually from the use of glycerol-based skin preparations. However, rarely, glycerol is also an important biomarker for the diagnosis of glycerol kinase deficiency, an X-linked genetic defect that presents with hypoglycemia. Subtle elevations in urine glycerol excretion may be predicted in females with partial

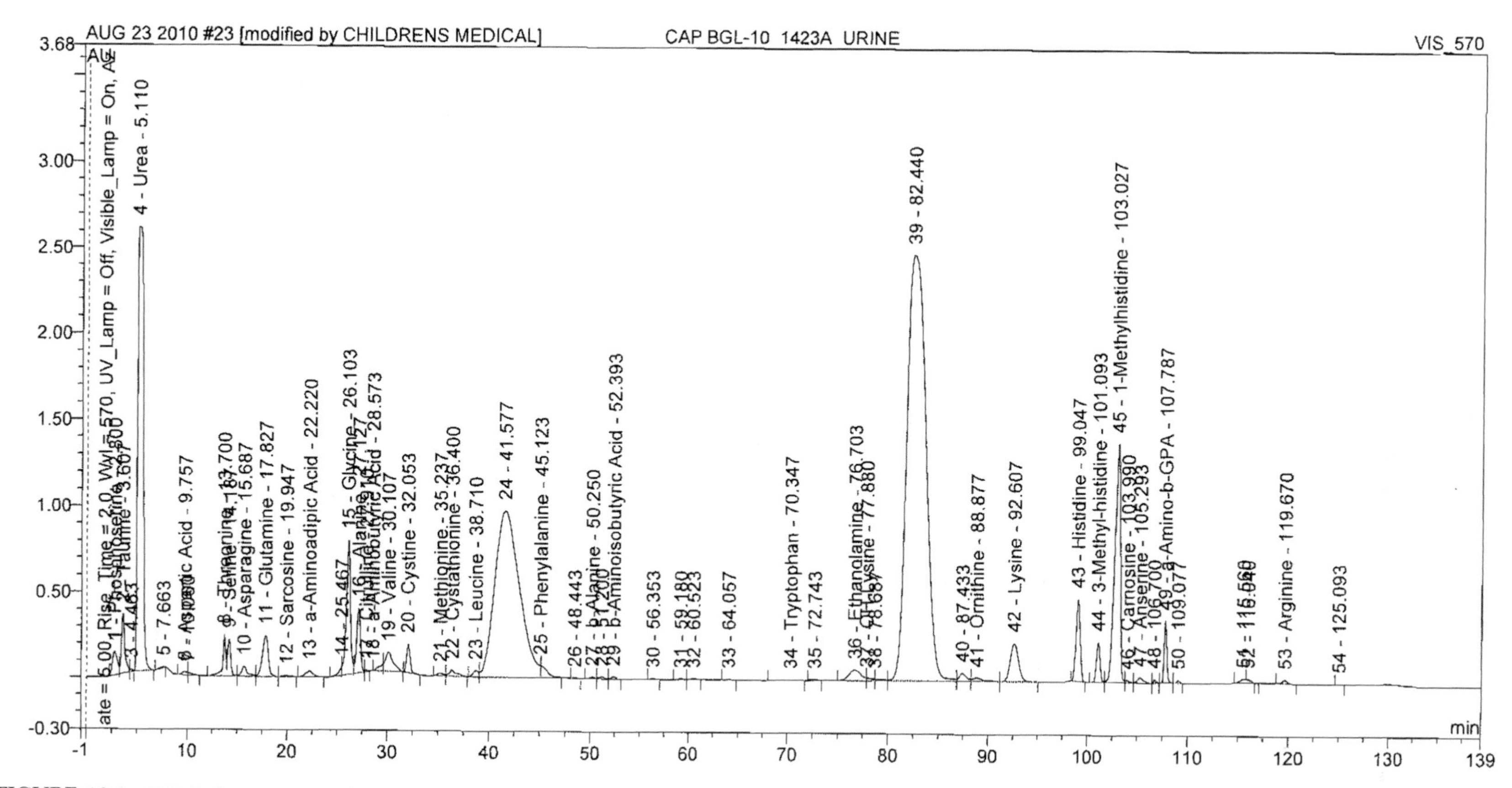


FIGURE 10.1 **HPLC chromatogram of urinary amino acids from a patient receiving ampicillin.** The large peak at approximately 42 min is an ampicillin metabolite that has co-eluted completely with tyrosine and also overlaps with phenylalanine and leucine, interfering with the measurement of all three amino acids. Source: *Courtesy of Dr. Patti Jones, Children's Medical Center of Dallas.*

X-chromosome inactivation, and this diagnosis is severely compromised as a result of the presence of endogenous glycerol. Thus, there may be multiple sources for some organic acids that would not be readily distinguishable on purely analytical terms for which additional clinical and pre-analytical information would be required for optimal interpretation. Table 10.2 lists a number of exogenous compounds that can interfere with interpretation of organic acid chromatograms (Figure 10.3).

**TABLE 10.1** Important Ninhydrin-Positive Interferences in Amino Acid Analysis[a]

| |
|---|
| Penicillin derivatives, particularly ampicillin |
| Cephalosporins |
| Gabapentin—in patients with seizure disorders |
| Glutathione—in hemolyzed blood samples |
| L-Dopa, methyldopa |
| Small peptides—may be important for diagnosing dipeptidase defects |
| Ammonia |
| Urea |

[a]*Interference will be highly dependent on the separation system used.*

## ANALYSIS OF ACYLCARNITINES

Acylcarnitine analysis is also a process that is dependent on positive identification of biomarkers through mass spectrometry. Most clinical laboratories performing routine acylcarnitine analysis use flow injection tandem mass spectrometry and multiple reaction monitoring of parent compounds or specific fragments for compound detection. This process does not differentiate isobaric compounds and does have certain interferences due to the presence of acylcarnitines with the same mass and also from a number of exogenous interfering sources [15–17]. Most notable interferences due to the presence of isobaric compounds are shown in Table 10.3 and include the following:

1. C4 acylcarnitine, which could be derived from butyrylcarnitine (biomarker for short-chain acyl-CoA dehydrogenase deficiency), isobutyrylcarnitine (isobutyryl-CoA dehydrogenase deficiency),

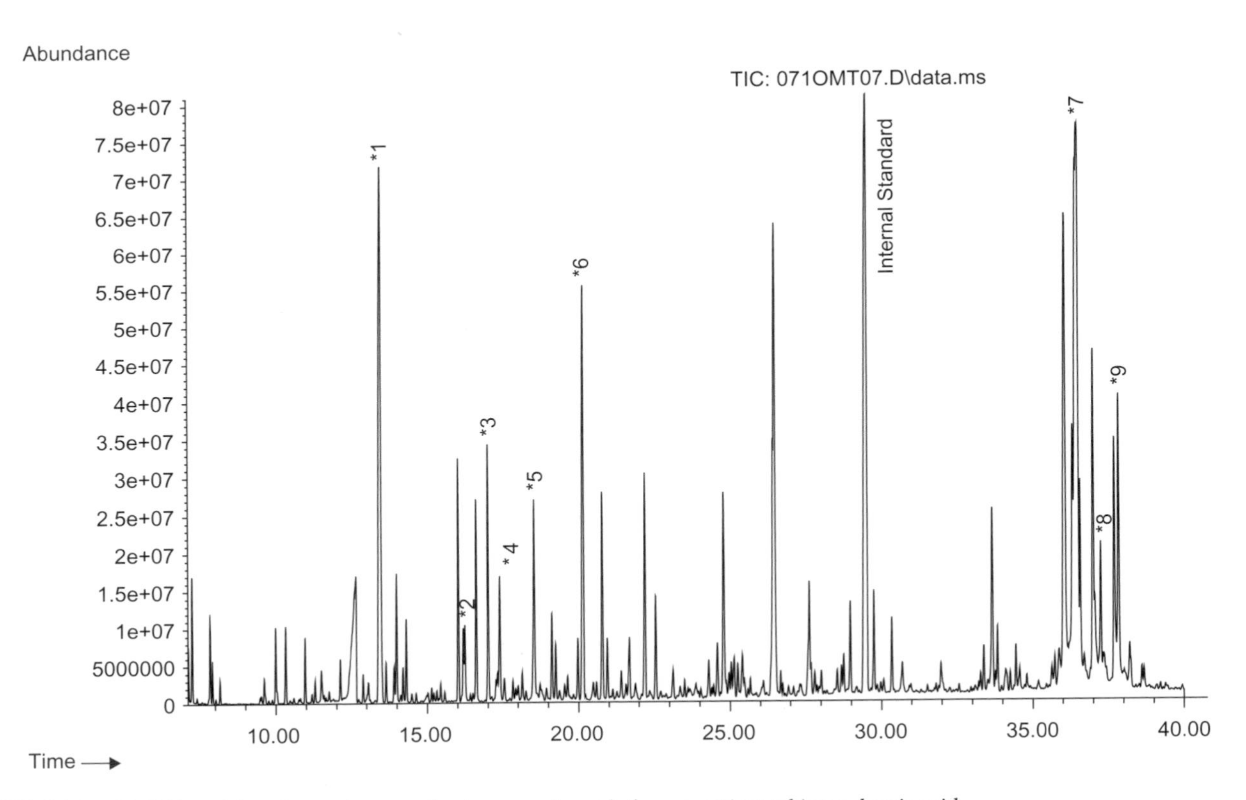


**FIGURE 10.2** GC-MS total ion chromatogram of urine organic acids from a patient taking valproic acid. Peak 1 is free valproic acid, peaks 2–6 are metabolites of valproic acid, and peaks 7–10 are glucuronide conjugates of valproic acid and its metabolites.

and formiminoglutamic acid (glutamate formiminotransferase deficiency)

2. C5 acylcarnitine, which could be derived from isovalerylcarnitine (isovaleric acidemia), 2-methylbutyrylcarnitine (2-methylbutyryl-CoA dehydrogenase deficiency), or pivaloylcarnitine (therapy with pivalate-containing medications)
3. C8 acylcarnitine, which could be derived from octanoylcarnitine (medium-chain acyl-CoA dehydrogenase deficiency or medium-chain

**TABLE 10.2** Important Interferences in Urine Organic Acid Analysis[a]

| |
|---|
| Valproic acid—has multiple metabolites, also interferes with mitochondrial metabolism |
| Salicylates |
| Ibuprofen |
| Phenobarbital |
| Dopamine—is converted to homovanillic acid, a biomarker for neuroblastoma |
| Tylenol (acetaminophen) |
| Carbamazepine |
| Ethosuccimide |
| Lactate—may be endogenous (L-lactate) or exogenous (D-lactate); these isomers are not separated |
| Succinate—may be endogenous (mitochondrial) or exogenous (bacterial) |
| 4-Hydroxyphenylacetic acid—may be endogenous (tyrosinemia) or exogenous (bacterial) |
| Medium-chain triglycerides—converted to medium-chain dicarboxylic acids-biomarkers for fatty acid oxidation defects |
| Glycerol—can be derived from glycerol-based skin products or glycerol kinase deficiency |

*[a]Identification of interferences rarely poses a problem in this assay, but the interfering compounds can influence data interpretation.*

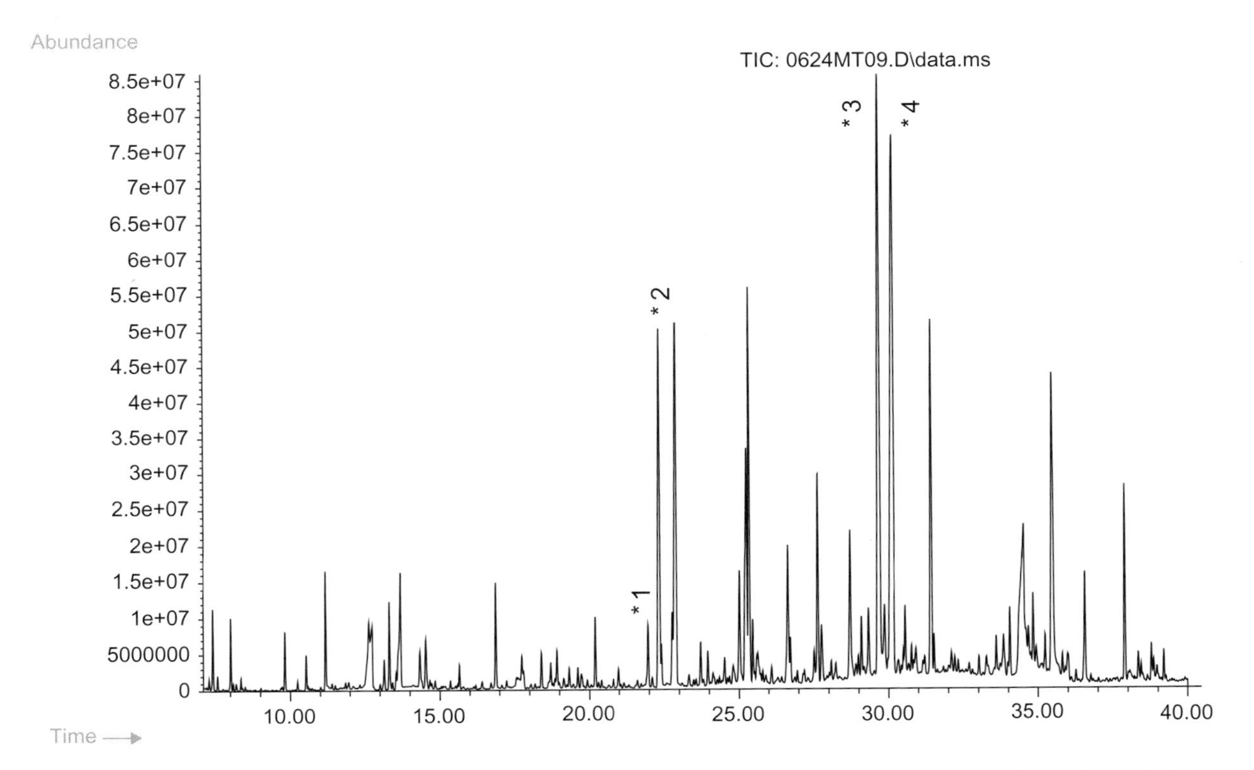


**FIGURE 10.3 GC-MS total ion chromatogram of urine organic acids from a patient taking acetaminophen, ibuprofen, and phenobarbital.** Peak 1 is free ibuprofen; peak 2 is acetaminophen; peak 3 is phenobarbital, which in this system interferes with the internal standard (undecanedioic acid); and peak 4 is carboxyibuprofen [19].

**TABLE 10.3** Isobaric Compounds That Cannot Be Isolated Using Flow-Injection Tandem Mass Spectrometry and Multiple-Reaction Monitoring for Acylcarnitines and Amino Acids

| |
|---|
| C4 acylcarnitine—may be butyrylcarnitine, isobutyrylcarnitine, formiminoglutamic acid |
| C5 acylcarnitine—may be isovalerylcarnitine, 3-methylbutyrylcarnitine, pivaloylcarnitine |
| C8 acylcarnitine—may be octanoylcarnitine, valproylcarnitine |
| C16:1 acylcarnitine—most likely a cefotaxime metabolite |
| Leucine, isoleucine, alloisoleucine, and hydroxyproline |

triglyceride therapy) or from valproylcarnitine (valproic acid therapy)

4. C16:1-hydroxy acylcarnitine, for which there is no defined metabolic disease as an isolated biomarker but is interfered by a cefataxime metabolite.

Many of these interferences can be evaluated by examining alternative disease biomarkers by organic acid analysis or by including a separation step into the procedure [18]. Also, many of the disorders described here have additional acylcarnitine abnormalities that would help define a true disorder from an artifact due to interference.

## CONCLUSIONS

Analysis of amino acids, organic acids, and acylcarnitine is important for detection of inborn errors of metabolism. In general, amino acid analysis using conventional HPLC is subject to more interferences than the other analytes as analysis of organic acids that utilizes more analytically specific GC-MS and acylcarnitine using tandem mass spectrometry have positive identification. Nevertheless, attention must be paid to eliminate any issue of interferences in these assays because clinicians rely on accurate results for proper diagnosis of inborn errors of metabolism.

## References

[1] Slocum RH, Cummings JG. Amino acid analysis of physiological samples. In: Hommes FA, editor. Techniques in diagnostic human biochemical genetics. New York: Wiley-Liss; 1991. p. 87–126.

[2] Walker V, Mills GA. Quantitative methods for amino acid analysis in biological fluids. Ann Clin Biochem 1995;32:28–57.

[3] Irvine GB. Amino acid analysis. Methods Mol Biol 1997;64:257–65.

[4] Narayan SB, Ditewig-Meyers G, Graham KS, Scott R, Bennett MJ. Measurement of plasma amino acids by ultraperformance liquid chromatography. Clin Chem Lab Med 2011;49:1177–85.

[5] Dietzen DJ, Weindel AL. Comprehensive determination of amino acids for diagnosis of inborn errors of metabolism. Methods Mol Biol 2010;603:27–36.

[6] Christenson RH, Azzazy HME, Amino acids. In: Burtis CA, Ashwood ER editors. Tietz fundamentals of clinical chemistry. 5th ed. Philadelphia: Saunders; p 300–24.

[7] Rehfeld SJ, Loken H, Korte WD. Interference by antibiotics in amino acid assays. Clin Chem 1974;20:1477.

[8] Lemons JA, Teng C, Naughton MA. Removal of glutathione interference in blood amino acid analysis. Biochem Med 1976;15:282–8.

[9] Hoover-Fong JE, Geraghty MT, Raymond GV, Thomas GH. Gabapentin interference with urine histidine as measured by the Beckman amino acid analyser. J Inherit Metab Dis 2001;24:415–16.

[10] Schulze A, Matern D, Hoffmann GF. Newborn screening. In: Sarafoglou K, Hoffmann GF, Roth KS, editors. Pediatric endocrinology and inborn errors of metabolism. New York: McGraw Hill; 2009. p. 17–32.

[11] Kumps A, Duez P, Martens Y. Metabolic, nutritional, iatrogenic, and artifactual sources of urinary organic acids: a comprehensive table. Clin Chem 2002;48:708–17.

[12] Wishart DS, Tzur D, Knox C, et al. HMDB: the human metabolome database. Nucleic Acids Res 2007;35:D521–526.

[13] Bennett MJ. The laboratory diagnosis of inborn errors of mitochondrial fatty acid oxidation. Ann Clin Biochem 1990;27:519–31.

[14] Bennett MJ, Ragni MC, Hood I, Hale DE. Azelaic and pimelic acids: metabolic intermediates or artifacts. J Inherit Metab Dis 1992;15:220–3.

[15] Millington DS. Tandem mass spectrometry in clinical diagnosis. In: Blau N, Duran M, Blaskovics ME, Gibson KM, editors. Physicians guide to the laboratory diagnosis of metabolic diseases. 2nd ed. Berlin: Springer; 2003. p. 57–75.

[16] Abdenur JE, Chamoles NA, Guinle AF, et al. Diagnosis of isovaleric academia by tandem mass spectrometry: false positive result due to pivaloylcarnitine in a newborn screening programme. J Inherit Metab Dis 1998;21:624–30.

[17] Vianey-Sabin C, Boyer S, Levrat V, et al. Interference of cefotaxime in plasma acylcarnitine profile mimicking an increase of 3-hydroxypalmitoleylcarnitine (C16:1-OH) using butyl esters. J Inherit Metab Dis 2004;27(Suppl. 1):94.

[18] Minkler PE, Stoll MS, Ingalls ST, et al. Quantification of carnitine and acylcarnitines in biological matrices by HPLC electrospray ionization mass spectrometry. Clin Chem 2008;54:1451–62.

[19] Bennett MJ, Sherwood WG, Bhala A, Hale DE. Identification of urinary metabolites of (±)-2-(p-isobutylphenyl) propionic acid (Ibuprofen) by routine organic acid screening. Clin Chim Acta 1992;210:55–62.

CHAPTER

# 11

# Challenges in Endocrinology Testing

*Lindsay A.L. Bazydlo, Neil S. Harris, William E. Winter*

University of Florida College of Medicine, Gainesville, Florida

## INTRODUCTION

The proper interpretation of endocrine hormone measurements requires consideration of the hormone concentration in the plasma, the sensitivity of the responding cells and tissues, and the clinical symptoms of the patient [1]. For example, in states of thyroid hormone resistance, the concentrations of the thyroid hormones and thyroid-stimulating hormone (TSH) are typically elevated, but the affected patients are usually euthyroid or even possibly hypothyroid [2]. In another example, in type 2 diabetes, the absolute insulin concentration may be elevated because of end-organ resistance. In such a case, hyperinsulinism reflects a lack of insulin responsiveness and not a primary excess of insulin secretion [3]. Fortunately, neither insulin nor connecting peptide (C-peptide) concentrations are used in the diagnosis of diabetes [4].

A variety of interferences can perturb the results of endocrine testing [5]. In the introduction to this chapter, we highlight pre-analytic considerations, sample collection and processing assay, specificity, high-dose hook effect, macrocomplexes, human anti-mouse antibodies, rheumatoid factor, and heterophile antibodies.

### Pre-Analytical Conditions

Similar to any analytical determination, the pre-analytical status of the sample is crucial [6]. The sample must be drawn on the right patient, at the right time, in the correct volume, properly identified, placed into the correct tube, and properly processed. Some hormones vary considerably in concentration over the course of 24 hr but a single reference interval is applied (e.g., thyrotropin, which is also known as TSH, and testosterone). For other hormones with a more clearly evident diurnal variation (e.g., cortisol), reference intervals apply to the time of the day of sampling (e.g., a.m. vs. p.m.). Some hormones exhibit ultradian variation (e.g., luteinizing hormone and follicle-stimulating hormone) or infradian variation (e.g., estradiol and progesterone in menstruating women). Such variations must be taken into consideration when the laboratorian or clinician interprets the significance of a hormone concentration in an individual.

Even the location of the phlebotomy can affect the measurement. For example, parathyroid hormone (PTH) measurements taken downstream of (i.e., proximal to) a forearm-implanted parathyroid gland can display higher PTH measurements than when the PTH is drawn from the contralateral arm [7]. For certain immunoassays or mass spectroscopy methods, sample preparation (e.g., extraction) is also a key factor for proper analysis [8–11].

### Sample Collection and Processing

Proper selection of the collection tube and post-phlebotomy sample handling are very important for many hormones. For example, glucagon degrades quickly. To avoid this, plasma is prepared using an ethylenediaminetetraacetic acid (EDTA) tube containing aprotinin (e.g., 0.04 mL of aprotinin added to each 1 mL of whole blood), followed by centrifugation (within 2 hr) and freezing of the plasma. Aprotinin is a bovine pancreatic trypsin inhibitor that is typically purified from bovine lung. Vasoactive intestinal polypeptide (VIP) is similarly collected. For adrenocorticotropic hormone (ACTH), blood is drawn into an EDTA tube that must be plastic or siliconized glass. For antidiuretic hormone (e.g., arginine vasopressin), the EDTA sample should be centrifuged, separated, and frozen within 2 hr of collection. Regarding

Accurate Results in the Clinical Laboratory.
DOI: http://dx.doi.org/10.1016/B978-0-12-415783-5.00011-6

TSH, in one study approximately 2% of TSH measurements were elevated in lithium–heparin plasma separator tubes compared to rapid serum tubes [12].

The clinical laboratory must inform the medical and nursing staff of the pre-analytical requirements of sample acquisition. Currently, most facilities provide this information online.

## Assays for Hormone Analysis

Usually, immunoassays are used in clinical laboratories for analysis of various hormones. Many factors affect accuracy of results obtained using these immunoassays, including assay specificity and various other factors that may interfere with assay technology. These factors are discussed next.

### *Assay Specificity*

In selecting the proper analytical assay, the laboratorian should advise the clinician concerning the specificity of the assay performance. For example, many steroid immunoassays are not specific for a single steroid but may identify several members of a class of related steroids [13]. Estradiol assays detect a variety of estrogens [14], and cortisol immunoassays detect a variety of glucocorticoids [15]. In addition, in the Advia Centaur cortisol immunoassay [16], the following selected percentage cross-reactivities are reported: corticosterone, 5.3%; cortisone, 31%; 11-deoxycortisol, 23%; 6β-hydroxycortisol, 6.8%; prednisolone, 109%; 6-methyl-prednisolone, 26%; and prednisone, 34%. There may also be negative interferences in immunoassays as reported for unconjugated estriol lowering the measured estradiol concentration [17]. In addition to specificity problems, immunoassays may lack low-end analytical sensitivity [18].

### *High-Dose Hook Effect*

Laboratorians must be aware of the possibility of the high-dose hook effect in immunoassays, in which samples with very high concentrations of an analyte produce a falsely low result [19]. In competitive immunoassays, one hypothesis explaining the high-dose hook effect is that the analyte self-aggregates and the analyte is thus underdetected. In double-antibody assays, high-dose hook effect might result from such a great excess of analyte that there is insufficient capture antibody and detection antibody to detect the analyte.

The high-dose hook effect is probably of greatest importance for prolactin (PRL) [20,21]. Prolactinomas can produce PRL levels (in ng/mL) in the hundreds to tens of thousands. However, damage to the hypothalamus or hypothalamic–pituitary portal system can also raise PRL at least into the hundreds. The difference between these two situations is critically important: Prolactinomas can be treated with dopamine agonists, whereas other large anterior pituitary adenomas are treated surgically. Therefore, dilution and re-assay may be required in certain clinical situations to exclude the high-dose hook effect where the PRL result is falsely low or not as high as it is *in vivo* (Table 11.1).

### *Macrocomplexes*

Autoantibodies against hormones can raise the circulating concentration of the hormone without producing clinical symptomatology of hormone excess because the free (unbound) hormone remains within the reference interval. The phenomenon of macrocomplexes is most widely observed for amylase [22]. For hormones, this is best appreciated for prolactin. Treatment of the patient's serum or plasma with polyethylene glycol (PEG) will precipitate antigen–antibody complexes, and the resulting supernatant is then assayed. If the decline in measured analyte is 50% or more, macroprolactin is present. However, the astute laboratorian (and clinician) will recognize that macroprolactinemia does not exclude true hyperprolactinemia of any cause. Other causes of hyperprolactinemia include dopamine, cholinergic, and serotonergic antagonists; estrogens and pregnancy; neurogenic stimulation (nursing, chest wall disease, and spinal cord injury); hypothyroidism with thyrotropin-releasing hormone (TRH) releasing PRL; chronic renal disease; and cirrhosis.

Anti-hormone autoantibodies that raise a hormone's concentration have also been reported for TSH [23,24], estradiol [25], thyroxine (T4) and triiodothyronine (T3) [26–28], insulin [29], and ACTH [30].

**TABLE 11.1** The Differential Diagnosis of Hyperprolactinemia

| |
|---|
| Central causes: prolactin release-inhibiting hormone (dopamine) deficiency |
| Hypothalamic disease |
| Interruption in the hypothalamic–pituitary portal system |
| Medications |
| Dopamine antagonists |
| Cholinergic antagonists |
| Serotonergic antagonists |
| Endocrine causes |
| Estrogen use |
| Pregnancy, postpartum |
| Hypothyroidism (pathologically elevated thyrotropin-releasing hormone can release prolactin) |
| Neurogenic |
| Nursing (nipple stimulation) |
| Chest wall disease |
| Spinal cord injury |
| Other disorders |
| Cirrhosis |
| Chronic renal disease |

### *Human Anti-Mouse Antibodies, Rheumatoid Factor, and Heterophile Antibodies*

Human anti-mouse antibodies (HAMA), rheumatoid factor (RF), and heterophile antibodies are responsible for causing false-positive immunoassay results. Heterophile antibodies represent naturally occurring antibodies and autoantibodies directed against heterogeneous and ill-described antigens [31]. For more details, see Chapter 6 and also the excellent review by Tate and Ward [32].

Rheumatoid factor is an IgM antibody that binds to the Fc portion of an IgG molecule. In a report from France, heterophile antibodies that raised the measured TSH concentration led to overtreatment of a patient with suspected hypothyroidism [33]. Surprisingly, heterophile autoantibodies have even been reported to lower hormone concentrations [34].

Many noncompetitive immunoassays use capture and detection monoclonal antibodies that are made in mice. Even without outright exposure to mice (e.g., somebody who works in a mouse colony or handles mice), the general population is exposed to mouse proteins and immunoglobulins (e.g., exposure of grain products to mouse urine). Individuals treated with therapeutic mouse monoclonal antibodies can develop HAMA (Table 11.2). Responses to mouse immunoglobulins can therefore produce HAMA. Manufacturers attempt to minimize the interference from HAMA by adding excess mouse immunoglobulin to their assay buffers. Thus, HAMA would bind to the free mouse immunoglobulin and would not bind to the mouse monoclonal immunoglobulins used in the assay. Some manufacturers have replaced mouse antibodies with antibodies from another species such as rabbit. For a discussion of the effect of HAMA on immunoassays, see Chapter 6 and the excellent paper by Kricka *et al* [35].

**TABLE 11.2** Examples of Mouse Monoclonal Autoantibodies in Therapeutic Use

| Monoclonal Antibody | Activity |
|---|---|
| Abciximab | Anti-GPIIbIIIa |
| Adalimumab | Anti-tumor necrosis factor-α |
| Bevacizumab | Anti-vascular endothelial growth factor (VEGF) |
| Cetuximab | Anti-epidermal growth factor receptor |
| Daclizumab | Anti-CD25 |
| Infliximab | Anti-tumor necrosis factor-α |
| Muromonab-CD3 | Anti-CD3 (Orthoclone) |
| Ofatumumab | Anti-CD20 cytolytic |
| Rituximab | Anti-CD20 |
| Situximab | Anti-IL-6 |
| Trastuzumab | Anti-human epidermal growth factor receptor-2 (HER2) |

The best-known false-positive results from HAMA involve those in human chorionic gonadotropin (hCG) testing. This can lead to the errant diagnosis of choriocarcinoma. In at least one case, a woman with a false-positive hCG result was treated with chemotherapy for choriocarcinoma followed by total bilateral oophorectomy and hysterosalpingectomy. HAMA interferences have also been reported in the measurement of B-type natriuretic peptide [36] and TSH [37–39].

When HAMA interferences are suspected, the assay should be rerun on an immunoassay platform of another manufacturer. Serial dilution linearity should also be sought. With serial dilution, if the analyte recovery is greater than expected, HAMA is likely present. However, serial dilutions that yield the expected recoveries do not exclude HAMA. Failure to detect hCG in the urine when hCG is elevated in the plasma is supportive of HAMA. As IgG antibodies, HAMA do not enter the urine, and thus serum or plasma immunoassays can be falsely positive but the urine immunoassays are not interfered with and are correctly negative. Assays for HAMA are available in the evaluation of suspected HAMA interferences [40].

HAMA should also be considered in nonhormone immunoassays. For example, a HAMA interference in digoxin measurements has been reported [41]. Other analytes with reported HAMA interferences include CA-125 [42–44] and hepatitis B surface antigen [45].

Although the release of hormones from plasma binding proteins is not an issue of interference, total hormone methods should release all protein-bound hormones for proper analysis. In addition, significant variations in binding protein concentrations are not interferences; however, such aberrations can produce clinically important changes in the total concentrations of hormones in the bloodstream. For example, total T4 and total T3 are affected by concentration changes in thyroxine-binding globulin (TBG) and thyroxine-binding prealbumin (transthyretin) [46]. Total T3 is also affected by concentration changes in TBG. Abnormal affinity for T4 raises total T4 in dysalbuminemic hyperthyroxinemia and familial euthyroid thyroxine excess [47,48]. Nevertheless, free hormone level measurements are not immune from the effects of different binding protein concentrations [49].

The recognition and prevention of errors is a major goal of medicine [50]. Some interferences can develop that are rare and/or unexplained [51]. The laboratorian must be able to assist the clinician when the laboratory findings and the patient's history and physical examination findings do not match. In such circumstances, analytical errors or interferences may be present, the clinical diagnosis may be incorrect, or the blood sample may have been misidentified. As laboratorians, we likely spend much of our time concerned with the analyte's measurement; however, analytical errors may account for only 10% of all laboratory "mistakes" [52].

## CHALLENGES IN TESTING OF HORMONES SECRETED BY PITUITARY

The pituitary gland in the human brain is responsible for secretion of many hormones, and accurate measurement of these hormones is crucial for diagnosis of many illnesses. In this section, assays for these hormones are discussed, with an emphasis on sources of errors in their measurement and steps that can be taken to avoid such errors.

### Growth Hormone

The measurement of growth hormone (GH) can be helpful in the diagnosis of disorders involving either GH excess or deficiency. Due to the pulsatile nature of GH release, measurements are typically performed during dynamic testing in an attempt to assist in the interpretation of the results. Aside from the pulsatile release of GH, a number of other interferences, related to both the patient's lifestyle and the laboratory's ability to measure the analyte, can alter the measurement.

The release of GH from the pituitary is controlled and stimulated by a number of processes. GH release is normally stimulated by growth hormone-releasing hormone (GHRH) and inhibited by somatotropin release-inhibitory hormone (SRIH), both of which are released from the hypothalamus under the control of a number of different stimuli. SRIH is also known as somatostatin, which is also secreted by the delta cells of the islets of Langerhans. It has been well documented that following the onset of exercise, GH levels will rise in the plasma [53,54]. GH increases can also be seen due to stress, hypoglycemia, sleep, and various pharmacologic agents such as glucagon, clonidine, and L-dopa; likewise, GH suppression is seen with somatostatin, cortisol, obesity, and hyperglycemia.

A number of variables can influence the measurement of GH in the laboratory. Much of the difficulty in measuring GH lies in the inherent heterogeneity of the analyte (Table 11.3). GH is released primarily from two different sites, the somatotroph cells of the pituitary gland and the placenta. Placental GH appears between 15 and 20 weeks of gestation and is present until term, completely replacing the pituitary GH such that it is undetectable [55]. Although this form only appears during pregnancy, to what degree it cross-reacts with GH immunoassays is uncertain, and interpretation should be performed with caution. It would be an extremely unusual clinical case if GH measurements were required during pregnancy. Due to alternative splicing of the gene encoding GH, two isoforms exist—the 22- and 20-kDa forms. The 22-kDa isoform is known to be the more biologically active molecule, and it is typically detected in all GH immunoassays. Efforts have been made to specifically detect the 20-kDa form through the development of a "non-22K assay" in which monoclonal antibodies against the 22-kDa isoform are used to complex and remove this isoform and the remaining GH is measured using a polyclonal immunoassay. An alternative method involves raising monoclonal antibodies specifically against the 20-kDa isoform [56]. Not only are there two GH isoforms but also GH can exist as homodimer, heterodimers, and multimer [57].

**TABLE 11.3** Circulating Forms of Growth Hormone (GH)

| |
|---|
| 22-kDa GH |
| 20-kDa GH |
| GH–GH (dimmer; a.k.a., "big" GH) |
| GH–GH binding protein (GHBP; "big-big" GH) |

Clinical measurement of GH is typically achieved by the use of an immunoassay, and because of the heterogeneity of GH, commutability of the assays is a problem. If monoclonal antibodies are used for capture, then only a subset of the GH molecules will be measured. Compared to an assay employing polyclonal antibodies, the measured concentrations can be higher due to an increase in the forms of GH detected. When immunoassays were first introduced for the detection of GH, commutability was less of a problem; however, as the assays increasingly utilize specific monoclonal antibodies, the differences among assays have become more pronounced. Another problem with commutability of the assays is the lack of a reference standard with which to standardize the assays. Various circulating GHs are listed in Table 11.3.

Indirect interferences in the assay also exist, adding more layers to the difficulty of GH measurement. In the body, up to 50% of GH is bound to GH binding protein. Depending on the epitope that a monoclonal antibody binds to on GH, steric hindrance (the inability of the epitope to be accessed by the antibody) can occur, thereby falsely decreasing the results. Another cause of falsely altered results can be a consequence of the use of therapeutic GH analogs. This was reported in the use of pegvisomant, a modified GH analog that is used as a GH receptor antagonist [58]. Patients taking this drug had falsely low GH concentrations in most of the assays tested. Because the concentration of pegvisomant was 1000-fold that of GH, the authors hypothesized that the drug was binding the capture antibody in the assay while blocking the formation of the sandwich with the second antibody, resulting in decreased levels.

In athletics, detection of GH abuse can be afforded by specifically measuring the 20- and 22-kDa GH isoforms separately. Both forms should be found in the plasma of normal individuals. However, the detection of exclusively the 22-kDa form would suggest the administration of exogenous GH because recombinant DNA-derived GH consists of only this isoform and lacks the 20-kDa isoform.

## Adrenocorticotrophic Hormone

The measurement of ACTH can be helpful in determining the etiology of decreased cortisol. ACTH is secreted from corticotrophic cells of the anterior pituitary in pulsatile bursts creating fluctuations in ACTH blood levels at timescales ranging from minutes to days. Corticotrophin-releasing hormone (CRH) mediates control of ACTH secretion based on cortisol levels. A negative feedback system operates between cortisol and CRH: As cortisol levels decrease, CRH is increased. CRH directly stimulates ACTH, and ACTH directly stimulates cortisol. As cortisol levels increase, CRH release is decreased. In addition to the feedback loop mentioned previously, the neurotransmitters serotonin and acetylcholine are also known to stimulate ACTH, whereas GABA inhibits its secretion. A number of other endogenous factors (fever, hypoglycemia, cytokines, anti-diuretic hormone, etc.) can affect the release of ACTH.

Aside from the endogenous regulation and variation, ACTH measurement is very sensitive to preanalytical variables. It has been shown that as little as 0.1% hemolysis can cause significant loss of ACTH immunoreactivity [59]. Storage temperature and time to centrifugation also play important roles in the stability of the ACTH prior to measurement. In evaluation of blood samples taken from 10 healthy volunteers, the storage temperature of the blood prior to centrifugation was found to affect stability [60]. A more recent study highlighted that the more critical factor affecting stability is the amount of time expired prior to centrifugation [61]. For samples that were kept at room temperature and stored at variable time periods prior to centrifugation, the authors saw no relevant change at 4 hr and a mean decrease of 10% in the ACTH concentration after 24 hr as measured with a manual radioactive assay. Based on their study, the authors recommended that centrifugation be performed within 4hr of the sample being drawn.

The methodology used to measure ACTH in the clinical laboratory has evolved from highly complex bioassays [62] and two-site immunoradiometric assays to automated immunoassays offering high throughput and fast turnaround times. A number of different assays are available, and similar to other antibody-based assays, standardization and agreeability are problematic. A 2011 publication compared 25 fresh-frozen plasma samples assayed by seven different ACTH assays in 35 different laboratories [63]. The authors reported a wide range of inter- and intra-assay coefficients of variation across these assays, ranging from less than 10% to greater than 20%. Not only did these assays show substantial differences with respect to imprecision but also they demonstrated variable agreement among the methods. Although the assays were able to correctly classify most patients with high or normal ACTH values, only 60% of measurements of samples with low ACTH concentrations were correctly classified. The conclusion from this study is one found widely across multiple analytes within the endocrine field: ACTH assays should be standardized.

## Thyroid-Stimulating Hormone

With the number of cases of thyroid disease on the rise, the measurement of TSH continues to be an important part of laboratory medicine. TSH measurements are frequently ordered by primary care physicians and endocrinologists. In addition, TSH measurements may be requested on a stat basis by emergency department physicians for suspected cases of thyroid storm, myxedema coma, or myxedematous heart failure. TSH concentrations in the serum can vary based on the time of day, with maximum levels achieved at 2:00–4:00 a.m. and minimum levels achieved between 6:00 and 10:00 p.m. The minimum level can be decreased by as much as 50% compared to the maximum level, which can make interpretation difficult; in euthyroid subjects, however, the TSH maximum and minimum will remain within the reference interval. Although TSH is a routinely measured analyte, it is still prone to many of the interferences that plague all immunoassays. A number of endogenous interferences have been reported, such as rheumatoid factor [33], heterophilic antibodies [64], and paraproteins [65]. Another confounding factor is that when exogenous T4 is ingested, the T4 (or free T4 (FT4)) measurements can become discordant from the TSH measurements. For example, suppose a patient has primary hypothyroidism with a low FT4 level and an elevated TSH level. With oral administration of thyroxine, the FT4 can return to the reference interval far more quickly than can the TSH, which may not return to the reference interval for several weeks.

TSH is typically measured by a two-site immunoassay, and interference from other endogenous antibodies is an ongoing problem. There have been reports of macro-TSH, in which TSH has complexed with an immunoglobulin, which is believed to have a decreased clearance resulting in the accumulation of these molecules, falsely increasing the TSH level. This is typically identified due to an unusually increased TSH level with normal free T3 and free T4 levels and no symptoms. Identification can be through the use of protein A beads to remove IgG-complexed TSH molecules [23], or it can be a diagnosis of exclusion due to negative thyroid studies with a persistently elevated TSH level [66]. There was an interesting case of macro-TSH in a neonate who was identified with increased TSH levels, but normal T4 and T3, on neonatal screening [24]. Both newborn and mother had elevated TSH levels that were due to macro-TSH, as shown by gel filtration chromatography. In the case of the newborn, the macro-TSH slowly decreased, only to completely disappear after 8 months of life, consistent with the rate of elimination of maternal IgG. TSH is unique in that the identification of an elevated TSH level in an otherwise euthyroid patient does not immediately identify a macro-TSH. A persistently elevated TSH in a patient with free T4 and T3 levels within the reference intervals may indicate subclinical hypothyroidism [67]. TSH with impaired biologic activity has a similar presentation to macro-TSH: an elevated TSH with a normal free T3 and T4 along with the presence of a higher molecular-weight TSH by chromatography [68,69]. Mutations in the TSH $\beta$ subunit have been reported [70]. A distinguishing characteristic for patients whose TSH is inactive is the reduction in TSH after the administration of exogenous free T3; if the TSH was elevated purely because of binding with IgG, exogenous T3 should not be able to influence the rate of clearance of TSH.

## Luteinizing Hormone/Follicle-Stimulating Hormone

Luteinizing hormone (LH) and follicle-stimulating hormone (FSH) are both gonadotropins that control the functional activity of the gonads. The release of LH and FSH from the anterior pituitary is stimulated by gonadotropin-releasing hormone. The release of LH/FSH is episodic. The levels of LH and FSH vary during the course of the menstrual cycle. Postmenopausal women have persistent elevations in LH and particularly FSH.

Initial LH measurements were performed using radioimmunoassay (RIA), which included the use of polyclonal antibodies. One limitation of the use of polyclonal antibodies in the RIA was cross-reactivity with hCG and an increased nonspecific background compared to the use of some monoclonal antibodies [71]. As this methodology was replaced by more easily automated immunoassays, monoclonal antibodies were incorporated into these assays. Although the performance of these assays was considered to be acceptable, the results obtained across platforms were not identical. Studies were initiated in an effort to better understand the differences among assays and there were multiple conclusions. For 11 different commercial companies manufacturing LH immunoassays, there were a total of 55 different monoclonal antibodies used [72]. These antibodies further differed in the LH epitopes that they recognized, including antibodies specific for the $\alpha$ subunit, the $\beta$ subunit, and the holomolecule (anti-$\alpha\beta$). The variety of antibodies used across differing vendors not only contributes to a lack of standardization but also can lead to very different concentrations measured by an assay, based on the characteristics of the antibodies used. Studies with these kits were undertaken using an international standard for LH (IS 80/552), and the authors were able to determine that the calibration curve differed among kits, adding to the discrepancy between the kits [73]. Similarly, these same problems were encountered when the RIA for FSH was converted to an immunometric assay. The immunometric assay gave results that were an average of 17% higher than the RIA results, whereas the LH results were 33% lower for the assays evaluated in this study [71]. This presented a unique problem in that LH:FSH ratios were used when evaluating a patient for polycystic ovarian syndrome (PCOS). In a study of 78 women presenting with polycystic ovaries on pelvic ultrasound and 59 controls with normal ovaries by ultrasound, by comparison of both the RIA and two different monoclonal immunoassays, the cutoff ratio for PCOS used would need to be re-established using the new monoclonal assays. The diagnostic ratio cutoffs determined using the RIA would not be relevant for these new assays [74]. Evaluation of different vendor assays in different patient populations has also further confirmed the variation seen between assays [75].

Similar to other hormone assay interferences, a macro-FSH has also been reported in the literature [76]. In this case, the presence of this macrocomplex resulted in a false-positive interference in a sandwich immunoassay. With regard to LH, a patient with a macrocomplex was also identified for this hormone in 2003 [77]. In this case report, the patient showed a high LH value, whereas other hormone studies were normal except for high concentrations of thyroperoxidase antibodies. The authors identified a macrocomplex through gel filtration and chromatography.

## Prolactin

Prolactin is a hormone released from the anterior pituitary. Its physiologic role is in lactation, where suckling increases PRL concentrations. Levels of this hormone are at their lowest at midday and peak at night just after the onset of deep sleep and can be stimulated in a number of events such as stress. Increases in estrogen can also stimulate secretion and gene transcription of PRL [78]. Multiple drug classes can have effects on PRL release. Antipsychotics are the most common drugs to cause increased PRL levels. Increases can also be seen with morphine and morphine analog use, whereas lithium can decrease PRL levels by 40% [79,80]. Very high levels of PRL can generally fall into two categories, either prolactinoma or macroprolactin. Prolactinoma describes a tumor in the anterior pituitary that secretes large amounts of PRL. Because of the high levels of PRL that are produced by this tumor, laboratory measurement of PRL in these patients can be challenging due to the high-dose hook effect. Frieze *et al.* [20]. describe a male patient who initially presented with headaches, personality changes, and bulging eyes. His initial PRL levels were slightly elevated at 164.5 ng/mL (reference interval, 1.6–18.8 ng/mL). A biopsy from a 10-cm mass in his anterior skull base and cranial fossa confirmed the presence of a prolactinoma. Due to this discrepancy in lab values with the clinical diagnosis of a large tumor, the original serum was retested, with serial dilutions eventually revealing a PRL of 26,000 ng/mL.

Macroprolactin, also known as big big PRL, has a molecular weight greater than 100 kDa, typically due to complex formation with an immunoglobulin. The characteristics of macroprolactin are similar to those of other macrocomplexes that result from immunoglobulins, typically IgG, binding to PRL, contributing to the larger molecular mass and increasing its persistence in the circulation. Classically, the presence of a macroprolactin does not contribute any diagnostic information because PRL in this form is not biologically active; however, macroprolactin can pose a serious problem when interpreting laboratory results. The high PRL results typically seen in patients with prolactinomas can also be encountered in patients with macroprolactin, potentially leading to misdiagnosis. In an effort to correctly classify patients, routine screening of hyperprolactinemic samples (females >500 mU/L and males >290 mU/L) for macroprolactin is recommended [81].

According to the literature, macroprolactin can account for up to 26% of hyperprolactinemia cases reported, depending on the commercial immunoassay used [81,82]. Smith *et al.* [83] identified 10 patient samples with elevated PRL levels due to macroprolactin. They then analyzed these samples at 18 different clinical laboratories using a total of nine different assay systems. Across the range of assay systems evaluated, they reported differences in PRL ranging from 2.3- to 7.8-fold, with the Roche Elecsys users reporting the highest PRL levels. These authors utilized the technique of gel chromatography to establish the presence of macroprolactin; however, the more frequently used technique was complex precipitation using PEG. This method is not without problems because there are reported interferences of PEG with the PRL immunoassay on a variety of analyzers [83]. In an effort to circumvent this issue, reference intervals have been determined for PRL values of PEG-treated samples. Evaluating samples from 146 healthy volunteers, reference intervals were established for Abbott Architect, ADVIA Centaur, Siemens Immulite, Beckman Coulter Access, Roche Elecsys, and Tosoh A1A [83]. The intervals were evaluated using samples with known macroprolactins and non-macroprolactin, as established by gel filtration chromatography. Another interference of PRL with PEG precipitation is increased concentrations of serum globulins, as seen in cases of hypergammaglobulinemia from IgG myeloma and HIV infection [84]. In both cases, it was concluded that monomeric PRL was being precipitated along with the immunoglobulins, leading to a falsely low level of PRL after PEG treatment.

# CHALLENGES IN MEASURING HUMAN CHORIONIC GONADOTROPIN

The measurement of hCG is predominantly used in the determination of pregnancy. Depending on the assay, the specimen may be either urine or blood and the measurement either qualitative (as in point-of-care) or quantitative. One of the major challenges in the measurement of hCG is the presence of different biologic forms of hCG. There exist at least six hCG variations, including intact hCG (nicked and non-nicked), the free $\beta$ subunit (hCG-$\beta$; nicked and non-nicked), and the free hCG $\alpha$ subunit (regular and hyperglycosylated). These different forms can be released in varied ratios based on disease state, and assays measuring this hormone can vary widely with respect to the different forms that are measured. Of the six forms listed previously, most assays focus on measuring either intact hCG alone or hCG-β plus intact hCG.

Urine pregnancy tests play an important role in both emergency departments (EDs) and radiology suites to detect the pregnancy status of patients where it may be unknown or to rule out pregnancy prior to radiation. The high-dose hook effect can have a serious effect on these tests, producing false negatives and potentially leading the physician in an incorrect direction. The hook effect has been reported to contribute to false-negative results in the ED in patients with molar pregnancies [85,86]. The presence of a hydatidiform mole can result in hCG concentrations greater than 1 million mIU/mL, and these cases are at high risk of false negatives due to the hook effect [86]. In these cases, dilutions of samples and reanalysis for hCG can help determine if the hook effect is contributing to a false-negative result. The sensitivity of a particular assay to the hook effect varies greatly depending on the immunoassay method used. To better understand the extent of this variation, a study evaluated six different commercial assays for hCG measurement [87]. Using a sample with a total hCG concentration of approximately 3,600,000 mIU/mL, the authors measured this sample on the six different platforms. Out of the six analyzers evaluated, four demonstrated the hook effect, whereas the remaining two did not.

In addition to the well-known high-dose hook effect, another cause of a false-negative result is a "hook-like effect" due to an increase in hCG-β core fragment (hCG-βcf). This fragment is formed by proteolytic degradation of hCG in the kidneys and is then excreted. The amount of hCG-βcf in the urine is first detected in the later stages of early pregnancy (5–8 weeks of gestation) until term, and it is hypothesized to bind and saturate the primary antibody in a sandwich immunoassay, essentially blocking any binding for other forms of hCG [88,89]. The second antibody in these methods does not recognize hCG-βcf because they were optimized for early pregnancy detection, and the result is a false negative [90]. This theory was proven by Gronowski *et al.* [92] through the evaluation of three urine samples with apparent false-negative results using a point-of-care test. In this study, the authors used specific hCG antibody assays to measure the different forms of hCG present in the urine samples. They found that in these false-negative samples, the hCG-βcf was in molar excess of the intact hCG (64–78% of all hCG immunoreactivity). They further showed that this was the fragment causing the false-negative result by adding 1 million pmol/L hCG-βcf to urine that had been previously determined to have a concentration of 17,800 IU/L hCG. The point-of-care urine hCG result changed from positive prior to the addition to negative after the hCG-βcf was added in two of the three point-of-care assays evaluated. This work was expanded to examine a fourth urine point-of-care pregnancy device, and it was found that false negatives due to the addition of hCG-βcf to urine varied with device lot number [91]. The interference from hCG-βcf also translated to the serum assay, where investigators found that in two of the nine assays evaluated, values with greater than a 50% reduction were observed in the presence of elevated hCG-βcf [92]. Although the issues concerning measurement of hCG-βcf are fairly well documented, the hCG hormone can exist in a number of different forms that can also make interpretation of hCG assays difficult. A case describes a discrepancy between results obtained using urine and plasma samples from the same patient [93]. A 41-year-old female presented with vaginal blood loss and was hospitalized. Both qualitative and quantitative assays were used to test two successive urine samples for hCG, with all results positive, whereas the corresponding plasma samples resulted in no hCG present. Through hCG isoform analysis, the authors were able to identify that the majority of hCG being produced was a degraded form of the β subunit that lacked the C-terminal peptide. This variant was thought to have a fast clearance, as a possible explanation for why it was undetectable in the plasma but measured repeatedly in the urine.

Although the occurrence of false-positive results is less frequent than that of false negatives, false-positive results are still encountered in the measurement of hCG. Specific to the hCG assay, false positives can be classified into two categories: a falsely increased hCG level or a positive pregnancy test when the patient is not pregnant. Analytically, the serum/plasma hCG assay is susceptible to many of the same interferences as any traditional immunoassay, such as heterophile antibodies [94]. There is even a report of macro-hCG involving IgM [95]. However, because these molecules are too large to be filtered by the glomeruli, the urine hCG test typically does not encounter these interferences. There have been cases of a true analytical interference causing a false-positive urine test [96], although these appear to be less frequent than false-positive serum hCG tests. In this particular case, the patient had a tubo-ovarian abscess from an intrauterine conception device with a positive enzyme-linked immunosorbent assay (ELISA)-based urine pregnancy kit. Pathologic evaluation showed a bacterial colony consistent with *Actinomyces* with no notation regarding the presence of trophoblastic tissue and a negative serum hCG result. The actual mechanism of interference in this case is still unresolved.

## CHALLENGES IN THYROID TUMOR MARKER TESTS

As noted previously, numerous interferences in thyroid immunoassays have been identified, including macro-TSH and T4 and T3 autoantibodies [97,98]. These interferences will not be recounted here. However, interferences in thyroglobulin (Tg) and calcitonin immunoassays were not mentioned previously and will be examined here [99].

### Thyroglobulin

Papillary and follicular carcinomas are differentiated thyroid cancers. Tg can be measured as a tumor marker of such thyroid cancers following surgical thyroidectomy [100]. Preoperative Tg measurements have no role in the diagnosis of thyroid cancer. However, preoperative confirmation that a differentiated thyroid cancer produces Tg would allow Tg to be used as a tumor marker in such cases postoperatively. Uncommonly, differentiated thyroid cancers will not produce Tg and therefore Tg production must be confirmed before it is used as a tumor marker [101]. Tg measurements do not have a role in the management of anaplastic thyroid carcinoma or medullary thyroid carcinoma (in which calcitonin is a tumor marker).

If a patient has had a complete thyroidectomy and there are no metastases, Tg should not be detectable several weeks postoperatively [102]. To ensure that any remaining functional thyroid tissue is maximally stimulated, thyroxine replacement is discontinued until the TSH is clearly elevated (which takes ~2 weeks) or recombinant DNA-produced TSH (thyrogen-$\alpha$) is injected. If the Tg remains undetectable, the prognosis is good and the patient can be retested for Tg in the future [103]. Many experts believe that such Tg measurements have replaced the need for repeated thyroid scans for the detection of thyroid tissue and metastases [104].

An important interference in Tg assays is the presence of thyroglobulin autoantibodies (TGA) [105]. Whereas the frequency of TGA in the general population is approximately 5%, for unclear reasons, TGA can be found in approximately 10% of patients with differentiated thyroid carcinomas. TGA interferences can be more significant for double-antibody (sandwich) assays than competitive immunoassays, but TGA can interfere in either assay format.

Although it might be expected that TGA would lower Tg measurements, there is no way to predict if the TGA interference will be a positive interference or a negative interference. Therefore, the best practice is to not report Tg measurements if TGA are present. This implies that every Tg measurement requires that TGA also be measured (which is a best practice) [106]. Recovery studies are not reliable to determine the "true" Tg level when TGA are present.

Although Tg should not be used to assess postoperative status in the presence of TGA, absolute TGA levels may substitute as a tumor marker. Here, the hypothesis is as follows: If TGA are detected in a postoperative patient with a history of differentiated thyroid cancer, the TGA would only be present if there were thyroid tissue remaining (either normal thyroid or thyroid cancer). Therefore, higher TGA levels would correlate with a higher postoperative tumor burden [107,108].

### Calcitonin

The tumor marker used in cases of medullary thyroid carcinoma (MTC) is calcitonin (CT). Like Tg, the CT value does not distinguish benign from malignant tumors. CT is followed postoperatively to monitor the completeness of surgical removal of the MTC and the possibility of local tumor recurrence or metastases. Because basal CT levels may not be detectable in MTC, CT should be measured after stimulation with intravenous calcium (pentagastrin is not currently available in the United States) [109,110]. Elevated calcitonin levels due to the presence of heterophile antibodies have been reported in approximately 1% of patients with thyroid nodules [111].

## ADRENAL FUNCTION TESTS

Cortisol is probably the most measured analyte for evaluating adrenal function, although aldosterone and rennin are also ordered frequently. In this section, pitfalls in testing these analytes are addressed.

### Cortisol

The levels of endogenous cortisol can be very sensitive to drugs that the patient may be taking. Such drugs can include various glucocorticoids or drugs that impair the synthesis of cortisol. Cortisol levels can be either suppressed or elevated depending on the specific drug. Ambrogio *et al.* [114] provide a fairly comprehensive review of the effects of drugs on the hypothalamic–pituitary–adrenal axis. In addition, there have been reports of assay interference in the patient population taking the drug metyrapone, which is used in the management of Cushing's syndrome. For patients taking this drug, the current guidelines recommend determining the dose of metyrapone based on the serum cortisol concentration. The mechanism of action for this drug is inhibition of the enzyme 11β-hydroxylase, which can lead to the accumulation of cortisol precursor steroids such as 11-deoxycortisol. In one study, the authors were able to identify that the increased cortisol precursor concentrations caused a positive interference in the cortisol immunoassay in two patients taking the drug [112]. They were able to identify the increased results from immunoassay by simultaneously measuring these samples by liquid chromatography–tandem mass spectrometry (LC-MS). Further study compared cortisol measurements made in three different groups: patients receiving metyrapone, patients diagnosed with Cushing not receiving any treatment, and patients with no adrenal pathology and no medication known to interfere with the cortisol immunoassay [113]. In these three different patient populations, they measured cortisol by both immunoassay and LC-MS. The two control groups not receiving metyrapone showed good agreement between immunoassay and LC-MS; in contrast, the metyrapone group showed a positive bias of 23%, and the difference between the two assays positively correlated with drug dose and 11-deoxycortisol concentration.

The immunoassay used to measure cortisol can also be affected by heterophile antibodies [114]. In one case, the interference led to a falsely low cortisol result, and the patient was given cortisol replacement for suspected adrenal insufficiency [115]. Due to the clinical doubt surrounding the patient's diagnosis, further workup with the samples was performed, which included measurement of cortisol with a different immunoassay and incubation with a heterophile antibody blocking tube. The authors were able to identify the source of the problem and stopped the patient's glucocorticoid replacement therapy.

Heterophile antibodies are just one contributing factor to the controversy surrounding the use of immunoassays [116]. There has been investigation into other ways to measure cortisol, aside from the use of immunoassays. The use of LC-MS has the potential to avoid some of the traditional pitfalls of using immunoassays for analyte measurements, such as interferences caused by endogenous immunoglobulins or the high-dose hook effect [117]. The success of an LC-MS assay relies heavily on the development and validation of the method. The mechanisms of interference when using mass spectrometry are very different from immunoassays. A source of potential interference in LC-MS is other analytes. These can be isobaric compounds that have the same molecular weight, such as cortisol and tetrahydro prednisolone, or the A + 2 isotopes of their unsaturated analogs, such as cortisone and cortisol [118]. There is also a report in which fenofibrate, which is not chemically or structurally related to cortisol, causes a falsely elevated urine free cortisol result by an LC-MS method [119]. This interference overlapped with the major transition of cortisol; however, a second mass transition was free from the interference and allowed for accurate quantitation. Another potential source of interference specific to LC-MS methods includes false elevations/depressions due to ion enhancements/suppression [120]. Ion suppression affects the amount of analyte that is introduced into the mass spectrometer. This can happen due to the presence of nonvolatile or less volatile solutes in the sample matrix, such as salts, ion pairing agents, and endogenous compounds [121]. This can also happen if other molecules being ionized have a higher mass or are more polar, which may decrease the ability of the analyte of interest to be ionized [122,123]. The use of tandem mass spectrometry (LC-MS/MS) and isotopically labeled analogous internal standards can minimize such interferences.

Measurement of urinary free cortisol (UFC) is a screening test used in cases of suspected Cushing syndrome. Although the use of LC-MS is increasing, traditional measurement techniques have often used immunoassay. The urine provides a different matrix with different potential interferences compared to the serum assay. One problem specific to urine is that the hydrophobic conjugate derivatives of cortisol are present in large excess concentrations when compared to free cortisol, providing an opportunity for cross-reactivity [124]. A study was performed comparing cortisol measurements by gas chromatography–mass spectrometry (GC-MS), LC-MS, and two commercial immunoassays [11]. The two immunoassays that were evaluated both overestimated the UFC concentrations by approximately 1.6- to 1.9-fold compared with GC-MS. Both ring-A dihydro- and tetrahydrometabolites contributed to this overestimation, and the positive interference was not seen in the GC-MS or LC-MS methods. Quantitation of UFC by immunoassay typically requires an extraction step with a solvent to release free cortisol from conjugated steroids, but unconjugated steroids can still be a potential source of interference [125]. A study evaluated the ability of a different sample preparation technique—solid phase extraction—to remove the excess steroids [126]. The authors were able to show that solid phase extraction was effective in removal of 6β-hydroxycortisol from urine samples, thereby eliminating this as a possible interference in treated urine samples.

## Aldosterone and Renin

Primary aldosteronism is being increasingly investigated as a possible secondary cause of hypertension in adults. From the laboratory perspective, this usually involves measuring both aldosterone and renin and calculating an aldosterone-to-renin ratio (ARR). Aldosterone is typically measured using either a radioimmunoassay or chemiluminescence immunoassay; however, LC-MS applications have also been developed for the measurement of this analyte [127]. The renin value in the ARR can be generated by measuring either the plasma renin activity (PRA) or the direct renin concentration (DRC). The actual method used to measure the effect of renin in the ARR can also impact the results; it has been reported that PRA is preferable to DRC when calculating the ARR [128]. The PRA is measured by allowing renin in plasma to convert endogenous angiotensinogen to angiotensin I and then measuring angiotensin I by immunoassay. DRC assays are usually measured using antibody-based testing, such as radioimmunoassay and chemiluminescent immunoassays.

Regarding endogenous variables affecting this assay, a number of factors can affect the ARR, such as posture, time of day, age, gender, sodium intake, potassium level, and renal dysfunction. A woman's menstrual cycle can also have an effect, and this has led to a study investigating whether women are more at risk for false positives in primary aldosteronism screening [128]. The authors concluded that ARRs generated using DRC in normal women during the luteal phase can lead to a higher incidence of false-positive ARRs in women than in men.

Multiple drugs can change the endogenous levels of aldosterone and plasma renin activity. Angiotensin-converting enzyme inhibitors, angiotensin receptor antagonists, and diuretics can all contribute to potential false elevations in PRA and should be interpreted with caution in these individuals [129]. Due to interference in the interpretation of the ARR, mineralocorticoid receptor antagonists and high-dose amiloride must be discontinued at least 6 weeks prior to testing. In the case of the other drugs, a detectable low PRA level does not exclude primary aldosteronism and an undetectably low PRA in patients taking these drugs should be considered highly suspicious. Adrenergic inhibitors also suppress renin secretion, which can then lead to suppressed aldosterone secretion in normal patients [130]. A 2002 study evaluated the bias introduced by a number of antihypertensive drugs on the ARR [131]. Measurements were made in samples from patients taking the drugs and the percentage changes from the control group were as follows: amlodipine, $-17 \pm 32\%$; atenolol, $62 \pm 82\%$; doxazosin, $-5 \pm 26\%$; fosinopril, $-30 \pm 24\%$; and irbesartan, $-43 \pm 27\%$. The measurements for PRA and aldosterone concentrations in this work utilized radioimmunoassays. Whereas drugs typically act to change endogenous levels of renin and/or aldosterone, steroids and other metabolites can cause false results due to cross-reactions with the antibodies used in the assays. It has been hypothesized that increases in 21-deoxyaldosterone and other steroids in patients with congenital adrenal hyperplasia due to 21-hydroxylase deficiency could be responsible for false increases in aldosterone measurements when compared with LC-MS [127]. Another study compared an automated chemiluminescence immunoassay with a radioimmunoassay both with and without extraction steps in samples obtained after saline infusion [132]. Although the vendors of the assays used in this study do provide a list of steroids that interfere with their assays, the authors speculate that other interfering substances may be playing a role in their discrepant results.

## TESTING OF PARATHYROID FUNCTION

Usually, parathyroid hormones and 25-hydroxyvitamin D are tested for evaluation of parathyroid function. In this section, the challenges in measuring these analytes are addressed.

### Parathyroid Hormone

Parathyroid hormone plays an important role in mineral and bone metabolism. PTH is secreted from the four parathyroid glands in response to decreased ionized calcium levels. The PTH secreted is often referred to as "intact PTH," and it refers to the form of the hormone that is 1–84 amino acids. With secretion, 1–84 amino acid PTH can be rapidly degraded to 7–84 amino acid PTH, which lacks biologic activity. Due to the existence of various forms of PTH with varying biological activity, the measurement of PTH by sandwich immunoassay techniques is not straightforward. The majority of assays in use today are either second- or third-generation assays. Both utilize antibodies directed at the C-terminal end of the hormone, but the second-generation assay has increased interference from the N-terminal truncated fragment (7–84 amino acids) [133] due to the location of binding of the C-terminal-directed antibody. The ratio of intact PTH (1–84) and the N-terminal truncated fragment (7–84) is approximately 1:1 in a healthy population but can vary based on the presence of disease [134]. In an effort to circumvent this interference, third-generation assays were developed so that the N-terminal antibody recognizes approximately the first 4 N-terminal amino acids, which are absent in the truncated 7–84 amino acid fragment. "True-intact," "whole," "bio-intact," and "cyclase activating" PTH are terms applied to the 1–84 amino acid PTH assays. There is still debate regarding whether the third-generation assays offer any improvement in sensitivity or specificity over the second-generation assays [135,136].

Similar to many other hormones measured by immunoassay, heterophile antibodies and rheumatoid factor interference have been a problem for PTH assays [137,138]. Tumors can be a source of excess PTH production, increasing the PTH level and making interpretation of results difficult. Although adenomas, hyperplasia, and carcinoma of the parathyroid are all causes of hyperparathyroidism, there have also been cases of nonparathyroid tumors producing PTH [139]. Surgery is the usual treatment for parathyroid masses, and intraoperative measurement of PTH can assist the surgeon in determining if all of the mass has been removed.

Exogenous factors can also have an effect on PTH, producing misleading interpretations of laboratory results. A study examined the effect of three different anesthetic techniques on intraoperative PTH levels [140]. Compared to preinduction levels, the intraoperative PTH levels measured at 3 min postinduction increased by 28% with monitored anesthetic care, 45% with general anesthesia with laryngeal mask airway, and 65% with laryngeal mask airway. Another interfering substance that can increase levels of PTH was identified as cinacalcet HCl, a calcimimetic. *In vivo*, the release of PTH is regulated by free ionized calcium in the blood, which interacts with the calcium-sensing receptor on the parathyroid gland. In the case of hyperparathyroidism, cinacalcet HCl is used to increase the sensitivity of the calcium-sensing receptor to ionized calcium, thereby reducing PTH levels [141]. A case study evaluated the dynamic changes in PTH and PTH fragments in a patient receiving cinacalcet HCl for secondary hyperparathyroidism [142]. This study utilized three different immunoassays for PTH: the intact PTH assay, which detects full-length PTH (1–84); the intact PTH assay, which detects both full-length (1–84) and non-full-length fragments; and an assay using an antibody specific to the C-terminal region of PTH. The authors found that the relationships between the PTH assays did not remain linear during therapy and that the PTH response after therapy was assay dependent.

## Assays for 25-Hydroxyvitamin D

The popularity of measuring 25-hydroxyvitamin D (25-OHD) has vastly increased since the introduction of the first assay to measure this analyte in 1971. The first assay was a competitive protein binding assay using vitamin D binding protein as the binding agent [143]. Since then, efforts have been made to create new assays that are easier to incorporate into the clinical laboratory. Today, multiple different methods are employed for the measurement of 25-OHD, including immunoassays, liquid chromatography, and LC-MS approaches. Immunoassays are currently the dominant choice among clinical laboratories for the measurement of 25-OHD, most likely due to their compatibility with the high volume demands of clinical laboratories. However, these assays are prone to the same interferences as all immunoassays, such as heterophile antibodies [144], and they have their own unique issues as well. In one study, investigators evaluated their 25-OHD assay by retesting stored samples that had been previously tested for 25-OHD [145]. Using the DiaSorin Liason for both the initial and repeat testing, they found that 5% of their results were discrepant from the original results. They further investigated the source of the discrepancy by using heterophile blocking tubes to assess the presence of heterophile antibodies and found that the samples did not differ significantly after treatment.

One of the largest interferences reported for immunoassay measurement of 25-OHD is vitamin D binding protein (VDBP). 25-OHD is hydrophobic, and therefore 25-OHD is tightly bound to VDBP in the circulation. Releasing 25-OHD from VDBP was traditionally done by using organic extraction protocols, but these solvents are not compatible with the automated immunoassays. One study compared five automated immunoassays, a radioimmunoassay, and two independent LC-MS methods [146]. The authors found excellent concordance across the LC-MS assays and outstanding performance by the DiaSorin radioimmunoassay. The automated immunoassays did not fare nearly as well, and it is speculated that the extraction step without aggressive solvents may be a contributing factor to this lack of comparability. It has also been discovered that inaccuracies found in immunoassays are VDBP concentration dependent [147]. The inability to effectively and reproducibly remove 25-OHD from VDBP often contributes to a negative bias, thereby falsely decreasing the total 25-OHD.

One way to resolve the problems encountered when immunoassays are used to measure 25-OHD is to use a more direct measurement approach such as LC-MS. The major analytes contributing to interference using this technique are from the 25-OHD$_3$ epimer and other isobaric compounds. The epimer, 3-epi-25-OHD$_3$, is the most abundant vitamin D metabolite, and it has the same molecular weight as vitamin $D_3$; it also forms the same mass-to-charge parent and product ions when ionized. There is an inverse relationship between patient age and epimer percentage, and in 172 children tested, the epimers contributed 8.7−61.1% of the total measured 25-OHD [148]. A newer LC-MS method is able to separate the epimers and other isobaric compounds, such as 3-epi-25-$OHD_3$, 1-α-$OHD_3$, and the bile precursor 7-α-hydroxy-4-cholesten-3-one, from 25-$OHD_2$ and 25-$OHD_3$ [149]. Interferences have also been noted using LC-MS when samples are collected in serum separator tubes [150]. Further investigation identified that only 25-$OHD_2$ was falsely increased upon storage of the samples in the serum separator tubes, and the authors were able to circumvent this problem by using an alternative transition for monitoring 25-$OHD_2$.

## GONADAL AND REPRODUCTIVE INTERFERENCES

The most important endocrine immunoassay interferences in the field of reproductive medicine are false-positive hCG results, as discussed previously [151]. HAMA, rheumatoid factor, and heterophile antibodies can affect LH and FSH measurements [71,76,152]. Assay specificity is an important issue for sex steroid measurements. Many experts in this field are utilizing tandem mass spectroscopy as the definitive method for steroid analysis [153].

Prior to immunoassays of the past approximately 20 years, some older LH assays cross-reacted with TSH. For example, in cases of severe primary hypothyroidism with a greatly elevated TSH level, there could be a false elevation in LH. If a woman of reproductive age with untreated or undertreated primary hypothyroidism had menstrual irregularity, primary hypogonadism might be mistakenly diagnosed on the basis of an elevated LH that really resulted from cross-reactivity with a massively elevated TSH.

Specificity of sex steroid measurements is very important in patient care [154−158], and possibly even more important in the analysis of sex steroids in athletes [159−161]. The laboratorian must be aware of all such possibilities and investigate possible interferences when an inquiry is made by a clinician [162,163].

Although not interferences, proper interpretation of sex steroid and gonadotropin results requires that the proper reference intervals be used. For women of reproductive age, this can be challenging because the LH, FSH, estradiol, and progesterone levels are constantly changing throughout the menstrual cycle. There is diurnal variation in testosterone, although in a healthy male these variations do not fall outside the reference interval [164]. Of interest, testosterone appears to rise with stress [165].

## PROSTATE

Possibly the best tumor marker in widespread clinical use is prostate-specific antigen (PSA) [166]. Total and free PSA can be measured [167]. Nevertheless, the use of PSA testing remains controversial [168].

PSA is measured by immunoassay [169]. Rarely (in 0.3% of samples), heterophile antibodies falsely raise PSA, although in the study reporting this finding, no adverse clinical consequences were identified [170]. In another report, heterophile antibodies raised the PSA concentration, leading to inappropriate therapy [171]. In post-prostatectomy patients treated for prostate cancer, if PSA rises without any clinical evidence of metastases or worsening metastases, heterophile antibodies should be considered and tested for [172,173]. PSA assays may be improved using F(ab')2 antibody fragments [174].

## TESTING FOR INSULIN-LIKE GROWTH FACTOR 1

Insulin-like growth factor 1 (IGF-1) is often measured in cases of GH disorders. IGF-1 is often preferred over GH due to its stability, and it has been shown to be useful in screening for GH deficiency. IGF-1 is a small molecule and is carried through the circulation attached to larger proteins, known as IGF-1 binding proteins, specifically IGF binding protein-3. Release of IGF-1 from the binding proteins is necessary for measurement, and incomplete liberation can cause discrepancies among results. Acid gel filtration chromatography is the best method available for accomplishing separation of IGF-1 from its binding proteins; however, this approach is not compatible with the typical clinical laboratory setting. An alternative method has been developed that includes dissociation by low pH, precipitation of the binding proteins by ethanol, and separation of free IGF-1 with centrifugation [54]. This method has two main problems: Not all the binding proteins are precipitated with ethanol, and precipitation of some of the IGF-1 falsely decreases the IGF-1 values [175−177]. To circumvent these problems, excess IGF-2 is introduced into the sample after the acid/ethanol precipitation to occupy the binding protein sites that would still remain [178]. Similar procedures have been employed when measuring IGF-2, with the exception of IGF-1 addition to saturate the binding protein [179]. Sandwich immunoassays have also been developed for the measurement of IGF-1 [180]. Heterophilic antibodies have also been shown to interfere and falsely decrease IGF-1 levels measured by the Immulite 2000 [34]. In this case, the low IGF-1 values suggested GH deficiency, but neither the clinical signs nor the GH stimulation test results supported this diagnosis. After treatment of the sample with a heterophile blocking tube, the IGF-1 results were within the reference interval.

Although IGF-1 is not as variable as GH, there are still multiple factors that can impact circulating levels. Age and sex are both related to IGF-1 levels to the extent that reference intervals change based on these demographics. A multicenter study evaluated 3961 healthy subjects aged 1 month to 88 years to describe age- and sex-specific reference intervals [181]. Using an automated chemiluminescence immunoassay system, it was found that in people younger than 20 years, females had higher serum IGF-1 levels with a mean peak of 410 μg/L at age 14 years, whereas males had a mean peak of 382 μg/L at age 16 years. The age dependence slowed with increasing age, and after 25 years males had higher IGF-1 levels than females. Diet, lifestyle, and ethnicity have been shown to correlate with IGF-1 levels [182,183]. Whereas GH levels can change significantly in the circulation over short periods of time, IGF-1 levels are relatively stable during the day [54] and the timing of sample collection generally does not impact IGF-1 results.

## MEASUREMENT OF OTHER HORMONES INCLUDING INSULIN

The normal endocrine products of the islets of Langerhans include insulin, glucagon, somatostatin, and pancreatic polypeptide [184]. The only situation in clinical medicine in which glucagon might be measured is in cases of suspected glucagonoma [185]. Glucagonoma can present with secretory diarrhea, glucose intolerance, or frank diabetes and rash (necrolytic migratory erythema). In the absence of longstanding diabetes, glucagon deficiency is rare [186]. The pre-analytical concerns in measuring glucagon were discussed previously in the Introduction. A macroglucagon (IgG bound to glucagon) has been reported to cause hyperglucagonemia [187]. Somatostatinomas may also present with diabetes and secretory diarrhea as well as cholelithiasis [188]. Somatostatin should be drawn into a prechilled EDTA tube with separation of the plasma from the cells within 2 hr of collection. The sample is shipped frozen. In pathologic circumstances, VIP-secreting islet tumors develop (i.e., VIPomas) that produce secretory diarrhea and flushing [189]. The collection of VIP has been discussed in the Introduction. Pancreatic polypeptidomas are not endocrinologically active because they do not secrete pancreatic polypeptide nor do they cause symptoms [190].

An important clinical interference in insulin assays is the common development of insulin antibodies in insulin-treated patients [191,192]. These antibodies may raise total insulin concentrations, making such measurements not interpretable in assessing endogenous insulin secretion in insulin-treated patients. Attempts have been made to measure "free insulin" after polyethylene glycol precipitation of the antibody-bound insulin complexes; however, free insulin has not proven to be a reliable measure of the biologically active insulin, and such measurements are not recommended [193].

For each insulin molecule released by the pancreas into the splenic vein, one molecule of C-peptide will also be released. Because injected insulin preparations lack C-peptide, C-peptide antibodies do not develop. Therefore, C-peptide can be measured as an index of endogenous insulin production in insulin-treated patients. An important caveat is that approximately 50% of the insulin secreted from the pancreas is cleared by the liver on its first pass. This reflects the fact that hepatocytes are a major target of insulin action. Because insulin clearance by the liver is greater than C-peptide clearance by the liver, in peripheral blood the molar ratio of C-peptide to insulin is greater than 1. Nevertheless, assuming that the hepatic clearances of insulin and C-peptide are comparable among individuals, C-peptide concentrations can be used as a proxy for endogenous insulin secretion.

Insulin autoantibodies can be found in individuals at increased risk for the development of type 1 diabetes as well as at the onset of type 1 diabetes [194]. Insulin autoantibodies arise spontaneously and are not the consequence of exposure to exogenous insulin injections. These autoantibodies are usually of very low titer compared to the insulin antibodies that develop in response to exogenous insulin administration [29]. Within 7 days of the initiation of insulin treatment of an insulin-naive individual, insulin antibodies can appear. Autoantibodies to proinsulin have also been described [195]. The insulin antibody assays do not discriminate autoantibodies from antibodies induced by insulin exposure. Note that insulin antibodies develop in response to both animal and human insulins.

The analytical specificity of insulin and C-peptide assays is an important consideration. The insulin immunoassay should not detect C-peptide, proinsulin, or partially cleaved proinsulin. Unfortunately, in both type 1 and type 2 diabetes, relative elevations in the secretion of proinsulin occur that likely reflect the failing nature of the β cell and/or β cells that are extremely stimulated [196]. Similarly, the C-peptide assay should not detect insulin, proinsulin, or partially cleaved proinsulin. Proinsulin-specific immunoassays have been reported [197,198]. Hyperproinsulinopathies are rare inborn errors that result from incomplete cleavage of the proinsulin to insulin plus C-peptide [199].

Another issue affecting insulin immunoassays is the detection of recombinant DNA-derived insulins whose sequences have been altered compared to human insulin. Laboratorians need to educate clinicians about the specificity of their insulin immunoassay. For example, the Roche insulin immunoassay does not detect glargine [200]. A 1999 publication reported on the development of an immunoassay specific for lispro insulin [201].

As with other hormone assays, standardization of insulin measurements across platforms has not been achieved but is an important goal for laboratory scientists [202,203]. Such work is underway [204–207]. Likewise, C-peptide assays are not standardized across platforms [208–210].

## PRENATAL TESTING

A variety of hormones and proteins are measured in second-trimester maternal serum in screening for neural tube defects (e.g., anencephaly and spina bifida), Down syndrome (trisomy 21), and Edward syndrome (trisomy 18). α-Fetoprotein, hCG, and unconjugated estriol constitute the triple screen, whereas the quadruple screen is constituted with the addition of dimeric inhibin B measurements [211,212]. The quadruple screen has a sensitivity of 80% and specificity of 95% for the detection of Down syndrome [213,214]. First-trimester prenatal screening for Down syndrome and Edward syndrome includes ultrasound-detected nuchal translucency and measurements of pregnancy-associated polypeptide-A (PAPP-A) and hCG [215].

A PubMed search (February 2012) revealed no analytical interferences for PAPP-A or α-fetoprotein. Estriol measurements should suffer from similar specificity problems as estradiol measurements and other steroids [216]. False-positive hCG measurements due to HAMA have been discussed. In a patient with testicular cancer, heterophile antibodies have been reported to raise hCG [217]. Heterophile antibodies producing a false elevation in inhibin A have been described [218]. In an ELISA format assay to measure inhibin A, a catalase inhibitor was recommended to prevent a decline in the measured inhibin A level in cases of delayed separation of serum from the whole blood [219].

## CONCLUSIONS

As discussed throughout this chapter, measuring hormones by immunoassay in the clinical laboratory can be technically demanding. Proper pre-analytical management of the sample is critical, especially for hormones that are particularly labile. The collection tubes, position of patient during phlebotomy, patient preparation, and the time of the day can all alter the results. Sample processing can also impact the results, as seen with ACTH, where the timing of centrifugation and storage of these samples can affect the results.

The analytical challenges in hormone immunoassays are multiple: hormones are typically present in the plasma in low concentrations, the biologically active hormone may only be the free (or unbound) fraction, hormones often circulate attached to binding proteins, hormones may exist in a variety of forms (e.g., the 22- and 20-kDa forms of GH), hormone metabolites may cross-react with the parent compound, and cross-reactivity between hormones—especially steroid hormones—may exist. In addition to these challenges are the analytical interferences discussed previously: even if the measurement of the hormone by immunoassay is theoretically straightforward, the opportunity exists for interference from HAMA or heterophile antibodies.

Even the best-validated assays may encounter some of the interferences discussed in this chapter. As laboratorians, we may not be able to prevent these interferences, but we can be prepared to recognize the possibility of such interferences when there is discordance between the laboratory result and clinical findings in the patient. In this way, the laboratorian can assist the clinician in the proper interpretation of the result. Further studies may be indicated, for example, to prove or disprove the possibility of HAMA interference. A number of tools for identifying the presence of interference, such as heterophile blocking tubes or PEG precipitation, are available. In the field of immunoassays, it is wise for both the clinician and the laboratorian to remember the Latin phrase "Caveat emptor," which translates as "Buyer beware."

### References

[1] Mantovani G, Spada A. Mutations in the Gs alpha gene causing hormone resistance. Best Pract Res Clin Endocrinol Metab 2006;20(4):501–13.

[2] Agrawal NK, Goyal R, Rastogi A, Naik D, Singh SK. Thyroid hormone resistance. Postgrad Med J 2008;84(995):473–7.

[3] Gallagher EJ, Leroith D, Karnieli E. The metabolic syndrome–from insulin resistance to obesity and diabetes. Med Clin North Am 2011;95(5):855–73.

[4] American Diabetes A. Diagnosis and classification of diabetes mellitus. Diabetes Care 2012;35(Suppl. 1):S64–71.

[5] Dimeski G. Interference testing. Clin Biochem Rev 2008;29 (Suppl. 1):S43–8.

[6] Raff H, Sluss PM. Pre-analytical issues for testosterone and estradiol assays. Steroids 2008;73(13):1297–304.

[7] Lightowler C, Carroll MJ, Chesser AMS, et al. Identification of Auto-Transplanted parathyroid tissue by Tc-99m Methoxy Isobutyl Isonitrile Scintigraphy. Nephrol Dial Transpl 1995;10 (8):1372–5.

[8] Ray JA, Kushnir MM, Rockwood AL, Meikle AW. Analysis of cortisol, cortisone and dexamethasone in human serum using liquid chromatography tandem mass spectrometry and assessment of cortisol: cortisone ratios in patients with impaired kidney function. Clin Chim Acta 2011;412(13–14):1221–8.

[9] Stanczyk FZ, Cho MM, Endres DB, Morrison JL, Patel S, Paulson RJ. Limitations of direct estradiol and testosterone immunoassay kits. Steroids 2003;68(14):1173–8.

[10] Turpeinen U, Hamalainen E, Haanpaa M, Dunkel L. Determination of salivary testosterone and androstendione by liquid chromatography-tandem mass spectrometry. Clin Chim Acta 2012;413(5–6):594–9.

[11] Wood L, Ducroq DH, Fraser HL, et al. Measurement of urinary free cortisol by tandem mass spectrometry and comparison with results obtained by gas chromatography-mass spectrometry and two commercial immunoassays. Ann Clin Biochem 2008;45(Pt 4):380–8.

[12] Strathmann FG, Ka MM, Rainey PM, Baird GS. Use of the BD vacutainer rapid serum tube reduces false-positive results for selected beckman coulter Unicel DxI immunoassays. Am J Clin Pathol 2011;136(2):325–9.

[13] Couchman L, Vincent RP, Ghataore L, Moniz CF, Taylor NF. Challenges and benefits of endogenous steroid analysis by LC-MS/MS. Bioanalysis 2011;3(22):2549–72.

[14] Blair IA. Analysis of estrogens in serum and plasma from postmenopausal women: past present, and future. Steroids 2010;75 (4–5):297–306.

[15] Rossi C, Calton L, Hammond G, et al. Serum steroid profiling for congenital adrenal hyperplasia using liquid chromatography-tandem mass spectrometry. Clin Chim Acta 2010;411(3–4):222–8.

[16] <http://labmed.ucsf.edu/labmanual/db/resource/Centaur_Cortisol.pdf>. [accessed 30.05.2012.

[17] Cao Z, Swift TA, West CA, Rosano TG, Rej R. Immunoassay of estradiol: unanticipated suppression by unconjugated estriol. Clin Chem 2004;50(1):160–5.

[18] Taieb J, Benattar C, Birr AS, Lindenbaum A. Limitations of steroid determination by direct immunoassay. Clin Chem 2002;48(3):583–5.

[19] Klee GG. Interferences in hormone immunoassays. Clin Lab Med 2004;24(1):1–18.

[20] Frieze TW, Mong DP, Koops MK. "Hook effect" in prolactinomas: case report and review of literature. Endocr Pract 2002;8 (4):296–303.

[21] Schofl C, Schofl-Siegert B, Karstens JH, et al. Falsely low serum prolactin in two cases of invasive macroprolactinoma. Pituitary 2002;5(4):261–5.

[22] Klonoff DC. Macroamylasemia and other immunoglobulin-complexed enzyme disorders. West J Med 1980;133(5):392–407.

[23] Sakai H, Fukuda G, Suzuki N, Watanabe C, Odawara M. Falsely elevated thyroid-stimulating hormone (TSH) level due to macro-TSH. Endocr J 2009;56(3):435–40.

[24] Rix M, Laurberg P, Porzig C, Kristensen SR. Elevated thyroid-stimulating hormone level in a euthyroid neonate caused by macro thyrotropin-IgG complex. Acta Paediatr 2011;100(9): e135–7.

[25] Gordon DL, Holmes E, Kovacs EJ, Brooks MH. A spurious markedly increased serum estradiol level due to an IgA lambda. Endocr Pract 1999;5(2):80–3.

[26] Pietras SM, Safer JD. Diagnostic confusion attributable to spurious elevation of both total thyroid hormone and thyroid hormone uptake measurements in the setting of autoantibodies: case report and review of related literature. Endocr Pract 2008;14(6):738–42.

[27] Massart C, Elbadii S, Gibassier J, Coignard V, Rasandratana A. Anti-thyroxine and anti-triiodothyronine antibody interferences in one-step free triiodothyronine and free thyroxine immunoassays. Clin Chim Acta 2009;401(1–2):175–6.

[28] Lai LC, Day JA, Clark F, Peaston RT. Spuriously high free thyroxine with the Amerlite MAB FT4 assay. J Clin Pathol 1994;47(2):181–2.

[29] Palmer JP, Asplin CM, Clemons P, et al. Insulin antibodies in insulin-dependent diabetics before insulin treatment. Science 1983;222(4630):1337–9.

[30] Wheatland R. Chronic ACTH autoantibodies are a significant pathological factor in the disruption of the hypothalamic-pituitary-adrenal axis in chronic fatigue syndrome, anorexia nervosa and major depression. Med Hypotheses 2005;65 (2):287–95.

[31] Fiad TM, Duffy J, McKenna TJ. Multiple spuriously abnormal thyroid function indices due to heterophilic antibodies. Clin Endocrinol 1994;41(3):391–5.

[32] Tate J, Ward G. Interferences in immunoassay. Clin Biochem Rev 2004;25(2):105–20.

[33] Georges A, Charrie A, Raynaud S, Lombard C, Corcuff JB. Thyroxin overdose due to rheumatoid factor interferences in thyroid-stimulating hormone assays. Clin Chem Lab Med 2011;49(5):873–5.

[34] Brugts MP, Luermans JG, Lentjes EG, et al. Heterophilic antibodies may be a cause of falsely low total IGF1 levels. Eur J Endocrinol 2009;161(4):561–5.

[35] Kricka LJ, Schmerfeld-Pruss D, Senior M, Goodman DB, Kaladas P. Interference by human anti-mouse antibody in two-site immunoassays. Clin Chem 1990;36(6):892–4.

[36] Pan XH, Zhang SZ, Chen HQ, Xiang MX, Wang JA. Spuriously high B-type natriuretic peptide level caused by human anti-mouse antibodies. Ann Intern Med 2011;155(6):407–8.

[37] John R, Henley R, Barron N. Antibody interference in a two-site immunometric assay for thyrotrophin. Ann Clin Biochem 1989;26(Pt 4):346–52.

[38] Santhana Krishnan SG, Pathalapati R, Kaplan L, Cobbs RK. Falsely raised TSH levels due to human anti-mouse antibody interfering with thyrotropin assay. Postgrad Med J 2006;82(973):e27.

[39] Wood JM, Gordon DL, Rudinger AN, Brooks MM. Artifactual elevation of thyroid-stimulating hormone. Am J Med 1991;90 (2):261–2.

[40] Labus JM, Petersen BH. Quantitation of human anti-mouse antibody in serum by flow cytometry. Cytometry 1992;13 (3):275–81.

[41] Ingels M, Rangan C, Morfin JP, Williams SR, Clark RF. Falsely elevated digoxin level of 45.9 ng/mL due to interference from human antimouse antibody. J Toxicol Clin Toxicol 2000;38 (3):343–5.

[42] Koper NP, Massuger LF, Thomas CM, Beyer C, Crooy MJ. An illustration of the clinical relevance of detecting human antimouse antibody interference by affinity chromatography. Eur J Obstet Gynecol Reprod Biol 1999;86(2):203–5.

[43] Morton BA, O'Connor-Tressel M, Beatty BG, Shively JE, Beatty JD. Artifactual CEA elevation due to human anti-mouse antibodies. Arch Surg 1988;123(10):1242–6.
[44] Oei AL, Sweep FC, Massuger LF, Olthaar AJ, Thomas CM. Transient human anti-mouse antibodies (HAMA) interference in CA 125 measurements during monitoring of ovarian cancer patients treated with murine monoclonal antibody. Gynecol Oncol 2008;109(2):199–202.
[45] Prince AM, Brotman B, Jass D, Ikram H. Specificity of the direct solid-phase radioimmunoassay for detection of hepatitis-B antigen. Lancet 1973;1(7816):1346–50.
[46] Schussler GC. The thyroxine-binding proteins. Thyroid 2000;10 (2):141–9.
[47] Bartalena L, Bogazzi F, Brogioni S, Burelli A, Scarcello G, Martino E. Measurement of serum free thyroid hormone concentrations: an essential tool for the diagnosis of thyroid dysfunction. Horm Res 1996;45(3-5):142–7.
[48] Ruiz M, Rajatanavin R, Young RA, et al. Familial dysalbuminemic hyperthyroxinemia: a syndrome that can be confused with thyrotoxicosis. N Engl J Med 1982;306(11):635–9.
[49] Ross HA, de Rijke YB, Sweep FC. Spuriously high free thyroxine values in familial dysalbuminemic hyperthyroxinemia. Clin Chem 2011;57(3):524–5.
[50] Kohn LT, Corrigan JM, Donaldson MS. To err is human: building a safer health system. Washington, D.C.: National Academy Press; 1999.
[51] Sakai H, Matsumoto K, Sugiyama M, et al. A case of factitious adrenal insufficiency after vascular graft surgery caused by spurious immunometric assays. Endocr J 2006;53(3):415–19.
[52] Plebani M, Carraro P. Mistakes in a stat laboratory: types and frequency. Clin Chem 1997;43(8 Pt 1):1348–51.
[53] Frystyk J. Exercise and the growth hormone-insulin-like growth factor axis. Med Sci Sports Exerc 2010;42(1):58–66.
[54] Frystyk J, Freda P, Clemmons DR. The current status of IGF-I assays–a 2009 update. Growth Horm IGF Res 2010;20(1):8–18.
[55] Alsat E, Guibourdenche J, Couturier A, Evain-Brion D. Physiological role of human placental growth hormone. Mol Cell Endocrinol 1998;140(1–2):121–7.
[56] Hashimoto Y, Ikeda I, Ikeda M, et al. Construction of a specific and sensitive sandwich enzyme immunoassay for 20 kDa human growth hormone. J Immunol Methods 1998;221(1–2):77–85.
[57] Baumann G, Stolar MW, Buchanan TA. The metabolic clearance, distribution, and degradation of dimeric and monomeric growth hormone (GH): implications for the pattern of circulating GH forms. Endocrinology 1986;119(4):1497–501.
[58] Paisley AN, Hayden K, Ellis A, Anderson J, Wieringa G, Trainer PJ. Pegvisomant interference in GH assays results in underestimation of GH levels. Eur J Endocrinol 2007;156 (3):315–19.
[59] Livesey JH, Dolamore B. Stability of plasma adrenocorticotrophic hormone (ACTH): influence of hemolysis, rapid chilling, time, and the addition of a maleimide. Clin Biochem 2010;43 (18):1478–80.
[60] Jane Ellis M, Livesey JH, Evans MJ. Hormone stability in human whole blood. Clin Biochem 2003;36(2):109–12.
[61] Reisch N, Reincke M, Bidlingmaier M. Preanalytical stability of adrenocorticotropic hormone depends on time to centrifugation rather than temperature. Clin Chem 2007;53(2):358–9.
[62] Liotta A, Krieger DT. A sensitive bioassay for the determination of human plasma ACTH levels. J Clin Endocrinol Metab 1975;40(2): 268–7
[63] Pecori Giraldi F, Saccani A, Cavagnini F. Study group on the hypothalamo-pituitary-adrenal axis of the italian society of E. Assessment of ACTH assay variability: a multicenter study. Eur J Endocrinol 2011;164(4):505–12.
[64] Ross HA, Menheere PP, Endocrinology Section of S, et al. Interference from heterophilic antibodies in seven current TSH assays. Ann Clin Biochem 2008;45(Pt 6):616.
[65] Imperiali M, Jelmini P, Ferraro B, et al. Interference in thyroid-stimulating hormone determination. Eur J Clin Invest 2010;40 (8):756–8.
[66] Mendoza H, Connacher A, Srivastava R. Unexplained high thyroid stimulating hormone: a "BIG" problem. BMJ Case Rep 2009.
[67] Khandelwal D, Tandon N. Overt and subclinical hypothyroidism: who to treat and how. Drugs 2012;72(1):17–33.
[68] Spitz IM, Le Roith D, Hirsch H, et al. Increased high-molecular-weight thyrotropin with impaired biologic activity in a euthyroid man. N Engl J Med 1981;304(5):278–82.
[69] Winter WE, Signorino MR. Review: molecular thyroidology. Ann Clin Lab Sci 2001;31(3):221–44.
[70] Partsch CJ, Riepe FG, Krone N, Sippell WG, Pohlenz J. Initially elevated TSH and congenital central hypothyroidism due to a homozygous mutation of the TSH beta subunit gene: case report and review of the literature. Exp Clin Endocrinol Diabetes 2006;114(5):227–34.
[71] Seth J, Hanning I, Bacon RR, Hunter WM. Progress and problems in immunoassays for serum pituitary gonadotrophins: evidence from the UK external quality assessment schemes, (EQAS) 1980–1988. Clin Chim Acta 1989;186(1):67–82.
[72] Costagliola S, Niccoli P, Florentino M, Carayon P. European collaborative study of luteinizing hormone assay: 1. Epitope specificity of luteinizing hormone monoclonal antibodies and surface mapping of pituitary and urinary luteinizing hormone. J Endocrinol Invest 1994;17(6):397–406.
[73] Costagliola S, Niccoli P, Florentino M, Carayon P. European collaborative study on luteinizing hormone assay: 2. Discrepancy among assay kits is related to variation both in standard curve calibration and epitope specificity of kit monoclonal antibodies. J Endocrinol Invest 1994;17(6):407–16.
[74] Milsom SR, Sowter MC, Carter MA, Knox BS, Gunn AJ. LH levels in women with polycystic ovarian syndrome: have modern assays made them irrelevant? BJOG 2003;110(8):760–4.
[75] Iwasa T, Matsuzaki T, Tanaka N, et al. Comparison and problems of measured values of LH, FSH, and PRL among measurement systems. Endocr J 2006;53(1):101–9.
[76] Webster R, Fahie-Wilson M, Barker P, Chatterjee VK, Halsall DJ. Immunoglobulin interference in serum follicle-stimulating hormone assays: autoimmune and heterophilic antibody interference. Ann Clin Biochem 2010;47(Pt 4):386–9.
[77] Vieira JG, Nishida SK, Faria De Camargo MT, Hauache OM, Monteiro De Barros Maciel R, Guimaraes V. "Macro LH": anomalous molecular form that behaves as a complex of luteinizing hormone (LH) and IgG in a patient with unexpectedly high LH values. Clin Chem 2003;49(12):2104–5.
[78] Winter WE, Jialal I, Vance ML, Bertholf RL. In: Burtis CA, Ashwood ER, Bruns DE, editors. Tietz textbook of clin chem and molecular diagnostics. USA: Elsevier; 1822.
[79] Chahal J, Schlechte J. Hyperprolactinemia. Pituitary 2008; 11(2):141–6.
[80] Basturk M, Karaaslan F, Esel E, Sofuoglu S, Tutus A, Yabanoglu I. Effects of short and long-term lithium treatment on serum prolactin levels in patients with bipolar affective disorder. Prog Neuropsychopharmacol Biol Psychiatry 2001;25 (2):315–22.
[81] Gibney J, Smith TP, McKenna TJ. The impact on clinical practice of routine screening for macroprolactin. J Clin Endocrinol Metab 2005;90(7):3927–32.
[82] Suliman AM, Smith TP, Gibney J, McKenna TJ. Frequent misdiagnosis and mismanagement of hyperprolactinemic patients

before the introduction of macroprolactin screening: application of a new strict laboratory definition of macroprolactinemia. Clin Chem 2003;49(9):1504–9.
[83] Beltran L, Fahie-Wilson MN, McKenna TJ, Kavanagh L, Smith TP. Serum total prolactin and monomeric prolactin reference intervals determined by precipitation with polyethylene glycol: evaluation and validation on common immunoassay platforms. Clin Chem 2008;54(10):1673–81.
[84] Ram S, Harris B, Fernando JJ, Gama R, Fahie-Wilson M. False-positive polyethylene glycol precipitation tests for macroprolactin due to increased serum globulins. Ann Clin Biochem 2008;45(Pt 3):256–9.
[85] Pang YP, Rajesh H, Tan LK. Molar pregnancy with false negative urine hCG: the hook effect. Singapore Med J 2010;51(3): e58–61.
[86] Hunter CL, Ladde J. Molar pregnancy with false negative beta-hCG urine in the emergency department. West J Emerg Med 2011;12(2):213–15.
[87] Al-Mahdili HA, Jones GR. High-dose hook effect in six automated human chorionic gonadotrophin assays. Ann Clin Biochem 2010;47(Pt 4):383–5.
[88] Kato Y, Braunstein GD. Beta-core fragment is a major form of immunoreactive urinary chorionic gonadotropin in human pregnancy. J Clin Endocrinol Metab 1988;66(6):1197–201.
[89] Stenman UH, Tiitinen A, Alfthan H, Valmu L. The classification, functions and clinical use of different isoforms of HCG. Hum Reprod Update 2006;12(6):769–84.
[90] Griffey RT, Trent CJ, Bavolek RA, Keeperman JB, Sampson C, Poirier RF. "Hook-like Effect" causes false-negative point-of-care urine pregnancy testing in emergency patients. J Emerg Med 2011.
[91] Gronowski AM, Powers M, Stenman UH, Ashby L, Scott MG. False-negative results from point-of-care qualitative human chorionic gonadotropin (hCG) devices caused by excess hCGbeta core fragment vary with device lot number. Clin Chem 2009;55(10):1885–6.
[92] Grenache DG, Greene DN, Dighe AS, et al. Falsely decreased human chorionic gonadotropin (hCG) results due to increased concentrations of the free beta subunit and the beta core fragment in quantitative hCG assays. Clin Chem 2010;56 (12):1839–44.
[93] Albersen A, Kemper-Proper E, Thelen MH, Kianmanesh Rad NA, Hoedemaeker RF, Boesten LS. A case of consistent discrepancies between urine and blood human chorionic gonadotropin measurements. Clin Chem Lab Med 2011;49(6):1029–32.
[94] Ballieux BE, Weijl NI, Gelderblom H, van Pelt J, Osanto S. False-positive serum human chorionic gonadotropin (HCG) in a male patient with a malignant germ cell tumor of the testis: a case report and review of the literature. Oncologist 2008;13 (11):1149–54.
[95] Heijboer AC, Martens F, Mulder SD, Schats R, Blankenstein MA. Interference in human chorionic gonadotropin (hCG) analysis by macro-hCG. Clin Chim Acta 2011;412(23–24):2349–50.
[96] Levsky ME, Handler JA, Suarez RD, Esrig ET. False-positive urine beta-HCG in a woman with a tubo-ovarian abscess. J Emerg Med 2001;21(4):407–9.
[97] Despres N, Grant AM. Antibody interference in thyroid assays: a potential for clinical misinformation. Clin Chem 1998;44 (3):440–54.
[98] Wu AH. Quality specifications in thyroid diseases. Clin Chim Acta 2004;346(1):73–7.
[99] Iervasi A, Iervasi G, Carpi A, Zucchelli GC. Serum thyroglobulin measurement: clinical background and main methodological aspects with clinical impact. Biomed Pharmacother 2006;60 (8):414–24.
[100] Harish K. Thyroglobulin: current status in differentiated thyroid carcinoma (review). Endocr Regul 2006;40(2):53–67.
[101] Nascimento C, Borget I, Al Ghuzlan A, et al. Persistent disease and recurrence in differentiated thyroid cancer patients with undetectable postoperative stimulated thyroglobulin level. Endocr Relat Cancer 2011;18(2):R29–40.
[102] Kloos RT. Papillary thyroid cancer: medical management and follow-up. Curr Treat Options Oncol 2005;6(4):323–38.
[103] Pagano L, Klain M, Pulcrano M, et al. Follow-up of differentiated thyroid carcinoma. Minerva Endocrinol 2004;29 (4):161–74.
[104] Mazzaferri EL, Massoll N. Management of papillary and follicular (differentiated) thyroid cancer: new paradigms using recombinant human thyrotropin. Endocr Relat Cancer 2002;9 (4):227–47.
[105] Spencer CA, Lopresti JS. Measuring thyroglobulin and thyroglobulin autoantibody in patients with differentiated thyroid cancer. Nat Clin Pract Endocrinol Metab 2008;4(4):223–33.
[106] Preissner CM, O'Kane DJ, Singh RJ, Morris JC, Grebe SK. Phantoms in the assay tube: heterophile antibody interferences in serum thyroglobulin assays. J Clin Endocrinol Metab 2003;88(7):3069–74.
[107] Spencer CA. Clinical review: clinical utility of thyroglobulin antibody (TgAb) measurements for patients with differentiated thyroid cancers (DTC). J Clin Endocrinol Metab 2011;96 (12):3615–27.
[108] Feldt-Rasmussen U, Rasmussen AK. Autoimmunity in differentiated thyroid cancer: significance and related clinical problems. Hormones 2010;9(2):109–17.
[109] Daniels GH. Screening for medullary thyroid carcinoma with serum calcitonin measurements in patients with thyroid nodules in the United States and Canada. Thyroid 2011;21 (11):1199–207.
[110] Colombo C, Verga U, Mian C, et al. Comparison of calcium and pentagastrin tests for the diagnosis and follow-up of medullary thyroid cancer. J Clin Endocrinol Metab 2012;97 (3):905–13.
[111] Giovanella L, Suriano S. Spurious hypercalcitoninemia and heterophilic antibodies in patients with thyroid nodules. Head Neck 2011;33(1):95–7.
[112] Owen LJ, Halsall DJ, Keevil BG. Cortisol measurement in patients receiving metyrapone therapy. Ann Clin Biochem 2010;47(Pt 6):573–5.
[113] Monaghan PJ, Owen LJ, Trainer PJ, Brabant G, Keevil BG, Darby D. Comparison of serum cortisol measurement by immunoassay and liquid chromatography-tandem mass spectrometry in patients receiving the 11beta-hydroxylase inhibitor metyrapone. Ann Clin Biochem 2011;48(Pt 5):441–6.
[114] Bolland MJ, Chiu WW, Davidson JS, Croxson MS. Heterophile antibodies may cause falsely lowered serum cortisol values. J Endocrinol Invest 2005;28(7):643–5.
[115] Saleem M, Lewis JG, Florkowski CM, Mulligan GP, George PM, Hale P. A patient with pseudo-Addison's disease and falsely elevated thyroxine due to interference in serum cortisol and free thyroxine immunoassays by two different mechanisms. Ann Clin Biochem 2009;46(Pt 2):172–5.
[116] Kricka LJ. Interferences in immunoassay–still a threat. Clin Chem 2000;46(8 Pt 1):1037–8.
[117] Hoofnagle AN, Wener MH. The fundamental flaws of immunoassays and potential solutions using tandem mass spectrometry. J Immunol Methods 2009;347(1–2):3–11.
[118] Kushnir MM, Rockwood AL, Roberts WL, Yue B, Bergquist J, Meikle AW. Liquid chromatography tandem mass spectrometry for analysis of steroids in clinical laboratories. Clin Biochem 2011;44(1):77–88.

[119] Meikle AW, Findling J, Kushnir MM, Rockwood AL, Nelson GJ, Terry AH. Pseudo-Cushing syndrome caused by fenofibrate interference with urinary cortisol assayed by high-performance liquid chromatography. J Clin Endocrinol Metab 2003;88(8):3521–4.

[120] Annesley TM. Ion suppression in mass spectrometry. Clin Chem 2003;49(7):1041–4.

[121] King R, Bonfiglio R, Fernandez-Metzler C, Miller-Stein C, Olah T. Mechanistic investigation of ionization suppression in electrospray ionization. J Am Soc Mass Spectrom 2000;11(11):942–50.

[122] Sterner JL, Johnston MV, Nicol GR, Ridge DP. Signal suppression in electrospray ionization Fourier transform mass spectrometry of multi-component samples. J Mass Spectrom 2000;35(3):385–91.

[123] Bonfiglio R, King RC, Olah TV, Merkle K. The effects of sample preparation methods on the variability of the electrospray ionization response for model drug compounds. Rapid Commun Mass Spectrom 1999;13(12):1175–85.

[124] Vogeser M, Parhofer KG. Liquid chromatography tandem-mass spectrometry (LC-MS/MS)–technique and applications in endocrinology. Exp Clin Endocrinol Diabetes 2007;115 (9):559–70.

[125] Lee C, Goeger DE. Interference of 6 beta-hydroxycortisol in the quantitation of urinary free cortisol by immunoassay and its elimination by solid phase extraction. Clin Biochem 1998;31 (4):229–33.

[126] Stowasser M, Ahmed AH, Pimenta E, Taylor PJ, Gordon RD. Factors affecting the aldosterone/renin ratio. Horm Metab Res 2012;44(3):170–6.

[127] Taylor PJ, Cooper DP, Gordon RD, Stowasser M. Measurement of aldosterone in human plasma by semiautomated HPLC-tandem mass spectrometry. Clin Chem 2009;55 (6):1155–62.

[128] Ahmed AH, Gordon RD, Taylor PJ, Ward G, Pimenta E, Stowasser M. Are women more at risk of false-positive primary aldosteronism screening and unnecessary suppression testing than men?. J Clin Endocrinol Metab 2011;96(2):E340–6.

[129] Young WF. Primary aldosteronism: renaissance of a syndrome. Clin Endocrinol 2007;66(5):607–18.

[130] Seifarth C, Trenkel S, Schobel H, Hahn EG, Hensen J. Influence of antihypertensive medication on aldosterone and renin concentration in the differential diagnosis of essential hypertension and primary aldosteronism. Clin Endocrinol 2002;57(4):457–65.

[131] Mulatero P, Rabbia F, Milan A, et al. Drug effects on aldosterone/plasma renin activity ratio in primary aldosteronism. Hypertension 2002;40(6):897–902.

[132] Schirpenbach C, Seiler L, Maser-Gluth C, Beuschlein F, Reincke M, Bidlingmaier M. Automated chemiluminescence-immunoassay for aldosterone during dynamic testing: comparison to radioimmunoassays with and without extraction steps. Clin Chem 2006;52(9):1749–55.

[133] John MR, Goodman WG, Gao P, Cantor TL, Salusky IB, Juppner H. A novel immunoradiometric assay detects full-length human PTH but not amino-terminally truncated fragments: implications for PTH measurements in renal failure. J Clin Endocrinol Metab 1999;84(11):4287–90.

[134] Winter WE, Harris NS. Calcium biology and disorders. In: Clarke W, editor. Contemporary practice in clinical chemistry. 2nd ed. Washington, D.C.: AACC Press; 2011.

[135] Boudou P, Ibrahim F, Cormier C, Chabas A, Sarfati E, Souberbielle JC. Third- or second-generation parathyroid hormone assays: a remaining debate in the diagnosis of primary hyperparathyroidism. J Clin Endocrinol Metab 2005;90 (12):6370–2.

[136] Silverberg SJ, Gao P, Brown I, LoGerfo P, Cantor TL, Bilezikian JP. Clinical utility of an immunoradiometric assay for parathyroid hormone (1–84) in primary hyperparathyroidism. J Clin Endocrinol Metab 2003;88(10):4725–30.

[137] Cavalier E, Carlisi A, Chapelle JP, Delanaye P. False positive PTH results: an easy strategy to test and detect analytical interferences in routine practice. Clin Chim Acta 2008;387(1–2):150–2.

[138] Levin O, Morris LF, Wah DT, Butch AW, Yeh MW. Falsely elevated plasma parathyroid hormone level mimicking tertiary hyperparathyroidism. Endocr Pract 2011;17(2):e8–11.

[139] Kandil E, Noureldine S, Khalek MA, Daroca P, Friedlander P. Ectopic secretion of parathyroid hormone in a neuroendocrine tumor: a case report and review of the literature. Int J Clin Exp Med 2011;4(3):234–40.

[140] Hong JC, Morris LF, Park EJ, Ituarte PH, Lee CH, Yeh MW. Transient increases in intraoperative parathyroid levels related to anesthetic technique. Surgery 2011;150(6):1069–75.

[141] Barman Balfour JA, Scott LJ. Cinacalcet hydrochloride. Drugs 2005;65(2):271–81.

[142] Henrich LM, Rogol AD, D'Amour P, Levine MA, Hanks JB, Bruns DE. Persistent hypercalcemia after parathyroidectomy in an adolescent and effect of treatment with cinacalcet HCl. Clin Chem 2006;52(12):2286–93.

[143] Haddad JG, Chyu KJ. Competitive protein-binding radioassay for 25-hydroxycholecalciferol. J Clin Endocrinol Metab 1971;33 (6):992–5.

[144] Holmes EW, Garbincius J, McKenna KM. Non-linear analytical recovery in the DiaSorin Liaison immunoassay for 25-hydroxy vitamin D. Clin Chim Acta 2011;412(23–24):2355–6.

[145] Becker N, McClellan AC, Gronowski AM, Scott MG. Inaccurate 25-hydroxyvitamin d results from a common immunoassay. Clin Chem 2012;58(5):948–50.

[146] Farrell CJ, Martin S, McWhinney B, Straub I, Williams P, Herrmann M. State-of-the-art vitamin D assays: a comparison of automated immunoassays with liquid chromatography-tandem mass spectrometry methods. Clin Chem 2012;58(3):531–42.

[147] Heijboer AC, Blankenstein MA, Kema IP, Buijs MM. Accuracy of 6 routine 25-hydroxyvitamin D assays: influence of vitamin D binding protein concentration. Clin Chem 2012;58(3):543–8.

[148] Singh RJ, Taylor RL, Reddy GS, Grebe SK. C-3 epimers can account for a significant proportion of total circulating 25-hydroxyvitamin D in infants, complicating accurate measurement and interpretation of vitamin D status. J Clin Endocrinol Metab 2006;91(8):3055–61.

[149] Shah I, James R, Barker J, Petroczi A, Naughton DP. Misleading measures in Vitamin D analysis: a novel LC-MS/MS assay to account for epimers and isobars. Nutr J 2011;10:46.

[150] Elder PA, Lewis JG, King RI, Florkowski CM. An anomalous result from gel tubes for vitamin D. Clin Chim Acta 2009;410 (1–2):95.

[151] Jones GR, Giannopoulos P. A method for routine detection of heterophile antibody interference in the Abbott AxSYM hCG assay. Pathology 2004;36(4):364–6.

[152] Segal DG, DiMeglio LA, Ryder KW, Vollmer PA, Pescovitz OH. Assay interference leading to misdiagnosis of central precocious puberty. Endocrine 2003;20(3):195–9.

[153] Honour JW. Steroid assays in paediatric endocrinology. J Clin Res Pediatr Endocrinol 2010;2(1):1–16.

[154] Wang C, Shiraishi S, Leung A, et al. Validation of a testosterone and dihydrotestosterone liquid chromatography tandem mass spectrometry assay: interference and comparison with established methods. Steroids 2008;73(13):1345–52.

[155] Sharp AM, Fraser IS, Caterson ID. Further studies on danazol interference in testosterone radioimmunoassays. Clin Chem 1983;29(1):141–3.

[156] Owen WE, Rawlins ML, Roberts WL. Selected performance characteristics of the Roche Elecsys testosterone II assay on the modular analytics E 170 analyzer. Clin Chim Acta 2010;411 (15−16):1073−9.

[157] Kane J, Middle J, Cawood M. Measurement of serum testosterone in women; what should we do? Ann Clin Biochem 2007;44(Pt 1):5−15.

[158] Cummings EA, Salisbury SR, Givner ML, Rittmaster RS. Testolactone-associated high androgen levels, a pharmacologic effect or a laboratory artifact? J Clin Endocrinol Metab 1998;83 (3):784−7.

[159] Shackleton C. Steroid analysis and doping control 1960−1980: scientific developments and personal anecdotes. Steroids 2009;74(3):288−95.

[160] Saudan C, Baume N, Robinson N, Avois L, Mangin P, Saugy M. Testosterone and doping control. Br J Sports Med 2006;40 (Suppl. 1):i21−4.

[161] Mulcahey MK, Schiller JR, Hulstyn MJ. Anabolic steroid use in adolescents: identification of those at risk and strategies for prevention. Phys Sportsmed 2010;38(3):105−13.

[162] Heald AH, Butterworth A, Kane JW, et al. Investigation into possible causes of interference in serum testosterone measurement in women. Ann Clin Biochem 2006;43(Pt 3):189−95.

[163] Warner MH, Kane JW, Atkin SL, Kilpatrick ES. Dehydroepiandrosterone sulphate interferes with the Abbott Architect direct immunoassay for testosterone. Ann Clin Biochem 2006;43(Pt 3):196−9.

[164] Collier CP, Morales A, Clark A, Lam M, Wynne-Edwards K, Black A. The significance of biological variation in the diagnosis of testosterone deficiency, and consideration of the relevance of total, free and bioavailable testosterone determinations. J Urol 2010;183(6):2294−9.

[165] Brennan PA, Herd MK, Puxeddu R, et al. Serum testosterone levels in surgeons during major head and neck cancer surgery: a suppositional study. Br J Oral Maxillofac Surg 2011;49 (3):190−3.

[166] Gomella LG, Liu XS, Trabulsi EJ, et al. Screening for prostate cancer: the current evidence and guidelines controversy. Can J Urol 2011;18(5):5875−83.

[167] Zhu L, Leinonen J, Zhang WM, Finne P, Stenman UH. Dual-label immunoassay for simultaneous measurement of prostate-specific antigen (PSA)-alpha1-antichymotrypsin complex together with free or total PSA. Clin Chem 2003;49 (1):97−103.

[168] Slomski A. USPSTF finds little evidence to support advising PSA screening in any man. JAMA 2011;306(23):2549−51.

[169] Datta P, Dasgupta A. Evaluation of an automated chemiluminescent immunoassay for complexed PSA on the Bayer ACS:180 system. J Clin Lab Anal 2003;17(5):174−8.

[170] Anderson CB, Pyle AL, Woodworth A, Cookson MS, Smith Jr JA, Barocas DA. Spurious elevation of serum PSA after curative treatment for prostate cancer: clinical consequences and the role of heterophilic antibodies. Prostate Cancer Prostatic Dis 2011.

[171] Henry N, Sebe P, Cussenot O. Inappropriate treatment of prostate cancer caused by heterophilic antibody interference. Nat Clin Pract Urol 2009;6(3):164−7.

[172] Park S, Wians Jr FH, Cadeddu JA. Spurious prostate-specific antigen (PSA) recurrence after radical prostatectomy: interference by human antimouse heterophile antibodies. Int J Urol 2007;14(3):251−3.

[173] Fritz BE, Hauke RJ, Stickle DF. New onset of heterophilic antibody interference in prostate-specific antigen measurement occurring during the period of post-prostatectomy prostate-specific antigen monitoring. Ann Clin Biochem 2009;46(Pt 3):253−6.

[174] Vaisanen V, Peltola MT, Lilja H, Nurmi M, Pettersson K. Intact free prostate-specific antigen and free and total human glandular kallikrein 2. Elimination of assay interference by enzymatic digestion of antibodies to F(ab')2 fragments. Anal Chem 2006;78(22):7809−15.

[175] Mesiano S, Young IR, Browne CA, Thorburn GD. Failure of acid-ethanol treatment to prevent interference by binding proteins in radioligand assays for the insulin-like growth factors. J Endocrinol 1988;119(3):453−60.

[176] Mohan S, Baylink DJ. Development of a simple valid method for the complete removal of insulin-like growth factor (IGF)-binding proteins from IGFs in human serum and other biological fluids: comparison with acid-ethanol treatment and C18 Sep-Pak separation. J Clin Endocrinol Metab 1995;80 (2):637−47.

[177] Clemmons DR. IGF-I assays: current assay methodologies and their limitations. Pituitary 2007;10(2):121−8.

[178] Blum WF, Breier BH. Radioimmunoassays for IGFs and IGFBPs. Growth Regul 1994;4(Suppl. 1):11−19.

[179] Blum WF, Ranke MB, Bierich JR. A specific radioimmunoassay for insulin-like growth factor II: the interference of IGF binding proteins can be blocked by excess IGF-I. Acta Endocrinol 1988;118(3):374−80.

[180] Khosravi MJ, Diamandi A, Mistry J, Lee PD. Noncompetitive ELISA for human serum insulin-like growth factor-I. Clin Chem 1996;42(8 Pt 1):1147−54.

[181] Brabant G, von zur Muhlen A, Wuster C, et al. Serum insulin-like growth factor I reference values for an automated chemiluminescence immunoassay system: results from a multicenter study. Horm Res 2003;60(2):53−60.

[182] DeLellis K, Rinaldi S, Kaaks RJ, Kolonel LN, Henderson B, Le Marchand L. Dietary and lifestyle correlates of plasma insulin-like growth factor-I (IGF-I) and IGF binding protein-3 (IGFBP-3): the multiethnic cohort. Cancer Epidemiol Biomarkers Prev 2004;13(9):1444−51.

[183] Juul A, Bang P, Hertel NT, et al. Serum insulin-like growth factor-I in 1030 healthy children, adolescents, and adults: relation to age, sex, stage of puberty, testicular size, and body mass index. J Clin Endocrinol Metab 1994;78(3):744−52.

[184] Goke B. Islet cell function: alpha and beta cells–partners towards normoglycaemia. Int J Clin Pract Suppl 2008;159:2−7.

[185] Chastain MA. The glucagonoma syndrome: a review of its features and discussion of new perspectives. Am J Med Sci 2001;321(5):306−20.

[186] Vidnes J, Oyasaeter S. Glucagon deficiency causing severe neonatal hypoglycemia in a patient with normal insulin secretion. Pediatr Res 1977;11(9 Pt 1):943−9.

[187] Yamamoto Y, Yamagishi S, Noto Y, et al. The first case of insulin-dependent diabetes mellitus with prominent spurious hyperglucagonemia due to interference of immunoglobulin G in glucagon radioimmunoassay (OAL-123) system. Horm Res 1996;45(6):295−9.

[188] Williamson JM, Thorn CC, Spalding D, Williamson RC. Pancreatic and peripancreatic somatostatinomas. Ann R Coll Surg Engl 2011;93(5):356−60.

[189] O'Dorisio TM, Mekhjian HS, Gaginella TS. Medical therapy of VIPomas. Endocrinol Metab Clin North Am 1989;18 (2):545−56.

[190] Klemanski DLO. Pancreatic neuroendocrine cancer. Armen Med Netw 2011;: Accessed 04.03.2012 from <http://www.health.am/cr/more/pancreatic-neuroendocrine-cancer/>.

[191] Agin A, Jeandidier N, Gasser F, Grucker D, Sapin R. Use of insulin immunoassays in clinical studies involving rapid-acting insulin analogues: Bi-insulin IRMA preliminary assessment. Clin Chem Lab Med 2006;44(11):1379−82.

[192] Sapin R. The interference of insulin antibodies in insulin immunometric assays. Clin Chem Lab Med 2002;40(7):705–8.

[193] Hanning I, Home PD, Alberti KG. Measurement of free insulin concentrations: the influence of the timing of extraction of insulin antibodies. Diabetologia 1985;28(11):831–5.

[194] Winter WE, Schatz DA. Autoimmune markers in diabetes. Clin Chem 2011;57(2):168–75.

[195] Kuglin B, Gries FA, Kolb H. Evidence of IgG autoantibodies against human proinsulin in patients with IDDM before insulin treatment. Diabetes 1988;37(1):130–2.

[196] Rhodes CJ, Alarcon C. What beta-cell defect could lead to hyperproinsulinemia in NIDDM? Some clues from recent advances made in understanding the proinsulin-processing mechanism. Diabetes 1994;43(4):511–17.

[197] Chevenne D, Deghmoun S, Coric L, Nicolas M, Levy-Marchal C. Evaluation of an ELISA assay for total proinsulin and establishment of reference values during an oral glucose tolerance test in a healthy population. Clin Biochem 2011;44(16):1349–51.

[198] Wu TJ, Lin CL, Taylor RL, Kao PC. Proinsulin level in diabetes mellitus measured by a new immunochemiluminometric assay. Ann Clin Lab Sci 1995;25(6):467–74.

[199] Roder ME, Vissing H, Nauck MA. Hyperproinsulinemia in a three-generation Caucasian family due to mutant proinsulin (Arg65-His) not associated with imparied glucose tolerance: the contribution of mutant proinsulin to insulin bioactivity. J Clin Endocrinol Metab 1996;81(4):1634–40.

[200] Kim S, Yun YM, Hur M, Moon HW, Kim JQ. The effects of anti-insulin antibodies and cross-reactivity with human recombinant insulin analogues in the E170 insulin immunometric assay. Korean J Lab Med 2011;31(1):22–9.

[201] Bowsher RR, Lynch RA, Brown-Augsburger P, et al. Sensitive RIA for the specific determination of insulin lispro. Clin Chem 1999;45(1):104–10.

[202] Sapin R. Insulin assays: previously known and new analytical features. Clin Lab 2003;49(3–4):113–21.

[203] Staten MA, Stern MP, Miller WG, Steffes MW, Campbell SE, Insulin Standardization W. Insulin assay standardization: leading to measures of insulin sensitivity and secretion for practical clinical care. Diabetes Care 2010;33(1):205–6.

[204] Marcovina S, Bowsher RR, Miller WG, et al. Standardization of insulin immunoassays: report of the American diabetes association workgroup. Clin Chem 2007;53(4):711–16.

[205] Miller WG, Thienpont LM, Van Uytfanghe K, et al. Toward standardization of insulin immunoassays. Clin Chem 2009;55 (5):1011–18.

[206] Robbins DC, Andersen L, Bowsher R, et al. Report of the American diabetes association's task force on standardization of the insulin assay. Diabetes 1996;45(2):242–56.

[207] Rodriguez-Cabaleiro D, Van Uytfanghe K, Stove V, Fiers T, Thienpont LM. Pilot study for the standardization of insulin immunoassays with isotope dilution liquid chromatography/ tandem mass spectrometry. Clin Chem 2007;53(8):1462–9.

[208] Cabaleiro DR, Stockl D, Kaufman JM, Fiers T, Thienpont LM. Feasibility of standardization of serum C-peptide immunoassays with isotope-dilution liquid chromatography-tandem mass spectrometry. Clin Chem 2006;52(6):1193–6.

[209] Koskinen P. Nontransferability of C-peptide measurements with various commercial radioimmunoassay reagents. Clin Chem 1988;34(8):1575–8.

[210] Little RR, Rohlfing CL, Tennill AL, et al. Standardization of C-peptide measurements. Clin Chem 2008;54(6):1023–6.

[211] Fang YM, Benn P, Campbell W, Bolnick J, Prabulos AM, Egan JF. Down syndrome screening in the United States in 2001 and 2007: a survey of maternal-fetal medicine specialists. Am J Obstet Gynecol 2009;201(1):97, e1–5.

[212] Reynolds T. The triple test as a screening technique for down syndrome: reliability and relevance. Int J Womens Health 2010;2:83–8.

[213] Kazerouni NN, Currier B, Malm L, et al. Triple-marker prenatal screening program for chromosomal defects. Obstet Gynecol 2009;114(1):50–8.

[214] Rawlins ML, La'ulu SL, Erickson JA, Roberts WL. Performance characteristics of the Access Inhibin A assay. Clin Chim Acta 2008;397(1–2):32–5.

[215] Knutsen-Larson S, Flanagan JD, Van Eerden P, Stein QP. The first-trimester screen in clinical practice. S D Med 2009;62 (10):389 92–93

[216] Schwarz S, Boyd J. Interference of danazol with the radioimmunoassay of steroid hormones. J Steroid Biochem 1982;16 (6):823–6.

[217] Gallagher DJ, Riches J, Bajorin DF. False elevation of human chorionic gonadotropin in a patient with testicular cancer. Nat Rev Urol 2010;7(4):230–3.

[218] Lambert-Messerlian G, Bandera C, Eklund E, Neuhauser A, Canick J. Very high inhibin A concentration attributed to heterophilic antibody interference. Clin Chem 2007; 53(4):800–1.

[219] Thirunavukarasu PP, Wallace EM. Measurement of inhibin A: a modification to an enzyme-linked immunosorbent assay. Prenat Diagn 2001;21(8):638–41.

CHAPTER

# 12

# Pitfalls in Tumor Markers Testing

*Alyaa Al-Ibraheemi, Amitava Dasgupta, Amer Wahed*

University of Texas Health Sciences Center at Houston, Houston, Texas

## INTRODUCTION

Tumor markers are molecules that are produced by cancer cells or by other cells of the body in response to cancer or under certain benign (noncancerous) conditions. Most tumor markers are produced by normal cells as well as by cancer cells, but in the process of developing cancer, concentrations of these markers are elevated manyfold compared to very low concentrations of these markers observed in blood under noncancerous conditions [1]. In addition to blood, these markers are also found in urine, stool, or bodily fluids of patients with cancer. More than 20 different tumor markers have been characterized and are used clinically for diagnosis and monitoring of treatment. Some tumor markers are elevated with only one type of cancer, whereas others are associated with two or more cancer types. There is no "universal" tumor marker for detection of a particular type of cancer. Common tumor markers are summarized in Table 12.1.

Although theoretically any type of biological molecule can act as a tumor marker, in practice, most markers are either proteins or glycoproteins. However, low-molecular-weight substances such as vanillylmandelic acid and homovanillic acid in neuroblastoma are also used, and nucleic acids (both DNA and RNA) are currently being evaluated as possible tumor markers. Patterns of gene expression and changes to DNA are also under intense investigation for use as tumor markers. These types of markers are measured specifically in tumor tissues [1]. Most of the traditionally used markers are probably not involved in tumorigenesis but are likely to be by-products of malignant transformation. In the future, however, molecules involved in carcinogenesis or cancer progression are likely to find increasing use as markers [2].

## CLINICAL USES OF TUMOR MARKERS

Tumor markers can be used for one of five purposes [3]:

1. Screening a healthy population or a high-risk population for the probable presence of cancer
2. Diagnosis of cancer or of a specific type of cancer
3. Evaluating prognosis in a patient
4. Predicting potential response of a patient to therapy
5. Monitoring recovery of a patient who has undergone surgery, radiation, or chemotherapy.

### Screening and Early Detection of Cancer

Tumor markers were first developed to test for cancer in people without symptoms, but very few markers are effective in achieving this goal. Today, the most widely used tumor marker in the clinical setting is prostate-specific antigen (PSA). In addition, only a few markers that are available have clinically useful predictive values for cancer at an early stage, when patients at high risk are tested.

### Diagnosing Cancer

Tumor markers are not the gold standard for diagnosis of a cancer. In most cases, a suspected cancer can only be diagnosed by a biopsy. $\alpha$-Fetoprotein (AFP) is an example of a tumor marker that can be used to aid in diagnosis of cancer, especially hepatocellular carcinoma (HCC). Although the level of AFP can also be increased in some liver diseases, when it reaches a certain threshold, it is usually indicative of HCC.

*Accurate Results in the Clinical Laboratory.*
DOI: http://dx.doi.org/10.1016/B978-0-12-415783-5.00012-8

TABLE 12.1 Commonly Used Tumor Markers

| Marker | Uses |
|---|---|
| Prostate-specific antigen (PSA) | Prostate carcinoma |
| CA-125 | Ovarian and fallopian carcinoma |
| α-Fetoprotein | Hepatocellular carcinoma and germ cell tumors |
| Carcinoembryonic antigen (CEA) | Colorectal, gastric, pancreatic, lung, and breast carcinomas |
| CA-19-9 | Pancreatic carcinoma and cholangiocarcinoma |
| $\beta_2$-Microglobin (B2M) | Multiple myeloma and lymphoma |
| β-Human chorionic gonadotropin (β-hCG) | Choriocarcinoma and testicular carcinoma |

### Determining the Prognosis for Certain Cancers

Some types of cancer grow and spread faster than others, whereas some cancers also respond well to various therapies. Sometimes, the level of a tumor marker can be useful in predicting the behavior and outcome for certain cancers. For example, in testicular cancer, very high levels of a tumor marker such as human chorionic gonadotropin (hCG) or AFP may indicate an aggressive cancer with poor survival outcome. Patients with these high levels may require very aggressive therapy even at the initiation of cancer therapy.

### Response to Therapy

Certain markers found in cancer cells can be used to predict whether a certain treatment is likely to produce a favorable outcome or not. For example, in breast and stomach cancer, if the cells have too much of a protein termed human epidermal growth factor receptor 2 (HER2), drugs such as trastuzumab (Herceptin) can be helpful if used during chemotherapy. If there is normal expression of HER2, these drugs may not produce expected therapeutic benefits. Therefore, expression of HER2 is routinely ordered prior to chemotherapy.

### Detecting Recurrent Cancer

Tumor markers are also used to identify recurrence of certain tumors after successful therapy. Certain tumor markers may be useful for further evaluation of a patient after completion of the treatment when there is no obvious sign of cancer in the body. The following are commonly measured tumor markers in clinical laboratories:

- PSA for prostate cancer
- hCG for gestational trophoblastic tumors and some germ cell tumors
- AFP for certain germ cell tumors and HCC
- CA-125 (carbohydrate antigen or cancer antigen 125) for ovarian cancer
- CA-19-9 (carbohydrate antigen 19-9) for pancreatic and gastrointestinal cancers
- Carcinoembryonic antigen (CEA) for colon and rectal cancers.

Less commonly monitored tumor markers include the following:

- CA-15-3 (cancer antigen 15-3), a marker for breast cancer
- CA-72-4 (cancer antigen 72-4), a marker for colorectal cancer
- CYFRA 21-1 (cytokeratin fragment), a marker for lung cancer
- Squamous cell carcinoma antigen, a marker of squamous cell lung cancer
- Neuron-specific enolase, a marker for lung cancer
- Chromogranin A, a marker for neuroendocrine tumor.

In this chapter, emphasis is placed on discussion of common tumor markers, including pitfalls in measuring these markers.

## PROSTATE-SPECIFIC ANTIGEN

PSA is a serine protease belonging to the kallikrein family and is also a single-chain glycoprotein containing 237 amino acid and four carbohydrate side chains (molecular weight, 28,430 Da). PSA is expressed by both normal and neoplastic prostate tissue.

### PSA Expression and Processing

Under normal conditions, PSA is produced as a proantigen (proPSA) by the secretory cells that line the prostate glands and secreted into the lumen, where the propeptide moiety is removed to generate active PSA. The active PSA can then undergo proteolysis to generate inactive PSA, of which a small portion then enters the bloodstream and circulates in an unbound state (free PSA). Alternatively, active PSA can diffuse directly into the circulation, where it is rapidly bound to protease inhibitors, including $\alpha_1$-antichymotrypsin (ACT) and $\alpha_2$-macroglobulin [4].

Although generating less PSA per cell than normal tissue, prostate cancer tissues lack basal cells, resulting in the disruption of the basement membrane and normal lumen architecture. As a result, the secreted proPSA and several truncated forms can be released into the circulation directly resulting in more PSA "leaked" into the blood. In addition, a larger fraction of the PSA produced by malignant tissue escapes proteolytic processing. The end result is that more PSA is found in blood in patients with prostate cancer compared to patients with no cancer.

In men with a normal prostate, the majority of free PSA in the serum reflects the mature protein that has been inactivated by internal proteolytic cleavage. In contrast, this cleaved fraction is relatively decreased in patients with prostate cancer. Thus, the percentage of free or unbound PSA is lower in the serum of men with prostate cancer and, conversely, the amount of complexed PSA (PSA complexed to $\alpha_1$-antichymotrypsin) is higher compared with those of men who have a normal prostate or benign prostatic hyperplasia (BPH) [5,6]. This finding has been used in the use of the ratio of free to total PSA and complexed PSA (cPSA) as a means of distinguishing between prostate cancer and BPH as a cause of an elevated PSA. Causes of elevated PSA include the following:

- BPH
- Prostate cancer
- Prostatic inflammation/infection
- Perineal trauma.

## Benign Prostatic Hyperplasia

The most common explanation for an elevated serum PSA is BPH due to the high prevalence of this condition in men older than the age of 50 years. Serum PSA levels overlap considerably in men with BPH and those with prostate cancer. For example, one published report retrospectively examined preoperative serum PSA in 187 men with a histologic diagnosis of BPH on a transurethral resection of the prostate (TURP) specimen and 198 men with organ-confined prostate cancer as determined by step-section analysis of a radical prostatectomy specimen [7]. The median serum PSA concentrations were 3.9 ng/mL (range, 0.2–55 ng/mL) and 5.9 ng/mL (range, 0.4–58 ng/mL), respectively. Although this difference was statistically significant, the distribution of serum PSA values in both groups overlapped considerably with a clustering of PSA values below 10.0 ng/mL (90 and 73%, respectively). A potentially confounding problem is that medical treatment for BPH can also reduce serum PSA concentrations.

## Elevated PSA in Prostate Cancer and Other Conditions

Studies in the 1980s confirmed that serum total PSA could be used as a screening tool to identify men with prostate cancer because elevated serum PSA is clearly a more sensitive marker than digital rectal examination. Although the majority of prostate cancers express PSA, between 20 and 50% of men with newly diagnosed prostate cancers in the United States have serum PSA values below 4.0 ng/mL (upper end of normal is usually considered as 4.0 ng/mL), indicating that PSA lacks specificity as a tumor marker. In general, patients with PSA below 4.0 ng/mL are more likely to have prostate cancer that is confined to the organ. These patients have a better prognosis than patients with prostate cancer who show levels above 4.0 ng/mL [8].

Prostatitis with or without active infection is an important cause of elevated PSA, and levels as high as 75 ng/mL have been reported in the literature. Thus, many physicians often initially treat a man with an isolated elevated serum PSA with antibiotics for a presumed diagnosis of prostatitis and then obtain a repeat serum PSA for further clinical evaluation. The percentage of free PSA may be less affected by the presence of inflammation, particularly when the total serum PSA is less than 10 ng/mL. However, at least one study suggested that the free-to-total ratio of serum PSA is unable to distinguish chronic inflammation from prostate cancer because both conditions may lower the percentage of free PSA. This would be expected because inflammation leads to elevated serum PSA in a similar manner as prostate cancer through disruption of the basal membrane and increased leakage of "immature" PSA into the bloodstream [9].

Any perineal trauma can also increase the serum PSA. Prostate massage and digital rectal examination may cause minor transient elevations that may be clinically insignificant. For example, in one study, 2750 healthy men and patients older than the age of 40 years undergoing digital rectal examination were divided into four groups based on their initial serum PSA [10]. The two groups with the lowest initial serum PSA values (0.1–4 and 4.1–10 ng/mL) had statistically insignificant changes in the serum PSA after digital rectal examination, whereas PSA increased in the group with an initial serum PSA of 10.1–20 ng/mL, indicating a trend toward statistical significance. In contrast, patients with an initial serum PSA greater than 20 ng/mL had statistically significant increases in PSA after digital rectal examination. Thus, it is reasonable to perform PSA testing without regard to whether a patient has had a recent digital rectal examination.

Mechanical manipulation of the prostate by cystoscopy, prostate biopsy, or TURP can more significantly affect the serum PSA. Vigorous bicycle riding has been reported to cause substantial elevations in serum PSA, but this is not a consistent finding [11]. Sexual activity can minimally elevate the PSA (usually in the 0.4 to 0.5 ng/mL range) for approximately 48–72 hr after ejaculation.

## PSA Testing

Emerging concepts regarding PSA testing that may help refine the interpretation of an elevated concentration include the following:

- PSA density
- PSA velocity
- Free versus complexed or bound PSA.

These modifications would presumably be most useful for prostate cancer screening when the total PSA is between 2.5 and 10.0 ng/mL, the range for which decisions regarding further diagnostic testing are most difficult. To more directly compensate for BPH and prostate size, transrectal ultrasound (TRUS) has been used to measure prostate volume. Serum PSA is then divided by prostate volume to obtain PSA density, with higher PSA density values (>0.15 ng/mL/cc) being more suggestive of prostate cancer and lower values more suggestive of BPH. Although an early study suggested that PSA density was a promising method for distinguishing patients with benign and malignant prostate disease, subsequent reports have found considerable overlap in PSA densities between these groups. One multicenter study that compared PSA density versus PSA for the early detection of prostate cancer found that almost half of the cancers would have been missed using 0.15 ng/mL/cc as a cutoff for biopsy. There are also inherent difficulties in measuring PSA density, which include errors of prostatic volume measurement with TRUS and an intrapatient variation of up to 15% in PSA density with repeated measurements [12]. Furthermore, although PSA density can reliably differentiate between large groups of patients with BPH and prostate cancer, the ability to extrapolate these data to the individual patient is quite limited. Most important, the requirement for TRUS substantially increases cost and discomfort, and it is impractical for widespread screening purposes.

Another approach has been to assess the rate of PSA change over time (the PSA velocity). An elevated serum PSA that continues to rise over time is more likely to reflect prostate cancer than one that is consistently stable. For practical purposes, the clinical usefulness of PSA velocity is in part limited by intrapatient variability in the serum PSA; at least three consecutive measurements should be performed [13]. A longer time over which values are continuously measured can be useful in reducing the general variation in the PSA measurements. Depending on the magnitude of the abnormal serum PSA, it is likely that a patient with an initially abnormal level would undergo prostate biopsy before waiting for a second measurement to be performed 1 year later. Therefore, a rapid PSA velocity should quickly result in an abnormal PSA level, which would be further evaluated due to the PSA elevation alone even in the absence of a rapid velocity. Furthermore, men with cancer often have a PSA velocity less than 0.75 ng/mL per year, especially those with lower PSA levels. Multivariate and receiver operator characteristic analyses as well as a systematic review suggest that calculation of PSA velocity and PSA doubling time (the number of months for a certain level of PSA to increase by a factor of 2) are of limited value in screening, although some studies suggest that they may be useful in detecting the most aggressive cancers.

### *Serum-Free and Bound PSA*

Prostate cancer is associated with a lower percentage of free PSA in the serum compared to PSA values observed in benign conditions. The percentage of free PSA has been used to improve the sensitivity of cancer detection when total PSA is in the normal range (<4 ng/mL) and also to increase the specificity of cancer detection when total PSA is in the "gray zone" (4.1–10 ng/mL). In this latter group (PSA between 4.1 and 10 ng/mL), the lower the value of free PSA, the greater the likelihood that an elevated PSA represents cancer rather than BPH. For example, in one study involving men with PSA values in the range of 4.1–10 ng/mL, the probability of cancer with a free-to-total PSA ratio below 10% was 56%, compared with only 8% when the ratio was greater than 25% [14]. As with PSA, there is no absolute free/total cutoff that can completely differentiate prostate cancer from BPH. The optimal cutoff value for free PSA is unclear and depends on whether optimal sensitivity or specificity is sought. The higher the cutoff value, the greater the sensitivity but the lower the specificity.

Free PSA could be useful for risk stratification in men with prostate cancer. A lower percentage of free to total PSA may be associated with a more aggressive form of prostate cancer. This was illustrated in a study that examined banked frozen serum from 20 men with prostate cancer [15]. Ten years before the diagnosis of cancer, at a time when the total PSA was not different between those who ultimately developed aggressive (defined by clinical stage T3 disease, nodal or bone metastases, pathologically positive margins, or a Gleason score of 7 or greater) versus nonaggressive tumors, there was a statistically significant difference in the percentage of free PSA between the two groups. All 8 men with aggressive cancers had a free PSA of 14% or greater compared with only 2 of 6 (33%) with nonaggressive cancer.

### *Complexed PSA*

Assays for ACT cPSA have been implemented that could theoretically provide a similar enhanced degree of specificity compared to free-to-total PSA ratio but require measurement of a single analyte. Most, but not all, reports suggest that cPSA outperforms both total PSA and the ratio of free to total PSA, with similar sensitivity but a higher specificity [16]. According to one study, for men with total PSA in the diagnostic gray zone (4.0–10.0 ng/mL), the use of cPSA alone would have missed only 1 of the 36 men with cancer who would be diagnosed with prostate cancer using both total PSA and biopsies. Interestingly, free-to-total PSA alone would also have missed one cancer but eliminated biopsy in only 20 men compared to 34 men for whom biopsy could be eliminated using cPSA assay alone. The utility of cPSA in men with a lower total PSA (2–4 ng/mL) is under investigation because there are conflicting data regarding whether cPSA improves specificity compared with free-to-total PSA ration [17]. Complexed PSA has been approved for the monitoring of men with prostatic carcinoma. The utility of complexed PSA for screening is uncertain, and it is not routinely used for clinical practice on a regular basis.

### *Percentage [−2]proPSA*

PSA is initially produced as proPSA, and this form can preferentially leak into the bloodstream in men with prostate cancer. One specific isoform of proPSA is [−2]proPSA, which is unbound and potentially higher in concentration in men with prostate cancer. Based on this observation, there has been growing interest in using the ratio of [−2]proPSA to free PSA (expressed as percent [−2]proPSA or %[−2]proPSA) for screening of prostate cancer. In one multicenter study involving 566 men undergoing biopsy, it was observed that %[−2]proPSA significantly outperformed both total PSA and percent free PSA [18]. At 80% sensitivity, %[−2]proPSA had 52% specificity, compared to 30% for PSA and 29% for percent free PSA. P%[−2]proPSA is currently approved by the European Union for prostate cancer detection and is being evaluated in the United States by the U.S. Food and Drug Administration.

### *False-Positive and Unexpected PSA Results*

False-positive PSA test results may be encountered, causing confusion regarding the diagnosis of prostate cancer. Based on screening of 61,604 men in Europe, Kilpelainen *et al.* [19] observed 17.8% false-positive PSA results. However, men who tested false positive with one PSA screening test were more prone to be diagnosed with prostate cancer in the future. Nevertheless, the major cause of false elevation of PSA is the presence of heterophilic antibody in the serum [20]. In fact, the presence of heterophilic antibody in the specimen may cause not only false elevation of PSA but also false elevation of other tumor markers. This important topic is addressed in detail later in this chapter and also in Chapter 6.

Condoms are commonly submitted for testing in crime laboratories in cases of sexual assault and rape because the presence of seminal fluid in the condom can be used as important evidence of sexual assault. Although microscopic examination of the seminal fluid and identification of human spermatozoa is desirable, sometimes due to degradation and other reasons, direct microscopic examination of seminal fluid is not possible. In this case, chemical testing of the seminal fluid can be conducted, and the presence of PSA can be indicative of seminal fluid. However, false-positive tests may be observed using the Seratec PSA SemiQuant Cassette test or other similar PSA test mostly due to the presence of spermicide nonoxynol-9, which is also a strong detergent [21].

Using assays capable of measuring very low levels of PSA in serum, it has been demonstrated that PSA may be produced in other sites in addition to prostate glands. Very low levels of PSA may be present in breast milk, and detectable PSA levels in the serum may be observed in women with breast cancer as well as breast cysts. Elevated PSA may also be observed in male patients with breast cancer, which is a rare form of cancer in males [22,23]. In addition, both PSA and free PSA have been detected in women with pancreatitis and pancreatic cancer [24].

## CA-125

CA-125, also known as mucin 16 or MUC16, is a glycoprotein. CA-125 in humans is encoded by the *MUC16* gene. CA-125 is used as a tumor marker because CA-125 concentrations may be elevated in the blood of some patients with specific types of cancers such as ovarian cancer and other benign conditions. CA-125 levels in serum are elevated in approximately 50% of women with early stage disease and in more than 80% of women with advanced ovarian cancer [25]. Monitoring CA-125 serum levels is also useful for determining response of a patient to ovarian cancer therapy as well as for predicting a patient's prognosis after treatment. In general, persistence of high levels of CA-125 during therapy is associated with poor survival rates in patients. Also, an increase in CA-125 levels in a patient during remission is a strong predictor of recurrence of ovarian cancer.

In April 2011, the United Kingdom's National Institute for Health and Clinical Excellence recommended that women with symptoms indicative of ovarian cancer should be offered a CA-125 blood test [26]. The aim of this guideline is to help diagnose the disease at an earlier stage when treatment is more likely to be successful. Women with higher levels of CA-125 in their blood would then be offered an ultrasound scan to determine whether they need any further testings. However, the specificity of CA-125 is limited. CA-125 levels are elevated in approximately 1% of healthy women and fluctuate during the menstrual cycle. CA-125 is also increased in a variety of benign and malignant conditions, including the following:

- Endometriosis
- Uterine leiomyoma
- Cirrhosis with or without ascites
- Pelvic inflammatory disease
- Cancers of the endometrium, breast, lung, and pancreas
- Pleural or peritoneal fluid inflammation due to any cause.

Monitoring of serum CA-125 levels also has value in patients with fallopian tube cancer, a relatively rare disease. The pretreatment serum CA-125 concentration appears to have prognostic value in patients with fallopian tube cancer, and following initial treatment it is a sensitive marker for monitoring of recurrence and response to chemotherapy. This was illustrated in a series of 53 patients with fallopian tube cancer, of whom 43 (81%) had an elevated pretreatment serum CA-125 concentration. In a univariate Cox regression model, preoperative serum CA-125 was independently associated with disease-free and overall survival outcome. The level of CA-125 was also correlated with response to therapy, decreasing in all 20 patients who responded to chemotherapy and increasing in both patients who demonstrated progressive disease. Postoperatively, in 90% of patients, an increase in serum CA-125 preceded the clinical or radiologic diagnosis of recurrent disease, with a median lead time of 3 months [27].

## False-Positive and False-Negative CA-125

Meigs' syndrome is the association of ovarian fibroma, pleural effusion, and ascites that may also cause marked elevation of CA-125. Timmerman *et al.* [28] reported two cases of Meigs' syndrome for which the authors observed elevated CA-125 and commented that a pelvic neoplasm in a woman presenting with hydrothorax ascites and elevated CA-125 levels may indicate a benign condition that can be rapidly resolved after surgical correction. Neither ultrasound nor computed tomography can offer a reliable preoperative diagnosis.

*CASE REPORT* A 46-year-old female with right pleural effusion and ascites caused by an ovarian tumor showed an elevated CA-125 level of 1808 U/mL. Computed tomography showed ascites and a bilobate pelvic tumor approximately 25 cm in size. The diagnosis of advanced epithelial ovarian cancer was considered, and the patient was treated with chemotherapy. After three cycles of chemotherapy, the CA-125 level was reduced to 90 U/mL. After that, surgery was performed on the patient in order to remove 25 cm of the left ovarian tumor with intact capsule. Histopathological examination revealed only fibroma. However, her CA-125 was further reduced to 11 U/mL postoperatively. The authors concluded that although the association between ovarian tumor, pleural effusion, and ascites with marked elevation of CA-125 is highly indicative of epithelial ovarian cancer, Meigs' syndrome must also be considered in these patients, as such minimally invasive surgery or biopsy collection must be considered [29].

Abnormally high values of both CA-125 and CA-19-9 have been reported in women with benign tumors. Nagata *et al.* [30] reported that three patients with endometriosis also showed elevated levels of CA-125 and two patients with dermoid cyst showed elevated levels of CA-19-9. Therefore, the authors recommended that tumor marker values should be considered along with bimanual examination, ultrasound, and computed tomography scan for diagnosis of ovarian tumors. Sometimes, F(ab')2 fragments of the murine monoclonal antibody OC-125 are administered to patients with ovarian cancer because OC-125 is directed against the CA-125 antigen present on the surface of human ovarian cancers. Exposure to such antibody may lead to development of an immune response causing the presence of human anti-mouse monoclonal antibody (HAMA; also broadly termed as heterophilic antibody), which may interfere in an unpredictable manner with the determination of CA-125 using serum specimens in such patients [31]. Reinsberg and Nocke [32] reported falsely low CA-125 values as well as one case of a falsely high value in patients who developed anti-idiotypic antibodies and nonspecific HAMA after exposure to [$^{131}$I] F(ab')2 fragments of the OC-125 antibody. However, such interference can be eliminated after removal of serum immunoglobulins, particularly IgG. Whereas false-positive results can also be corrected by adding nonspecific murine IgG to the serum, false-negative results cannot be corrected by such treatment. The false-negative results were due to the presence of anti-idiotypic antibody, whereas the false-positive result was due to the presence of HAMA [32]. Falsely elevated CA-125 due to the presence of heterophilic antibodies in the specimen is a serious concern; for case studies, see the heterophilic antibody section of this chapter.

Measurable CA-125 concentrations can also be observed in patients without any cancer. CA-125 concentrations are known to rise in patients with severe congestive heart failure, and the elevations correlate with the severity of disease and elevations of a specific marker of heart failure, B-type natriuretic peptide [33]. In the menstrual phase of the cycle in women, CA-125 values may be elevated, causing false-positive test results [34]. CA-125 may also increase after abdominal surgery, chronic obstructive pulmonary disease, active tuberculosis, and lupus erythematous. During pregnancy, CA-125 concentrations increase 10 weeks after gestation and remain high throughout the pregnancy. During the terminal phase of pregnancy, the CA-125 concentration may be as high as twice the upper limit of the reference range [35].

## α-FETAL PROTEIN

AFP, sometimes called $\alpha_1$-fetoprotein or α-fetoglobulin, is a protein encoded in humans by the *AFP* gene, which is located on the q arm of chromosome 4 (4q25) [36]. AFP is a major plasma protein produced by the yolk sac and the liver during fetal development and is considered as the fetal form of albumin. AFP binds to copper, nickel, fatty acids, and bilirubin. AFP is found in monomeric, dimeric, and trimeric forms [36]. The half-life of AFP is approximately 5–7 days. Following effective cancer therapy, normalization of the serum AFP concentration over 25–30 days is indicative of an appropriate decline. However, it is essentially undetectable in the serum in normal men [37]. The upper limit of normal serum AFP concentration is less than 10–15 μg/L. Many tissues regain the ability to produce this oncofetal protein while undergoing malignant degeneration, but serum AFP concentrations above 10,000 μg/L are most commonly observed in patients with nonseminomatous germ cell tumors (NSGCTs) or hepatocellular carcinoma.

In men with NSGCTs, AFP is produced by yolk sac (endodermal sinus) tumors and less often due to embryonal carcinomas. As with β-hCG, the frequency of an elevated serum AFP increases with advancing clinical stage of the tumor, from 10–20% in men with stage I tumors to 40–60% in those with disseminated NSGCTs [37].

By definition, pure seminomas do not cause an elevated serum AFP. However, molecular studies have demonstrated AFP mRNA in minute quantities in pure seminoma and several case reports have documented pure seminoma with borderline elevations in serum AFP (10.4 to 16 ng/mL). Higher serum AFP concentrations are considered diagnostic of a nonseminomatous component of the tumor (especially yolk sac elements) or hepatic metastases [38]. If the presence of an elevated serum AFP is confirmed, patients should be treated as if they had an NSGCT.

Serum AFP is the most commonly used marker for diagnosis of HCC because it is often elevated in patients with HCC. Serum levels of AFP do not correlate well with other clinical features of HCC, such as size, stage, or prognosis. Elevated serum AFP may also be seen in patients without HCC, such as those with acute or chronic viral hepatitis. AFP may be slightly elevated in patients with liver cirrhosis due to chronic hepatitis C infection. A significant rise in serum AFP in a patient with cirrhosis should raise concern that HCC may have developed. It is generally accepted that serum levels greater than 500 μg/L (upper limit of normal in most laboratories is between 10 and 20 μg/L) in a high-risk patient is diagnostic of HCC. However, HCC is often diagnosed at a lower AFP level in patients undergoing screening.

Not all tumors secrete AFP, and serum AFP concentrations could be normal in up to 40% of patients with small HCCs. Furthermore, an elevated AFP may be more likely in patients with HCC due to viral hepatitis compared to alcoholic liver disease. In a study of 357 patients with hepatitis C and without HCC, 23% had an AFP greater than 10.0 μg/L [39]. Elevated levels were associated with the presence of stage III or IV fibrosis, an elevated international normalized ratio, and an elevated serum aspartate aminotransferase level.

However, AFP levels are normal in the majority of patients with fibrolamellar carcinoma, a variant of HCC [40]. The sensitivity, specificity, and predictive value for the serum AFP in the diagnosis of HCC depend on the characteristics of the population under study, the cutoff value chosen for establishing the diagnosis, and the gold standard used to confirm the diagnosis. A number of studies have described test characteristics in different settings. The following estimates were based on a cutoff value greater than 20 μg/L in a systematic review that included five studies [41]:

- Sensitivity: 41–65%
- Specificity: 80–94%
- Positive likelihood ratio: 3.1–6.8
- Negative likelihood ratio: 0.4–0.6.

At a prevalence of HCC of 5%, a serum AFP of 20 μg/L or greater had a positive and negative predictive value of 25 and 98%, respectively. At a prevalence of 20%, these percentages were 61 and 90%, respectively. The low positive predictive values indicate the limitation of using the serum AFP as a screening test for HCC [42]. Despite the issues inherent in using AFP for the diagnosis of HCC, it has emerged as an important prognostic marker, especially in patients being considered for liver transplantation. Patients with AFP levels greater than 1000 μg/L have an extremely high risk of recurrent disease following the transplant, irrespective of the tumor size seen on imaging.

### False-Positive AFP

False-positive elevations of serum AFP can occur from tumors of the gastrointestinal tract, particularly hepatocellular carcinoma, or from liver damage (e.g., cirrhosis, hepatitis, or drug or alcohol abuse) [37]. Lysis of tumor cells during the initiation of chemotherapy may result in a transient increase in serum AFP. Elevated serum AFP occurs in pregnancy with tumors of gonadal origin (both germ cell and non-germ cell) and in a variety of other malignancies, of which gastric cancer is the most common [43]. German *et al.* [44] commented that elevated α-fetoprotein concentrations can also be due to liver damage secondary to drugs (chemotherapy, anesthetics, or antiepileptics), virus, or alcoholism without any clinical evidence of malignant tumor. Johnson *et al.* [45] reported the case of a patient with elevated maternal serum α-fetoprotein in which an ultrasound scan of the fetus showed a lemon sign, a banana sign, an effaced cisterna magna, and splayed lumbar vertebrae. After pregnancy termination, no spinal abnormality was observed during autopsy, although X-rays of fetal spine showed narrowing in the thoracic spine. The sonographic cranial findings and elevated α-fetoprotein suggestive of neural tube defect were misleading in this case, although karyotype of the fetus was 69,XXY [45].

As expected, heterophilic antibodies, if present in the specimen, can also cause falsely elevated α-fetoprotein concentration. Values up to 140 times the upper reference range have been observed in cases of hereditary tyrosinemia type 1, with 71% of patients showing levels twice the upper references range [46]. The concentration of α-fetoprotein may also be increased in pregnant women with systemic lupus erythematous [47].

## CARCINOEMBRYONIC ANTIGEN

CEA is a glycoprotein involved in cell adhesion that is normally produced during fetal development, but the production of CEA stops before birth. Therefore, it is not usually present in the blood of healthy adults, although levels are raised in heavy smokers. CEA is a glycosylphosphatidylinositol cell surface anchored glycoprotein whose specialized sialofucosylated glycoforms serve as functional colon carcinoma L-selectin and E-selectin ligands, which may be critical to the metastatic dissemination of colon carcinoma cells [48]. It is found in sera of patients with colorectal carcinoma (CRC) [49], gastric carcinoma, pancreatic carcinoma, lung carcinoma, and breast carcinoma. Patients with medullary thyroid carcinoma also have higher levels of CEA compared to healthy individuals (>2.5 ng/mL). However, CEA blood test is not reliable for diagnosing cancer or as a screening test for early detection of cancer. Most types of cancer do not produce a high CEA. Elevated CEA levels should return to normal after successful surgical resection or within 6 weeks of starting treatment if cancer treatment is successful. Because it lacks both sensitivity and specificity, serum CEA is not a useful screening tool for CRC. However, in patients with established disease, the absolute level of the serum CEA correlates with disease burden and is of prognostic value [50]. Furthermore, elevated preoperative levels of CEA should return to baseline after complete resection; residual disease should be suspected if they do not.

### Serum CEA and Colorectal Carcinoma

Serum levels of the tumor marker CEA should be routinely measured preoperatively in patients undergoing potentially curative resections for CRC for two reasons:

- Elevated preoperative CEA levels that do not normalize following surgical resection imply the presence of persistent disease and the need for further evaluation.
- Preoperative CEA values are of prognostic significance. CEA levels 5.0 ng/mL or greater are associated with an adverse impact on survival that is independent of tumor stage [51].

The 2010 TNM staging criteria for malignant tumor (T describes the size of tumor, N describes the lymph nodes involved, and M describes the distant metastatics) do not include serum CEA in stage assignment but recommend that the information be collected for prognostic value [52]. The data also support the view that decisions regarding adjuvant chemotherapy should include consideration of preoperative CEA levels. However, although a patient with node-negative cancer and a high preoperative CEA could be considered at higher than average risk for recurrence after surgery alone, and this might influence the decision to administer adjuvant chemotherapy, particularly if other risk factors (e.g., obstruction and perforation) are present, there are no data that directly support benefit for adjuvant chemotherapy in this particular setting based on CEA concentration alone. An expert panel convened by the American Society of Clinical Oncology (ASCO) in 2006 concluded that the data were insufficient to support the use of CEA to determine whether or not to treat a patient with adjuvant therapy [53]. Serial measurement of CEA can detect disease recurrence even among patients with an initially normal CEA level, although the sensitivity is lower (between 27 and 50% in four separate studies).

Thus, a postoperative CEA elevation indicates recurrence with high probability, but a normal postoperative CEA level (even if it was initially elevated) is not useful for excluding a disease recurrence.

## Arguments against Serial CEA Testing

- Thirty to forty percent of all cases of CRC recurrences are not associated with measurable elevations in serum CEA [54].
- The benefit of CEA monitoring is limited to a small number of patients with recurrent CRC and is not cost-effective.
- There are no data showing that CEA testing improves quality of life.

## Frequency of Testing

The optimal timing of CEA measurements is unclear. Although one study suggests better disease-free survival in patients who have a CEA level checked every 1 or 2 months compared to less frequent intervals, these results have not been confirmed by others, and no study has shown a survival benefit for any CEA testing frequency. ASCO guidelines suggest that postoperative serum CEA testing be performed every 3 months for at least 3 years after initial therapy in patients with stage II or III disease, if the patient is a potential candidate for surgery or systemic therapy [55]. Guidelines from the National Comprehensive Cancer Network differ slightly, suggesting postoperative CEA testing every 3–6 months for 2 years and then every 6 months for a total of 5 years [45]. These guidelines suggest that serial testing be limited to patients who have T2 or higher stage and who would be considered potential candidates for resection of isolated metastases.

Adjuvant therapy with a 5-fluorouracil-based regimen may falsely elevate the serum CEA level, possibly because of treatment-related changes in liver function. Thus, waiting until adjuvant treatment is completed to initiate CEA surveillance is advisable [55].

## CEA in Cholangiocarcinoma

As a single analyte, serum levels of CEA are neither sufficiently sensitive nor specific to diagnose cholangiocarcinoma. Many conditions other than cholangiocarcinoma can increase serum levels of CEA. Non-cancer-related causes of an elevated CEA include gastritis, peptic ulcer disease, diverticulitis, liver disease, chronic obstructive pulmonary disease, diabetes, and any acute or chronic inflammatory state. One large study evaluated serum CEA levels in 333 patients with primary sclerosing cholangitis (PSC), of whom 44 (13%) were diagnosed with cholangiocarcinoma either by histologic confirmation or after at least 1 year of clinical follow-up. A serum CEA level greater than 5.2 ng/mL had a sensitivity and specificity of 68% (95% confidence interval (CI), 48–84%) and 82% (95% CI, 74–88%), respectively [56]. Biliary levels of CEA have also been evaluated. In one report, the biliary CEA concentration was elevated approximately fivefold in patients with cholangiocarcinoma compared to those with benign strictures; patients with PSC or choledochal cysts without malignancy had intermediate values [57].

## False-Positive CEA

As expected, false-positive CEA test results can occur due to the presence of heterophilic antibodies in the specimen (see discussion on heterophilic antibodies at the end of this chapter and also Chapter 6). However, CEA concentrations can also be elevated in non-neoplastic conditions. Renal failure and fulminant hepatitis can falsely increase CEA values. CEA concentrations may be also have elevated in patients receiving hemodialysis. Patients with hypothyroidism may also have elevated levels of CEA, correlated with the duration of hypothyroidism. CEA levels may also be raised in some non-neoplastic conditions like ulcerative colitis, pancreatitis, cirrhosis, chronic obstructive pulmonary disease (COPD), Crohn's disease, as well as in smokers [35].

*CASE REPORT* A 66-year-old female was admitted to the hospital for treatment of uterine and rectal prolapse, pleural and pericardial effusion, as well as ascites. She complained about general fatigue, constipation, and anal pain. Blood testing reveled normocytic anemia (hemoglobin, 9.5 g/dL), and thyroid function tests showed severe thyroid dysfunction with a thyroid-stimulating hormone (TSH) concentration of 47.09 μU/mL (normal, 0.5–3.0 μU/mL), free thyroxin 0.1 ng/dL (normal, 0.8–1.5 ng/dL), and free triiodothyronine 0.6 pg/mL (normal, 2.1–3.8 pg/mL). Her thyroid microsome antibody titer was elevated, and thyroid ultrasonography demonstrated diffuse swelling of the thyroid. Therefore, she was diagnosed with primary hypothyroidism. Interestingly, her serum CEA was elevated to 16.1 ng/mL (normal, <5 ng/mL) and her CA-125 was 193 U/mL (normal, <35 U/mL). These results suggested malignancy, but further investigations ruled out any malignancy. After treating the patient with oral levothyroxine, her CEA and CA-125 levels returned to normal values [58].

## CA-19-9

CA-19-9, also called cancer antigen 19-9 or sialylated Lewis (a) antigen, is a tumor marker used primarily in the management of pancreatic cancer [59]. Guidelines from ASCO discourage the use of CA-19-9 as a screening test for cancer, particularly pancreatic cancer. The reason is that the test may be falsely negative in many cases or abnormally elevated in people with no cancer at all (false positive). However, in individuals with pancreatic masses, CA-19-9 can be useful in distinguishing between cancer and other pathology of the gland [59,60]. The reported sensitivity and specificity of CA-19-9 for pancreatic cancer are 80 and 90%, respectively [61]. However, these values are closely related to tumor size. The accuracy of CA-19-9 to identify patients with small surgically resectable cancers is limited [61–63]. The specificity of CA-19-9 is limited because CA-19-9 is frequently elevated in patients with cancers other than pancreatic cancer and also in various benign pancreaticobiliary disorders [64]. As a result of all these issues, CA-19-9 is not recommended as a screening test for pancreatic cancer.

The degree of elevation of CA-19-9 (both at initial presentation and in the postoperative setting) is associated with long-term prognosis [65,66]. Furthermore, in patients who appear to have potentially resectable disease, the magnitude of the CA-19-9 level can also be useful in predicting the presence of radiographically occult metastatic disease [66,67]. The rates of unresectable disease among all patients with a CA-19-9 level $\geq$ 130 units/mL versus $<$130 units/mL were 26 and 11%, respectively. Among patients with tumors in the body/tail of the pancreas, more than one-third of those who had a CA-19-9 level $\geq$ 130 units/mL had unresectable disease. Although high levels of CA-19-9 may help surgeons to better select patients for staging laparoscopy, an expert panel convened by ASCO recommended against the use of CA-19-9 alone as an indicator of operability.

Serial monitoring of CA-19-9 levels (once every 1–3 months) is useful for further monitoring of patients after potentially curative surgery and for those who are receiving chemotherapy for advanced disease. Elevated CA-19-9 levels usually precede the radiographic appearance of recurrent disease, but confirmation of disease progression should be pursued with imaging studies and/or biopsy.

CA-19-9 can be elevated in many types of gastrointestinal cancer, such as colorectal cancer, esophageal cancer, and hepatocellular carcinoma. Apart from cancer, elevated levels may also occur in pancreatitis, cirrhosis, and diseases of the bile ducts. It can be elevated in people with obstruction of the bile duct [60]. In patients who lack the Lewis antigen (a blood-type protein on red blood cells), which is approximately 10% of the Caucasian population; CA-19-9 is not expressed even in those with large tumors. This is because of a deficiency of the fucosyltransferase enzyme that is needed to produce CA-19-9 as well as the Lewis antigen [64,68].

### Combined CEA and CA-19-9

The use of a combined index of serum CA-19-9 and CEA (CA-19-9 + (CEA $\times$ 40)) has also been proposed for screening of cholangiocarcinoma. In a study of 72 patients with primary sclerosing cholangitis, the use of CA-19-9 alone (cutoff value $\geq$ 37 U/mL) was 63% sensitive for detecting cholangiocarcinoma, whereas the sensitivity of the combined CA-19-9/CEA index was only 33% [69]. Determination of serum levels of CA-19-9 and CEA may aid in the differentiation of invasive from noninvasive intraductal papillary mucinous neoplasm of the pancreas. The tests had a sensitivity of 80% and a specificity of 82%.

### Pitfalls in Measuring CA-19-9

Interference of heterophilic antibodies causing false-positive CA-19-9 results has been documented, and usually treating the specimen with heterophilic antibody blocking agents eliminates such interference. Monaghan *et al.* [70] described a case in which a false-positive CA-19-9 result obtained by the ADVIA Centaur CA-19-9 assay (Siemens Diagnostics) was not resolved after treating the specimen with a heterophilic antibody blocking agent. By performing a gel filtration study, the authors speculated that the interfering substance was a low-molecular-weight (approximately 100 kDa) compound. Interestingly, when the specimen was reanalyzed using Roche Modular Analytics E170 and Brahms KRYPTOR analyzer, CA-19-9 values were within reference limits, indicating that the interfering substance was only affecting the CA-19-9 assay on the ADVIA Centaur analyzer.

*CASE REPORT* A 76-year-old female patient with symptoms of shaking, chills, fever, abdominal pain, pruritus, fatigue, and dark urine was admitted to the hospital. Her liver enzymes were elevated, with an aspartate aminotransferase level of 201 U/L, alanine aminotransferase level of 228 U/L, $\gamma$-glutamyl transpeptidase level of 296 U/L, alkaline phosphatase level of 866 U/L, and total bilirubin level of 10 mg/dL. Interestingly, her serum CA-19-9 was highly elevated to 9586 U/mL. Ultrasonographic examination of liver and biliary system revealed cholelithiasis, dilation of common bile duct, and portal hilar lymph nodes. Abdominal computerized tomography showed a stone in the common bile duct. Endoscopic examination demonstrated dilatation of common bile duct, choledocholithiasis, and multiple stones in the gall bladder. However, no cancer was detected by any investigation. She was treated with ceftriaxone 2 g/day. Sphincterotomy and extraction of biliary stones was performed, and her symptoms gradually improved over 10 days. Interestingly, not only did her liver enzymes return to normal values but also her CA-19-9 level was significantly reduced to 50 U/L [71].

Patients with acute or chronic pancreatitis may also have elevated levels of CA-19-9. In addition, pulmonary diseases may elevate CA-19-9 levels. Liver cirrhosis, Crohn's disease, and benign gastrointestinal diseases may also increase CA-19-9 levels [35].

*CASE REPORT* A 52-year-old female had a history of epigastric pain and anorexia for 2 months and also lost weight. She was a nonsmoker and a nondrinker but reported overconsumption of black tea (1.5–2 L) every day. Her physical examination was unremarkable. Her electrolytes, liver function tests, lipid profile, and fasting blood glucose were all within acceptable limits, but her serum C-19-9 was significantly elevated to 1432 U/mL (normal, <37 U/mL). Abdominal ultrasonography and computed tomography showed mild enlargement of the pancreas, but further investigation with endoscopic ultrasonography showed no pancreatic malignancy or biliary abnormality. The patient was advised to stop tea consumption, and 4 weeks later her symptoms resolved and her CA-19-9 level had almost returned to a normal value of 42 U/mL [72].

## $\beta_2$-MICROGLOBULIN

Also known as B2M, $\beta_2$-microglobulin is a component of major histocompatibility complex class I molecules, present on all nucleated cells (excludes red blood cells). In humans, the $\beta_2$-microglobulin B2M protein is encoded by the *B2M* gene [73,74]. For the diagnosis of multiple myeloma, the serum B2M level is one of the prognostic factors incorporated into the International Staging System. The serum B2M level is elevated (i.e., >2.7 mg/L) in 75% of patients at the time of diagnosis. Patients with high values have inferior survival [75]. The prognostic value of serum B2M levels in myeloma is probably due to two factors:

- High levels are associated with greater tumor burden.
- High levels are associated with renal failure, which carries an unfavorable prognosis.

In lymphoma, B2M levels usually correlate with disease stage and tumor burden in patients with chronic lymphocytic leukemia (CLL), with increasing levels associated with a poorer prognosis. B2M may be regulated, at least in part, by exogenous cytokines. The source of these elevated cytokines in CLL is unclear, although interleukin-6, which inhibits apoptosis in CLL cells, may be released from vascular endothelium [76].

However, B2M levels also rise with worsening renal dysfunction, leading some investigators to suggest a measure of B2M adjusted for the glomerular filtration rate (GFR-adjusted B2M) [77]. This GFR-adjusted B2M requires validation in prospective confirmatory studies. The plasma B2M concentration is increased in dialyzed patients, with a level ranging from 30 to 50 mg/L—much higher than the normal value of 0.8–3.0 mg/L. Infection with the AIDS virus, hepatitis, and active tuberculosis may also elevate the level of B2M [35].

## HUMAN CHORIONIC GONADOTROPIN

hCG is a hormone composed of $\alpha$ and $\beta$ subunits, and the $\beta$ subunit is specific for hCG ($\beta$-hCG) and provides functional specificity. $\beta$-hCG is synthesized in large amounts by placental trophoblastic tissue and in much smaller amounts by the hypophysis and other organs such as testicles, liver, and colon. Therefore, elevated levels of hCG are observed during pregnancy, produced by the developing placenta after conception and later by the placental component syncytiotrophoblast [78]. Laboratory tests for hCG are essentially very sensitive and specific for diagnosis of trophoblast-related conditions including pregnancy and the gestational trophoblastic diseases. Rarely, very low levels of hCG are detected in the absence of one of these conditions. However, hCG exists in many forms in serum, including intact molecule, $\beta$-hCG, and hyperglycated forms, and other forms including C-terminal peptide. Therefore, an assay capable of measuring all forms of hCG is desirable to resolve low values and ensure that they are indeed low values. Although all assays detect regular hCG, they do not necessarily detect all hCG variants [79]. For example, many over-the-counter pregnancy tests do not measure hyperglycosylated hCG, which accounts for most of the total hCG at the time of missed menses. Clinical tests for pregnancy may only detect total hCG levels $\geq$ 20 mIU/mL. Therefore, when following hCG levels to negative (<1 mIU/mL) in women with gestational trophoblastic disease, it is important to use a sensitive hCG test that detects both regular and other forms of hCG.

At levels of hCG above 500,000 mIU/mL, a "hook effect" can occur resulting in an artifactually low value for hCG (i.e., 1–100 mIU/mL) [80]. This is because the sensitivity of most hCG tests is set to the pregnancy hCG range (i.e., 27,300–233,000 mIU/mL at 8–11 weeks of gestation); therefore, when an extremely high hCG concentration is present, both the capture and the tracer antibodies used in assays become saturated, preventing the binding of the two to create a sandwich [81]. For this reason, a suspected diagnosis of gestational trophoblastic disease must be communicated to the laboratory so that the hCG assay can also be performed at 1:1000 dilution.

## Causes and Evaluation of Persistent Low Level of hCG

Determining the clinical value of a low level of hCG can be challenging [78]. It is important to determine if the hCG represents an actual early pregnancy (intrauterine or ectopic), active gestational trophoblastic disease (complete or partial mole, invasive mole, choriocarcinoma, or placental site trophoblastic tumor), quiescent gestational trophoblastic disease, a laboratory false positive (also called phantom hCG), or a physiologic artifact (pituitary hCG). For example, a false-positive hCG test result or pituitary hCG is commonly found in women who also have a history of gestational trophoblastic disease. Unless tumor is evident, it is essential to exclude these possibilities before initiating chemotherapy for assumed persistence of disease. Persistent low-level positive hCG results can be defined as hCG levels varying by no more than twofold over at least a 3-month period in the absence of tumor on imaging studies. The hCG level is less than 1000 mIU/mL.

## False-Positive hCG

The capture and tracer antibodies used for hCG testing may be goat, sheep, or rabbit polyclonal antibodies or mouse, goat, or sheep monoclonal antibodies. Humans extensively exposed to animals or certain animal by-products can develop human antibodies against animal antibodies (HAAA). Humans with recent exposure to mononucleosis are prone to develop HAAA; those with IgA deficiency syndrome also often have heterophilic antibodies [82]. False-positive hCG test due to the presence of heterophilic antibody in the serum specimen has been well documented in the literature. Such false-positive results in the absence of pregnancy have led to many men and women being misdiagnosed with cancer, confusion and misunderstanding, and needless surgery and chemotherapy [83–85]. There are two main methods for identifying false-positive hCG:

- The most readily available approach is to show the absence of hCG in the patient's urine because large molecules such as an immobilized capture antibody–heterophilic antibody/HAAA–tracer antibody sandwich fail to cross the glomerular basement membrane. A true hCG elevation should be present in both serum and urine [86].
- A second useful way of identifying a false-positive serum hCG result is to send the serum to two laboratories using different commercial assays. If the assay results vary greatly or are negative in one or both alternative tests, then a false-positive hCG can be presumed.

Patients who have false-positive hCG test results are at risk for recurrent false-positive hCG assay results [87]. They are also at risk for other false positives, such as CA-125 and thyroid antibodies [88]. They should inform future health care providers of this problem, and it should be noted in their medical records.

*CASE REPORT* A 39-year-old male patient underwent a unilateral orchiectomy due to a right testicular mass. He showed an elevated serum hCG level of 30 U/L (normal, <5 U/L). Histological examination revealed embryonal carcinoma. He underwent three cycles of chemotherapy with bleomycin, etoposide, and cisplatin. His serum hCG after 3 weeks of chemotherapy was undetectable. Two months later, his serum hCG was increased to 55 U/L, suggesting relapse of the cancer. However, further investigations revealed no relapse of cancer, hCG was undetectable in cerebrospinal fluid, and examination of lymph node did not show any malignancy. His serum hCG was increased further to 280 U/L and after chemotherapy was reduced to 55 U/L. Unfortunately, serum hCG was increased again to 341 U/L, and after chemotherapy it was reduced to 145 U/L. Interestingly, in the following month, his serum hCG returned to normal level, which coincided with changing serum hCG assay to Roche Modular analyzer E170. Previously, the Abbott's assay was used on the AxSYM analyzer for measurement of serum hCG. Retesting high hCG results using the Roche analyzer showed remarkably low levels, indicating interference in the Abbott's serum hCG assay. However, reanalysis of these samples using Abbott's Architect analyzer and Immulite analyzer (Siemens Diagnostics) also eliminated this interference. Unfortunately, this male patient with testicular cancer received two additional unnecessary chemotherapy cycles due to falsely elevated serum hCG levels as a result of analytical interference in the assay [89].

*CASE REPORT* A 44-year-old HIV-positive male presented with a painless swelling of his left testicle. He underwent left radical orchiectomy for a pathological stage T1 NSGCT. However, after the procedure, his serum hCG was persistently elevated and he went through four cycles of chemotherapy with etoposide and cisplatin. Despite chemotherapy, his serum hCG values did not return to normal level. However, further investigations revealed that the patient was cancer-free. Suspecting false-positive serum hCG levels, the authors reanalyzed the sample after adding a heterophilic antibody blocking agent and serum hCG levels became undetectable, indicating the presence of a heterophilic antibody in the serum specimen causing false-positive serum hCG values in this patient. Other tumor markers including α-fetoprotein were not elevated [90].

## INTERFERENCE FROM HETEROPHILIC ANTIBODIES IN LABORATORY TESTING OF TUMOR MARKERS

Heterophilic antibodies are human antibodies that interact with assay antibodies causing false-positive or -negative results. For a complete discussion of heterophilic antibodies, see Chapter 6. In this section, case reports showing interferences from heterophilic antibodies on various tumor marker measurements are presented. Although among tumor markers, serum hCG assay is affected most by the presence of heterophilic antibody, false-positive test results occur with other tumor markers, including PSA, CA-125, CA-19-9, CEA, α-fetoprotein, and even B2M. In fact, interference of heterophilic antibody is the major problem in assays of various cancer markers. Even IgM-λ antibody to *Escherichia coli* can produce false-positive test results with determination of various tumor markers as well as troponin I.

*CASE REPORT* A 56-year-old male presented to the hospital with a history of benign prostatic hypertrophy and a urinary tract infection probably related to urethral obstruction. The patient did not improve with intravenous ampicillin/sulbactam and cefepime therapy, and several days later *E. coli* was identified in the blood culture. The patient was managed by intravenous cefepime and oral ciprofloxacin therapy. On the third day of hospitalization, the patient complained about jaw pain and his cardiac troponin I was 8.7 ng/mL, which was further elevated to 12.7 ng/mL 6 hr later, and the patient was admitted to the coronary care unit. However, further investigation found no evidence of myocardial obstruction or damage. The troponin I value was elevated even further to 220 ng/mL on day 7. Interestingly, his creatine kinase-MB and myoglobin were within normal limits, indicating no myocardial infarction. More strikingly, his CA-125, α-fetoprotein, and thyrotropin were all elevated, as measured by the Abbott assays on an AxSYM analyzer. Suspecting false-positive results for all these analytes, the authors treated the specimens with protein A–Sepharose and murine mAB-conjugated Sepharose. Treating with protein A–Sepharose did not eliminate the interference, but treatment with mAB-conjugated Sepharose did eliminate the interference, indicating that interference was due to heterophilic antibody of the IgM class. The authors speculated that IgM antibody production in this patient was induced by *E. coli* infection [91].

*CASE REPORT* A 62-year-old male underwent radical prostatectomy for localized adenocarcinoma of the prostate. His preoperative PSA level was 12.2 ng/mL. Six weeks after the procedure, during a routine visit, his PSA was 7.48 ng/mL. To rule out specimen error, the test was repeated 1 week later using a different sample, and the PSA was 7.94 ng/mL. In order to eliminate the possibility of false-positive PSA levels, the serum specimen was treated with a heterophilic antibody blocking agent and the PSA value decreased below detection level, indicating that the false-positive postoperative PSA in this patient was due to the presence of a heterophilic antibody [92].

*CASE REPORT* A 55-year-old female underwent surgical treatment for ovarian cancer in 1995 and was free from cancer. In July 2001, a blood specimen for CA-125 showed a CA-125 level of 11 U/mL. The test was repeated 1 month later, and the value was still slightly elevated to 14 U/mL. When aliquots of the serum samples were sent to a different laboratory, CA-125 values were 156 and 187 U/mL. Suspecting interference, specimens were treated with heterophilic antibody blocking agent, and values were returned to near normal, indicating falsely elevated CA-125 in sera due to the presence of interfering heterophilic antibodies [93].

*CASE REPORT* A 76-year-old male was referred to the authors' hospital, complaining of pain in his right upper quadrant. He had complained about that pain previously. Ultrasound revealed hepatomegaly and biliary polyps, whereas computed tomography showed pancreatic polycysts and a cyst on his left kidney. Although CA-125, CEA, and α-fetoprotein concentrations were within normal ranges, his CA-19-9 was elevated to 1047.4 U/mL (normal, <37 U/mL) as measured by the Abbott's assay using the AxSYM analyzer. Interestingly, when CA-19-9 concentration was measured using a different assay (Roche assay on Modular E170 analyzer), the value was 11.9 U/mL, which was within the normal limit. Suspecting high rheumatoid factor present in the patient as the cause of interference with the Abbott assay, based on previous reports of interference of rheumatoid factors in

**TABLE 12.2** Common Causes of Elevation of Levels of Various Tumor Markers in the Absence of Neoplasia

| Tumor Marker | Common Causes of Elevated Levels |
|---|---|
| Prostate-specific antigen | Prostatitis/benign prostatic hyperplasia, breast cancer/cyst, heterophilic antibody |
| $\alpha$-Fetoprotein | Hepatobiliary disease, pneumonia, pregnancy, autoimmune disease, heterophilic antibody |
| CA-125 | Hepatobiliary disease, pulmonary disease, renal failure, hypothyroidism, endometriosis, pregnancy, autoimmune disease, skin disease, cardiovascular disease, heterophilic antibody |
| Carcinoembryonic antigen (CEA) | Hepatobiliary disease, renal failure, hypothyroidism, gastrointestinal disease, pancreatitis, endometriosis, autoimmune disease, heterophilic antibody |
| CA-19-9 | Hepatobiliary disease, renal failure, pulmonary disease, pancreatitis, gastrointestinal disease, endometriosis, heterophilic antibody |
| CA-15-3 | Vitamin $B_{12}$ deficiency, effusion, renal failure |
| $\beta_2$-Microglobin | Renal failure, autoimmune disease, cerebral lesion |
| hCG or $\beta$-hCG | Renal failure, pregnancy, autoimmune disease, heterophilic antibody |
| CA-72-4 | Hepatobiliary disease, renal failure, effusion, pancreatitis, gastrointestinal disorder |
| CYFRA 21-1 | Hepatobiliary disease, renal failure, effusion, pulmonary disease |
| Chromogranin A | Cardiovascular disease, viral infection, prostatitis/benign prostatic hyperplasia, gastrointestinal disease, heterophilic antibody |

CA-19-9 assay, the authors analyzed 18 specimens with high rheumatoid factors using the Abbott assay but did not observe any false-positive CA-19-9 results. Therefore, the authors ruled out a rheumatoid factor as the interfering agent in this patient. Because the patient lived on a farm where mice were present, the authors suspected a heterophilic antibody (HAMA) as the cause of interference. After incubating the specimen with 50% mouse serum followed by treatment with polyethylene glycol 6000, the CA-19-9 value decreased from 1047.4 to 19.8 U/mL, indicating that the cause of false-positive CA-19-9 in this patient using the Abbott assay was HAMA [94].

## ANALYSIS OF LESS FREQUENTLY ASSAYED TUMOR MARKERS

CA-15-3 antigen is a marker for breast cancer, although in the absence of cancer CA-15-3 concentration can be significantly elevated in patients with vitamin $B_{12}$ deficiency. Increases up to twice the upper reference range have been observed in some healthy women during the menstrual cycle. Pancreatitis, Crohn's disease, ulcerative colitis, and benign gastrointestinal disease may also cause some elevation of CA-15-3 [35].

CA-72-4 is a marker for colorectal cancer. In general, approximately 5% women show CA-72-4 values higher than the reference range. In addition, 3–7% of patients with pancreatitis show CA-72-4 values higher than the normal range. Interestingly, approximately 50% of patients suffering from Mediterranean fever demonstrate CA-72-4 values higher than the reference range [35].

CYFRA 21-1, squamous cell carcinoma antigen, and CEA are the most useful markers for diagnosis of non-small cell lung cancer, whereas neuron-specific enolase is a marker for small cell lung cancer [95]. However, patients with pulmonary disease without cancer may show elevated levels of CYFRA 21-1 antigen. Patients receiving dialysis, patients with renal failure, and patients with liver disease may also show elevated levels of this antigen [20]. Various dermatological diseases, such as psoriasis, eczema, epidermitis, erythrodermia, and atopic dermatitis, may cause false elevations of squamous cell carcinoma antigen levels. Patients with chronic renal failure may also show elevation of this marker [35].

Chromogranin A is a marker for various neuroendocrine diseases such as gastroenteropancreatic tumor, endocrine tumors, and bronchial carcinoid. It may also be useful in the diagnosis of pheochromocytoma along with urine/plasma metanephrine determination. However, the assay for chromogranin A may be affected by the presence of heterophilic antibodies in the serum [96].

## CONCLUSION

Currently, tumor markers are widely used. However, there are some limitations with the interpretation of tumor marker measurements. All the markers in current use can be synthesized by normal tissues, and levels in serum can be elevated in certain benign diseases. Frequently, both benign disorder and malignant transformation of the same organ can give rise to elevations in the same marker; for example, PSA can be increased in both BPH and prostate carcinoma. In most situations, however, marker levels are higher in malignancy (especially when distant metastases are present) than in benign disease.

Markers are only rarely elevated in serum in patients with early malignancy or malignancy confined to the primary site. For example, CEA is present at high concentrations in less than 10% of patients with Duke's A colorectal cancer and is only frequently elevated when distant metastases are present.

Markers are generally elevated in only a proportion of patients (at most 70–80%) with a particular tumor type, even in the presence of advanced disease—for example, CA-125 in ovarian cancer, CA-19-9 in pancreatic cancer, and AFP in hepatocellular cancer. In contrast, hCG is present at high concentrations in almost all patients with trophoblastic malignancies.

Apart from PSA, which is almost prostate specific, no other marker in clinical use is organ specific. Many markers, such as CEA, CA-125, CA-15-3, CA-19-9, and tissue polypeptide antigen, can be elevated in serum with most types of advanced adenocarcinoma. AFP, however, is relatively specific for hepatocellular carcinoma and nonseminomatous germ cell tumors, whereas HCG is relatively specific for trophoblastic malignancies. Although heterophilic antibodies can also cause significant interference in tumor marker immunoassays, various heterophilic blocking reagents, steps to remove immunoglobulins, and serial dilutions may correct the problem. There are also various non-neoplastic conditions that may elevate values of various tumor markers as discussed throughout the chapter. Table 12.2 summarizes common non-neoplastic causes of elevation of various tumor markers.

## References


[1] National Cancer Institute. Tumor Markers. Available at: <http://www.cancer.gov/cancertopics/factsheet/Detection/tumor-markers>.

[2] Informa. Critical Reviews in Clinical Laboratory Sciences. Retrieved June 2012 from: <http://informahealthcare.com> by HAM/TMC Library.

[3] American Cancer Society. Tumor Markers. Available at: <http://www.cancer.org/docroot/PED/content/PED_2_3X_Tumor_Markers.asp?sitearea = PED>; April 2012 [accessed].

[4] Lilja H, Christensson A, Dahlén U, Matikainen MT, et al. Prostate-specific antigen in serum occurs predominantly in complex with alpha 1-antichymotrypsin. Clin Chem 1991;37 (9):1618–25.

[5] Björk T, Piironen T, Pettersson K, Lövgren T, et al. Comparison of analysis of the different prostate-specific antigen forms in serum for detection of clinically localized prostate cancer. Urology 1996;48(6):882–8.

[6] Partin AW, Hanks GE, Klein EA, Moul JW, et al. Prostate-specific antigen as a marker of disease activity in prostate cancer. Oncology 2002;16(8):1024–38.

[7] Sershon PD, Barry MJ, Oesterling JE. Serum prostate-specific antigen discriminates weakly between men with benign prostatic hyperplasia and patients with organ-confined prostate cancer. Eur Urol 1994;25(4):281–7.

[8] Hudson MA, Bahnson RR, Catalona WJ. Clinical use of prostate specific antigen in patients with prostate cancer [Abstract]. J Urol 1989;142(4):1011–17.

[9] Jung K, Meyer A, Lein M, Rudolph B, et al. Ratio of free-to-total prostate specific antigen in serum cannot distinguish patients with prostate cancer from those with chronic inflammation of the prostate. J Urol 1998;159(5):1595–8.

[10] Crawford ED, Schutz MJ, Clejan S, Drago J, et al. The effect of digital rectal examination on prostate-specific antigen levels. JAMA 1992;267(16):2227–8.

[11] Luboldt HJ, Peck KD, Oberpenning F, Schmid HP, et al. Bicycle riding has no important impact on total and free prostate-specific antigen serum levels in older men. Urology 2003;61 (6):1177–80.

[12] Brawer MK, Aramburu EA, Chen GL, Preston SD, et al. The inability of prostate specific antigen index to enhance the predictive the value of prostate specific antigen in the diagnosis of prostatic carcinoma. J Urol 1993;150(2 Pt 1): 369–73.

[13] Riehmann M, Rhodes PR, Cook TD, Grose GS, et al. Analysis of variation in prostate-specific antigen values. Urology 1993;42 (4):390–7.

[14] Catalona WJ, Partin AW, Slawin KM, Brawer MK, et al. Use of the percentage of free prostate-specific antigen to enhance differentiation of prostate cancer from benign prostatic disease: a prospective multicenter clinical trial. JAMA 1998;279 (19):1542–7.

[15] Carter HB, Partin AW, Luderer AA, Metter EJ, et al. Percentage of free prostate-specific antigen in sera predicts aggressiveness of prostate cancer a decade before diagnosis. Urology 1997;49 (3):379–84.

[16] Brawer MK, Meyer GE, Letran JL, Bankson DD, et al. Measurement of complexed PSA improves specificity for early detection of prostate cancer. Urology 1998;52(3):372–8.

[17] Tanguay S, Bégin LR, Elhilali MM, Behlouli H, et al. Comparative evaluation of total PSA, free/total PSA, and complexed PSA in prostate cancer detection. Urology 2002;59 (2):261–5.

[18] Sokoll LJ, Sanda MG, Feng Z, Kagan J, et al. A prospective, multicenter, national cancer institute early detection research network study of [-2]proPSA: improving prostate cancer detection and correlating with cancer aggressiveness. Cancer Epidemiol Biomarkers Prev 2010;19(5):1193–200.

[19] Kilpelainen TP, Tammela TL, Roobol M, Hugosson J, et al. False positive screening results in the European randomized study of screening for prostate cancer. Eur J Cancer 2011;47:2698–705.

[20] Morgan BR, Tarter TH. Serum heterophilic antibodies interfere with prostate specific antigen test and results in over treatment in a patient with prostate cancer. J Urol 2001;166:2311–12.

[21] Bitner SE. False positives observed on the seratec PSA semi-quant cassette test with condom lubricants. J Forensic Sci 2012; [Epub ahead of print]

[22] Yu H, Diamandis EP. Prostate specific antigen in milk of lactating women. Clin Chem 1995;41:54–8.

[23] Yu H, Diamandis EP, Sutherland DJ. Immunoreactive prostate specific antigen in female and male breast tumors and its association with steroid hormone receptors and patients age. Clin Biochem 1994;27:75–9.
[24] Pezzilli R, Wirnsberger C, Billi P, Zanarini L, et al. Serum prostate specific antigen in pancreatic disease. Ital J Gastroenterol Hepatol 1999;31:580–3.
[25] Carlson KJ, Skates SJ, Singer DE. Screening for ovarian cancer. Ann Intern Med 1994;121(2):124–32.
[26] United Kingdom National Institute for Health and Clinical Excellence. Women should be offered a blood test for ovarian cancer. NICE guidance. 2011-04-27. Available at: <http://www.nice.org.uk/newsroom/news/WomenShouldBeOfferedBloodTest.jsp>. Retrieved April 2012.
[27] Hefler LA, Rosen AC, Graf AH, Lahousen M, et al. The clinical value of serum concentrations of cancer antigen-125 in patients with primary fallopian tube carcinoma: a multicenter study. Cancer 2000;89:1555–60.
[28] Timmerman D, Moerman P, Vergote I. Meigs' syndrome with elevated serum CA-125 levels: two case reports and review of literature. Gynecol Oncol 1995;59:405–8.
[29] Moran-Mendoza A, Alvarado-Luna G, Calderillo-Ruiz G, Serrano-Olvera, et al. Elevated CA-125 level associated with Meigs syndrome: case report and review of the literature. Unt J Gynecol Cancer 2006;16(Suppl. 1):315–18.
[30] Nagata H, Takahashi K, Yamane Y, Yoshino K, et al. Abnormally high values of CA-125 and CA 19-9 in women with benign tumors. Gynecol Obstet Invest 1989;28:156–68.
[31] Maher VE, Drukman SJ, Kinders RJ, Hunter RE, et al. Human antibody response to the intravenous and intraperitoneal administration of F(ab') fragment of OC 125 murine monoclonal antibody. J Immunother 1991;11:56–66.
[32] Reinsberg J, Nocke W. Falsely low results in CA-125 determination due to anti-idiotypic antibodies induced by infusion of [131]F(ab')2 fragments of the OC 125 antibody. Eur J Clin Chem Clin Biochem 1993;31:323–7.
[33] Nunez J, Nunez E, Consuegra L, Sanchis J, et al. Carbohydrate enzyme 125: an emerging prognostic risk factor in acute heart failure? Heart 2007;93:716–21.
[34] Bon GG, Kenemans P, Dekker JJ, Hompes PG, et al. Fluctuations in CA-125 and CA-15-3 serum concentrations during spontaneous ovulatory cycle. J Human Reprod 1999;14:566–70.
[35] Trape J, Filella X, Alsina-Donadeu M, Juan-Pereira L, et al. Increased plasma concentrations of tumor markers in the absence of neoplasia. Clin Chem Lab Med 2011;49:1605–20.
[36] Harper ME, Dugaiczyk A. Linkage of the evolutionarily-related serum albumin and alpha-fetoprotein genes within q11–22 of human chromosome 4. Am J Human Genetics 1983;35(4):565–72.
[37] Gilligan TD, Seidenfeld J, Basch EM, Einhorn LH, et al. American society of clinical oncology. american society of clinical oncology clinical practice guideline on uses of serum tumor markers in adult males with germ cell tumors. J Clin Oncol 2010;28(20):3388–404.
[38] Javadpour N. Significance of elevated serum alpha-fetoprotein (AFP) in seminoma. Cancer 1980;45(8):2166–8.
[39] Hu KQ, Kyulo NL, Lim N, Elhazin B, et al. Clinical significance of elevated alpha-fetoprotein (AFP) in patients with chronic hepatitis C, but not hepatocellular carcinoma. Am J Gastroenterol 2004;99(5):860–4.
[40] Soreide O, Czerniak A, Bradpiece H, Bloom S, et al. Characteristics of fibrolamellar hepatocellular carcinoma: a study of nine cases and a review of the literature. Am J Surg 1986;151(4):518–23.
[41] Gupta S, Bent S, Kohlwes J. Test characteristics of alpha-fetoprotein for detecting hepatocellular carcinoma in patients with hepatitis C: a systematic review and critical analysis. Ann Intern Med 2003;139(1):46–50.
[42] Pomfret EA, Washburn K, Wald C, Nalesnik MA, et al. Report of a national conference on liver allocation in patients with hepatocellular carcinoma in the United States. Liver Transpl 2010;16(3):262–78.
[43] Liu X, Cheng Y, Sheng W, Lu H, et al. Clinicopathologic features and prognostic factors in alpha-fetoprotein-producing gastric cancers: analysis of 104 cases. J Surg Oncol 2010;102(3):249–55.
[44] German JR, Llanos M, Tabernero M, Mora J. False positive elevations of alpha-fetoprotein associated with liver dysfunction in germ cell tumors. Cancer 1993;72:2491–4.
[45] Johnson DD, Nager CW, Budorick NE. False-positive diagnosis of spina bifida in a fetus with triploidy. Obstet Gynecol 1997;89(5 Pt 2):809–11.
[46] Phaneuf D, Lambert M, Laframboise R, Mitchell G, et al. Type 1 hereditary tyrosinemia: evidence for molecular heterogeneity and identification of a casual mutation in a French Canadian patient. J Clin Invest 1992;90:1185–92.
[47] Petri M, Jo AC, Patel J, Demers F, et al. Elevation of maternal alpha-fetoprotein in systematic lupus erythematosus: a controlled study. J Rheumatol 1995;22:1365–8.
[48] Konstantopoulos K, Thomas SN. Cancer cells in transit: the vascular interactions of tumor cells. Annu Rev Biomed Eng 2009;11:177–202.
[49] Cancer Diagnosis Information about Cancer-Stanford Cancer.
[50] Clinical practice guidelines for the use of tumor markers in breast and colorectal cancer. Adopted on May 17, 1996 by the American Society of Clinical Oncology. J Clin Oncol 1996;14(10):2843-2877.
[51] Meling GI, Rognum TO, Clausen OP, Børmer O, et al. Serum carcinoembryonic antigen in relation to survival, DNA ploidy pattern, and recurrent disease in 406 colorectal carcinoma patients. Scand J Gastroenterol 1992;27(12):1061–8.
[52] Edge SB, Byrd DR, Compton CC, et al. American joint committee on Cancer. Cancer staging manual. 7th ed. New York: Springer; 2010. p. 143
[53] Locker GY, Hamilton S, Harris J, Jessup JM, et al. ASCO 2006 update of recommendations for the use of tumor markers in gastrointestinal cancer. J Clin Oncol 2006;24(33):5313–27.
[54] Benson 3rd AB, Desch CE, Flynn PJ, Krause C, et al. 2000 update of American Society of Clinical Oncology colorectal cancer surveillance guidelines. J Clin Oncol 2000;18(20):3586–8.
[55] National Comprehensive Cancer Network. NCCN guidelines. Available at: <www.nccn.org>.
[56] Zeng Z, Cohen AM, Urmacher C. Usefulness of carcinoembryonic antigen monitoring despite normal preoperative values in node-positive colon cancer patients. Dis Colon Rectum 1993;36(11):1063–8.
[57] Nakeeb A, Lipsett PA, Lillemoe KD, Fox-Talbot MK, et al. Biliary carcinoembryonic antigen levels are a marker for cholangiocarcinoma. Am J Surg 1996;171(1):147–52.
[58] Takahasi N, Shimada T, Ishibashi Y, Oyake N, et al. Transient elevation of serum tumor markers in a patient with hypothyroidism. Am J Med Sci 2007;333:387–9.
[59] Perkins G, Slater E, Sanders G, Prichard J. Serum tumor markers. Am family Phys 2003;68(6):1075–82.
[60] Goonetilleke KS, Siriwardena AK. Systematic review of carbohydrate antigen (CA 19-9) as a biochemical marker in the diagnosis of pancreatic cancer. Eur J Surg Oncol 2007;33(3):266–70.
[61] European Group on Tumour Markers. Tumour markers in gastrointestinal cancers—EGTM recommendations. Anticancer Res 1999;19(4A):2811–15.

[62] Cwik G, Wallner G, Skoczylas T, Ciechanski A, et al. Cancer antigens 19-9 and 125 in the differential diagnosis of pancreatic mass lesions. Arch Surg 2006;141(10):968–73.

[63] DiMagno EP, Reber HA, Tempero MA. AGA technical review on the epidemiology, diagnosis, and treatment of pancreatic ductal adenocarcinoma; American gastroenterological association. Gastroenterology 1999;117(6):1464–84.

[64] Kim HJ, Kim MH, Myung SJ, Lim BC, et al. A new strategy for the application of CA-19-9 in the differentiation of pancreaticobiliary cancer: analysis using a receiver operating characteristic curve. Am J Gastroenterol 1999;94(7):1941–6.

[65] Maisey NR, Norman AR, Hill A, Massey A, Oates J, et al. CA-19-9 as a prognostic factor in inoperable pancreatic cancer: the implication for clinical trial. Br J Cancer 2005;93(7):740–3.

[66] Kondo N, Murakami Y, Uemura K, Hayashidani Y, et al. Prognostic impact of perioperative serum CA 19-9 levels in patients with resectable pancreatic cancer. Ann Surg Oncol 2010;17(9):2321–9.

[67] Maithel SK, Maloney S, Winston C, Gönen M, et al. Preoperative CA 19-9 and the yield of staging laparoscopy in patients with radiographically resectable pancreatic adenocarcinoma. Ann Surg Oncol 2008;15(12):3512–20.

[68] Tempero MA, Uchida E, Takasaki H, Burnett DA, et al. Relationship of carbohydrate antigen 19-9 and Lewis antigens in pancreatic cancer. Cancer Res 1987;47(20):5501–3.

[69] Björnsson E, Kilander A, Olsson R. CA 19-9 and CEA are unreliable markers for cholangiocarcinoma in patients with primary sclerosing cholangitis. Liver 1999;19(6):501–8.

[70] Monaghan PJ, Leonard MB, Neithercut WD, Raraty MG, et al. False positive carbohydrate antigen 19-9 (CA-19-9) results due to a low molecular weight interference in an apparently healthy male. Clin Chim Acta 2009;406:41–4.

[71] Kormaz M, Unal H, Selcuk H, Yilmaz S. Extraordinary elevated serum levels of CA-19-9 and rapid decrease after successful therapy: a case report and review of literature. Turk J Gastroenterol 2010;21:461–3.

[72] Howaizi M, Abboura M, Krespine C, Sbai-Idrissi MS, et al. A new case of CA-19-9 elevation: heavy tea consumption. Gut 2003;52:913–14.

[73] Güssow D, Rein R, Ginjaar I, Hochstenbach F, et al. The human beta 2-microglobulin gene: primary structure and definition of the transcriptional unit. J Immunol 1987;139(9):3132–8.

[74] Suggs SV, Wallace RB, Hirose T, Kawashima EH, et al. Use of synthetic oligonucleotides as hybridization probes: isolation of cloned cDNA sequences for human beta 2-microglobulin. Proc Natl Acad Sci USA 1981;78(11):6613–17.

[75] Rossi D, Fangazio M, De Paoli L, Puma A, et al. Beta-2-microglobulin is an independent predictor of progression in asymptomatic multiple myeloma. Cancer 2010;116(9):2188–200.

[76] Moreno A, Villar ML, Cámara C, Luque R, et al. Interleukin-6 dimers produced by endothelial cells inhibit apoptosis of B-chronic lymphocytic leukemia cells. Blood 2001;97(1):242–9.

[77] Delgado J, Pratt G, Phillips N, Briones J, Fegan, et al. Beta2-microglobulin is a better predictor of treatment-free survival in patients with chronic lymphocytic leukemia if adjusted according to glomerular filtration rate. Br J Haematol 2009;145 (6):801–5.

[78] Gregory JJ, Finlay JL. Alpha-fetoprotein and beta-human chorionic gonadotropin: their clinical significance as tumour markers. Drugs 1999;57(4):463–7.

[79] Muller CY, Cole LA. The quagmire of hCG and hCG testing in gynecologic oncology. Gynecol Oncol 2009;112(3):663–72.

[80] Cole LA. Immunoassay of human chorionic gonadotropin, its free subunits, and metabolites. Clin Chem 1997;43(12):2233–43.

[81] Flam F, Hambraeus-Jonzon K, Hansson LO, Kjaeldgaard A. Hydatidiform mole with non-metastatic pulmonary complications and a false low level of hCG. Eur J Obstet Gynecol Reprod Biol 1998;77(2):235–7.

[82] Knight AK, Bingemann T, Cole L, Cunningham-Rundles C. Frequent false positive beta human chorionic gonadotropin tests in immunoglobulin a deficiency. Clin Exp Immunol 2005;141(2):333–7.

[83] Baulieu JL, Lepape J, Baulieu F, Besnard JC. Falsely elevated results of radioimmunoassays using double antibody method: arguments for a third anti-rabbit IgG antibody present in certain human sera. Eur J Nucl Med 1982;7(3):121–6.

[84] Vladutiu AO, Sulewski JM, Pudlak KA, Stull CG. Heterophilic antibodies interfering with radioimmunoassay: a false-positive pregnancy test. JAMA 1982;248(19):2489–90.

[85] Hussa RO, Rinke ML, Schweitzer PG. Discordant human chorionic gonadotropin results: causes and solutions. Obstet Gynecol 1985;65(2):211–19.

[86] Cole LA, Khanlian SA, Sutton JM, Davies S, et al. Hyperglycosylated hCG (invasive trophoblast antigen, ITA) a key antigen for early pregnancy detection. Clin Biochem 2003;36(8):647–55.

[87] Committee on Gynecologic Practice, American College of Obstetricians and Gynecologists. Committee opinion: Number 278, November 2002. Avoiding inappropriate clinical decisions based on false-positive human chorionic gonadotropin test results. Obstet Gynecol 2002;100(5 Pt 1):1057–59.

[88] Cole LA, Khanlian SA. Inappropriate management of women with persistent low hCG results. J Reprod Med 2004;49(6):423–32.

[89] Ballieux BE, Weiji NI, Gelderblom H, van Pelt J, et al. False positive serum human chorionic gonadotropin (hCG) in a male patient with a malignant germ cell tumor of the testis: a case report and review of the literature. Oncologist 2008;13:1149–54.

[90] Gallagher DJ, Riches J, Bajorin DF. False positive elevation of human chorionic gonadotropin in a patient with testicular cancer. Nat Rev Urol 2010;7:230–3.

[91] Covinsky M, Laterza O, Pfeifer JD, Farkas-Szallasi T, et al. An IgM λ antibody to *Escherichia coli* produces false positive results in multiple immunometric assays. Clin Chem 2000;46:1157–61.

[92] Poyet C, Hof D, Sulser T, Muntener M. Artificial prostate specific antigen persistence after radical prostatectomy. J Clin Oncol 2012;30:e62–3.

[93] Bertholf R, Johannsen L, Benrubi G. False elevation of serum CA-125 levels caused by human anti-mouse antibodies. Annals Clin Lab Sci 2002;32:414–18.

[94] Liang Y, Yang Z, Ye W, Yang J, et al. falsely elevated carbohydrate antigen 19-9 level due to heterophilic antibody interference but not rheumatoid factor: a case report. Clin Chem Lab Med 2009;47:116–17.

[95] Kulpa J, Wojcik E, Reinfuss M, Kolodziejski L. Carcinoembryonic antigen, squamous cell carcinoma antigen, VYFRA 21-1 and neuron specific enolase in squamous cell lung cancer patients. Clin Chem 2002;48L:1931–7.

[96] Giovanella L, Ceriani L. Spurious increase in serum chromogranin A: the role of heterophilic antibody. Clin Chem Lab Med 2010;48:1497–9.

CHAPTER

# 13

# Issues of Interferences in Therapeutic Drug Monitoring

*Gwendolyn A. McMillin, Kamisha L. Johnson-Davis*
University of Utah School of Medicine and ARUP Laboratories, Salt Lake City, Utah

## INTRODUCTION

Therapeutic drug management (TDM) is a cornerstone of personalized medicine, designed to select and optimize drug therapy for individual patients. By optimizing therapy, the desirable effects of a drug (efficacy) can be maximized and the undesirable effects of a drug (toxicity) minimized. TDM is also employed to evaluate patient compliance, identify drug–drug interactions or other changes in pharmacokinetics, and both characterize and manage acute intoxications or emergent overdose situations. TDM is most useful when there is a good correlation between drug concentration in blood and effect (therapeutic and toxic). It is particularly useful for drugs that have a narrow therapeutic window, prodrugs, and drugs that are highly bound to plasma proteins. TDM is not widely available for most drugs due in part to limitations in analytical methods.

Inaccuracies in TDM results will impact patient care by contributing to unnecessary or inappropriate dose adjustments. Inappropriate increases in dose could lead to dose-dependent adverse drug reactions (ADRs). ADRs are reported to account for 41% of all hospital admissions, due to inappropriate dose or prescription, allergic reactions, and drug–drug interactions. The consequences of ADRs are costly and can be fatal [1]. Errors in TDM results could also cause unnecessary dose reductions and minimize the therapeutic effect. Therapeutic failure due to inappropriately low dosing could contribute to unnecessary testing and procedures, and for time-sensitive conditions such as organ transplantation, cancer, uncontrolled seizures, and cardiac disturbances, it could be fatal. The laboratory has an important opportunity to promote clinically useful TDM through involvement in appropriate specimen collection, sample handling, sample processing, selection of appropriate analytical methods, and interpretation of results. As such, the laboratory must be prepared to consult regarding therapeutic ranges, toxic thresholds, as well as strengths and limitations of available analytical methods. Inaccuracies in TDM may be a result of pre-analytical factors surrounding the specimen and patient or analytical factors surrounding selection and performance characteristics of the analytical method (Figure 13.1). Sources of interference may arise from endogenous or exogenous sources and may impact specific technologies in a direct or indirect manner. However, by using good laboratory practice, interferences can be identified, managed, and/or prevented.

This chapter describes sources of pre-analytical and analytical interferences that have been reported to adversely affect the accuracy of TDM results. Specific drug, technology, and case examples are provided to illustrate these sources in more detail and guide appropriate utilization and interpretation of TDM from the clinical laboratory perspective.

## SOURCES OF PRE-ANALYTICAL INTERFERENCES IN TDM

Pre-analytical sources of inaccuracy in laboratory test results are discussed in detail elsewhere in this book (see Chapters 1–4). However, there are many pre-analytical factors important to TDM that are unique from other types of laboratory testing, and these are briefly emphasized here. In particular, the coordination of pharmacokinetic factors for the specific drug formulation(s) of interest, with the clinical status

*Accurate Results in the Clinical Laboratory.*
DOI: http://dx.doi.org/10.1016/B978-0-12-415783-5.00013-X

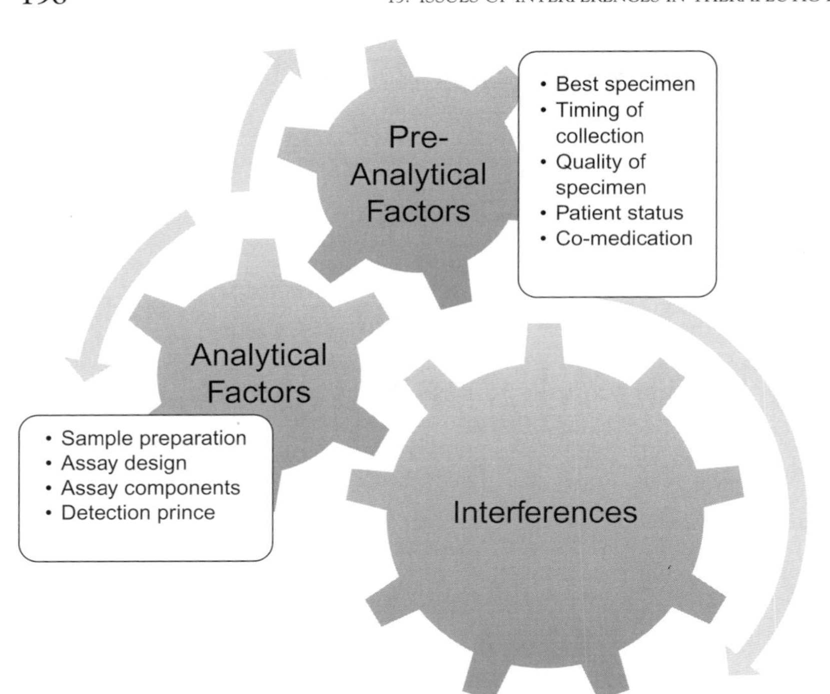


**FIGURE 13.1** Factors that affect accuracy of a TDM result.

and co-medications of the patient, will dictate the most appropriate specimen to collect, at the most appropriate time. In addition, specimen collection using the wrong anticoagulant and environmental factors such as heat and light can affect drug stability and alter the concentration of therapeutic drugs. TDM measurements are typically collected after a patient is predicted to have achieved steady-state concentration from predose (trough) collections, which occur before the next scheduled dose. Steady state is achieved when the concentration of drug in the body is in equilibrium with the rate of dose administered and the rate of elimination (after initiation of therapy, the time equivalent to at least five half-lives of the drug is needed for establishment of steady state). A random collection may reflect steady state for some drugs, particularly those with long elimination half-lives. Random collections are also important when evaluating therapeutic failure, toxicity, or overdose situation. Consideration of inaccuracies and appropriate interpretation of results depends on understanding the pre-analytical circumstances surrounding a request for TDM.

Specimen collection containers can alter TDM results by affecting drug concentration and proportional stability. Failure to separate cells promptly from serum or plasma may lead to *in vitro* metabolism, as has been observed with fosphenytoin and mycophenolic acid acyl glucuronide [2,3]. Serum versus plasma, variation among tube preservatives, and gel separator tubes are relevant for accurate recovery of certain drugs or fractions of drugs. For example, heparin collection tubes can increase the concentration of free (not bound to proteins) drug concentration by activating lipoprotein lipase and fatty acid concentration, which will displace protein-bound drug from albumin and $\alpha_1$ glycoprotein [4]. Free fraction of phenobarbital, phenytoin, and valproic acid was shown to be elevated in serum versus plasma, whereas the free fraction of carbamazepine and theophylline was lower in serum versus plasma [5]. The proportion of free valproic acid was found to be time-sensitive, and it decreased significantly after 96 hr of storage at ambient temperature, most likely due to degradation of binding proteins. Blood specimens that are exposed to extreme temperatures can lead to plasma protein degradation as well and increase the free fraction of drugs that are highly bound to protein (e.g., phenytoin and valproic acid). Inappropriate dose adjustments can be made when based on free drug concentrations determined with suboptimal specimens. In addition, citrate collection tubes were associated with a decrease in the total concentration of valproic acid [6]. More dramatically, gel separator tubes have the potential to decrease total drug recovery through adsorption of the drug to the gel material. Separator tubes were reported to decrease the recovery of cardiac drugs, tricyclic antidepressants, anticonvulsants, and antipsychotics. As such, gel separator tubes should not be used as collection tubes for TDM unless validated to be appropriate [4].

Drug stability can also be influenced by sensitivity to light or heat and may lead to erroneous TDM results if specimens are stored under inappropriate conditions. Amiodarone, methotrexate, chlordiazepoxide, carbamazepine, chlorpromazine, haloperidol, and fluoxetine are examples of light-sensitive drugs. Rapid *in vitro* degradation is also recognized for bupropion, busulfan, carbamazepine, lithium, and olanzapine. The consequences of *in vitro* degradation include falsely low results, which could potentially contribute to inappropriate dose adjustments. Specimens containing unstable drugs should be stored at the appropriate temperature (i.e., frozen) to preserve integrity and prevent falsely low results. In addition, the impact of repeat freeze–thaw cycles on drug integrity should be carefully evaluated [4].

TDM results can also be affected by a patient's clinical status and lifestyle factors such as diet, smoking, alcohol use, co-morbidities, pharmacogenetics, and polydrug therapy. Food and fluid intake can alter the pharmacokinetics (absorption, distribution, metabolism, and elimination) of a drug by impacting gastric pH and emptying time, which can affect drug absorption. Drugs that are administered orally are absorbed into the bloodstream through the gastrointestinal tract by passive diffusion if the drugs are lipid soluble or non-ionized. Factors that affect intestinal motility, hepatic blood flow, and bile flow will also have an impact on the pharmacokinetics of a drug. For example, protein intake can affect drug–protein binding for drugs and alter drug clearance. In 1987, Fagan *et al.* demonstrated that high-protein diets can increase the clearance of propranolol and theophylline [7].

Food–drug interactions will influence drug bioavailability and drug metabolism by inhibiting or inducing drug-metabolizing enzymes. For example, grapefruit juice can inhibit the cytochrome P450 isozyme 3A4 (CYP3A4) in the small intestine and increase bioavailability of several drugs [8]. This isozyme, part of the superfamily of CYP enzymes associated with drug metabolism, is involved in metabolism of approximately half of all drugs. Increased drug bioavailability can lead to enhanced drug activity and possible toxicity. Drug classes that are affected by this food–drug interaction include calcium channel blockers (e.g., nifedipine), antiarrhythmic drugs (e.g., amiodarone), benzodiazepines (e.g., diazepam), antiepileptic drugs (e.g., carbamazepine), antibiotics (e.g., erythromycin), antiretrovirals (e.g., indinavir), immunosuppressants (e.g., cyclosporine A), and cholesterol-lowering drugs (e.g., simvastatin). In addition, cruciferous vegetables such as cabbage and cauliflower and charbroil foods can induce metabolism of drugs metabolized by the CYP1A2 isozyme. Substrates of CYP1A2 include theophylline, clozapine, and olanzapine. Alcohol consumption can also induce the CYP2E1 isozyme to increase metabolism of certain drugs, in addition to the well-recognized synergistic effect with the pharmacodynamics of depressant drugs. Cigarette smokers tend to require high doses of many therapeutic drugs to obtain optimal therapy, in part due to the fact that nicotine induces isozymes CYP1A1, CYP1A2, and CYP2E1.

Genetic polymorphisms in the genes that code for drug-metabolizing enzymes can also affect an individual's ability to biotransform drugs for activation or elimination (pharmacogenetics). For example, genetic polymorphisms can cause a patient to be a poor or slow drug metabolizer. However, the overall impact of induction or inhibition of drug-metabolizing enzymes, whether due to interacting substances or genetic predispositions, depends on whether the drug substrate is activated or inactivated by the affected isozyme(s). For drugs that are inactivated by metabolism (e.g., tricyclic antidepressants and warfarin), poor metabolizers are at risk of accumulating drug and are at risk of drug-induced toxicity if standard dosing is administered. Rapid metabolizers are at risk of therapeutic failure due to suboptimal dosing and will require higher doses to achieve optimal therapy. The opposite is true for drugs that are activated by metabolism (e.g., clopidogrel and codeine). Accurate TDM is an important tool for optimizing dosing under conditions of unpredictable pharmacokinetics, such as pharmacogenetic variants and food–drug and drug–drug interactions.

Another source of unpredictable pharmacokinetics that requires TDM to optimize dosing is patients who are critically ill. Of particular concern are patients with impaired renal, hepatic, gastrointestinal, or cardiovascular function. Renal disease will reduce the glomerular filtration rate and reduce the clearance of drugs that are eliminated via the kidney. Renal disease will also impact drug–protein binding in patients with uremia, due to uremia toxins competing for drug-binding sites to albumin, and increase the concentration of non-protein-bound (free) drug and therapeutic effect. Liver toxicity or disease will decrease production of albumin and other proteins that bind to drugs. Hypoalbuminemia will increase the fraction of pharmacologically active drugs and may increase the risk of toxicity. Decreased liver function will reduce first-pass metabolism and also impact expression of CYP450 isozymes in general. Gastrointestinal disease or a history of bariatric surgery, malabsorptive disorders, or intestinal disease may impact absorption of drugs [9]. Cardiovascular disease will decrease cardiac output, tissue perfusion, drug disposition, and absorption. Reduced blood flow to the liver will decrease drug metabolism.

TDM is also necessary in pregnant or nursing women to accommodate changes in maternal physiology and to protect the fetus or infant from toxicity. Pregnant women have increased body fat, total body water, and plasma volume, which will decrease plasma protein concentration and drug–protein binding. Lipophilic drugs will distribute and accumulate in fat and decrease bioavailability. Drugs that are hydrophilic will have a lower volume of distribution in the body and may have increased clearance and decreased bioavailability in patients with excessive body fat content. Cardiac output in pregnant women is increased, resulting in enhanced drug metabolism. A general recommendation is to carefully monitor pregnant or nursing women who require drug therapy to ensure efficacy and safety for both mother and her unborn child [10].

## SOURCES OF ANALYTICAL INTERFERENCES IN TDM

The vulnerability of an analytical method to interferences is dependent on the sample matrix, the sample preparation method, the assay design, assay components, and detection principle (see Figure 13.1). Sources of analytical interferences may be endogenous or exogenous. Laboratories must consider common sources of interferences, such as hemolysis, during development and validation of analytical methods, and they must investigate suspected interferences that arise with analytical methods in routine use. Unfortunately, analytical interferences may go unrecognized by laboratories unless TDM results are questioned by clinicians who have carefully evaluated results in the context of a specific patient scenario. Examples of analytical interference sources affecting common technologies used to support TDM are listed in Table 13.1.

Endogenous interferences (matrix components) known to affect TDM include classical biological interferences such as hemolysis, bilirubinemia, hematocrit, and lipemia. For example, hemolysis is known to compromise lithium quantitation due to a dilution effect. Both hemolysis and bilirubinemia may interfere with spectral detection methods, particularly for automated homogeneous immunoassays [11]. Also, the tacrolimus II MEIA (microparticle enzyme immunoassay) has been shown to be affected by hematocrit, wherein a tacrolimus result is biased if the blood sample hematocrit falls outside 30–40% [12]. Electrolytes and serum protein affect ion-transfer voltammetry methods, such as in quantification of propranolol, or ion-selective electrodes used for lithium quantitation [13]. Heterophilic antibodies, steroids, and other endogenous substances provide

**TABLE 13.1** Example Sites and Sources of Analytical Interferences

| Analytical Method | Major Vulnerabilities to Interference | Example Sources of Interference |
|---|---|---|
| Immunoassay | Antibodies (species, clone, epitopes) | Drug metabolites |
| | Assay design (homogeneous vs. heterogeneous, specific steps, timing of reactions, reagents, detection) | Similar drugs in same drug class |
| | | Any structurally similar compounds<br>Heterophilic antibodies<br>Matrix components |
| Chromatography | Sample preparation (extraction, dilution, protein precipitation)<br>Assay design (chemistry of phases, flow rates, temperature)<br>Detection technology | Any co-eluting compounds<br>Compounds with similar detection characteristics<br>Matrix components<br>Carryover |
| Mass spectrometry | Sample preparation | Isobaric or cross-talking compounds |
| | Assay design (SIM, MRM, ions detected) | Impurities in reagents |
| | Instrument (mass resolution, ionization, voltages, data collection parameters, signal/noise) | Matrix components |
| | | Carryover |

MRM, multiple reaction monitoring; SIM, selected ion monitoring.

interesting case reports for isolated, patient-specific interferences with certain technologies. Examples are provided later in this chapter. It can be stated that biological samples are very complex and essentially all endogenous compounds have the potential to introduce inaccuracies through interferences, depending on the unique characteristics of an individual patient specimen, coupled with the analytical method employed.

Exogenous interferences are introduced from outside the patient and add another level of sometimes unpredictable complexity. Such interferences may be qualified as direct or indirect. A direct exogenous interference is related to the drug of interest, including drug metabolites, or drug impurities. Direct interferences can and should be carefully evaluated during assay development and validation. Compounds that interfere with a TDM technology but are unrelated to the specific drug of interest could be qualified as indirect exogenous interferences and may be less intuitive.

Examples of indirect interferences include co-medications, herbal medicines, vitamin and mineral supplements, social drugs (e.g., nicotine and ethanol), components of the diet, or components of the patient's environment (e.g., water and air sources). Indirect interferences may be predicted and characterized during assay development and validation, but they are more often recognized and described after an assay is in routine use.

Analytical methods should be designed to detect, minimize, or compensate for predictable interferences. Some interfering substances may be minimized through sample preparation methods such as dilution, protein precipitation, liquid–liquid or solid-phase extraction. Physical separation techniques (e.g., chromatography, ultrafiltration, and ultracentrifugation) may also help minimize matrix effects. Most automated immunoassays do not incorporate any sample preparation methods and are consequently at higher risk from interferences. However, if the source of a potential interference is defined and is measureable (e.g., hematocrit), patient specimens could be qualified for testing in advance. Samples that exceed established thresholds for a known interference could be disqualified for testing. To avoid inaccuracies in TDM results, alternate methods could be made available for disqualified specimens.

All TDM results should be interpreted within the context of the clinical, pre-analytical, and analytical scenario. Any result that is inconsistent with expectations should be investigated. Figure 13.2 illustrates a possible algorithm for investigating a discordant TDM result, with intent to guide identification or characterization of the interference. Thus, an investigation should begin by considering pre-analytical and analytical variables. If suspicions surround the pre-analytical components of testing, a new specimen should be collected. If suspicions are aligned better with analytical components of testing, an alternate method should be sought. Repeat testing of the original sample with the original technology is also an important preliminary step for investigating the possibility of an analytical interference. Comparing results between technically distinct analytical approaches is very informative. Additional tools that may help characterize or resolve interference include dilution studies, to potentially reduce the impact of an interference sufficiently to restore accuracy of results; removing plasma components and washing red cells for assays based on detection of erythrocyte-bound drug; and ultrafiltration or protein precipitation for physically removing large-sized or protein-based interfering substances or sample extraction methods for lipid-based interfering substances. Solid-phase extraction, liquid–liquid extraction, or other sample cleanup steps should be aligned with the specimen type and the drug(s) of interest [14,15]. Many other approaches to characterizing, eliminating, and preventing interferences may be

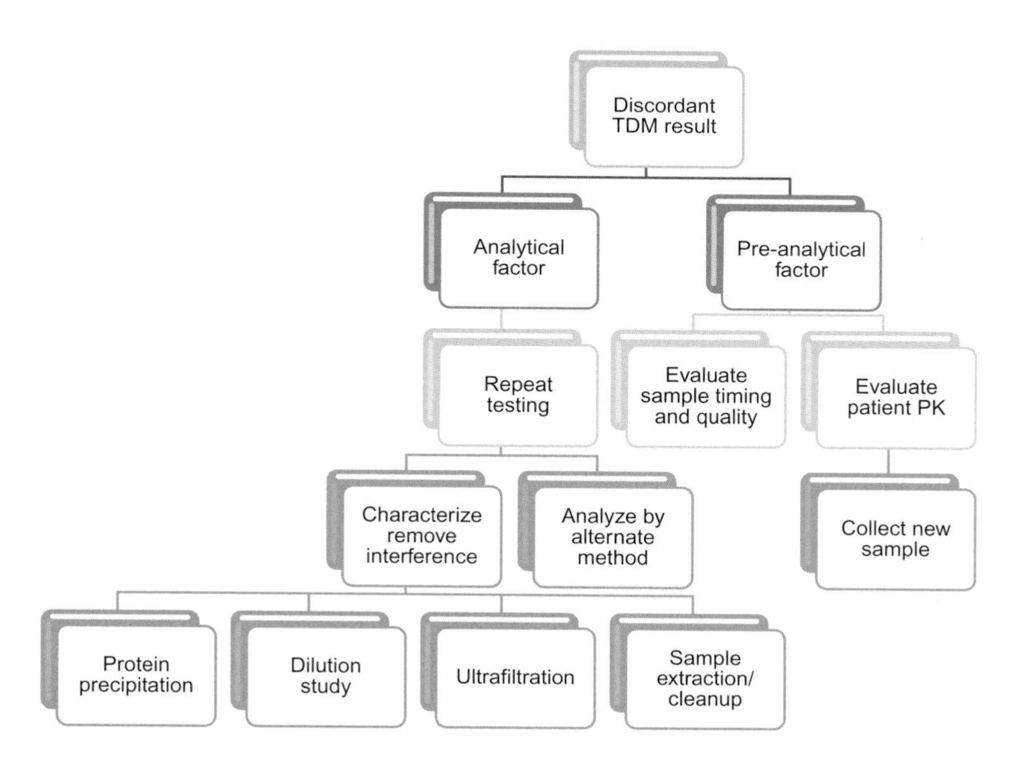


**FIGURE 13.2** Algorithm for evaluating a discordant TDM result.

appropriate based on the actual source of interference and the available analytical technology.

## MECHANISMS OF ANALYTICAL INTERFERENCES IN TDM

Most well-described analytical interferences that affect TDM are associated with commercially available immunoassays. Among immunoassay technologies, the source of the interference may relate to the antibodies on which the test is built or to the assay design. Immunoassays are available in homogeneous and heterogeneous formats (see Chapter 6), which have unique vulnerabilities to interferences based on timing or various assay steps (e.g., washes and incubations), chemistry (e.g., pH), temperatures, reactions (e.g., read times), and detection technology (e.g., spectral). All immunoassay formats employ a capture antibody and are typically designed to detect a single representative chemical structure against which the capture antibody is raised. Sometimes, immunoassays employ more than one antibody, and the antibodies may have originated from several different species (e.g., mouse and rabbit). Actual performance characteristics and vulnerabilities to specific sources of analytical interferences vary tremendously among commercial products, based in part on the antibodies that are unique to each product. In general, immunoassays that utilize polyclonal antibodies are more likely to be associated with analytical interferences than are those that utilize monoclonal antibodies due to the fact that polyclonal antibodies have a larger number of epitopes available for binding drug antigens with similar chemical structures. Also, in general, heterogeneous assay formats are less vulnerable to nonspecific interferences than are homogeneous assay formats.

For drug classes that contain many structurally similar compounds, cross-reactivity across the drug class inherently compromises specificity for an individual drug [16]. For example, immunoassays designed to detect tricyclic antidepressants generally detect several common tricyclic antidepressant drugs at similar concentrations. Because an acute intoxication with a tricyclic antidepressant is likely to be managed based on the drug class and clinical presentation, rather than based on the individual drug(s) responsible, analytical interference across the class of drugs is desirable and can be useful clinically for drug detection, such as in the emergency department, or for a psychiatric clinic that wants to evaluate compliance with therapy. However, an immunoassay for tricyclic antidepressants is generally not useful clinically for TDM due to the fact that other tricyclic structures are known to cross-react with common immunoassays. Such assays are also not useful for supporting TDM or pharmacokinetic evaluations or whenever concentrations of individual drugs and drug metabolites are clinically relevant. In the case of tricyclic antidepressants, the preferred technology for analysis is liquid chromatography or liquid chromatography coupled to tandem mass spectrometry (LC-MS/MS). In any case, review of the cross-reactivity profile for any immunoassay product is required to evaluate which drugs and other compounds are likely to be detected and with what degree of sensitivity/specificity. See the discussion of tricyclic antidepressants and other specific drug examples for more information.

Examples of specific immunoassay designs that are associated with analytical interferences known to adversely affect TDM (exemplified later in this chapter) include fluorescent polarization immunoassay (FPIA), enzyme multiplied immunoassay (EMIT), microparticle enzyme immunoassay (MEIA), and chemiluminescence immunoassay (CLIA).

## CHROMATOGRAPHY AND MASS SPECTROMETRY

Chromatography has long been recognized to impart specificity to TDM because of the inherent ability to separate components of the matrix and, hence, to separate drugs and drug metabolites from one another. As such, chromatographic techniques physically separate the sample and are frequently referred to as physical techniques. Chromatographic methods are typically developed by individual laboratories and vary tremendously relative to sample preparation methods, chemistry of related solid (e.g., column) and mobile (e.g., gas or liquid) phases, flow rates, temperatures, instrument platforms and associated capabilities, detection technologies, methods of data collection, and extent of data analysis. Chromatographic methods are extremely vulnerable to interferences from co-eluting compounds that may or may not be structurally similar to the compounds of interest. The relevance of a co-eluting substance depends on whether that substance interferes with the analyte signal that is generated, which is largely based on the signal-to-noise ratio, and on the detection method. Common detection methods include spectral (e.g., ultraviolet and fluorescent), electrochemical, and mass spectrometry. Despite possible limitations, chromatographic methods, particularly coupled with mass spectrometric detection, are recognized to resolve many of the interferences that are encountered with immunoassay methods and are generally regarded as gold standard platforms for TDM.

Mass spectrometry is a highly selective detection technology that detects mass-to-charge ratios of ionized molecules and fragments. This detection technology is frequently coupled to gas or liquid chromatography. Output of data can be quantitative, qualitative for targeted masses, compared to a library of data to identify a particular compound, or used to determine element composition and structure. Assays can be designed using selected ion monitoring, which is a mode that scans a limited mass-to-charge range; multiple reaction monitoring, in which an ion mass is selected in the first mass spectrometer and the fragment ion mass is selected in the second mass spectrometer; or full mass spectrum, which scans a wide mass-to-charge ratio range.

Although mass spectrometric methods are very specific, the technology is not exempt from interferences. Ion suppression or enhancement, which alters ion intensity/abundances and signal-to-noise ratios, have become well recognized in mass spectrometric methods but may go unrecognized when isolated to a single patient specimen. For example, by affecting analyte ionization, ion suppression can occur in the presence of compounds that are less volatile than the analyte of interest. Salts, anions, ion-pairing agents (e.g., trifluoroacetic acid), drugs/metabolites, uncharacterized sample matrix components and co-eluting compounds, and reagent impurities are known to produce ion suppression or enhancement. In electrospray ionization, the primary source of ion suppression is due to changes in spray droplet formation in the presence of nonvolatile compounds [17,18]. Another source of interference in mass spectrometry that may be overlooked includes compounds that are isobaric or isometric to the analyte of interest. An example of isobaric interferences was observed with endogenous nucleosides and ribavirin in an LC-MS/MS method [19]. Plasticizers, reagents, lipids, and phospholipids in particular are well recognized to introduce matrix effects, manifested as either suppression or enhancement of analytical response [20]. Consequently, internal standards, such as stable isotope-labeled compounds, are used to minimize the effects of ion suppression and improve the analytical accuracy of mass spectrometry [21]. Lastly, cross talk can cause interference when different parent ions fragment and have identical product ion masses [22].

## EXAMPLES OF INTERFERENCES THAT AFFECT TDM

In this section, specific examples are given regarding interferences affecting the measurement of various therapeutic drugs.

### Interferences in Digoxin Measurement

Digoxin (Lanoxin) is a cardiac glycoside with positive inotropic effects that is used to treat cardiac arrhythmias and congestive heart failure. Effects are dose-related, and TDM is required because the clinical symptoms of inadequate and excessive dose are similar. The traditional therapeutic range for digoxin, 0.5–2.0 ng/mL, is narrow among drugs that require TDM. It is also noteworthy that the upper limit of the therapeutic range overlaps somewhat with the threshold for toxicity. In congestive heart failure, an even more narrow range has been proposed (0.5–0.8 ng/mL), and patient outcomes begin to decline at concentrations greater than 0.9 ng/mL. The clinical expectations for maintaining a very narrow circulating concentration of digoxin raise questions about whether analytical techniques available for TDM of digoxin can perform with adequate accuracy and precision [23]. A very narrow therapeutic range also demands that pre-analytical variables be recognized and managed.

#### *Pre-Analytical Variables*

Digoxin has a high volume of distribution and requires several hours to distribute after dosing. A common cause of an unexpectedly high digoxin is therefore based on inappropriate timing of specimen collection. Blood for digoxin analysis should be collected at least 8 hr after a dose is administered.

#### *Analytical Variables*

Both positive and negative interferences have been described for immunoassays designed for TDM of digoxin. It is usually assumed that negative interference is most dangerous to the patient management because it may contribute to inappropriate dose increases. Sources of analytical interferences that are recognized include digoxin metabolites; endogenous digoxin-like immunoreactive factor or substance (DLIF or DLIS); and exogenous drugs, including anti-digoxin Fab fragments used to treat digoxin overdose, aldosterone antagonists, other cardiac glycosides, and several herbal medicines. Some forms of interference can be effectively removed through physical separation, accomplished by ultrafiltration or chromatography. Analysis of ultrafiltrate, chromatographically distinct fractions, or direct analysis of independent specimen components such as via mass spectrometric detection, further reduces the potential for analytical interferences [24].

Tables 13.2 and 13.3 summarize examples of interferences described for common immunoassay technologies in the absence or presence of digoxin, respectively. The type of interference described (positive, negative, or none) for FPIA, EMIT, MEIA-II, and

**TABLE 13.2** Interferences with Digoxin Immunoassays in the Absence of Digoxin

| Interferent | FPIA | EMIT | MEIA-II | CLIA | Resolved by Ultrafiltration |
|---|---|---|---|---|---|
| DLIF | + | + | | | Yes |
| Drugs | | | | | |
| Digitoxin | + | + | + | + | Partial |
| Anti-digoxin Fab | | + | + | + | Yes |
| Aldosterone antagonists | + | + | + | + | No |
| Herbals | | | | | |
| Chan su | + | + | + | | Partial |
| Danshen | + | | | | Yes |

**TABLE 13.3** Interferences with Digoxin Immunoassays in the Presence of Digoxin

| Interferent | FPIA | EMIT | MEIA-II | CLIA | Resolved by Ultrafiltration |
|---|---|---|---|---|---|
| DLIF | + | + | - | | Yes |
| Drugs | | | | | |
| Digitoxin | + | | +/− | | Partial |
| Anti-digoxin Fab | | +/− | +/− | +/− | Yes |
| Aldosterone antagonists | + | + | − | + | Partial |
| Herbals | | | | | |
| Chan su | + | + | − | | Partial |
| Danshen | + | | − | | Yes |

CLIA, chemiluminescence immunoassay; EMIT, enzyme multiplied immunoassay; FPIA, fluorescent polarization immunoassay; MEIA, microparticle enzyme immunoassay.

CLIA is indicated, along with whether or not ultrafiltration will resolve the interference. The data presented in these tables are intended for example purposes and are inherently simplified. For example, the tables do not reflect that the presence and magnitude of interferences is closely related to the concentration of interfering substances and/or digoxin, and that specific reagents that utilize these technologies may differ. The package insert for a specific assay of interest should be consulted, and independent validation studies should be performed as needed, to fully characterize any interference and its potential impact to a unique clinical setting. In the following sections, each major source of interferences is described in more detail.

## DIGOXIN METABOLITES

Digoxin is extensively metabolized. Cross-reactivity of the metabolites with antibodies used to develop digoxin immunoassays was historically a source of analytical interference, such as with radioimmunoassays. In particular, the hydrolysis metabolites digoxigenin, digoxigenin monodigitoxoside, and digoxigenin bisdigitoxoside exhibited near equal cross-reactivity with digoxin in some assays. Because these metabolites exhibit physiological activity, it was proposed that this cross-reactivity could represent a total "bioactive" digoxin concentration [25,26]. Newer assays exhibit little cross-reactivity to metabolites, suggesting that current assays are not subject to the same cross-reactivity. The concentration of active metabolites is also not thought to be sufficiently high to contribute significantly to the efficacy or toxicity of digoxin in most patients. Nonetheless, awareness of cross-reactivity to digoxin metabolites may be relevant for application of immunoassay testing to patients who exhibit poor elimination kinetics, who may be at risk for accumulation of clinically relevant drug metabolites. Because metabolites do accumulate in urine, the analytical method selected for testing urine for digoxin should be evaluated carefully.

## DIGOXIN-LIKE IMMUNOREACTIVE FACTORS

The chemical structure of digoxin is steroid-like and therefore exhibits high potential for endogenous counterparts. Indeed, a source of endogenous interferences in digoxin assays has been recognized in select populations in the absence of digoxin exposure or administration. This steroid-like substance was called DLIF and DLIS in the mid-1980s and was shown to cross-react with digoxin immunoassays. DLIF has been recognized most for patients with conditions of volume expansion. Vulnerable populations include infants and young children, pregnant women, patients with impaired renal or hepatic function, and persons who have exercised vigorously. It has also been reported to be elevated immediately after a myocardial infarction and in the critically ill [25].

It is now thought that there are at least two classes of DLIF that may or may not cross-react with digoxin immunoassays and exert activity at endogenous sodium−potassium ATPase [27]. The degree and type of interference may change in the presence of digoxin. In the absence of digoxin, clinically significant concentrations of apparent digoxin (no actual digoxin present) have been observed in samples collected from newborns and pregnant women and analyzed by FPIA [28]. In the presence of digoxin, DLIF may lead to positive or negative interference, based on the specific analytical technology. Ultrafiltration or chromatographic separation remains the best means of resolving this type of interference, when suspected. Due to the extensive characterization of DLIF and improvements in immunoassay design, currently available immunoassays may exhibit minimal or no clinically significant interference from DLIF [29,30].

### ANTI-DIGOXIN IMMUNE FRAGMENTS

Digibind and DigiFab are examples of pharmaceutical immune (Fab) products that bind digoxin with high affinity, and they are used as antidotes to treat digoxin overdose. These Fab products are responsible for variable amounts of analytical interference with digoxin assays, both in the absence and in the presence of digoxin. Technologies such as CLIA, MEIA, and EMIT demonstrate concentration-dependent positive interference in the absence of digoxin, and concentration-dependent negative interference is observed in the presence of digoxin, to the point of molar equivalency (neutralization). As might be expected, the presence of digoxin under conditions of Fab excess results in positive interference [31].

Patients undergoing Fab-based decontamination treatment can be monitored by measuring free digoxin in ultrafiltrate or by incorporating a pre-analytical protein-precipitation step to measure total digoxin (e.g., FPIA). Monitoring free digoxin concentrations in plasma ultrafiltrate is the most accurate and accepted approach. It is important to employ a method of analysis such as FPIA or MEIA that is free of matrix effects because ultrafiltrate exhibits different chemical/physical properties than plasma or serum. Ultrafiltrate is not an appropriate specimen for methods that depend on sample fluid dynamics, such as slide technologies (e.g., Vitros) [32].

### CARDIAC GLYCOSIDES

The digitalis family is composed of several cardiac glycosides with similar chemical structures, including the once popular drug digitoxin. Whereas it is not currently used in the United States, digitoxin is used in European, Scandinavian, and other countries. Although it would not be expected that digoxin would be co-administered with digitoxin, a patient may be transitioned from one drug to another or may be using undisclosed medication obtained outside the primary clinic. The therapeutic dose and corresponding therapeutic concentration range of digitoxin is significantly higher (10- to 15-fold) than that of digoxin. Due to the higher therapeutic range, and structural similarity, even low cross-reactivity with a digoxin immunoassay could produce clinically significant interference. Only the CLIA digoxin assay has sufficiently low cross-reactivity with digitoxin to avoid analytical interferences. FPIA assays are particularly vulnerable to positive interference from digitoxin in the absence of digoxin. MEIA assays are moderately susceptible to positive interference from digitoxin in the absence of digoxin. Of note, negative interference is observed with MEIA when digoxin and digitoxin are present. Other cardiac glycosides, including oleandrin (used in folk medicines), may exhibit the same patterns of positive interference in the absence of digoxin and either positive or negative interference in the presence of digoxin. Interferences from nondigoxin cardiac glycosides may be at least partially resolved by measuring free digoxin in ultrafiltrate [33,34].

### ALDOSTERONE ANTAGONISTS

Spironolactone and potassium canrenoate are used to treat hypertension and congestive heart failure, and they are often co-prescribed with digoxin. These co-medications, along with their common active metabolite canrenone, are structurally similar to digoxin and have been described to produce positive interference in FPIA and CLIA digoxin assays. Negative interference has also been described for MEIA digoxin assays [35]. Newer-generation MEIA III assays that employ monoclonal antibodies have improved resistance to the negative analytical interferences described. However, clinically significant positive analytical interference has been described for the MEIA III assay in patients co-medicated with higher-dose spironolactone (e.g., 100 mg/day). Similar positive interferences are described with EMIT digoxin assays. The impact of the analytical interferences on dosing can be significant, easily exceeding the therapeutic range of digoxin. One study demonstrated clinically significant differences in approximately 45% of patients co-medicated with digoxin and either spironolactone or canrenoate. Such interferences were not eliminated by measurement of ultrafiltrate [36]. As such, digoxin monitoring for patients co-medicated with aldosterone antagonists is best pursued with chromatographic techniques.

### HERBAL MEDICINES

Herbal medicines are a cornerstone of traditional Eastern medicine and lifestyle, and they are readily available without prescription. These medicines are not well controlled in terms of potency or purity, and they may not be reported by patients who use them. Several case examples and *in vitro* studies have demonstrated that select herbal medicines produce bidirectional analytical interferences in digoxin immunoassays. For example, positive and negative interferences have been described extensively for Chan su and Danshen. Thus, Chan su produced concentration-dependent positive interference in FPIA, EMIT, and MEIA digoxin immunoassays in the absence of digoxin. In the presence of digoxin, Chan su produced positive interference in FPIA and EMIT and negative interference in MEIA. Chan su did not affect CLIA with or without digoxin present. Interferences with Chan su were only partially resolved by analysis of ultrafiltrate. Positive interference from Danshen has been reported in the absence of digoxin when analyzed by FPIA but not by EMIT, MEIA, or CLIA. As with Chan su, positive interference has been observed with FPIA and EMIT, and negative interference with MEIA, when Danshen is found in the presence of digoxin; no effect was observed with CLIA. Danshen interferences were largely resolved by analysis of ultrafiltrate [37–39].

*CASE REPORT*[1] A 19-year-old Taiwanese female presented to an emergency department with nausea, vomiting, dizziness, weakness, and numbness in the mouth. Noteworthy was her admission to having ingested cooked toad eggs approximately 1 hr prior. Vitals and neurological exam were normal, but an electrocardiogram revealed a first-degree atrioventricular block. Serum digoxin concentration was measured by immunoassay. Despite no history of digoxin therapy, the concentration was 1.9 ng/mL; potassium concentration was 7.1 mmol/L. The patient was administered anti-digoxin Fab and admitted to the intensive care unit. Both the apparent digoxin concentration and potassium concentrations declined, and the patient was released a few days later. The most likely cause of the digoxin cross-reactivity is that the toad eggs ingested contained Chan su, a traditional Chinese aphrodisiac commonly made from toad venom.

## Interferences in Carbamazepine Measurement

Carbamazepine (Tegretol and Carbatrol) is a tricyclic anticonvulsant drug used for the treatment of partial and tonic–clonic seizures, trigeminal neuralgia, and manic depression. The therapeutic range for carbamazepine is 4–12 μg/mL; signs of toxicity, such as stupor, seizures, and respiratory depression, may occur at concentrations exceeding 15 μg/mL. Carbamazepine is metabolized primarily by cytochrome P450 isozyme (CYP) 3A4 to the active metabolite, carbamazepine-10,11-epoxide. The 10,11-epoxide metabolite has similar activity as the parent drug, and it may contribute significantly to efficacy of the drug in populations that accumulate the metabolite, such as children. The concentration of the metabolite may be higher than the parent drug concentrations in situations of carbamazepine overdose and in patients with renal failure (uremia); therefore, monitoring drug ratios of parent and metabolite may be useful for evaluating compliance and drug–drug interactions.

[1]From Kuo *et al.* [40].

### *Pre-Analytical Variables*

Drug–drug interactions are common with carbamazepine due to relatively high protein binding (65–80%) and involvement of drug-metabolizing enzymes, including CYP2C19 and CYP3A4. Carbamazepine is eliminated primarily via the kidneys; thus, renal failure or insufficiency can lead to elevated results and increase the risk for toxicity.

### *Analytical Variables*

Major sources of analytical interference for carbamazepine include cross-reactivity from the 10,11-epoxide metabolite and structurally similar drugs. Cross-reactivity of carbamazepine and its metabolite can vary in commercial immunoassays, from 0% for the Vitros assay to approximately 93.6% on the Dade Dimension. Other assays, such as MEIA (AxSYM), have moderate cross-reactivity (22%) [41]. This cross-reactivity can be utilized to calculate the total amount of epoxide present, but chromatographic methods are preferred [42]. Analytical interferences with carbamazepine immunoassays have been described for drugs that are structurally similar to carbamazepine, such as oxcarbazepine, which produce positive interference in carbamazepine analysis by EMIT [43]. Cross-reactivity with carbamazepine analogs that are currently in development, such as eslicarbazepine, is not yet characterized. Because this new drug is a purified isomer, it is important to recognize that nonchiral chromatographic assays would not be able to distinguish between use of eslicarbazepine and use of racemic licarbazepine. Noncarbamazepine analogs may also cross-react with carbamazepine assays. For example, Parant *et al.* reported that the antihistamine drug hydroxyzine and its metabolite, cetirizine, produced a false-positive result for carbamazepine using the particle-enhanced turbidimetric inhibition immunoassay (PETINIA) but not with EMIT or turbidimetry (ADVIA Centaur) [44,45].

## Interferences in Phenytoin Measurement

Phenytoin (Dilantin) is administered to manage generalized tonic–clonic seizures, partial or complex–partial seizures, and status epilepticus. There is a good correlation between plasma concentration of phenytoin and its clinical effect, and a well-accepted therapeutic range is 10–20 μg/mL. Fosphenytoin (Cerebyx) is a prodrug that is rapidly converted by hydrolysis to phenytoin. Total plasma phenytoin concentrations greater than 20 μg/mL do not enhance seizure control and are associated with adverse effects such as nystagmus, ataxia, psychosis, hyperglycemia, nystagmus, and hematological disorders; concentrations greater than 35 μg/mL have been shown to precipitate seizure activity.

### *Pre-Analytical Variables*

Drug–drug interactions are common with phenytoin due to high protein binding (>90%) and involvement of drug-metabolizing enzymes, including CYP2C9 and CYP2C19. Phenytoin is primarily eliminated via the kidneys; thus, renal failure or insufficiency can lead to elevated results and increase the risk for toxicity due to accumulation of the parent drug. Samples collected after administration of fosphenytoin may lead to falsely elevated results due to *in vitro* metabolism of fosphenytoin, particularly in patients with elevated alkaline phosphatase activity [2].

### *Analytical Variables*

A major source of analytical interference for phenytoin is the primary metabolite 5-(*p*-hydroxyphenyl)-5-phenylhydantoin (HPPH). Several phenytoin immunoassays cross-react with the metabolite HPPH, leading to falsely elevated results [46,47]. The glucuronide form of HPPH is eliminated in urine. Analytical interference due to HPPH is of particular concern for uremic patients as well as any patients who have poor renal function and may thereby accumulate the metabolite [48].

Other sources of analytical interference reported for phenytoin include the nonsteroidal anti-inflammatory drug oxaprozin (Daypro), which can exhibit cross-reactivity with TDx phenytoin assays to produce a falsely elevated result [49,50]. Fosphenytoin cross-reacts directly with several immunoassay formats, including FPIA, EMIT, and CLIA, to produce falsely elevated results. Of interest, the extent of cross-reactivity increases in the presence of phenytoin for FPIA (TDx) but is independent of phenytoin concentration for CLIA (ACS:180) [51]. False-negative results are also possible, as demonstrated by Brauchli *et al.* in a patient with monoclonal IgM-λ and testing performed by an automated particle-enhanced turbidimetric inhibition immunoassay (see case report) [52].

*CASE REPORT*[2] A 73-year-old female was admitted to a hospital due to confusion and recurrent convulsive facial contractions. The convulsions increased and extended to the right side of her body. An EEG identified epileptic zones in the occipital lobe, and treatment by phenytoin infusion was initiated. Despite consistent daily dosing, total serum phenytoin concentration was undetectable (<0.4 ng/mL) by PETINIA (Dimension Analyzer, Siemens Diagnostics) in multiple samples. Other medications administered to the patient were detected in the same samples, eliminating the possibility of sample mix-up. The possibility of drug–drug interactions was also eliminated. Because of the unexpected result, testing by alternate methods was pursued. Phenytoin was detected and concentrations were reasonable when analysis was performed by FPIA or by chromatographic methods. Testing with PETINIA was repeated after a protein precipitation step, and phenytoin was detected as expected. It was therefore hypothesized that the false-negative results were due to an interfering protein. A monoclonal IgM-λ was detected in the patient serum by immunofixation, and this was likely responsible for the analytical interference.

[2]From Brauchli *et al.* [52].

## Interferences in Measurement of Immunosuppressant Drugs

Immunosuppressant drugs (cyclosporine A, tacrolimus, sirolimus, everolimus, and mycophenolate mofetil) function to suppress the immune system by inhibiting T cell activation and proliferation. Immunosuppressant drugs are used to treat autoimmune disease, allergies, multiple myeloma, and chronic nephritis and to prevent rejection after organ transplantation. TDM for immunosuppressant drugs is routinely performed in the transplant setting, and dose adjustments are made based on results. Blood concentrations below the therapeutic range can lead to underdosing, which can put a patient at risk for graft rejection because the immune system is not suppressed. Blood concentrations that exceed the therapeutic range can lead to oversuppression of the immune system and increase the risk of opportunistic infections, malignancy, and organ toxicity.

Cyclosporine (Sandimmune and Neoral) and tacrolimus (Prograf) are both calcineurin inhibitors and have serious adverse effects, such as renal dysfunction and toxicity, hypertension, and hyperlipidemia, with excess exposure. The general therapeutic range is 100–400 ng/mL. The therapeutic range for tacrolimus is 5–20 ng/mL. Sirolimus (Rapamune) and everolimus (Zortress and Afinitor) are inhibitors of the mammalian target of rapamycin (mTOR) and function to inhibit T-lymphocyte activation and proliferation. They are also associated with serious adverse effects, such as anemia, leucopenia, thrombocytopenia, hyperlipidemia, and gastrointestinal effects. The therapeutic range for sirolimus is 12–20 ng/mL, and it is 3–8 ng/mL for everolimus. Mycophenolate mofetil (CellCept) is a prodrug that is rapidly metabolized to mycophenolic acid (MPA) in the liver. MPA is a reversible and noncompetitive inhibitor of inosine monophosphate dehydrogenase, which is important for guanine nucleotide synthesis for T cell proliferation. MPA is associated with adverse effects such as leukopenia, diarrhea, and vomiting. The therapeutic range for predose MPA is 1.0–3.5 mg/L when combined with cyclosporine.

### *Pre-Analytical Variables*

Cyclosporine, tacrolimus, sirolimus, and everolimus are highly bound to red blood cells and proteins; therefore, TDM is best performed with whole blood. MPA is not bound to red blood cells but highly bound to plasma proteins; therefore, TDM is performed with plasma. Changes in hematocrit may affect accuracy of whole blood testing results.

Bioavailability is extremely variable with most immunosuppressant drugs. These drugs are extensively metabolized and eliminated primarily in the urine. As such, patients with impaired gastrointestinal, hepatic, or renal function are subject to unpredictable pharmacokinetics. For example, liver transplant patients with hyperbilirubinemia will accumulate tacrolimus metabolites leading to a high positive bias in tacrolimus results by immunoassay [11]. Drug–drug interactions that lead to elevated or reduced concentrations of immunosuppressant drugs are also a significant concern.

### *Analytical Variables*

Immunosuppressant drugs can be monitored by immunoassays and chromatographic methods (high-performance liquid chromatography (HPLC) and LC-MS/MS). LC-MS/MS methods are now commonly used to create multianalyte panels and to increase specificity compared to immunoassay methods [53]. These methods are not free of interferences, and they should be evaluated with the same scrutiny as an immunoassay or other analytical method [20]. Nonetheless, immunoassays remain a convenient means of testing, particularly in a hospital setting. Sources of analytical interferences for immunoassays include drug metabolites and endogenous factors such as heterophilic antibodies and hematocrit. In addition, the structural similarity between sirolimus and everolimus leads to near equivalent cross-reactivity with sirolimus immunoassays. Everolimus can be monitored by sirolimus immunoassay, but if a patient is co-medicated with both drugs, distinguishing everolimus and sirolimus requires a chromatographic approach [54–56]. Functional assays designed to monitor and optimize immunosuppressant therapy are available as well, particularly for MPA, but are not discussed here [57,58]. See Table 13.4 for a summary of drugs and drug metabolites that are reported to interfere with immunoassays for immunosuppressant drugs.

**TABLE 13.4** Drugs and Drug Metabolites That May Interfere with Immunoassays for Immunosuppressant Drugs

| Drug | Metabolites | Drugs |
|---|---|---|
| Cyclosporine (Sandimmune) | 31 (inactive)<br>Hydroxylated cyclosporine (10% activity) | Amikacin<br>Amphotericin B<br>Azathioprine<br>Carbamazepine<br>Chloramphenicol<br>Cimetidine<br>Digoxin<br>Disopyramide<br>Erythromycin<br>Furosemide<br>Gentamicin<br>Kanamycin A<br>Lidocaine<br>Mycophenolic acid<br>Phenobarbital<br>Phenytoin<br>Prazosin<br>Prednisone |
| Tacrolimus (Prograf) | 8 (inactive)<br>31-*O*-desmethyl tacrolimus (active) | Quinidine<br>Rifampin |
| Sirolimus (Rapamune) | Hydroxy-sirolimus (unknown activity)<br>Desmethyl sirolimus (unknown activity)<br>Sirolimus (ring-open form; unknown activity) | Tobramycin<br>Verapamil |
| Everolimus (Zortress, Affinitor) | Hydroxy-everolimus (unknown activity)<br>Dihydroxy-everolimus (unknown activity)<br>Desmethyl everolimus (unknown activity)<br>Everolimus (ring-open form; unknown activity) | |
| Mycophenolate mofetil (CellCept, Myfortic) | 7-*O*-glucuronide (MPAG) (inactive)<br>Acyl-glucuronide (AcMPAG) (active) | |

## DRUG METABOLITES

Cyclosporine has 31 metabolites; one of the major metabolites has approximately 10% of the immunosuppressive activity of the parent compound, whereas the rest of the metabolites are inactive. Tacrolimus has 1 active metabolite, 31-*O*-desmethyl tacrolimus, and 8 metabolites that are inactive. Sirolimus has 7 inactive metabolites, and everolimus has at least 20 inactive metabolites. The primary metabolite of MPA is MPA-glucuronide (MPAG), which is pharmacologically inactive.

Several immunoassay methods are known to cross-react with inactive metabolites to produce results that are 20–60% higher than those obtained by chromatographic techniques such as HPLC or LC-MS/MS [59–61]. Immunoassay methodologies that utilize polyclonal antibodies exhibit the highest degree of positive bias compared to immunoassays based on monoclonal antibodies; however, LC-MS/MS is the preferred technology. For MPA, immunoassay results may be overestimated by cross-reactivity with the MPAG metabolite, as well as the prodrug mycophenolate mofetil [62,63].

#### ENDOGENOUS INTERFERENTS

Heterophilic antibodies may interfere with immunoassays of any sort, and they can be identified with various dilution and pretreatment methods. Falsely elevated tacrolimus and cyclosporine results were reported in patients receiving these drugs (see case report), most likely due to endogenous anti-β-galactosidase antibodies [64]. Hematocrit values have been shown to correlate with response in an MEIA assay for tacrolimus as well. Specifically, low hematocrit values are associated with overestimated tacrolimus concentrations by MEIA; the overestimation can be accounted for by correcting for hematocrit [12].

*CASE REPORT*[3] A 53-year-old male who received a kidney transplant was treated with tacrolimus and corticosteroids. Tacrolimus was monitored using the antibody conjugated magnetic immunoassay (ACMIA)-tacrolimus (Siemens Dimension). The tacrolimus results obtained for the renal transplant patient were consistent with expectations for the first 3 weeks post-transplant (< 12 ng/mL). An unexpected tacrolimus result of 21.5 ng/mL was then observed, and dosing was suspended for a few days. During the period in which tacrolimus was discontinued, blood tacrolimus concentrations persisted at greater than 10 ng/mL. When tacrolimus was reintroduced, it was measured by MEIA-tacrolimus (IMx), and results were consistent with clinical expectations. For comparison, samples were analyzed in parallel with ACMIA (on Dimension analyzer: Siemens Diagnostics) and results were approximately twice the MEIA concentrations. In the automated ACMIA-tacrolimus method (Dimension), whole blood is lysed and mixed with anti-tacrolimus–β-galactosidase antibody conjugate and magnetic particles to which unlabeled tacrolimus is bound. Tacrolimus in the patient blood competes with the tacrolimus bound to particles for binding with the conjugate. The magnetic particles are separated from the reaction mixture, and the tacrolimus-bound conjugate is detected in the presence of enzyme substrate. In the case report, a dilution study was performed that showed nonlinearity. In addition, the interference was resolved completely by treatment with polyethylene glycol and by testing washed erythrocytes, suggesting that the interferent was a plasma component. Positive interference was observed with ACMIA-cyclosporine as well, despite the fact that the patient never received cyclosporine, which further suggests that the cause of the interference was endogenous anti-β-galactosidase antibodies.

[3]From D'Alessandro *et al.* [64].

## Interferences in Measurement of Antidepressant and Mood-Stabilizer Drugs

### *Tricyclic Antidepressants*

Tricyclic antidepressants (TCAs), such as amitriptyline (Elavil), clomipramine (Anafranil), doxepin (Sinequan, Prudoxin, Zonalon, and Silenor), imipramine (Tofranil), trimipramine (Surmontil), nortriptyline (Pamelor), and desipramine (Norpramin), are named for their three-ring structure and used to treat various forms of depression, anxiety disorders, eating disorders, attention deficit hyperactivity disorder, enuresis in children, and chronic and neuropathic pain. Plasma concentrations of TCAs have a positive correlation with clinical improvement and toxicity. Serum TCA concentrations greater than 500 ng/mL can produce adverse anticholinergic symptoms, including dry mouth, sedation, blurred vision, fever, urinary retention, agitation, confusion, and seizures. TCA toxicity from overdose can be life-threatening and involve cardiovascular complications such as hypotension, tachycardia, and cardiac arrhythmias. Serum concentrations greater than 1000 ng/mL can be fatal. Chronically administered TCAs can contribute to cardiac toxicity by accumulating in cardiac tissues [13,65].

### *Pre-Analytical Variables*

Several TCAs have active metabolites that may be important to identify and quantitate because their concentrations contribute to the therapeutic efficacy of TCAs. Some active metabolites are available as independent drugs. For example, nortriptyline is the active metabolite for amitriptyline, and desipramine is the active metabolite for imipramine. Drug–drug or drug–herb interactions involving cytochrome P450 isozymes, such as CYP2D6 and CYP2C19, commonly contribute to accumulation of active drug or drug metabolites that may lead to concentration-related toxicity [13,66].

### *Analytical Variables*

Immunoassays designed to detect TCAs are prone to interferences from other TCA drugs and drug metabolites, as well as structurally similar compounds from different drug classes. Table 13.5 provides a list of drugs that have been reported to interfere with immunoassays designed to detect TCAs. For example, cyclobenzaprine, a skeletal muscle relaxant, has a similar structure to TCA and can cause a false-positive result for TCA by immunoassay due to antibody cross-reactivity [67]. Phenothiazine antipsychotic drugs, such as thioridazine, have three-ring structures and can also cross-react with TCA antibodies and produce false-positive results. Other three-ring structured compounds can produce false-positive TCA results via cross-reactivity with TCA immunoassays, including phenothiazine antipsychotic drugs such as thioridazine [68]; antiepileptic drugs such as carbamazepine and oxcarbazepine [69]; antihistamines such as cetirizine, hydroxyzine, and diphenhydramine [70]; and the antipsychotic drug quetiapine [71].

HPLC methods are well-established for TCA testing; however, this methodology is prone to interferences in peak separation and retention time, which can occur in patients on multidrug therapy. Thioridazine, an antipsychotic drug for schizophrenia therapy, can interfere with imipramine metabolism and analysis by HPLC [68]. Mass spectrometric methods provide high specificity and minimize interferences; thus, they are considered the method of choice for identification, differentiation, and quantitation of TCA drugs [72].

*CASE REPORT*[4] A 17-year-old female with a history of bulimia was admitted to an emergency department after unknown drug ingestion. The patient was drowsy with dilated, reactive pupils, slurred speech, and ataxia. Electrolytes and liver function test results were normal. Urine toxicology testing was negative, but a serum toxicology screen was positive for TCAs by FPIA. An electrocardiograph was pursued that revealed only sinus tachycardia. Targeted chromatographic testing failed to identify any specific TCA, and the patient's condition improved rapidly. Subsequent testing revealed carbamazepine at 18.6 μg/mL in a blood sample collected approximately 12 hr postingestion, which decreased to 10 μg/mL in a sample collected at approximately 18 hr postingestion, but no TCAs, demonstrating that the initial screening result was a false positive.

[4]From Matos *et al.* [73].

**TABLE 13.5** Drug Analytes That May Interfere with Tricyclic Antidepressant Immunoassays

| | |
|---|---|
| Carbamazepine | Maprotiline |
| Carbamazepine-10,11-epoxide | Oxcarbazepine |
| Cetirizine | Perphenazine |
| Chlorpromazine | Phencyclidine |
| Cyclobenzaprine | Prochlorperazine |
| Fluphenazine | Quetiapine fumarate |
| Hydroxyzine | Thioridazine |

### *Lithium*

Lithium carbonate (Eskalith and Lithane) is a monovalent cation that is used as a mood-stabilizing agent for treatment of bipolar disorder, acute manic episode, and depression. Lithium TDM is recommended due to its narrow therapeutic range (0.6–1.2 mmol/L), unpredictable serum concentrations, and concentration-dependent toxicity. Concentrations of lithium greater than 1.5 mmol/L are associated with lethargy, muscle weakness, tremors, and speech difficulties. Lithium concentrations greater than 2.5 mmol/L can produce muscle rigidity, mental confusion, seizures, coma, cardiac arrhythmias, and death [13].

### *Pre-Analytical Variables*

Patient specimens should not be collected in lithium heparin tubes because the additive can increase the concentration of lithium in patient samples, causing a falsely elevated result. For this reason, serum is preferred over plasma. In addition, specimens that are grossly hemolyzed may cause a dilution effect on lithium concentrations and should not be used for lithium analysis. Co-administration of diuretics and nonsteroidal anti-inflammatory agents can cause drug–drug interactions and affect lithium concentrations in serum by altering lithium excretion. Lithium is excreted primarily in urine, where it is actively absorbed by the kidneys; hence, serum concentrations of lithium are affected by the glomerular filtration rate [13].

### *Analytical Variables*

Lithium is most frequently measured by ion-selective electrodes (ISE). Automated colorimetric, photometric, and enzymatic methods, as well as elemental analysis using flame atomic absorption spectrometry, flame atomic emission spectrophotometry, or inductively coupled plasma mass spectrometry methods, are also available and could be used to resolve a suspected interference with ISE [74]. Sources of analytical interferences are based on the technology. For ISE, sources of interference include ions originating from other drugs and elements. Drugs such as quinidine, procainamide, and its metabolite, *N*-acetylprocainamide, can produce a positive interference with lithium results by ISE, particularly with the Beckman analyzer. The interference from the three drugs is additive. Quinidine can also produce a negative bias in lithium results by colorimetric methods, and valproic acid can produce a positive bias in lithium results by the same method. A positive bias in lithium results is also caused by drug interferences from the cardiac drug lidocaine and the anticonvulsant drug carbamazepine by ISE. Calcium concentrations greater than 8.9 mmol/L can cause a positive bias interference on ISE methods and a negative bias interference by colorimetric methods. Potassium and sodium can also produce a negative and positive bias interference, respectively, by colorimetry [75].

## CONCLUSIONS

TDM is a critical tool for optimization of drug therapy. Accuracy of laboratory testing to support TDM depends on a deliberate coordination of both pre-analytical and analytical sources of interferences. Timing of specimen collection, performance characteristics of analytical techniques, and patient history guide interpretation of a TDM result. Any result that is inconsistent with clinical expectations should be critically evaluated before dose adjustments are made.

## References


[1] Bennett CL, Nebeker JR, Lyons EA, Samore MH, Feldman MD, McKoy JM, et al. The Research on Adverse Drug Events and Reports (RADAR) project. JAMA 2005;293(17):2131–40.

[2] Dasgupta A, Schlette E. Rapid *in vitro* conversion of fosphenytoin into phenytoin in sera of patients with liver disease: Role of alkaline phosphatase. J Clin Lab Anal 2001;15(5):244–50.

[3] Shipkova M, Schutz E, Armstrong VW, Niedmann PD, Oellerich M, Wieland E. Determination of the acyl glucuronide metabolite of mycophenolic acid in human plasma by HPLC and Emit. Clin Chem 2000;46(3):365–72.

[4] Hammett-Stabler CA, Pesce AJ, Warner A. Standards of laboratory practice. Washington, DC: National Academy of Clinical Biochemistry; 1999.

[5] Ohshima T, Hasegawa T, Johno I, Kitazawa S. Variations in protein binding of drugs in plasma and serum. Clin Chem 1989;35(8):1722–5.

[6] Tarasidis CG, Garnett WR, Kline BJ, Pellock JM. Influence of tube type, storage time, and temperature on the total and free concentration of valproic acid. Ther Drug Monit 1986;8(3): 373–6.

[7] Fagan TC, Walle T, Oexmann MJ, Walle UK, Bai SA, Gaffney TE. Increased clearance of propranolol and theophylline by high-protein compared with high-carbohydrate diet. Clin Pharmacol Ther 1987;41(4):402–6.

[8] Hanley MJ, Cancalon P, Widmer WW, Greenblatt DJ. The effect of grapefruit juice on drug disposition. Expert Opin Drug Metab Toxicol 2011;7(3):267–86.

[9] Edwards A, Ensom MH. Pharmacokinetic effects of bariatric surgery. Ann Pharmacother 2012;46(1):130–6.

[10] Best BM, Ċapparelli EV. Implications of gender and pregnancy for antiretroviral drug dosing. Curr Opin HIV AIDS 2008;3 (3):277–82.

[11] Gonschior AK, Christians U, Winkler M, Linck A, Baumann J, Sewing KF. Tacrolimus (FK506) metabolite patterns in blood from liver and kidney transplant patients. Clinical chemistry 1996;42(9):1426–32.

[12] Doki K, Homma M, Hori T, Tomita T, Hasegawa Y, Ito S, et al. Difference in blood tacrolimus concentration between ACMIA and MEIA in samples with low haematocrit values. J Pharm Pharmacol 2010;62(9):1185–8.

[13] Mitchell PB. Therapeutic drug monitoring of psychotropic medications. Br J Clin Pharmacol 2001;52(Suppl. 1):45S–54S.

[14] Juhascik MP, Jenkins AJ. Comparison of liquid/liquid and solid-phase extraction for alkaline drugs. J Chromatogr Sci 2009;47(7):553–7.

[15] Scheurer J, Moore CM. Solid-phase extraction of drugs from biological tissues—A review. J Anal Toxicol 1992;16(4):264–9.

[16] Krasowski MD, Pizon AF, Siam MG, Giannoutsos S, Iyer M, Ekins S. Using molecular similarity to highlight the challenges of routine immunoassay-based drug of abuse/toxicology screening in emergency medicine. BMC Emerg Med 2009;9(5): 1–18.

[17] Guo X, Lankmayr E. Phospholipid-based matrix effects in LC-MS bioanalysis. Bioanalysis 2011;3(4):349–52.

[18] Honour JW. Development and validation of a quantitative assay based on tandem mass spectrometry. Ann Clin Biochem 2011;48(Pt 2):97–111.

[19] Danso D, Langman LJ, Snozek CL. LC-MS/MS quantitation of ribavirin in serum and identification of endogenous isobaric interferences. Clin Chim Acta 2011;412(23-24):2332–5.

[20] Annesley TM. Methanol-associated matrix effects in electrospray ionization tandem mass spectrometry. Clin Chem 2007; 53(10):1827–34.

[21] Keller BO, Sui J, Young AB, Whittal RM. Interferences and contaminants encountered in modern mass spectrometry. Anal Chim Acta 2008;627(1):71–81.

[22] Vogeser M, Seger C. Pitfalls associated with the use of liquid chromatography-tandem mass spectrometry in the clinical laboratory. Clin Chem 2010;56(8):1234–44.

[23] Rathore SS, Curtis JP, Wang Y, Bristow MR, Krumholz HM. Association of serum digoxin concentration and outcomes in patients with heart failure. JAMA 2003;289(7):871–8.

[24] Srinivas NR, Ramesh M. Digoxin—A therapeutic agent and mechanistic probe: review of liquid chromatographic mass spectrometric methods and recent nuances in the clinical pharmacology attributes of digoxin. Bioanalysis 2009;1(1):97–113.

[25] Jortani SA, Valdes Jr. R. Digoxin and its related endogenous factors. Crit Rev Clin Lab Sci 1997;34(3):225–74.

[26] Miller JJ, Straub Jr. RW, Valdes Jr. R. Digoxin immunoassay with cross-reactivity of digoxin metabolites proportional to their biological activity. Clin Chem 1994;40(10):1898–903.

[27] Dasgupta A. Endogenous and exogenous digoxin-like immunoreactive substances: Impact on therapeutic drug monitoring of digoxin. Am J Clin Pathol 2002;118(1):132–40.

[28] Sanchez BM, De Cos MA, Peralta FG, Arribas C, Armijo JA. Syva Emit 2000 and Roche "on line" subject to less interference by digoxin-like factors than Abbott TDx FPIA in newborns and pregnant women. Clin Chem 1996;42(6 Pt 1):974–6.

[29] Chicella M, Branim B, Lee KR, Phelps SJ. Comparison of microparticle enzyme and fluorescence polarization immunoassays in pediatric patients not receiving digoxin. Ther Drug Monit 1998;20(3):347–51.

[30] Azzazy HM, Duh SH, Maturen A, Schaller E, Shaw L, Grimaldi R, et al. Multicenter study of Abbott AxSYM Digoxin II assay and comparison with 6 methods for susceptibility to digoxin-like immunoreactive factors. Clin Chem 1997;43(9):1635–40.

[31] McMillin GA, Owen WE, Lambert TL, De BK, Frank EL, Bach PR, et al. Comparable effects of DIGIBIND and DigiFab in thirteen digoxin immunoassays. Clin Chem 2002;48(9):1580–4.

[32] Jortani SA, Pinar A, Johnson NA, Valdes Jr. R. Validity of unbound digoxin measurements by immunoassays in presence of antidote (Digibind). Clin Chim Acta 1999;283(1-2):159–69.

[33] Dasgupta A, Scott J. Unexpected suppression of total digoxin concentrations by cross-reactants in the microparticle enzyme immunoassay: elimination of interference by monitoring free digoxin concentration. Am J Clin Pathol 1998;110(1):78–82.

[34] Datta P, Dasgupta A. Bidirectional (positive/negative) interference in a digoxin immunoassay: importance of antibody specificity. Ther Drug Monit 1998;20(3):352–7.

[35] Steimer W, Muller C, Eber B. Digoxin assays: Frequent, substantial, and potentially dangerous interference by spironolactone, canrenone, and other steroids. Clin Chem 2002;48(3): 507–16.

[36] Cobo A, Martin-Suarez A, Calvo MV, Dominguez-Gil A, de Gatta MM. Clinical repercussions of analytical interferences due to aldosterone antagonists in digoxin immunoassays: an assessment. Ther Drug Monit 2010;32(2):169–76.

[37] Dasgupta A, Biddle DA, Wells A, Datta P. Positive and negative interference of the Chinese medicine Chan Su in serum digoxin measurement: elimination of interference by using a monoclonal chemiluminescent digoxin assay or monitoring free digoxin concentration. Am J Clin Pathol 2000;114(2): 174–9.

[38] Datta P, Dasgupta A. Effect of Chinese medicines Chan Su and Danshen on EMIT 2000 and Randox digoxin immunoassays: wide variation in digoxin-like immunoreactivity and magnitude of interference in digoxin measurement by different brands of the same product. Ther Drug Monit 2002;24(5): 637–44.

[39] Wahed A, Dasgupta A. Positive and negative *in vitro* interference of Chinese medicine dan shen in serum digoxin measurement: elimination of interference by monitoring free digoxin concentration. Am J Clin Pathol 2001;116(3):403–8.

[40] Kuo HY, Hsu CW, Chen JH, Wu YL, Shen YS. Life-threatening episode after ingestion of toad eggs: a case report with literature review. Emerg Med J 2007;24(3):215–16.

[41] Hermida J, Tutor JC. How suitable are currently used carbamazepine immunoassays for quantifying carbamazepine-10,11-epoxide in serum samples? Ther Drug Monit 2003;25(3):384–8.

[42] McMillin GA, Juenke JM, Tso G, Dasgupta A. Estimation of carbamazepine and carbamazepine-10,11-epoxide concentrations in plasma using mathematical equations generated with two carbamazepine immunoassays. Am J Clin Pathol 2010;133 (5):728–36.

[43] Kumps A, Mardens Y. Cross-reactivity assessment of oxcarbazepine and its metabolites in the EMIT assay of carbamazepine plasma levels. Ther Drug Monit 1986;8(1):95–7.

[44] Parant F, Bossu H, Gagnieu MC, Lardet G, Moulsma M. Cross-reactivity assessment of carbamazepine-10,11-epoxide, oxcarbazepine, and 10-hydroxy-carbazepine in two automated carbamazepine immunoassays: PETINIA and EMIT 2000. Ther Drug Monit 2003;25(1):41–5.

[45] Dasgupta A, Tso G, Johnson M, Chow L. Hydroxyzine and cetirizine interfere with the PENTINA carbamazepine assay but not with the ADVIA CENTEUR carbamazepine assay. Ther Drug Monit 2010;32(1):112–15.

[46] Soldin SJ, Wang E, Verjee Z, Elin RJ. Phenytoin overview—Metabolite interference in some immunoassays could be clinically important: results of a college of American pathologists study. Arch Pathol Lab Med 2003;127(12):1623–5.

[47] Roberts WL, Rainey PM. Phenytoin overview—Metabolite interference in some immunoassays could be clinically important. Arch Pathol Lab Med 2004;128(7):734 author reply 735.

[48] Tutor-Crespo MJ, Hermida J, Tutor JC. Phenytoin immunoassay measurements in serum samples from patients with renal insufficiency: comparison with high-performance liquid chromatography. J Clin Lab Anal 2007;21(2):119–23.

[49] Rainey PM, Rogers KE, Roberts WL. Metabolite and matrix interference in phenytoin immunoassays. Clin Chem 1996; 42(10):1645–53.

[50] Oxaprozin Datta P. and 5-(*p*-hydroxyphenyl)-5-phenylhydantoin interference in phenytoin immunoassays. Clin Chem 1997;43(8 Pt 1):1468–9.

[51] Datta P, Dasgupta A. Cross-reactivity of fosphenytoin in four phenytoin immunoassays. Clin Chem 1998;44(3):696–7.

[52] Brauchli YB, Scholer A, Schwietert M, Krahenbuhl S. Undetectable phenytoin serum levels by an automated particle-enhanced turbidimetric inhibition immunoassay in a patient with monoclonal IgM lambda. Clin Chim Acta 2008;389(1-2): 174–6.

[53] Taylor PJ. Therapeutic drug monitoring of immunosuppressant drugs by high-performance liquid chromatography-mass spectrometry. Ther Drug Monit 2004;26(2):215–19.

[54] Dasgupta A, Moreno V, Balark S, Smith A, Sonilal M, Tejpal N, et al. Rapid estimation of whole blood everolimus concentrations using architect sirolimus immunoassay and mathematical equations: comparison with everolimus values determined by liquid chromatography/mass spectrometry. J Clin Lab Anal 2011;25(3):207–11.

[55] Bouzas L, Tutor JC. Determination of blood everolimus concentrations in kidney and liver transplant recipients using the sirolimus antibody conjugated magnetic immunoassay (ACMIA). Clin Lab 2011;57(5-6):403–6.

[56] Johnson-Davis KL, De S, Jimenez E, McMillin GA, De BK. Evaluation of the Abbott ARCHITECT i2000 sirolimus assay and comparison with the Abbott IMx sirolimus assay and an established liquid chromatography-tandem mass spectrometry method. Ther Drug Monit 2011;33(4):453–9.

[57] Maiguma T, Yosida T, Otsubo K, Okabe Y, Sugitani A, Tanaka M, et al. Evaluation of inosin-5′-monophosphate dehydrogenase activity during maintenance therapy with tacrolimus. J Clin Pharm Ther 2010;35(1):79–85.

[58] Glander P, Hambach P, Liefeldt L, Budde K. Inosine 5′-monophosphate dehydrogenase activity as a biomarker in the field of transplantation. Clin Chim Acta 2011;413:1391–7.

[59] Sallustio BC, Noll BD, Morris RG. Comparison of blood sirolimus, tacrolimus and everolimus concentrations measured by LC-MS/MS, HPLC-UV and immunoassay methods. Clin Biochem 2011;44 (2-3):231–6.

[60] Brown NW, Franklin ME, Einarsdottir EN, Gonde CE, Pires M, Taylor PJ, et al. An investigation into the bias between liquid chromatography-tandem mass spectrometry and an enzyme multiplied immunoassay technique for the measurement of mycophenolic acid. Ther Drug Monit 2010;32(4):420–6.

[61] Wallemacq P, Goffinet JS, O'Morchoe S, Rosiere T, Maine GT, Labalette M, et al. Multi-site analytical evaluation of the Abbott ARCHITECT tacrolimus assay. Ther Drug Monit 2009;31(2): 198–204.

[62] Rebollo N, Calvo MV, Martin-Suarez A, Dominguez-Gil A. Modification of the EMIT immunoassay for the measurement of

unbound mycophenolic acid in plasma. Clin Biochem 2011; 44(2-3):260–3.

[63] Shipkova M, Schutz E, Besenthal I, Fraunberger P, Wieland E. Investigation of the crossreactivity of mycophenolic acid glucuronide metabolites and of mycophenolate mofetil in the Cedia MPA assay. Ther Drug Monit 2010;32(1):79–85.

[64] D'Alessandro M, Mariani P, Mennini G, Severi D, Berloco P, Bachetoni A. Falsely elevated tacrolimus concentrations measured using the ACMIA method due to circulating endogenous antibodies in a kidney transplant recipient. Clin Chim Acta 2011;412(3-4):245–8.

[65] Linde K, Schumann I, Meissner K, Jamil S, Kriston L, Rucker G, et al. Treatment of depressive disorders in primary care—Protocol of a multiple treatment systematic review of randomized controlled trials. BMC Fam Pract 2011;12:127.

[66] Linder MW, Keck Jr. PE. Standards of laboratory practice: antidepressant drug monitoring. National Academy of Clinical Biochemistry. Clin Chem 1998;44(5):1073–84.

[67] Van Hoey NM. Effect of cyclobenzaprine on tricyclic antidepressant assays. Ann Pharmacother 2005;39(7-8):1314–17.

[68] Maynard GL, Soni P. Thioridazine interferences with imipramine metabolism and measurement. Ther Drug Monit 1996; 18(6):729–31.

[69] Saidinejad M, Law T, Ewald MB. Interference by carbamazepine and oxcarbazepine with serum- and urine-screening assays for tricyclic antidepressants. Pediatrics 2007;120(3):e504–9.

[70] Dasgupta A, Wells A, Datta P. False-positive serum tricyclic antidepressant concentrations using fluorescence polarization immunoassay due to the presence of hydroxyzine and cetirizine. Ther Drug Monit 2007;29(1):134–9.

[71] Caravati EM, Juenke JM, Crouch BI, Anderson KT. Quetiapine cross-reactivity with plasma tricyclic antidepressant immunoassays. Ann Pharmacother 2005;39(9):1446–9.

[72] Breaud AR, Harlan R, Di Bussolo JM, McMillin GA, Clarke W. A rapid and fully-automated method for the quantitation of tricyclic antidepressants in serum using turbulent-flow liquid chromatography-tandem mass spectrometry. Clin Chim Acta 2010;411(11-12):825–32.

[73] Matos ME, Burns MM, Shannon MW. False-positive tricyclic antidepressant drug screen results leading to the diagnosis of carbamazepine intoxication. Pediatrics 2000;105(5):E66.

[74] Dou C, Aleshin O, Datta A, Yuan C. Automated enzymatic assay for measurement of lithium ions in human serum. Clin Chem 2005;51(10):1989–91.

[75] Sampson M, Ruddel M, Elin RJ. Lithium determinations evaluated in eight analyzers. Clin Chem 1994;40(6):869–72.

CHAPTER

# 14

# Limitations of Drugs of Abuse Testing

*Amitava Dasgupta*
University of Texas Health Sciences Center at Houston, Houston, Texas

## INTRODUCTION

Substance abuse is a serious public health and safety issue not only in the United States but also worldwide. According to the *2010 National Survey on Drug Use and Health* by the Office of Applied Science of the Substance Abuse and Mental Health Services Administration (SAMHSA), an estimated 22.6 million Americans 12 years or older used an illicit drug during the month prior to the survey interview (Figure 14.1). This represented 8.9% of the U.S. population aged 12 years or older. Marijuana was the most commonly used drug, with an estimated 17.4 million people using it in the month prior to the survey. In addition, there were 1.5 million cocaine users, and an alarming 7.0 million people abused prescription medications. Illicit drug abuse was more common among unemployed people (17.5% among unemployed people vs. 8.4% among employed people). Another alarming finding was that an estimated 10.6 million people aged 12 years or older reported driving under the influence of an illicit drug [1]. Thus, December has been designated National Impaired Driving Prevention Month.

Because of widespread use of drugs and alcohol, drug testing using urine specimens and blood alcohol testing are commonly performed in emergency room patients who are involved in accidents, especially car accidents, as well as in anyone with clinical symptoms indicating drug or alcohol overdose or both. In many hospital laboratories, blood alcohol testing is performed using enzymatic assays that can be easily adopted on automated analyzers. However, such enzymatic methods are subject to interference most notably from elevated blood lactate and lactate dehydrogenase. This topic is discussed in Chapter 16. This chapter focuses on limitations of drugs of abuse testing conducted in clinical laboratories, with a brief discussion on the challenges of legal drug testing, especially the issue of urine adulteration.

## DRUGS OF ABUSE TESTING: MEDICAL VERSUS LEGAL

Whereas medical drug testing has been practiced for a long time, especially for patients admitted to emergency departments, legal drug testing was initiated by President Reagan, who issued Executive Order 12564 on September 15, 1986. This executive order directed drug testing for all federal employees who are involved in law enforcement, national security, protection of life and property, public health, and other services requiring a high degree of public trust. Following this executive order, the National Institute of Drug of Abuse (NIDA) was given the responsibility of developing guidelines for federal drug testing. Currently, SAMHSA, an agency under the U.S. Department of Health and Human Services, is responsible for providing mandatory guidelines for federal workplace drug testing. Although Reagan's executive order was not intended for private employers, currently a majority of *Fortune* 500 companies have policies for workplace drug testing. Drug abuse costs American industry and the public an estimated $100 billion a year, but workplace drug testing can contribute to a better work environment, improves morale of employees, and is very effective in preventing work-related accidents. Therefore, developing a cost-effective corporate workplace drug testing program that meets federal guidelines, capable of standing legal challenge and also accepted by employees, is the objective of all workplace drug testing programs [2]. Several publications have established large negative

*Accurate Results in the Clinical Laboratory.*
DOI: http://dx.doi.org/10.1016/B978-0-12-415783-5.00014-1

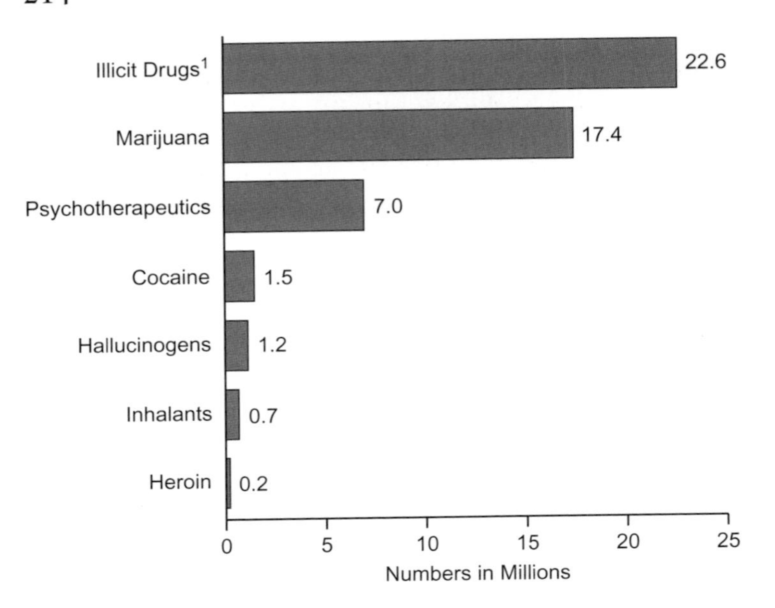


**FIGURE 14.1 Abuse of various illicit drugs and inhalants in the United States.** Illicit drugs include marijuana/hashish, cocaine (including crack), heroin, hallucinogens, inhalants, and prescription-type psychotherapeutics used nonmedically. Source: *Data from the 2010 National Survey on Drug Use and Health: National Findings. U.S. Department of Health and Human Services, Substance Abuse and Mental Health Services Administration (SAMHSA), Office of Applied Studies.*

correlations between workplace drug testing and employee substance abuse [3].

In medical drug testing, informed consent may not be obtained from a patient. An overdosed patient admitted to the emergency department may not be able to grant an informed consent anyway. In contrast, in a workplace or any other legal drug testing program, obtaining an informed consent prior to testing is mandatory. Another major difference between medical and legal drug testing is that in medical testing, an initial positive screening result obtained by using immunoassays may not be confirmed by using gas chromatography combined with mass spectrometry (GC-MS), but in legal drug testing, GC-MS confirmation is mandatory. In addition, a chain of custody must be maintained in legal drug testing indicating all personnel who have possession of the specimen from the time of collection to the time of reporting results. Chain of custody is not usually initiated in medical drug testing. Therefore, a medical drug testing result may not stand a legal challenge.

## DRUGS TESTED IN DRUGS OF ABUSE TESTING PROTOCOLS

The fourth edition of the *Diagnostic and Statistical Manual of Mental Disorders* distinguishes 11 categories of abused substances, including amphetamines, cocaine, marijuana, hallucinogens, inhalants, opioids, phencyclidine (PCP), sedative hypnotics, anxiolytics, and alcohol. Despite this guideline, many drugs of abuse programs may screen only five drugs, also known as NIDA-5 or SAMHSA-5 drugs. The five SAMHSA-mandated drugs for federal workplace drug testing are amphetamine, cocaine (tested as benzoylecgonine, the inactive metabolite), opiates, PCP, and marijuana (tested as 11-nor-9-carboxy-Δ-9-tetrahydrocannabinol, the inactive metabolite). Although these drugs are more frequently tested, there is no industry standard and terms such as "routine drug screen" or "comprehensive drug screen" are used by different laboratories, but the drugs included in each protocol may vary from one laboratory to another [4]. Some private employers may test for additional drugs in their workplace drug testing protocols, and such comprehensive drug panel may include barbiturates, benzodiazepines, oxycodone, methadone, methaqualone, and propoxyphene.

## DRUG TESTING METHODOLOGIES

Immunoassays are widely used as the first step of drug screening in both medical and workplace drug testing programs. Immunoassays can be easily automated, several drugs can be analyzed using one specimen, and results can be directly downloaded in the laboratory information system. The main component of the immunoassay reagents is the analyte-specific antibody, which can be either polyclonal or monoclonal in nature. In general, monoclonal antibody-based immunoassays are more specific to the target analyte than are the polyclonal antibody-based assays. All immunoassay methods used for drugs of abuse screening require no specimen pretreatment. The assays use very small amounts of sample volumes (most $<100\ \mu L$), reagents are stored in the analyzer, and most assays employ stored calibration curves in the automated analyzer. With respect to assay design, there are two formats of immunoassays: competition and immunometric (commonly referred to as "sandwich"). Competition immunoassays work best for drugs that are small molecules such as therapeutic drugs and drugs of abuse requiring a single analyte-specific antibody. In the competition format, the analyte molecules in the specimen compete with analyte (or its analogs), labeled with a suitable tag and provided in the reagent, for a limited number of binding sites provided by an analyte-specific antibody (also provided in the reagent). Thus, in these types of assays, the higher the analyte concentration in the sample, the less label can bind to the antibody to form the conjugate. If the bound label provides the signal, such as in fluorescence polarization immunoassay (FPIA), the analyte concentration in the specimen is inversely proportional to the signal produced. On the other hand, if the signal is generated by the free label, then the signal is proportional to the concentration of the abused drug in the specimen. The signals are mostly optical, such as absorbance, fluorescence, or chemiluminescence.

There are several variations in this basic format. Although the assays can be homogeneous or heterogeneous, most drugs of abuse assays use the homogeneous format, in which the bound label has different properties than the free label and physical separation between bound and free label is unnecessary before measuring the signal. For example, in FPIA, the free label has different Brownian motion than when the label is complexed to a large antibody (146 kDa). This results in differences in the fluorescence polarization properties of the label, where the bound label is capable of producing the signal [5]. In another type of homogeneous immunoassay, an enzyme is the label, whose activity is modulated differently in the free versus the antibody-bound conditions of the label. This principle is used in the enzyme multiplied immunoassay technique (EMIT) and cloned enzyme donor immunoassay (CEDIA) technologies [6,7]. In the EMIT method, the label enzyme, glucose 6-phosphodehydrogenase (G6PDH), is active unless in the antigen–antibody complex. The active enzyme reduces nicotinamide adenine dinucleotide (NAD) to NADH, and the absorbance is monitored at 340 nm (NAD has no signal at 340 nm, whereas NADH absorbs at 340 nm). To guard against interference from a specimen's native G6PDH, the newer assays use recombinant bacterial enzymes whose activity conditions are different from those of the human enzyme. Similarly, in the CEDIA method, two genetically engineered inactive fragments of the enzyme β-galactosidase are coupled to the antigen and the antibody reagents. When they combine, the active enzyme is produced, and the substrate—a chromogenic galactoside derivative—produces the assay signal. In a third commonly used format of homogeneous immunoassay (turbidimetric immunoassay), analyte (antigen) or its analogs are coupled to colloidal particles of latex, for example [8]. Because antibodies are bivalent, the latex particles agglutinate in the presence of the antibody. However, in the presence of free analytes in the specimen, there is less agglutination. In a spectrophotometer, the resulting turbidity can be monitored as end point or as rate. In the kinetic interaction of microparticles in solution (KIMS) assay method, in the absence of drug molecules, free antibodies bind to drug microparticle conjugate forming particle aggregates, and an increase in absorption is observed. When drug molecules are present in urine specimen, these molecules bind with free antibody molecules and thus prevent formation of particle aggregates and diminish absorbance in proportion to drug concentration. The ONLINE drugs of abuse testings immunoassays marketed by Roche Diagnostics (Indianapolis, IN) are based on the KIMS format.

In the heterogeneous immunoassay format, the bound label is physically separated from the unbound labels before signal is measured. The separation is often done magnetically, where the reagent analyte (or its analog) is provided as coupled to paramagnetic particles (PMPs), and the antibody is labeled. Conversely, the antibody may also be provided as conjugated to the PMPs, and the reagent analyte may carry the label. After separation and wash, the bound label is reacted with other reagents to generate the signal. This is the mechanism in many chemiluminescent immunoassays, in which the label may be a small molecule that generates chemiluminescent signal [9]. The label may also be an enzyme (enzyme immunoassay (EIA) or enzyme-linked immunosorbent assay (ELISA)) that generates chemiluminescent, fluorometric, or colorimetric signal. Another type of heterogeneous immunoassay uses polystyrene particles. If the particles are micro sizes, this type of assay is called microparticle-enhanced immunoassay [10].

Usually, multiple calibrators (four to six levels) are recommended for accurate measurements of the analyte across the entire assay range in an immunoassay, although 2-point calibrations are also used. Most automated assay systems can store a calibration curve depending on the assay stability of the systems. Therefore, when a sample is analyzed during the period denoted by calibration stability, the assay signal is automatically converted into analyte concentration via the stored calibration curve. Drugs of abuse assays are often used to report "qualitative" results—that is, positive or negative with respect to a certain analyte concentration (the "cutoff" level). Thus, many of the assays are in qualitative or quantitative formats, and in most cases such formats are defined by assay protocol and calibration. In qualitative formats, the calibration can be simplified to only one or two calibrators, centering on the cutoff point, thus providing the most accuracy around that point. The algorithm compares the signal observed with a sample with that of the cutoff calibrator and reports the result as positive or negative. Semiquantitative results can be reported with a calibration curve containing a minimum of three calibrator samples; often, the combination of the zero-calibrator, together with two or more calibrators at or near cutoff level, is used to generate a calibration curve. Obviously, such assay formats will have increased inaccuracy at analyte concentrations much higher than the cutoff concentration.

Immunoassays used for drugs of abuse testing have high sensitivity, but specificity may vary widely between assays designed for individual drugs. In general, there are certain cutoff concentrations for immunoassays used for drugs of abuse testing. These cutoffs are set by SAMHSA guidelines. Therefore, if the concentration of a target drug or metabolite is below the cutoff concentration, the immunoassay result should be negative, and a negative test result may be indicative of no drug present in the specimen or a drug concentration lower than the cutoff concentration. The GC-MS confirmation cutoff as proposed by SAMHSA guidelines may be the same as the cutoff concentration recommended for immunoassay screening or may be lower than the immunoassay screening threshold. Immunoassay and GC-MS cutoff concentrations of various SAMHSA drugs are summarized in Table 14.1. Immunoassay and GC-MS confirmation cutoffs of other commonly abused drugs (in the workplace drug testing protocols of private employers; non-SAHMSA drugs) are given in Table 14.2. Drugs or metabolites can be detected for a limited time in urine after abuse. Detection windows of various drugs in urine are summarized in Table 14.3.

In legal drug testing, a person may intentionally dilute their urine specimen with water in order to avoid a positive test result. Many investigators have explored the possibility of detecting abused drugs in urine specimens using lower cutoff concentrations than recommended by SAMHSA guidelines in order to identify more subjects who may be abusing drugs. Unfortunately, in the United States, such practice cannot stand legal challenges except for opiates, for which some private employers are still using the 300 ng/mL cutoff concentration that was proposed in the original drug testing guidelines by NIDA. However, the Correctional Services of Canada (CSC) incorporates lower screening and confirmation cutoff for drug/metabolites if needed for diluted urine. These guidelines include the following: all amphetamine screening cutoff, 100 ng/mL; confirmation cutoff, 100 ng/mL; benzodiazepines screening and confirmation cutoff, 50 ng/mL; benzoylecgonine screening and confirmation cutoff, 15 ng/mL; opiates screening and confirmation cutoff, 120 ng/mL; phencyclidine screening and confirmation cutoff, 5 ng/mL; and cannabinoids screening cutoff, 20 ng/mL, but confirmation cutoff of 3 ng/mL. Fraser and Zamecnik [11] reported that between 2000 and 2002, 7912 urine specimens collected by the CSC were dilute, and 26% of these screened positive using SAMHSA cutoff values. When lower values for cutoff and confirmation were adopted, 1100 specimens tested positive for one or more illicit drugs. The drug most often confirmed positive in a diluted specimen was marijuana. Soldin *et al.* [12] reported a more than 100% increase

**TABLE 14.1** Original Cutoff Values and New Cutoff Values (Effective October 1, 2010) of SAMHSA-Mandated Drug Testing

| Drug or Drug Class | Immunoassay Cutoff (ng/mL) | | GC-MS Confirmation (ng/mL) |
|---|---|---|---|
| | Original Value | New Value | New Value |
| Amphetamine/methamphetamine | 1000 | 500 | 500 |
| MDMA | Not applicable | 500 | 250 |
| Cannabinoids | 50 | 50 | 15 |
| Cocaine metabolites | 300 | 150 | 100 |
| Opiates (codeine/morphine)[a] | 2000 | 2000 | 2000 |
| 6-Acetylmorphine[b] | 10 | 10 | 10 |
| Phencyclidine | 25 | 25 | 25 |

[a]*In the first guidelines, the published cutoff concentration for opiate screening was 300 ng/mL, but the value was increased to 2000 ng/mL in 1998. Some private employers may still use immunoassays of opiate with a cutoff concentration of 300 ng/mL.*
[b]*Testing is recommended if opiate test is positive (>2000 ng/mL).*

**TABLE 14.2** Immunoassay Screening and GC-MS Confirmation Cutoffs for Non-SAMHSA Drugs

| Drug | Immunoassay Cutoff (ng/mL) | GC-MS Confirmation (ng/mL) |
|---|---|---|
| Barbiturates | 200 or 300 | 200 |
| Benzodiazepines | 200 or 300 | 200 |
| Methadone | 300 | 300 |
| Methaqualone | 300 | 300 |
| Propoxyphene | 300 | 300 |
| Oxycodone | 100 or 300 | 100 |
| 6-Acetylmorphine | 10 | 10 |
| LSD | 0.5 | 0.5 |

TABLE 14.3 Typical Window of Detection of Various Drugs in Urine Specimens Using SAMSHA Cutoffs[a]

| Drug | Window of Detection |
|---|---|
| Amphetamine | 2 days |
| Methamphetamine | 2 days |
| 3,4-Methylenedioxyamphetamine | 1–2 days |
| Short-acting barbiturates | 1–2 days |
| Long-acting barbiturates | 14–21 days |
| Short-acting benzodiazepine | 3 days |
| Long-acting benzodiazepine | 14–21 days |
| Cocaine | 2–3 days |
| Morphine | 2 days |
| Codeine | 2 days |
| Heroin | >1 day for detecting 6-acetylmorphine |
| | 2 days for detecting morphine |
| Oxycodone | 2–4 days |
| Methadone | 3 days |
| Propoxyphene | 2–14 days |
| Marijuana | 2–3 days after acute use |
| | 5–7 days for moderate use |
| | 30 days or more in chronic users |
| Methaqualone | 14 days |
| Phencyclidine | 8 days |
| LSD | 2–3 days |

[a]*Detectability depends on drug amounts, frequency and chronicity of abuse, and individual variables related to drug metabolism and excretion.*

TABLE 14.4 Antibody Specificity of Various Immunoassays Used for Drugs of Abuse Testing

| Immunoassay for the Drug | Antibody Target |
|---|---|
| Amphetamine/methamphetamine assays | Methamphetamine or amphetamine |
| 3,4-Methylenedioxymethamphetamine assay | 3,4-Methylenedioxymethamphetamine |
| Cocaine assay | Benzoylecgonine |
| Opiate assays | Morphine |
| Oxycodone assay | Oxycodone |
| Heroin assay | 6-Acetylmorphine |
| Methadone assay | Methadone or EDDP metabolite |
| Marijuana assay | 11-nor-9-carboxy-Δ-9-tetrahydrocannabinol |
| PCP assay | Phencyclidine |
| Benzodiazepine assays | Commonly oxazepam, but nordiazepam, nitrazepam, or lormetazepam may be used |
| Barbiturates | Commonly secobarbital |
| LSD assay | LSD |
| Methaqualone assay | Methaqualone |

EDDP, 2-ethyldene-1,5-dimethyl-3,3-diphenpyrrolodide.

in cocaine-positive specimens when the cutoff was lowered from 300 to 80 ng/mL in a pediatric population because neonates are not capable of concentrating urine to the same extent as adults. Luzzi *et al.* [13] investigated the analytic performance criteria of three immunoassay systems (EMIT, Beckman EIA, and Abbott FPIA) for detecting abused drugs below established cutoff values. The authors concluded that these drugs can be screened at concentrations much lower than established SAMHSA cutoff values. For example, the authors proposed a marijuana metabolite cutoff of 35 ng/mL using EMIT and 14 ng/mL for the Beckman EIA and the Abbott FPIA assays. The proposed cutoff values were based on imprecision studies in which the coefficient of variation was less than 20%. Such lowering of cutoff values increased the number of positive specimens in the screening tests by 15.6%. A 7.8% increase was also observed in the confirmation stage of drugs of abuse testing [13].

In general, immunoassays used for drugs of abuse testing target a specific drug or a metabolite. For example, immunoassays designed for detecting cocaine in urine target benzoylecgonine, the active metabolite of cocaine. In contrast, an antibody used in the immunoassay for screening barbiturates may target secobarbital. Antibody targets of various immunoassays used in drugs of abuse testing are listed in Table 14.4. The major limitation of immunoassays is that an antibody may cross-react with a structurally similar drug, causing false-positive test results. Therefore, initial drug screening should be confirmed by GC-MS. Most mass spectrometers used for drugs of abuse testing are operated in electron ionization mode, although mass spectrometers can also be operated in chemical ionization mode. Confirmation methods for drug testing using GC-MS can be found in books devoted to drugs of abuse testing [14]. Liquid chromatography combined with mass spectrometry (LC-MS) and liquid chromatography combined with tandem mass spectrometry (LC-MS/MS) are gaining popularity as alternatives to GC-MS for drug confirmation [15]. Eichhorst *et al.* [16] proposed the use of LC-MS for rapid screening and confirmation of the presence of abused drugs in urine specimen as an alternative to immunoassay screening.

## CHALLENGES IN TESTING FOR AMPHETAMINES

In general, immunoassays for amphetamines, in addition to detecting the presence of amphetamine and methamphetamine, can also detect the widely abused designer drugs 3,4-methylenedioxyamphetamine (MDA) and 3,4-methylenedioxymethamphetamine (MDMA). However, certain amphetamine immunoassays may have lower capability of detecting MDMA and MDA due to poor cross-reactivity of the methamphetamine-specific antibody. For example, the Abuscreen ONLINE amphetamine assay (Roche Diagnostics) has only 0.1% cross-reactivity with MDMA and 38% cross-reactivity with MDA. The Neogen amphetamine assay has only 2.2% cross-reactivity with MDMA and 0.4% cross-reactivity with MDA. Kunsman *et al.* [17] reported that the typical concentration of MDMA in urine specimens varied widely from 380 to 96,200 ng/mL (mean, 13,400 ng/mL), whereas the typical concentration of MDA varied from 150 to 8600 ng/mL (mean, 1600 ng/mL). The presence of MDA in urine specimens at a concentration of approximately 10–15% of MDMA concentration is consistent with MDMA metabolism that may be indicative of MDMA abuse only [17]. Therefore, an immunoassay for amphetamine with low cross-reactivity with MDMA and MDA may produce false-negative test results if a low concentration of MDMA or MDA is present in the urine specimen. However, specific immunoassays for detecting the presence of MDMA in urine are commercially available. Stout *et al.* [18] compared DRI methamphetamine, DRI ecstasy (MDMA), and Abuscreen ONLINE amphetamine assays at a cutoff of 500 ng/mL and observed that the DRI ecstasy assay performed the best, as expected, with a GC-MS confirmation rate of 90% of the specimens screened positive. In contrast, the DRI amphetamine assay had poor capability of detecting the presence of MDMA in urine because only 6% of specimens were confirmed positive by GC-MS. Only 20% of specimens screened positive by Abuscreen ONLINE amphetamine assay were confirmed by GC-MS. In another study, the authors commented that DRI and CEDIA amphetamine assays do not have good sensitivity in identifying urine specimen containing MDMA [19]. Poklis *et al.* [20] reported that the EMIT d.a.u. monoclonal amphetamine/methamphetamine immunoassay has a cutoff concentration of 3000 ng/mL for racemic MDMA but only 800 ng/mL for MDA. The assay had higher sensitivity for detecting the S(+) isomer of both MDMA and MDA. The authors found EMIT d.a.u. monoclonal amphetamine/methamphetamine assay to be vastly superior to EMIT d.a.u. polyclonal amphetamine/methamphetamine assay for detecting MDMA and MDA.

The FPIA amphetamine/methamphetamine assay for application on the AxSYM analyzer is capable of detecting abuse of paramethoxyamphetamine (PMA) and paramethoxymethamphetamine if the cutoff concentration is set at 300 ng/mL [21]. However, amphetamine/methamphetamine immunoassays are not suitable for detection of the majority of designer drugs structurally related to amphetamine/methamphetamine. Kerrigan *et al.* [22] evaluated cross-reactivities of 11 designer drugs with nine commercially available immunoassays. The designer drugs included in the study were 2,5-dimethoxy-4-bromophenylethylamine (2C-B); 2,5-dimethoxyphenethylamine (2C-H); 2,5-dimethoxy-4-iodophenethylamine (2C-I); 2,5-dimethoxy-4-ethylthiophenethylamine (2C-T-2); 2,5-dimethoxy-4-isopropylthiophenethylamine (2C-T-4); 2,5-dimethoxy-4-propylthiophenethylamine (2C-T-7); 2,5-dimethoxy-4-bromo-amphetamine (DOB); 2,5-dimethoxy-4-ethylamphetamine (DOET); 2,5-dimethoxy-4-iodoamphetamine (DOI); 2,5-dimethoxy-4-methylamphetamine (DOM); and 4-methylthioamphetamine (4-MTA). Cross-reactivities of these designer drugs with immunoassays studied were less than 0.4%, and even at a concentration of 50,000 ng/mL, these designer drugs were insufficient to produce a positive response, indicating that amphetamine/methamphetamine immunoassays are not capable of detecting the presence of most of these drugs in urine. However, 4-MTA was the only drug that demonstrated 5% cross-reactivity with the Neogen amphetamine ELISA assay (Lexington, KY) but a significant 200% cross-reactivity with the Immunalysis amphetamine ELISA assay (Pomona, CA). Apollonio *et al.* [23] reported that 4-MTA had 280% cross-reactivity with the Bio-Quant Direct ELISA assay for amphetamine. However, LC-MS or GC-MS methods are capable of detecting the presence of these drugs or their metabolites in urine specimens. Kerrigan *et al.* [24] described a GC-MS protocol for analysis of 2C-B, 2C-H, 2C-I, 2C-T-2, 2C-T-7, 4-MTA, DOB, DOET, DOI, and DOM in urine specimens. Ewald *et al.* [25] analyzed designer drugs and 2,5-diemethoxy-4-bromo-methamphetamine using GC-MS. Takahashi *et al.* [26] created a psychoactive drug data library by performing analysis using LC with photodiode array spectrophotometry as well as GC-MS. This library has data on 104 drugs with the potential for abuse.

***CASE REPORTS*** Two men experimented with capsules filled with white powder containing an unknown substance in order to experience hallucination. Within 15 min of taking such capsules orally, they experienced intense hallucinations followed by vomiting, and eventually they became unconscious. After an unknown period of time, both men were admitted to the emergency department. The first patient, a 28-year-old, survived but experienced serious convulsion. The second patient, a 29-year-old, died in the hospital 6 days after admission. The initial drug screens of both patients using CEDIA immunoassays were negative for amphetamines but positive for cannabinoid metabolite for both men. The presence of DOB was confirmed in urine specimens using GC-MS. The concentration of DOB in the serum of the deceased patient was 19 ng/mL, whereas that of the patient who survived was 13 ng/mL. Although DOB is structurally related to amphetamine, the CEDIA amphetamine immunoassay is unable to detect its presence in the urine [27].

Immunoassays used for amphetamine/methamphetamine usually have good sensitivity but poor specificity due to cross-reactivity of structurally related drugs causing false-positive test results. Therefore, GC-MS confirmation is essential for many amphetamine-positive specimens even in a medical drug testing setting, especially if a patient denies any such drug abuse. In one report, the authors investigated the sensitivity and specificity of two immunoassays for detecting amphetamines and reported that the DRI amphetamine assay (Thermo Scientific) identified 1104 presumptive positive urine specimens (out of 27,400 randomly collected specimens), but the presence of amphetamine, methamphetamine, MDA, or MDMA was confirmed by GC-MS in only 1.99% of these presumptive positive specimens. The presence of ephedrine, pseudoephedrine, or phenylpropanolamine was confirmed in 833 urine specimens, which were presumptive positive for amphetamine using the DRI assay. However, the Abuscreen ONLINE assay identified only 317 presumptive positive amphetamine specimens, and the presence of amphetamine, methamphetamine, MDA, or MDMA was confirmed in 7.94% of specimens by GC-MS [28].

Interference of various sympathomimetic amines found in over-the-counter (OTC) cold medications in amphetamine immunoassays is well recognized. However, the GC-MS confirmation test must be able to identify the sympathomimetic amine causing the false-positive test result. Ephedrine or pseudoephedrine, which is present in many OTC cold medications, is responsible for the majority of false-positive results in amphetamine immunoassay screening tests. These sympathomimetic amines may be present in large amounts in urine specimens that are initially tested positive by amphetamine immunoassays. Commonly encountered medications that interfere with amphetamine screening assays are listed in Table 14.5.

Amphetamine and methamphetamine have optical isomers designated *d* (or +) for dextrorotatory and *l* (or −) for levorotatory. The *d* isomers, the more physiologically active compounds, are the intended targets of immunoassays because *d* isomers are abused. Ingestion of medications containing the *l* isomer can cause false-positive results. For example, Vicks inhaler contains the active ingredient *l*-methamphetamine, and extensive use of this product may cause false-positive results for immunoassay screening. Specific isomer resolution procedures must be performed to differentiate the *d* and *l* isomers because routine confirmation by GC-MS does not determine isomer composition. Poklis *et al.* [29] reported relatively high concentrations of *l*-methamphetamine observed in two subjects (1390 and 740 ng/mL, respectively) after extensively inhaling Vicks inhaler every hour for several hours. However, urine specimens tested negative by the EMIT II amphetamine/methamphetamine assay (Dade Behring), even after such extensive use of the Vicks inhaler.

Ranitidine is an $H_2$ receptor blocking agent (antihistamine) that reduces acid production by the stomach and is available OTC without any prescription. Dietzen *et al.* [30] reported that ranitidine, if present in urine at a concentration greater than 43 μg/mL, may cause a false-positive amphetamine screen test result using Beckman Synchron immunoassay reagents (Beckman Diagnostics, Brea, CA). These concentrations of ranitidine are expected in patients taking ranitidine at the recommended dosage. Trazodone interferes with both amphetamine and MDMA assay. A series of patients who tested positive for MDMA (using

**TABLE 14.5** Drugs that Interfere with Amphetamine Immunoassays

| | |
|---|---|
| Structurally similar amines | Brompheniramine, benzphetamine, ephedrine, isometheptene, mephentermine, methylphenidate, pseudoephedrine, phentermine, propylhexedrine, phenylephrine, phenmetrazine, tyramine |
| Antidepressant/antipsychotic | Chlorpromazine, bupropion, desipramine, doxepin, fluoxetine, perazine, thioridazine, trimipramine, trazodone |
| Antihistamine | Ranitidine |
| Antimalarial | Chloroquine |
| Antispasmodic | Mebeverine |
| β-Blocker | Labetalol |
| Cardioactive | *N*-acetyl procainamide (metabolite of procainamide), Mexiletine |
| Narcotic analgesic | Fentanyl |
| Nonsteroidal anti-inflammatory | Tolmetin |
| Tocolytic agent | Ritodrine |
| Antiviral | Amantadine |
| Herbal supplement | Dimethylamine |

Ecstasy EMIT II assay) did not show any presence of MDMA in urine when confirmed by a specific LC-MS/MS method. However, all specimens showed the presence of trazodone and its metabolite meta-chlorophenylpiperazine [31]. Baron *et al.* [32] demonstrated that the trazodone metabolite meta-chlorophenylpiperazine is responsible for the interference.

Labetalol, a β-blocker commonly used for control of hypertension in pregnancy, can cause false-positive amphetamine screen results using an immunoassay. A labetalol metabolite is structurally similar to amphetamine and methamphetamine, thus causing interference in the assay [33]. Casey *et al.* [34] reported that bupropion, a monocyclic antidepressant and an aid for smoking cessation, may cause false-positive screen results using the EMIT II amphetamine immunoassay. Vidal and Skripuletz [35] reported the case of a 50-year-old male who showed positive amphetamine and lysergic acid diethylamide (LSD) in his urine specimen analyzed by the CEDIA immunoassay. However, GC-MS confirmation failed to confirm either drug. The authors identified bupropion as the cause of false-positive immunoassay screening results for both amphetamine and LSD. The antidepressant desipramine and the antiviral agent amantadine also interfere with amphetamine immunoassays [36].

Mebeverine, an N-substituted ethylamphetamine, is an antispasmodic drug that is metabolized to mebeverine-alcohol, veratric acid, methoxyethylamphetamine, hydroxyethylamphetamine, and PMA. Kraemer *et al.* [37] reported that an FPIA amphetamine assay showed positive response in urine specimens following oral ingestion of a 405-mg mebeverine tablet by volunteers. The authors concluded that positive amphetamine immunoassay test results were due to the presence of methoxyethylamphetamine, hydroxyethylamphetamine, and PMA in the urine specimens. Confirmation of these compounds can be achieved by using GC-MS [37]. Vorce *et al.* [38] showed that dimethylamylamine (DMAA) may cause false-positive test results with a KIMS amphetamine assay and the EMIT II Plus amphetamine assay if present at a concentration of 6900 ng/mL due to structural similarity of DMAA with amphetamine. DMMA is an aliphatic amine naturally found in geranium flowers but also used in body building natural supplements such as Jack3d and OxyELITE Pro. It has been promoted as a safe alternative to ephedrine. The authors further analyzed 134 urine specimens that were tested false positive for amphetamine but confirmed negative by GC-MS and did not contain any known drugs that may cause false-positive amphetamine test results. They observed the presence of DMAA in 92.3% of specimens, with concentrations varying from 2500 to 67,000 ng/mL [38].

## CHALLENGES IN TESTING OF COCAINE METABOLITE

Cocaine abuse is usually confirmed by detecting the presence of benzoylecgonine, an inactive metabolite of cocaine, in the urine. In general, the antibodies used in cocaine immunoassays are specific for benzoylecgonine, and these assays, unlike amphetamine immunoassays, demonstrate good specificity as well as sensitivity. Armbruster *et al.* [39] reported that EMIT II and Abuscreen ONLINE cocaine immunoassays had good sensitivity and specificity because 99% of specimens screened positive by these assays were confirmed for the presence of benzoylecgonine by GC-MS. However, life-threatening acute cocaine overdose may not be identified by urine toxicological screen because sufficient concentration of benzoylecgonine may not be present in the urine due to the fact that not enough time has passed after cocaine ingestion for benzoylecgonine to accumulate in the urine. In one case report, in which the person died from cocaine overdose, the urine drug test was negative for cocaine using the EMIT assay. Later GC-MS analysis confirmed that the concentration of benzoylecgonine was only 75 ng/mL, whereas the concentration of cocaine in the urine was only 55 ng/mL in the urine specimen. The immunoassay cutoff concentration of 300 ng/mL cannot detect such a low level of benzoylecgonine. Moreover, cocaine has poor cross-reactivity with the antibody specific for benzoylecgonine (reported cross-reactivity of cocaine with the EMIT assay was 25,000 ng/mL). However, the heart blood concentration of cocaine was 18,330 ng/mL, thus explaining the cause of death as cocaine overdose [40].

Although fluconazole, an antifungal agent, does not produce false-negative test results with cocaine immunoassays, it may cause false-negative results in the GC-MS confirmation step using trimethylsilyl derivative because derivatized fluconazole elutes with derivatized benzoylecgonine. However, such interference can be eliminated by using a pentafluoropropionyl derivative of benzoylecgonine because derivatized benzoylecgonine elutes before derivatized fluconazole [41].

Importantly, positive cocaine test results, both by immunoassay screening and by GC-MS confirmation method, may occur after drinking coca tea, Mazor *et al.* [42] reported that when five healthy subjects drank a cup of coca tea, all urine specimens tested positive for cocaine (as benzoylecgonine) 2 hr after ingestion, and three of five participants' urine specimens remained positive up to 36 hr postingestion. The reason is that coca tea may still contain cocaine. Mean urinary benzoylecgonine concentration in all specimens was 1777 ng/mL.

## CHALLENGES IN TESTING FOR OPIATES

In order to circumvent false-positive test results due to ingestion of poppy seed-containing food in opiate assays, the current SAMHSA guideline recommends a cutoff of 2000 ng/mL for opiate screening tests. However, private employers may still use the old cutoff concentration of 300 ng/mL for opiate, and consumption of poppy seed-containing food can easily results in a positive screening as well as confirmation of codeine and morphine by GC-MS. In contrast, 6-monoacetylmorphine (also known as 6-acetylmorphine), which is a specific metabolite of heroin, is not consistent with ingestion of poppy seed-containing food. Antibodies used in most immunoassays target morphine because it is the common metabolite of both codeine and heroin. In general, antibodies specific for opiates have poor cross-reactivity with oxycodone and other 6-keto opioids such as oxymorphone and hydrocodone. Smith *et al.* [43] commented that, in general, immunoassays for opiates displayed substantially lower sensitivity for detecting 6-keto opioids, and urine specimens containing low to moderate amounts of hydromorphone, hydrocodone, oxymorphone, and oxycodone will likely be undetected by opiate immunoassays. Detecting oxycodone is a practical challenge for opiate immunoassays due to the low cross-reactivity of oxycodone with opiate immunoassays. For example, based on information provided in package inserts, the Abuscreen ONLINE opiate assay has only 0.3% cross-reactivity with oxycodone, whereas the CEDIA opiate assay has only 3.1% cross-reactivity with oxycodone. To circumvent such problems, several diagnostic companies have marketed immunoassays that are specifically capable of detecting oxycodone. In one report, the authors analyzed 17,069 urine specimens for the presence of oxycodone using the DRI oxycodone immunoassay during a 4-month period. The DRI oxycodone assay demonstrated 97.7% sensitivity and 100% specificity at an oxycodone cutoff concentration of 300 ng/mL [44]. Gingras *et al.* [45] evaluated the performance of CEDIA and DRI immunoassays for oxycodone (both from Microgenics) using the Hitachi 917 analyzer and concluded that a combination of both assays provided the best performance (98% sensitivity and specificity) when results were compared with those of the GC-MS confirmation method, in contrast to using only one assay. Opiate immunoassays are also not capable of detecting fentanyl and methadone, which also belong to the opioid class. Immunoassays are commercially available for screening for the presence of methadone or its metabolites in urine. A fentanyl homogeneous enzyme immunoassay is now commercially available from Immunalysis Corporation. Snyder *et al.* [46] reported that this assay has 97% sensitivity and 99% specificity in comparison to LC-MS/MS assay for fentanyl. The authors concluded that this assay is a rapid way of accurately detecting fentanyl in urine.

*CASE REPORT* A 2-year-old female was brought to the hospital after exhibiting signs of rubbing mouth and staggering. A urine toxicology screen performed in the hospital laboratory was negative. She was eventually discharged, but she was brought to the hospital the next morning because she was unresponsive. She experienced severe cardiopulmonary arrest and was pronounced dead on arrival to the hospital. Toxicological analysis of postmortem specimens by GC-MS showed the presence of oxycodone at a concentration of 1360 ng/mL in heart blood, 47,230 ng/mL in urine, and 222,340 ng/mL in the gastric content. The cause of death was oxycodone poisoning. This case report indicates the limitation of hospital urine toxicology screening tests using opiate immunoassays that do not significantly detect oxycodone [47].

Heroin is first metabolized to 6-acetylmorphine and then further transformed into morphine by a liver enzyme. The presence of 6-monoacetylmorphine is considered as confirmation of heroin abuse. Opiate immunoassays may not be capable of detecting the presence of 6-monoacetylmorphine in the urine as effectively as morphine. A specific immunoassay for detecting 6-acetylmorphine is available from Microgenics. Holler *et al.* [48] compared the performance of this CEDIA 6-acetylmorphine assay and the Roche Abuscreen ONLINE opiate assay for detecting the presence of 6-monoacetylmorphine in the urine. They observed that out of 37,713 urine specimens analyzed, 3 specimens screened positive for 6-monoacetylmorphine at a cutoff concentration of 10 ng/mL using the CEDIA assay, whereas the presence of 6-acetylmorphine was confirmed in only 1 specimen using the GC-MS confirmation method. However, when 87 urine specimens in which the presence of 6-acetylmorphine was previously confirmed by GC-MS were re-analyzed using CEDIA assay, all specimens screened positive. However, 12 specimens containing 6-monoacetylmorphine were screened negative by the Abuscreen ONLINE opiate assay. The authors concluded that urine specimens containing predominately 6-monoacetylmorphine may screen negative (false negative) using the opiate immunoassay. In 2010, the new SAMHSA guidelines implemented for drugs of abuse testing recommended screening 6-monoacetylmorphine in urine at the cutoff concentration of 10 ng/mL.

*CASE REPORT* A 3-year-old female was brought to the medical center after being found shivering and unattended outside a public shopping center in winter. An intravenous bolus of naloxone was administered in the emergency room. However, toxicological analysis of her urine upon arrival to the emergency room was negative. In contrast, when a urine toxicology screen was ordered the next day, it was positive for opiates at the 300 ng/mL cutoff concentration. However, GC-MS confirmation failed to identify any opioids, including codeine, hydrocodone, oxycodone, morphine, hydromorphone, and oxymorphone. The CEDIA opiate assay used in the hospital laboratory cross-reacts with naloxone at a concentration of 6000 ng/mL, and the cause of the opiate-positive second specimen was established to be due to the presence of naloxone in the specimen because the naloxone was administered minutes before the first urine specimen was collected. It was absent in the first specimen, which was negative for opiate and other illicit drugs [49].

Buprenorphine is a morphine-based semisynthetic opioid that is a partial antagonist of mu-opioid receptors in the brain. This drug is used in addiction treatment and also has analgesic properties. However, the morphine antibody used in opiate immunoassay does not detect the presence of buprenorphine. In order to monitor compliance of buprenorphine therapy, a specific immunoassay that is designed for detecting buprenorphine in urine must be used. Hull *et al.* [50] reported that using a 5 ng/mL cutoff concentration, there was 97.9% agreement between the results obtained using the CEDIA buprenorphine immunoassay and those obtained using LC-MS/MS.

Opiate immunoassays, like other immunoassays, suffer from providing false-positive test results due to the presence of various cross-reacting substances other than opioid in urine specimens. Certain quinolone antibiotics may cause false-positive test results with opiate immunoassay screening. Baden *et al.* [51] evaluated potential interference of 13 commonly used quinolones (levofloxacin, ofloxacin, pefloxacin, enoxacin, moxifloxacin, gatifloxacin, trovafloxacin, sparfloxacin, lomefloxacin, ciprofloxacin, clinafloxacin, norfloxacin, and nalidixic acid) with various opiate immunoassays (at 300 ng/mL cutoff concentration) and observed that levofloxacin and ofloxacin may cause false-positive opiate test results with Abbot Laboratories assays manufactured for application on the AxSYM analyzer, as well as with CEDIA, EMIT II, and Abuscreen ONLINE assays. In addition, pefloxacin administration may cause false positives with CEDIA, EMIT II, and Abuscreen ONLINE assays; gatifloxacin administration may cause false positives with CEDIA and EMIT II assays; and lomefloxacin, moxifloxacin, ciprofloxacin, and norfloxacin administration may cause false positives with the Abuscreen ONLINE assay [51]. Straley *et al.* [52] reported a case of a 48-year-old male participating in a residential treatment program who tested positive for opiate during a routine urine drug screen but the GC-MS confirmation was negative for any opiate. The patient received gatifloxacin for treating a urinary tract infection. The authors concluded that the presence of gatifloxacin in the urine specimen was responsible for the false-positive opiate test result. Rifampicin is used in treating tuberculosis and may cause false-positive test results with opiate immunoassays such as the KIMS assay on the Cobas Integra analyzer (Roche Diagnostics). A false-positive result may be observed even after 18 hr of administration of a single oral dose of 600 mg of rifampicin [53].

False-positive test results with methadone immunoassays due to the presence of interfering substance in the urine have also been reported. In one report, the authors observed false-positive methadone test results using the Cobas Integra Methadone II test kit (Roche Diagnostics) in three schizophrenia patients treated with quetiapine monotherapy. The authors used a 300 ng/mL cutoff concentration of methadone in urine specimens for their screening of urine specimens. However, no methadone was detected in the plasma specimen of any patient using LC-MS [54]. Rogers *et al.* [55] reported positive methadone urine drug test results in a patient using the One Step Multi-Drug, Multi-Line Screen Test Device (ACON Laboratories, San Diego, CA), a point-of-care device for urine drug screen. The patient had no history of methadone exposure but ingested diphenhydramine. The GC-MS confirmatory test failed to detect any methadone in the urine specimen, confirming that the presumptive methadone test result was a false positive. When drug-free urine specimens were supplemented with diphenhydramine, false-positive methadone tests were also observed using the point-of-care device. Doxylamine intoxication may cause false-positive results with both EMIT d.a.u. opiate and methadone assays. The urine doxylamine concentration needed to cause a positive test result was 50 μg/mL for methadone and 800 μg/mL for opiate [56].

# CHALLENGES IN TESTING FOR MARIJUANA METABOLITES

Immunoassays for screening of marijuana in urine usually target 11-nor-9-carboxy-Δ-9-tetrahydrocannabinol, the major metabolite of marijuana. In general, marijuana immunoassays show good sensitivity and specificity. For example, only 2 or 3% of urine specimens that test positive for marijuana metabolite by EMIT d.a.u. marijuana assay cannot be confirmed by GC-MS [57]. Passive inhalation of marijuana, however, cannot be detected by marijuana immunoassays because the concentration of the metabolite is substantially lower than the cutoff concentration of the assay. In one study, after eight volunteers were exposed to passive inhalation of marijuana in a coffee shop for 3 hr, none of the urine specimens collected from the volunteers tested positive by marijuana immunoassays even at a cut-off concentration of 25 ng/mL (usual cutoff concentration, 50 ng/mL) because the concentration of the metabolite was up to 7.8 ng/mL after hydrolysis of the conjugated metabolite as determined by GC-MS [58]. Similarly, use of hemp oil should not produce positive marijuana test results because hemp seeds are washed with water prior to extraction of oil, a procedure that removes traces of marijuana from the seed hull. However, prescription use of synthetic marijuana (Marinol) should cause positive marijuana test results.

A number of new designer drugs have emerged in the market known as "legal highs" or "herbal highs." These drugs include both herbal substances and synthetic designer drugs that can be purchased through Internet sites and include synthetic cannabinoids such as JWH-018, JWH-073, JWH-250, HU-210, and CP-47,497 and its homologs. These compounds are lipid soluble, typically containing 20–26 carbon atoms, are more potent than marijuana, and are called "spice." JWH-018 was the first compound of this class reported in 2008, and it is an effective cannabinoid receptor (CB1) agonist [59]. HU-20 is a synthetic agonist analog of marijuana [60]. Unfortunately, the presence of these compounds in urine cannot be determined using marijuana immunoassays because these compounds do not cross-react with the antibody. Therefore, GC-MS or LC-MS/MS must be used for detecting the presence of these compounds in urine.

Although uncommon, false-positive marijuana test results may occur during the screening step due to cross-reactivity from other compounds that are not illicit drugs. Boucher *et al.* [61] described a case of a 3-year-old female who was hospitalized because of behavioral disturbance of unknown cause. The only remarkable finding in her medication history was suppositories of neflumic acid, which was initiated 5 days before hospitalization. After admission, her urinary toxicology screen was positive for the presence of marijuana metabolite, but her parents strongly denied such exposure. Further analysis of the specimen using chromatography failed to confirm the presence of marijuana metabolite, but niflumic acid was detected in the specimen. The authors concluded that the false-positive marijuana test result was due to the presence of niflumic acid in the urine specimen. The antiviral agent efavirenz is known to cross-react with marijuana immunoassays. In one study, the authors analyzed 30 urine specimens collected from patients receiving efavirenz using the Rapid Response Drugs of Abuse Test Strips, the Synchron marijuana immunoassay (Beckman Coulter, Brea, CA), and the Cannabinoid II assay (Roche Diagnostics). Only the Rapid Response test strips demonstrated positive marijuana test results in 28 of 30 specimens, whereas the two other immunoassays did not show any interference from efavirenz. As expected, GC-MS confirmation failed to demonstrate the presence of marijuana metabolite in any of the 30 specimens analyzed [62].

## CHALLENGES IN TESTING FOR PHENCYCLIDINE

False-positive test results may occur in phencyclidine (PCP) immunoassays due to cross-reactivity of several drugs with various commercially available immunoassays. Dextromethorphan is an antitussive agent found in many over-the-counter cough and cold medications. Ingesting large amounts of dextromethorphan (>30 mg) may result in positive false-positive test results with opiate and PCP immunoassays. In one report, the authors observed three false-positive PCP tests in pediatric urine specimens using an on-site testing device (Instant-View multi-test drugs of abuse panel; Alka Scientific Designs, Poway, CA). The authors concluded that false-positive PCP tests were due to the cross-reactivities of ibuprofen, metamizol, dextromethorphan, and their metabolites with the PCP assay [63]. Thioridazine is known to cause false-positive PCP test with both EMIT d.a.u. and EMIT II phencyclidine immunoassays [64]. In our experience, most screened positive PCP urine specimens are false positive because the prevalence of PCP abuse is currently low in the United States.

***CASE REPORT*** A 13-year-old female taking venlafaxine regularly for depression was overdosed with 48 tablets of 150 mg venlafaxine. Her other medications were topical Benzamycin and pyridoxine 50 mg daily for acne. On admission, her urine drug screen using Abbott assays on an AxSYM analyzer was positive for PCP, but GC-MS failed to confirm the presence of PCP in urine. A serum specimen obtained 3 hr after overdose showed a venlafaxine concentration of 3930 ng/mL as measured by GC-MS. The therapeutic concentration of venlafaxine along with its *O*-desmethylvenlafaxine metabolite is 250–750 ng/mL. When a urine specimen was supplemented with venlafaxine and its metabolite, the authors observed a positive phencyclidine test using the same immunoassay, thus confirming the positive test result due to interference of venlafaxine [65].

## CHALLENGES IN TESTING FOR BENZODIAZEPINES

Currently, more than 14 various benzodiazepines are approved for use in the United States, and usually antibodies used in benzodiazepine immunoassays target oxazepam and have variable cross-reactivity to other benzodiazepines. However, less commonly, antibodies may target other drugs in the benzodiazepine class (see Table 14.4). Unfortunately, benzodiazepine immunoassays suffer from producing false-negative results for two reasons:

1. The antibody may have poor cross-reactivity with a particular drug, such as lorazepam.
2. The drug may be present in concentrations lower than 200 ng/mL, the usual cutoff for benzodiazepine immunoassays. A drug such as clonazepam may be present at a concentration much lower than 200 ng/mL after therapeutic use.

Therefore, benzodiazepine immunoassay may not be appropriate for monitoring compliance of a patient with a drug in the benzodiazepine class. Clonazepam is a common benzodiazepine used in treating panic disorder and also in controlling seizure. However, detecting clonazepam in urine after therapeutic use using benzodiazepine immunoassays is a challenge due to the low concentration of the drug in urine specimens. Clonazepam is metabolized to 7-aminoclonazepam. West *et al.* [66] reported that when urine specimens collected from subjects taking clonazepam were tested using the DRI benzodiazepine immunoassay at 200 ng/mL cutoff, 38 specimens out of 180 tested positive by the immunoassay (21% positive). However, using LC-MS/MS, 126 out of 180 specimens tested positive (70% positive) when the detection limit of the LC-MS/MS assay was set at 200 ng/mL, the same cutoff used by the DRI benzodiazepine assay, indicating poor detection capability of the benzodiazepine immunoassay for clonazepam and its metabolite in urine. When the authors used a lower cutoff (40 ng/mL), 157 out of 180 specimens tested positive, indicating that the 200 ng/mL cutoff is too high to monitor compliance of patients with clonazepam therapy.

Flunitrazepam (Rohypnol) is used in date rape and is not approved for clinical use in the United States. However, despite the fact that it is not available in the United States, patients may have access to the drug, particularly those living in southern states bordering Mexico. Although flunitrazepam is a benzodiazepine drug, some commercial immunoassays for benzodiazepines using an antibody-targeting oxazepam may have relatively low cross-reactivity with flunitrazepam. In addition, urine concentration of flunitrazepam may not be adequate, and it is a challenge for routinely used benzodiazepine assays to detect the presence of flunitrazepam in urine. Forsman *et al.* [67], using a CEDIA benzodiazepine assay at a cutoff of 300 ng/mL, failed to obtain a positive result in the urine of volunteers after they received a single dose of 0.5 mg flunitrazepam. In addition, only 22 of 102 urine specimens collected from volunteers after receiving the highest dose of flunitrazepam (2 mg) showed positive screening test results using the CEDIA benzodiazepine assay. Kurisaki *et al.* [68] reported that the Triage benzodiazepine assay has low sensitivity in detecting estazolam, brotizolam, and clotiazepam. Therefore, a negative result in a Triage test may not mean the absence of these drugs in the urine specimen. In another report, the authors demonstrated that the Abuscreen ONLINE benzodiazepine assay has 96% specificity but only 36% sensitivity because all urine specimens containing lorazepam and lormetazepam (as confirmed by GC-MS) tested negative by the immunoassays due to poor cross-reactivities of these drugs with the antibody used in the assay [69].

*CASE REPORT* A 58-year-old divorced white female with familial manic–depressive disorder who was socially isolated and unable to cope with her problem ingested her prescription medication lorazepam along with alcohol despite a physician's warning that such practice may cause severe respiratory depression. She was brought to the hospital by her sister, and it was determined that she had ingested 12 lorazepam tablets (2 mg) in the past 24 hr. Surprisingly, her urinary toxicology screen was negative for benzodiazepine using the Cobas Integra benzodiazepine assay (KIMS assay). Other drugs of abuse tests were also negative. Another urine specimen collected 4 hr after admission was also negative. However, GC-MS analysis of both urine specimens showed concentrations of lorazepam greater than 20,000 ng/mL. The author further investigated why the urine screen using the KIMS benzodiazepine assay was negative despite such a high concentration of lorazepam. On dilution of the original urine specimen (20 and 40 ×), positive benzodiazepine test results by the immunoassay were observed. The author concluded that the false-negative benzodiazepine test using immunoassay was either due to antigen excess (lorazepam level too high) causing immune complex formation or due to the presence of an inhibitor in the specimen that may change the expected microparticle aggregation-generating signal [70].

In general, benzodiazepine assays have good sensitivity, but specificity may vary widely among various assays. Lum *et al.* [71] reported that with the Multigent benzodiazepine assay for application on the Architect chemistry analyzer (Abbott Laboratories), 615 urine specimens out of 2447 screened between June and July 2007 tested positive for the benzodiazepines. However, the presence of benzodiazepines was confirmed in 457 specimens using high-performance liquid chromatography (HPLC), indicating that 25.1% of the specimens screened positive by the immunoassays were false positives. The authors randomly selected 50 false-positive urine specimens (which were tested negative by HPLC) and determined after reviewing medical records that 16 false-positive specimens were obtained from patients receiving sertraline (Zoloft). Of these 50 specimens, 47 screened negative by the Syva EMIT assay.

Interference of oxaprozin, a nonsteroidal anti-inflammatory drug (NSAID), in immunoassays for benzodiazepines has been reported. In one study, the authors investigated potential interference of oxaprozin with FPIA, CEDIA, and EMIT d.a.u. benzodiazepine immunoassays. In their study, 36 urine specimens collected from 12 subjects after each subject received a single dose of 1200 mg of oxaprozin were analyzed by all three immunoassays of benzodiazepines using a cutoff concentration of 200 ng/mL. All 36 urine specimens showed positive results using both CEDIA and EMIT d.a.u. assays, but the FPIA assay showed 35 of 36 specimens positive for benzodiazepines. The authors concluded that a single dose of oxaprozin may cause false-positive benzodiazepine test results using immunoassays [72].

## CHALLENGES IN DETERMINING OTHER DRUGS BY IMMUNOASSAYS

Barbiturates that are abused are usually short or intermediate acting, such as amobarbital, pentobarbital, secobarbital, and butalbital. Long-acting barbiturates such as phenobarbital, which is also an anticonvulsant, are rarely abused. In general, the specificity for detecting individual barbiturates varies with the immunoassays. Propoxyphene is used for treating mild to severe pain. Although both propoxyphene and its metabolites are found in urine, in general, antibodies used in propoxyphene immunoassays target the parent drug but may have variable cross-reactivity with norpropoxyphene, a metabolite of propoxyphene. McNally *et al.* [73] concluded that the ONLINE propoxyphene assay (Roche Diagnostics) has better sensitivity that the EMIT propoxyphene assay for detecting the presence of propoxyphene in urine because the antibody used in the ONLINE assay has 77% cross-reactivity with norpropoxyphene, whereas the EMIT assay showed only 7% cross-reactivity with norpropoxyphene. Another report indicates that diphenhydramine (Benadryl) interferes with the EMIT propoxyphene immunoassay [74].

Methaqualone is metabolized to 2′-hydroxy and 3′-hydroxy metabolites, which are then conjugated and excreted in urine as glucuronide. Brenner *et al.* [75] reported that both the Roche ONLINE methaqualone immunoassay and the EMIT II methaqualone immunoassay have high cross-reactivity toward both 2- and 3-hydroxy metabolites of methaqualone as well as their conjugated form and are useful for screening of methaqualone in urine specimens. When volunteers received 200 mg of methaqualone, all urine specimens tested highly positive (300 ng/mL cutoff) for 72 hr. When the specimens were analyzed by GC-MS without hydrolysis of glucuronide conjugates, low levels of methaqualone and metabolites were detected. However, when urine specimens were hydrolyzed with β-glucuronidase and then analyzed again by GC-MS, high concentrations of metabolites were found. Therefore, authors recommend hydrolysis of the urine specimen prior to GC-MS analysis.

Studies indicate that NSAIDs may interfere with results in multiple immunoassays screening for the presence of drugs of abuse in urine specimens. Joseph *et al.* [76] studied 14 NSAIDs for potential interference with EMIT and FPIA assays for various drugs of abuse and observed that tolmetin interferes with EMIT immunoassays at high concentrations (1800 μg/mL and higher) because of high molar absorptivity at 340 nm, the wavelength used for detection in the EMIT technology. Samples containing cannabinoid and benzoylecgonine tested negative in the presence of tolmetin, but there was no effect on the FPIA assay because the detection wavelength was 525 nm. Rollins *et al.* [77] commented that although the frequency of false-positive test results with immunoassays is low with acute or chronic ibuprofen use, chronic use of naproxen even at therapeutic dosage may cause false-positive test results using the EMIT d.a.u. marijuana assay or the FPIA barbiturate assay (Abbott Laboratories).

## ADULTERANTS AND DRUGS OF ABUSE TESTING

Adulteration of a specimen is not an issue in medical drug testing because the specimen is collected by a heath care professional in an overdosed patient admitted to the emergency department. However, adulteration of a urine specimen is possible in workplace drug testing, in which a drug abuser may want to cheat. Many detoxifying agents for beating drug tests are available through the Internet. However, contrary to claims, such agents usually contain high amounts of caffeine or a diuretic such as hydrochlorothiazide and cannot flush out a drug from the system. The manufacturers usually recommend a subject drink excess water along with these detoxifying agents, and the end result is production of diluted urine in which the drug concentration may be reduced. However, in workplace drug testing, diluted urine can be easily identified with observation of low creatinine and specific gravity, and further analysis may not be conducted. In this case, a person may be denied a job, or an employee may be fired.

Common household chemicals such as laundry bleach, table salt, toilet bowl cleaner, hand soap, and vinegar have been used for many years as adulterants of urine specimens in an attempt to avoid a positive drug test. Common adulterants used to invalidate drug testing include table salt, vinegar, liquid laundry bleach, lemon juice, and Visine eye drops [78,79]. Household vinegar and concentrated lemon juice make urine acidic and can be easily detected by checking the pH of the specimen. Table salt increases specific gravity of urine. However, the presence of Visine eye drops in urine cannot be detected by the usual specimen integrity tests. Both the collection site and the laboratory have a number of mechanisms to detect potentially invalid specimens. For example, the temperature should be 90.5-98.9°F. The specific gravity should be between 1.005 and 1.030, and pH should be between 4.0 and 10.0. The creatinine concentration should be 20–400 mg/dL. However, some drug testing laboratories consider a creatinine concentration of 15 mg/dL as the lower end of the cutoff concentration. Adulteration with sodium chloride at a concentration necessary to produce a false-negative result always produces a specific gravity greater than 1.035.

However, there are urinary adulterants available through the Internet that are effective in producing false-negative test results during immunoassay screening tests. Because GC-MS confirmation may not be performed if the screening immunoassay test is negative, a person might effectively beat drug testing by adulteration of their urine specimen using these agents. Unfortunately, the presence of these compounds in the specimen cannot be determined using routine specimen integrity testing, and special tests must be performed during the pre-analytical stage to determine the presence of these adulterants in the specimen. Because adulterating a specimen is equivalent to refusal to test, the person may be denied employment. In addition, in certain states, adulteration of a urine specimen submitted for legal drug testing is a violation of the state law and the person can be prosecuted. One such adulterant to mask drug screening, Stealth, is available through the Internet. The reagent pack contains a powdered catalyst, which should be added to a urine sample cup before voiding. Then a liquid activator reagent should be added to the specimen. The combination of reagents successfully masks urine drug screening by both EMIT and FPIA. Unfortunately, the color of the urine does not change after adding these reagents. However, these reagents contain some powerful reducing substances and show a strong positive glucose urine dipstick result. In addition, Stealth can be detected in urine by using a simple spot test utilizing a stock solution containing 2% potassium dichromate in distilled water and 2N hydrochloric acid. When a few drops of potassium dichromate solution were added to five drops of urine in a test tube followed by addition of two drops of 2N hydrochloric acid, an intense blue color developed that became colorless after approximately 2 min (Dasgupta, unpublished data). No color change was observed if no Stealth was present in the urine specimen.

Wu *et al.* [80] reported that the active ingredient of another Internet-based urinary adulterant, Urine Luck, is 200 mmol/L of pyridinium chlorochromate (PCC). The authors reported a decrease in the response rate for all EMIT II drug screens and for the Abuscreen morphine and marijuana assays. In contrast, the Abuscreen amphetamine assay produced a higher response rate, whereas no effect was observed on the results of benzoylecgonine and PCP. This adulteration of urine did not alter GC-MS confirmation of methamphetamine, benzoylecgonine, and phencyclidine. However, apparent concentrations of opiates and marijuana metabolite were reduced. Wu *et al.* also described the protocol for detection of PCC in urine using spot tests. The indicator solution contains 10 g/L of 1,5-diphenylcarbazide in methanol. The indicator detects the presence of chromium ions and is colorless when prepared. Two drops of indicator solution are added to 1.0 mL of urine. If a reddish-purple color develops, the test is positive. In this author's experience, addition of a few drops of 3% hydrogen peroxide causes a dark brown precipitation if PCC is present. In the absence of PCC, the light yellow color of urine is bleached to almost colorless.

Glutaraldehyde has also been used as an adulterant to mask urine drug test. This product is available under the trade name UrinAid. A 10% solution of glutaraldehyde is available from pharmacies as over-the-counter medication for the treatment of warts. Glutaraldehyde at a concentration of 0.75–2% by volume can lead to false-negative screening results on EMIT II drugs of abuse screening assays. The assay for cocaine (as benzoylecgonine) was mostly affected [81].

Klear is available through various Internet sites, and the manufacturer claims that it can mask all positive drug test results. The Klear product consists of two microtubes of white crystalline material, with each tube containing approximately 500 mg. This product readily dissolves in urine with no change of color or temperature of urine, and it may cause false-negative GC-MS confirmation of marijuana metabolite. ElSohly *et al.* [82] first reported this product as potassium nitrite, and they provided evidence that nitrite leads to decomposition of ions of marijuana metabolite. The authors further reported that using a bisulfite treatment step at the beginning of sample preparation could eliminate this interference. Nitrite in urine may arise *in vivo* and is found in urine in low concentrations. Patients receiving medications such as nitroglycerine, isosorbide dinitrate, nitroprusside, and ranitidine may increase nitrite levels in blood. However, concentrations of nitrite were below 36 μg/mL in specimens cultured positive for microorganisms, and nitrite concentrations were below 6 μg/mL in patients receiving medications that are metabolized to nitrite. On the other hand, nitrite concentrations were 1910–12,200 μg/mL in urine specimens adulterated with nitrite [83]. In our experience, if present in a urine specimen as an adulterant, nitrite can be detected by adding a few drops of 2% potassium permanganate solution to five drops of urine followed by addition of two drops of 2N hydrochloric acid. The pink color of the solution immediately becomes colorless if nitrite is present in the urine [84].

AdultaCheck 4 and AdultaCheck 6 test strips can be used to detect common adulterants in urine. AdultaCheck 4 consists of four individual tests, whereas AdultaCheck 6 detects creatinine, oxidants, nitrite, glutaraldehyde, pH, and chromate. The Intect7Check test strip for checking adulteration in urine is composed of seven different pads to test for creatinine, nitrite, glutaraldehyde, pH, specific gravity, bleach, and PCC [85]. SAMSHA guidelines require additional tests for urine specimens with abnormal physical characteristics or ones that show characteristics of an adulterated specimen during initial screening or confirmatory tests (nonrecovery of internal standard, unusual response, etc.) [86]. A pH less than 3 or greater than 11 and nitrite concentrations greater than 500 mg/mL indicate the presence of adulterants. A nitrite colorimetric test or a general oxidant colorimetric test can be performed to identify nitrite. The presence of chromium in a urine specimen can be confirmed by a chromium colorimetric test or a general test for the presence of oxidant. A confirmatory test can be performed using multi-wavelength spectrophotometry, ion chromatography, atomic absorption spectrophotometry, capillary electrophoresis, or inductively coupled plasma mass spectrometry. Halogens such as fluorine, chlorine, bromine, and iodine are found in nature, and these halide salts (e.g., sodium chloride) are also found in urine. However, elemental halogens (e.g., pure bromine or iodine) can be used as adulterants. The presence of these elemental halogens should be confirmed by a halogen colorimetric test or a general test for the presence of oxidants. The presence of glutaraldehyde should be detected by a general aldehyde test or the characteristic immunoassay response in one or more drug immunoassay tests for initial screening. The presence of PCC should be confirmed by using a general test for the presence of oxidant and a GC-MS confirmatory test. The presence of a surfactant should be verified by using a surfactant colorimetric test with a dodecylbenzene sulfonate equivalent cutoff of 100 mg/mL or greater.

## OTHER DRUGS NOT DETECTED BY ROUTINE TOXICOLOGY SCREENS

Ketamine is used at rave parties, but this drug is not tested routinely in toxicological screen. As mentioned previously, designer drugs related to the structure of amphetamine and marijuana may not be detected by routine toxicology screens performed in most hospital laboratories for diagnosis of drug overdoses. In addition, drugs such as LSD and methaqualone are infrequently abused; therefore, routine testing may be unnecessary [87]. Nevertheless, immunoassays are available for both drugs. Wiegand *et al.* [88] compared EMIT II, CEDIA, and DPC RIA assays for detecting LSD in forensic urine specimens and commented that at 500 pg/mL LSD cutoff, of 221 forensic urine specimens that screened positive by the EMIT II assay, only 11 tested positive by the CEDIA assay and 3 with the RIA assay, indicating a high false-positive rate with the EMIT II assay for LSD. However, each assay correctly identified 23 of 24 urine specimens that had previously been found to contain LSD by GC-MS at a cutoff of 200 pg/mL. The authors concluded that the CEDIA assay demonstrated superior precision, accuracy, and decreased cross-reactivity to compounds other than LSD compared with the EMIT II assay and does not require handling of radioactive compounds. The chemical structure of ketamine is given in Figure 14.2.

*CASE REPORT* A 31-year-old male with severe end-stage cardiomyopathy secondary to rheumatic heart disease and crack cocaine use called emergency medical services for shortness of breath. He was diagnosed with cardiogenic shock secondary to sepsis and was admitted to the hospital. His initial urine drug

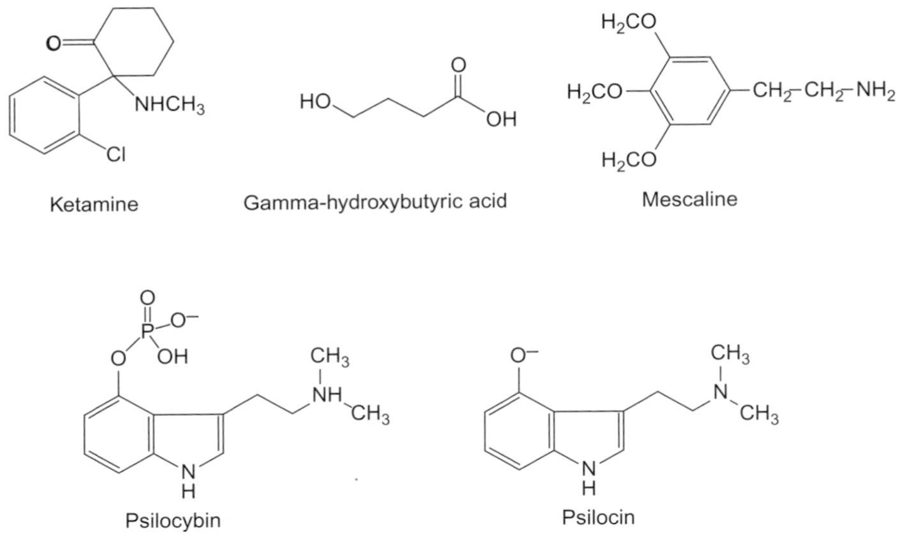

Ketamine  Gamma-hydroxybutyric acid  Mescaline

Psilocybin  Psilocin

**FIGURE 14.2** Chemical structures of ketamine, GHB, mescaline, psilocybin, and psilocin.

screen was negative. However, his girlfriend, who visited him regularly, had suspicious behavior, and the patient became incoherent and began hallucinating. One urine specimen collected at that time was positive for LSD using both CEDIA and EMIT assays. Another specimen collected 3.5 hr later was also positive for LSD, but LC-MS failed to show the presence of LSD or its metabolite, 2-oxo-3-hydroxy LSD, in both specimens. Examination of the medical record by the authors showed that the patient received fentanyl 24 hr prior to each false-positive LSD specimen. GC-MS analysis revealed the presence of fentanyl in both urine specimens (0.67 μg/mL in the first specimen and 0.7 μg/mL in the second specimen). The authors concluded that fentanyl may cause false-positive test results with LSD immunoassays [89].

γ-Hydroxybutyric acid (GHB) is often used at rave parties, and especially in date rape situations, because this compound is tasteless and colorless and can be easily mixed with a drink to make the victim unconscious. Currently, there is no immunoassay for routine screening of GHB in urine or any other biological matrix. Unfortunately, GHB cannot be detected by routine drugs of abuse testing protocols. In the case of suspected overdose of GHB, a more sophisticated analytical technique such as GC-MS should be employed for confirming the presence of GHB in blood or urine. GHB in blood can be determined using GC-MS after liquid–liquid extraction and di-trimethylsilyl derivatization [90]. The chemical structure of GHB is given in Figure 14.2.

The active component of peyote cactus is mescaline. The chemical structure of mescaline is given in Figure 14.2. Native Americans sometimes use peyote cactus for religious ceremonies. There is no commercially available immunoassay for determining the presence of mescaline in urine, and only chromatographic methods are available for determination of mescaline concentration in biological fluids after suspected overdose. Although uncommonly encountered, abuse of peyote cactus may cause clinically significant symptoms requiring hospitalization. In one study, the authors identified 31 cases of peyote cactus abuse in the California Poison Control System database between 1997 and 2008 [91]. Severe toxicity and even death from mescaline overdose have been reported. One person who died under the influence of mescaline showed 9.7 μg/mL of drug in serum and 1163 μg/mL of drug in urine [92].

Magic mushrooms (psychoactive fungi), which grow in the United States, Mexico, South America, and many other areas of the world, contain the hallucinogenic compounds psilocybin and psilocin. Psilocybin and psilocin, along with other compounds in the "tryptamine" class of drugs, are classified as Class I controlled substances with no known medical use but have a high abuse potential. Chemical structures of psilocybin and psilocin are given in Figure 14.2. Unlawful possession of a Class I controlled substance is a felony by law in the United States. Although not commonly abused, and not routinely tested due to lack of availability of immunoassays, magic mushroom abuse may cause serious medical complications and even death. After ingestion of magic mushroom, psilocybin, often the major component of magic mushroom, is rapidly converted by dephosphorylation into psilocin, which has psychoactive effects similar to those of LSD. Although the presence of psilocybin and psilocin in biological fluids can only be determined by chromatographic methods, Tiscione and Miller [93] identified psilocin in a urine specimen during a routine investigation for driving under the influence of drugs using FPIA for screening for

TABLE 14.6 Drugs Not Usually Detected by Routine Toxicology Screen

| Drug | Comments |
|---|---|
| Designer drugs related to amphetamine | Other than MDMA and MDA, most drugs structurally related to amphetamine or methamphetamine cannot be detected by amphetamine/methamphetamine assays |
| Flunitrazepam | Date rape drug flunitrazepam (Rohypnol) may not be detected by benzodiazepine assays due to low concentration in urine |
| Clonazepam and lorazepam | May not be detected due to low levels |
| Oxycodone, methadone, fentanyl | Opiate assay does not detect these drugs. Specific assays must be used |
| Hydrocodone, oxymorphone, hydromorphone | May have low cross-reactivity with certain opiate immunoassays |
| Designer drugs such as "spice" | These designer drugs related to structure of THC may not cross-react with marijuana immunoassays |
| Ketamine | No immunoassay available |
| Magic mushroom abuse (psilocybin) | No immunoassay available |
| Peyote cactus abuse (mescaline) | No immunoassay available |

THC, tetrahydrocannabinol, the active component of marijuana.

amphetamine/methamphetamine in urine. The authors determined that at a concentration of 50 μg/mL, the cross-reactivity of psilocin with the amphetamine immunoassay is 1.3%. In contrast, McClintock *et al.* [94] reported a case of a 28-year-old male with a history of alcohol and drug abuse who had three emergency room visits and three admissions to the hospital, including one in the intensive care unit, in the past 2 months of the study. All laboratory toxicology studies, including GC-MS analysis of urine specimens, were negative. The patient admitted using magic mushroom to a nurse, and the authors concluded that his symptoms were consistent with magic mushroom abuse. This case illustrates the difficulty of diagnosing magic mushroom poisoning using routine toxicological analysis. Drugs that are not detected by routine toxicology screen are listed in Table 14.6.

## CONCLUSIONS

Drug of abuse testing in urine specimens is most common, although for legal drug testing, alternative specimens such as hair and oral fluids are gaining popularity. Usually, for both medical and legal drug testing, initial screening of urine specimens is conducted using commercially available immunoassays. If the initial screening is positive, then the individual drug or drug class must be confirmed by an alternative method, most commonly GC-MS for all legal drug testing. For medical drug testing, GC-MS confirmation may or may not be performed depending on the physician's request. Although the initial screening of specimens using immunoassays is a fast and effective way for determining the presence of a drug or drug class in the specimens, immunoassays suffer from cross-reactivity to structurally related compounds and false-positive drug testing is common with immunoassays. Moreover, due to poor cross-reactivity with the morphine antibody used in opiate immunoassays, opioids such as oxycodone, methadone, fentanyl, propoxyphene, and, to a certain extent, oxymorphone, hydrocodone, and hydromorphone may not be detected during routine toxicological screen. Therefore, specific immunoassays must be used for detecting oxycodone, methadone, propoxyphene, and fentanyl. Brahm *et al.* [95] reviewed the effects of commonly prescribed drugs causing false-positive test results with immunoassays, and Tenore [96] reviewed challenges in urine toxicology screening. Interested readers should refer to these two articles for more in-depth information on this topic.

## References


[1] 2010 National Survey on Drug Use and Health. Substance Abuse and Mental Health Services Administration, Rockville, MD. Available at: http://www.oas.samhsa.gov/nhsda.htm.

[2] Montoya ID, Elwood WN. Fostering a drug free workplace. Health Care Superv 1995;14:1–3.

[3] Carpenter CS. Workplace drug testing and worker drug use. Health Serv Res 2007;42L:795–810.

[4] Jaffee WB, Truccp E, Teter C, Levy S, et al. Focus on alcohol and drug abuse: ensuring validity in urine drug testing. Psychiatr Serv 2008;59:140–2.

[5] Jolley ME, Stroupe SD, Schwenzer KS, et al. Fluorescence polarization immunoassay III: an automated system for therapeutic drug determination. Clin Chem 1981;27:1575–9.

[6] Hawks RL, Chian CN, editors. Urine testing for drugs of abuse. Rockville, MD: National Institute of Drug Abuse. Department of Health and Human Services; 1986 [NIDA research monograph 73].

[7] Jeon SI, Yang X, Andrade JD. Modeling of homogeneous cloned enzyme donor immunoassay. Anal Biochem 2004;333:136–47.

[8] Datta P, Dasgupta A. A new turbidimetric digoxin immunoassay on the ADVIA 1650 Analyzer is free from interference by spironolactone, potassium canrenoate, and their common metabolite canrenone. Ther Drug Monit 2003;25:478–82.

[9] Dai JL, Sokoll LJ, Chan DW. Automated chemiluminescent immunoassay analyzers. J Clin Ligand Assay 1998;21:377–85.

[10] Montagne P, Varcin P, Cuilliere ML, Duheille J. Microparticle-enhanced nephelometric immunoassay with microsphere-antigen conjugate. Bioconjugate Chem 1992;3:187–93.

[11] Fraser AD, Zamecnik J. Impact of lowering the screening and confirmation cutoff values for urine drug testing based on dilution indicators. Ther Drug Monit 2003;25:723–7.

[12] Soldin SJ, Morales AJ, D'Angelo LJ, Bogema SC, Hicks JC. The importance of lowering the cut-off concentrations of urine screening and confirmatory tests for benzoylecgonine/cocaine [Abstract]. Clin Chem 1991;37:993.

[13] Luzzi VI, Saunders AN, Koenig JW, Turk J, et al. Analytical performance of immunoassays for drugs of abuse below established cutoff values. Clin Chem 2004;50:717–22.

[14] Paul B, Past MR. Confirmation methods in drug testing: an overview. In: Dasgupta A, editor. Critical issues in alcohol and drugs of abuse testing. Washington, DC: AACC Press; 2009. p. 125–61.

[15] Maralikova B, Weinmann W. Confirmatory analysis for drugs of abuse in plasma and urine by high performance liquid chromatography-tandem mass spectrometry with respect to criteria for compound identification. J Chromatogr B Analyt Technol Biomed Life Sci 2004;811:21–30.

[16] Eichhorst JC, Etter ML, Rousseaux N, Lehotay DC. Drugs of abuse testing by tandem mass spectrometry: a rapid simple method to replace immunoassay. Clin Biochem 2009;42:1531–42.

[17] Kunsman GW, Levine B, Kuhlman JJ, Jones RL, et al. MDA-MDMA concentrations in urine specimens. J Anal Toxicol 1996;20:517–21.

[18] Stout PR, Klette KL, Wiegand R. Comparison and evaluation of DRI methamphetamine, DRI ecstasy, Abuscreen ONLINE amphetamine, and a modified Abuscreen ONLINE amphetamine screening immunoassays for the detection of amphetamine, methamphetamine, 3,4-methylenedioxyamphetamine (MDMA) and 3,4-methylenedioxyamphetamine (MDA) in human urine. J Anal Toxicol 2003;27:265–9.

[19] Huang MK, Dai YS, Lee CH, Liu C, et al. Performance characteristics of DRI, CEDIA and REMEDi system for preliminary tests for amphetamines and opiates in urine. J Anal Toxicol 2006;30:61–4.

[20] Poklis A, Fitzgerald RL, Hall KV, Saddy JJ. EMIT-d.a.u monoclonal amphetamine/methamphetamine assay II: detection of methylenedioxyamphetamine (MDA) and methylenedioxymethamphetamine (MDMA). Forensic Sci Int 1993;59:63–70.

[21] Lin DL, Liu HC, Tin HL. Recent paramethoxymethamphetamine (PMMA) deaths in Taiwan. J Anal Toxicol 2007;31:109–13.

[22] Kerrigan S, Mellon MB, Banuelos S, Arndt C. Evaluation of commercial enzyme-linked immunosorbent assays to identify psychedelic phenethylamine. J Anal Toxicol 2011;25:444–51.

[23] Apollonio LG, Whittall IR, Pianca DJ, Kyd JM, et al. Matrix effect and cross-reactivity of select amphetamine-type substances, designer analogues, and putrefactive amines using the Bio-Quant direct ELISA presumptive assays for amphetamines and methamphetamines. J Anal Toxicol 2007;31:208–13.

[24] Kerrigan S, Banuelos S, Perrella L, Hardy B. Simultaneous detection of ten psychedelic phenethylamine in urine by gas chromatography-mass spectrometry. J Anal Toxicol 2011;35:459–69.

[25] Ewald AH, Fritschi G, Bork WR, Maurer HH. Designer drugs 2,5-dimethoxy-4-bromo-amphetamine (DOB) and 2,5-diemthoxy-4-bromo-methamaphetamine (MDOB): studies of their metabolite and toxicological detection in rat urine using gas chromatography/mass spectrometric techniques. J Mass Spectrom 2006;41:487–98.

[26] Takahashi M, Nagashima M, Suzuki J, Seto T, et al. Creation and application of psychoactive designer drugs data library using liquid chromatography with photodiode array spectrometry detector and gas chromatography-mass spectrometry. Talanta 2009;77:1245–72.

[27] Balikova M. Nonfatal and fatal DOB (2,5-dimethoxy-4-bromoamphetmaine) overdose. Forensic Sci Int 2005;153:85–91.

[28] Stout PR, Klette KL, Horn CK. Evaluation of ephedrine, pseudoephedrine, and phenylpropanolamine concentrations in human urine specimens and a comparison of the specificity of DRI amphetamines and Abuscreen online (KIMS) amphetamines screening immunoassays. J Foresnic Sci 2004;49:160–4.

[29] Poklis A, Jortani WSA, Brown CS, Crooks CR. Response of the EMIT II amphetamine/methamphetamine assay to specimens collected following use of Vicks inhalers. J Anal Toxicol 1993;17:284–6.

[30] Dietzen DJ, Ecos K, Friedman D, Beason S. Positive predictive values of abused drug immunoassays on the Beckman SYNCHRON in a Veteran population. J Anal Toxicol 2001;25:174–8.

[31] Logan BK, Costantino AG, Rieders EF, Sanders D. Trazodone, meta-chlorophenylpiperazine (an hallucinogenic drug and trazodone metabolite) and the hallucinogen trifluoromethylphenylpiperazine cross-react with EMITII ecstasy immunoassay in urine. J Anal Toxicol 2010;34:587–9.

[32] Baron JM, Griggs DA, Nixon AL, Long WH, et al. The trazodone metabolite meta-chlorophenylpiperazine can cause false positive urine amphetamine immunoassay result. J Anal Toxicol 2011;35:364–8.

[33] Yee LM, Wu D. False positive amphetamine toxicology screen results in three pregnant women using labetalol. Obstet Gynecol 2011;117(2 Pt 2):503–6.

[34] Casey ER, Scott MG, Tang S, Mullins ME. Frequency of false positive amphetamine screens due to bupropion using the Syva EMIT II immunoassay. J Med Toxicol 2011;7:105–8.

[35] Vidal C, Skripuletz T. Bupropion interference with immunoassays for amphetamines and LSD. Ther Drug Monit 2007;29:373–5.

[36] Merigian KS, Beowning RG. Desipramine and amantadine causing false positive urine test for amphetamine. Ann Emerg Med 1993;22:1927–8.

[37] Kraemer T, Wenning R, Maurer HH. The antispasmodic drug mebeverine leads to positive amphetamine results by fluorescence polarization immunoassays (FPIA) studies on the toxicological analysis of urine by FPIA and GC-MS. J Anal Toxicol 2001;25:333–8.

[38] Vorce SP, Holler JM, Cawrse BM, Magluilo J. Dimethylamine: a drug causing positive immunoassay results for amphetamines. J Anal Toxicol 2011;35:183–7.

[39] Armbruster DA, Schwarzhoff RH, Hubster EC, Liserio MK. Enzyme immunoassay, kinetic microparticle immunoassay, radioimmunoassay and fluorescence polarization immunoassay compared for drugs of abuse screening. Clin Chem 1993;39:2137–46.

[40] Baker JE, Jenkins AJ. Screening for cocaine metabolite fails to detect an intoxication. Am J Forensic Med Pathol 2008;29:141–4.

[41] Dasgupta A, Mahle C, McLemore J. Elimination of fluconazole interference in gas chromatography/mass spectrometric confirmation of benzoylecgonine, the major metabolite of cocaine using pentafluoropropionyl derivative. J Forensic Sci 1996;41:511–13.

[42] Mazor SS, Mycyk MB, Wills BK, Brace LD, et al. Coca tea consumption causes positive urine cocaine assay. Eur J Emerg Med 2006;13:340–1.

[43] Smith ML, Hughes RO, Levine B, Dickerson S, et al. Forensic drug testing for opiates: VI. Urine testing for hydromorphone, hydrocodone, oxymorphone, and oxycodone with commercial opiate immunoassays and gas chromatography-mass spectrometry. J Anal Toxicol 1995;19:18–26.

[44] Abadie JM, Allison KH, Black DA, Garbin J, et al. Can an immunoassay become a standard technique in detecting oxycodone and its metabolites? J Anal Toxicol 2005;29:825–9.

[45] Gingras M, Laberge MH, Lafebvre M. Evaluation of the usefulness of an oxycodone immunoassay in combination with a traditional opiate immunoassay for the screening of opiates in urine. J Anal Toxicol 2010;34:78–83.

[46] Snyder ML, Jarolim P, Melanson SE. A new automated urine fentanyl immunoassay: technical performance and clinical utility for monitoring fentanyl compliance. Clin Chim Acta 2011;412:946–51.

[47] Armstrong EJ, Jenkins AJ, Sebrosky GF, Balraj EK. An unusual fatality in a child due to oxycodone. Am J Forensic Med Pathol 2004;25:338–41.

[48] Holler JM, Bosy TZ, Klette KL, Wiegand R, et al. Comparison of the microgenics CEDIA heroin metabolite (6-AM) and the Roche Abuscreen ONLINE opiate immunoassays for detection of heroin use in forensic urine samples. J Anal Toxicol 2004;28:489–93.

[49] Straseski JA, Stolbach A, Clarke W. Opiate positive immunoassay screen in a pediatric patient. Clin Chem 2010;56:1220–5.

[50] Hull MJ, Bierer MF, Griggs DA, Long WH, et al. Urinary buprenorphine concentration in patients treated with Suboxone as determined by liquid chromatography-mass spectrometry and CEDIA immunoassay. J Anal Toxicol 2008;32:516–21.

[51] Baden LR, Horowitz G, Jacoby H, Eliopoulos GM. Quinolones and false positive urine screening for opiates by immunoassay technology. JAMA 2001;286:3115–19.

[52] Straley CM, Cecil EJ, Herriman MP. Gatifloxacin interfere with opiate urine drug screen. Pharmacotherapy 2006;26:435–9.

[53] De Paula M, Saiz LC, Gonzalez-Revalderia J, Pascual T, et al. Rifampicin causes false positive immunoassay results for opiates. Clin Chem Lab Med 1998;36:241–3.

[54] Widschwendter CG, Zernig G, Hofer A. Quetiapine cross-reactivity with urine methadone immunoassays [Letter to the Editor]. Am J Psychiatry 2007;164:172.

[55] Rogers SC, Pruitt CW, Crouch DJ, Caravati EM. Rapid urine drug screens: diphenhydramine and methadone cross-reactivity. Pediatr Emerg Care 2010;26:665–6.

[56] Hausmann E, Kohl B, von Boehmer H, Wellhoner HH. False positive EMIT indication for opiates and methadone in doxylamine intoxication. J Clin Chem Clin Biochem 1983;21:599–600.

[57] Frederick DL, Green J, Fowler MW. Comparison of six cannabinoid metabolite assays. J Anal Toxicol 1985;9:116–20.

[58] Rohrich J, Schimmel I, Zorntlein S, Becker J, et al. Concentrations of delta-9-tetrahydrocannabinol and 11-nor 9-carboxytetrahydrocannabinol in blood and urine after passive exposure to cannabis smoke in a coffee shop. J Anal Toxicol 2010;34:196–203.

[59] Zawilska JB. "Legal highs"—New players in the old drama. Curr Drug Abuse Rev 2011;4:122–30.

[60] Vardakou I, Pistos C, Spiliopoulou CH. Spice drugs as a new trend: mode of action, identification and legislation. Toxicol Lett 2010;197:157–62.

[61] Boucher A, Vilette P, Crassard N, Bernard N, et al. Urinary toxicological screening: analytical interference between niflumic acid and cannabis. Arch Pediatr 2009;16:1457–60 [In French].

[62] Oosthuizen NM, Laurens JB. Efavirenz interference in urine screening immunoassays for tetrahydrocannabinol. Ann Clin Biochem 2011;49:194–6.

[63] Marchei E, Pellegrini M, Pichini S, Martin I, et al. Are false positive phencyclidine immunoassay instant-view multi test results caused by overdose concentrations of ibuprofen, metamizol and dextromethorphan?. Ther Drug Monit 2007;29:671–3.

[64] Long C, Crifasi J, Maginn D. Interference of thioridazine (Mellaril) in identification of phencyclidine. Clin Chem 1996;42:1885–6 [Letter to the Editor].

[65] Bond GR, Steele PE, Uges DR. Massive venlafaxine overdose resulted in a false positive Abbott AxSYM urine immunoassay for phencyclidine. J Toxicol Clin Toxicol 2003;41:999–1002.

[66] West R, Pesce A, West C, Crews B, et al. Comparison of clonazepam compliance by measurement of urinary concentration by immunoassay and LC-MS/MS in patient management. Pain Physician 2010;13:71–8.

[67] Forsman M, Nystrom I, Roman M, Berglund L, et al. Urinary detection times and excretion patterns of flunitrazepam and its metabolites after a single oral dose. J Anal Toxicol 2009;33:491–501.

[68] Kurisaki E, Hayashida M, Nihira M, Ohno Y, et al. Diagnosis performance of Triage for benzodiazepines: urine analysis of the dose of therapeutic cases. J Anal Toxicol 2005;29:539–43.

[69] Augsburger M, Rivier L, Mangin P. Comparison of different immunoassays and GC-MS screening of benzodiazepines in urine. J Pharm Biomed Anal 1998;18:681–7.

[70] Wenk RE. False negative urine immunoassay after lorazepam overdose. Arch Pathol Lab Med 2006;130:1600–1 [Letter to the Editor].

[71] Lum G, Mushlin B, Farney L. False positive rates for the qualitative analysis of urine benzodiazepines and metabolites with the reformulated Abbott Multigent reagents. Clin Chem 2008;54:220–1 [Letter to the Editor].

[72] Fraser AD, Howell P. Oxaprozin cross-reactivity in three commercial immunoassays for benzodiazepines in urine. J Anal Toxicol 1998;22:50–4.

[73] McNally AJ, Pilcher I, Wu R, Salamone SJ, et al. Evaluation of the online immunoassay for propoxyphene: comparison to EMIT II and GC-MS. J Anal Toxicol 1996;20:537–40.

[74] Schneider S, Wennig R. Interference of diphenhydramine with the EMIT II immunoassay for propoxyphene. J Anal Toxicol 1999;23:637–8.

[75] Brenner C, Hui R, Passarelli J, Wu R, et al. Comparison of methaqualone excretion patterns using Abuscreen ONLINE and EMIT II immunoassay and GC/MS. Forensic Sci Int 1996;79:31–41.

[76] Joseph R, Dickerson S, Willis R, Frankenfield D, et al. Interference by nonsteroidal antiinflammatory drugs in EMIT and TDx assays for drugs of abuse. J Anal Toxicol 1995;19:1–7.

[77] Rollins DE, Jennison TA, Jones G. Investigation of interference by non-steroidal antiinflammatory drugs in urine tests for abused drugs. Clin Chem 1990;36:602–6.

[78] Mikkelsen SL, Ash O. Adulterants causing false negative in illicit drug test. Clin Chem 1988;34:2333–6.

[79] Warner A. Interference of household chemicals in immunoassay methods for drugs of abuse. Clin Chem 1989;35:648–51.

[80] Wu A, Bristol B, Sexton K, Cassella-McLane G, Holtman V, Hill DW. Adulteration of urine by Urine Luck. Clin Chem 1999;45:1051–7.

[81] George S, Braithwaite RA. The effect of glutaraldehyde adulteration of urine specimens on Syva EMIT II drugs of abuse assay. J Anal Toxicol 1996;20:195–6.

[82] ElSohly MA, Feng S, Kopycki WJ, Murphy TP, Jones AB, Davis A, et al. A procedure to overcome interference's caused by adulterant "Klear" in the GC-MS analysis of 11-nor-Δ9-THC-9-COOH. J Anal Toxicol 1997;20:240–2.

[83] Urry F, Komaromy-Hiller G, Staley B, Crockett D, Kushnir M, Nelson G, et al. Nitrite adulteration of workplace drug testing specimens: sources and associated concentrations of nitrite and distinction between natural sources and adulteration. J Anal Toxicol 1998;22:89–95.

[84] Dasgupta A, Wahed A, Wells A. Rapid spot tests for detecting adulterants in urine specimens submitted for drug testing. Am J Clin Pathol 2002;117:325–9.

[85] Dasgupta A, Chughtai O, Hannah C, Davis B, Wells A. Comparison of spot test and AdultaCheck 6 and Intect 7 urine test strips for detecting the presence of adulterants in urine specimens. Clin Chim Acta 2004;348:19–25.
[86] Bush DM. The US mandatory guidelines for federal workplace drug testing programs: current status and future considerations. Forensic Sci Int 2008;174:111–19.
[87] Melanson SE, Baskin L, Magnani B, Kwong TC, et al. Interpretation and utility of abuse immunoassays: lessons from laboratory drug testing surveys. Arch Pathol Lab Med 2010;134:735–9.
[88] Wiegand RF, Klette KL, Stout PR, Gehlhausen JM. Comparison of EMIT II, CEDIA and DPC RIA assays for the detection of lysergic acid diethylamide in forensic urine samples. J Anal Toxicol 2002;26:519–23.
[89] Gagajewski A, David GK, Poch GK, Anderson CJ, et al. False positive lysergic acid diethylamide immunoassay screen associated with fentanyl medication. Clin Chem 2002;48:205–6 [Letter to the Editor].
[90] Elian AA. GC-MS determination of gamma-hydroxybutyric acid (GHB) in blood. Forensic Sci Int 2001;122:43–7.
[91] Carstairs SD, Cantrill FL. Peyote and mescaline exposures: a 12 year review of a statewide poison center database. Clin Toxicol 2010;48:350–3.
[92] Reynolds PC, Jindrich EJ. A mescaline associated fatality. J Anal Toxicol 1985;9:183–4.
[93] Tiscione NB, Miller MI. Psilocin identified in a DUID investigation. J Anal Toxicol 2006;30:342–5.
[94] McClintock RL, Watts DJ, Melanson S. Unrecognized magic mushroom abuse in a 28 year old man. Am J Emerg Med 2008;26(972):e3–4.
[95] Brahm NC, Yeager LL, Fox MD, Farmer KC, et al. Commonly prescribed medications and potential false-positive urine drug screens. Am J Health Syst Pharm 2010;67:1344–50.
[96] Tenore PL. Advanced urine toxicology testing. J Addictive Dis 2010;29:436–48.

CHAPTER

# 15

# Challenges in Confirmation Testing for Drugs of Abuse

*Larry A. Broussard*

**Louisiana State University Health Sciences Center, New Orleans, Louisiana**

## INTRODUCTION: FACTORS TO CONSIDER WHEN INTERPRETING DRUG TESTING RESULTS

In order to accurately interpret drug testing results, more information than the result itself (typically positive or negative) is required. Some of the factors influencing interpretation are sample tested, purpose of test request, scope of testing performed (i.e., which drugs are included in the drug panel), knowledge of components and analytical parameters of the initial and confirmation methods used, the presence of other chemicals in the sample, physiological state of the donor, and metabolism of the drug(s). Several of these issues were addressed in Chapter 14, which focused on the initial (often called screening) tests for drugs of abuse using immunoassay techniques. This chapter focuses on the confirmatory testing typically performed after the immunoassay screening described in Chapter 14 and challenges in interpreting such results, including positive, negative, or, in some cases, questionable results.

Although urine is the sample most commonly used for drugs of abuse testing, a variety of other samples, including serum/plasma/blood, hair, oral fluid, and sweat, can also be used. Advantages of urine specimens include noninvasive collection, relatively easy, cost-effective proven technology, and a moderate window of detection for most of the drugs of interest. A primary disadvantage of urine is the lack of correlation between drug concentration and impairment. In situations in which correlation between drug concentrations and pharmacological effects or impairment is desirable, the appropriate sample is serum, plasma, or blood. The use of oral fluid offers the advantage of ease of collection, but the window of detection for most drugs of interest differs from those reported in urine. Hair offers the advantage of the longest window of detection because the presence of a drug in hair reflects use or exposure of several weeks, even months, past. A disadvantage of hair as a sample is that testing methods are fairly cumbersome, often requiring significant sample preparation. Although this chapter focuses on the interpretation of results from testing urine samples, some of the information may be applied to results obtained using other samples. When interpreting results from samples other than urine, additional factors to consider include analytes detected (parent drug or metabolite(s)), cutoff concentrations, windows of detection, methods of analysis, and sample handling techniques [1].

The purpose of testing is another factor to consider when interpreting results. Drug testing may be performed for different reasons, including the setting and possible medical or legal implications. For example, in the clinical emergency setting, testing is usually requested to assist in determining whether the symptoms are caused by one or more drugs versus trauma or disease. In the pain management setting, testing may be used to monitor patient compliance and/or identify use of other drugs. Another example is the use by social services of newborn meconium drug screening results in investigations involving parental custody. The purpose of testing may dictate whether confirmation testing by a more sensitive and specific technique should be performed or not. In workplace and forensic drug testing, confirmation is mandated either by law or by consensus within the forensic testing profession. In the clinical setting, confirmation may not be required, particularly if treatment decisions are

*Accurate Results in the Clinical Laboratory.*
DOI: http://dx.doi.org/10.1016/B978-0-12-415783-5.00015-3

based on a combination of patient presentation and initial screening results. When confirmation is required, it is typically performed only for positive screening results; however, in some situations, such as pain management monitoring, negative results may require confirmation [2].

The scope of a drug test—that is, which drugs are detected/reported—is affected by several factors, including the intended purpose of the testing. The Substance Abuse and Mental Health Services Administration (SAMHSA) in the Department of Health and Human Services (HHS) administers the regulation of the federal workplace drug testing program. The drugs/metabolites to be tested in this program are established by law (see Chapter 14, Table 14.1). For clinical samples, factors determining the scope of testing include prevalence of local usage, clinician requests, complexity and availability of testing methods and instrumentation, and personnel requirements.

Specific components of the confirmation testing procedure may affect drug testing results and should be considered when interpreting results. Knowledge of analytical parameters of the confirmation procedure such as specificity, sensitivity (limit of detection (LOD) and limit of quantitation (LOQ)), upper limit of linearity, and interferences aids the interpretative process. Laboratories should determine these parameters as part of the initial validation of a method and the periodic revalidation typically required by regulatory and accreditation agencies. Knowledge of the sensitivity and specificity of the assay is essential for interpretation of results and implementation of follow-up action because false-positive (incorrectly reporting the presence of a drug) and false-negative (failure to detect the presence of a drug) results have serious consequences. The specificity of immunoassays is crucial for screening procedures and was discussed in Chapter 14. Confirmation procedures are typically very specific for targeted compounds but may be subject to interference from compounds with very similar structures. The specificity of screening procedures may depend on the intended use of the test. For example, in the clinical setting, it may be desirable to detect all of the compounds of a given class, but in the workplace setting detection of only the compounds specified by law or contract is allowed. The practical sensitivity for immunoassays is the cutoff concentration, although they have the ability to detect drugs below these levels. The cutoff concentrations for confirmation of many drugs are listed in Chapter 14 (Tables 14.1 and 14.2), but laboratories have the ability to detect these drugs at lower concentrations. HHS-certified (also known as SAMHSA-certified) laboratories are required to have a minimum sensitivity (expressed as the LOQ) of 40% of the cutoff concentration, and although laboratories can often detect lower concentrations, some choose to maintain the LOQ at this concentration. For reconfirmation of positive results, these laboratories are allowed to use the LOD (i.e., the lowest level at which a compound can be detected but not accurately quantitated) because of the possibility of deterioration of the drug during storage between time of initial reporting and subsequent reconfirmation. Regardless of the sensitivity of an assay it is important to recognize that neither immunoassay- nor mass spectrometry-based methods can measure to zero [1].

Deliberate attempts to mask or destroy drugs present in urine obviously affect the interpretation of drug testing results. Detection and specific effects of certain adulterants on immunoassay screening results are discussed in Chapter 14. The tests performed to detect dilution, substitution, or adulterations are referred to as specimen validity tests (SVTs), and interpretation of these results is discussed later in this chapter.

The physiological state of the donor may affect interpretation of drug testing results. Any condition or disease affecting the absorption, metabolism, or excretion of drugs can impact drug testing results. Because the primary site of drug metabolism for most drugs is the liver and the primary mode of excretion is most often urinary, hepatic or renal conditions affect drug testing results. Some drugs are excreted unchanged, whereas others undergo chemical modification such as demethylation, oxidation, or reduction. Often, drugs or metabolites have glucuronide or sulfate molecules attached via the process of conjugation to form a more water-soluble compound excreted in urine.

Most drugs of interest appear in urine as a combination of unconjugated (free) and conjugated parent drug and metabolites. Each of these has a different excretion pattern, with some detectable for a few hours and others for days to weeks. Thus, the choice of compounds (parent drug and/or metabolite(s)) in the drug panel can influence the results obtained [2].

## CONFIRMATION TESTING PROCESS

In the workplace and forensic settings, confirmation is usually performed using gas chromatography–mass spectrometry (GC-MS) or liquid chromatography–tandem mass spectrometry (LC-MS/MS). Although confirmation procedures are specific for each drug/metabolite, there are common features that are discussed later. The conditions of collection and transport to the laboratory can affect the ultimate drug test results, but these actions may not be under the laboratory's control. For regulated workplace drug testing, the laboratory may take action including rejection of the specimen based on errors such as sample misidentification, improper or broken security seal, specimen leakage, and incomplete chain of custody documentation.

The testing process for GC-MS confirmation procedures typically includes the following steps: sample pretreatment, extraction, derivatization, chromatographic separation, and mass spectrometric detection. LC-MS/MS procedures typically require much less sample extraction and may consist of simple dilution of the sample without any prior extraction followed by injection onto the column ("dilute and shoot"). Sample pretreatment includes addition of internal standard (IS) and may include sample dilution and chemical treatment such as hydrolysis, periodate oxidation, or oxime derivatization. Errors in any of these processes can lead to inaccurate results. Typically, pipetting of the donor, standard(s), and control(s) samples is the initial step, followed by addition of IS to all samples. The most commonly used ISs are compounds with physical–chemical properties similar to those of the drug and that are deuterated compounds in which deuterium replaces hydrogen atoms. Ideally, an IS has an extraction recovery, retention time, and ionization response similar to those of the target analyte and only a slight difference in mass. Internal standards are used to determine drug analyte concentrations (using ratios of drug to IS response) and to account for any conditions that can adversely affect drug recovery. Because the IS and drug of interest will be affected the same during the pre-analytical and analytical processes, the ratio of drug analyte to IS remains constant even when conditions cause absolute drug loss. Factors affecting recovery include extraction efficiency, pipetting variance, gas or liquid flow rates, column variance, and sample injection volume. Addition of the same amount of IS to every sample in the batch is critical. HHS-certified laboratories are required to monitor IS recovery, with acceptable limits being 50–200% of the IS area of calibrator(s) or control (s). Use of deuterated IS solution contaminated with nondeuterated analyte can lead to inaccurate concentrations and potential false results. IS solutions should be monitored for the presence of the nondeuterated analyte of interest. Error in addition of IS solution is another potential source of inaccurate results.

Urinary excretion of drugs and metabolites as conjugated compounds enhances their water solubility but increases the difficulty of extraction using an organic solvent. Hydrolysis of the conjugate prior to extraction is necessary for these drug analytes. The three most common modes of hydrolysis are enzymatic, such as β-glucuronidase; acidic, such as using hydrochloric acid (HCl); and basic, such as using sodium or potassium hydroxide (NaOH and KOH). Hydrolysis using a strong acid or base and higher temperatures is usually faster but can destroy the drug analyte or produce interfering artifacts. A control containing the conjugated drug should be included in the batch to demonstrate the effectiveness of the hydrolysis step. Specific drug analytes requiring a hydrolysis step or other chemical pretreatment such as periodate oxidation or oxime derivatization are discussed later [3].

After sample pretreatment, the next step in confirmation procedures is extraction of the drug analyte from the urine and subsequent concentration by drying and reconstitution. The two most common techniques used for extraction of drug analyte(s) from urine are liquid–liquid and solid-phase extraction columns. Analytical factors to be considered for liquid–liquid extraction procedures include the choice of solvent, pH, and buffer. The drug analyte is extracted into an organic solvent after adjustment of pH to maximize the extraction efficiency for that particular analyte. Some procedures include back extraction from the organic solvent into an aqueous solution and then re-extraction of the analyte into an organic solvent. The solvent extract can be concentrated by evaporation for subsequent reconstitution, possible derivatization, and then GC-MS analysis.

The second type of procedure commonly used for drug analyte extraction from urine is the use of solid-phase extraction (SPE) columns of various types and phases. Columns are used to extract drug analyte(s) from urine through a process of adsorption onto the column matrix, removal of other compounds during a wash procedure, and elution into an organic solvent. Three types of SPE columns are normal-phase, reverse-phase, and ion-exchange. Reverse-phase and ion-exchange columns are widely used in forensic toxicology laboratories. Most SPE procedures include common steps of pH adjustment of the sample and conditioning of the column in order to maximize retention of the drug analyte onto the column. The sample is applied to the column and potential interfering compounds are washed through, followed by elution using an organic solvent. Often, the SPE columns used for drug testing are mixed mode, containing both ion-exchange and reverse-phase properties. In this case, the elution solvent is often methanol containing a 2–5% solution of acid or base depending on the properties (acidic or basic) of the analyte drug(s).

After a compound (drug or metabolite) has been extracted from urine, the extract is typically dried and reconstituted for injection into the gas chromatograph. Frequently, it is necessary to convert the drug to a more volatile compound, and this is accomplished by reconstitution with a derivatizing reagent. Derivatized compounds are typically more volatile, more stable to thermal decomposition, and less likely to adsorb to gas chromatograph column walls or solid supports. If the derivatizing reagent is optically active, it can be used to separate *d*- and *l*-stereoisomers of a drug such as methamphetamine. The most popular derivatization method is silylation, in which active hydrogens on the target analyte are replaced with a trimethylsilyl (TMS) group. *N*,*O*-bismethylsilyltrifluroacetamide (BSTFA) and *N*-(*tert*-butyldimethylsilyl)-*N*-methyltrifluoroacetamide (MTBSTFA) are two silylating reagents. Another derivatization method is acylation, which converts amino, hydroxyl, and thiol groups to esters, thioesters, and amines, respectively. Commonly used acylation agents are acetic anhydride, trifluoroacetic anhydride (TFAA), heptafluorobutyric anhydride (HFBA), and pentafluoropropionic acid anhydride (PFPA). A third derivatization method is alkylation, in which the active hydrogen in a carboxylic acid, hydroxyl, thiol, or amino functional group is replaced with an alkyl group that decreases polarity of drugs. Alkylation reagents include trimethylanilinium hydroxide (TMAH) and pentafluorobenzyl bromide (PFBBr) [3].

## CONFIRMATION OF AMPHETAMINES

Federally regulated workplace drug testing has always included testing for amphetamine and methamphetamine, and in October 2010, screening for the synthetic stimulant 3,4-methylenedioxymethamphetamine (MDMA; Ecstasy) with confirmation testing for MDMA, 3,4-methylenedioxyamphetamine (MDA), and methylenedioxyethylamphetamine (MDEA) was added. The current cutoffs for amphetamines (amphetamine and methamphetamine) and MDMA for federal workplace testing are 500 ng/mL for screening and 250 ng/mL for confirmation (Chapter 14, Table 14.1) [4]. These compounds and others with similar structures and properties, such as pseudoephedrine, ephedrine, propylhexedrine, phenylephrine, phenmetrazine, and phentermine available by prescription or as over-the-counter (OTC) treatments for nasal congestion and appetite suppression, are known as sympathomimetic amines. Table 14.5 in Chapter 14 lists prescription medications or their metabolites and ingredients in OTC drugs that have been reported to interfere with amphetamine screening immunoassays.

Several approaches have been taken to minimize the occurrence of false-positive results with amphetamine immunoassays [5]. These approaches include performing a second immunoassay screen from a different manufacturer using an immunoassay with different cross-reactivity or pretreatment of samples with sodium periodate to eliminate interference from sympathomimetic amines. Another method for differentiating true positives from false positives is neutralization of the signal in a true-positive sample by the addition of antibody to the target analyte [6]. Using serial dilution testing and optimal slope cutoffs (determined by receiver operating characteristic analysis) enabled Woodworth *et al.* [7] to differentiate samples containing amphetamine/methamphetamine from those containing cross-reacting compounds and to increase the positive predictive value of the immunoassay.

This chapter focuses on results obtained by GC-MS confirmation procedures following positive screening results using immunoassays targeting amphetamines and/or MDMA. Confirmatory testing of amphetamines can be performed in a single method that includes all five analytes (amphetamine, methamphetamine, MDMA, MDA, and MDEA) or as separate methods for amphetamine/methamphetamine and MDMA/MDA/MDEA. A review in 2011 of the amphetamine confirmatory method parameters reported by HHS-certified laboratories showed that 11 laboratories used one assay for all amphetamine analytes, whereas 27 laboratories used two separate assays—one for amphetamine and methamphetamine and one for MDMA, MDA, and MDEA [3].

Reports of positive amphetamines or MDMA results typically include a positive result for the screening class (amphetamines or MDMA) and identification of the compound(s) that was confirmed—amphetamine and/or methamphetamine for positive amphetamine results and MDMA, MDA, and MDEA for positive MDMA results. Interpretation of these results should consider several factors, including possible further testing for optical isomers and investigation of medications and supplements ingested. Amphetamine and methamphetamine have optical isomers designated *d* (or +) for dextrorotatory and *l* (or −) for levorotatory, and the *d* isomers—the more physiologically active compounds—are the intended targets of immunoassays. Because GC-MS procedures using non-chiral derivatives and non-chiral columns do not differentiate the *d* or *l* isomers of amphetamine and methamphetamine, it is necessary to perform isomer resolution to determine that a positive result is due to the presence of the *d* isomer. Primary examples are the presence of *l*-methamphetamine in Vicks inhaler that cannot be distinguished from use of illicit methamphetamine (*d* isomer or racemic mixture depending on the method of production), the excretion of *l*-methamphetamine and *l*-amphetamine by patients taking selegiline for Parkinson's disease, and the excretion of *d*- and *l*-amphetamine and methamphetamine by patients taking the analgesic famprofazone [8]. Studies have shown that heavy use of the inhaler can cause false-positive results but that when the inhaler was used as directed, no false-positive results were obtained [9]. Isomer resolution can be accomplished using a chiral, optically active column or chiral derivatizing reagents [10]. Use of chiral derivatizing reagents allows analysis on instrument/column systems used for other routine analyses but has the potential disadvantage of possibly obtaining four isomers instead of two if the derivatizing agent is not optically pure. *N*-trifluoroacetyl-*l*-propyl chloride (TPC) and (−)-methylchloroformate can be used to distinguish methamphetamine enantiomers with GC-MS without a chiral column. Chiral columns such as *N*-3,5-(dinitrobenzoyl)phenyl-glycine or β-cyclodextrin products can differentiate *d* and *l* isomers, or enantiomers, of amphetamines for GC and LC methods [3]. The generally accepted interpretation of isomer resolution results is that greater than 80% of the *l* isomer is considered consistent with use of legitimate medication or, conversely, greater than 20% of the *d* isomer (and total concentration above the cutoff) is considered evidence of illicit use [9]. Positive amphetamine results can be obtained due to ingestion of medications containing amphetamine, methamphetamine, or compounds metabolized to these compounds.

Amphetamine (Dexedrine (*d*-amphetamine), Adderall (*d*- and *l*-amphetamine), etc.) and methamphetamine (Desoxyn (*d*-methamphetamine)) are the active compounds of medications prescribed for appetite suppression, narcolepsy, and attention deficit disorder, and ingestion of these drugs will result in true-positive results due to excretion of these compounds in the urine. In addition, compounds known to be metabolized to amphetamine and/or methamphetamine include selegiline, amphetaminil, benzphetamine, clobenzorex, dimethylamphetamine, ethylamphetamine, famprofazone, fencamine, fenethylline, fenproporex, furfenorex, mefenorex, and prenylamine [11]. Some weight loss or nutritional supplements contain fenproporex, and use of these supplements has resulted in detection of *d*-amphetamine in the urine of users [12]. In January 2006, the U.S. Food and Drug Administration warned consumers that Brazilian dietary supplements Emagrece Sim and Herbathin contain active drug ingredients [9,13].

Another important consideration when interpreting confirmation results for amphetamines is the possibility of false-negative results due to destruction of the compounds or false-positive results due to generation of the compounds during the procedure [9]. All GC-MS confirmation procedures for amphetamines and related compounds should include preventative measures to avoid loss of these volatile compounds during the evaporation step of extraction or during analysis. Procedures to reduce/eliminate loss of amphetamines during evaporation include lowering the temperature for evaporation, performing incomplete evaporation, or adding methanolic HCl prior to evaporation in order to produce more stable hydrochloride salts [11,14,15]. Derivatization using compounds such as HFBA, PFPA, TFAA, 4-carboxyhexafluorobutyryl chloride (4-CB), (MTBSTFA, TPC, 2,2,2-trichloroethyl chloroformate, and propylchloroformate decreases the volatility of amphetamines in addition to improving chromatography and quantitation and forming higher molecular weight fragments yielding different mass ions and ion ratios from those of potentially interfering compounds [9]. Contaminants in the heptafluorobutyryl derivatizing reagent have been shown to give methamphetamine peak interferences when ephedrine is present in the sample [16]. In 1993, it was demonstrated that concentrations of methamphetamine less than 50 ng/mL can be generated from high levels of pseudoephedrine or ephedrine in injection ports at a temperature of 300 °C after derivatization with 4-CB, HFBA, and TPC [17]. False-positive results in the federal drug testing program due to generation of methamphetamine resulted in the implementation of the requirement that the metabolite amphetamine must be present at a concentration of 100 ng/mL (250 ng/mL prior to October 2010) or higher in order to report a positive methamphetamine [4]. Lowering the injector temperature and periodate pretreatment of samples are two procedures used to prevent generation of methamphetamine. Periodate treatment (0.35 M sodium periodate for 10 min at room temperature) eliminates formation of methamphetamine by selective oxidation of pseudoephedrine, ephedrine, phenylpropanolamine, and norpseudoephedrine at concentrations of 1,000,000 ng/mL in the presence of amphetamine and methamphetamine [18]. Inclusion of a quality control sample containing a high concentration of sympathomimetic amines in each confirmation batch can be used to monitor the effectiveness of the periodate oxidation procedure [9].

As previously noted, there is a HHS requirement that the metabolite amphetamine must be present at a concentration of 100 ng/mL in order to report a positive methamphetamine. Jemionek *et al.* [19] reported the detection of low concentrations of methamphetamine in urine specimens containing high concentrations of amphetamine, suggesting that the methamphetamine was either a minor by-product of the amphetamine manufacture or the product of a minor metabolic product of amphetamine elimination.

Compounds with structures similar to those of amphetamine and methamphetamine could potentially interfere with GC-MS confirmation procedures. Laboratories that are HHS-certified must perform interference studies for amphetamine confirmation assays by analyzing samples containing interferents (phentermine at 50,000 ng/mL and phenylpropanolamine, ephedrine, and pseudoephedrine at 1 mg/mL) in the presence of and without amphetamine and methamphetamine at 40% of the cutoff [20]. Hydroxynorephedrine, norephedrine, norpseudoephedrine, phenylephrine, and propylhexedrine are other compounds with structures similar to those of amphetamine and methamphetamine that may be tested for interference [21]. When testing for designer amphetamines such as MDMA and MDA, the laboratory must be aware of the metabolism and excretion patterns for these drugs because some of the metabolites (3,4-dihydroxymethamphetamine (HHMA); 3,4-dihydroxyamphetamine (HHA); 4-hydroxy-3-methoxymethamphetamine (HMMA); and 4-hydroxy-3-methoxyamphetamine (HMA)) are excreted as glucuronide and sulfate conjugates. In order to obtain adequate recovery of these metabolites, a hydrolysis procedure should be included as part of the confirmation testing [9,21,22].

## CONFIRMATION OF COCAINE METABOLITE BENZOYLECGONINE

Cocaine, a naturally occurring alkaloid obtained from the South American shrub *Erythroxylum coca*, is rapidly metabolized by cholinesterases to ecgonine methyl ester and benzoylecgonine, which are excreted in urine. Both immunoassay screening and GC-MS confirmation procedures target benzoylecgonine. Because no compounds causing false-positive benzoylecgonine confirmation results have been reported, the interpretation of positive results should focus on explanations of cocaine introduction into the body. Two possible unintended sources to consider are use of cocaine as a local anesthetic in ENT surgery and ingestion of tea prepared from coca leaves. A positive result due to use of a cocaine solution as a local anesthetic can be verified by the physician performing the procedure. Although teas made from coca leaves are illegal in the United States, they could be obtained unintentionally during travel in South America or purchased in ethnic or other stores selling herbal teas such as Health Inca and Mate De Coca, two coca-leaf teas previously removed from the market.

In October 2010, the confirmation cutoff concentration used in federally regulated workplace drug testing programs was lowered to 100 ng/mL for GC-MS confirmation methods. Much lower cutoffs are often found in clinical and nonregulated workplace settings. Several LC-MS/MS methods have been described for measurement of cocaine and/or benzoylecgonine in plasma, oral fluid, and hair [23–28].

False-negative benzoylecgonine results could be caused by intentional destruction from a chemical adulterant or hydrolysis due to nonspecific esterases or spontaneously occurring at pH above 7. Destruction by hydrolysis can be slowed by freezing the specimen [2]. In order to avoid benzoylecgonine loss, sample preparation techniques for confirmation testing should not be performed at a high pH. A typical extraction is performed at pH 6.0. Liquid–liquid extractions using organic solvents such as diethyl ether, chloroform, or a chloroform–isopropanol mixture can be used, but many laboratories use SPE. A review of the benzoylecgonine GC-MS confirmatory method parameters reported by HHS-certified laboratories in 2011 shows that the majority of labs use SPE, trideuterated benzoylecgonine IS, and anhydride- and trifluoroacetamide-type derivatizing reagents [3].

## CONFIRMATION OF OPIATES

Opiates and opioids have important therapeutic uses in pain management and other purposes. The term *opiates* refers to naturally occurring alkaloids of the poppy plant, *Papaver somniferum*, notably morphine and codeine, and semisynthetic alkaloids with similar structures. The semisynthetic alkaloids of this class include heroin and the prescriptive medications buprenorphine (Buprenex), hydrocodone (Vicodin), hydromorphone (Dilaudid), oxycodone (Percodan and OxyContin), and oxymorphone (Opana). The term *opioids* refers to compounds that have affinity toward opioid receptors and exhibit pharmacological properties similar to morphine but may be structurally unrelated to morphine. These include fentanyl (Duragesic and Sublimaze), meperidine (Demerol), methadone, pentazocine (Talwin), propoxyphene (Darvon), and tramadol (Ultram). Thus, it can be said that all opiates are opioids, but not all opioids are opiates [1].

As discussed in Chapter 14, the immunoassay screens for opiates are designed to detect morphine and codeine and have varying sensitivity to other related members of this class. These compounds were targeted because of the desire to detect heroin abuse. With the emergence of pain management testing and the abuse of other opiates, there is now a need for more sensitive and specific methods to detect these other opiate compounds and their metabolites. Immunoassays to detect methadone and propoxyphene have been available for decades, and several immunoassays specific to oxycodone have recently been developed. Similarly, an assay for buprenorphine was developed when it was introduced as a treatment option for heroin dependency, and immunoassays for fentanyl and its metabolites are also now available. Clinicians often mistakenly think that opiate assays (screening and confirmatory) also detect any opiates, including methadone, oxycodone, and fentanyl, and do not realize that the tests have to be ordered separately when detection of these drugs is desired [2].

Knowledge and understanding of the complex metabolism of opiates is essential for the interpretation of opiate results. Figure 15.1 illustrates the metabolism

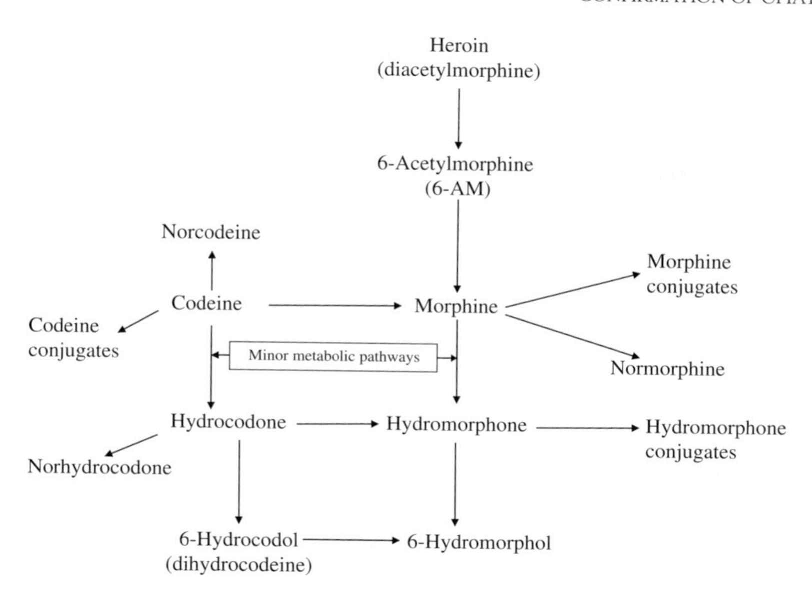


**FIGURE 15.1** Metabolism of heroin, morphine, codeine, hydrocodone, and hydromorphone.

and relationships of several, but not all, opiates and metabolites. Metabolism of opiates typically includes a combination of conjugation via glucuronidation and sulfation and/or oxidation or reduction prior to excretion in urine. The metabolism of several opiates is discussed next in order to provide information necessary for the interpretation of test results.

## Heroin

The primary concern in workplace drug testing of opiates has traditionally been the detection of heroin. Heroin (diacetylmorphine) is rapidly metabolized to 6-acetylmorphine, also called 6-monoacetylmorphine (6-AM or 6-MAM), and subsequently to morphine. Testing for 6-AM in addition to morphine and codeine is now required in the federally regulated drug testing system.

## Morphine

Approximately 85% of a morphine dose is excreted as parent morphine (75% morphine-3-glucuronide and 10% unconjugated (free) morphine), with another 5% undergoing demethylation to normorphine (4% conjugated normorphine and 1% free normorphine) [29]. Minor metabolites that have been identified include morphine-6-glucuronide, morphine-3-ethereal sulfate, and morphine-3,6-diglucuronide [30]. Hydromorphone has been identified as a minor metabolite in patients chronically treated with morphine with ratios of hydromorphone to morphine from 0.2 to 2.2% [31]. Oxymorphone was not detected in urine from high-dose morphine and high-dose hydromorphone patients, indicating that oxymorphone is not a metabolite of morphine or hydromorphone [32,33].

## Codeine

Codeine is demethylated to morphine and norcodeine, with some individuals (approximately 7% of Caucasians and 1–3% of those in other ethnic groups) being poor metabolizers because of a CYP2D6 deficiency [34]. The urinary excretion pattern for a dose of codeine includes the following approximate percentages: 32–46% conjugated codeine, 5–17% free codeine, 10–21% conjugated norcodeine, 5–13% conjugated morphine, and trace amounts of free norcodeine and morphine [29]. Oyler *et al.* [33,35] reported that hydrocodone can be produced as a minor metabolite of codeine and may be excreted in urine at concentrations as high as 11% of the codeine concentration.

## Hydrocodone

Hydrocodone is initially metabolized to hydromorphone and norhydrocodone, with the urinary excretion pattern of a hydrocodone dose giving these approximate levels: 12% free hydrocodone, 5% norhydrocodone, 4% conjugated hydromorphone, 3% 6-hydrocodol, and 0.1% as conjugated 6-hydromorphol [29].

## Hydromorphone

Approximately 36% of a dose of hydromorphone (30% conjugated hydromorphone and 6% free hydromorphone) is excreted without undergoing oxidation or reduction. Cone *et al.* [36] reported metabolism of hydromorphone via reduction to trace amounts of conjugated 6α- and 6β-hydromorphol.

## Oxycodone

Oxycodone is extensively metabolized, with the following approximate levels reported from a single dose: 8% free oxycodone, 23.1% free noroxycodone, 10.4% conjugated oxymorphone, 0.3% free oxymorphone, 8.6% conjugated noroxymorphone, 5.6% free noroxymorphone, 6.0% α-oxycodol, 1.9% β-oxycodol, 6.8% α-noroxycodol, and 1% β-noroxycodol [37].

## Oxymorphone

The urinary excretion pattern of oxymorphone includes approximately 44% as conjugated oxymorphone, 1.9% free oxymorphone, and 3% as 6-oxymorphol (2.9% free and conjugated 6β-oxymorphol and 0.1% 6α-oxymorphol) [29].

## Methadone

Methadone, a synthetic opioid used as a treatment for heroin (and other opioids) dependence and for chronic pain, is subject to wide variations in inter- and intraindividual pharmacokinetics. The primary methadone metabolite is 2-ethylidene-1,5-dimethyl-3,3-diphenylpyrrolidine (EDDP), which is inactive. Approximately 5–50% of a dose is eliminated in the urine as methadone and 3–25% as EDDP. Detection of both methadone and its primary metabolite, EDDP, should be included when using urine drug testing to monitor methadone treatment for addiction or pain management. When immunoassay is used as the testing technique, two separate assays are required because the immunoassays for methadone and EDDP demonstrate little or no cross-reactivity to each other [1].

## Interpretation of Opiate Results

Because of the complex metabolic relationships described previously (see Figure 15.1), interpretation of opiate results is not always straightforward. One may not be able to definitively determine ingestion of certain compounds versus the presence of these compounds as metabolites of ingestion of other compounds. Some factors to be considered when interpreting opiate results are discussed here.

Heroin use can be confirmed by the detection of morphine and 6-AM, the metabolite unique to heroin. The absence of 6-AM does not confirm non-use of heroin because it has a shorter half-life than morphine. More difficult to explain is a positive 6-AM in the absence of morphine, an occurrence reported in some samples since 6-AM screening became part of the federal testing program in October 2010. Speculative explanations include the existence of a unique metabolic pathway in some individuals or the simultaneous introduction of small amounts of morphine and generation of 6-AM by the laboratory during the testing process (discussed later). In addition to heroin or morphine use, there are other possible considerations for the interpretation of positive morphine results. Poppy seeds contain morphine naturally, and their ingestion may cause a positive opiates result regardless of the testing method (immunoassay or chromatography). In an effort to eliminate or reduce the number of positive results due to poppy seed ingestion, the U.S. Department of Defense raised the opiate cutoff levels from 300 to 2000 ng/mL in 1994. In December 1998, federally mandated workplace programs followed suit.

Lower cutoffs continue to be used in clinical and nonregulated settings. In the pain management setting, testing is performed not only to confirm use of the medication as prescribed but also to determine if the patient is taking other opiates. Sometimes the presence of a compound not previously reported as a metabolite is interpreted as use of the compound. For example, before hydromorphone was identified as a minor metabolite in patients chronically treated with morphine, the presence of hydromorphone in the urine of a patient treated with morphine could be interpreted as ingestion of hydromorphone [31]. Attempts have been made to use ratios of parent and metabolite concentrations to interpret opiate results, but this is difficult [38].

The variable pharmacokinetics and potential drug–drug interactions associated with methadone contribute to a range of methadone and EDDP concentrations in the urine of patients being monitored. Compliant patients are expected to be positive for both methadone and EDDP, but a sample that is negative for methadone and positive for EDDP could indicate that the patient is a rapid metabolizer. Negative results for both methadone and EDDP typically indicate that methadone was not ingested and could result from bingeing. These results could also occur if levels of both methadone and EDDP are below the cutoff concentrations as would occur in a dilute urine sample. A technique known as spiking has been used by patients to mimic compliance when diverting methadone for sale. Because this involves the addition of a small amount of the dose to the urine after collection (before testing), this yields a strongly positive result for methadone but a negative EDDP. Note that this is also attempted with other drugs (e.g., buprenorphine) used in rehabilitation programs [1].

Because of the complex metabolism of opiates and the similarity in structures, another factor to consider when interpreting opiate results is the potential for interferences in the GC-MS confirmation procedure. HHS-certified laboratories must perform interference studies for opiate confirmation assays by analyzing samples containing interferents (hydrocodone, hydromorphone, oxycodone, oxymorphone, and norcodeine at 5000 ng/mL) in the presence and absence of morphine and codeine at 40% of the cutoff [20]. Similarly, interference studies for 6-acetylmorphine assays must test free morphine, codeine, hydrocodone, hydromorphone, oxycodone, oxymorphone, and norcodeine at 5000 ng/mL as potential interferents [20]. Several of these potential interferents are keto opiates, and a dual-derivatization procedure (oxime derivatization step using reagents such as hydroxylamine and methoxylamine followed by TMS derivatization) produces compounds that separate chromatographically from codeine and morphine and prevent interference [39]. Similar procedures eliminate potential interferences with the measurement of 6-AM [40,41].

As previously discussed, opiates such as morphine and codeine are excreted primarily as conjugated parent compound or metabolites. Incomplete recovery of these compounds during the confirmation procedure can lead to falsely decreased concentrations or even false-negative results. Hydrolysis to convert the conjugate to free drug prior to extraction is necessary when testing for opiates. Wang *et al.* [42,43] reported that acid hydrolysis, which liberated greater than 90% of morphine and hydromorphone glucuronides, was preferable to hydrolysis using β-glucuronidase from various sources. This report of inefficient enzyme hydrolysis is in disagreement with other reports, perhaps due to the fact that many factors, including buffer pH and composition, ionic strength of the solution, preparation in silylized glassware, temperature, and length of incubation, can affect enzymatic hydrolysis [39,44–52]. It is important to note that acid hydrolysis will lead to the degradation of 6-AM, and its extraction is typically carried out separately from the extraction of morphine and codeine. To prevent loss of 6-AM, extraction is typically performed using silanized glassware. Furthermore, once extracted, the analytes are subject to additional losses and should be analyzed immediately.

As of 2011, all HHS-certified laboratories were using GC-MS for codeine and morphine analysis; however, for 6-AM analysis, one laboratory was using GC-MS/MS, and two laboratories were using LC-MS/MS as an alternative method of analysis [3]. As previously mentioned, one factor to consider if a sample tests positive for 6-AM is the possibility that the result was due to the generation of the compound during the testing process.

The National Laboratory Certification Program reported two instances of a laboratory generating 6-AM when an acetyl donor such as ethyl acetate or acetic acid was used during extraction, derivatization, and/or confirmatory analysis of specimens containing morphine [53]. One sample was a proficiency testing sample containing 5000 ng/mL free morphine, and the other sample was a donor specimen containing more than 200,000 ng/mL morphine. Laboratories were advised not to use compounds that could serve as acetyl donors during 6-AM confirmatory procedures. In order to demonstrate the absence of generation of 6-AM in samples containing high concentrations of morphine, laboratories should include a control containing approximately 90% conjugated and 10% free morphine. Because this situation would not occur in the absence of morphine, it would not explain a result of positive 6-AM and negative morphine unless morphine was also introduced during the testing process (i.e., as a contaminant in the deuterated morphine IS).

## CONFIRMATION OF MARIJUANA METABOLITE

Cannabinoids are a group of compounds present in marijuana, a recreational drug that originates from the dried leaves and flowers of *Cannabis sativa*. The principal psychoactive cannabinoid is $\Delta^9$-tetrahydrocannabinol (THC). More than 100 THC metabolites have been identified, and more of the dose is excreted in feces than in urine. The major urinary metabolites are 11-nor-tetrahydrocannabinol-9-carboxylic acid (THCA) and its glucuronide conjugate. Screening immunoassays for cannabinoids detect multiple metabolites of marijuana, with a cutoff concentration of 50 ng/mL used for regulated drug testing and as low as 20 ng/mL in the clinical and nonregulated setting. GC-MS confirmation identifies and quantifies THCA in urine using a cutoff of 15 ng/mL in regulated drug testing. GC-MS confirmation procedures include a basic hydrolysis step of the glucuronide conjugate, using a reagent such as NaOH.

Interpretation of a positive marijuana metabolite result using a cutoff of 50 ng/mL indicates use, but it is difficult to determine the time of use. Due to the lipophilicity of THC and its metabolites, marijuana may be detected in urine for several weeks postexposure, particularly with chronic use. Chronic marijuana users may produce positive results for longer periods of time because of accumulation of cannabinoid metabolites in fatty tissue followed by slow release. This variable release from tissue and differences in hydration status can lead to a false assumption of new marijuana use if a negative result is followed by a positive result. Creatinine normalization (dividing the THCA result by the creatinine result) has been used to address this problem. For less-than-daily users, an increase of greater than 50% in the normalized THCA concentration is considered indicative of new marijuana use. For chronic users, the excretion pattern is more complicated because they have smaller decreases in excretion later in the elimination phase. An increase of greater than 150% for these users has been suggested as an indicator of new use. Marijuana smoke can be inhaled by a nonsmoking individual, but studies have shown that it is virtually impossible for this passive inhalation to cause a positive result when using a screening cutoff of 50 ng/mL [1].

## CONFIRMATION OF PHENCYCLIDINE

Although phencyclidine (l-phenylcyclohexylpiperidine (PCP)) is classified as a dissociative anesthetic, it also exhibits hallucinogenic, stimulant, depressant, and analgesic properties.

The cutoff concentration for immunoassay screening and GC-MS confirmation of PCP in regulated drug testing is 25 ng/mL. Compounds causing false-positive immunoassay screening results are discussed in Chapter 14, and no compounds causing false-positive GC-MS results have been reported.

## CONFIRMATION OF BENZODIAZEPINES

Benzodiazepines, among the most frequently prescribed drugs in the United States, include alprazolam (Xanax), chlordiazepoxide (Librium), clonazepam (Klonopin), diazepam (Valium), estazolam (ProSom), flunitrazepam (Rohypnol), flurazepam (Dalmane), lorazepam (Ativan), midazolam (Versed), oxazepam (Serax), prazepam (Centrax), quazepam (Doral), temazepam (Restoril), and triazolam (Halcion). With the exception of flunitrazepam, those listed are available for use in the United States. Note that a number of others are available in other countries, and occasionally these are brought into the United States illegally.

After oral administration, benzodiazepines are well absorbed and distributed throughout the body. They are extensively metabolized in the liver via dealkylation, reduction, and hydroxylation, followed by conjugation. The major urinary products of benzodiazepines are the glucuronide metabolite conjugates. Interpretation of urinary metabolite results is difficult because the metabolism of different benzodiazepines often results in common metabolites such as oxazepam, temazepam, and nordiazepam. The cytochrome P450's 3A3 and 3A4 mediate the hydroxylation and dealkylation reactions, so co-administration of drugs that inhibit or induce these enzymes affects benzodiazepine metabolism. Similarly, any drugs that alter glucuronyl transferase activity may affect benzodiazepine metabolism.

Benzodiazepine immunoassays target oxazepam or nordiazepam, and the variable cross-reactivity to other benzodiazepines was discussed in Chapter 14. Benzodiazepines that are not metabolized to oxazepam or nordiazepam, such as flurazepam, alprazolam, lorazepam, clonazepam, and triazolam, may not be detected by immunoassays. To date, no immunoassay detects every benzodiazepine nor all of the metabolites; as a result, false-negatives occur. Because metabolites are excreted as glucuronide conjugates, the sensitivity of immunoassays may be improved by using a hydrolysis step prior to immunoassay testing. This is cumbersome and not realistic in many settings. It is thus important that the laboratory and clinician clearly understand which benzodiazepines and metabolites are detected.

Methods for confirmation testing of benzodiazepines include GC and LC with MS detection. Confirmation methods using GC-MS generally report cutoff concentrations of 50 ng/mL. Most procedures detect oxazepam, nordiazepam, temazepam, α-hydroxyalprazolam, and diazepam at a minimum. The first step in confirmation testing is hydrolysis of the conjugated metabolites using either glucuronidase or HCl. HCl hydrolysis cleaves the benzodiazepine ring to form benzophenone, which can be separated and identified. Disadvantages of benzophenone-producing methods include inability to identify specific benzodiazepines present and the inability to detect triazolobenzodiazepines such as alprazolam. For this reason, some laboratories have turned to LC-MS/MS, and some of these methods permit the identification of multiple benzodiazepines and metabolites at low concentrations. As mentioned previously, interpretation of results to determine benzodiazepines ingested is challenging because several of these drugs are metabolized to the same metabolites. Depending on the method and drug ingested, it may not be possible to determine exactly which drug was ingested.

Flunitrazepam (Rohypnol) is a potent sedative hypnotic that is banned in the United States but is prescribed in other countries for the treatment of insomnia. It is important to remember that benzodiazepine immunoassays vary considerably in their ability to detect flunitrazepam, and many have very little cross-reactivity with the drug or its metabolites. However, there are specific immunoassays designed specifically for flunitrazepam detection. Confirmation testing using GC-MS or LC-MS/MS typically targets detection of flunitrazepam and its principal metabolite, 7-aminoflunitrazepam [1].

## CONFIRMATION OF BARBITURATES

Barbiturates are typically classified by their duration of action. Common barbiturates include phenobarbital (long acting), butalbital (intermediate acting), pentobarbital and secobarbital (both short acting), and thiopental (ultra-short acting). Abuse of barbiturates decreased significantly as the use of benzodiazepines increased.

All barbiturates are strong inducers of hepatic microsomal enzymes; as such, they are often found to influence the metabolism of co-administered drugs. For example, phenobarbital induces uridine diphosphate glucuronyltransferase enzymes and the CYP2C and CYP3A subfamilies of cytochrome P450. Other drugs metabolized by these enzymes would be metabolized more rapidly when co-administered with phenobarbital. Similarly, secobarbital is a strong CYP2A6 and CYP2C8/9 inducer. Thus, use of barbiturates could conceivably cause false-negative results for other drugs due to rapid metabolism of the drug(s).

The immunoassays for barbiturates are designed to detect secobarbital, usually with a screening cutoff of 200 or 300 ng/mL. Cross-reactivity to other barbiturates varies but is generally sufficient to allow their detection. Acceptable confirmation methodologies include GC, high-performance liquid chromatography, GC-MS, and LC-MS/MS. The most common barbiturates included in a confirmation panel are amobarbital, butalbital, pentobarbital, phenobarbital, secobarbital, and, less often, butabarbital. Interpretation of the results is simpler than that of opiates and benzodiazepines.

## SPECIMEN VALIDITY TESTING

Drug abusers try to avoid detection by drug testing in many ways, including use of products designed to prevent detection of drugs present in a urine sample. Some products advertised to "beat a drug test" require ingestion (usually in conjunction with large amounts of water) prior to submission of a urine sample, whereas other products are to be added to or substituted for the urine sample itself. Several of these products were discussed in Chapter 14. Collection procedures have been designed to prevent such substitution, dilution, or adulteration of the specimen. These procedures include collection site preparation (i.e., no hot water, addition of a coloring agent to the toilet, no coats or purses, etc.) and use of collection cup temperature monitoring devices. Tests of specimen validity can be performed in an attempt to detect tampering with the samples.

HHS-certified laboratories are now required to perform specimen validity testing to identify dilute, substituted, adulterated, or invalid samples submitted for regulated drug testing (Table 15.1). This testing includes creatinine, specific gravity (when the creatinine is less than 20 ng/mL), pH, and nitrite measurements. Testing for known adulterants such as glutaraldehyde, pyridinium chlorochromate, or oxidizing chemicals as a class is optional. Criteria for the testing process are very similar to those of testing for

**TABLE 15.1** Reporting Specimen Validity Test Results by HHS-Certified Laboratories

| Test(s) | Result(s) | Reported as[a] |
|---|---|---|
| Ph | <3.0 or ≥11.0 | Adulterated |
| | ≥3.0 and <4.5 | Invalid |
| | ≥9.0 and <11.0 | Invalid |
| Nitrite | ≥500 μg/mL | Adulterated |
| | ≥200 and ≤500 μg/mL | Invalid |
| Creatinine | ≥2.0 and <20 mg/dL | |
| and specific gravity | >1.0010 and <1.0030 | Dilute |
| Creatinine | <2.0 mg/dL | |
| and specific gravity | ≤1.0010 or ≥1.0200 | Substituted |
| Creatinine | <2.0 mg/dL | |
| and specific gravity | Acceptable (1.0011–1.0199) | Invalid |
| Creatinine | ≥2.0 mg/dL | |
| and specific gravity | ≤1.0010 | Invalid |

Source: *National Laboratory Certification Program Manual for Laboratories and Inspectors, October 1, 2010.*

[a]*Definitions:*

*Adulterated specimen*: A urine specimen that has been altered, as evidenced by test results showing either a substance that is not a normal constituent or an abnormal concentration of an endogenous substance. Program cutoffs have been established for pH, nitrite, and other adulterants to distinguish between acceptable and adulterated urine specimens.

*Substituted specimen*: A specimen that has been submitted for testing in place of the donor's urine, as evidenced by creatinine and specific gravity values that are outside the physiologically producible ranges of human urine. Program cutoffs have been established for specific gravity and creatinine to distinguish between acceptable, dilute, and substituted specimens.

*Dilute specimen*: A urine specimen with creatinine and specific gravity values that are lower than expected but are still within the physiologically producible ranges of human urine. Donors may deliberately dilute specimens by consuming large amounts of water or other liquid or by adding liquid to their urine specimens. A dilute finding is *not* direct evidence of specimen tampering.

*Invalid result*: Refers to the result reported by a laboratory for a urine specimen that contains an unidentified adulterant, contains an unidentified interfering substance, has an abnormal physical characteristic, or has an endogenous substance at an abnormal concentration that prevents the laboratory from completing testing or obtaining a valid drug test result.

drugs—that is, the results must be obtained using approved initial and confirmatory tests on two separate aliquots of urine. A sample is reported as adulterated when test results show the presence of a substance that is not a normal constituent or an endogenous substance present in an abnormal concentration. A sample is reported as substituted when creatinine and specific gravity values are outside the physiologically possible ranges of human urine, indicating submission of a non-urine specimen. A result of *substituted* or *adulterated* indicates intention to circumvent testing for drugs and, in some settings, is treated the same as a positive drug result. A result is reported *dilute* when the creatinine and specific gravity values are lower than expected but are still within the physiologically producible ranges of human urine. A dilute specimen does not necessarily indicate specimen tampering because it can result from consumption of large amounts of fluid or by addition of liquid to the urine specimen.

The result of *invalid* can apply to a number of situations related to the ability to perform all tests satisfactorily. A result is reported as invalid if the laboratory is unable to complete testing or obtain a valid drug test result due to the specimen containing an unidentified adulterant or interfering substance or having an abnormal physical characteristic or endogenous substance at an abnormal concentration. Reporting the specimen validity test results (including an invalid result) does not preclude reporting the drug test results. Thus, a result of dilute, adulterated, substituted, or invalid can be reported in conjunction with positive results for specific drugs if the criteria for identification and reporting of the positive result are met. To reiterate, selected drugs may be reported positive even if the laboratory is unable to complete testing for all of the drugs in the panel. For example, "THC positive, invalid result" means that the laboratory was able to confirm the presence of THC (and hence legally defend this result) but was unable to complete the testing for one or more of the other drugs in the panel for one of the reasons previously mentioned. Regulated samples reported as positive for any drug, substituted, adulterated, and/or invalid must be retained in secured frozen storage for a minimum of 1 year [1].

## CONCLUSIONS

In order to accurately interpret drug testing results, more information than the result itself (typically positive or negative) is required. A positive result for a drug or a class of drugs does not automatically indicate intentional use of an illegal substance or nonprescribed ingestion of medication. Some of the factors influencing interpretation are the sample tested, purpose of test request, scope of testing performed (i.e., which drugs are included in the drug panel), knowledge of components and analytical parameters of the initial and confirmation methods used, the presence of other chemicals in the sample, physiological state of the donor, and metabolism of the drug(s). All these factors should be considered when decisions involving interpretations of results are made, particularly considering the possible subsequent consequences.

## References

[1] Broussard LA, Hammett-Stabler CA. Introduction to drugs of abuse. In: Dasgupta A, Langman LJ, editors. Pharmacogenomics of alcohol and drugs of abuse. Boca Raton FL: CRC Press; 2012. p. 93–128.

[2] Hammett-Stabler CA, Broussard LA. Toxicology and the clinical laboratory. In: Clarke W, editor. Contemporary practice in clinical chemistry. 2nd ed. Washington, DC: AACC Press; 2011. p. 587–600.

[3] Analytical Methods in Workplace Drug Testing online course. National Laboratory Certification Program Training by RTI Center for Forensic Sciences. 2012 Course registration available at https://www.forensiced.org/training/NLCP/?csection = NLCP %20Training. Online program information available at https://www.forensiced.org/index.cfm.

[4] Substance Abuse and Mental Health Services Administration: Federal Workplace Drug Testing Program; Final Notice of Revisions to the Mandatory Guidelines for Federal Workplace Drug Testing Programs (Revisions to Mandatory Guidelines). Fed Regist 2008; 73: 71858–907.

[5] Broussard L. Critical issues when testing for amphetamines: pitfalls of immunoassay screening and GC-MS confirmation for amphetamines and methamphetamines. In: Dasgupta A, editor. Critical issues in alcohol and drugs of abuse testing. Washington, DC: AACC Press; 2009. p. 245–52.

[6] Shindelman J, Mahal J, Hemphill G, Pizzo P, Coty WA. Development and evaluation of an improved method for screening of amphetamines. J Anal Toxicol 1999;23:506–10.

[7] Woodworth A, Saunders AN, Koenig JW, Moyer TP, Turk J, Dietzen DJ. Differentiation of amphetamine/methamphetamine and other cross-immunoreactive sympathomimetic amines in urine samples by serial dilution testing. Clin Chem 2006;52:743–6.

[8] Greenhill B, Valtier S, Dody JT. Metabolic profile of amphetamine and methamphetamine following administration of the drug famprofazone. J Anal Toxicol 2003;27:479–84.

[9] Broussard L. Interpretation of amphetamines screening and confirmation testing. In: Dasgupta A, editor. Handbook of Drug Monitoring Methods. Totowa, NJ: Humana Press; 2008. p. 379–93.

[10] Moore KA. Amphetamines/sympathomimetic amines. In: Levine B, editor. Principles of Forensic Toxicology. Washington, DC: AACC Press; 2003. p. 341–8.

[11] Kraemer T, Maurer HH. Determination of amphetamine, methamphetamine and amphetamine-derived designer drugs or medicaments in blood and urine. J Chromatog B 1998;713:163–87.

[12] Jemionek J, Bosy TJ, Jacobs A, Holler J, Magluli J, Dunkley C. Five cases of *d*-amphetamine positive urines resulting from ingestion of "Brazilian nutritional supplements" containing fenproporex have been reported. ToxTalk SOFT Newsl 2006;30(2):11.

[13] FDA News P06-07, January 13, 2006.

[14] Holler JM, Vorce SP, Bosy TZ, Jacobs A. Quantitative and isomeric determination of amphetamine and methamphetamine from urine using a nonprotic elution solvent and R (-)-α-methoxy-α-trifluoromethylphenylacetic acid chloride derivatization. J Anal Toxicol 2005;29:652–7.

[15] Blandford DE, Desjardins PRE. Detection and identification of amphetamine and methamphetamine in urine by GC/MS. Clin Chem 1994;40:145–7.

[16] Wu AHB, Wong SS, Johnson KG, Ballatore, Seifert WE. The conversion of ephedrine to methamphetamine and methamphetamine-like compounds during and prior to gas chromatographic/mass spectrometric analysis of CB and HFB derivatives. Biol Mass Spectrom 1992;21:278–84.

[17] Hornbeck CL, Carrig JE, Czarny RJ. Detection of a GC/MS artifact peak as methamphetamine. J Anal Toxicol 1993;17:257–63.

[18] ElSohly MA, Stanford DF, Sherman D, Shah H, Bernot D, Turner CE. A procedure for eliminating interferences from ephedrine and related compounds in the GC/MS analysis of amphetamine and methamphetamine. J Anal Toxicol 1992;16:109–11.

[19] Jemionek JF, Addison J, Past MR. Low concentrations of methamphetamine detectable in urine in the presence of high concentrations of amphetamine. J Anal Toxicol 2009;33:170–3.

[20] National Laboratory Certification Program Manual for Laboratories and Inspectors. October 1, 2010.

[21] Goldberger BA, Cone EJ. Confirmatory tests for drugs in the workplace by gas chromatography–mass spectrometry. J Chromatog A 1994;674:73–86.

[22] Butler D, Guilbault GG. Analytical techniques for Ecstasy. Anal Lett 2004;37:2003–30.

[23] Sergi M, Bafile E, Compagnone D, Curini R, D'Ascenzo G, Romolo FS. Multiclass analysis of illicit drugs in plasma and oral fluids by LC-MS/MS. Anal Bioanal Chem 2009;393:709–18.

[24] Concheiro M, Gray TR, Shakleya DM, Huestis MA. High-throughput simultaneous analysis of buprenorphine, methadone, cocaine, opiates, nicotine, and metabolites in oral fluid by liquid chromatography tandem mass spectrometry. Anal Bioanal Chem 2010;398(2):915–24.

[25] Fritch D, Blum K, Nonnemacher S, Haggerty BJ, Sullivan MP, Cone EJ. Identification and quantitation of amphetamines, cocaine, opiates, and phencyclidine in oral fluid by liquid chromatography–tandem mass spectrometry. J Anal Toxicol 2009;33(9):569–77.

[26] Kala SV, Harris SE, Freijo TD, Gerlich S. Validation of analysis of amphetamines, opiates, phencyclidine, cocaine, and benzoylecgonine in oral fluids by liquid chromatography–tandem mass spectrometry. J Anal Toxicol 2008;32(8):605–11.

[27] Miller EI, Wylie FM, Oliver JS. Simultaneous detection and quantification of amphetamines, diazepam and its metabolites, cocaine and its metabolites, and opiates in hair by LC-ESI-MS-MS using a single extraction method. J Anal Toxicol 2008;32(7):457–69.

[28] Lopez P, Martello S, Bermejo AM, DeVincenzi E, Tabernero MJ, Chiarotti M. Validation of ELISA screening and LC-MS/MS confirmation methods for cocaine in hair after simple extraction. Anal Bioanal Chem 2010;397(4):1539–48.

[29] Baselt RC. Disposition of toxic drugs and chemicals in Man. 9th ed. Seal Beach, CA: Biomedical Publications; 2011.

[30] Yeh SY, Gorodetzky, Krebs HA. Isolation and identification of morphine 3- and 6-glucuronides, morphine 3,6-diglucuronide, morphine 3-ethereal sulfate, normorphine, and normorphine 6-glucuronide as morphine metabolites in human. J Pharm Sci 1977;66:1288–93.

[31] Cone EJ, Heit HA, Caplan YH, Gourlay D. Evidence of morphine metabolism to hydromorphone in pain patients chronically treated with morphine. J Anal Toxicol 2006;30:1–5.

[32] Cone EJ, Caplan YH, Moser F, Robert T, Black D. Evidence that morphine is metabolized to hydromorphone but not to oxymorphone. J Anal Toxicol 2008;32:319–23.

[33] Bourland J. Opiates metabolism. National Laboratory Certification Program Drug Testing Matters; 2012; Part 3 of 4: 1–7. Available at http://www.datia.org/eNews/2011/NLCP_DTM_Bourland_Opiates_Part1_12Dec2011.pdf

[34] de Leon J, Susce MT, Murray-Carmichael E. The AmpliChip CYPP450 genotyping test: integrating a new clinical tool. Mol Diag Ther 2006;10(3):135–51.

[35] Oyler JM, Cone EJ, Joseph RE, Huestis MA. Identification of hydrocodone in human urine following controlled codeine administration. J Anal Toxicol 2000;24:530–5.

[36] Cone EJ, Phelps BA, Gorodetzky CW. Urinary excretion of hydromorphone and metabolites in humans, rats, dogs, guinea pigs and rabbits. J Pharm Sci 1977;66:1709–13.

[37] Lavolic B, Kharasch E, Hoffer C, Risser L, Lie-Chen L-Y, Shen DD. Pharmacokinetics and pharmacodynamics of oral oxycodone in healthy human subjects: role of circulating active metabolites. Clin Pharmacol Ther 2006;79(5):461–79.

[38] Chang BL, Huang MK. Urinary excretion of codeine and morphine following the administration of codeine-containing cold syrup. J Anal Toxicol 2000;24:133–9.

[39] Broussard LA, Presley LC, Pittman T, Clouette R, Wimbish GH. Simultaneous identification and quantitation of codeine, morphine, hydrocodone, and hydromorphone in urine as trimethylsilyl and oxime derivatives by gas chromatography–mass spectrometry. Clin Chem 1997;43:1029–32.

[40] Singh J, Burke RE, Mertens LE. Elimination of the interferences by keto-opiates in the GC-MS analysis of 6-monoacetylmorphine. J Anal Toxicol 2000;24:27–31.

[41] Ropero-Miller JD, Lambing MK, Winecker RE. Simultaneous quantitation of opioids in blood by GC-EI-MS analysis following deproteination, detautomerization of keto analytes, solid-phase extraction, and trimethylsilyl derivatization. J Anal Toxicol 2002;26:524–8.

[42] Wang P, Stone JA, Chen KH, Gross SF, Haller CA, Wu AHB. Incomplete recovery of prescription opioids in urine using enzymatic hydrolysis of glucuronide metabolites. J Anal Toxicol 2006;30:570–5.

[43] Duer WC, McFarland S. Comments on "Incomplete recovery of prescription opioids in urine using enzymatic hydrolysis of glucuronide metabolites" [Letter to the Editor]. J Anal Toxicol 2007;31:419–21.

[44] Meatherall R. GC-MS confirmation of codeine, morphine, 6-acetylmorphine, hydrocodone, hydromorphone, oxycodone, and oxymorphone in urine. J Anal Toxicol 1999;23:177–86.

[45] Jennison TA, Wozniak E, Nelson G, Urry FM. The quantitative conversion of morphine 3-β-D glucuronide to morphine using β-glucuronidase obtained from *Patella vulgate* as compared to acid hydrolysis. J Anal Toxicol 1993;17:208–10.

[46] Romberg RW, Lee L. Comparison of the hydrolysis rates of morphine-3-glucuronide and morphine-6-glucuronide with acid and β-glucuronidase. J Anal Toxicol 1995;19:157–62.

[47] Combie J, Blake TE, Nugent TE, Tobin T. Morphine glucuronide hydrolysis: superiority of β-glucuronidase from *Patella vulgate*. Clin Chem 1982;28:83–6.

[48] Hackett LP, Dusci LJ, Ilett KF, Chiswell GM. Optimizing the hydrolysis of codeine and morphine glucuronides in urine. Ther Drug Moni 2002;24:652–7.

[49] Cremese M, Wu AH, Cassella G, O'Connor E, Rymut K, Hill DW. Improved GC/MS analysis of opioids with use of oxime-TMS derivatives. J Forensic Sci 1998;43:1220–4.

[50] Delbeke FT, Debackere M. Influence of hydrolysis procedures on the urinary concentrations of codeine and morphine in relation to doping analysis. J Pharm Biomed Anal 1993;11:339–43.

[51] Solans A, de la Torre R, Segura J. Determination of morphine and codeine in urine by gas chromatography–mass spectrometry. J Pharm Biomed Anal 1990;8:905–9.

[52] Zezulak M, Snyder JJ, Needleman SB. Simultaneous analysis of codeine, morphine and heroin after β-glucuronidase hydrolysis. J Forensic Sci 1993;38(6):1275–85.

[53] Possible acetylation of morphine to 6-acetylmorphine(6-AM). NLCP Notice to HHS-certified laboratories, applicant laboratories, and NLCP inspectors, May 4, 2011.

CHAPTER

# 16

# Alcohol Determination Using Automated Analyzers

## Limitations and Pitfalls

*Sheila Dawling*

**Vanderbilt University Medical Center, Nashville, Tennessee**

## INTRODUCTION

Ethanol is currently the most expensive and prevalent addiction, affecting an estimated 140 million people worldwide in 2003 [1,2]. The European Union currently has the highest rates of illness and premature death from alcohol-related problems. The European Information System on Alcohol and Health [3] and the National Institute on Alcohol Abuse and Alcoholism [4] and the Centers for Disease Control and Prevention [5] in the United States provide a wide range of health statistics focused on production and availability, levels of consumption, patterns of consumption, harms and consequences, economic aspects, alcohol control policies, prevention and treatment, and comparative risk assessment.

In the United States, the annual cost of ethanol addiction is estimated at a staggering $166 billion, with smoking second at $157 billion and drugs third at $110 billion [1]. In 2011, approximately $18 billion was spent on the direct costs of alcohol and drug treatment—1.3% of all health care spending [4]. There is increasing concern regarding binge drinking, classified as five or more drinks (or 50 g of alcohol) at least once a week for men and four for women [6,7]. More than one-fifth of the European population aged 15 years or older report binge drinking, which is widespread across all demographics and geographical areas.

There is a paucity of recent data on the incidence of ethanol in the emergency setting, but uncomplicated ethanol intoxication is estimated to account for approximately 600,000 emergency department (ED) visits annually in the United States [8]. Some studies suggest almost half of all trauma beds are occupied by patients who were self-injured (or injured by someone else) while under the influence of alcohol [9,10]. It is important to have accurate knowledge of the patient's alcohol status because of its potential to interact with co-ingested substances, increasing their potency and/or confounding or masking their symptoms. Substance abuse (including ethanol) is more prevalent in health-related professionals than in the general population: An estimated 8–12% of physicians develop a substance-use problem during their lifetime, with anesthesiologists having the highest risk, [11] and 32% of nurses report some type of substance-use problem, with those in the ED having the highest rates recorded in 1998 [12].

Hospital patients who use ethanol chronically incur a two to four times higher morbidity and mortality during anesthesia and during their postoperative course have a heightened susceptibility to infection, cardiopulmonary insufficiency, and bleeding disorders. Many patients do not readily admit to alcohol dependence, and withdrawal becomes evident approximately 6–24 hr after their last drink. This development can change a normal postoperative course into a life-threatening situation, and it escalates the severity tier of care and the concomitant expense of treatment [13].

Because of its extensive availability, ethanol is the most frequently encountered toxic substance in both clinical and forensic analytical settings. Ethanol analysis is often requested in life-threatening settings to evaluate neurological status, for monitoring patients undergoing ethanol therapy for methanol or ethylene glycol toxicity, monitoring those enrolled in alcohol and other drug treatment programs, and when evaluating patients' suitability for organ transplant. Clinical

*Accurate Results in the Clinical Laboratory.*
DOI: http://dx.doi.org/10.1016/B978-0-12-415783-5.00016-5

laboratories therefore need rapid and reliable methods for detection and quantitation of ethanol in biological fluids, usually plasma or serum and urine. Unfortunately, the convenience of having ethanol testing available on automated chemistry analyzers has been traded for specificity, and interferences do occur, some of which can have significant clinical and legal implications if not fully appreciated. Laboratory personnel and those served by the laboratory need to have a clear understanding of the limitations of these tests and have protocols in place for determining which results might be suspect.

Laboratory practice guidelines for use of drug tests in patients treated in the ED have been prepared by expert panels of clinical toxicologists and ED physicians. The three most recent are those of the U.S. National Academy of Clinical Biochemistry [14], the UK National Poisons Information Service and Association of Clinical Biochemists [15], and the Alberta Medical Association in Canada [16]. Although for the most part they are in agreement that the use of emergency drug screens is not warranted, all endorse the provision of STAT (within 1 hr) quantitative ethanol tests.

Because there is no concentration that defines clinical intoxication, as opposed to the legal setting, the laboratory or ED has choices regarding specimens and testing methods that best satisfy its needs. As such, many hospitals have implemented alternative sample testing, such as urine, saliva, or breath, which are easier or safer to obtain and more reliable to process and interpret. Any difference in ethanol content between blood and these matrices is without clinical significance. Breath analyzers as routinely used in law enforcement are accurate, precise, legally defensible, inexpensive, and perhaps, equally important, are perceived to be so by the general public. Testing methodology is not flawed in the way that enzymatic tests are flawed. However, there is clearly inadequate supervision of quality because breath testing is not performed under the guidance of the laboratory personnel nor are tests regulated by the Clinical Laboratory Improvement Act (CLIA) or the College of American Pathologists (CAP) or similar agencies in other countries.

## DEVELOPMENT OF LABORATORY METHODS FOR ETHANOL MEASUREMENT

Because of the widespread availability and toxicity of ethanol, and its relatively large concentration in body fluids, it naturally became a target for early analytical toxicologists. Just as today, their methods were based on the reducing properties of ethanol: Addition of a strong oxidizing agent (potassium dichromate in concentrated sulfuric acid) induced a color change (from yellow/orange to blue/green chromic sulfate) that was easily monitored. Widmark [17] introduced a distillation step to vaporize the ethanol from the biological fluid prior to the reaction, and the ethanol vapor captured by reagent impregnated into filter paper placed above the boiling solution. Later, a microdiffusion well was used for the same purpose. These methods are now considered obsolete because of their lack of specificity (methanol and isopropanol are both strong enough reducing agents to produce a positive reaction), their tediousness, and their inapplicability to automation.

Volatile alcohols are extremely well suited to analysis by gas chromatography (GC); mixtures could be separated by molecular weight and volatility on a column packed with nothing more than activated charcoal. The relatively high concentrations ensured adequate sensitivity with even the most basic flame ionization detector introduced in the 1960s [18]. With the advent of capillary columns and programmable ovens, quantitative analysis of alcohols by GC became one of the easiest and most robust tests available in the clinical laboratory, and it was soon established as the gold standard for both alcohols and glycols [19]. Despite its obvious advantages in specificity, and the ability to identify and quantitate several toxic alcohols simultaneously, GC is not widely employed in clinical laboratories, especially those that do not serve a major trauma or poison treatment center. The equipment is relatively expensive (approximately \$30,000–\$40,000), and its use requires specialized training. The need to test large numbers of samples for ethanol necessarily requires a method that can be adapted to an automated chemistry platform.

To this end, enzymatic assays were investigated based either on alcohol dehydrogenase (ADH) or on alcohol oxidase (AOD). Again, both these tests are based on the ability of ethanol to act as a reducing agent and produce acetaldehyde plus a reduced product. Of these enzymatic methods, ADH proved to be more easily automated because of its good stability in solution, and methods for the most part use spectrophotometric detection of the NADH (nicotinamide adenine dinucleotide reduced form) product as the detection mechanism. Point-of-care methods for ethanol, however, mostly use AOD because although the enzyme is highly unstable in solution; it has good stability when formulated onto a dry dipstick pad or solid support. A wide variety of test kits are available for use with urine, saliva, and breast milk, but these are discrete single-use tests and not suitable for automated analyzers. AOD methods are currently of interest because their oxygen consumption and peroxide formation make them more amenable to conversion to electrical signals that can be measured with an electrode sensor. Alternatively, the hydrogen peroxide product can be used to make a visible color reaction using added peroxidase. The reaction mechanisms for both AOD and ADH assays are discussed in more detail as the different methods are evaluated later in this chapter.

## CURRENT AUTOMATED METHODOLOGIES

Figure 16.1 summarizes the many different strategies for measurement of ethanol by enzymatic assays. All the major clinical chemistry automated instrument manufacturers currently market alcohol test kits. Several diagnostic reagent manufacturers sell kits suitable for use on a number of different automated chemistry platforms. All are based on the enzymatic oxidization of ethanol to acetaldehyde, using the enzyme ADH (EC 1.1.1.1., ethanol:$NAD^+$ oxidoreductase). However, the specific application of the test may differ between manufacturers, and it is naive to assume that all ADH-based ethanol tests are created equal, will have the same performance characteristics, and will be subject to the same limitations and interferences. Variability between the different tests arises from differences in the following:

- Species source of ADH
- Reagent concentration of substrate, co-factors, and enzyme
- Source contamination of $NAD^+$
- Use of $NAD^+$ or an $NAD^+$ analog
- Use of a trapping reagent to drive the reaction to completion
- Sample and reagent addition volumes, incubation timings, and blanking (preincubation)
- Nature of any coupled indicator reaction and its reagent formulation
- Detection and blanking spectrophotometer wavelength
- Measurement timings and mode (end point or kinetic).

Table 16.1 shows the specifics for each of the most commonly used kits currently available, as extracted from manufacturer-provided package inserts. Manufacturers' addresses are included in the footnote to this table. The list is intended to be fairly comprehensive, but the exclusion of a particular product is not deliberate

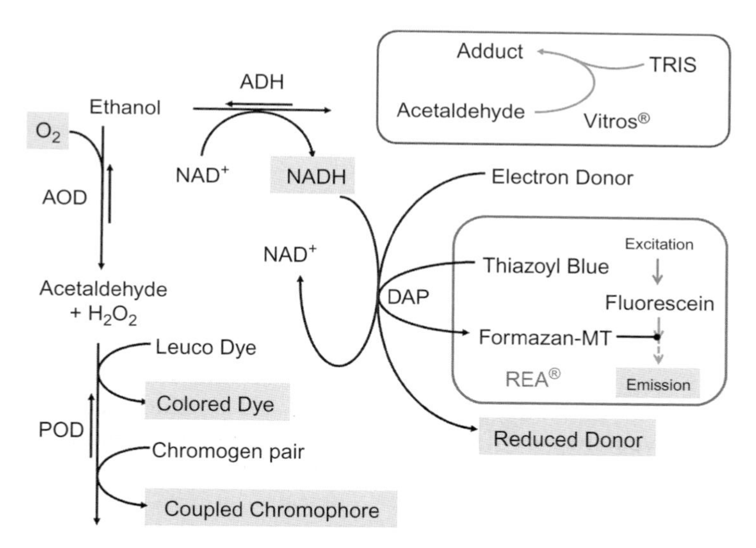


**FIGURE 16.1 Enzymatic ethanol assays.** Ethanol, the substrate and starting point of the assays, is shown in the top left of the figure. Alcohol oxidase (AOD) assays are shown down the left-hand side, and alcohol dehydrogenase (ADH) assays are shown horizontally and down the right-hand side. Both of these reactions produce acetaldehyde, but this product is not tracked or utilized further. Although both reactions are reversible, the direction required for ethanol quantitation is shown with the longer arrow. Compounds in gray boxes are products that can be measured to follow the reaction rate, and with the exception of molecular oxygen, these are all reaction products. Oxygen consumption in AOD assays can be measured with a standard oxygen-sensing electrode. The secondary coupled enzymatic reaction for ADH assays with diaphorase (DAP; EC 1.6.99.1 NADPH:(acceptor) oxidoreductase) using NADH as a substrate is shown with a generic electron donor substrate (e.g., compounds such as tetrazolium salts that form colored formazan dyes when reduced). The secondary coupled enzymatic reaction for AOD assays with peroxidase (POD) using hydrogen peroxide as substrate is shown with two generic chromogens: Common chromogen pairs are 4-aminopyrine with either chromotropic acid (4,5-dihydroxynaphthalene-2,7-disulfonic acid) or DHBS (3,5-dichloro-2-hydroxybenzenesulfonic acid), which form blue and red chromophores, respectively. A common leuco dye is TMB (tetramethylbenzidine), which forms blue TMB diimine. Reactions that are unique to a specific manufacturer (Vitros and Abbott's REA) are sequestered into boxes for illustrative purposes.

TABLE 16.1 Kit Details

| Instrument/ Method | Enzyme | pH | Reaction Time | NAD | Wavelength (nm) | Measurement Mode | Calibrators (mg/dL) | Assay Range (mg/dL) | Secondary Reaction |
|---|---|---|---|---|---|---|---|---|---|
| Abbott Architect[a] | ADH <4 mg/mL | 8.75 | | NAD <7.6 mM | 340/416 | Kinetic | 0, 100 | 10–600 | None |
| Abbott REA[a] | Yeast ADH | | | NAD | 565 | Kinetic | 0, 25, 50, 100, 200, 300 | 13–300 | Yes |
| Beckman platforms[b] | ADH | | | NAD | 340 | Kinetic | 0, 100 | 10–600 | None |
| Microgenics DRI multiplatform[c] | ADH | | | NAD | 340 | Kinetic | 0, 100 | 10–600 | None |
| Olympus EMIT II Plus[b] | ADH 908 U/mL | 8.9 | | NAD 18 mM | 340 | Kinetic | 0, 100 | 10–600 | None |
| Ortho Vitros[d] | ADH | 8.75 | 5 min | NAD | 340 reflectance | End point | 2, 100, 300 | 10–300 | Yes |
| Randox[e] | ADH | 8.8 | | NAD | 340 | Kinetic | 0, 100 | 5–500 | None |
| Roche platforms[f] | Yeast ADH >15 U/mL | | 8 min | Yeast NAD >1.25 mM | 340/659 | Kinetic | 0, 220 | 10–500 | None |
| Siemens Advia[g] | Yeast ADH >203 U/mL in well | 7.3 | | NAD analog 1.35 mM | 340/505 | Kinetic | 0, 100 | 10–500 | None |
| Siemens Dimension Vista[g] | Yeast ADH 525 U/ml in well | | 4 min | NAD 18 mM | 340/383 | Kinetic | 0, 300 | 3–300 | None |

[a]*Abbott Diagnostics, Abbott Park, IL.*
[b]*Beckman Coulter, Brea, CA (incorporating Olympus).*
[c]*Microgenics Corporation, Fremont, CA.*
[d]*Ortho-Clinical Diagnostics, Rochester, NY.*
[e]*Randox Laboratories, Crumlin, Co. Antrim, UK.*
[f]*Roche Diagnostics, Indianapolis, IN.*
[g]*Siemens Healthcare Diagnostics, Tarrytown, NY.*

and does not indicate disapproval of its use. Likewise, inclusion does not signify an endorsement of the product.

Most assays are calibrated with a 2-point calibration curve, using a 0 and a 100 mg/dL ethanol calibrator (or 300 mg/dL Siemens Dimension). Vitros uses three calibrators at 2, 100, and 300 mg/dL; Abbott REA uses multiple levels (0, 25, 50, 100, 200, and 300 mg/dL) because the concentration–response relationship is nonlinear. In contrast to peptide assays, there should be, and is, excellent gravimetric agreement between different sources of ethanol used for calibration. When the information is included in the package insert, there is generally little or no interference by hemolysis, icterus, or lipemia (except if this exceeds the absorbance maximum of the individual instrument photometer). The specimen pH does not affect the reaction over the range 3–11, an important consideration for urine testing.

## ADH with Ultraviolet Detection

This is the most straightforward of the enzymatic methods: A single enzymatic reaction is required, and one of the products is measured directly. Both the oxidized co-factor $NAD^+$ and its reduced product NADH strongly absorb ultraviolet (UV) light attributable to the adenine moiety. The peak absorption of $NAD^+$ is at 259 nm, with a molar extinction coefficient of 16,900 $M^{-1}cm^{-1}$. Many automated clinical chemistry methods take advantage of the fact that NADH (but not $NAD^+$) also absorbs at higher wavelengths, with a second peak in UV absorption at 339 nm with an extinction coefficient of 6220 $M^{-1}cm^{-1}$ [20]. The reaction obeys Beer's law, and the concentration of ethanol in a specimen can be inferred directly from the increase in absorbance at 340 nm as NADH is produced.

The Vitros application is a variant of this method. The patient's sample applied to the slide is evenly distributed by the spreading layer (containing $BaSO_4$ and $NAD^+$) onto the underlying reagent layer (containing ADH and Tris at pH 8.75). The Tris buffer (tris(hydroxymethyl) aminomethane) in the reaction layer traps the acetaldehyde produced by the reaction of ethanol with ADH and drives it to completion. The reflection density of the reduced NADH is measured and is proportional to the concentration of ethanol in the sample. The filtering of the specimen through the spreading layer reduces potential interference in many assays from myeloma proteins, and it also reduces the interference from elevated concentrations of lactate dehydrogenase. Other trapping reagents for acetaldehyde have been used in other applications, such as semicarbazide.

### ADH with Radiative Energy Attenuation

In this assay, ethanol substrate is used in a coupled catalytic reaction with ADH and a second enzyme, diaphorase, to generate a color change in a dye. The method was originally designed for the Abbott TDx/FLx platforms, and it is currently available on the Abbott AxSYM. radiative energy attenuation (REA) measures the degree of inhibition of the fluorescence of fluorescein dye resulting from the production of a colored product. In the original version of the assay, [21,22] the ADH reaction with ethanol and $NAD^+$ was coupled to a second reaction between the NADH produced and the chromogen iodonitrotetrazolium violet dye (INT). This diaphorase-catalyzed reaction results in reoxidation of NADH to $NAD^+$, along with the generation of a red chromophore formazan-INT. This product has an absorbance peak at 492 nm, which overlaps the excitation and emission spectral profile of the fluorescein included in the reaction mixture. Subsequent reformulation of the kit eliminated the need for a time-consuming probe wash by replacing the INT with thiazoyl blue (monotetrazolium or methylthiotetrazolium (MTT)). The detection strategy is unchanged with the reduced MTT (formazan-MT) yielding a purple color with an absorbance maximum at 565 nm. The logarithm of the fluorescent light emitted is inversely proportional to the chromophore present, the production of which is directly linked to the ethanol concentration in the sample.

### Alcohol Oxidase Methods

Although currently no automated methods use AOD, a brief description is appropriate because technological advances may soon result in the successful automation of AOD methods. The left portion of Figure 16.1 shows generic AOD methods. These are amenable to electrochemical detection methods and are available on the Analox GM series of benchtop instruments (Analox Instruments, London). They offer the potential for adaptation onto handheld devices such as iSTAT (Abbott Diagnostics, Abbott Park, IL) or blood gas analyzers. The main drawback of AOD methods is that the enzyme is not very specific for ethanol: All the common toxic alcohols are substrates for this enzyme, producing a signal that cannot be distinguished from that produced by ethanol. However, substrate preferences (as indicated by different $K_m$ values) very much depend on the biological origin of the AOD.

## PERFORMANCE OF AUTOMATED ETHANOL ASSAYS

Most vendors of assay kits include calibration materials in their kits or as separately purchased items. These are often matrix matched to give optimal performance on the specific platform. Calibration materials can also be obtained from independent vendors such as Cerilliant Corporation (Round Rock, TX) and UTAK Laboratories (Valencia, CA). These can also be used for chromatographic methods, although many laboratories performing ethanol analysis by GC prefer to make their own multicomponent calibrators and verify them with independent controls. All vendors of commercial test kits provide quality control materials for assessing method performance prior to testing of patient specimens. Automated test methods are usually designated as applicable to more than one body fluid, typically including both serum/plasma and urine. To fulfill regulatory requirements for using matrix-matched control materials, commercially available controls in blood, serum, saliva, and urine are widely available (UTAK, Bio-Rad Laboratories (Hercules, CA), and Microgenics (Freemont, CA)). In addition to these internal controls, a laboratory's performance of all clinically important analytes is monitored by their participation in an external proficiency testing program, such as that offered by CAP. In the United States, CAP is required to submit proficiency test results to Centers for Medicare and Medicaid Services for all laboratories that have provided a CLIA identification number.

CAP interprets the CLIA '88 regulations (42 CFR Part 493, Section 493.937) for acceptable performance as obtaining a value within ±25% of the all-methods mean for ethanol and as a value between 0 and 9 mg/dL for alcohol-free samples. Results are issued for each sample provided at least 80% of participating laboratories have returned results. If the result is in doubt, there is concern regarding the quality of the specimen, or there is no clear consensus, it is ungraded. Returns are evaluated by each of the four major analytical method groups: GC, ADH with UV-visible (UV-Vis) detection, ADH with REA, and the Vitros method.

During 2010 and 2011, six survey sets of five challenges each were distributed by CAP. Almost 3200 laboratories participated in the CAP alcohol survey (AL2), and the vast majority of these (approximately 98%) used automated methods for measurement. Of the participants, 80% used ADH UV-Vis methods, 16% Vitros, 2.6% GC (84 labs), and 1.3% Abbott ADH with REA.

Of the 30 survey samples circulated during 2010 and 2011, 3 were negative for ethanol (<10 mg/dL). Fifteen challenges contained one or more additional volatile compounds (4 each methanol, acetone, and isopropanol; and 6 ethylene glycol). Twelve challenges contained only ethanol, with concentrations between 24 and 240 mg/dL as determined by GC. The CAP performance goal is 8% error or 2 mg/dL (whichever is the larger), with a total error of 16%.

Method imprecision (as defined by the coefficient of variation) for all four method groups was similar, usually ranging from 4 to 6% across the full range of ethanol concentrations tested. Both GC and the enzymatic methods performed equally in this regard. Accuracy was assessed by comparison of the returned result to the gold standard GC results.

Whereas ADH REA and ADH UV-Vis methods compared very favorably with GC across the ethanol range measured (Table 16.2), the Vitros method consistently underestimated the true ethanol content by approximately 16%. This bias is believed to be explained by a matrix effect because the Vitros method compares favorably with other methods when testing authentic patient materials [23]. Differences in test results in preserved quality control materials, or those designed to remain liquid at −20°C by the addition of ethylene or another glycol, are well documented for a number of analytes across a wide range of analytical platforms. With the configuration of the Vitros dry slide technology, such additives will clearly affect the surface tension of the specimen as it migrates on the spreading layer.

In addition, CAP provides linearity/calibration verification surveys twice yearly that consist of seven different concentrations of ethanol from 10 to 500 mg/dL (LN11). No other volatiles are added to these specimens. Only 850 laboratories participated—approximately one in four of those in the AL2 survey—with a small percentage of participants using GC and Abbott REA (1.5% each); 15% using Vitros, and the remaining 82% using ADH UV-Vis. Participation in the CAP (or a similar) survey is required only if the assay is not routinely calibrated using a full range of calibrators, and patient results are reported that measure above the highest-level calibrator. The group composition is therefore skewed compared to the spot challenge survey. CAP evaluates responses as precise or imprecise, linear or nonlinear, with calibration either verified or different. Summary failure data are shown in Table 16.3 for five challenges through 2010–2012. The peer group for each is its own method mean. All assays performed well in terms of producing a linear response, with Vitros producing the only failure. Imprecision was observed in all methods except ADH UV-Vis (ADH REA 4%, GC 3.4%, and Vitros 1.2%). In terms of calibration slope, ADH UV-Vis performed best (only 4.8% of labs were different from their peers), and ADH REA and Vitros each had 15% bias. The same gravimetric bias of Vitros compared to other methods noted previously was also evident here. The relatively large percentage of laboratories using GC

**TABLE 16.2** Comparison of Three Automated ADH-Based Methods for Ethanol to Gas Chromatography

| Method | Slope | Intercept (mg/dL) | Linear Correlation ($r^2$) | No. of Laboratories Reporting |
|---|---|---|---|---|
| Abbott ADH REA | 1.0155 | 1.0197 | 0.9997 | 41 |
| ADH UV-Vis all manufacturers | 1.0071 | 1.1739 | 0.9998 | 2548 |
| Ortho Vitros | 0.838 | 2.175 | 0.9998 | 516 |

Source: *Data from CAP Alcohol Survey (AL2) samples for 2010 and 2011* (N = 30). *Gas chromatography performed by 84 laboratories. Linear regression analysis* $y = mx + c$, $r^2$ *correlation coefficient.*

**TABLE 16.3** Calibration Failure Rates for Ethanol with Three Automated ADH-Based Methods and Gas Chromatography

| Reagent Kit Failure | Percentage of Peer Group Participants | | | |
|---|---|---|---|---|
| | GC | Abbott ADH REA | ADH UV-Vis, all Methodologies | Ortho Vitros |
| Nonlinear | 0 | 0 | 0 | 1.8 |
| Imprecise | 3.4 | 4 | 0 | 1.2 |
| Different | 11.6 | 15.2 | 4.8 | 15.4 |

Source: *Data from CAP Alcohol Linearity Survey Samples for 2010 and 2011 (four LN11 challenges).*

that had a different calibration from their peers (11.6%) may be a reflection of the high incidence of home-brew calibration materials used.

## PROBLEMS WITH CURRENT METHODOLOGIES

Assays based on the measurement of enzyme activity offer many advantages over GC methods because of their amenability to automation and consequent speed and widespread use, and they have undoubtedly contributed to saving many lives. However, these methods are prone to inaccuracy due to (1) cross-reactivity of the enzyme with alternate substrates such as toxic alcohols other than ethanol and (2) interference from elevated plasma concentrations of substrates and/or enzymes that could also generate NADH, the indicator compound used to monitor the reaction. Although industrial propriety often prevents the end user from knowing the precise ingredients of the kit, appreciation of enzyme reaction principles at least gives the operator some clues regarding the appropriate starting point in troubleshooting wayward results.

At the outset of this discussion, it is important that the reader appreciate that the ADH present in human liver and other tissues has quite different characteristics from the *Saccharomyces cerevisiae* (yeast) enzyme present in all of the ADH-based ethanol kits [24]. Active yeast ADH is a homotetramer with a molecular mass of approximately 145 kDa. The active site of each subunit contains a zinc atom, two reactive sulfhydryl groups, and a histidine residue. Its pH optimum is alkaline, 8.6–9.0. Inhibitors are compounds that react with free sulfhydryl, chelate zinc, or are alternate substrates such as NAD analogs, purine and pyrimidine derivatives, and halogenated alcohols [25]. Yeast ADH is most reactive with ethanol, its preferred substrate, and less so with methanol, and then its activity decreases with increasing size of the alcohol substrate molecule. Branch-chain and secondary alcohols (e.g., isopropanol) have very low reactivity [24,26].

Enzyme reactions are characterized in terms of two parameters: the velocity (rate of change of substrate concentration with time) and the Michaelis constant, $K_m$ (the substrate concentration that produces half the maximum velocity). A typical enzyme rate curve is shown in Figure 16.2. Velocity is typically expressed in amount (moles or milligrams) of substrate turned over to product in a given time, and it is sometimes factored to the amount of enzyme present. The maximum reaction velocity possible, $V_{max}$, is a function of the reactivity of the enzyme, including any polymorphic forms, as well as the prevailing reaction conditions (amount of enzyme in the reaction mixture, temperature, co-factor concentration, pH, etc.). First-order reaction kinetics is followed when the exclusive determinant of the enzyme rate is the substrate concentration available (see notation in Figure 16.2). Therefore, during this time, the reaction rate can be used to calculate the substrate concentration, and this is the basis for a large number of clinical enzymatic assays for substrates (glucose, urea, creatinine, etc.; see Chapter 8). At higher substrate concentrations, the reaction kinetics might conform more closely to zero-order (see notation in Figure 16.2). Here, the enzyme is substrate saturated (all available substrate binding sites on the enzyme are occupied), and the reaction proceeds at a fixed rate ($V_{max}$), which cannot be exceeded regardless of any increase in substrate concentration that might occur. Whereas ADH kinetics is zero order at relatively low ethanol concentrations in humans, yeast ADH has a much higher $K_m$, and first-order kinetics prevails under the operating conditions of most test methods.

All enzyme reactions are reversible, and ADH is no exception. The reaction is as follows, with ethanol being oxidized to acetaldehyde and the reaction monitored spectrophotometrically at 340 nm by the production of NADH:

$$\text{Ethanol} + \text{NAD}^+ + \text{H}^+ \xrightleftharpoons{\text{ADH}} \text{acetaldehyde} + \text{NADH}$$

The position of the equilibrium of an enzymatic reaction cannot be changed by altering the amount of enzyme present in the reaction mixture. For this reaction, the equilibrium position lies far to the left, and the time required for the assay to reach equilibrium may be from 10 to 60 min depending on prevailing conditions of temperature, substrate or co-factor

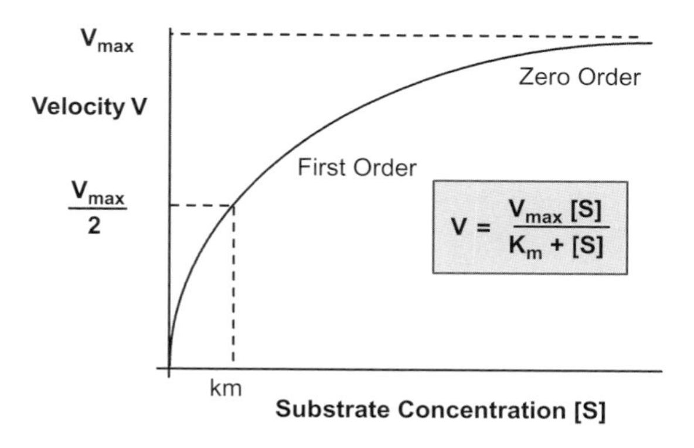


**FIGURE 16.2 Typical Michaelis–Menten enzyme plot of reaction velocity against substrate concentration.** The equation is shown in the gray box, and regions corresponding to first-order and zero-order kinetics are indicated. Velocity = rate of change in substrate concentration with time; $V_{max}$, maximum velocity. $K_m$ is the Michaelis constant, the substrate concentration at half maximum velocity.

concentration, etc. Because equilibrium (or end point) methods are relatively insensitive to small changes in the reaction conditions, this approach was initially preferred, but it lost favor once technological advances made it possible to finely control instrument conditions. However, the reaction can be driven toward the right, primarily by increasing the reaction pH or by adding a trapping reagent to remove the acetaldehyde product, such as semicarbazide or tris(hydroxymethyl)aminomethane. Coupling one of the ADH reaction products to a second enzymatic reaction also accomplishes the same end, but this strategy is usually employed for the purpose of modifying the measurement wavelength to avoid potential interferences and is a more expensive option because other substrates/co-factors and enzyme must be added to the reagent kit.

The analysis time can also be shortened by using a kinetic measurement to monitor the reaction rather than the equilibrium end point. However, this requires that the substrate concentration ([$S$]) be kept sufficiently low ([$S$] $<<<$ Michaelis constant, $K_m$) to ensure that first-order reaction conditions prevail so that the amount of product formed is dependent only on the amount of substrate present. This originally necessitated pre-dilution of the sample because ethanol concentrations are relatively large, but modern automated pipetters can now dispense specimen volumes of 2–5 μL with excellent precision (<2%). Addition of extra ADH enzyme could also be used to increase the amount of substrate consumed, but this can be expensive, so an alternative strategy is to add a competitive enzyme inhibitor such as pyrazole [27]. Once the $K_m$ is increased by the competitive inhibitor, the assay response is linear over the required range of ethanol concentration to cover most clinical eventualities.

Automated ADH assays are liable to two potential sources of interference that can lead to inaccuracy in some clinical settings. The first is having a degree of cross-reactivity with other toxic alcohols, and the second is being subject to positive bias introduced by elevated concentrations of lactate and/or lactate dehydrogenase (LDH). Interestingly, toxic alcohol interference is better known but occurs less frequently, whereas the lactate/LDH interference is less predictable and may be more prevalent but is less well recognized.

## Cross-Reactivity with other Alcohols

Most enzymes can use a number of different substrates, albeit with different preferences. Substrates that are metabolized most quickly or selectively have low values for $K_m$, the substrate concentration at which the reaction proceeds at half its maximum velocity (see Figure 16.2). ADH is no exception, and a number of different alcohols can be converted into their respective aldehydes with the concomitant use of $NAD^+$ and production of NADH. Thus, NADH production, which is usually monitored to specifically determine the ethanol concentration, is not specific and gives no indication of which substrate has been utilized. A large concentration of an alcohol such as methanol with a high $K_m$ value produces NADH at a relatively low rate, and this might lead the technologist to believe there is a small amount of ethanol present in the sample.

However, when ethanol is present in addition to another alcohol, the situation is more complicated because both alcohols can be metabolized by ADH. The term *competitive inhibitor* refers to a compound that directly competes with ethanol for substrate binding site on the enzyme and occupies it, thus reducing the amount of ethanol that can bind and be converted to acetaldehyde. The magnitude of the effect is determined by the relative proportions of ethanol to inhibitor that are present, and once the inhibitor has been metabolized or excreted, the effect is dissipated. Some competitive inhibitors are also alternate substrates for the enzyme, and their binding to the active site results in their metabolism into products (and thus NADH production). Most toxic alcohols are alternate substrates for ADH and can therefore inhibit the metabolism of other alcohols that have higher $K_m$ values. Indeed, this very principle—preventing the metabolism of methanol or ethylene glycol by addition of ethanol—was taken advantage of for many years in the treatment of poisoning from toxic alcohols with intravenous or oral ethanol until it was replaced by 4-methyl pyrazole (fomepizole, Antizol), another competitive inhibitor of ADH. However, for the yeast ADH used in these automated assays, the $K_m$ for ethanol, the preferred substrate, is only 2.1 mM, whereas that for methanol is very much larger at 130 mM, and that for isopropanol is 140 mM [28]. The potential for interference is therefore more theoretical than practical because a serum methanol of 130 mg/dL gives an apparent ethanol of only 2 mg/dL. Likewise, to obtain a positive ethanol result (10 mg/dL), the methanol concentration must be 650 mg/dL, which requires a massive ingestion.

CAP endeavors to assess and monitor toxic alcohol interference in ethanol tests by incorporating a number of other alcohols into some of its proficiency survey challenges. Reviewing the data from the 30 2010 and 2011 CAP alcohol challenges shows there were 15 ethanol-only samples and 15 "mixed" samples. However, toxic alcohol concentrations were relatively low—methanol, 48–61 mg/dL; acetone, 30–44 mg/dL; isopropanol, 38–72 mg/dL; and ethylene glycol, 46–152 mg/dL—but three samples were included containing only a toxic alcohol in the absence of ethanol. These data (summarized in Table 16.4) clearly show that there is no interference in ethanol measurements at these values. It is interesting to note that CAP also appears satisfied because these were all in the 2010 challenges, and none of the 2011 challenges were formulated in this way. Comparing performance of automated ADH alcohol methods for the mixed alcohol challenges again shows very little interference compared to the gold standard GC values (data not shown). Propylene glycol (70 mg/dL) was added for the first time to one of the challenges in the 2012 AL2-A set. This sample also contained 130 mg/dL ethanol, and although there was no added inaccuracy in the measured ethanol concentrations for this sample, 14% of respondents reporting a quantitative result incorrectly identified the compound as ethylene glycol. Failures resulted from both GC and enzymatic methods, which is rather disturbing, and this proves that although GC may be the gold standard analytical technique, inattention to detail can negate any added advantage offered. Furthermore, 10% of labs reporting a qualitative result also misidentified ethylene glycol.

These studies show that the presence of a toxic alcohol, even at large concentrations, is unlikely to produce a false-positive ethanol result by any of the automated ADH methods in current usage. The downside of this, however, is that toxic alcohol ingestions are not going to be detected by automated ADH ethanol tests, and other strategies must be employed to avoid missed diagnoses and missed treatment opportunities.

Nearly all kit manufacturers include interference information for methanol, acetone, isopropanol, and ethylene glycol in their package inserts, and these data are compared in Table 16.5. Manufacturers' interference data can sometimes be misleading and induce a false sense of security (see Case Reports). However, authentic patient specimens represent a complex mix of endogenous and drug-related metabolites that can produce a quite different pattern of interference than when alcohols are spiked into clean samples *in vitro*. Data on interfering substances other than toxic alcohols are not always available and sometimes only come to light in case reports. Alcohol oxidase ethanol methods are much more susceptible to false-positive interference in samples containing other toxic alcohols. This, coupled with the instability of AOD in solution, led to the overwhelming preference for the use of ADH in automated ethanol test kits.

## Interference by Elevated Lactate and LDH Concentrations

As outlined previously, the production of NADH by ADH is not specific for the presence of ethanol, but the mere presence of large concentrations of toxic alcohols is unlikely to produce a false-positive ethanol result. However, there may be other enzymes and/or substrates in the patient specimen that could result in the generation of NADH from the reduction of $NAD^+$. It has been known for more than 20 years that high concentrations of LDH and/or lactate interfere with ADH-based ethanol assays, giving false-positive results, [29,30] and this has also been observed in the glycerol dehydrogenase-based ethylene glycol assays [31–33]. This appears in some quarters to have been forgotten with the passage of time or sidelined in the belief that this will only be applicable to postmortem tests and not relevant to clinical testing [34,35]. CAP attempted to address this knowledge gap by incorporating a dry case study into the 2011 AL-2C proficiency set.

The ADH and LDH (E.C. 1.1.1.27 L-lactate: $NAD^+$ oxidoreductase) reactions are outlined here. When both enzymes and their substrates are present in a specimen, there are clearly two sources of NADH production:

$$\text{Ethanol} + NAD^+ + H^+ \overset{\text{ADH}}{\rightleftharpoons} \text{acetaldehyde} + \text{NADH}$$

$$\text{Lactate} + NAD^+ + H^+ \overset{\text{LDH}}{\rightleftharpoons} \text{pyruvate} + \text{NADH}$$

**TABLE 16.4** Comparison of Measured Ethanol Concentrations in Three Samples Containing Only a Nonethanol Toxic Alcohol

| Sample | Mean Measured Ethanol (mg/dL) | | | | Toxic Alcohol by GC (mg/dL) | | | |
|---|---|---|---|---|---|---|---|---|
| | GC | Abbott ADH REA | ADH UV Vis, all Methodologies | Ortho Vitros | Methanol | Acetone | Isopropanol | Ethylene Glycol |
| 2010 AL2-10 | 0.75 | 0.19 | 0.57 | 0.37 | 48.4 | | | |
| 2010 AL2-01 | 0 | 0.33 | 0.18 | 0.14 | | 30.4 | 71.6 | |
| 2010 AL2-11 | 0 | 0.18 | 0.17 | 0.04 | | | | 152.5 |
| No. of labs reporting | 84 | 41 | 2548 | 516 | | | | |

Source: *Data from CAP Alcohol Survey (AL2) Samples for 2010 and 2011.*

**TABLE 16.5** Toxic Alcohol Interference Data as Reported in Manufacturer Package Inserts[a]

| Method used | Concentration of Alcohol Tested (Apparent Ethanol Concentration), mg/dL | | | | | | | |
|---|---|---|---|---|---|---|---|---|
| | Methanol | Acetone | Isopropanol | Acetaldehyde | Ethylene Glycol | Propylene Glycol | *n*-Propanol | *n*-Butanol |
| Abbott Architect[b] | 2000 (0) | 2000 (0) | 2000 (0) | 2000 (0) | 2000 (0) | No data | 2000 (214) | 2000 (4) |
| Abbott REA[b] | 1000 (<10) | 1000 (<10) | 1000 (<50) | No data | 1000 (<10) | 1000 (<10) | 100 (64) | 1000 (274) |
| Microgenics DRI multiplatform[c] | 2000 (0) | 2000 (0) | 2000 (9) | 2000 (0) | 2000 (0) | 2000 (0) | 1500 (213) | 1000 (37) |
| Olympus EMIT II Plus[d] | 2000 (0) | 2000 (0) | 2000 (0) | 2000 (0) | 2000 (0) | No data | 2000 (214) | 2000 (1.7) |
| Ortho Vitros[e] | 700 (0) | 450 (0) | 200 (20) | 10 (0) | 200 (4) | 200 (0) | 200 (108) | 200 (58) |
| Randox[f] | 3000 (0) | 3000 (0) | 3000 (10) | 3000 (0) | 3000 (0) | 3000 (0) | 3000 (321) | 3000 (40) |
| Roche platforms[g] | 2000 (0) | 2000 (0) | 2000 (6) | 2000 (−3) | 2000 (2) | No data | 2000 (120) | 2000 (34) |
| Siemens Dimension Vista[h] | 2000 (2) | 2000 (2) | 2000 (4) | 2000 (2) | 2000 (0) | 2000 (0) | 2000 (346) | 500 (10) |

[a]*Beckman and Siemens Advia package inserts do not list interferences.*
[b]*Abbott Diagnostics, Abbott Park, IL.*
[c]*Microgenics Corporation, Fremont, CA.*
[d]*Beckman Coulter, Brea, CA (incorporating Olympus).*
[e]*Ortho-Clinical Diagnostics, Rochester, NY.*
[f]*Randox Laboratories, Crumlin, Co. Antrim, UK.*
[g]*Roche Diagnostics, Indianapolis, IN.*
[h]*Siemens Healthcare Diagnostics, Tarrytown, NY.*

Both ADH and LDH reactions are equilibrium reactions, and if both reactions are occurring within a single specimen on the analyzer, the enzymes not only compete for $NAD^+$ but also produce NADH at rates consistent with their concentrations and the appropriate substrate concentrations in a kinetic assay. Thus, an apparently large amount of ethanol might be determined to be present in the sample, which may in fact be completely untrue. Furthermore, because there is no mechanism in the test reaction mixture (as opposed to *in vivo*) to regenerate $NAD^+$ from NADH, $NAD^+$ may become the limiting factor in the ADH and/or the LDH reaction.

In 2010, the CAP Toxicology Resource Committee sponsored a pilot study that demonstrated the continued occurrence of false positives in automated ADH ethanol test kits. In this minisurvey, 50 laboratories agreed to test a serum specimen containing 120 mmol/L sodium plus 25 mmol/L sodium lactate and also 5000 U/L human LDH (mixed isoenzymes not characterized). Summarized results for the seven different test systems targeted are shown in Table 16.6. Using a cutoff of 10 mg/dL for a positive ethanol result, the Dade Dimension, Abbott REA, and Vitros platforms gave the correct negative ethanol result for the sample. The Roche platform assay results were positive but close to 30 mg/dL, and the other three assays (Beckman, Abbott Aeroset, and Microgenics DRI) gave ethanol results between 70 and 106 mg/dL, representing significant inebriation. Elevated lactate/LDH is mentioned in five of the manufacturer package inserts: The Abbott Aeroset, Roche, and Microgenics DRI methods give a cautionary note about lactate and LDH in postmortem samples but do not give specific details, whereas Abbott ADH REA, Randox, and Vitros give specific data only for lactate (tested at 25, 90, and 20 mmol/L, respectively, producing <10% interference), which is in line with the CAP experimental data. Reference ranges are somewhat method dependent, but for serum lactate it is typically below 2 mmol/L, and for LDH it is less than 300 U/L.

It may be confusing to observe that in alcohol determination seemingly similar assays may perform differently. Although informative, manufacturers' interference data can be misleading. Authentic patient specimens represent a complex component mix composed of endogenous metabolites and drug components, whereas manufacturers' data are generated by spiking good-quality specimens with purified pharmaceuticals and chemicals. Having a high serum LDH activity is usually not sufficient for a significant false positive to occur in the ADH-based ethanol assay because a high substrate (lactate) concentration is also required. Blood (plasma or serum) is therefore more likely to be affected than urine because although there may be a high concentration of lactate in urine, it is unlikely that high-molecular-weight LDH will also be present in the urine specimen.

It is difficult to identify a precise LDH activity at which interference occurs in each ethanol assay kit. Most LDH assays (and hence the value for LDH

**TABLE 16.6** Lactate and LDH Interference in Ethanol Assays

| Test Method | No. of Labs | Mean Measured Ethanol (mg/dL) |
|---|---|---|
| Abbott Aeroset | 6 | 102.5 |
| Abbott AxSYM REA | 2 | 9 |
| Beckman (LX & DX) | 14 | 73.2 |
| Microgenics DRI | 2 | 106 |
| Ortho Vitros | 7 | 9.8 |
| Roche Platforms | 7 | 29.3 |
| Siemens Dade Dimension | 10 | 4.2 |

Source: *Data from CAP Toxicology Resource Committee 2010 Pilot Survey, reported with 2011 AL2-C.*

attributed to the offending sample) employ the International Federation of Clinical Chemistry recommended lactate-to-pyruvate (L→P) reaction direction because it is less dependent on the $NAD^+$ and lactate concentrations, and there is less contamination of $NAD^+$ with inhibiting products during manufacture. The reference range is up to 225 U/L for L→P assays, but P→L methods give rather higher results (normal, <450 U/L). More important, the LDH present in human serum comprises five isoenzymes (LD1–LD5), composed of heterotetramers of H- and M-type subunits (4H = LD1 through to 4M = LD5). The typical percentage composition in a healthy person's serum is approximately 22% LD1, 35% LD2, 23% LD3, and 10% each of LD4 and LD5. The conditions favorable for the P→L reaction direction are a more acidic pH (7.0 optimum), a high concentration of pyruvate, a high ratio of NADH to $NAD^+$, and a preponderance of LD5 and LD4. Converse conditions are favorable for the L→P together with a preponderance of LD1 and LD2.

LDH is highly specific for the L(+)-isomer of lactate that is produced metabolically in humans. Although β-hydroxybutyrate (produced in excess in ketosis) is considered a poor substrate for LDH, the isomers do not behave equally in this regard. β-Hydroxybutyrate is in fact the preferred substrate for LD1 (resulting in LD1 being known as hydroxybutyrate dehydrogenase (HBD)), the serum activity of which can be measured, and the ratio of HBD to LDH is used to indicate the tissue-specific source of LDH in a more efficient manner than electrophoretic separation of LDH isomers.

It would be tempting to think that by contrast, the determination of serum lactate would be straightforward. Most assays for lactate use the enzyme L-lactate oxidase (EC 1.13.12.4; (S)-lactate:oxygen oxidoreductase) to oxidize L-lactate to pyruvate with the concomitant production of hydrogen peroxide. Whole blood analyzers (e.g., blood gas analyzers) measure the hydrogen peroxide amperometrically. These methods show significant interference from glycolic acid, the toxic metabolite of ethylene glycol [36], and may give values falsely elevated by up to 60 mM [37]. In most plasma lactate methods on chemistry analyzers, the hydrogen peroxide is consumed in a second enzymatic reaction with peroxidase and a chromogenic substrate to form a colored product that is measured spectrophotometrically. Typically, these are less subject to glycolate interference [38] but may return an apparent lactate at approximately twice the glycolate concentration. This can sometimes result in a high absorbance flag and a suppressed lactate result. The Roche methods do not use lactate oxidase but, rather, LDH to measure lactate concentration, and they are subject to less (20%) glycolate interference [39]. Therefore, these are more reliable in ethylene glycol intoxication.

*CASE REPORT* Gharapetian *et al.* [40] presented three patients with virtually identical apparent serum ethanol values (33, 33, and 34 mmol/L (152–156 mg/dL)) measured using the Siemens ADH-based ethanol assay on a Advia 1650 analyzer. Following calculation of very abnormal negative osmolal gaps (−23, −42, and −44), the samples were retested for ethanol on the Siemens (Dade Behring) Dimension RXL Flex reagent and also by GC and found to be negative (<2 mmol/L or <9.2 mg/dL). Interestingly, all three patients had marked elevations in serum LDH (between 5000 and 10,000 U/L), caused by acetaminophen hepatotoxicity, but only cases 1 and 3 had elevated serum lactate concentrations (method of measurement not given) at 22.5 and 5.1 mmol/L, respectively; patient 2 had a lactate of only 1.2 mmol/L. However, Siemens reported to the authors that the LDH would need to be greater than 100,000 U/L (by L→P assay), with a concomitantly elevated lactate, to produce an apparent ethanol concentration of 17.4 mmol/L. The patients' damaged hepatocytes will undoubtedly have spilled a number of substrates and other enzymes besides LDH into their serum, resulting in very high circulating levels of transaminases, other DHs, amino acids, ammonia, etc. This highlights the importance of disassociating manufacturer's spiked interference data from clinical samples because clearly additional factors are at play here in producing the erroneous ethanol results. The Siemens Advia assay uses a $NAD^+$ analog with high oxidizing potential rather than $NAD^+$ as the enzyme co-factor.

In most instances, the amount of either lactate or LDH present is not sufficient to make a significant contribution to the measured ethanol result. This is partly because compared to other toxic substances, serum ethanol concentrations are relatively large (10–100 mg/dL corresponds to 2.17–21.7 mmol/L). In contrast, lactate has a reference interval of approximately 0.5–2.2 mmol/L in venous blood.

# STRATEGIES FOR REMOVING INTERFERENCES

When endogenous interference in an enzymatic assay is problematic, blanking and preincubation can help reduce interference. This additional step increases both cost and analytical time. Unless a large number of samples are affected, this modification is not justified. Instrument parameters for U.S. Food and Drug Administration (FDA)-approved reagent kits are predefined within the instrument application and not adjustable by the end user. However, these can sometimes be altered if the analyzer has "open-channel" or "user-defined" applications. It is hoped that the re-awareness of this problem and the legal ramifications of an unreliable result will prompt the manufacturers to reformulate the poorly performing kits. Further research is required to pursue $NAD^+$ analogs that react almost exclusively with ADH and not LDH or do so at vastly differing speeds so that the two reactions can be isolated from each other. Alternate strategies could focus on side reactions that make colorless or non-interfering products. Coupling of the acetaldehyde product to another reaction (peroxidase to hydrogen peroxide) with judicious choice of chromogen offers promise of a visible color reaction that might be distinguished from the LDH assay.

LDH can be removed from serum with PEG (polyethylene glycol) 3000, which osmotically sequesters water from the sample so the proteins precipitate out. After centrifugation, the clear supernatant is analyzed. This technique is used extensively in clinical chemistry to remove interference from immunoglobulin-bound "macroproteins" such as prolactin, creatine kinase, and amylase [41,42]. Care should be taken to establish that added chemicals do not interfere with subsequent reactions, and that any specimen dilution is accounted for in the final result calculation. Alternatively, a protein-free filtrate of the sample can be prepared by centrifugation in a centrifugal filter device to obtain a protein-free ultrafiltrate [43]; a 30 kDa cutoff is appropriate for LDH. Such specimen manipulation must be carefully controlled by the processing of quality control samples in parallel to the patient sample. Clearly, it is not practical to process all samples by this extended assay, and identification of problem samples would require simultaneous testing for LDH, which again is not practical. Specimen filtration also does not remove excess lactate from the sample, but once the LDH is removed, lactate is probably no longer a significant problem.

## Determining Whether a Problem Exists: Checking for Concordance among Data

When presented with many hundreds or thousands of test results in a day, it is difficult for the clinical chemist to detect those few that might be in error. Modern laboratory information systems (LIS) may be able to run algorithms combining a number of different results for one patient and then flag or withhold for checking those identified as outliers. However, when ethanol is the only test requested, identifying potential errors becomes even more challenging. Some additional checks are presented here, which can be divided into two broad strategies. First, is the ethanol result accurate or could there have been some interference due to the patient's clinical condition? Second, if the ethanol result is accurate, did the patient actually ingest ethanol, or are there other co-ingestants, particularly other toxic alcohol metabolites, that complicate the clinical picture?

Figure 16.3 sketches an algorithm for evaluating an ethanol result for possible error. For the sake of clarity, only the main connections are shown, although in practice there is considerable overlap between the different routes. One must, however, be cognizant of the legal ramifications of reporting results from (or charging for) tests not requested by the physician but run by the laboratory solely for the purpose of validating the result of the requested test. Running serum indices for potassium is an obvious and well-accepted example, and many instrument manufacturers provide these reagents at no additional cost.

## Is the Ethanol Result Consistent with the Patient's Presentation?

Many labs performing testing for their own patients have access to a regularly updated electronic medical record through their hospital information system. Laboratories in smaller hospitals may not be so fortunate, but their LIS or their automated instrument middleware should be able to capture results of all tests performed on the patients. However, the interpretation is more difficult if the testing is not being done at the site of the patient's admission because not all records of tests done may even be available. In these latter situations, it is even more important that testing personnel or those designated to interpret the tests have a firm understanding of the limitations and are able to explain them concisely to personnel at the site where the patient is located.

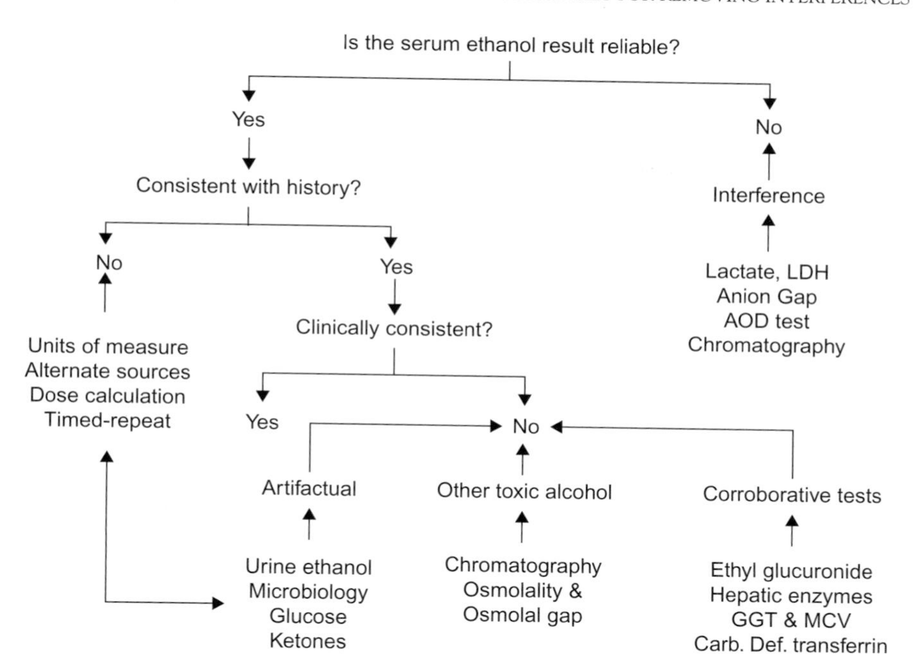


**FIGURE 16.3 Algorithm for evaluating the reliability of an ethanol result.** For clarity, only the main connections are shown, although in practice there is considerable overlap between the different routes.

Unfortunately, because of the sheer volume of results generated for each patient on a daily basis, many results are taken at face value and accepted as being 100% reliable without question provided that there are no instrument error flags.

Because of the development of tolerance in habituated individuals, and individual differences in susceptibility, there is no clear relationship between ethanol concentration and clinical effect. However, in children and young teens, and in unhabituated adults, there is general consensus on symptomatology, as shown in Table 16.7. The presence of other drugs can complicate the clinical picture, either because ethanol can potentiate the central nervous system (CNS) effects of other drugs or because it might mask their stimulatory effects. Many patients who make suicidal gestures or even those who take accidental overdoses have ethanol in their system at the time of hospitalization, and the use of ethanol increases susceptibility to accidents, especially motor vehicle accidents.

Because it is beyond the scope of the clinical chemist's responsibility to determine the precise cause of the patient's altered mental status, the information presented here suffices for most eventualities. Clinical chemists affiliated with regional poison treatment centers may be called on for more in-depth knowledge. In contrast to ethanol, many drugs are difficult, if not impossible, to detect by rapid tests because screening tests are available for only a handful of drugs, mostly illicit rather than clinically relevant. Unless screening of the patient's urine or blood by GC–mass spectrometry or liquid chromatography–mass spectrometry is available, the presence of other drugs is based almost exclusively on patient history rather than on any concrete analytical findings.

National Academy of Clinical Biochemistry (NACB) and other professional guidelines recommend the development of regional laboratories for such advanced drug testing and emphasize the benefit of collaborative work between clinical chemists and physicians. Debate continues over whether such testing could be completed in a relevant time span to affect more than a handful of patients [44].

## Are the Units of Measure Correct?

Most international toxicology publications use grams/liter (g/L) as the standard unit of measure for ethanol, whereas clinical chemists prefer millimoles/liter (mmol/L). In the United States, milligrams/deciliter (mg/dL) is commonplace, but to add further confusion, ethanol in the legal arena is expressed simply as percentage (%), and clinicians often drop the units completely. Good laboratory practice mandates a reference range appended to any result, reducing the potential for misunderstanding. The following are the most common interconversions between these different units of ethanol measurement:

- To convert g/L into mg/dL, multiply result by 100.
- To convert mg/dL into mmol/L, multiply result by 0.217.
- To convert mmol/L into mg/dL, multiply by 4.6.
- To convert % into mg/dL, multiply by 1000.

TABLE 16.7 Relationship of Serum Ethanol Concentration and Clinical Effect in Children and Nonhabituated Adults

| Serum Ethanol (mg/dL) | Typical Clinical Picture |
|---|---|
| <25 | Feeling of warmth and well-being |
| | Talkativeness and mild disinhibition |
| 25–50 | Euphoria |
| | Poor judgment |
| | Diminished fine motor coordination |
| 50–100 | Decreased sensorium |
| | Impaired coordination, balance, and gait |
| | Slowed reflexes and increased reaction time |
| | Emotional lability |
| 100–250 | CNS dysfunction—ataxia, diplopia, slurred speech, visual impairment, and nystagmus |
| | Confusion, stupor, difficulty sitting upright without assistance |
| | Nausea, vomiting |
| 250–400 | Stupor, coma, little response to stimuli |
| | Incontinence |
| | Respiratory depression |
| >400 | Respiratory paralysis |
| | Loss of gag and other protective reflexes |
| | Hypothermia |
| | Death |

## Is the Measured Ethanol Concentration Appropriate for the Ethanol Dose Ingested if Known?

Estimating dosages for exposures is notoriously difficult. It is tempting to envisage the worst possible scenario, in children especially, and surmise that if the bottle is now half full, then the patient must have ingested the other half. Small children do not often ingest a sufficient amount to seriously harm themselves, and because they have no better aim with tablets or alcohol than they have with food, the scene often looks far more dramatic than its reality.

The following is a relatively simple calculation (Widmark's formula):

$$Dose = Vd \times BW \times Cp \text{ or } Cp = \frac{Dose}{Vd \times BW} \quad (16.1)$$

where the dose of pure ethanol is estimated in grams; $V_d$, the distribution volume of ethanol, is 0.55 L/kg in females and 0.68 L/kg in males; BW is the body weight in kilograms; and $C_p$ is the blood ethanol concentration in grams/liter.

It is then only a short step to calculate the volume of beverage consumed by multiplying by the ethanol content of the formulation, remembering that "proof" is twice the percentage composition (approximate values are as follows: beer, 3 g/100 mL; wine, 8 g/100 mL; and spirits, 35 g/100 mL). As a rule, one "unit" of ethanol is approximately 12 g.

If the ethanol concentration is not appropriate for the ingested dose, the patient history is in error, the ethanol result may not be correct, or there may have been an alternate source of ethanol, such as those described in the next section.

Occasionally, a falsely low serum ethanol concentration is discovered. This most often occurs when a severely traumatized patient has been resuscitated emergently with large amounts of transfused (ethanol-negative) blood, red blood cells, or colloid at the time the samples are being drawn for analysis. Dilution of the sample directly from the line fluid by poor phlebotomy technique is also possible, and this is usually discovered because the hematocrit is very low, glucose is exceptionally high, or the electrolytes are out of balance [45].

## Can the Presence of Ethanol in a Specimen be Explained by Something other Than Ingestion by the Donor?

If the serum or urine ethanol result is valid, there are a number of alternative explanations for its presence other than ingestion of beverage alcohol. These include, but are not limited to, direct or indirect contamination. "Auto-brewery syndrome" has been invoked as a legal defense against DUI, but this *in vivo* generation of ethanol by the gastrointestinal tract has rarely been proven, and then not outside of Japan. It is believed to be the result of a combination of low mucosal and hepatic ADH activity, high intake of rice, and concomitant yeast infection [46].

Non-beverage ethanol is present in many common household items: Perfumes and aftershaves are 60–95% ethanol, hand sanitizer 65%, food essences 35%, mouthwash 15–35%, and NyQuil 10% ethanol. There is a trend for young teens to abuse ethanol in products such as beverages with caffeine and in canned whipped cream, ethanol-laced Jell-O and gummy bears, and even ethanol-soaked tampons [47–49].

Dermal absorption of alcohols is generally insignificant, [50] but once the epidermis is damaged, absorption may be increased 1000-fold. Bandages soaked with ethanol and applied for preoperative preparation for plastic surgery in a 2-year-old girl resulted in significant toxicity: 12 hr later, the child was comatose, with serum ethanol 80 mg/dL and glucose 15 mg/dL [51]. Immature skin is also highly alcohol-permeable, and there are case reports of premature infants with systemic toxicity following application of surgical spirit to cleanse skin prior to surgical procedures. For example, a serum ethanol of 259 mg/dL was reported 18 hr postprocedure in a 27-week-old neonate, with methanol of 26 mg/dL [52]. Surgical spirit (95% ethanol and 5% methanol) is the equivalent of rubbing alcohol, which in the United States contains isopropanol.

Adulteration of urine samples submitted for drug testing has been a problem since testing first began. A number of adulterant tests are used in legal testing, but these are not commonly applied in the clinical setting. Because testing proceeds to confirmation only if the primary screen is positive, adulteration aims to make the screen test negative by denaturing the antibody or preventing it from binding to its target drug. Ethanol or other alcohols are sometimes used for this purpose, but adulterers are often rather generous with the amount of additive, and the contamination is often obvious as soon as the top is taken off the tube.

*CASE REPORT* A 46-year-old patient with a long psychiatric history was receiving treatment for overuse of prescription pain medications. On her first visit, her urine screened positive for opiates, oxycodone, and acetaminophen; acetone was detected by GC at greater than 400 mg/dL, but with no trace of isopropanol (<10 mg/dL) or evidence of ketone bodies. The unusual alcohol result was noted in the patient's chart as suggestive of contamination rather than ingestion. Two weeks later, her urine screened positive for oxycodone, benzodiazepines, and salicylates, and it was found to contain isopropanol at 6500 mg/dL; again a comment on suspected contamination was entered in the chart. On the third visit, the drugs of abuse panel was negative, but the urine appeared very dilute with a measured ethanol of 98,050 mg/dL (creatinine < 0.02 mg/dL and blood urea nitrogen (BUN) < 1 mg/dL). After the physician was alerted, the patient was challenged and elected to seek medical care elsewhere.

Any specimen with an ethanol content greater than 15,000 mg/dL is almost certainly contaminated regardless of the symptomatology. In the case presented by Foley [53], a specimen submitted for drug testing from a heroin abuser was determined to have an ethanol content of 125 g/L (12,500 mg/dL) by headspace GC. A concomitant blood ethanol was negative, and it was determined that the ethanol had been added to the urine by a laboratory technician who misread the ethanol order for an instruction to add ethanol as a fixative for cytology testing. In addition to these less common sources of ethanol exposure, generation of ethanol *in vitro* is a real cause for concern, and its prevention is addressed next.

## Has the Specimen Been Preserved Correctly?

Ideally, specimens for ethanol testing (and lactate) should be collected into fluoride oxalate to inhibit glycolysis and prevent generation of ethanol by certain bacteria or yeasts within the specimen. The desire to prevent therapeutic anemia results in specimen consolidation, and because fluoride oxalate is unsuitable for many clinical tests (the salt adds approximately 150 mOsm/kg to the osmolality) and preservation is considered unnecessary when samples are to be processed emergently, ethanol tests are usually performed on serum, heparinized plasma, or unpreserved urine. Microbial contamination could be endogenous if the patient is septic or has a urinary tract infection (UTI), or it might have been introduced by aseptic collection techniques. The amount of required substrates in urine from healthy donors is usually too low to be of importance, but if the patient has ketosis—whether diabetic (hyperglycemic) or alcoholic (hypoglycemic)—ethanol production may be sufficient, if the correct microbes are present, to suggest severe toxicity [54]. Indeed, a small amount of fermentation may actually take place within the bladder if there is urinary stasis with a UTI such as *Candida albicans*.

When dealing with nonbiological specimens submitted for ethanol analysis, the same contamination considerations apply, as highlighted by the following case.

*CASE REPORT* A 2-year-old child was admitted repeatedly with unexplained ethanol intoxication (serum ethanol 95–148 mg/dL) and hypoglycemia (glucose 30–42 mg/dL). Because of the allegations of abuse and/or neglect, the parents were interviewed. Several items were confiscated from the patient's room, including a part-full bottle of orange juice. The juice tested positive for ethanol (2530 mg/dL), and the cause of the poisoning was believed to have been found. However, the finding that "still" orange juice was submitted for analysis disproved this theory because the juice contained not only ethanol but also considerable gas by the time it was tested. Fermentation had obviously occurred.

## Does the Ethanol Concentration Decline at the Expected Rate? The Twenty-Per-Hour Rule

In suspected intoxication, the differential diagnosis is broad, and additional tests, including imaging, are warranted if:

- the reported presence of ethanol was completely unexpected

- the concentration is of extremely high proportion
- the concentration does not decline at a rate of approximately 20 mg/dL per hour
- the patient has not woken up in the expected time or
- the patient is a child or young adolescent.

Ethanol "dosing" typically occurs over an extended time span compared to conventional oral medications. Its bioavailability ranges from less than 10% at 10 mg/dL to 90% at 100 mg/dL due to saturable first-pass metabolism of ADH in the duodenal/gastric mucosa and liver [55], with peak blood concentrations approximately 60 min after ingestion; the effect of delayed gastric emptying is less pronounced than with solid drug formulations.

The majority of postabsorptive ethanol metabolism occurs in the liver at a rate dependent on the activity of hepatic ADH, which converts approximately 95% of the dose to acetaldehyde. CYP2E1 also oxidizes ethanol to acetaldehyde once the blood concentration exceeds 30 mg/dL (its $K_m = 60-80$ mg/dL). In contrast to ADH, this enzyme is inducible by ethanol, and alcoholics may route up to 30% of the dose via this pathway [56]. Regardless of the route of formation, acetaldehyde is toxic both to the liver and systemically as it forms adducts with cellular components—DNA, proteins, enzymes, and lipids. Acetaldehyde degradation is rapid through mitochondrial aldehyde dehydrogenase (ALDH) to produce acetate [57]. The reoxidation of NADH determines the redox potential of the cell and thus determines many aspects of hepatocyte homeostasis.

Ethanol kinetics differs between males and females: Distribution volume (total body water) is lower in females (0.55 vs. 0.68 L/kg) due to their higher percentage of fat, and gastric ADH is expressed at lower activity [58]. Human ADH (in contrast to the yeast enzyme) is substrate saturated at low ethanol concentrations ($K_m$ is only 5–10 mg/dL), and metabolism proceeds by zero-order kinetics when serum ethanol is greater than 25 mg/dL. The rate is usually expressed in terms of a decline in serum ethanol, and a range between 13 and 25 mg/dL/hr is often quoted—hence the mean "twenty per hour" rule, although this is sometimes wider in ED patients [59,60]. Although hepatic function is severely compromised in cirrhosis, there is no compelling evidence that the rate of ethanol metabolism is altered [61,62].

Both ADH and ALDH in humans have important isoenzymes and polymorphisms. Human ADH is a heterodimeric protein, encoded by at least seven genes, producing more than 20 dimers with different rates of ethanol metabolism [63,64]. ALDH2 is the most important of the nine different subtypes for ethanol metabolism. The polymorphism $^{487}$Glu → Lys, associated with almost complete loss of function, is carried by approximately 40% of Asians, and the accumulation of acetaldehyde is responsible for the flushing, hypotonia, and tachycardia seen in these individuals after only minute doses of ethanol [65].

## Are the Osmolality and Osmolal Gap Appropriate?

Osmolality is a colligative property of solutions: All dissolved compounds contribute equally based on their number of particles present. Serum osmolality can be measured directly by osmometry (true osmolality; OSM), or it can be calculated (OSMc) by measuring the major osmotically active substances present. In a healthy person, these are sodium, glucose, and urea (BUN). Once these values are substituted into a formula, OSMc is obtained (Eqs. 16.2a–16.2d). Because the presence of ethanol is so common, a number of formulae have been derived based on studies of healthy individuals or patients with known exposure to ethanol alone. The one most commonly used in medicine is that of Smithline and Gardner [66] modified to include ethanol by Purssell *et al.* [67], noting that all measurements are made in mmol/L:

$$\text{OSMc} = 2(\text{sodium}) + \text{BUN} + \text{glucose} \tag{16.2a}$$

$$\text{OSMc} = 2(\text{sodium}) + \text{BUN} + \text{glucose} + 1.25(\text{ethanol}) \tag{16.2b}$$

These are more familiarly seen in the United States as follows:

$$\begin{aligned}\text{OSMc} = 2(\text{sodium}) &+ (\text{BUN mg/dL} \div 2.8) \\ &+ (\text{glucose mg/dL} \div 18)\end{aligned} \tag{16.2c}$$

$$\begin{aligned}\text{OSMc} = 2(\text{sodium}) + (\text{BUN mg/dL} \div 2.8) \\ + (\text{glucose mg/dL} \div 18) + (\text{ethanol mg/dL} \div 3.7)\end{aligned} \tag{16.2d}$$

Using the 1.25 molar multiplier (3.7 traditional unit divisors) as opposed to the theoretical 1.0 and 4.6 in part accounts for the body's response to the osmotic effect of ethanol by releasing cellular osmolyte and metabolic derangements such as elevated serum amino acids that result from the effects of ethanol on metabolism [68]. In addition, ethanol alters the dissociation behavior of other analytes within the serum, thus distorting the osmolality [69].

The difference between the measured and the calculated osmolality is termed the osmole gap (OG = OSM − OSMc). The practical value of the OG from the laboratory's perspective is to ensure that a measured ethanol value is consistent with the measured osmolality. From the physician's perspective, the clinical value of the OG is to determine whether other pathological substances are in the patient's serum contributing to OSM that are not accounted for in the OSMc value. Specifically, in practical terms, the OG is helpful in deducing the presence of other toxic alcohols, such as methanol, isopropanol, or ethylene and propylene glycols, especially in situations in which direct measurements of these are not available in a time-sensitive manner. Therefore, if the ethanol measurement is inaccurate, then an OG may be attributed erroneously to the presence of a toxic alcohol, or conversely one may be assumed not to be present when in fact it is, thus missing potential treatment options. Also note that normal gap values can vary considerably with instrumentation, so what constitutes normal must be taken in the context of each laboratory or patient group. For example, in one study, [70] OGs in healthy patients increased by approximately 12 mOsm/kg during the period 1996 to 2004.

There is considerable interest in using the rearranged form of the OG equation to calculate the ethanol concentration present [71]:

$$\text{Ethanol conc. mg/dL} = \text{OG} \times 3.7 \qquad (16.3)$$

However, actual ethanol concentrations could not be predicted with sufficient accuracy when testing patient specimens [72]. This is partly because the wide range for a normal OG makes it difficult to decide how much of the OG to attribute to the ethanol; at higher ethanol concentrations, investigators consistently report overestimation of the ethanol concentration by the OG, sometimes by as much as 30% [73]. Taking this notion one step further, even in the presence of ethanol, the concentration of another toxic alcohol could theoretically be estimated from its molecular weight. The osmotic activity attributed to 1 mg/dL of each is as follows: methanol, 0.34 mOsm/kg; isopropanol, 0.17 mOsm/kg; ethylene glycol, 0.16 mOsm/kg; and propylene glycol, 0.13 mOsm/kg. OGs in children are even more problematic to interpret because a normal OG may reach 22 mOsm/kg, which is unfortunate because young children are one of the patient populations most at risk of an unknown ingestion who would benefit from an early diagnosis [74].

Despite these caveats, the OG remains a particularly useful screening test for toxic alcohols in asymptomatic patients due to the fact that the latency period to symptoms can be long because it is the production of metabolites that is responsible for the majority of the toxicity.

In the three cases reported by Gharapetian *et al.* [40] described earlier in this chapter, the OSMs were 306, 286, and 278 mOsm/kg, respectively. The calculated OGs if ethanol was absent were relatively normal for such sick patients at 18, −1, and −1 mOsm/kg, respectively, but if ethanol was indeed positive at the concentrations measured, the OGs recalculated to −23, −42, and −44 mOsm/kg, respectively. Such grossly abnormal results should immediately raise suspicion that one of the parameters measured is in error.

The method by which serum osmolality is measured significantly affects the reliability of the result in the presence of volatile components. Because of the frequency with which ethanol is responsible for hyperosmolar samples, osmolality must be measured by freezing point depression (FDP) rather than by vapor pressure dew point depression (VPD). Because there is a linear relationship between the ethanol concentration and ΔOSM (the difference between FPD and VPD osmometry results), this has been used as a tool for estimating serum ethanol [75]. In clinical practice, it would be unusual to have both types of osmometry available because FPD is increasingly becoming the technique of choice.

***CASE REPORT*** Walker *et al.* [76] described a patient in whom methanol toxicity was strongly suspected based on history, visual impairment, and severe AG metabolic acidosis. No OG was present, but the patient was dialyzed and treated with ethanol infusion with good result. Later, the serum methanol was reported to be greater than 100 mg/dL. Investigation revealed that osmolality had been measured using VPD, and subsequent *in vitro* testing confirmed the insensitivity of VPD to toxic values of both methanol and ethanol. Thus, it is important to be cognizant of the technique used by the laboratory to avoid missing a potentially life-threatening but treatable intoxication.

## Are the Ethanol Ratios Appropriate in Different Fluids?

Table 16.8 shows the average values for blood ethanol compared to other fluids. There is considerable variation between individuals and with time since ingestion. The presence of certain bacteria, either endogenous or exogenously introduced, adds further variability. Postmortem considerations for obtaining aseptic samples are not applicable to clinical testing, and vitreous humor is included only for illustrative purposes.

With the exception of the first 90 min after ingestion, urine ethanol always exceeds that of the serum, with a ratio typically of 1.2–1.3 urine:serum in the early postabsorptive phase, increasing to 1.4–1.5 later. In the terminal stages after exposure, the ratio increases appreciably because ethanol is present in

**TABLE 16.8** Comparison of Ethanol Content of Different Fluids Relative to Whole Blood

| Fluid | Fluid:Whole Blood Ratio |
|---|---|
| Whole blood | 1.00 |
| Serum or plasma | 1.16 |
| Urine | 1.3 at equilibrium |
| Vitreous humor | 1.2 |
| Bile | 1.0 |
| Cerebrospinal fluid | 1.1 |
| Kidney | 0.7 |
| Liver | 0.6 |
| Brain | 0.8 |
| Breath | 1/2100 = 0.000476 |

bladder urine hours after the serum alcohol is no longer detectable [77]. However, during the absorption phase, the plasma is higher than the urine, although it takes only 10 min for ingested ethanol to appear in "primary" (i.e., ureter) urine [78]. Thus, the ratio of the ethanol content of the two fluids might be used not just to determine the historical time course of events but also to check the validity of results. For example, a serum ethanol measured at 350 mg/dL with a simultaneous urine ethanol of only 20 mg/dL must indicate that the patient ingested a large amount of ethanol in the minutes prior to sampling, that the serum ethanol was grossly overestimated due to an interfering substance, or that the blood was contaminated during phlebotomy. Although the urine may have been measured using the same assay as the serum, active LDH will not be present in the urine and so will not produce a false-positive ethanol result in the way that the serum might.

When attempting to reconcile results between different fluids, the following caveats should be noted:

- The effect of shock on metabolism and excretion of ethanol
- How blood loss affects the dynamics of distribution
- Administration of blood product, fluids, or colloid.

The validity of measuring ethanol in breath is based on the relationship of ethanol in end-expired (alveolar) air and that in the alveolar blood supply described by Henry's law. Thus, 1 mL of blood contains the same amount of ethanol as 2100 mL of breath at 34°C, the temperature of end-exhaled air. Currently, a number of breath ethanol analyzers are approved for use by the FDA [79] for workplace testing and ED use. These use infrared spectrometry; the primary wavelengths used for ethanol are 3.4 and 9.5 μm, corresponding to the C–H and C–O bond vibrational stretching, respectively. Loss in infrared energy across the sample represents the amount of ethanol present. Current instruments use five different wavelengths to enhance selectivity and abolish interference from acetone [80]. Many devices are calibrated to display in blood ethanol equivalent rather than the absolute breath ethanol concentration. The 15 min delay in sampling eliminates contamination from residual beverage in the mouth.

Saliva or, more correctly, oral fluid is becoming increasingly popular for all types of drug testing because of its noninvasive nature and relative immunity from adulteration [81]. This fluid can be tested by any of the automated ADH-based tests described in this chapter. However, for point-of-care testing, AOD methods predominate because these are most easily converted to visual signals (see Figure 16.1). Semiquantitative test strips are available that show good correlation between saliva and serum ethanol concentrations [82,83]. (Example products include DrugCheck Alcohol Rapid Test and DrugCheck SalivaScan, Express Diagnostics International, Blue Earth, MN; Alco-Screen, Craig Medical Distribution, Vista, CA; Saliva Alcohol Test Strip, Dipromed, Weigelsdorf, Austria; OratectPlus, Branan Medical, Irvine, CA; and Q.E.D. Saliva Alcohol Test, OraSure Technologies, Bethlehem, PA.)

## Could there be Interference from Lactate Dehydrogenase?

The properties of LDH and the reaction by which it causes interference in ADH-based assays that rely on NADH production were described previously. Although LDH assays are widely available and inexpensive on automated chemistry platforms, the test is not routinely ordered because it is not tissue specific. Its main clinical use is to monitor hemolysis or cell lysis following chemotherapy or progression of skeletal muscle disorders. LDH is found in high concentrations in all cells that utilize glucose for energy. Because erythrocytes contain more than 500 times the LDH of serum, even a small amount of hemolysis can substantially elevate the serum LDH. LDH is elevated in many disorders, as shown in Table 16.9; however, this list is by no means exhaustive, and the amount of elevation is quite variable, although the hemolytic causes often give higher results.

Common sense dictates that an isolated elevation in serum LDH, although necessary, is not sufficient to produce interference; an abnormally large substrate concentration (i.e., elevated serum lactate) should also be a prerequisite. The $K_m$ of hepatic LDH for lactate is 14 mM [84], whereas physiologic concentrations are less than 2 mM. However, case studies highlight that

TABLE 16.9 Causes of Elevated Lactate Dehydrogenase

| | |
|---|---|
| Hemolytic | Megaloblastic anemia ($B_{12}$/folate deficiency) |
| | RBC metabolic disorders (G6PD deficiency) |
| | Hemoglobinopathy (sickle disease, thalassemia) |
| | Autoimmune |
| | Physical destruction (prosthetic valves, burn) |
| | Poor phlebotomy technique |
| Cancers | Leukemia |
| | Lymphoma |
| | Testicular cancer |
| | Tumor lysis syndrome |
| Infective | Malaria |
| | Infective mononucleosis |
| | Hepatitis |
| | Influenza |
| Tissue trauma | Myocardial infarction |
| | Stroke |
| | Intestinal and pulmonary infarction |
| | Pancreatitis |
| | Alcoholic hepatitis |
| | Chronic liver disease |
| | Gunshot wound, laceration |
| | Seizures (neurologic, metabolic, or drug-induced) |
| | Rhabdomyolysis |
| | Compartment syndrome |
| | Crush injury |
| Muscle disease | Metabolic (glycogen storage disorders) |
| | Structural muscle disease (muscular dystrophy) |
| Excretion | Kidney disease |

additional unidentified chemical reactions are contributory because interference can occur at normal lactate concentrations [40]. A concentration of 1000 U/L of hepatic LDH uses 1 mmol lactate/min, furnishing 1 mmol NADH/min. The same NADH production rate from ADH corresponds to approximately 5 mg/dL ethanol.

If the result of an ADH-based ethanol result is in doubt, it is quick and inexpensive to test the urine specimen with a point-of-care testing semiquantitative alcohol dipstick based on AOD detection. Because the two methodologies have different common interferents, it is unlikely that the results will be concordant unless ethanol is indeed present.

## Does the Patient have an Anion Gap Acidosis or Ketosis?

Metabolic acidosis results from increased production of metabolic acids or lack of metabolic base (bicarbonate). Those with a normal anion gap typically result from gastrointestinal or renal disorders of electrolyte balance. An elevated anion gap acidosis, particularly in the intensive care unit setting, is commonly caused by the presence of lactic acid. The differential diagnosis of lactic acidosis is long (Table 16.10) and beyond the scope of this chapter. In hypoxic states, or where there is mitochondrial dysfunction inhibiting the tricarboxylic acid cycle, the end product of glycolysis must be converted into lactate, which subsequently accumulates.

Ketosis occurs when cells contain insufficient glucose to sustain ATP demand. Counterregulatory hormones ensure lipolysis, and fatty acids are oxidized for energy. Accumulated hepatic acetyl CoA is converted to acetoacetate and then to acetone or β-hydroxybutyrate. Ketonemia arises when hepatic ketone production outstrips the capacity of extrahepatic tissues to oxidize them. The organic anions acetoacetate and β-hydroxybutyrate result in anion gap acidosis. The two main causes of ketoacidosis are diabetic ketoacidosis (DKA) and alcoholic ketoacidosis (AKA); starvation ketosis is similar to AKA. The urine screens positive for ketones in both presentations, and the serum hydroxybutyrate is elevated (more so in DKA than in AKA or starvation), but the two can easily be distinguished by simple and inexpensive tests: hyperglycemia (>500 mg/dL) and positive urine glucose in DKA, and hypo- or normoglycemia and an absence of glucose in the urine in AKA. DKA was discussed previously as a potential cause of factitious urine ethanol.

Determination of the anion gap ($AG = [Na + K] - [Cl + HCO_3]$) usually involves no additional cost to the laboratory because electrolyte measurements are standard of care. The anion gap calculation can be performed automatically for all electrolyte orders in the lab's middleware or LIS. Lactate is also commonly measured in the emergency or critical care setting, and it is available on most blood gas analyzers and confirms the AG calculation. It was previously noted that L-lactate assays, especially those using electrochemistry, may be unreliable in the presence of the ethylene glycol metabolite glycolate. As with any laboratory test, it is important to understand the performance characteristics of the assay used to measure lactate.

## Are there other Indicators or Markers of Ethanol Ingestion?

Markers of ingestion may be aimed at detecting either acute (ethyl glucuronide) or chronic exposure (hepatic enzymes, hematological markers, and carbohydrate-deficient transferrin). Although chronic markers are not essential for verification of ethanol ingestion, they can help identify alcohol abusers because physicians may otherwise recognize only 20–50% of patients with alcoholism [85]. For example, chronic alcoholics may have an unusually mild intoxication for the amount of ethanol measured, and they are also more likely to use other toxic alcohols for heightened effects or for economic reasons. No single laboratory test is reliable enough on its own to support a diagnosis of alcoholism [86,87].

### *Ethyl Glucuronide*

Ethyl glucuronide is a minor metabolite of ethanol, accounting for less than 0.1% of the dose, produced by enzymatic conjugation of ethanol with glucuronic acid. In contrast to ethanol, ethyl glucuronide is nonvolatile, water-soluble, and stable to storage—properties that give it significant advantages as a routine test for ethanol use. Its presence is highly specific for ethanol, and it can be found in urine, serum, and oral fluid for 3–5 days after a single 50 g dose of ethanol. This is a favorable detection span compared to that of ethanol, which is typically 10–24 hr in body fluids. Currently, the test is not widely used clinically, although enzyme immunoassay kits are available for some automated platforms [88]. This screening test has the potential to provide a rapid cross-check on a serum ethanol that is inconsistent with the history or symptoms, and it may be performed in the routine chemistry lab on the same automated analyzer as that used for the original ethanol test [89].

**TABLE 16.10** Causes of Elevated Lactate Concentration

| | |
|---|---|
| Type A: | Tissue hypoperfusion and hypoxia |
| | Hemodynamic shock: hemolysis, bleeding |
| | Low oxygenation: hypoxia, respiratory failure, anemia, dyshemoglobinemia, carbon monoxide |
| | Tissue hypoperfusion: cardiovascular instability, heart failure, fluid overload, drug-induced hypotension |
| | Compartment syndrome, seizures, severe exercise |
| | Septic shock |
| Type B: | Without clinically apparent tissue hypoxia |
| | Underlying diseases: diabetes mellitus, uremia, liver disease, infection, malignancy |
| | Toxic alcohol metabolism: ethanol, methanol, ethylene glycol, metformin, isoniazid |
| | Uncouplers of oxidative phosphorylation: carbon monoxide, aspirin, cyanide |
| | Inborn errors of metabolism: pyruvate dehydrogenase deficiency, glycogen storage disease, pyruvate carboxylase deficiency, etc. |
| | D-Lactic acidosis: Gram-positive bowel overgrowth, short bowel syndrome, propylene glycol |

### *Hepatic Transaminases*

Assays for the transaminases AST and ALT are universally available on automated chemistry analyzers and are inexpensive to perform. Both enzymes are elevated in chronic liver disease but usually less than 1000 U/L; higher values reflect acute hepatic disturbance. The De Rittis ratio (AST:ALT) is a reliable marker to distinguish between alcoholic (>2) and non-alcoholic (<1) liver disease (see Chapter 9) [90].

### *γ-Glutamyl Transferase and Mean Corpuscular Volume*

The activity of γ-glutamyl transferase (GGT) and the red blood cell mean corpuscular volume (MCV) have been the traditional markers of chronic ethanol use: Serum GGT greater than 500 U/L with an MCV greater than 100 fL in the absence of another condition should raise suspicion of ethanol abuse [91,92]. Both tests are readily available in most laboratories. Care should be taken to rule out megaloblastic anemia from $B_{12}$ and/or folate deficiency as the cause of an increased MCV. GGT lacks specificity because numerous disorders and drugs can elevate GGT, including diabetes, hyperthyroidism, cardiac and renal disease, smoking, barbiturates, phenytoin, carbamazepine, and rifampicin [93,94].

### *Carbohydrate-Deficient Transferrin*

In habitual drinkers, the proportion of transferrin with none, one, or two sialic acid chains (collectively termed carbohydrate deficient transferrins (CDT)) is greater than 6%, whereas healthy individuals with no or low alcohol consumption will have a CDT less than 2%. CDT analysis is somewhat specialized (high-performance liquid chromatography, capillary electrophoresis, and isoelectric focusing) [95–97], but an immunoassay has become available on automated chemistry platforms [98]. However, genetic transferrin variants, and some non-alcoholic hepatic diseases such as hepatocellular carcinoma, chronic hepatitis C, and primary biliary cirrhosis, give false positives, and the main strength of the test lies in detection of relapse in chronic alcoholics [99,100].

## Does the Clinical Picture Suggest Exposure to other Toxic Alcohols, and are there Simple Tests for their Detection?

No discussion of laboratory testing for ethanol is complete without mentioning toxic alcohols. Often, ethanol (with the OG) is measured as a means to determine if another alcohol is present because these compounds are notoriously difficult for the average clinical laboratory to measure. Despite its shortcomings, the OG remains a useful screening test for toxic alcohols in asymptomatic patients because the metabolites are primarily responsible for the toxicity. Performance of the OG for such use was investigated by Lynd *et al.* [101] using rigorous Standards for Reporting Diagnostic Accuracy criteria. Samples from 20 patients with analytically proven exposure to ethylene glycol (10), methanol (9), or both (1) were used to determine the validity of Eq. (16.2b). Using a diagnostic threshold of OG $\geq$ 10 mOsm/kg for patients requiring antidotal therapy resulted in sensitivity of 0.9 and specificity of 0.5; sensitivity was 1.0 when identifying patients requiring dialysis.

Toxic alcohols are in common usage and often taken in overdose, either deliberately or unintentionally, particularly in children. It is therefore important to have a method that can distinguish between these and identify them correctly both in the presence and in the absence of ethanol. The salient laboratory tests and distinguishing clinical features of exposure to toxic alcohols are presented in Table 16.11, which is intended to be illustrative rather than exhaustive.

The NACB places the onus on diagnostic reagent manufacturers to provide reliable tests for these toxic alcohols and glycols rather than have the laboratory make a "diagnosis by exclusion" or rely on surrogate testing methods. GC should be available for this purpose, but only rarely is this the case. Clinical laboratories should provide direct measurements for methanol and ethylene glycol in serum or plasma because toxicity can occur without clinical signs of inebriation and so surrogate markers could be useful [102].

Currently, enzymatic procedures for toxic alcohols can only be used on automated analyzers that have "open" applications and are not FDA-approved for this purpose. None of these are widely adapted or well controlled, characterized, or standardized. For example, methanol is converted to formaldehyde by AOD and then to formic acid by formaldehyde dehydrogenase (FDH), following the reaction by NADH production [103]. Because the latter reaction is highly specific for formate, the entity that correlates directly with methanol toxicity, it may be more appropriate to measure formate directly with FDH [104]. Another simple assay utilizes periodate to oxidize ethylene glycol to formaldehyde, which is then reacted with a chromogen. This commercially available veterinary product is reported to show good performance with human samples [105]. A similar test generates a blue chromophore, which also shows positive for propylene glycol,

**TABLE 16.11** Uses, Toxic Metabolites, and Clinical and Laboratory Features of Toxic Alcohols

| Compound (Common Name) | Sources | Toxic Metabolite(s) | Distinctive Clinical and Lab Features |
|---|---|---|---|
| Ethanol (sugar alcohol) | Surgical spirit, Perfumes, Beverages, Mouthwash, Hand sanitizer, Food flavors, Cough syrups | Acetaldehyde | Increased OSM and OG, Anion gap metabolic acidosis, Hypoglycemia, Ketonuria if chronic, Tests for chronic use, Inebriation CNS depression, Vomiting |
| Methanol (wood alcohol) | Windshield washer fluid, Gasoline antifreeze, Paint remover, Photocopying fluid, Surgical spirit, Moonshine | Formaldehyde Formate | Increased OSM and OG, Profound anion gap metabolic acidosis, Lactic academia, Hypoglycemia, Transient inebriation, Nausea/vomiting, Blindness |
| Isopropanol (rubbing alcohol) | Rubbing alcohol, Windshield washer fluid, Glue remover, Cleaners and degreasers, Skin lotions | Acetone (nontoxic) | Increased OSM and OG, Normal anion gap and pH, Normoglycemia, Ketonuria, no glucose, Profound inebriation and CNS depression, Tachycardia, hypertension, vasodilation Gastritis, hematemesis |
| Ethylene glycol (antifreeze) | Radiator fluid, Antifreeze | Glycolate Oxalate | Increased OSM and OG, Anion gap metabolic acidosis, Lactic acidemia (may be falsely increased), Hypoglycemia, Hypocalcemia, Muscle dysfunction, seizures, Acute renal failure, Oxalate crystalluria, Transient inebriation |
| Propylene glycol | Intravenous drug preparations | D-Lactate | Increased OSM and OG, Anion gap metabolic acidosis, Lactic acidemia (may be falsely low), Acute renal failure, Hippuric aciduria |

OG, osmole gap; OSM, osmolality.

but has some cross-reactivity with ethanol (>50 mg/dL) [106]. Serum oxalate may also be measured by converting it to formate with oxalate decarboxylase and then analyzing the formate as described previously [107].

Failing the provision of such tests, an alternate strategy is to use the "gap" between the results of two non-ideal tests in the hope of providing a good result. Several examples illustrate this notion.

When added to serum at relevant concentrations, acetone, ethanol, isopropanol, and methanol produce the expected linear increase in FPD osmolality, whereas VPD osmolality is unaffected: Only ethylene glycol produces a linear increase in both VPD and FPD osmolality [108]. Therefore, the relationship between the difference in FPD and VPD osmometry measurements (ΔOSM) might be used to infer the isopropanol or methanol concentration. In clinical practice, however, it would be unusual to have both types of osmometry available because FPD is increasingly replacing VPD as the technique of choice.

The presence of a significant difference between the osmometry-calculated alcohol value and ADH-based ethanol result has been used to validate the presence of other toxic alcohols [109]. A similar strategy comparing the sample response between a semiquantitative AOD dipstick and an ADH-based "specific" ethanol assay could be used to gauge the presence of methanol, which reacts strongly with AOD but not ADH [110].

As described previously, the lactate gap (amperometric value minus a glycolate-unaffected assay value) can be used as a surrogate to identify ethylene glycol exposure and thus to estimate the glycolate concentration [111].

Ethylene glycol might be inferred from the "triglyceride gap" between the results of two different automated analyzer kits, one containing ethylene glycol-reactive glycerol dehydrogenase and the other nonreactive glycerol dehydrogenase. This again showed some cross-reactivity with propylene glycol (<15%) [112].

## FURTHER CASE STUDIES

The cases presented in this section highlight the importance of assessing all the available information and not focusing exclusively on the result of one test in preference to another. The use of ethanol concentration and osmolal and anion gaps is vital to arriving at the correct diagnosis and treatment.

*CASE REPORT* A 4-year-old male was hospitalized for 6 weeks following a viral illness during which time attempts to control status epilepticus required numerous medications. He became profoundly hypotensive (75/35 mmHg) despite extensive pressor support, and there was some concern that morphine might have been administered, so a point-of-care drug screen was performed, which showed a positive AOD ethanol test. He was transferred to our institution unresponsive. He had a partially compensated metabolic acidosis on ventilatory support (pH 7.26) and elevated anion gap (36 mmol/L). Most remarkably, his OSM was 611 mOsm/kg. Volatile analysis by GC revealed a propylene glycol concentration of 1548 mg/dL. Lactate was measured at 7.1 mmol/L, which although elevated was probably a 90% underestimate of the true lactate because propylene glycol is metabolized to both L- and D-lactate, and the body has very limited capacity to metabolize the latter, whereas only the L-isomer is detected by laboratory tests. Only after all these entities were taken into account could the OSM be reconciled.

The patient had five plasmapheresis treatments and recovered after an episode of hypotensive/gentamicin-induced acute renal failure. Propylene glycol is present in large concentrations as a solubilizing agent in a number of intravenous medications, [113,114] and his estimated dose was greater than 16 g/day once his medications were reconciled (lorazepam 85%, phenobarbital 75%, and phenytoin 40%).

*CASE REPORT* Vasiliades *et al.* [115] reported a 45-year-old male admitted after having fallen from a ladder. He had a sweet odor on his breath but no evidence of head trauma. He was hypotensive (blood pressure 80/60 mmHg), his pulse was 90 beats/min, and he had shallow respirations at 14 min. Arterial blood gas showed respiratory acidosis (pH 7.25, $pO_2$ 70 mmHg, and $pCO_2$ 60 mmHg) and was ventilated. Serum ethanol by ADH assay was 186 mg/dL, but his OG (65 mOsm/kg) was too large for the amount of ethanol present. Subsequent GC analysis showed the ethanol to be only 90 mg/dL, with isopropanol at 186 mg/dL and acetone metabolite 18 mg/dL. As expected, urine ketones were negative (because this reacts with β-hydroxybutyrate and only poorly with acetone). The nonspecific ADH assay is dangerously misleading when more than one alcohol is present, and the fact that isopropanol is a profound respiratory depressant complicates the differential with a known head trauma.

*CASE REPORT* Platteborze *et al.* [116] presented an equally confusing case of a 32-year-old female with a 72 hr history of abdominal pain and nausea who had ingested only ginger ale and homemade liquor during this time. She had a partially compensated metabolic acidosis; an elevated anion gap (27 mmol/L) explained by an elevated β-hydroxybutyrate and lactate; and an OG of 56 mOsm/kg, which could only be fully accounted for once the ethanol (128 mg/dL), isopropanol (14 mg/dL), and acetone metabolite (37 mg/dL) were taken into account.

*CASE REPORT* A previously healthy obese 13-year-old male awoke with visual loss. He was somnolent and mildly confused, and he had been vomiting since the previous afternoon.

On presentation to the ED, he was found to have tachycardia (121 min) and tachypnea (38 min); pulse oximetry showed 98% saturation. Initial labs showed severe metabolic acidosis: pH 7.18, $pCO_2$ 10 mmHg, and $HCO_3$ 3.7 mmol/L. He was dehydrated (protein, creatinine, and BUN elevated). Urinalysis showed protein 2 + and ketones 1 + with occasional red blood cells; plasma glucose was 196 mg/dL. Serum acetaminophen and salicylate were negative, and urine ethanol by dipstick (AOD) was approximately 100 mg/dL. By history, the previous day he drank half a bottle of Gatorade after playing outside, and soon afterward he developed nausea and emesis that persisted overnight. There was concern for methanol ingestion, but because the ethanol was already positive, and the hospital was in a "dry" county, he was transferred to a tertiary care hospital for treatment without any further intervention.

Toxicological analysis on transfer 40 min later showed a negative ethanol in serum and urine by GC, but methanol was detected in urine, and the serum concentration was 31 mg/dL approximately 22 hr post-ingestion. Ethylene glycol was negative by GC, and the serum calcium was normal. OSM was measured at 302 mOsm/kg, and OSMc was 253 mOsm/kg, giving an OG of 48 mOsm/kg. His AG was elevated at 25 mmol/L, and lactate was 11 mmol/L. The patient was treated with Fomepizole and remarkably recovered full eyesight.

This case highlights the potential danger of methanol cross-reactivity in the ethanol test, leading to the erroneous belief that the patient is already protected by antidote, but attention to the size of the OG shows it to be too large to be explained solely by the ethanol.

# FUTURE DIRECTIONS

A number of lines of development offer promise in improving automated ethanol assays. Novel fluorophores are under development as potential reaction indicators with high sensitivity. Advances in carbon nanotube technology have enabled the construction of highly stable ADH-based electrodes that are being investigated for use as biosensors. The same nanotechnology has enabled AOD incorporation into electrodes with vastly improved stability, including co-bonding into polymers with dyes for visual reaction detection [117]. Despite this, electrodes are still only reliable for a few hundred assays each [118], they become saturated at low ethanol concentrations, and the requirement for repeated dilutions negates the advantages of automation. The brewing industry has instigated many important advances in laboratory medicine throughout the years, including Kjeldahl's contribution to protein assays. The industry's scientists have manufactured point-mutated AODs with reduced activity to methanol (wild-type $K_m$ 0.62 mM, and mutant up to 2.48 mM) but without loss of velocity, and they have successfully incorporated them into amperometric detectors and biosensors with improved linear response to ethanol [119].

# CONCLUSIONS

There are a number of possibilities to be explored for improving automated tests for ethanol to increase their specificity. Manufacturers need to focus on the following five points:

1. Eliminate interference in ethanol assays from NADH generated by non-ADH reactions.
2. Eliminate toxic alcohol interference in AOD assays.
3. Find mechanisms to stabilize AOD in solution.
4. Provide rapid and reliable automated assays for:
   a. ethyl glucuronide
   b. methanol/formate
   c. isopropanol/acetone
   d. ethylene glycol/glycolate/oxalate and
   e. propylene glycol/L-lactate.
5. Continue to "open up" analyzers to permit customer modifications or use of non-branded reagents, and offer support to customers in these endeavors.

Automated ethanol assays were a remarkable advance when they were introduced three decades ago. They doubtless improved treatment outcomes and likely saved many lives, but there are many reasons why their performance should be improved. Laboratories and clinical chemists must collaborate with manufacturers by providing clinically relevant samples for method development and providing education on what is clinically desirable. Only when this happens will the current problems be resolved.

## References

[1] Van Riper T. The most expensive addictions. Forbes Business 10/3/2006. Available at <http://www.forbes.com/2006/10/02/addictions-most-expensive-biz-cx_tvr_1003addictions.html> [accessed March 2012].

[2] Lieber CS. Medical disorders of alcoholism. N Engl J Med 1995;333:1058–65.
[3] European Information System on Alcohol and Health. Available at <http://www.euro.who.int/en/what-we-do/health-topics/disease-prevention/alcohol-use/facts-and-figures> [accessed March 2012].
[4] National Institute on Alcohol Abuse and Alcoholism. <http://www.niaaa.nih.gov/publications> [accessed March 2012].
[5] Centers for Disease Control and Prevention. Alcohol-Related Disease Impact (ARDI). Atlanta, GA: CDC. <http://www.cdc.gov/alcohol/about.htm> [accessed March 2012].
[6] National Survey on Drug use and Health. SAMHSA publication Available at 2010 <http://www.oas.samhsa.gov/nhsda.htm>.
[7] Tenore PL. Advanced urine toxicology testing. J Addictive Dis 2010;29:436–48.
[8] Pletcher MJ, Maselli J, Gonzales R. Uncomplicated alcohol intoxication in the emergency department: an analysis of the National hospital ambulatory medical care survey. Am J Med 2004;117:863–7.
[9] Gentilello LM, Donovan DM, Dunn CW, Rivara FP. Alcohol interventions in trauma centers. JAMA 1995;274:1043–8.
[10] Spies C, Neuner B, Newmann T, et al. Intercurrent complications in patients admitted to the intensive care unit following trauma. Intensive Care Med 1996;22:286–93.
[11] Cicala RS. Substance abuse among physicians: what you need to know. Hosp Physician 2003;39:39–46.
[12] Trinkoff AM, Storr CL. Substance use among nurses: differences between specialties. Am J Public Health 1998;88:581–5.
[13] Spies C, Rommelspacher H. Alcohol withdrawal in the surgical patient: prevention and treatment. Anesth Analg 1999;88:946–54.
[14] Wu AH, McKay C, Broussard LA, Hoffman RS, Kwong TC, Moyer TP, et al. National academy of clinical biochemistry laboratory medicine practice guidelines: recommendations for the use of laboratory tests to support poisoned patients who present to the emergency department. Clin Chem 2003;49:357–79.
[15] Watson I. Laboratory analyses for poisoned patients: joint position paper. National poisons information service, association of clinical biochemists. Ann Clin Biochem 2002;39:328–39.
[16] Toxicology Working Group of the Alberta Clinical Practice Guideline Program. Laboratory guideline for the investigation of the poisoned patient. 2006. Available at <http://www.topalbertadoctors.org/download/316/poisoned_patient_guideline.pdf> [accessed May 2012].
[17] Widmark EMP. Eine Mikromethode zur Bestimmung von Äthylalkohol im Blut. Biochem Z 1922;181:473–5.
[18] Tangerman AT. Highly sensitive gas chromatographic analysis of ethanol in whole blood, serum, urine, and fecal supernatants by the direct injection method. Clin Chem 1997;43:1003–9.
[19] Penton Z. Gas-chromatographic determination of ethanol in blood with 0.53 mm fused-silica open tubular columns. Clin Chem 1987;33:2094–5.
[20] Dawson RB. Data for biochemical research. 3rd ed. Oxford: Clarendon; 1985. p. 122
[21] Yost DA, Boehnlein L, Shaffer M. A novel assay to determine ethanol in whole blood on the abbott TDx. Clin Chem 1984;30:1029A.
[22] Schaffar M, Stroupe SD. A general method for routine clinical chemistry in the abbott TDx analyzer. Clin Chem 1883;29:1251A.
[23] Ortho-Vitros Ethanol package insert. Available at <http://www.cmmc.org/cmmclab/IFU/ETOH_060.PDF> [accessed May 2012].
[24] Pietruszko R, Crawford K, Lester D. Comparison of substrate specificity of alcohol dehydrogenases from human liver, horse liver, and yeast towards saturated and 2-enoic alcohols and aldehydes. Arch Biochem Biophys 1973;159:50–60.
[25] Heitz JR, Anderson CD, Anderson BM. Inactivation of yeast alcohol dehydrogenase by *N*-alkylmaleimides. Arch Biochem Biophys 1968;127:627–36.
[26] Green DW, Sun HW, Plapp BV. Inversion of the substrate specificity of yeast alcohol dehydrogenase. J Biol Chem 1993;268:7792–8.
[27] Young E, Rafter-Tadgell.. Use of a competitive inhibitor in a kinetic enzymatic method for measuring ethanol in serum. Clin Chem 1987;33:2296–8.
[28] Green DW, Sun HW, Plapp BV. Inversion of the substrate specificity of yeast alcohol dehydrogenase. J Biol Chem 1993;268:7792–8.
[29] Nine JS, Moraca M, Virji M, Rao K. Serum-ethanol determination: comparison of lactate and lactate dehydrogenase interference in three enzymatic assays. J Anal Tox 1995;19:192–6.
[30] Winek CL, Wahba WW. A response to "Serum-ethanol determination: comparison of lactate and lactate dehydrogenase interference in three enzymatic assays". J Anal Tox 1996;20:211–12.
[31] Eder AF, McGrath CM, Dowdy YG, Tomaszewski JE, Rosenberg FM, Wilson RB, et al. Ethylene glycol poisoning: toxicokinetic and analytical factors affecting laboratory diagnosis. Clin Chem 1998;44:168–77.
[32] Wax P, Branton T, Cobaugh D, Kwong T. False positive ethylene glycol determination by enzyme assay in patients with chronic acetaminophen hepatotoxicity [Abstract]. J Toxicol Clin Toxicol 1999;37:604.
[33] Eder AF, Dowdy YG, Gardiner JA, Wolf BA, Shaw LM. Serum lactate and lactate dehydrogenase in high concentrations interfere in enzymatic assay of ethylene glycol. Clin Chem 1996;42:1489–91.
[34] Badcock NR, O'Reilly DA. False-positive EMIT-st ethanol screen with post-mortem infant plasma. Clin Chem 1992;38:434.
[35] Powers R, Dean D. Evaluation of potential lactate/lactate dehydrogenase interference with an enzymatic alcohol analysis. J Anal Tox 2009;33:561–3.
[36] Morgan TJ, Vlark C, Clague A. Artifactual elevation of measured plasma L-lactate concentration in the presence of glycolate. Crit Care Med 1999;27:2177–9.
[37] Castanares-Zapatero D, Filèe C, Philippe M, Hantson P. Survival with extreme lactic acidosis following ethylene glycol poisoning. Can J Anaesth 2008;55:318–19.
[38] Meng QH, Adeli K, Zello GA, Porter WH, Krahn J. Elevated lactate in ethylene glycol poisoning: true or false? Clin Chim Acta 2010;411:601–4.
[39] Manini AF, Hoffmann RS, Martin KE, Nelson LS. Relationship between serum glycolate and falsely elevated lactate in severe ethylene glycol poisoning. J Anal Toxicol 2009;33:174–6.
[40] Gharapetian A, Holmes DT, Urquhart N, Rosenberg F. Dehydrogenase interference with enzymatic ethanol assays: forgotten but not gone. Clin Chem 2008;54:1251–2.
[41] Kavanagh L, McKenna TJ, Fahie-Wilson MN, Gibney J, Smith TP. Specificity and clinical utility of methods for the detection of macroprolactin. Clin Chem 2006;52:1366–72.
[42] Wyness SP, Hunsaker JJ, La'ulu SL, Rao LV, Roberts WL. Detection of macro-creatine kinase and macro-amylase by polyethylene glycol precipitation and ultrafiltration methods. Clin Chim Acta 2011;412:2052–7.
[43] Thompson WC, Malhotra D, Schammel DP, Blackwell W, Ward ME, Dasgupta A. False positive ethanol by enzymatic assay: elimination of interference by measuring alcohol in protein-free ultra-filtrate. Clin Chem 1994;40:1594–5.
[44] Wu AH, McKay C, Broussard LA, Hoffman RS, Kwong TC, Moyer TP, et al. National academy of clinical biochemistry laboratory medicine practice guidelines: recommendations for the use of laboratory tests to support poisoned patients who present to the emergency department. Clin Chem 2003;49:357–79.

[45] Riley D, Wigmore JG, Yen B. Dilution of blood collected for medicolegal alcohol analysis by intravenous fluids. J Analyt Toxicol 1996;20:330–1.

[46] Logan BJ, Jones AW. Endogenous ethanol "auto-brewery" syndrome. Med Sci Law 2000;40:206–15.

[47] Binakonsky J, Giga N, Ross C, Siegel M. Jello shot consumption among older adolescents: a pilot study of a newly identified public health problem. Subst Use Misuse 2011;46:828–35.

[48] PRWeb. Spiked gummy candy offers teens a hiding place for alcohol. Available at <http://www.prweb.com/releases/2012/1/prweb9120244.htm>.

[49] Teen Alcohol Abuse. Vodka soaked tampons? Hoax? Available at <http://www.teenalcoholabuse.us/content/vodka-soaked-tampons.html>.

[50] Jones AW, Rajs J. Appreciable blood-ethanol concentration after washing abraised and lacerated shin with surgical spirit. J Anal Toxicol 1997;21:587–8.

[51] Püschel K. Percutaneous alcohol intoxication. Eur J Pediatr 1981;136:317–18.

[52] Harpin V, Rutter N. Percutaneous alcohol absorption and skin necrosis in a preterm infant. Arch Dis Child 1982;57:477–9.

[53] Foley KF. A 25-proof urine. Clin Chem 2011;57:142–3.

[54] Alexander WD, Wills PD, Eldred N. Urinary ethanol and diabetes mellitus. Diabet Med 1988;5:463–4.

[55] Norberg A, Jones AW, Hahn RG, Gabrielsson JL. Role of variability in explaining ethanol pharmacokinetics: research and forensic applications. Clin Pharmacokinet 2003;42:1–31.

[56] Caro AA, Cederbaum AI. Oxidative stress, toxicology, and pharmacology of CYP2E1. Annu Rev Pharmacol Toxicol. 2004;44:22–42.

[57] Agarwal DP. Genetic polymorphisms of alcohol metabolizing enzymes. Path Biol 2001;49:703–9.

[58] Freeza M, Di Padova C, Pozzato G, Terpin M, Baraona E, Lieber CS. High blood alcohol levels in women: the role of decreased gastric alcohol dehydrogenase activity and first-pass metabolism. N Engl J Med 1990;32:95–9.

[59] Brennan DF, Betzelos S, Reed R. Ethanol elimination rates in an ED population. Am J Emerg Med 1995;13:276–80.

[60] Gershman H, Steeper J. Rate of clearance of ethanol from the blood of intoxicated patients in the emergency department. J Emerg Med 1991;9:307–11.

[61] Jones AW. Ethanol metabolism in patients with liver cirrhosis. J Clin Forens Med 2000;7:48–51.

[62] Pitzele HZ, Tolia VM. Twenty per hour: altered mental state due to ethanol abuse and withdrawal. Emerg Med Clin N Am 2010;28:683–705.

[63] Agarwal DP. Genetic polymorphisms of alcohol metabolizing enzymes. Path Biol 2001;49:703–9.

[64] Sultatos LG, Pastino GM, Rosenfeld CA, Flynn EJ. Incorporation of the genetic control of alcohol dehydrogenase into a physiologically based pharmacokinetic model for ethanol in humans. Toxicological Sci 2004;78:20–31.

[65] Eriksson CJP, Fukunaga T, Sarkola T, Chen WJ, Chen CC, Ju JM, et al. Functional relevance of human ADH polymorphism. Alcohol Clin Exp Res 2001;25:157S–63S.

[66] Smithline N, Gardner Jr. KD. Gaps—Anionic and osmolal. JAMA 1976;236:1594–7.

[67] Purssell RA, Pudek M, Brubacher J, Abu-Laban RB. Derivation and validation of a formula to calculate the contribution of ethanol to the osmolal gap. Ann Emerg Med 2001;38:653–9.

[68] Griveas I, Gompou A, Kyristis I, Papatheodorou G, Agroyannis I, Tsakoniatis M, et al. Osmolal gap in hemodialyzed uremic patients. Artificial Organs 2011;36:16–20.

[69] Lund ME, Banner Jr W, Finley PR, Burnham L, Dye JA. Effect of alcohols and selected solvents on serum osmolality measurements. J Toxicol Clin Toxicol 1983;20:115–32.

[70] Krahn J, Khajuria A. Osmolality gaps: diagnostic accuracy and long-term variability. Clin Chem 2006;52:737–9.

[71] Purssell RA, Pudek M, Brubacher J, Abu-Laban RB. Derivation and validation of a formula to calculate the contribution of ethanol to the osmolal gap. Ann Emerg Med 2001;38:653–9.

[72] Lund ME, Banner Jr W, Finley PR, Burnham L, Dye JA. Effect of alcohols and selected solvents on serum osmolality measurements. J Toxicol Clin Toxicol 1983;20:115–32.

[73] Bhagat CI, Beilby JP, Garcia-Webb P, Dusci LJ. Errors in estimating ethanol concentration in plasma using the "osmolal gap". Clin Chem 1985;31:647–8.

[74] McQuillen KK, Anderson AC. Osmol gaps in the pediatric population. Acad Emerg Med 1999;6:27–30.

[75] Draviam EJ, Custer EM, Schoen I. Vapor pressure and freezing point osmolality measurements applied to a volatile screen. Am J Clin Pathol 1984;82:706–9.

[76] Walker JA, Schwartzbard A, Krauss EA, Sherman RA, Eisinger RP. The missing gap: a pitfall in the diagnosis of alcohol intoxication by osmometry. Arch Intern Med 1986;146:1843–4.

[77] Jones AW. Ethanol distribution ratios between urine and capillary blood in controlled experiments and in apprehended drinking drivers. J Forens Sci 1992;37:2–34.

[78] Jones AW. Urine as a biological specimen for forensic analysis of alcohol and variability in the urine-to-blood relationship. Toxicol Rev 2006;25:15–35.

[79] Federal Register Vol. 72, No. 241/Monday, December 17, 2007/Notices.

[80] Jones AW, Pounder DJ. Measuring blood alcohol concentrations for clinical and forensic purposes. In: Karch SB, editor. Drug Abuse Handbook. Boca Raton FL: CRC Press; 1988.

[81] Huestis MA, Verstraete A, Kwong TC, Morland J, Vincent MJ, de la Torre R. Oral fluid testing: promises and pitfalls. Clin Chem 2011;57:805–10.

[82] Schwartz RH, O'Donnell RM, Thorne MM, Getson PR, Hicks JM. Evaluation of colorimetric dipstick test to detect alcohol in saliva: a pilot study. Ann Emerg Med 1989;18:1001–3.

[83] Christopher TA, Zaccardi JA. Evaluation of the QED saliva alcohol test: a new, rapid, accurate device for measuring ethanol in saliva. Ann Emerg Med 1992;21:1135–7.

[84] Nakae Y, Stoward PJ. The initial reaction velocities of lactate dehydrogenase in various cell types. Histochem J 1994;26:283–91.

[85] Solomon J, Vanga N, Morgan JP, Joseph P. Emergency room physicians: recognition of alcohol misuse. J Stud Alcohol 1980;41:583–6.

[86] Sharpe PC. Biochemical detection and monitoring of alcohol abuse and abstinence. Ann Clin Biochem 2001;38:652–64.

[87] Hannuksela ML, Liianantti MK, Nissinen AE, Savolainen MJ. Biochemical markers of alcoholism. Clin Chem Lab Med 2007;45:953–61.

[88] Böttcher M, Beck O, Heelander A. Evaluation of a new immunoassay for urinary ethyl glucuronide testing. Alcohol Alcohol 2008;43:46–8.

[89] Neumann T, Helander A, Dahl H, et al. Value of ethyl glucuronide in plasma as a biomarker of recent alcohol consumption in the emergency room. Alcohol Alcohol 2008;43:431–5.

[90] Sharpe PC, McBride R, Archbold GPR. Biochemical markers of alcohol abuse. Q J Med 1996;89:137–44.

[91] Chick J, Kreitman N, Plant M. Mean cell volume and gammaglutamyl-transferase as markers of drinking in working men. Lancet 1981;1(8232):1249–51.

[92] Bernadt MW, Mumford J, Taylor C, Smith B, Murray RM. Comparison of questionnaire and laboratory tests in the detection of excessive drinking and alcoholism. Lancet 1982;1 (8267):325–8.

[93] Rosalki SB, Rau D. Serum gamma-glutamyl transpeptidase activity in alcoholism. Clin Chim Acta 1972;39:41–7.

[94] Rosman AS, Lieber CS. Biochemical markers of alcohol consumption. Alcohol Health Res World 1990;14:210–18.

[95] Chronic alcohol abuse. Chromsystems Dialog. 2006. Available at <http://www.chromsystems.com/news-en/dialog/dialog_2006_1_e.pdf> [accessed May 2012].

[96] Schellenberg F, Wielders JP. Evaluation of capillary electrophoresis for CDT on SEBIA's capillarys system: intra and interlaboratory precision, reference interval and cut-off. Clin Chim Acta 2010;411:1888–93.

[97] Hacker R, Torsten A, Helwig-Rolig A, Kropf J, Steinmetz A, Schaefer JR. Investigation by isoelectric focusing of the initial carbohydrate-deficient transferrin (CDT) and non-CDT transferrin isoform fractionation step involved in determination of CDT by the ChronAlcoI.D. assay. Clin Chem 2000;46:483–92.

[98] Schwarz MJ, Domke I, Helander A, Janssens PMW, van Pelt J, Springer B, et al. Multicenter evaluation of a new assay for the determination of carbohydrate deficient transferrin. Alcohol Alcohol 2003;38:270–5.

[99] Huseby NE, Bjordal E, Nilssen O, Barth T. Utility of biological markers during outpatient treatment of alcohol-dependent subjects: carbohydrate-deficient transferrin responds to moderate changes in alcohol consumption. Alcohol Clin Res 1997;21:1343–6.

[100] Reynaud M, Hourcade F, Planche F, Albuisson E, Meunier MN, Planche R. Usefulness of carbohydrate-deficient transferrin in alcoholic patients with normal γ-glutamyltransferase. Alcohol Clin Exp Res 1998;22:615–18.

[101] Lynd DL, Richardson KJ, Pursse RA, Abu-Laban RB, Brubacher JR, Lepik KJ, et al. An evaluation of the osmole gap as a screening test for toxic alcohol poisoning. BMC Emerg Med 2008;8:5–14.

[102] Wu AH, McKay C, Broussard LA, Hoffman RS, Kwong TC, Moyer TP, et al. National academy of clinical biochemistry laboratory medicine practice guidelines: recommendations for the use of laboratory tests to support poisoned patients who present to the emergency department. Clin Chem 2003;49:357–79.

[103] Vinet B. An enzymic assay for the specific determination of methanol in serum. Clin Chem 1987;33:2204–8.

[104] Hovda KE, Urdal P, Jacobsen D. Case report: increased serum formate in the diagnosis of methanol poisoning. J Analyt Toxicol 2005;29:586–8.

[105] Long H, Nelson LS, Hoffman RS. A rapid qualitative test for suspected ethylene glycol poisoning. Acad Emerg Med 2008;15:688–90.

[106] Shin JM, Sachs G, Kraut JA. Simple diagnostic tests to detect toxic alcohol intoxications. Transl Res 2008;152:194–201.

[107] Urdal P. Enzymatic assay for oxalate in unprocessed urine, as adapted for a centrifugal analyzer. Clin Chem 1984;30:911–13.

[108] Lund ME, Banner Jr W, Finley PR, Burnham L, Dye JA. Effect of alcohols and selected solvents on serum osmolality measurements. J Toxicol Clin Toxicol 1983;20:115–32.

[109] Draviam EJ, Custer EM, Schoen I. Vapor pressure and freezing point osmolality measurements applied to a volatile screen. Am J Clin Pathol 1984;82:706–9.

[110] Hack JB, Chiang WK, Howland MA, Patel H, Goldfrank L. The utility of an alcohol oxidase reaction test to expedite the detection of toxic alcohol exposures. Acad Emerg Med 2000;7:294–7.

[111] Shirey T, Silvetti M. Reaction of lactate electrodes to glycolate. Crit Care Med 1999;27:2305–7.

[112] Blandford SE, Desjardins RP. A rapid method for measurement of ethylene glycol. Clin Biochem 1994;27:25–30.

[113] Yaucher NE, Fish JT, Smith HL, Wells JA. Propylene glycol-associated renal toxicity from lorazepam infusion. Pharmacother 2003;23:1094–9.

[114] Hayman M, Seidl EC, Ali M, Malik K. Acute tubular necrosis associated with propylene glycol from concomitant administration of intravenous lorazepam and trimethoprim-sulfamethoxazole. Pharmacotherapy 2003;23(9):1190–4.

[115] Vasiliades J, Pollock J, Robinson CA. Pitfalls of the alcohol dehydrogenase procedure for the emergency detection of alcohol: a case study of isopropanol overdose. Clin Chem 1978;24:383–5.

[116] Platteborze PL, Rainey PM, Baird GS. Ketoacidosis with unexpected serum isopropyl alcohol. Clin Chem 2011;57:1361–5.

[117] Ho Y-H, Periasamy AP, Shen-Ming Chen S-M. Amperometric ethanol biosensor based on alcohol dehydrogenase immobilized at poly-L-lysine coated carminic acid functionalized multiwalled carbon nanotube film. Int J Electrochem Sci 2011;6:3922–37.

[118] Guilbault GG, Danielsson B, Mandenius CF, Mosback K. Enzyme electrode and thermistor probes for the determination of alcohols with alcohol oxidase. Anal Chem 1983;55:1582–5.

[119] Dmytruk KV, Smutok OV, Ryabova OB, Gayda GZ, Sibirny VA, Schuhmann W, et al. Isolation and characterization of mutated alcohol oxidases from the yeast *Hansenula polymorpha* with decreased activity toward substrates and their use as selective elements in an amperometric biosensor. BMC Biotechnol 2007;7:33–9.

CHAPTER

# 17

# Pre-Analytical Issues and Interferences in Transfusion Medicine Tests

*Elena Nedelcu*

University of Texas Health Sciences Center at Houston, Houston, Texas

## INTRODUCTION

Blood transfusion risks are generally categorized as infectious and noninfectious. Infectious risks associated with the likelihood to acquire transfusion-transmitted infectious diseases (TTID) are prevented in two steps. Initially, prospective blood donors are screened with a predonation questionnaire, and individuals identified at risk for TTID are deferred from blood donation. If no TTID risk is identified in this screening phase, individuals are accepted for donation. Donated blood is then tested for major TTID as well as for ABO/Rh typing. Testing for known TTID is performed in two steps—a screening and a confirmatory test. Screening for major TTID such as hepatitis B and C, HIV-1 and -2, HTLV-I and -II, *Treponema pallidum*, *Trypanosoma cruzi*, and West Nile virus is routinely performed. If positive in the screening phase, confirmatory tests are performed, but regardless of their results, the donor status is deferred and the blood discarded. If screening test results are negative, the blood is released for further transfusion. Noninfectious risks of blood transfusions are associated with adverse reactions occurring after administration of blood products. Some of the acute adverse reactions (acute hemolytic transfusion reaction, bacterial contamination or sepsis, and transfusion-related lung injury) and transfusion-related acute graft-versus-host disease, a delayed adverse event of transfusion, are potentially fatal [1,2]. Transfusion-related immunomodulation, another delayed adverse event of blood transfusion, has been associated with worse clinical outcome in transfused patients compared with that of nontransfused. For these reasons, prevention of transfusion risks begins with ensuring that no transfusions are performed unless necessary, a step in the whole transfusion process that involves ordering.

Ordering a transfusion medicine test or a blood product is initiated, as in the case of most tests initiated, by a clinician. Most routine orders are "blood type," "type and screen" (T&S), and "type and cross" (T&C). Blood type is performed to determine the ABO/Rh(D) typing of a patient. T&S is ordered usually when a blood transfusion is likely, but not definite, and includes the ABO/Rh(D) typing and antibody screen performed on blood samples of the recipient. T&C order indicates that a transfusion is imminent, so complete pre-transfusion tests, including ABO/Rh(D) typing, antibody screen, and crossmatch, are done for the number of units requested.

The importance of the next step, blood sample collection from the right patient, cannot be emphasized enough. This step is crucial in transfusion medicine perhaps even more than in any other area of laboratory medicine because errors occurring within this phase may result in issuing the wrong type of blood. Blood sample collection should be performed directly by the phlebotomist after correct patient identification by full name and hospital identification number (attached wristband) and matching with physician's order. It is paramount that the collected blood sample is labeled at the bedside and not elsewhere or prelabeled due to the high potential of mislabeling or sample mix-up. Correct labeling includes patient's full name, hospital number, ordering physician's name, date and time of collection, and phlebotomist's initials. Blood is collected in siliconized plain tubes with no additives or in tubes with potassium ethylenediaminotetraacetic acid (K2-EDTA) as

*Accurate Results in the Clinical Laboratory.*
DOI: http://dx.doi.org/10.1016/B978-0-12-415783-5.00017-7

anticoagulant. If the sample is drawn from an intravenous (i.v.) line, the i.v. infusion should be stopped 5–10 min prior to blood drawing and the first 10 mL discarded to prevent blood contamination with fluids that may cause hemolysis. A tube should be collected full to ensure an appropriate blood:anticoagulant ratio, inverted 3–10 times for proper mixing, and transported immediately to the laboratory. Fewer inversions may lead to clotted samples, whereas excessive shaking, exposure to excessive heat or cold temperature, as well as delays in transportation or even routine transportation via pneumatic tube system may all lead to sample rejection. Samples should be collected within less than 72 hr from a scheduled transfusion; otherwise, complement-dependent antibodies may be missed due to complement becoming unstable. New blood samples from the recipient are needed for repeat pre-transfusion testing every 72 hr if new transfusion orders are made. The age of the samples may be extended to up to 1 month in certain clinical settings, such as preoperative evaluation of elective surgery patients if they have a negative antibody screen and no red blood cell (RBC) exposure via pregnancy or blood transfusions within the past 3 months. Blood samples and segments from the transfused units are stored at 1–6 °C for 7–10 days post-transfusion [2,3]. Correctly labeled samples accompanied by requisition form are sent to the blood bank, and a careful clerical check is performed to confirm that the information on the label and on the transfusion request (requisition form) is identical. Patient history is reviewed for previous blood transfusion, immunohematology testing, or transfusion reactions. Blood samples are rejected based on specific criteria if deemed improperly labeled or inadequate for testing. The presence of hemolysis is perhaps the most common reason for sample rejection in the blood bank because it cannot be distinguished from the antibody-mediated hemolysis, which is considered as positive reaction in most immunohematology assays. A hemolyzed or grossly lipemic sample may give false-positive results. If both clerical check and sample are acceptable, the specimen is processed by immediate centrifugation, and the RBCs and supernatant are separated. Documentation of all steps from receipt of specimen to testing and result interpretation is performed manually or electronically, and records are kept confidential for a period of time in accordance with requirements of federal, state, and accrediting agencies.

## BASICS OF IMMUNOHEMATOLOGY

Immunohematology is the study of RBC antigens and antibodies associated with blood transfusions. There are more than 230 types of antigens present on the surface of RBCs that, based on their chemical structure, can be grouped into two major categories—carbohydrates and polypeptides. The RBC antigen formation is encoded by specific genes inherited from parents and categorized in *blood group systems* if genes are known and found on closely located loci and in *blood group collections* if the genes responsible for their formation have not yet been discovered.

The ABO blood group system was the first one discovered at the beginning of the 20th century, and it is still considered the most important antigenic system mostly because the ABO mismatch is potentially fatal. Other major blood group systems are Rh, Kell, Kidd, Duffy, Lutheran, and MNS. The presence or absence on the RBC surface of specific antigens gives an individual antigenic profile or *RBC phenotype*. For routine transfusions, blood is typed only for ABO and Rh blood group systems. "Rh typing" is a misnomer because it does not involve phenotyping for all major antigens belonging to the Rh system but, rather, only for the D antigen, the most immunogenic of all. Therefore, Rh-positive or Rh-negative should read as D-positive or D-negative, respectively. Determination of the full RBC antigenic profile (RBC phenotype) is not routinely performed, but it is required for correct selection of blood products in certain situations in which antibodies are identified in recipient's plasma.

Transfusion of ABO-compatible blood is required at all times because the ABO blood group system has preformed antibodies capable of causing hemolysis with release of free hemoglobin and other intracellular components into the circulating plasma, leading to renal failure, activation of the coagulation system, and potentially death. Other preformed antibodies, also called "naturally occurring" or non-RBC stimulated, are formed against RBC antigens belonging to the carbohydrate group. These antibodies are not considered clinically significant because they do not usually react at body temperature. Unlike antibodies against RBC antigens belonging to the carbohydrate group, the antibodies against polypeptide RBC antigens are not preformed but require prior RBC exposure (sensitization) via prior transfusion or pregnancy for their formation. A second exposure to the same RBC antigen results in RBC coating with antibody and/or complement fractions, shortened RBC life span with hemolysis.

The principle of most immunohematology tests is the antigen–antibody reaction leading to RBC agglutination or hemolysis. If any of these occur, the reaction should be interpreted as positive. Hemolysis is considered the strongest positive reaction that can occur and indicates the presence of a potent, complement-fixing antibody, but it is not commonly seen. The RBC agglutination is most commonly seen as a sign of a positive reaction in transfusion medicine tests. RBC agglutination occurs in two phases. In the first phase, also known as RBC sensitization, a reversible reaction occurs between the paratope and the epitope held together by noncovalent attraction, whereas in the second phase, RBCs with bound antibodies form a stable lattice via antigen–antibody bridges bound to different RBCs. The formation of this lattice is naturally prevented by the net negative charge of the RBC membrane created by sialic acid. If suspended in an ionic medium, a charge difference or electrical potential forms between cations closest to the RBC membrane and the outer cations that move more freely in the solution. This is called the $\zeta$ potential, and it is responsible for keeping the RBCs in solution approximately 25 nm apart. Factors affecting the RBC agglutination include the inherent characteristics of the specific antigens and antibodies involved, the density and accessibility of specific antigens on the RBC membrane, the antibody isotype and concentration, as well as physical and/or chemical parameters of the reaction environment (temperature, incubation time, and ionic strength of the solution). The impact of the RBC antigen density is illustrated by the ease with which a reaction occurs with A or B antigens, which are present in hundreds of thousands to millions of molecules per RBC. Another example is illustrated by the so-called "dosage effect," meaning stronger antibody reaction with RBCs that are homozygous than with heterozygous RBCs for a particular antigen because homozygous RBCs express twice the number of antigen molecules per cell. The impact of the antibody isotype is mostly due to the difference in size of various classes of immunoglobulins. For example, because they are large molecules (35 nm), the IgM antibodies can easily reach two adjacent RBCs in solution; therefore, they have an intrinsic agglutinin ability and are thus referred to as *direct agglutinins*. In contrast, most IgG antibodies are smaller (14 nm) and unable to induce agglutination without an enhancement of the reaction, which is why they are referred to as *indirect agglutinins* [1–4]. In case of a negative immunohematology reaction, an extra quality control step is performed using reagent RBCs known to give a positive reaction (also called "check cells"). Check cells are commercially available reagent RBCs known to give a positive reaction and tested exactly as the patient's RBCs. A positive test result with check cells validates the patient's negative test result. If the check cells do not give a positive reaction, as expected, the patient's results cannot be considered truly negative, and the test should be repeated.

# METHODOLOGY USED FOR IMMUNOHEMATOLOGY TESTS

Routine blood bank testing has been dependent on the technology used for detection of antigen–antibody reactions. When interpreting the test results, it is important to know the advantages and limitations of each methodology, as well as the type of reagents and antibodies used (monoclonal vs. polyclonal) in the commercially available kits. If testing for plasma antibodies is performed in a test tube, reagents and patient sample are added to the tube and centrifuged, resulting in an RBC pellet at the bottom of the tube. The presence of hemolysis or RBC agglutination indicates a positive reaction. If present, agglutination is observed after gently dislodging the RBC pellet. The agglutination is graded according to the strength of the reaction from 0 to $4^+$. An example of grading based on the visible agglutination pattern for reactions performed in test tubes is illustrated in Table 17.1. Macroscopically negative reactions are reviewed under microscopic magnification to distinguish between true negative and microscopically positive ($m^+$) reactions. Appropriate grading systems are used for other testing methodologies.

Other methods used for detection of antigen–antibody reactions are column agglutination (gel technology) and solid-phase testing. Briefly, gel technology is the most widely used method and utilizes a dextran acrylamide gel microtube containing anti-human IgG to which plasma and reagent RBC of the antibody screen kit are added. If plasma contains antibodies reacting with any screen RBCs antigens, agglutinates form and are trapped at the top of the microtube, indicating a positive reaction. If antibodies are not present in a patient's plasma, agglutination does not occur and the RBCs pass through the gel layer and form a pellet at the bottom of the microtube, indicating a negative

**TABLE 17.1** Grading the Agglutination (or Reaction Strength) in Test Tubes

| Appearance | Grade |
|---|---|
| One solid RBC aggregate | 4+ |
| Several medium to large RBC aggregates | 3+ |
| Many small aggregates with a clear background | 2+ |
| Many small aggregates with a turbid background | 1+ |
| Few small aggregates with many unagglutinated RBCs | w+ |
| Small RBC aggregates visible only under microscopic magnification | m+ |
| Negative (absence of aggregates, dispersed RBCs) | 0 |

reaction. In solid-phase technology, the patient's plasma is added to a microplate coated with a monolayer of reagent RBCs. Following incubation and washing, indicator RBCs are added. If RBC agglutination occurs, there is a diffuse adherence of RBC to the well. If no reaction occurs, the indicator RBCs settle to the bottom of the well [4,5]. Knowledge of the type of reagents used for testing (polyclonal vs. monoclonal antisera) is also critical in interpretation [6]. Polyclonal sera may contain anti-A, anti-B and anti-A,B, anti-T or Tn antibodies and may give more false-positive results than monoclonal reagents. Unexpected reactivity may also be detected, especially if polyclonal antisera originate from multiparous women or immunized donors. The issues described with polyclonal reagents are less common today because most of the reagents used are monoclonal.

Overall, interpretation of transfusion medicine tests requires knowledge of the basics of immunohematology reactions and methodology used and also how other factors, such as potentiators and enzymes, may affect this reactivity. Recognition of false-positive reactions or pseudoagglutination, which can be misinterpreted as true agglutination, is important to avoid extended testing, which may delay or unnecessarily increase the cost of transfusion. The presence of fibrin clots or rouleaux formation mimics RBC agglutination and may also be misinterpreted as positive reaction. Knowledge of interferences occurring in blood testing is critical for issuing the correct blood in a timely manner. Within this process, the patient's medical history and blood bank history are essential factors for understanding test results and deciding if further testing is required because the same test results can occur in different clinical scenarios, but the recommendations for transfusion might be different.

## PRE-TRANSFUSION EVALUATION

Pre-transfusion evaluation includes a review of the patient's medical history, prior blood transfusions, pregnancies, and blood bank testing (history of antibody against RBC antigens), clinical diagnosis, and assessment of transfusion urgency. All this information variably modulates the selection decision regarding the blood type to be transfused. For emergent situations, such as life-threatening bleeding, when there is no time for performing tests, a decision is made to release universal RBC (O Rh-negative) or plasma (AB type) because the risk of bleeding outweighs the risk of hemolysis due to antibodies that might be present in the patient's plasma. Excluding these cases, immunohematology tests are performed in the blood bank as part of the pre-transfusion evaluation for each unit transfused.

Pre-transfusion testing includes ABO/Rh typing performed on blood samples of recipient and donor, antibody screen, and the *in vitro* compatibility testing (crossmatch). The aim of this test set is to ensure that the transfused blood is compatible with the recipient's blood and no harm is done to the patient. Results are compared with historical ones, if available, and if any discrepancies occur, they must be resolved before blood can be released for transfusion. In general, when the antibody screen is positive, further testing is performed to determine whether there is any specificity against major RBC antigens. Antibodies against RBC antigens are known to be clinically significant or insignificant based on their known propensity to produce immune hemolysis and to cause hemolytic transfusion reactions and hemolytic disease of the fetus and newborn (HDFN). If the alloantibodies against RBCs are clinically insignificant alloantibodies, crossmatch-compatible RBC units can be released [2–4,7]. When clinically significant alloantibodies are detected in a patient's plasma, antigen-negative RBC units should be selected and issued for that particular patient.

### Determination of ABO/Rh(D) Type

Prior to any transfusion, the patient's previous ABO/Rh(D) type is checked in the medical records. If not previously performed, two samples are required for blood type confirmation. If previously performed but not within the past 72 hr, another blood sample is required for confirmation. The results obtained with the new blood sample are checked against the previous results and, if they yield the same blood type result, are valid only for 72 hr, after which a new sample is required for testing. If the blood type does not match the one recorded in the past, an ABO discrepancy workup is initiated and should be resolved prior to blood release, unless there is an emergent need for transfusion, in which case universal blood products are issued. Other relevant blood bank history, such as prior detection of antibodies, is also critical because it is a known phenomenon that antibody levels may become undetectable with time. However, if a clinically significant antibody was once detected, the patient must continue to receive RBC units negative for the corresponding antigen even if current testing is negative in order to prevent delayed hemolytic transfusion reactions. The patient's clinical history is critical also because specific medical conditions require administration of special blood products, such as CMV-negative, leukoreduced, or irradiated units. The requirement of special blood products, however, is neither directly dependent on nor interferes with the immunohematology testing.

Determination of the ABO blood group is performed in two steps, a forward and a reverse reaction. In the forward reaction, the presence of the A or/and B antigens is determined by mixing the RBC to be tested with licensed anti-A and anti-B reagents. In the reverse reaction, the serum is mixed with reagent RBCs of A1 and B types to detect the presence of anti-A1 and anti-B antibodies. The reactivity strength between the AB antigens and corresponding antibodies is naturally strong and usually graded as 4+. If the forward and the reverse reactions lead to different conclusions regarding the ABO type, or if the reaction strength is weaker than expected, and/or the historical blood type does not match the current one, the cause of the discrepancy must be fully investigated for a final interpretation of the ABO blood group type. The forward and reverse ABO typing is performed on donated blood, and the forward reaction is confirmed prior to transfusion using blood contained in a segment attached to the donor unit. Both forward and reverse reactions are performed on recipient and donor blood as part of the pre-transfusion testing.

The Rh or D type is determined in a similar form as the forward ABO typing but using appropriate reagents for detection of D antigen. For D typing, a negative or positive reaction occurring in the reverse concurrent ABO reaction may serve as a negative or positive control, respectively. If D type appears positive in an AB patient who does not have anti-A or anti-B, the D typing should be repeated with an inert control reagent (e.g., bovine serum albumin) to confirm a true D-positive reaction. Blood donors who are type Rh-negative (D-negative) are further tested to detect the presence of so-called "weak D antigen" (Du) using more sensitive methods, such as the presence of anti-human globulin, which acts as an enhancer of the reaction between a D antigen that is present but not detected by the routine (direct) method. Direct typing for the D antigen is performed in test tubes at room temperature (RT) by the immediate spin technique after mixing one drop of a 3–5% suspension of the patient's RBCs with one drop of commercial anti-D. Anti-D reagents used in the past were pooled from human sera and associated with false-positive reactions due to high protein concentration. Current anti-D reagents used in tube testing contain a saline-reactive blend of polyclonal and murine monoclonal IgM. Other commercially available reagents used for both direct and weak D testing contain a blend of monoclonal IgM and monoclonal IgG anti-D from human/murine heterohybridoma cell lines or a blend containing murine monoclonal IgM anti-D and human polyclonal IgG anti-D. Determination of the weak D is required only for blood donors in order to establish the true D status and is performed by testing the donor or patient's RBCs with IgG anti-D in the anti-human globulin (AHG) phase. If the weak D testing is positive, the Rh(D) type is interpreted as Rh(D)-positive, and if it is negative, the Rh(D) type is interpreted as a true negative. A parallel test with a negative diluent control is always included for the weak D test. All RBC units labeled Rh(D)-negative are re-tested for the D antigen as part of the pre-transfusion testing using blood from a segment attached to the donor unit. Repeat testing of donor units by the transfusion service for weak D is not required by the AABB standards. The weak D status is neither required nor routinely determined in recipients because Rh(D)-negative blood is safe to be transfused regardless of the true Rh(D) status of the recipient. Weak D testing may be performed in certain instances, such as in fetuses/newborns of Rh(D)-negative mothers or when there is a suspicion that the patient possesses a partial D antigen and is at risk of alloimmunization. The presence of other antigens belonging to the Rh blood group is not determined for routine transfusions.

## Detection of Antibodies

Antibodies are initially detected by the antibody screen and confirmed by additional testing.

### *Antibody Screen*

The antibody screen is performed by mixing a patient's plasma with three reagent RBCs with a known phenotype present in a commercially available kit containing a table of the antigenic profiles of the RBCs used, called antigram. AABB requires that reagent RBCs present in the antibody screen should contain at least one homozygous RBC reagent positive for the following major RBC antigens: C, c, D, E, e, $Fy^a$, $Fy^b$, $Jk^a$, $Jk^b$, K, k, $Le^a$, $Le^b$, $P_1$, M, N, S, and s antigens. If the patient's plasma contains alloantibodies against major RBC antigens, the antibody screen becomes positive and further testing by an extended panel is required. The antibody screen and panel are both indirect anti-human globulin tests (IATs) performed with reagent RBCs prepared from donors of type O so that the naturally occurring anti-A or anti-B antibodies in the patient's plasma would not interfere with the testing. Other IATs are the crossmatch test, in which the patient's plasma is mixed with donor RBCs obtained from a segment attached to the unit to be transfused, and the RBC phenotype, in which the patient's RBCs are mixed with commercially available antisera against major RBC antigens in order to determine if they are

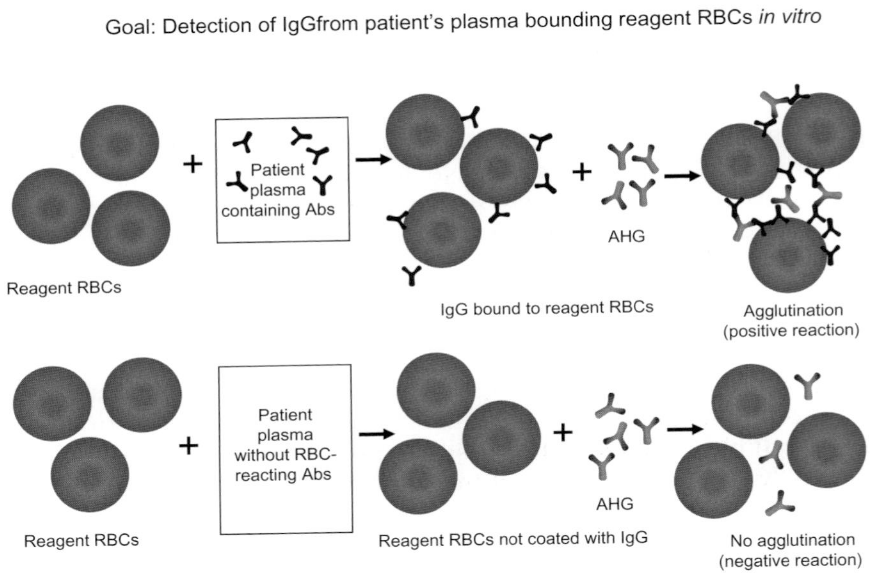

**FIGURE 17.1** Illustration of the principle of the indirect antiglobulin test.

present or not. The principle of any IAT is illustrated in Figure 17.1.

The plasma reactivity detected *in vitro* in the antibody screen and panel may be due to various causes, immune or nonimmune, and should be distinguished from false-positive reactions. The presence of false-negative reactions and technical errors should also be ruled out [8]. The reaction conditions leading to increased or decreased detection of antibodies play a significant role in blood bank testing and results interpretation [9–11]. Extended and additional testing is performed with the goal of identifying whether plasma reactivity is due to an antibody with specificity against a major RBC antigen. Alloantibodies against a certain RBC antigen may develop if individuals with antigen-negative RBCs are exposed to RBCs possessing that antigen via transfusion, pregnancy, or transplant.

### *Antibody Identification by Extended (Panel) Testing*

Positive antibody screens require further testing by testing the patient's plasma against a panel of 10–12 RBCs of varying phenotypes. The identification of antibody specificity is accomplished in definite steps. Briefly, in a first step, the antigens present on cells that do not react with patient's plasma are crossed out based on the assumption that if the corresponding antibody would have been present in patient's plasma, it would have resulted in a positive reaction. This is only an assumption; in reality, antibodies reactivity is not always typical, and sometimes testing additional cells is needed to achieve an acceptable level of confidence. If three homozygous RBCs do not react with a patient's plasma, there is a less than 5% chance that the plasma reactivity may be due to the corresponding antibody. After the crossing out is complete, a list of antibodies that cannot be ruled out is made and additional testing, if needed, is performed. In the next step, called "ruling in," the plasma reactivity pattern is compared with the profile of antigens across all cell lines. If at least three cell lines react with the patient's plasma, there is a 95% chance that the reactivity is due to an antibody corresponding to that antigen. Antibodies against major RBC antigens should be ruled out, and any additional reactivity should be explained, if possible. However, often, even if the major alloantibodies are ruled out, an extra-weak reactivity might still be present. The clinical significance of this weakly reactive nonspecific (WRNS) antibody is most likely limited if the patient does not have a history of prior RBC exposure, in which case crossmatch-compatible RBC units are provided for the patient. However, if the patient had prior RBC sensitization via pregnancy or recent transfusion, the clinical significance of RBC is indeterminate, meaning that a developing alloantibody with partial reactivity cannot be completely ruled out. In this case, additional testing, including determination of the patient's RBC phenotype, may be indicated, and the safest blood to be provided for that patient would match the patient's antigen profile. It is required by AABB that the antibody testing be performed in the presence of anti-human globulin, also called "in AHG phase." A negative test does not completely rule out the presence of antibodies against major RBC antigens due to two factors. One is the intrinsic limitation of the test, with a minimum of 100–500 IgG or C3 molecules bound per cell being required for detection of a positive reaction. The second is the reaction variability due to different subclasses of IgG.

When plasma reacts with all reagent RBCs tested (pan-reactivity), the process of ruling out antibodies against major RBC antigens cannot be performed because no negative reactions are found. If plasma also reacts with all reagent RBCs and the patient's own RBCs (positive autocontrol), an autoantibody is suspected. If the autocontrol is negative, an antibody against a high-incidence (HI) antigen is likely in a case with no recent history of blood transfusions. HI antigens are present on most, if not all, reagent RBCs supplied on the antibody screen and panel kits. In the case of autoantibodies, a determination of the antibody reactivity at various temperatures may be needed. Based on the reactivity pattern at RT, 37 °C, or AHG phase, a conclusion can usually be made about the autoantibody type. Briefly, cold autoantibodies (CAA) typically react $2^+$ or mixed-field (mf) at RT but not at 37 °C, whereas the warm autoantibodies (WAA) strongly react ($4^+$) at 37 °C and AHG phase. As previously highlighted, knowledge of the methodology used for the antibody identification is critical for a correct interpretation. If the antibody screen is performed in test tubes, the patient's plasma is admixed with screening reagent RBCs prepared in a 3–5% cell suspension and examined for hemolysis and agglutination immediately after centrifugation (immediate spin phase), after incubation at 37 °C and washing, or after saline wash and addition of AHG (AHG phase). Therefore, in general, antibodies against different types of RBC antigens are recognized by their characteristic pattern of reaction [4,12].

The most common method used for antibody detection is column agglutination technology. This platform (ID-Micro Typing System gel agglutination test, Ortho-Clinical Diagnostics, Raritan, NJ) utilizes card sets of six microtubes containing dextran acylamide infused with anti-IgG. They are used for antibody detection and identification, and also for ABO/Rh typing, crossmatching, and direct antiglobulin testing. Screening RBCs in 0.8% cell suspension and plasma to be tested are added to microcolumns, incubated at 37 °C and then centrifuged. If alloantibodies against RBC antigens are present in the patient's plasma, they will coat the screening RBCs positive for the corresponding antigen and the coated RBCs agglutinate and remain trapped in the AHG gel at the top of the column, which is interpreted as a positive (4+) reaction. If the RBCs are not coated, they will pass through the gel and pellet at the bottom of the column, interpreted as a negative reaction. Intermediate reaction strengths (3+ to 1+) are given by the migration of RBC agglutinates through the column visualized after centrifugation. This technology is equivalent or superior to low ionic strength solution (LISS) tube testing for detection of clinically significant alloantibodies, and it is less sensitive to clinically insignificant cold antibodies. Major advantages of this method are the ease of training and standardization of testing. A solid-phase adherence platform utilizes microplate wells with bound reagent RBCs at the bottom to which patient's plasma is added, followed by incubation at 37 °C. The microplate is washed to remove plasma, indicator RBCs coated with anti-human globulin are added, and the plate is then centrifuged. A positive reaction is indicated by diffuse binding of indicator RBCs at the bottom surface of the well, whereas a negative reaction is visualized by the presence of a distinct RBC pellet at the bottom of the well. The direct solid-phase test system is designed for ABO/Rh typing in which the microplate wells are coated with the anti-A, anti-B, and anti-D antibodies and the patient's RBCs are directly added without the need for indicator RBCs. After incubation and centrifugation, the wells are read in a similar manner as described previously. This technology has increased sensitivity to alloantibodies and less sensitivity to cold alloantibodies. However, more WRNS antibodies and weak WAA are detected, which may require additional workup to rule out clinically significant alloantibodies. Reactions performed with column and solid-phase technology are stable and can be read days later. Automated or semiautomated testing platforms are available for gel technology (ProVue, Ortho-Clinical Diagnostics, Rochester, NY) and solid-phase adherence (Galileo, Immucor, Norcross, GA).

## *In Vitro* Compatibility Testing

*In vitro* compatibility testing is historically known as the serologic crossmatch test. Compatibility testing in a general sense refers to a quality process that ensures that the safest blood products are issued and includes all the steps described in the Introduction, beginning with blood test or product ordering and ending with record keeping and documentation. Pre-transfusion evaluation and testing includes: (1) recipient testing (ABO/Rh typing on the recipient's blood specimen and review of previous records for clinically significant antibodies, adverse effects to blood transfusions, and special transfusion requirements); (2) donor testing (confirmation of the ABO typing on all units and of the Rh in the units labeled Rh-negative); (3) selection of ABO/Rh-compatible units from the blood bank inventory; (4) performance of electronic (computer) crossmatch, if the recipient antibody screen is negative and there is no history of clinically significant antibodies, or full (serologic, in AHG phase) crossmatch, if the recipient has such a history or the current antibody screen is positive; and (5) labeling the RBC units with at least two independent patient identifiers, donor unit number, and compatibility test results.

The serologic crossmatch tests the blood *in vitro* compatibility between recipient and potential donor. It also serves to reconfirm the ABO type of the donor; thus, it is a checkpoint for preventing ABO typing errors. In the *major crossmatch*, the patient's plasma is tested against donor RBCs obtained from a segment of the blood product. The major crossmatch aims to detect whether the transfused RBCs will survive in the patient's blood and to detect the presence of plasma reactivity that might not be detected by the antibody screen, such as antibodies to low-incidence (LI) antigens not present in the screen RBCs. In the *minor crossmatch*, patient's RBCs are tested against donor's plasma; this test was discontinued in 1976 when the antibody screen on donor blood became routine.

According to the AABB standards, the crossmatch "shall use methods that demonstrate ABO incompatibility and clinically significant antibodies to red cell antigens and shall include an antiglobulin phase." The method of performing the major crossmatch depends on the results of the antibody screen. If the antibody screen is negative and the patient does not have a history of clinically significant antibodies, the major crossmatch performed is only in immediate-spin (IS) phase—the so-called incomplete crossmatch. A "computer crossmatch" or electronic crossmatch wherein the ABO types of both donor and recipient are electronically confirmed and assessed for ABO compatibility by AABB and U.S. Food and Drug Administration (FDA)-approved computer software is also acceptable when certain regulatory conditions are met. The crossmatch performed at 37 °C and in AHG phase has been historically discontinued for cases with negative antibody screen due to significant savings in labor and cost. A negative antibody screen may miss cases with antibodies against low-incidence (LI) antigens because these antigens are not usually present on the screening RBCs. Therefore, alloimmunization, as well as immediate hemolytic reactions caused by clerical error that could have been prevented by AHG crossmatch, is still possible. If the antibody screen is positive, the major crossmatch is performed in AHG phase with selected antigen-negative RBC units.

## Direct Anti-Human Globulin Test

The direct antiglobulin test (DAT) or direct Coombs' test detects the presence of antibody or complement fraction bound to RBCs *in vivo*. The test is simply performed by adding AHG or Coombs' serum to washed patient's RBCs and is illustrated in Figure 17.2. The RBCs agglutinate if they are coated *in vivo* with IgG or the complement fractions, indicating a positive reaction. A positive DAT suggests decreased survival of RBC and/or destruction, but it is not always associated with a disease entity causing hemolysis. As many as 10% of hospital patients and between 1 in 1000 to 1 in 9000 blood donors can have a positive DAT [13]. When hemolysis is present, a positive DAT indicates an immune cause for RBC destruction. Positive DATs are seen in hemolytic transfusion reactions (recipient alloantibody coats transfused donor RBCs), in HDFN (maternal antibody crosses the placenta and coats fetal RBCs), autoimmune hemolytic anemia (the autoantibody coats the patient's own RBCs), drug-induced hemolytic anemia (drug or drug–antibody complex coats the RBCs), and passenger lymphocyte syndrome (antibodies produced by passenger lymphocytes from a transplanted organ coat recipient RBCs). Hypergammaglobulinemia may also be associated with a positive DAT due to nonspecific passive

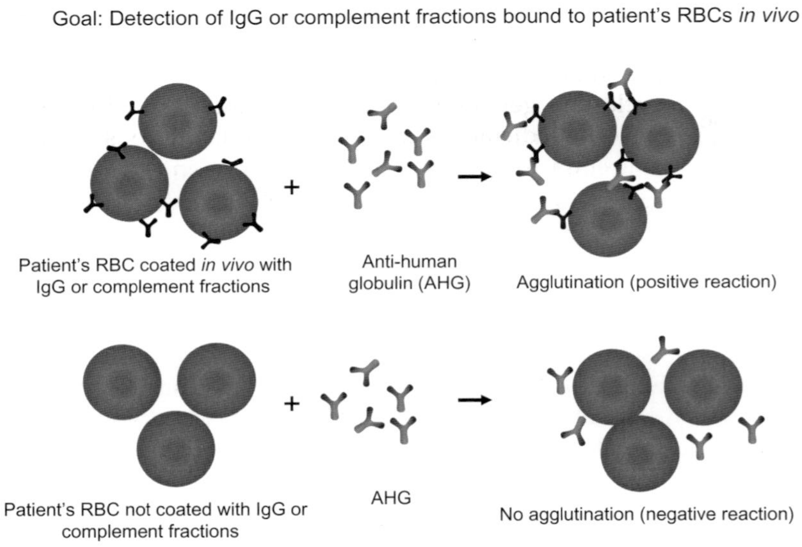

**FIGURE 17.2** Illustration of the principle of the direct antiglobulin test.

adsorption of immunoglobulins onto circulating RBCs. Due to limitations of the test, a negative DAT does not totally exclude the previous conditions. The routine DAT detects only anti-IgG and anticomplement fraction antibodies if more than 200 molecules are coating the RBCs, and antibodies belonging to other immunoglobulin subclasses (IgA, IgD, and IgE) are not detected. In the case of a positive DAT with a prior RBC exposure via transfusion or pregnancy, the antibodies bound to the patient's RBCs are detached by performing an RBC eluate and then tested against the antibody screening/panel to identify if any RBC antigen specificity is present. For correct interpretation, it is necessary to know the type of reagents used and to know that the use of polyclonal reagents has been reported to give false-positive DAT [14,15]. Similarly, knowledge of reaction condition is critical for a correct interpretation of positive RBC eluate [16,17]. Positive DAT and/or IAT may also be caused by certain drugs [18,19]. Regardless of the mechanisms and type of antibody bound, RBCs with a positive DAT have a decreased *in vivo* survival [20].

### Autocontrol

In case the antibody screen is positive, the patient's plasma is tested against the patient's own RBCs (autocontrol) in order to help differentiate whether the plasma reactivity is due to the presence of an autoantibody, alloantibody, or both. The autocontrol indicates the presence of antibody in the patient's plasma *in vitro*, and in that sense, it is a test similar to antibody screen (IAT); however, it also detects the *in vivo* coating of the patient's RBCs and in that regard is similar to DAT. A negative autocontrol implies that the reactivity detected in plasma is most likely due to an alloantibody, whereas a positive result suggests an autoantibody but does not rule out the presence of alloantibodies, especially if the patient has a history of recent transfusions. A positive autocontrol should prompt a DAT and, if positive, the RBC eluate should be tested against the antibody screen. Both auto- and alloantibodies might be present in plasma or bind the patient's RBCs, and their reactivity can overlap. This phenomenon is more often encountered in patients with a history of multiple cell transfusions who have a higher rate of alloimmunization than the general patient population [21].

### Determination of the Red Blood Cell Phenotype

The extended RBC phenotype is not performed for routine transfusions, but it is used when identifying the presence of antigens is essential to the antibody identification or selection of compatible blood, for frequently transfused patients, for detection of rare phenotypes in donors, or as part of paternity or forensic investigations. Ideally, the RBC phenotype of frequently transfused patients should be performed prior to the initiation of transfusion therapy because this category has a much higher risk of alloimmunization. The test is performed by mixing the patient's RBC with commercially available reagent antisera against common RBC antigens (CcEe, MNSs, Kell, Duffy, Kidd, etc.).

## INTERFERENCES IN BASIC TRANSFUSION MEDICINE TESTS

Interferences occurring in blood testing can result in potentially fatal errors or cause a delay in supplying blood for transfusion. The sources of interference in testing are generally categorized as pre-analytical, analytical, and post-analytical. In general, a large proportion of errors occur in the pre-analytical phase of the tests. Pre-analytic variables that can influence the test results can be grouped broadly into two categories: (1) those related to specimen collection, handling, and processing and (2) those related to physiologic factors and patient's endogenous variables, including diseases, circulating antibodies, and drug therapy. Physiologic variables include the effects of age, sex, time, season, altitude, conditions such as menstruation and pregnancy, and lifestyle. Among these, age, sex, and pregnancy status are very important when making recommendations in transfusion medicine. Other patient-related variables, such as clinical diagnosis, problem list, past and current medication, and medical or surgical procedures, are important and may be directly related to the testing results. For this reason, knowledge of medical history is required for correct interpretation of test results and release of the right type of blood. Unlike for other laboratory testing, the fasting or postprandial status, prior exercise, posture during blood drawing, or specific timing of blood sample collection in regard to circadian or physiologic cycles are generally not known to interfere with transfusion medicine assays. Factors related to specimen collection, such as effects of the duration of tourniquet application, the anticoagulant:blood ratio, specimen handling, and processing steps in the preparation of serum/plasma and RBC separation, can introduce an important pre-analytic variable. If they result in hemolysis or fibrin clots being present in the sample, this may be misinterpreted as a positive reaction.

At the initial time of interpretation, the true cause for test results is not known; thus, interferences and errors may look alike. Hemolysis resulting from incorrect drawing looks like hemolysis due to immune causes. If suspected, potential errors, including technical and clerical errors, have to be ruled out for correct test interpretation. Interferences should be distinguished from errors occurring in the process of blood testing and administration. A major potential source of error can occur when the protocols for patient identification or blood drawing are not respected and the patient is not properly identified (blood is drawn from a patient different from the one intended) or when the blood sample is collected from the right patient but mislabeled or blood samples are mixed up. In these instances, the wrong blood in the tube (WBIT) is tested. Although uncommon, WBIT does not represent a true interference but, rather, an error in the process that requires early recognition and further investigation to prevent such events from reoccurring. Two blood samples from the same patient are required for ABO/Rh determination prior to transfusion in order to prevent clerical error or detect possible WBIT. If WBIT is suspected, the test should be repeated on a different sample collected from the same patient. Case Study 1 demonstrates why a high level of alertness should be maintained concerning this rare but still possible source of error in blood bank testing. It also highlights that interpretation of blood tests is part of a process with multiple sources of potential errors and how one of the most critical errors can be initially presented to the clinical pathologist as a discrepancy.

Methodology used for testing plays a definite role in test interpretation. In general, microcolumn gel technology is known to be more sensitive than tube testing and less sensitive than solid-phase adherence technology for detection of clinically significant alloantibodies. The advantage of using a methodology with increased sensitivity is better identification of patterns of reactivity specific to an antibody that might not be elucidated by a less sensitive methodology. Studies showed that some clinically significant alloantibodies identified by solid-phase technology displayed a WRNS pattern in gel testing. Few studies reported, however, that gel technology is less sensitive than the tube test methodology for detecting ABO isohemagglutinins and expected anti-A or anti-B may be missed. The monoclonal anti-A and -B reagents used in the gel cards are also not known to react with B(A), acquired-B, or polyagglutinable RBCs. On the other hand, reagents and chemicals present in the reaction environment of various testing kits may also result in false-positive reactions and trigger extensive testing to rule out the presence of alloantibodies. Similarly, a patient's autoantibodies can react with reagent RBCs present in the antibody screen and identification panel and consequently mask detection of reactivity due to an alloantibody. Cold agglutinins and warm autoantibodies are known for giving a pan-reactivity pattern and masking possible underlying alloantibodies. Autoantibodies to reagents usually lead to false-negative reactions due to reagent consumption or less availability for detection of the true target.

Interferences, defined as alterations in the expected reactivity pattern potentially misleading the interpretation and correct identification of significant findings, can occur in various tests performed in the blood bank, but their correct interpretation requires an integrated review of the case.

## Interferences in ABO/Rh Typing

ABO discrepancies include disagreement between historical and current blood type, discrepancy between forward and reverse reactions, reactivity weaker than expected (4+) between the RBC antigens and corresponding antibodies, and detection of mf type of reactivity. ABO discrepancies detected in patients to be transfused must be resolved before any blood component is transfused unless blood is urgently needed, in which case group O RBCs or AB plasma are issued. Discrepancies occurring in donor samples must be resolved before the blood unit is labeled with a blood type. At the initial read of an ABO discrepancy, it is not known which of the typing reactions—the forward or the reverse—reflects the patient's true ABO type. Patient's age, clinical diagnosis, historical blood type, transfusion history, and the results of other tests are useful hints needed for final interpretation.

Clerical and technical errors must first be ruled out because they are the most common causes for ABO discrepancies. Clerical errors are responsible for more than 95% of fatal transfusion reactions and imply patient or specimen misidentification or blood sample mix-up. Causes of technical errors include failure to follow manufacturer's instruction, failure to add cells or reagents, incorrect test cell preparation so that the pipette dispenses a lower amount of A or B cells into the well, improper centrifugation or incubation conditions, use of contaminated reagents, or defective equipment [2,4,10,12]. Because clerical and technical errors are not uncommon, the best course of action is to repeat testing on a better washed RBC sample and use plasma from the original specimen to determine if the discrepancy persists. Repeat blood drawing might be needed, especially if sample mix-up is a concern.

True ABO discrepancies (not due to clerical and technical errors) are generally grouped into four categories:

1. Weak or loss of expected RBC antigen
2. The presence of unexpected RBC antigen-like reactivity
3. Weak or loss of expected antibody
4. The presence of unexpected antibody reactivity.

### *Weak or Absent Reactivity of Expected Antigen*

Weak A or B antigens are seen in subgroups of A and B blood types due to age, such as in newborns and the elderly, or disease process. Certain hematological malignancies (leukemia and Hodgkin's lymphoma) are associated with aberrant transferase formation leading to fewer antigens being formed on the red cell membrane, whereas solid organ cancers are associated with an excess of soluble substance similar to A or B antigen and can neutralize the typing reagents and result in a false-negative reaction. Chimerism, the presence of a dual population of cells, is routinely seen post ABO-mismatched stem cell transplant and, rarely, due to vascular anastomosis in fraternal twins. In chimerism, ABO/Rh testing is expected to be discrepant within approximately 1 month post-transplant. Transfusion of non-ABO-specific blood, such as massive transfusion of O Rh-negative RBC units in trauma patients, as well as fetal–maternal hemorrhage, also creates a chimeric state. Therefore, the resolution of an ABO discrepancy starts with checking the patient's age and clinical condition.

### *The Presence of Unexpected Red Blood Cell Antigen-Like Reactivity*

Detection of unexpected antigen may be due to misidentification of rouleaux for agglutination, polyagglutinable RBCs, or interference from other substances causing RBC agglutination, such as Wharton's jelly. Autoagglutination due to cold autoantibodies often causes unexpected reactivity (antigen-like in the ABO/Rh typing but also antibody-like in the antibody screen). Antibodies not specific to blood group antigens but reacting with chemicals or drugs present in the reaction microenvironment may also cause this type of discrepancy. Stem cell transplantation, B(A) phenomenon, and fetal–maternal hemorrhage are other known causes of unexpected antigen reactivity. Mixed-field agglutination is usually noted in chimeras, indicating a mixed RBC population (such as in post-transfusion, massive transfusion of another blood group, and bone marrow transplant). Polyagglutination is a phenomenon suspected when patient's RBCs cross reacts with anti-A reagent (mf). Polyagglutinable RBCs also react with most normal adult sera but not with autologous serum. This phenomenon is due to exposure of a cryptantigen on the RBC surface as a result of the action of an enzyme associated with a microorganism (acquired microbial polyagglutination), passive adsorption of microbial structures with antigenic structures similar to A, B, H, T, and Tn antigens or may be associated with inherited conditions or acquired due to a somatic stem cell mutation.

### *Weak or Loss of Expected Antibody*

The most frequent cause of this discrepancy is failure to detect anti-B in A or O plasma samples. This can readily be corrected by the traditional tube test, which is usually performed at RT and sometimes at 4 °C. Weak or absent antibodies are found in newborns, elderly, immunosuppressed patients, post stem cell transplant, and in severe immunodeficiencies.

### *The Presence of Unexpected Antibody Reactivity*

Detection of unexpected reactivity may be due to rouleaux formation, blood subgroups, passively acquired antibodies (via transfusions or administration of Rh immunoglobulins, intravenous immunoglobulins, or other drugs), and passenger lymphocytes syndrome. Individuals with an ABO subgroup can develop alloantibodies that react with reagent RBCs of the reverse typing. The classic example is the A2 subgroup with anti-A1 antibody. Positive reaction can also result from cold agglutinins and true alloantibodies reacting with antigens present on the surface RBCs used for testing (e.g., anti-M, anti-N, anti-P1, and anti-c). A citrate-dependent autoantibody causing errors in blood grouping has also been described [22].

## Interferences in the Detection of Antibodies

The antibody identification process is not always straightforward. If clerical and technical errors are ruled out, the reactivity observed *in vitro* might be due to false positives, real antigen–antibody reactions, and nonimmune interactions.

False-positive reactions or pseudoagglutination were described previously. Common causes of interferences in this category include the presence of fibrin clots, rouleaux formation, and agglutination due to albumin and plasma expanders. Real antigen–antibody reactions can result from simple or combined presence of allo- or autoantibodies either directed against specific RBC antigens or not related to them.

Autoantibodies typically react with the patient's own cells and all reagent RBCs present in the antibody screen and panel in a nonspecific manner. Due to their pan-reactivity, they may mask the detection of alloantibodies. Distinguishing between autoantibodies and alloantibodies is essential, but the answer is not always in the positive autocontrol because transfused patients with developing alloantibodies have a positive test. It is rather the reactivity pattern that suggests the type of autoantibody. Autoantibodies reacting only at cold temperatures, including RT (thus in *in vitro* testing) but not at 37 °C, are known as cold autoantibodies (CAA) and are generally considered clinically insignificant. They become clinically significant if their reactivity extends beyond 32 °C into the body temperature, such as in the case of cold agglutinin disease (CAD), or if the patient undergoes hypothermia, such as in cardiac surgery. The typical CAA is (2+) or (mf) equally pan-reactive with reagent RBCs present in the antibody screen and extended panel. Autoantibodies reacting at 37 °C or in the presence of AHG (AHG phase) are described as warm autoantibodies (WAA). The typical WAA is strong (3+ or 4+) and equally pan-reactive with all reagent RBCs. The clinical significance of WAA is related to their propensity to cause hemolysis. Because both CAA and WAA are typically pan-reactive, if present, they may mask or interfere with the detection of alloantibodies. In such cases, further workup is needed to rule out the presence of underlying alloantibodies against major RBC antigens. Interferences due to passively transfused antibodies occur in patients with a prior transfusion with plasma products containing alloantibodies (e.g., intravenous immunoglobulins (IVIg) or anti-D) or status post-organ or stem cell transplant (passenger lymphocyte syndrome).

Interferences due to antibodies present in the patient's plasma not due to RBC antigens have been described for a variety of chemicals, antibiotics, potentiators (LISS), or other substances (lactose, lactate, melibiose, phenol, sucrose, and thrombin) present in the testing environment or commercial antisera [23–25]. Chemicals that may lead to generation of antibodies reacting with RBCs are paraben, thimerosal, sugars, EDTA, inosine, citrate, acriflavine, sodium caprylate, yellow No. 5, and tartrazine. Antibiotics known to cause production of antibodies that are also capable of reacting with RBCs but not via blood group antigens are penicillin, chloramphenicol, neomycin sulfate, gentamicin, tetracycline, streptomycin, and vancomycin.

At least three mechanisms have been described for the reactivity of RBCs: (1) Antigen–antibody complexes may form in the test environment leading to RBC agglutination, (2) antibodies may adsorb onto RBC surface and bind to antigen, and (3) RBC agglutination may be enhanced by the presence of exogenous substances independent of an antigen–antibody reaction. Chemicals and drugs interfere with testing by either covalent (penicillin) or noncovalent link to RBCs. In the latter case, washing may remove the reactivity due to residual plasma or due to antibodies that are not covalently bound. Antibodies against neomycin, chloramphenicol, gentamicin, hydrocortisone, sugars, dyes (acriflavine, yellow No.5, and tartrazine), sodium azide, and thimerosal are noncovalently bound, thus the reactivity due to these antibodies is removed after RBC washing, whereas antibodies against penicillin, inosine or EDTA are not removed by washing and additional testing is required. Occasionally, these antibodies may display blood group antigen specificity. Other immune interactions were described for an RBC age-dependent antibody and antibodies to lower oxiranes (ethylene, propylene, and butylene oxides) used in the sterilization of polyvinyl chloride blood donor packs. Sterilization of the outside of the bag with propylene oxide sometimes caused the anticoagulant in the bag to acquire properties that could induce RBCs to acquire a new antigen, termed lower oxirane. If these antibodies are present, they can create problems in pre-transfusion testing and present anomalies in ABO, Rh grouping, and antibody detection.

# CASE STUDIES

The following case studies are meant to provide the basic understanding for most common interferences seen in transfusion medicine tests. The emphasis is on correlation with clinical summary and in building a stepwise approach to final interpretation.

## Case Study 1

### *Clinical Summary*

The patient is a 34-year-old male admitted 4 days prior to this testing for multiple head and abdominal wounds status post motor vehicle accident who underwent exploratory laparatomy with splenectomy for multiple splenic lacerations. At that time, the patient was designated in the emergency department (ED) as *Omega, Male*, and his blood type was typed as O Rh-positive. Three ABO and Rh-specific uncrossmatched RBC units were issued. As his hemoglobin level dropped to 6.8 mg/dL during the past 2 days, another RBC unit was requested.

### *Blood Bank Investigation*

Because the prior ABO testing was performed more than 72 hr from the request, a new blood sample was requested for testing. Results are as follows:

| Forward | | | Reverse | |
|---|---|---|---|---|
| Patient's RBCs Reaction with | | | Patient's Plasma Reaction with | |
| Anti-A | Anti-B | Anti-D | A1 RBCs | B RBCs |
| 4+ | 0 | 0 | 0 | 4+ |

His current blood type is A Rh-negative and discrepant with the prior ABO/Rh type determined as O Rh-positive 3 days prior to this testing. Further investigation revealed that this patient arrived to the ED via ambulance at the same time as a different patient involved in the same accident. Both men were unconscious on arrival and received "trauma" designations, *Alpha, Male* and *Omega, Male*. While both patients were being examined in the ED, the transfusion medicine service received two separate specimens labeled *Alpha, Male* and *Omega, Male* with appropriately completed requisitions. The physician was willing to wait for ABO determination for ABO type-specific uncrossmatched blood. The ABO/Rh results obtained on the initially submitted specimens are as follows:

| Case 1 | Forward | | | Reverse | |
|---|---|---|---|---|---|
| Initial Tests | Patient's RBCs Reaction with | | | Patient's Plasma Reaction with | |
| | Anti-A | Anti-B | Anti-D | A1 RBCs | B RBCs |
| Alpha, Male | 4+ | 0 | 0 | 0 | 4+ |
| Omega, Male | 0 | 0 | 3+ | 4+ | 4+ |

The results of antibody screen tests were negative on both patients.

Futher investigation revealed that the *Alpha, Male* was found to have blunt-force abdominal injury and possible pelvic fractures. Despite administration of crystalloids, he remained hypotensive, thus three units of group A Rh-negative RBCs were requested. The transfusions started while he was transported to the operating room. At that time, red urine draining from the patient's Foley catheter bag was noted, but it was attributed to possible renal or lower genitourinary trauma. Another three A Rh-negative units were transfused during the surgery. Profound microvascular hemorrhage from the wounds was noted, and the patient remained hypotensive despite fluids, blood, and pressors administered. He experienced cardiac arrest and could not be resuscitated. The other patient, initially designated as *Omega, Male*, underwent exploratory laparatomy with splenectomy and received three O Rh-positive RBCs. Retrospective crossmatches were found to be compatible in both patients.

### *Interpretation*

ABO discrepancies between the current and the past ABO/Rh types require in-depth investigation of the patient's clinical history and blood bank testing and, if not explained by typical scenarios, they are highly suspicious for WBIT or clerical errors. Repeat ABO/Rh and crossmatch testing on the blood samples collected at the time of arrival to the ED gave the same result. A repeat ABO/Rh testing on a new sample from the patient initially designated as *Omega, Male* confirmed that his blood type was A Rh-negative.

The explanation for serologic testing of the patients described in this case study is WBIT at initial testing. The true blood type of patient *Omega, Male* is A Rh-negative. He was transfused with three O Rh-positive RBC units. The ABO types of the donor and recipient were compatible, and the Rh mismatch may lead to formation of anti-D antibodies; overall, however, WBIT did not have a fatal effect on the patient. The true blood type of *Alpha, Male* was O Rh-positive, and he received six A Rh-negative RBC units. Visual inspection of the post-transfusion sample of this patient identified marked hemolysis. These findings are consistent with an acute hemolytic reaction due to ABO incompatibility, and it is highly likely that the ABO mismatch contributed, if not caused, the patient's fatal outcome. Such cases are FDA reportable, and a root-cause analysis should be performed to prevent similar occurrences.

## Case Study 2

### *Clinical Summary and Blood Bank Testing*

A request for Type and Cross for one RBC unit is received for a 74-year-old male with active gastrointestinal bleeding and a historical blood type known as A Rh-negative and a prior history of massive transfusion 6 years ago for multiple trauma status post pedestrian accident. The antibody screen identifies an anti-D alloantibody as presented here:

| Forward | | | Reverse | |
|---|---|---|---|---|
| Patient's RBCs Reaction with | | | Patient's Plasma Reaction with | |
| Anti-A | Anti-B | Anti-D | A1 RBCs | B RBCs |
| 3+ | 0 | 0 | 2+ | 4+ |

### *Interpretation*

The forward reaction shows a blood type A with a weaker than expected reaction with anti-A, and the reverse reaction shows a strong reaction with B reagent RBCs, indicating the presence of normal anti-B in patient's plasma and thus a blood type A. In the reverse reaction, a weak reactivity with A1 RBCs is detected. Correlation with the patient's known historical blood type, clinical history, antibody screen, and testing for cold antibodies is required in cases with extra antibody reactivity. Most likely, this is a case of A2 subgroup with anti-A1 antibody. This is proven by further testing the patient's RBCs with *Dolichos biflorus*, a lectin that agglutinates A1 antigen. In this case, the patient being A2 subgroup, no agglutination occurs with the anti-A lectin.

ABO subgroups usually present with weaker than expected reactivity in the forward typing. Most common A phenotypes are A1 (80%) and A2 (<20%). The $A^1$ and $A^2$ genes code for different A transferases, and the A2 transferase is less efficient in converting H to A substance, resulting in red cells that have approximately 20–25% less A antigen than A1 cells. In addition to this quantitative difference, A1 and A2 antigens have a different carbohydrate composition. This biochemical difference explains why 1–8% of A2 individuals and 22–35% of A2B individuals produce anti-A1 antibody. This antibody causes discrepancy between the forward and the reverse typing because A1 reagent RBCs are used for reverse testing. *Dolichos biflorus*, an anti-A1 lectin, agglutinates $A_1$ and $A_1B$ but not $A_2$ or $A_2B$ red cells. The anti-A1 is not usually clinically significant because it does not usually react at 37 °C; therefore, release of A blood group is safe for the patient. For such cases, however, a mini-thermal amplitude test should be performed to rule out cold antibodies that may give a similar ABO discrepancy and also prove that the anti-A1 does not react at body temperature.

In this case, the results of the mini-thermal amplitude tests showed that A1 reactivity is maintained at 37 °C and in AHG phase, as shown here:

| | IS | RT | 37 °C | AHG |
|---|---|---|---|---|
| A1 cells | 2+ | 2+ | 2+ | 2+ |
| A2 cells | 0 | 0 | 0 | 0 |
| B cells | 4+ | 4+ | 4+ | 4+ |
| I | 0 | 0 | 0 | 0 |
| II | 0 | 0 | 0 | 0 |
| III | 0 | 0 | 0 | 0 |
| Autocontrol | 0 | 0 | 0 | 0 |

The finding that the anti-A1 reactivity is carried at 37 °C and in AHG phase is critical and relatively uncommon. Unlike in usual cases of anti-A1 antibody, an anti-A1 reacting in these phases is potentially associated with hemolysis, and transfusion of A blood type should be avoided. Only crossmatch-compatible O Rh-negative units are safe for this patient. The lack of reactivity with the screening RBCs and patient's own RBCs rules out the presence of CAA.

## Case Study 3

### *Clinical Summary and Blood Bank Testing*

A request for type and screen is made for a 24-year-old pregnant female as part of her routine prenatal testing. The patient historical blood type is AB Rh-positive. The patient's current ABO typing results are presented here:

| Forward | | | Reverse | |
|---|---|---|---|---|
| Patient's RBCs Reaction with | | | Patient's Plasma Reaction with | |
| Anti-A | Anti-B | Anti-D | A1* RBCs | B RBCs |
| 4+ | 4+ | 4+ | 2+ | 0 |

Results of the thermal amplitude test for Case Study 3 are shown here:

| | IS | RT | 37 °C | AHG |
|---|---|---|---|---|
| A1 cells | 0 | 0 | 0 | 0 |
| A2* cells | 2+ | 2+ | 0 | 0 |
| B cells | 4+ | 4+ | 4+ | 4+ |
| I | 0 | 0 | 0 | 0 |
| II* | 2+ | 2+ | 0 | 0 |
| III | 0 | 0 | 0 | 0 |
| Cord RBCs | 0 | 0 | 0 | 0 |
| Autocontrol | 0 | 0 | 0 | 0 |

### *Initial Interpretation and Further Testing*

This patient's plasma reacts with A1 RBCs as in the A2 subtype cases described previously; however, this reaction is likely not due to an anti-A1 antibody because no reactivity is seen with other A1 cells. The patient's RBCs type agglutinate with *D. biflorus*, indicating that her true type is A1. In addition, the patient's plasma reacted with A2 cells and one screening cells only in IS and RT phases suggsting the presence of a cold antibody. As in any case of a positive antibody screen, further testing was performed by an extended panel and an anti-M alloantibody was identified. All the clinically significant alloantibodies were ruled out. To further prove that the reactivity seen in the ABO typing and mini-thermal amplitude test was due to the anti-M antibody, the reacting (asterisks) RBCs were tested for M antigen, and all of them were found to be positive. The results of the M antigen typing for Case Study 3 are presented here:

| Reagent RBC Tested | Patient's Plasma | Anti-M Reagent |
|---|---|---|
| A1 RBC* | 2+ | 4+ |
| A2* | 2+ | 4+ |
| II* | 2+ | 4+ |

This is an example of a cold allo-antibody interfering with the ABO testing.

## Case Study 4

### *Clinical Summary and Initial Blood Bank Testing*

Two RBC units are requested for a 54-year-old male admitted for melena who was found to have a hemoglobin level of 6 g/dL. His previous blood type is unknown. Results of the ABO/Rh typing for Case Study 4 are presented here:

| Forward | | | Reverse | |
|---|---|---|---|---|
| Patient's RBCs Reaction with | | | Patient's Plasma Reaction with | |
| Anti-A | Anti-B | Anti-D | A1 RBCs | B RBCs |
| 4+ | 3+ | 0 | 0 | 4+ |

What ABO type should the RBC units released for this patient be?

### *Initial Interpretation and Further Testing*

Current blood type reveals an ABO discrepancy between the forward and reverse reaction. The forward type appears to be AB with a weaker B antigen expression, and the reverse type reveals blood type A. Based on reactions pattern and strength, it is most likely that the blood type is A with unexpected B antigen present. This scenario has been described as acquired B phenomenon and occurs in certain clinical settings (patients with carcinoma of colon or rectum, intestinal obstruction, septicemia, or wound infections). The microorganisms involved in these conditions, usually *Proteus vulgaris*, *Clostridium*, and *Escherichia coli* 086, release bacterial enzymes that deacetylate the true A antigen (*N*-acetyl galactosamine) to galactosamine. This molecule reacts with the anti-B reagent as the true B antigen (galactose), but weaker. The discrepancy becomes evident in the reverse typing when strong anti-B antibodies are identified.

Distinction of true B versus acquired B requires knowledge of clinical conditions and extra testing. In the acquired B phenomenon, although the patient's own RBCs appear to type as B, they do not agglutinate with his own serum containing anti-B antibodies, whereas other reagent RBCs of B type give a positive reaction. Repeating the forward reaction after acidifying the anti-B reagent to pH 6 or adding galactosamine will maintain reactivity only in the case of true B antigens but not with the acquired B antigens. Other methods used to distinguish the true B antigen from the acquired B are typing with *Griffonia simplicifolia* (former *Bandeira simplicifolia*) or *Phaseolus lunatus*, which will agglutinate only the true B antigen. Genotyping for the B gene is also helpful, if available, as well as detection of the B antigen in secretions (with the limitation that only 80% of individuals have secretor status with expression of soluble AB antigens in secretions). A summary of tests performed to distinguish the true B antigen from acquired B phenomenon is presented in Table 17.2. In this case, acidification of the B typing showed no reactivity with patient's plasma, and further genotyping the patient revealed A blood type.

TABLE 17.2 Distinction between Acquired B Phenomenon and True B Antigen

| Test | Acquired B Phenomenon | True B Antigen |
|---|---|---|
| Patient's RBC reaction with | | |
| Anti-B | 2+ to 3+ | 4+ |
| Monoclonal anti-B | 0 | 4+ |
| *Griffonia simplicifolia* | 0 | + |
| Repeat reaction with anti-B after | | |
| Acidification | 0 | 4+ |
| Addition of galactosamine | 0 | 4+ |
| Other test results | | |
| Anti-B antibody detected in the reverse typing | Yes | No |
| Genotyping for B gene | Absent | Present |
| B antigen in secretions | Absent | Present |

### *Final Interpretation*

Initial ABO discrepancy is resolved. The patient's true blood type is A Rh-negative. Early recognition of the acquired B phenomenon is critical for cases with weak anti-B antibody (not this case) that might be mistaken as AB. Transfused with type-specific RBC unit in such cases may result in acute hemolytic reaction because the anti-B antibodies are hemolytic even if weaker than expected. As in any case of ABO discrepancy, if emergent RBC transfusion is needed, universal RBC units of O type should be issued.

## Case Study 5

### *Clinical Summary and Initial Blood Bank Testing*

One RBC unit is requested for a 27-year-old female with a history of dysfunctional uterine bleeding and symptoms of anemia who was found to have a hemoglobin level of 7 mg/dL. Her previous blood type is unknown, and she does not have a history of transfusions, pregnancies, or abortions. Her ABO/Rh typing results are illustrated here:

| Forward | | | Reverse | |
|---|---|---|---|---|
| Patient's RBCs Reaction with | | | Patient's Plasma Reaction with | |
| Anti-A | Anti-B | Anti-D | A1 RBCs | B RBCs |
| 4+ | 3+ | 0 | 0 | 4+ |

### *Initial Interpretation and Further Testing*

The patient's ABO typing results are similar to those described for Case Study 4; however, the clinical scenario is not suggestive of an acquired B phenomenon, and additional testing performed for ruling out this entity is inconclusive. Repeat ABO typing gave similar results. The antibody screen was negative, and crossmatches performed with segments from several B type-specific RBC units were all compatible. The patient's genotyping was ordered, but the results were not readily available at the time of transfusion request. It is likely that the patient's true blood type is A, and that she has antibodies to acriflavin, a dye present in the anti-B reagent. In such cases, the reactivity detected is due to the patient's residual plasma coating his or her own RBCs and binding the dye present in the anti-B reagent causing a false-positive reaction. In this case, repeat typing with washed RBCs or B typing with reagents not containing acriflavin did not show any reactivity of the patient's plasma.

### *Final Interpretation*

The ABO discrepancy is resolved. The patient's true blood type is A. Interference in the ABO testing of antibody against acriflavin caused the plasma reactivity seen in initial reverse ABO typing. This is a known phenomenon, and its early recognition of the presence of plasma reactivity due to antibodies against chemicals of reagents present in the testing kit may prevent unnecessary testing or delays in issuing blood transfusions.

## Case Study 6

### *Clinical Summary and Blood Bank Testing*

A 2-year-old female with acute intestinal obstruction and a hematocrit of 21% underwent emergent surgery to remove necrotic bowel. During the surgery, she received 250 mL fresh frozen plasma (FFP). Immediately postsurgery, she developed fever, hemoglobin/hematocrit drop, hemoglobinemia, and hemoglobinuria. A transfusion workup was initiated and showed that her pre-transfusion sample had a repeat negative antibody screen. Repeat crossmatch was found to be compatible, as it was prior to transfusion. The DAT of the pre-transfusion sample was negative; the DAT of the post-transfusion sample was positive (1 + ) with complement fraction detected on RBC surface. The RBC eluate was nonreactive with reagent RBCs present in the antibody screen.

### *Initial Interpretation and Further Testing*

This patient experienced DAT conversion to positive after transfusion of FFP. The RBC eluate is negative, and repeated antibody screen and crossmatch are negative. The clinical summary and blood bank testing are suggestive of polyagglutination due to T activation phenomenon (T from Thomsen).

T activation, also called Hubener–Thomsen–Friedenreich phenomenon, refers to alterations of the RBC membrane antigens usually by microbial enzymatic action with subsequent exposure of previously hidden antigens (cryptantigens) that will react with most adult normal sera. The enzyme responsible for the reaction is a neuraminidinase removing *N*-acetylneuraminic acid (sialic acid) residue from glycophorin A and B exposing the cryptantigen (T antigen). This is often accompanied by decreased expression of M and N antigens normally situated on glycophorin molecules. The exposed T antigen reacts with the anti-T antibody found in normal sera. The anti-T antibody is a naturally occurring IgM antibody formed as a result of stimulation with T-like antigens present on bacteria and viruses [26].

In the past, when pooled normal sera were used for typing, the presence of T activation was detected first as an ABO discrepancy with unexpected antigens. Today, T activation is usually detected by finding a positive DAT with a negative RBC eluate within certain clinical contexts or by incompatible crossmatch. Typical clinical scenarios include children with bacterial infections with *Streptococcus pneumonia*, *Clostridium perfringens*, *Vibrio cholera*, or influenza virus who experience DAT conversion and hemolysis often after exposure to plasma rather than RBCs. T activation is a transient phenomenon lasting weeks to months and accompanied by hemolysis. If suspected, T activation

**TABLE 17.3** Major Patterns of RBC Agglutination with Lectins

| Cryptantigen | T | Tk | Th | Tx | Tn |
|---|---|---|---|---|---|
| *Arachis hypogeal* | + | + | + | + | 0 |
| *Glycine soja* | + | 0 | 0 | 0 | + |
| *Vicia cretica* | + | 0 | + | 0 | 0 |
| *Vicia hyrcanica* | + | + | + | NT | 0 |
| *Medicago disciformis* | + | 0 | + | 0 | 0 |
| *Salvia sclarea* | 0 | 0 | 0 | 0 | + |
| *Salvia horminum* | 0 | 0 | 0 | 0 | + |
| *Griffonia simplicifolia* | 0 | + | 0 | 0 | 0 |

should be proven by further testing with a lectin panel. Lectins are extracts from plant seeds used in blood bank testing for their property to agglutinate certain RBC antigens in characteristic patterns (Table 17.3).

In this case, further testing revealed that the patient's RBCs strongly reacted with *Arachis hypogea* (peanut lectin) and *Glycine soja* (soya bean). No reactions were detected with *Salvia sclarea*, *Salvia horminum*, and *D. biflorus*.

#### Final Interpretation

Additional testing with specific lectins identified a T activation phenomenon. Recognition of this phenomenon is important to explain ABO discrepancies, incompatible crossmatches or DAT conversion with complement coating the patient's RBC, and post-transfusion hemolysis with negative workup because these patients undergo immune hemolysis, but not due to alloantibodies against specific RBC antigens [27–30]. Especially if detected within the context of a transfusion reaction workup, the presence of an acute hemolytic transfusion reaction, clerical error, and alloantibodies against RBC antigens have to be ruled out. The DAT is usually negative in patients with T-activated RBCs, but it may be positive with complement such as in this case. Because these patients have hemolysis and usually blood products are requested for them, early identification and testing is critical in order to recommend avoidance of transfusion with plasma because it may lead to significant hemolysis. If RBC transfusions are needed, they should be washed, and if platelets are transfused, plasma should be removed first and platelets resuspended in 5% albumin. Low-titer anti-T plasma may be administered, if needed and available.

## Case Study 7

#### Clinical Summary and Initial Blood Bank Testing

The patient is a 55-year-old female with a history of coronary artery disease who was admitted for fatigue and was found to have a creatinine of 2 mg/dL. Her hemoglobin level is 8 mg/dL, and she is currently asymptomatic; however, she complained of chest pain 1 day prior, and a myocardial infarction was ruled out. A request for one RBC unit is made for this patient. Her blood was tested 2 years ago and she was found to be A Rh-negative with a negative antibody screen. She does not have a history of blood transfusions or pregnancies. Results of her ABO/Rh typing are as follows:

| Forward | | | Reverse | |
|---|---|---|---|---|
| Patient's RBCs Reaction with | | | Patient's Plasma Reaction with | |
| Anti-A | Anti-B | Anti-D | A1 RBCs | B RBCs |
| 4+ | 0 | 0 | 1+ | 4+ |

The antibody screen and extended testing reveals pan-reactivity (1 +) with all reagent RBC tested. The autocontrol is positive.

#### Initial Interpretation and Further Testing

Pan-reactivity with mild reaction strength (1 +) may be due to multiple causes. Most important, when present, it interferes with detection of clinically significant alloantibodies. However, before embarking on additional time-consuming testing, a clinical history suggestive of multiple myeloma should promptly raise the question of pseudoagglutination due to rouleaux formation of RBCs, a phenomenon known to be occasionally mistaken as true agglutination. When viewed microscopically on peripheral blood smear or microscopic examination of reaction tube, the RBCs have the appearance of a "stack of coins." If rouleaux phenomenon is suspected, the first step is to repeat testing after washing the patient's RBCs. Washed RBCs yield a negative reaction, whereas in the case of true agglutination, the positive reactions persist. In our case, the repeated ABO typing after washing the patient's RBCs yielded the following results:

| Forward | | | Reverse | |
|---|---|---|---|---|
| Patient's RBCs Reaction with | | | Patient's Plasma Reaction with | |
| Anti-A | Anti-B | Anti-D | A1 RBCs | B RBCs |
| 4+ | 0 | 0 | 0 | 4+ |

A review of the patient's peripheral blood smear revealed the characteristic rouleaux formation. Plasma reactivity disappeared after addition of normal saline. Further serum protein electrophoresis and immunofixation studies revealed the presence of a monoclonal gammopathy of IgG-λ. The patient was scheduled for a bone marrow biopsy.

#### Final Interpretation

The ABO discrepancy is resolved. The patient's blood type is A Rh-negative. The (1 +) plasma pan-reactivity is due to her high level of immunoglobulins causing pseudoagglutination. This is a common occurrence for a category of patients with hypergammaglobulinemia with or without monoclonal gammopathies (multiple myeloma, macroglobulinemia, Waldenstrom's macroglobulinemia, hyperviscosity syndrome, and liver diseases). It may also be caused by hyperfibrinogenemia or exogeneous factors, including infusion of high-molecular-weight plasma expanders such as dextran, polyvinyl-pyrolidone, or some intravenous X-ray contrast materials that circumvent the natural influence of the $\zeta$ potential to keep RBCs apart. The rouleaux phenomenon may cause

ABO discrepancies, false positive antibody screens and extended panels, and incompatible crossmatches. It usually does not interfere in the AHG phase because the recipient's serum is washed away prior to AHG addition. In these cases, review of patient clinical and transfusion history is essential for interpretation.

Mild plasma pan-reactivity may also be caused by antibodies to cold agglutinins (CA) or HI RBC antigens, such as k, $Kp^b$, $Js^b$, and $Lu^b$. In the case of antibodies against HI antigens, the autocontrol is negative. Also, being true antigen–antibody interactions, the reactivity does not disappear with addition of normal saline. These antibodies are rare and difficult to identify due to lack of negative panel cells for HI antigens (difficult to rule out); thus, further testing in a reference blood bank laboratory is often needed. CA may be auto- or alloantibodies, and if pseudoagglutination is not likely or proven, a mini-thermal amplitude test is indicated to prove or rule out CAA.

## Case Study 8

### *Clinical Summary and Initial Blood Bank Testing*

A newborn baby boy was delivered to a 28-year-old mother via vaginal delivery at 39 weeks of gestation. The mother's blood type is O Rh-negative, with anti-D antibody detected in her serum early in the pregnancy. The baby has been monitored for hemolytic disease of the newborn, and he has not required any interventions, including intrauterine blood transfusions, throughout his fetal life. His bilirubin level is 12 mg/dL, and the forward typing shows the following results:

| Forward | | |
|---|---|---|
| Patient's RBCs Reaction with | | |
| Anti-A | Anti-B | Anti-D |
| 3+ | 4+ | 0 |

### *Initial Interpretation and Further Testing*

The baby forward typing suggests that the baby's RBCs are of AB type. Reverse typing is not performed on newborns because they do not produce antibodies until the age of 6 months at least. His Rh is negative; thus, HDFN due to anti-D antibody is ruled out. Because his mother is blood type O and she does not have other alloantibodies that can explain the baby's high bilirubin, a possible extra-reactivity is suspected in the baby's forward typing reaction likely due to sample contamination with Wharton's jelly. Spontaneous agglutination of samples contaminated with Wharton's jelly has been described in textbooks, but it not very common in practice. The positive reactions are caused by nonspecific agglutination due to hyaluronic acid and albumin found in cord blood. Repeated testing with the baby's RBCs washed three times with 0.9% saline yielded the following results:

| Forward | | |
|---|---|---|
| Patient's RBCs Reaction with | | |
| Anti-A | Anti-B | Anti-D |
| 0 | 4+ | 0 |

The direct anti-human globulin performed on the baby's RBC was positive (2 +), with IgG and no complement fixation noted. An anti-B antibody was eluted from the baby's RBCs.

### *Final Interpretation*

The baby's true blood type is B Rh-negative. Initial reactivity was most likely due to Wharton's jelly coating the patient's RBCs and interfering with the RBC testing. However, he has hemolytic disease of the newborn due to an IgG anti-B antibody present in his mother's blood that crossed the placenta. Phototherapy is indicated due to his high bilirubin, as well as monitoring for clinical and laboratory indicators of hemolysis.

## Case Study 9

### *Clinical Summary and Initial Blood Bank Testing*

The patient is a 36-year-old male who was admitted for multiple abdominal gunshot wounds and underwent a massive transfusion for life-threatening bleeding 4 days prior to the current testing. At that time, the patient received six O Rh-negative RBC units, six AB-negative FFP units, and one dose of AB platelets. His blood type was retrospectively found to be B Rh-positive. The patient underwent successful surgical repair, but he remained critically ill and his hemoglobin levels dropped from 9 to 7 mg/dL during the next few days. Two RBC units are requested in preparation of surgical re-exploration. Current ABO typing is shown here. His antibody screen is negative.

| Forward | | | Reverse | |
|---|---|---|---|---|
| Patient's RBCs Reaction with | | | Patient's Plasma Reaction with | |
| Anti-A | Anti-B | Anti-D | A1 RBCs | B RBCs |
| 0 | 2+ mf | 0 | 3+ | 0 |

### Interpretation

The patient's initial blood type was O Rh-negative. He now has an ABO discrepancy with a weaker than expected B antigen with mf reaction and a weaker than expected anti-A antibody. The history of massive transfusion is essential for interpretation, and it explains the current findings due to dilution of the patient's own B cells by transfused RBCs of O type. Additional testing is not required in this case. The patient's true blood type is B Rh-negative, and crossmatch and ABO-compatible or -specific RBC units can be released.

## Case Study 10

### Clinical Summary and Initial Blood Bank Testing

One RBC unit is requested for a 55-year-old female with a history of stem cell transplant (SCT) 3 weeks ago for recurrent acute myelogeneous leukemia. Her current ABO/Rh typing results are shown here. The patient was O Rh-negative, and the SCT donor was B Rh-negative.

| Forward | | | Reverse | |
|---|---|---|---|---|
| Patient's RBCs Reaction with | | | Patient's Plasma Reaction with | |
| Anti-A | Anti-B | Anti-D | A1 RBCs | B RBCs |
| 0 | 2+ mf | 0 | 3+ | 0 |

### Interpretation

This case is serologically similar to Case Study 9; however, the patient's clinical history is different and critical for recommendations regarding blood transfusions. ABO-mismatched SCTs are routinely performed since ABO, unlike HLA, is not the most important histocompatibility antigenic system for SCT. Within the engraftment period, which lasts approximately 1 month post-SCT, the recipient ABO type gradually switches to the donor type. During this time, ABO discrepancies are routinely noted. They are explained by the patient's history; no further testing is usually required. The patient is transitioning to a blood type B reflected by detection of a B antigen in the forward typing and the presence of anti-A antibody in the reverse typing. These findings are expected and reflect good SCT engraftment. Transfusion of blood products compatible with both recipient and donor blood types is necessary during this time. In this case, O Rh-negative RBC would be preferred, but R Rh-negative RBCs are also appropriate; if FFP or platelets are needed, the AB type is adequate.

## Case Study 11

### Clinical Summary and Initial Blood Bank Testing

A 29-year-old pregnant female is admitted for abdominal pain and light vaginal bleeding at 37 weeks of gestation. Her blood type is known to be AB Rh-negative, and she was administered Rh immunoglobulin at 28 weeks of gestation for HDFN prophylaxis. Her current blood type is also AB Rh-negative, and the antibody screen is positive. The extended panel detects a (2 + ) anti-D reactivity in her plasma. She does not have a history of prior transfusions.

### Interpretation

This case is typical. Based on the patient's clinical history of Rh immunoglobulin administration, the anti-D reactivity is interpreted as due to passive antibody transfer from the drug. Despite its simplicity, it is listed here as a reminder that clinical history is always critical in interpretation of test results. Passive transfer of alloantibodies can also be seen after administration of IVIg, such as described in Case Study 13.

## Case Study 12

### Clinical Summary and Initial Blood Bank Testing

Type and screen request is made for a 30-year-old pregnant woman as part of her prenatal care. The patient has another living child and a history of two abortions. She has never received Rh immunoglobulin or RBC transfusions. Her blood type is found to be A Rh-negative, and she has a positive antibody screen. Additional testing reveals the presence of (4 + ) anti-D reactivity. In addition, an anti-C antibody cannot be ruled out. The patient does not remember if she received Rh immunoglobulin during her first pregnancy.

### Initial Interpretation and Further Testing

The presence of anti-D reactivity in this patient's plasma is most likely due to a true anti-D antibody based on the reactivity strength and lack of history for Rh immunoglobulin administration at least during the current pregnancy. The presence of true anti-D makes a patient ineligible for HDFN prophylaxis. Because allo-anti-C could not be ruled out in initial testing, the patient's plasma was further tested and found to react with C-positive reagent RBCs in the presence of ficin. The patient's RBCs were negative for the C antigen.

When plasma reactivity is suggestive of both anti-D and anti-C, additional testing should be performed at a reference laboratory to rule out anti-G antibody. This is directed against the G antigen, which is part of the Rh blood group and is a compound antigen formed when both C and D antigens are present. Anti-G antibody cannot be detected by routine methods because reagent RBCs are not profiled for the G antigen, but it should be suspected when both anti-D and anti-C reactivity is detected. In this case, additional testing revealed the presence of an anti-G antibody in the patient's plasma. True anti-C and anti-D antibodies were not detected.

### *Final Interpretation and Recommendations*

The patient has an anti-G alloantibody present in her plasma. This antibody may cross the placenta and cause HDFN. Because she does not have a true anti-D alloantibody, she is still eligible for HDFN prophylaxis. Should she require further transfusions, RBC units negative for both C and D antigens should be issued. The same transfusion requirements are made if the anti-G is detected in a male's plasma except that knowledge of his true anti-D versus anti-G status is neither required nor necessary. This distinction is useful only to capture the true D-negative women for RhIg administration to prevent HDFN.

## Case Study 13

### *Clinical Summary and Initial Blood Bank Testing*

A 9-year-old boy with a history of bilateral renal dysplasia status post renal transplant was recently diagnosed with graft rejection based on a recent renal biopsy. His hemoglobin level is 7 mg/dL. One RBC unit was requested. The patient has a history of blood transfusion when he was 7 years old during the peritransplant period but not recently. His blood type is known to be O Rh-positive, and his antibody screen was negative at prior testing. Current ABO/Rh type is the same, but the antibody screen is positive and additional testing reveals the presence of anti-D and anti-c antibodies.

### *Interpretation*

A careful review of medical and drug history reveals that this patient was administered intravenous immunoglobulin for his renal rejection. His newly identified anti-D and anti-c alloantibodies are most likely due to passive transfer from the IVIg preparations. Passive transfer of alloantibodies is commonly seen after administration of IVIg [12, 31–33]. These preparations are extracted by cold fractionation from large pools of volunteer plasma donors and contain a broad spectrum of antibodies to viral, bacterial, and other microorganisms. Isohemagglutinins and alloantibodies against RBC antigens may be present in up to 17–50% of these preparations and continue to be present in the patient's plasma several weeks to months post IVIg administration. The reactivity detected in the patient's plasma is due to passive transfer of true alloantibodies. These antibodies may produce hemolysis, usually self-limited but occasionally clinically significant, especially after large doses. Early recognition of this typical clinical scenario may prevent additional unnecessary testing and transfusion delays.

## Case Study 14

### *Clinical Summary and Initial Blood Bank Testing*

A request for type and screen is made for a 56-year-old female with coronary artery disease who is scheduled for coronary artery bypass grafting. Her hemoglobin level is 10 mg/dL, and she is currently asymptomatic. Her ABO/Rh test results are shown here:

| Forward | | | Reverse | |
|---|---|---|---|---|
| Patient's RBCs Reaction with | | | Patient's Plasma Reaction with | |
| Anti-A | Anti-B | Anti-D | A1 RBCs | B RBCs |
| 4+ | 0 | 0 | 2+ | 4+ |

The antibody screen is equally pan-reactive (2 + ). The extended panel also showed (2 + ) pan-reactivity.

### *Initial Interpretation and Further Testing*

The initial presentation is identical to that of Case Study 2 (A2 subgroup with anti-A1 antibody) and Case Study 3 (anti-M antibody). Unlike Case Study 3, in which the antibody screen was positive with one cell, in this case (2 + ) pan-reactive plasma is seen. This pattern is typical of a cold agglutinin. Due to pan-reactivity, underlying alloantibodies against major RBC antigens cannot be ruled out.

Cold agglutinins are antibodies against I/i antigens of the Ii blood group collection. They are associated with CAD if the thermal amplitude extends up to the lower body temperature (35 °C). The presence of cold agglutinins may be demonstrated by cold autoadsorption or rabbit erythrocyte stromal test (RESt). In this test, the patient's plasma is incubated with rabbit erythrocytes, which are rich in I antigen and absorb anti-I antibody if present. The plasma incubated one or several times with RESt is then run against the usual antibody screen for detection of any leftover reactivity. A negative antibody screen after RESt is suggestive of anti-I, anti-IH, or anti-H cold agglutinins. The RESt technique can be used if the patient has recently been transfused, however, careful interpretation is necessary because RESt can also remove the reactivity of clinically significant alloantibodies, such as anti-B, anti-D, and anti-E. Re-testing after washing the cells three or four times with warm (37 °C) saline and/or performing serum testing at 37 °C may also be useful (prewarm technique). Cold agglutinins with anti-i specificity are suspected if patient's plasma reacts with cord blood cells (rich in little i antigen). Unusual cold agglutinins reacting strongly with cord cells and weakly with adult I cells have been described in lymphoproliferative diseases [34]. Alternatively, allogeneic cold adsorption may be used with phenotypically similar allogeneic cells if the patient's full RBC phenotype is known or with selected reagent RBCs (usually R1, R2, and rr) negative for K, Jka, Jkb, Fya, Fyb, and S so that potential clinically significant alloantibodies can be detected in the adsorbed plasma. If the patient has not been transfused within the past 3 months, a cold autoadsorption technique may be used. In this method, the patient's own RBCs treated with a proteolytic enzyme solution to remove antibody binding sites are subsequently exposed to the patient's serum/plasma at 4 °C. Results of the thermal amplitude test and RESt for Case Study 14 are shown here:

| Antibody Screen | IS | RT | 37 °C | AHG | CC | RESt × 2 |
|---|---|---|---|---|---|---|
| I | 2+ | 2+ | 0 | 0 | 4+ | 0 |
| II | 2+ | 2+ | 0 | 0 | 4+ | 0 |
| III | 2+ | 2+ | 0 | 0 | 4+ | 0 |
| Autocontrol | 2+ | 2+ | 0 | 0 | 4+ | 0 |
| Cord blood | 0 | 0 | 0 | 0 | 4+ | NT |

### *Final Interpretation*

Plasma reactivity is due to an anti-I cold agglutinin. A negative reaction with cord cells rules out the presence of anti-i cold agglutinins. These are not considered clinically significant unless there is exposure to cold temperature, such as in induced hypothermia during surgery. Once a cold agglutinin has been identified, measures to warm the blood transfused are required. ABO and crossmatch-compatible RBC units can be released for these patients.

If the autocontrol is negative and the antibody screen is positive, the patient's blood type has to be considered with regard to H antigen expression, knowing that the H antigen expression is highest in O RBCs (O > A2 > B > A2B > A1 > A1B). Blood groups with the least H expression, such as A1 and A1B, may have anti-H antibodies. Anti-H is a clinically insignificant cold antibody that would react with O reagent RBC present in the antibody screen and panel but not with autologous RBCs or at body temperature.

## Case Study 15

### *Clinical Summary and Initial Blood Bank Testing*

A 72-year-old female with known chronic lymphocytic lymphoma and an indolent course of disease not requiring chemotherapy is evaluated in the hematology clinic for fatigue. During initial investigations started 2 month ago, she was found to have a hemoglobin level of 7 mg/dL and received two RBC units at a different hospital. Her current hemoglobin level is 8 mg/dL, and she is still weak. An additional RBC unit is requested. The patient's ABO/Rh typing results show a B Rh-positive blood type. The antibody screen and extended panel show that her plasma reacts equally and strongly (4 + ) with all reagent RBCs tested. The DAT is positive with the same strength (4 + ) with IgG and mild complement fixation. Results of the ABO/Rh typing for Case Study 14 are shown here:

| Forward | | | Reverse | |
|---|---|---|---|---|
| Patient's RBCs Reaction with | | | Patient's Plasma Reaction with | |
| Anti-A | Anti-B | Anti-D | A1 RBCs | B RBCs |
| 0 | 4+ | 4+ | 4+ | 0 |

### *Initial Interpretation and Further Testing*

The presentation suggests the presence of a warm autoantibody; underlying clinically significant alloantibodies cannot be ruled out. Because the patient received two RBC units, her phenotype cannot be performed. The clinician was contacted to elucidate the urgency for blood, and he stated that the patient was clinically stable. The RBC phenotype determined by molecular methods revealed that the patient's RBC phenotype was negative for C, D, and Jk(b) antigens. Allogeneic adsorption with phenomatched cells showed that she had an anti-Jk(b) antibody in her plasma and RBC eluate.

### *Final Interpretation*

This patient has a WAA with an underlying anti-Jk (b) alloantibody in her plasma and RBC eluate. These findings are consistent with a delayed serologic/hemolytic reaction; hemolytic WAA is also possible. WAA are usually IgG antibodies without RBC specificity reacting strongly and equally with all reagent RBCs tested. However, WAA with apparent specificity for e, Kell, Kidd, MNSs, ABO, Vel, and LW antigens have also been reported. In the case of WAA, the autocontrol (and DAT) is positive and the RBC eluate shows a similar 4 + pan-reactive pattern. The clinical significance of WAA is related to their propensity to cause hemolysis, which cannot be determined from the immunohematology testing but from results of a hemolysis panel (hemoglobin, hematocrit, bilirubin, LDH, and haptoglobin). If RBC transfusion is necessary, least-incompatible phenomatched RBC units can be issued.

## CONCLUSIONS

The reactivity detected in blood bank testing can result from various causes. Discrimination of clinically significant alloantibodies is critical for issuing the correct type of RBCs and preventing hemolytic transfusion reactions. Clerical and technical errors, spurious results, and false-positive and false-negative reactions are discerned after careful consideration of the whole case, including the patient's clinical and blood bank history, type of test, reaction conditions, and technology used. During this process, typical clinical scenarios should be considered. A summary of the basic transfusion process, routine tests, and typical case studies were presented in this chapter with the goal of aiding the rapid recognition of some of the most common interferences encountered in transfusion medicine.

## References

[1] Harmening D. Modern blood banking & transfusion practices. 5th ed. Philadelphia: F. A. Davis; 2005.

[2] Roback JD, Combs MR, Grossman BJ, Hillyer CD. Technical manual and standards for blood banks and transfusion services on CD-ROM. 17th ed. Bethesda, MD: American Association of Blood Banks; 2011.

[3] Simon TL, Dzik WH, Snyder EL, Rossi EC, Stowell CP, Strauss RG. Rossi's principles of transfusion medicine. 3rd ed. Philadelphia: Lippincott Williams & Wilkins; 2002.

[4] Quinley ED. Immunohematology: principles and Practice. 3rd ed. Philadelphia: Lippincott Williams & Wilkins; 2010.

[5] Morelati F, Revelli N, Maffei LM, Poretti M, Santoro C, Parravicini A, et al. Evaluation of a new automated instrument for pretransfusion testing. Transfusion 1998;38(10):959–65.

[6] Milam JD. Laboratory medicine parameter: utilizing monospecific antihuman globulin to test blood-group compatibility. Am J Clin Pathol 1995;104(2):122–5.

[7] Klein HG, Anstee DJ. Mollison's blood transfusion in clinical medicine. 11th ed. Boston: Blackwell; 2005.

[8] Powers A, Chandrashekar S, Mohammed M, Uhl L. Identification and evaluation of false-negative antibody screens. Transfusion 2010;50(3):617–21.

[9] Rolih S, Thomas R, Fisher F, Talbot J. Antibody detection errors due to acidic or unbuffered saline. Immunohematology 1993;9(1):15–18.

[10] Bobryk S, Goossen L. Variation in pipetting may lead to the decreased detection of antibodies in manual gel testing. Clin Lab Sci 2011;24(3):161–6.

[11] Arndt P, Garratty G. Evaluation of the optimal incubation temperature for detecting certain IgG antibodies with potential clinical significance. Transfusion 1988;28(3):210–13.

[12] Issitt PD, Anstee DJ. Applied blood group serology. 4th ed. Durham, NC: Montgomery Scientific; 1998.

[13] Freedman J. False-positive antiglobulin tests in healthy subjects and in hospital patients. J Clin Pathol 1979;32(10):1014–18.

[14] Nasongkla M, Hummert J, Chaplin Jr. H. Weak "false positive" direct antiglobulin test reactions with polyspecific antiglobulin reagents: lack of correlation with red-blood-cell-bound C3d. Transfusion 1982;22(4):273–5.

[15] Bruce M, Watt AH, Hare W, Blue A, Mitchell R. A serious source of error in antiglobulin testing. Transfusion 1986;26(2):177–81.

[16] Leger RM, Arndt PA, Ciesielski DJ, Garratty G. False-positive eluate reactivity due to the low-ionic wash solution used with commercial acid-elution kits. Transfusion 1998;38(6):565–72.

[17] Dumaswala UJ, Sukati H, Greenwalt TJ. The protein composition of red cell eluates. Transfusion 1995;35(1):33–6.

[18] Arndt PA, Leger RM, Garratty G. Positive direct antiglobulin tests and haemolytic anaemia following therapy with the beta-lactamase inhibitor, tazobactam, may also be associated with non-immunologic adsorption of protein onto red blood cells. Vox Sang 2003;85(1):53.

[19] Fried MR, Scofield TL, Stroncek DF, Swanson JL. Chloramphenicol-dependent antibody: a case report. Transfusion 1996;36(2):187–90.

[20] Garratty G. Effect of cell-bound proteins on the in vivo survival of circulating blood cells. Gerontology 1991;37(1-3):68–94.

[21] Aygun B, Padmanabhan S, Paley C, Chandrasekaran V. Clinical significance of RBC alloantibodies and autoantibodies in sickle cell patients who received transfusions. Transfusion 2002;42(1):37–43.

[22] Joshi SR. Citrate-dependent auto-antibody causing error in blood grouping. Vox Sang 1997;72(4):229–32.

[23] Garratty G. Problems in pre-transfusion tests related to drugs and chemicals. Am J Med Technol 1976;42(6):209–19.

[24] Garratty G. Screening for RBC antibodies: what should we expect from antibody detection RBCs. Immunohematology 2002;18(3):71–7.

[25] Garratty G. The significance of complement in immunohematology. Crit Rev Clin Lab Sci 1984;20(1):25–56.

[26] Horn KD. The classification, recognition and significance of polyagglutination in transfusion medicine. Blood Rev 1999;13(1):36–44.

[27] Hall N, Ong EG, Ade-Ajayi N, et al. T cryptantigen activation is associated with advanced necrotizing enterocolitis. J Pediatr Surg 2002;37(5):791–3.

[28] Eder AF, Manno CS. Does red-cell T activation matter?. Br J Haematol 2001;114(1):25–30.

[29] Crookston KP, Reiner AP, Cooper LJ, et al. RBC T activation and hemolysis: implications for pediatric transfusion management. Transfusion 2000;40(7):801–12.

[30] Boralessa H, Modi N, Cockburn H, et al. RBC T activation and hemolysis in a neonatal intensive care population: implications for transfusion practice. Transfusion 2002;42(11):1428–34.

[31] Lichtiger B, Rogge K. Spurious serologic test results in patients receiving infusions of intravenous immune gammaglobulin. Arch Pathol Lab Med 1991;115:467–9.

[32] Lichtiger B. Laboratory Serologic Problems Associated with Administration of Intravenous IgG. Curr Issues Transfus Med 1994;April-June.

[33] Buckly RH, Schiff RI. The use of intravenous immunoglobulin in immunodeficiency diseases. N Engl J Med 1991;325:110–17.

[34] Leger RM, Lowder F, Dungo MC, Chen W, Mason HM, Garratty G. Clinical evaluation for lymphoproliferative disease prompted by finding of IgM warm autoanti-IT in two cases. Immunohematology 2009;25(2):60–2.

CHAPTER

# 18

# Issues with Immunology and Serology Testing

*Amer Wahed, Semyon Risin*
University of Texas Health Sciences Center at Houston, Houston, Texas

## INTRODUCTION

Clinicians depend on clinical laboratories for obtaining accurate results in immunology and serology testing, and inaccurate results may have tremendous impact not only on the diagnosis but also on the patient. For example, a false-positive HIV test result may have a devastating psychological impact on the patient. Various methods are used in immunology and serology testing, including electrophoresis, immunoassays, and various other analytical techniques. Even for a single analyte, multiple technologies may be available. As expected, no method is free from analytical errors. In this chapter, various sources of errors in immunology and serology testings are addressed. The emphasis is on how to minimize errors as well as eliminate them when possible.

## CHALLENGES IN HEMOGLOBINOPATHY DETECTION

Multiple methodologies exist for testing for hemoglobinopathies in the clinical laboratory (Table 18.1). The most common ones employed are conventional electrophoresis, capillary electrophoresis, and high-performance liquid chromatography (HPLC). In hemoglobin electrophoresis, red cell lysates are run in electric fields under alkaline (alkaline gel) and acidic (acid gel) pH. This can be carried out on filter paper, a cellulose acetate membrane, a starch gel, a citrate agar gel, or an agarose gel. Separation of different hemoglobins is largely but not solely dependent on the charge of the hemoglobin molecule. Change in the amino acid composition of the globin chains alters the charge of the hemoglobin molecule, resulting in a change in the speed of migration.

HPLC utilizes a weak cation exchange column system. A sample of a red blood cell (RBC) lysate in buffer is injected into the system. Hemoglobin molecules are adsorbed onto the column as they are charged molecules in the buffer system. An eluting buffer is then injected into the system. The hemoglobin fraction then elutes off the column. The time required for different hemoglobin molecules to elute is referred to as retention time. The eluted hemoglobin molecules are detected by light absorbance. HPLC permits the provisional identification of many more variant hemoglobins than can be distinguished by conventional gel electrophoresis.

In capillary electrophoresis, a thin capillary tube made of fused silica is used. When an electric field is applied, the buffer solution within the capillary generates an electroendosmotic flow that moves toward the cathode. Separation of individual hemoglobins takes place due to differences in overall charges. Other less commonly used methodologies include isoelectric focusing, DNA analysis, and mass spectrometry.

Any one of the previously discussed methods can be used for screening purposes. Detection of abnormal hemoglobin requires validation by a second method. In addition, relevant clinical history, review of the complete blood count (CBC), and peripheral smear provide important correlation in the pursuit of an accurate diagnosis.

### Hemoglobinopathy Diagnosis Errors

Blood specimens for hemoglobinopathy diagnosis may be sent to outside clinical laboratories requiring specimens to be transported by couriers. Sometimes the specimens are mailed. Unduly long transit times without refrigeration may result in artifactual bands on gel electrophoresis causing confusion in

*Accurate Results in the Clinical Laboratory.*
DOI: http://dx.doi.org/10.1016/B978-0-12-415783-5.00018-9

TABLE 18.1 Common Methodologies for Detection of Variant Hemoglobins

| Methodology | Advantages | Disadvantages |
|---|---|---|
| Agarose gel electrophoresis | Cheaper to perform | Labor-intensive and time-consuming; common variant hemoglobins distributed in the four major lanes |
| HPLC | Faster turnaround time; greater resolution of common variant hemoglobins | Expensive |
| Capillary electrophoresis | Faster turnaround time; common variant hemoglobins distributed in 15 zones | |
| Isoelectric focusing | Greater resolution | More difficult to interpret; particularly suitable for small samples including dried blood spots and is thus often used for neonatal screening |

TABLE 18.2 Common Hemoglobin Disorders

| Disorder | Defective Chain | Comments |
|---|---|---|
| Hb S | β chain defect | Most common variant hemoglobin seen in the United States; mostly seen in African Americans |
| Hb C | β chain defect | Implies ancestry is from Western Africa. HB SC is a sickling disorder |
| Hb E | β chain defect | Most common variant hemoglobin seen in Asia |
| Hb D | β chain defect | Seen in African, Indian, Pakistani, as well as English individuals. Hb SD is a sickling disorder |
| Hb G | α chain defect | Most common α chain variant. Found in African Americans and African Caribbeans. It is of no clinical significance |
| Hb A2′ | δ chain defect | Most common δ chain defect. Seen in 1–2% of the African American population |
| Hb O | β chain defect | Wide geographic distribution—from Africa to Middle East and eastern Europe. HB SO is a sickling disorder |

interpretation and delay in final diagnosis [1]. The method of preparing the red cell lysate can sometimes be important. One study demonstrated that for accurate quantification of hemoglobin H from electrophoresis on agarose gel at alkaline pH, the lysate carbon tetrachloride is required. If other lysates are used, much less hemoglobin H is detected, which could result in missing its (Hb H) detection [2]. When HPLC is used, a recognized problem is carryover of sample from one to the next [3]. For example, if the first sample belongs to a patient with sickle cell disease (Hb SS), then a small peak may be seen at the "S" window in the next sample. This can lead to diagnostic confusion as well as the sample needing to be re-run. Relevant to the final diagnosis is the transfusion history. Blood transfusion-acquired hemoglobinopathy is an established phenomenon [4]. This potential for transfusion-acquired hemoglobinopathy exists because heterozygous individuals show no significant abnormalities during the blood donor screening process [5]. The abnormal hemoglobin in recipients accounts for between 0.8 and 14% of the total hemoglobin [4]. Transfusion histories thus remain vital in explaining such findings.

Common hemoglobin disorders are Hb S, Hb E, Hb C, Hb D, Hb G, and Hb O (Table 18.2). Hb S is the most common hemoglobinopathy in the United States, whereas Hb E is the most common hemoglobinopathy in Southeast Asia. These two hemoglobinopathies account for the majority of all hemoglobinopathies detected worldwide. Although only a handful of hemoglobinopathies are dealt with in the clinical laboratory, more than 1000 hemoglobinopathies have been described, the vast majority of which are rare or clinically insignificant.

### *Hemoglobin A2*

In the diagnosis of β-thalassemia trait, the proportion of Hb A2 relative to the other hemoglobins is clinically important [6]. In certain cases, Hb A2 variants may also be present. In such cases, the total Hb A2 (Hb A2 and Hb A2 variant) needs to be considered for the diagnosis of β-thalassemia [6]. Hb A2′ is the most common of the known Hb A2 variants and has been reported in 1 or 2% of African Americans [7]. Hb A2′ has been detected in heterozygous and homozygous states and in combination with other Hb variants and thalassemia [8–11]. The major clinical significance of Hb A2′ is that for the diagnosis or exclusion of β-thalassemia minor, the sum of Hb A2 and Hb A2′ must be considered. When present, Hb A2′ accounts for a small percentage (1 or 2%) in heterozygotes and is difficult to detect by gel electrophoresis [12]. However, it is easily picked up by capillary electrophoresis and HPLC. By HPLC, Hb A2′ elutes in the "S" window. In Hb AS trait and HB SS disease, Hb A2′ will be masked by the presence of Hb S. In Hb AC trait and Hb CC disease, glycosylated Hb C will also elute in the "S" window. In these conditions, Hb A2′ will remain undetected. Conversely, sickle cell patients on chronic transfusion protocol or recent, efficient RBC exchange may have a very small percentage of Hb S, which the pathologist may misinterpret as Hb A2′.

It has been documented that the Hb A2 concentration may be raised in HIV during treatment [13]. Other causes of a raised Hb A2 level are thought to be very unusual [14]. Severe iron deficiency anemia can reduce Hb A2 levels [15,16], and in some reports this has been shown to interfere with the diagnosis of β-thalassemia trait.

### *Hemoglobin F*

An increase in the percentage of hemoglobin F is associated with multiple pathologic states, including β-thalassemia, δβ-thalassemia, and hereditary persistence of fetal hemoglobin (HPFH). The former is associated with high Hb A2, and the latter two states are associated with normal Hb A2 values. Hematologic malignancies are associated with increased hemoglobin F and include acute erythroid leukemia (M6) and juvenile myelomonocytic leukemia. Aplastic anemia is also associated with an increase in Hb F %. When elucidating the actual cause of high Hb F, it is important to consider the actual percentage of Hb F, Hb A2 values, as well as correlation with CBC and peripheral smear findings.

It is also important to note that drugs (hydroxyurea, sodium valproate, and erythropoietin) and stress erythropoiesis may also result in high Hb F. Hydroxyurea is used in sickle cell disease patients to increase the amount of Hb F, the presence of which may help to reduce the clinical effects of the disease. Measuring the level of Hb F may be useful for determining the appropriate dose of hydroxyurea. In 15–20% of pregnancies, Hb F may be raised as much as 5%.

Hb F quantification may be an issue when HPLC is used. Fast variants (e.g., Hb H or Hb Bart's) may not be quantified because they may elute off the column before the instrument begins to integrate in many systems designed for adult samples. This will affect the quantity of Hb F. If an α-globin variant separates from Hb A, often an Hb F variant separates from normal Hb F but may not separate from other hemoglobin adducts present so that the total Hb F will not be adequately quantified. Hb F variants may also be due to mutation of the γ-globin chain, and again this may result in a separate peak and incorrect quantification. Some β chain variants and/or their adducts may not separate from Hb F, leading to incorrect quantification [17]. Capillary zone electrophoresis has an advantage over HPLC in that hemoglobin adducts (glycated hemoglobins and the aging adduct Hb X1d) do not separate from the main hemoglobin peak so that interpretation is easier than with HPLC. If Hb F appears to be greater than 10% on HPLC, its nature should be confirmed by an alternative method to exclude misidentification of Hb N or Hb J as Hb F [18].

### *Hemoglobinopathy S*

Hemoglobin S hemoglobinopathy is the most common hemoglobinopathy detected in the United States. Possible diagnoses of patients with Hb S hemoglobinopathy include sickle cell trait (Hb AS), sickle cell disease (Hb SS), and sickle cell disease status post RBC transfusion/exchange. Patients with sickle cell trait may also have concomitant α-thalassemia, and the diagnosis of Hb S/β-thalassemia (0/ + / + +) is also occasionally made. Double heterozygous states of Hb SC, Hb SD, and Hb SO are important sickling states that should not be missed.

Patients with Hb SS disease may have increased Hb F. The distribution of Hb F among the haplotypes of Hb SS is as follows: Hb F, 5–7% in Bantu, Benin, or Cameroon; 7–10% in Senegal; and 10–25% in Arab/Indian types [19]. Hydroxyurea also causes an increase in Hb F. This is usually accompanied by macrocytosis. Hb F can also be increased in Hb S/HPFH.

Hb A2 values are typically increased in sickle cell disease and more so by HPLC. This is because the post-translational modification form of Hb S, Hb S1d, produces a peak in the A2 window. This elevated value of Hb A2 may produce diagnostic confusion with Hb SS disease and Hb S/β-thalassemia. It is important to remember that microcytosis is not a feature of Hb SS disease, and patients with Hb S/β-thalassemia typically exhibit microcytosis.

Hb SS patients and Hb S/β-0-thalassemia patients do not have any Hb A, unless the patient has been transfused or has undergone red cell exchange. Glycated Hb S has the same retention time (approximately 2.5 min) as Hb A in HPLC [19]. This will produce a small peak in the A window and raise the possibility of Hb S/β + -thalassemia.

Hb S/α-thalassemia is considered when the percentage of Hb S is lower than expected. Classical cases of sickle cell trait are 60% of Hb A and approximately 35–40% of Hb S. Cases of Hb S/α-thalassemia will have lower values of Hb S, typically below 30% with microcytosis. A similar picture will also be present in patients with sickle cell trait and iron deficiency. Common challenges in hemoglobinopathy detection are summarized in Table 18.3.

TABLE 18.3 Common Challenges in Hemoglobinopathy Detection

| Scenario | Cause of Challenge |
|---|---|
| HPLC | Carryover of sample from one to the next |
| Transfusion | Transfusion from donors who are, for example, Hb AS (S trait) or Hb AC (C trait) |
| Medication | Hydroxyurea raises Hb F levels |
| Iron deficiency | Lowers Hb A2 levels, thus masking diagnosis of β-thalassemia trait |
| HIV | Falsely increases Hb A2 levels |
| Agarose gel | Small percentages of abnormal hemoglobins (e.g., Hb A2′) may be undetected |
| HPLC | Hb A2′ elutes in the S window. Thus, in sickle cell disease or trait, Hb A2′ will be undetected |
| HPLC | Fast hemoglobin variants (e.g., Hb H and Hb Bart's) may not be quantified effectively. Hb F quantification will also be affected |

## DETECTION OF MONOCLONAL PROTEINS

A monoclonal protein (paraprotein or M protein) is a monoclonal immunoglobulin that is secreted by an abnormal clone of plasma cells [20]. The M protein can be an intact immunoglobulin, only light chains (light chain myeloma, light chain deposition disease, or AL amyloidosis), or rarely only heavy chains (heavy chain disease).

Serum protein electrophoresis (SPEP) is an inexpensive, easy-to-perform screening procedure for detection of monoclonal proteins [21]. It is usually done by the agarose gel method or by the capillary zone electrophoresis method. Monoclonal bands are usually seen in the γ zone but may be seen in proximity of the β band or rarely in the $\alpha_2$ area. Urine protein electrophoresis is analogous to SPEP and is used to detect monoclonal proteins in the urine. Ideally, it should be performed on a 24-hr urine sample. Common problems encountered in SPEP are listed in Table 18.4.

When a monoclonal band is identified on SPEP, serum immunofixation and 24-hr urine immunofixation is typically recommended. There are certain situations in which a band may be apparent that in reality is not a monoclonal band. Examples include the following:

1. Fibrinogen is seen as a discrete band when electrophoresis is performed on plasma. This fibrinogen band is seen between the β and γ regions. If the electrophoresis is repeated after the addition of thrombin, this band will disappear or serum immunofixation study will be negative.

TABLE 18.4 Common Problems Encountered in Serum Protein Electrophoresis

1. SPE performed on plasma will result in a band due to fibrinogen. Subsequent immunofixation will be negative
2. A band may be seen at the point of application. Typically, this band is present in all samples performed at the same time
3. If the concentration of transferrin is high (e.g., due to iron deficiency), this may result in a band in the β region
4. In nephrotic syndrome, prominent bands may be seen in $\alpha_2$ and β regions that are not due to monoclonal proteins
5. Hemoglobin–haptoglobin complexes (seen in intravascular hemolysis) may produce a band in the $\alpha_2$ region
6. M proteins may form dimers, pentamers, polymers, or aggregates with each other, resulting in a broad smear rather than a distinct band
7. In light chain myeloma, the light chains are rapidly excreted in the urine and thus SPE may fail to show a band

2. With intravascular hemolysis, the free hemoglobin binds to haptoglobin. The hemoglobin–haptoglobin complex may appear as a large band in the $\alpha_2$ area. Serum immunofixation studies should be negative in such cases.
3. In patients with iron deficiency anemia, concentrations of transferrin may be high. This may result in a band in the β region.
4. If electrophoresis is performed on a nephrotic syndrome patient, total protein and serum albumin is typically low. The condition also produces increases in $\alpha_2$ and β fractions. Bands in either of these regions may mimic a monoclonal band.
5. When performing gel electrophoresis, a band may be visible at the point of application. Typically, this band is present in all samples performed at the same time.

If the quantity of the M protein is low, this may not be detected by serum electrophoresis. There are also certain situations in which a false-negative interpretation may be made on serum electrophoresis:

1. A clear band is not seen in cases of α heavy chain disease (HCD). This is presumably due to the tendency of these chains to polymerize or due to their high carbohydrate content [22–25].
2. In μ-HCD, a localized band is found in only 40% of cases [26]. Panhypogammaglobulinemia is a prominent feature of such patients.
3. In occasional cases of γ-HCD, again a localized band may not be seen [24,27,28].
4. When an M protein forms dimers, pentamers, polymers, or aggregates with each other or when they form complexes with other plasma components, this may result in a broad smear rather than a discrete band.

5. Some patients produce only light chains. These light chains are rapidly excreted in the urine [29]. Serum electrophoresis fails to show any band. Urine studies are typically more fruitful. When the light chains cause nephropathy and result in renal insufficiency, excretion of the light chains is hampered. It is at this point that a band may be seen in serum electrophoresis.
6. In some patients with IgD myeloma, the M protein spike may be small enough to be disregarded.

## Hypogammaglobulinemia

Hypogammaglobulinemia may be congenital or acquired. Among the acquired causes are multiple myeloma and primary amyloidosis. Panhypogammaglobulinemia can occur in approximately 10% of cases of multiple myeloma. Most of these patients have a Bence–Jones protein in the urine but lack intact immunoglobulins in the serum [30,31]. Bence–Jones proteins are monoclonal-free $\kappa$ or $\lambda$ light chains in the urine. Panhypogammaglobulinemia can also be seen in 20% of cases of primary amyloidosis. It is important to recommend urine immunofixation studies when panhypogammaglobulinemia is present in serum protein electrophoresis.

## Immunofixation Studies

An apparent monoclonal protein on serum protein electrophoresis may or may not be a true monoclonal protein. Also, M proteins may not be apparent on electrophoresis. To confirm the former and because of the latter, if there is a high index of clinical suspicion, immunofixation studies are required. Immunofixation can be performed on both serum and urine specimens. It is preferable to perform urine immunofixation on a 24-hr urine sample. Immunofixation studies are more sensitive than regular electrophoresis and also determine the particular isotype of the monoclonal protein. However, they cannot estimate the quantity of the M protein (which the electrophoresis can do). One source of possible error in urine immunofixation study is the "step ladder" pattern. Here, multiple bands are seen in the $\kappa$ (more often) or $\lambda$ lanes and are indicative of polyclonal spillage rather than monoclonal spillage into the urine.

## Capillary Zone Electrophoresis

This is an alternative method of performing serum protein electrophoresis. Protein stains are not required. A point of application is not seen. It is considered to be faster and more sensitive compared to agarose gel electrophoresis. Classical cases of monoclonal gammopathy produce a peak, typically in the $\gamma$ zone. However, subtle changes in the $\gamma$ zones may also represent underlying monoclonal gammopathy. Interpretation can be subjective. Pathologists with a high index of suspicion will refer a high percentage of cases for ancillary studies such as immunofixation. Others, disregarding the subtle changes, may potentially miss positive cases.

## Free Light Chain Immunoassay

Patients with monoclonal gammopathy may have negative serum protein electrophoresis as well as serum immunofixation studies. Reasons include very low level of M proteins and light chain gammopathy, where the light chains are very rapidly cleared from the serum by the kidneys. Because of the latter, urine electrophoresis and urine immunofixation are part of the workup for cases in which monoclonal gammopathy is a clinical consideration. Urine electrophoresis and urine immunofixation studies are also performed to document the amount (if any) of potentially nephrotoxic light chains being excreted in the urine in the case of monoclonal gammopathy.

Quantitative serum assays for $\kappa$ and $\lambda$ free light chain (FLC) have increased the sensitivity of serum testing strategies for identifying monoclonal gammopathies, especially the light chain diseases [32–35]. Cases that appear as nonsecretory myeloma can actually be cases of light chain myeloma. FLC assays allow disease monitoring as well as provide prognostic information for monoclonal gammopathy of undetermined significance (MGUS) and smoldering myeloma.

The rapid clearance of light chains by the kidney is reduced in renal failure. Levels may be 20–30 times higher than normal in end-stage renal disease. In addition, the $\kappa$:$\lambda$ ratio may be as high as 3:1 in renal failure (normal, 0.26–1.65). Therefore, patients with renal failure may be misdiagnosed as having $\kappa$ light chain monoclonal gammopathy. If a patient has $\lambda$ light chain monoclonal gammopathy, with the relative increase in $\kappa$ light chain in renal failure, the ratio may become normal. Thus, a case of $\lambda$ light chain monoclonal gammopathy may be missed. It is also important to be aware that the presence of circulating M proteins may interfere with other laboratory tests. The most common errors that occur are falsely low high-density lipoprotein cholesterol, falsely high bilirubin, as well as altered values of inorganic phosphate. Other tests in which altered results may occur include low-density lipoprotein cholesterol, C-reactive protein, creatinine, glucose, urea nitrogen, and inorganic calcium.

There is a potential for inappropriate clinical decisions based on altered lab values.

### Cerebrospinal Fluid Electrophoresis

Qualitative assessment of cerebrospinal fluid (CSF) for oligoclonal bands is the most important diagnostic CSF study when determining a diagnosis of multiple sclerosis (MS). In MS, elevation of the CSF immunoglobulin level relative to other protein components occurs, suggesting intrathecal synthesis. The immunoglobulin increase is predominantly IgG, although the synthesis of IgM and IgA may also be increased.

Oligoclonal bands are defined as at least two bands seen in the CSF lane with noncorresponding band present in the serum lane. Thus, it is crucial to perform CSF and serum electrophoresis simultaneously. Oligoclonal bands may be found in 95% or more of patients with clinically definite MS [36]. However, they may also be seen in central nervous system infections (e.g., Lyme disease), autoimmune diseases, brain tumors, and lymphoproliferative disorders. Thus, it is important to realize that an oilgoclonal band is not equivalent to MS.

The first step when interpreting CSF electrophoresis is to establish that the sample is indeed a CSF sample. The presence of the prealbumin band and the band in the $\beta_2$ regions due to desialated transferin establishes that the sample is indeed a CSF sample. Both of these bands are not present in the serum lane. The presence of oligoclonal bands may then be confirmed by performing immunofixation studies.

An abnormality of CSF IgG production can be demonstrated in 90% of clinically definite MS patients [37]. There are various ways to document this. The CSF IgG level may be expressed as a percentage of the total protein, as a percentage of albumin, or by the use of the IgG index. It is important to correlate the findings of the electrophoresis with the IgG index, CSF study, as well as any pertinent magnetic resonance imaging and clinical findings.

In MS, CSF is grossly normal, and CSF pressure is normal. The total leukocyte count is normal in the majority of patients. If the CSF white blood cell count is elevated, it rarely exceeds 50 cells/μL [38]. Lymphocytes are the predominant cell type. CSF protein is also usually normal. If there is a systemic immune reaction or a monoclonal gammopathy, then bands will be seen in both the serum and the CSF lanes. These bands will correspond with each other, and they are not oligoclonal bands.

## CHALLENGES IN HIV TESTING

Testing for HIV can be broadly divided into screening tests and confirmatory tests. Screening tests include standard testing, rapid HIV testing, and combination HIV antibody and antigen testing. Confirmatory testing is performed by Western blot.

**TABLE 18.5** HIV Types and Groups

| HIV Type | Group | Description |
|---|---|---|
| I | | Related to viruses found in chimpanzees and gorillas |
| I | M | M denotes "major." Most common type of HIV; responsible for the AIDS pandemic |
| I | N | N denotes "non-M, non-O"; seen in Cameroon |
| I | O | O denotes "outlier"; most common in Cameroon; not usually seen outside west-central Africa |
| I | P | P denotes "pending the identification of further human cases"; the virus was isolated from a Cameroonian woman residing in France |
| II | | Related to viruses found in sooty mangabeys |

Standard screen is performed by enzyme immunoassay (EIA). The test is based on the detection of IgG antibody against HIV-1 antigens in the serum. These HIV antigens include p24, gp120, and gp41. Antibodies to gp41 and p24 are the first detectable serologic markers following HIV infection [39]. IgG antibodies appear 6–12 weeks following HIV infection in the majority of patients and generally persist for life. Assays for IgM antibodies are not used because they are relatively insensitive.

HIV viruses are categorized into the following groups: M, N, O, and P (Table 18.5). M is considered to be the pandemic strain and accounts for the vast majority of strains of HIV. Group O strains are from certain areas of Africa (Cameroon, Gabon, and Equatorial Guinea). Groups N and P are from Cameroon. Group M viruses are divided into 10 subtypes, A–J. Subtype B is the most common one found in the United States and Europe.

The two important issues regarding HIV screen tests are the ability of the test to detect non-M strain and the timing of the test post exposure. If the patient has not been seroconverted, as expected, the antibody is absent and the individual may have a false-negative test. There are also rare patients with HIV infection who become seronegative even though they showed seropositive results after exposure to HIV virus [40,41]. Other causes of false-negative results include the following:

- Fulminant HIV infection
- Immunosuppression or immune dysfunction
- Delay in seroconversion following early initiation of antiretroviral therapy.

False-positive serologic test for HIV is extremely rare. However, false-positive test results for HIV infection have been documented in individuals who have

TABLE 18.6 HIV Testing Methods

| Screening | Confirmatory |
|---|---|
| Standard testing (EIA) | Western blot |
| Rapid HIV testing | HIV RNA by PCR |
| Combination HIV antigen and antibody test | |

EIA, enzyme immunoassay; PCR, polymerase chain reaction.

received HIV vaccines in vaccine trials [42]. Some of these individuals who became HIV positive on screen tests also had a positive Western blot. HIV RNA testing is an approach that should be used to resolve such issues. HIV testing methods are listed in Table 18.6.

## Rapid HIV Antibody Tests

Rapid HIV antibody testing is available in the United States. Results are available within minutes because the tests can be done on-site and can be read by the provider. Rapid tests may not be as sensitive as second-generation EIAs. In one study of more than 14,000 specimens, rapid testing failed to detect 16 samples, which were positive by EIA and subsequently confirmed by Western blot [43]. A positive rapid test should be considered as "preliminary" and requires confirmation by Western blot. Reports of false positives do exist in the literature. A negative test may be reported as such and does not require further workup. However, early testing prior to seroconversion will naturally result in false-negative results.

## Combined Antibody Antigen Tests

Fourth-generation tests have the ability to detect both HIV antibody and the p24 antigen. Sensitivity and specificity of such tests are generally excellent. The primary advantage of these assays is the ability to detect HIV infection prior to seroconversion.

# HEPATITIS TESTING

Hepatitis infection is a worldwide problem. Hepatitis A is a problem in many developing countries because infection may spread from contaminated water (fecal–oral route). However, hepatitis A testing is straightforward; IgM anti-hepatitis A virus (HAV) denotes recent infection, and IgG anti-HAV appears in the convalescent phase of acute hepatitis. Hepatitis E virus (HEV) is also an enterically transmitted virus. HEV can also be transmitted by blood transfusion, particularly in endemic areas. Chronic hepatitis does not develop after acute HEV infection, except in the transplant setting and possibly in other settings of immunosuppression. Fulminant hepatitis can occur, resulting in an overall case fatality rate of 0.5–3%. For reasons as yet unclear, the mortality rate in pregnant women can be as high as 15–25%, especially in the third trimester. The diagnosis of HEV is based on the detection of HEV in serum or stool by polymerase chain reaction (PCR) or on the detection of IgM antibodies to HEV. Antibody tests against HEV alone are less than ideal because they have been associated with frequent false-positive and -negative results. The hepatitis D virus (HDV; also called the delta virus) is a defective pathogen that requires the presence of the hepatitis B virus (HBV) for infection. HDAg can elicit a specific immune response in the infected host, consisting of antibodies of the IgM and IgG class (anti-HDV). Hepatitis B and C infection serology, which is the most important component of hepatitis testing, is discussed in detail in the following sections.

TABLE 18.7 Hepatitis B Serology

| Antigen | |
|---|---|
| HBsAg | First detectable agent in acute infection |
| HBcAg | Not tested because it is not detectable in blood |
| HBeAg | Indicates virus is replicating and patient is highly infectious |
| **Antibody** | |
| Anti-HBc | First antibody to appear; should be positive when other tests for hepatitis B are negative during the window period (HBsAg is negative and anti-HBs is not yet detectable) |
| Anti-HBe | Virus is not replicating |
| Anti-HBs | Patient is immune |

## Serology for Hepatitis B

Serologic markers available for hepatitis B infection are HBsAg (hepatitis B surface antigen), HBeAg (hepatitis B e antigen), anti-HBc (antibody against hepatitis B core antigen; both IgG and IgM), anti-HBs (antibody against hepatitis B surface antigen), anti-HBe (antibody against hepatitis B e antigen), and HBV DNA (Table 18.7). HBsAg is the first marker to be positive after exposure to HBV. It can be detected even before the onset of symptoms. Most patients may clear the virus and HBsAg typically becomes undetectable within 4–6 months. Persistence of HBsAg for more than 6 months implies chronic infection. The disappearance of HBsAg is followed by the presence of anti-HBs. During the window period (after the disappearance of HBsAg and before the appearance of anti-HBs), evidence of infection is documented by the presence of anti-HBc (IgM).

Coexistence of HBsAg and anti-HBs has been documented in approximately 24% of HBsAg-positive individuals. It is thought that the antibodies fail to neutralize the virus particles. These individuals should be considered as carriers of HBV. A subset of patients have undetectable HBsAg and are positive for HBV DNA. Most of these patients have very low viral load with undetectable levels of HBsAg. Uncommon situations are infection with HBV variants that decrease HBsAg production or mutant strains that have altered epitopes normally used for detection of HBsAg.

Individuals with recent infection will develop anti-HBc, IgM antibodies. Individuals with chronic infection and individuals who have recovered from an infective episode will develop anti-HBc, IgG antibodies. However, anti-HBc, IgM antibodies may remain positive for up to 2 years after an acute infection. Levels may also increase and be detected during exacerbations of chronic hepatitis B. This may lead to diagnostic confusion.

Some individuals have isolated anti-HBc antibody positivity. The clinical significance of this finding is unclear. Some of these individuals have been found to have HBV DNA by PCR. This is true for samples from serum or liver. Transmission of hepatitis B has been reported from blood and organ donors who have had isolated anti-HBc antibody positivity. On the other hand, a certain percentage of individuals who have anti-HBc are false positive. The presence of HBeAg usually indicates that the HBV is replicating and the patient is infectious. Seroconversion to anti-HBe typically means the virus is no longer replicating. This is associated with a decrease in serum HBV DNA and clinical remission. In some patients, seroconversion is still associated with active liver disease. This may be due to low levels of wild-type HBV or HBV variants that prevent or decrease the production of HBeAg.

### Serology for Hepatitis C

The tests for hepatitis C virus (HCV) include serologic assays and molecular assays for HCV RNA. The commonly utilized screening assay to detect anti-HCV antibody is an EIA. The latest version is EIA-3, which has better sensitivity and specificity compared to EIA-1 or EIA-2. In addition, the mean time to detection of seroconversion has been reduced by 2 or 3 weeks. Rapid tests for detection of anti-HCV also exist. If an individual is positive for anti-HCV, then the logical next step is to test for HCV RNA. If there is true infection, both anti-HCV antibody and tests for HCV RNA should be positive. If the HCV RNA is negative, then the possibilities include the following:

1. A false-positive anti-HCV antibody test.
2. If the individual is a newborn, the anti-HCV may be that of the mother with transfer of antibodies across the placenta.
3. Intermittent viremia.
4. Past infection.

For establishing the diagnosis of past infection, another test, recombinant immunoblot assay (RIBA), may be undertaken. If the anti-HCV is a false-positive test, then RIBA will be negative. If it is a case of past infection, RIBA will be positive.

There may be individuals who have HCV RNA but anti-HCV antibody testing is negative. This can be seen in immunocompromised individuals or in the setting of early acute infection. HCV RNA tests are positive earlier than anti-HCV antibody tests.

## ANTI-NUCLEAR ANTIBODIES

Anti-nuclear antibodies (ANA) are serologic hallmarks of systemic autoimmune diseases [44]. A positive ANA may be seen in individuals with systemic or organ-specific autoimmune diseases and a variety of infections. They can also be found in otherwise normal individuals. Examples of organs implicated in organ-specific autoimmune diseases are thyroid, liver, and lung. Examples of infections in which ANA may be found to be positive include infectious mononucleosis, hepatitis C infection, subacute bacterial endocarditis, tuberculosis, and HIV infections [45–47]. If they are positive in otherwise normal individuals, these individuals are most likely women and elderly.

The ANA assay is performed by incubating human epithelial cell tumor line (Hep2) cells fixed with methanol and/or acetone with the patient's serum. Fluorochrome-labeled antihuman globulin is added, which binds to the antigen–antibody complex. The slide is then viewed through a fluorescent microscope. Antibodies present in the patient's serum will bind to the nuclear antigen and should also produce a pattern that is noted. The dilution of the serum at which the reaction pattern disappears is also noted.

The accurate interpretation of patterns requires considerable experience. Also, the pattern type has been recognized to have a relatively low sensitivity and specificity for different autoimmune disorders. As mentioned previously, false positives may be seen in normal individuals. The majority of these are present in low titer.

False negatives are also an established phenomenon due to technical and physical issues that include method of substrate fixation, solubility of the antigen in question, and localization of the antigen outside the nucleus. Once the ANA is positive, further testing is required. These tests include antibodies to double-stranded DNA, anti-histone antibodies, antibodies to chromatin, and antibodies to other nuclear proteins and RNA–protein complexes.

## CONCLUSIONS

Clinicians rely on accurate results from clinical laboratories for diagnosis of hemoglobinopathy, HIV, and hepatitis. A false-positive or false-negative test result has serious consequences for patient management. Moreover, a monoclonal band, if missed, can also be a serious patient safety concern. In this chapter, the pitfalls of immunology and serology testing were addresses, with emphasis on both false-positive and false-negative test results and how to eliminate some of these errors in the clinical laboratory.

## References


[1] Fairbanks VF. A pitfall in hemoglobin electrophoresis: artefactual minor unstable hemoglobin results from improper specimen handling. Am J Clin Pathol 1980;73(2):245–7.

[2] Lafferty J, Ali M, Carstairs K, Crwaford L. The effect of carbon tetrachloride on the detection of hemoglobin H using various commercially available electrophoresis products. Am J Clin Pathol 1998;109:651–2.

[3] Williams JS, Donahue SH, Gao H, Brummel CL. Universal LC-MS method for minimized carryover in a discovery bioanalytical setting. Bioanalysis 2012;4:1025–37.

[4] Lippi G, Mercadanti M, Alberta C, Franchini M. An unusual case of a spurious, transfusion-acquired haemoglobin S. Blood Transfus 2010;8:199–202.

[5] Suarez AA, Polski JM, Grossman BJ, Johnston MF. Blood transfusion-acquired hemoglobin C. Arch Pathol Lab Med 1999;123(7):642–3.

[6] Stephens AD, Angastiniotis M, Baysal E, Chan V, Fucharoen S, Giordano PC, et al. ICSH recommendations for the measurement of haemoglobin A2. Int J Lab Hematology 2012;34:1–13.

[7] Hoyer JD, Kroft SH. Color atlas of hemoglobin disorders. Northfield, IL: College of American Pathologists; 2003.

[8] Codrington JF, Li HW, Kutlar F, et al. Observations on the levels of Hb A2 in patients with different beta thalassemia mutations and a delta chain variant. Blood 1990;76:1246–9.

[9] Horton B, Payne RA, Bridges MT, et al. Studies on an abnormal minor hemoglobin component (Hb B2). Clin Chim Acta 1961;6:246–53.

[10] Pearson HA, Moore MM. Human hemoglobin gene linkage: report of a family with hemoglobin B2, hemoglobin S, and beta thalassemia, including a probable crossover between thalassemia and delta loci. Am J Hum Genet 1965;17:125–32.

[11] Vella F. Variation in HbA2. Hemoglobin 1977;1:619–50.

[12] Van Kirk R, Sandhaus LM, Hoyer JD. The detection and diagnosis of hemoglobin A2′ by high-performance liquid chromatography. Am J Clin Pathol 2005;123:657–61.

[13] Galacteros F, Amaudric F, Prehu C, Feingold N, Doucet-Populaire F, Sobel A, et al. Acquired unbalanced hemoglobin chain synthesis during HIV infection. Comptes Rendus de l'Academie des Sci Serie III, Sciences de la Vie 1993;316:437–40.

[14] Weatherall DJ, Clegg JB. The thalassemia syndrome. Blackwell, Oxford; 2001.

[15] Cartei G, Chisesi T, Cazzavillan M, Battista R, Barbui T, Dini E. Relationship between Hb and HbA2 concentrations in beta thalassemia trait and effect of iron deficiency anaemia. Biomedicine 1976;25:282–4.

[16] Steinberg MH. Case report: effects of iron deficiency and the −88C→T mutation on HbA2 levels in beta thalassemia. Am J Med Sci 1993;305:312–13.

[17] Stephens AD, Angastiniotis M, Baysal E, Chan V, Fucharoen S, Giordano PC, et al. ICSH recommendations for the measurement of haemoglobin F. Int J Natl Lab Hem 2012;34:14–20.

[18] Wild BJ, Banin BJ. Detection and quantification of variant haemoglobins: an analytical review. Ann Clin Biochem 2004;41:355–69.

[19] Baib BJ, Wild BJ, Stephens AD, Phelan L. Variant haemoglobins: a guide to identification. Hoboken, NJ: Wiley-Blackwell; p. 30–8.

[20] Bird J, Behrens J, Westin J, Turesson I, Drayson M, Beetham R, et al. UK myeloma forum (UKMF) and nordic myeloma study group. Guidelines for the investigation of newly detected M-proteins and the management of monoclonal gammopathy of undetermined significance (MGUS). Br J Haematol 2009;147(1):22.

[21] Katzmann JA, Kyle RA. Immunochemical characterization of immunoglobulins in serum, urine, and cerebrospinal fluid. In: Detrick B, editor. Manual of molecular and clinical laboratory immunology. 7th ed. Washington, DC: American Society for Microbiology Press; 2006.

[22] Seligmann M, Mihaesco E, Preud'homme JL, Danon F, Brouet JC. Heavy chain diseases: current findings and concepts. Immunol Rev 1979;48:145.

[23] Fermand JP, Brouet JC. Heavy-chain diseases. Hematol Oncol Clin North Am 1999;13(6):1281.

[24] Al-Saleem T, Al-Mondhiry H. Immunoproliferative small intestinal disease (IPSID): a model for mature B-cell neoplasms. Blood 2005;105(6):2274.

[25] Wahner-Roedler DL, Kyle RA. Mu-heavy chain disease: presentation as a benign monoclonal gammopathy. Am J Hematol 1992;40(1):56.

[26] Fermand JP, Brouet JC, Danon F, Seligmann M. Gamma heavy chain "disease": heterogeneity of the clinicopathologic features. Report of 16 cases and review of the literature. Medicine 1989;68(6):321.

[27] Wahner-Roedler DL, Witzig TE, Loehrer LL, Kyle RA. Gamma-heavy chain disease: review of 23 cases. Medicine 2003;82(4):236.

[28] Kyle RA, Greipp PR. "Idiopathic" bence jones proteinuria: long-term follow-up in seven patients. N Engl J Med 1982;306(10):564.

[29] Grosbois B, Jégo P, de Rosa H, Ruelland A, Lancien G, Gallou G, et al. Triclonal gammopathy and malignant immunoproliferative syndrome. Rev Med Interne 1997;18(6):470.

[30] Pola V, Tichý M. Bisalbuminemia: critical review and report of a case of an acquired form in a myeloma patient. Folia Haematol Int Mag Klin Morphol Blutforsch 1985;112(1):208.

[31] Rambaud JC, Halphen M, Galian A, Tsapis A. Immunoproliferative small intestinal disease (IPSID): relationships with alpha-chain disease and "Mediterranean" lymphomas. Springer Semin Immunopathol 1990;12(2-3):239.

[32] Bradwell AR, Carr-Smith HD, Mead GP, Tang LX, Showell PJ, Drayson MT, et al. Highly sensitive, automated immunoassay for immunoglobulin free light chains in serum and urine. Clin Chem 2001;47:673–80.

[33] Katzmann JA, Clark RJ, Abraham RS, Bryant S, Lymp JF, Bradwell AR, et al. Serum reference intervals and diagnostic

ranges for free kappa and lambda immunoglobulin light chains: relative sensitivity for detection of monoclonal light chains. Clin Chem 2002;48:1437–44.

[34] Drayson M, Tang LX, Drew R, Mead GP, Carr-Smith HD, Bradwell AR. Serum free light chain measurements for identifying and monitoring patients with nonsecretory myeloma. Blood 2001;97:2900–2.

[35] Lachmann H, Gallimore R, Gillmore JD, Carr-Smith HD, Bradwell AR, Pepys MB, et al. Outcome in systemic AL amyloidosis in relation to changes in concentration of circulating free immunoglobulin light chains following chemotherapy. Br J Haematol 2003;122:78–84.

[36] Freedman MS, Thompson EJ, Deisenhammer F, et al. Recommended standard of cerebrospinal fluid analysis in the diagnosis of multiple sclerosis: a consensus statement. Arch Neurol 2005;62:865.

[37] McLean BN, Luxton RW, Thompson EJ. A study of immunoglobulin G in the cerebrospinal fluid of 1007 patients with suspected neurological disease using isoelectric focusing and the Log IgG-index: a comparison and diagnostic applications. Brain 1990;113(Pt 5):1269.

[38] Rudick RA, Whitaker JN. Cerebrospinal fluid tests for multiple sclerosis. In: Scheinberg P, editor. Neurology and neurosurgery update series, vol. 7. Princeton, NJ: CPEC; 1987. p. 1.

[39] Allain JP, Laurian Y, Paul DA, Senn D. Serological markers in early stages of human immunodeficiency virus infection in haemophiliacs. Lancet 1986;2:1233.

[40] Farzadegan H, Polis MA, Wolinsky SM, et al. Loss of human immunodeficiency virus type 1 (HIV-1) antibodies with evidence of viral infection in asymptomatic homosexual men: a report from the multicenter AIDS cohort study. Ann Intern Med 1988;108:785.

[41] Sullivan JF, Kessler HA, Sha BE. False positive HIV test: implications for the patient. JAMA 1993;269:2847.

[42] Cooper CJ, Metch B, Dragavon J, et al. Vaccine induced HIV seropositivity/reactivity in noninfected HIV vaccine recipients. JAMA 2010;304:275.

[43] Stekler JD, Swenson PD, Coombs RW, et al. HIV testing in a high incidence population: is antibody testing alone good enough? Clin Infect Dis 2009;49:444.

[44] Reichlin M. Diagnosis criteria and serology. In: Schur PH, editor. Clinical management of systemic lupus erythematosus. New York: Grune & Stratton; 1983. p. 49.

[45] Kaplan ME, Tan EM. Antinuclear antibodies in infectious mononucleosis. Lancet 1968;1:561.

[46] Clifford BD, Donahue D, Smith L, et al. High prevalence of serological markers of autoimmunity in patients with chronic hepatitis C. Hepatology 1995;21:613.

[47] Bonnet F, Pineau JJ, Taupin JL, et al. Prevalence of cryoglobulinemia and serological markers of autoimmunity in human immunodeficiency virus infected individuals: a cross sectional study of 97 patients. J Rheumatol 2003;30:2005.

CHAPTER

# 19

# Sources of Errors in Hematology and Coagulation Testing

*Andy Nguyen, Amer Wahed*

University of Texas Health Sciences Center at Houston, Houston, Texas

## INTRODUCTION

This chapter is divided into two parts: The first part discusses sources of errors in hematology testing, and the second part addresses challenges in coagulation testing. To understand sources of errors in hematology, it is important to understand the steps involved in providing blood count values as well as interpretation of peripheral blood smears [1].

Blood for complete blood counts (CBC) is typically collected in vacuum tubes that contain the anticoagulant ethylenediaminetetraacetic acid (EDTA). The blood collected in the vacuum tube is analyzed on automated hematology analyzers for CBC results. These automated instruments have various channels. Different channels are used to obtain different counts. One channel is used for red blood cell (RBC) count and platelet count. Another channel is used to obtain the total white blood cell (WBC) count and hemoglobin level. In this channel, the red cells are lysed. Some instruments have a separate channel for hemoglobin. Other channels are for WBC differential count, reticulocyte count, and nucleated red cell count. Different methodologies exist to obtain the actual counts, including impedance (based on the measurement of changes of electrical resistance produced by a particle suspended in a conductive medium as it passes through an aperture of known dimension), conductivity measurement with high-frequency electromagnetic current, light scatter, and florescence-based (flow cytometric) methods. For the measurement of hemoglobin concentration, the red cells are lysed and hemoglobin (and also methemoglobin and carboxyhemoglobin) is converted to cyanmethemoglobin. The absorbance of light at 540 nm is measured to provide a hemoglobin level [2]. Each time a cell passes through the aperture, a pulse is produced. The pulse height is proportional to the cell volume. The distribution curves for the volume are separated from each other with a moving discriminator. Cells with a volume between 2 and 30 fL are counted as platelets. Cells with a volume of 40–250 fL are counted as red cells. In addition to the actual counts, RBC and platelet histograms are also provided. Each cell volume is measured directly, and the mean corpuscular volume (MCV) is calculated by averaging the volume of all the cells or by drawing a perpendicular line from the peak of the RBC histogram to the baseline. The red cell distribution width (RDW), which is a measure of anisocytosis, is calculated from the RBC histogram at 20% of peak height.

The lysing reagent causes WBCs to lose cytoplasm, and the cell membrane collapses around the nucleus. This allows differentiating the cells as the nuclear size differences are accentuated. WBCs are counted between the range of 30 and 300 fL. Typically, three peaks are seen in the WBC histograms. The first peak represents the lymphocytes, and the third peak represents the neutrophils. All other white cells are represented as the second peak.

As discussed previously, the automated instruments are actually measuring red cell counts, volumes, and hemoglobin levels. The RDW and MCV are calculated by the instrument from the red cell volume histogram. Values for hematocrit, mean corpuscular hemoglobin (MCH), and mean corpuscular hemoglobin concentration (MCHC) are calculated by the instrument as following:

$$\text{Hematocrit} = \text{MCV} \times \text{RBC count}$$
$$\text{MCH} = \text{hemoglobin/RBC count}$$
$$\text{MCHC} = \text{hemoglobin/hematocrit}$$

*Accurate Results in the Clinical Laboratory.*
DOI: http://dx.doi.org/10.1016/B978-0-12-415783-5.00019-0

## ERRORS IN HEMOGLOBIN MEASUREMENT AND RBC COUNT

Hemoglobin measurement is based on absorption of light at 540 nm. If the sample is turbid, this will produce higher hemoglobin levels. Examples of such state include hyperlipidemia [3], patients on parenteral nutrition [4], hypergammaglobulinemia, and cryoglobulinemia. Turbidity from very high WBC count can also falsely elevate hemoglobin levels. Smokers have high carboxyhemoglobin, which may falsely elevate the measured hemoglobin level.

Large platelets may be counted by some instruments as red cells. Also, red cell fragments greater than 40 fL will be counted as whole red cells. In both situations, the RBC count will be falsely high. Cold agglutinins will cause red cell agglutination *in vitro* and result in low RBC counts. If cold agglutinins are suspected, the sample should be warmed to obtain an accurate RBC count.

## ERRORS IN MCV AND RELATED MEASUREMENTS

If there is red cell agglutination, then red cell clumps will be counted as single red cells but the volume of the estimated cell will be much higher. This will result in falsely high MCV values. If large platelets are counted as red cells, then these platelets typically have less volume than a normal red cell. This will result in falsely low MCV values.

If the patient is in a state of high osmolarity, the cytoplasms of the red cells are also hyperosmolar. When diluents are added to the blood in the analyzer, water will move into the red cells, causing them to swell in size. MCV values will be higher than that in the *in vivo* state. Examples of hyperosmolar states are uncontrolled diabetes mellitus, hypernatremia, and dehydration [5]. The converse will occur in hypo-osmolar states.

Values for hematocrit, MCH, and MCHC are obtained by calculation using hemoglobin levels, RBC counts, and MCV values. If there is an error in any of these values, the calculated values will also be inaccurate.

## ERRORS IN WBC COUNTS AND WBC DIFFERENTIAL COUNTS

Falsely high WBC counts are more common than falsely low WBC counts. There are several situations in which the WBC count may be falsely elevated. One of the most frequent situations is high WBC count in the presence of a significant number of nucleated red blood cells (NRBCs). If an accurate WBC count is required, then a corrected WBC count needs to be performed. This can be done by some hematology analyzers by running the sample again in the "NRBC mode" or performing a manual count. Platelet aggregates and nonlysis of red cells are other causes of spuriously high WBC counts. If the high WBC count is due to nonlysis of red cells, this may be a tip-off for hemoglobinopathies. Target cells seen in hemoglobinopathies are typically resistant to lysis. Platelet aggregates may be due to EDTA, and redrawing blood in a citrate tube may be the solution in such cases. Erroneous WBC counts with spurious leukocytosis can be seen with the presence of cryoglobulins and microorganisms. Spurious leukopenia can be seen in cold agglutinins and EDTA-dependent leukoagglutination [6].

As discussed previously, WBC histograms have three peaks. The first peak represents lymphocytes, and it is during this peak that WBCs have the lowest cell volume. It is easy to understand that when there are giant platelets or nucleated red cells or red cells resistant to lysis, these may be counted as lymphocytes in some instruments, giving rise to a falsely high lymphocyte count. Hemoglobinopathies and target cells are important causes of nonlysis of red cells. The presence of malarial parasites in red cells has also been known to increase the lymphocyte count.

In myelodysplastic syndrome, if the myeloid series is affected, then hypolobated and hypogranular neutrophils can be present. Automated analyzers may no longer count these dysplastic neutrophils as such; instead, these neutrophils may be counted as lymphocytes.

Basophilia is typically seen in chronic myelogenous leukemia. Basophils are cells with coarse granules that may even obscure the nucleus. If the analyzer falsely recognizes all the dense granules of basophils as one single nucleus, then these cells could be counted as lymphocytes.

It is thus apparent that there can be multiple situations in which the lymphocyte count is inappropriately elevated. Whereas falsely low lymphocyte count is rare, falsely low neutrophils can be encountered more frequently. If there is an error in the neutrophil count, it is more likely to be a falsely low count than a high count. Neutrophil aggregation is a documented phenomenon and can result in low neutrophil count. Neutrophils have fine granules, whereas the granules of eosinophils are larger. Basophils have quite large granules. If neutrophils have hemosiderin granules, they may be counted as eosinophils. If eosinophils are hypogranular, they may be counted as neutrophils. Red cells infected by malarial parasites may contain malarial pigments. Malaria-infected red cells are resistant to lysis. These red cells with malarial pigments may be counted as eosinophils.

A key difference between lymphocytes and monocytes is that monocytes are significantly larger. Reactive (activated) lymphocytes typically have more abundant cytoplasm compared to nonreactive lymphocytes. Their size approaches that of a monocyte. These lymphocytes may thus be counted as monocytes.

Also, it has been reported that abnormal lymphocytes such as those seen in chronic lymphocytic leukemia, lymphoblasts, and leukemic or lymphoma cells can be miscounted as monocytes. When there is left shift in the WBC series, there is a tendency for slightly more immature cells such as bands and metamyelocytes to be seen. Cells that are more immature are naturally larger and may also be counted as monocytes. Storage of blood at room temperature and delay in running the sample on the analyzer may also contribute to inaccurate WBC differential values.

When differential counts obtained by automated analyzers are compared to differential counts performed manually, differences in results are relatively frequent. Most often, they are clinically inconsequential. However, it is important to correlate significantly abnormal results with morphological review of the peripheral smear.

## ERRORS IN PLATELET COUNT

In certain situations, hematology analyzers are known to provide a falsely low platelet count when the true platelet count is adequate. This may give rise to suboptimal clinical management.

Partial clotting of specimen or platelet activation during venipuncture may cause platelet aggregation. Both mechanisms may lead to low platelet counts. Checking the specimen for clots, analyzing the histograms, as well as reviewing the smear are all important steps to avoid misleading low platelet counts. There are various other mechanisms to explain falsely low platelet counts, otherwise referred to as pseudothrombocytopenia: anticoagulant-induced pseudothrombocytopenia, platelet satellitism, giant platelets, and cold agglutinin-induced platelet agglutination.

Falsely elevated platelet counts are much less common than falsely low counts. Fragmented red cells or white cell fragments may be counted as platelets, giving rise to high platelet counts. Fragmented red cells can be seen in states of microangiopathic hemolysis such as disseminated intravascular coagulation (DIC). White cell fragments can be seen in leukemic or lymphoma states. Patients with leukemia, especially acute leukemia, need supportive therapy in the form of blood component transfusions. If the platelet count is falsely elevated in a patient with acute leukemia, the decision to transfuse platelets may be delayed, with undesirable clinical consequences. Falsely high platelet counts may also be seen in the presence of cryoglobulins and microorganisms present in blood [7].

## ERRORS IN SPECIFIC HEMATOLOGY TESTING

In the following sections, specific selected test errors often seen in the hematology laboratory are discussed in more detail.

### Cold Agglutinins

Cold agglutinins are polyclonal or monoclonal autoantibodies directed against RBC i or I antigens and preferentially binding erythrocytes at cold temperatures [8]. These autoantibodies are typically immunoglobulin M subtype, which may be associated with malignant disorder (e.g., B cell neoplasm) or benign disorders (e.g., postinfection and collagen vascular disease) and can be manifested clinically as autoimmune hemolytic anemia [9]. In the hematology laboratory with automated analyzers, cold agglutinins typically present as a discrepancy between the RBC indexes [9,10]. The agglutinated erythrocytes may be recognized as single cells or may be too large to be counted as erythrocytes; subsequently measured mean corpuscular volume is falsely elevated and the RBC count is disproportionately low. Although the measured hemoglobin is correct due to its independence of cell count, the calculated indexes are incorrect: The hematocrit (red cell count × MCV) is low, whereas the MCH (hemoglobin/red cell count) and the MCHC (hemoglobin/hematocrit) are elevated. Hemagglutination may be grossly visible to the unaided eye [8], and microscopic examination of the peripheral blood smear would show erythrocyte clumping [9]. By rewarming the blood sample to 37°C, the erythrocyte agglutination is alleviated and correct values may be obtained [9]. More severe cases of cold agglutinin may require saline replacement technique if rewarming the sample fails to resolve the RBC index discrepancy.

Spurious leukopenia due to cold agglutinin is also occasionally encountered with automated hematology analyzers. The mechanism is postulated to be an IgM autoantibody directed against components of the granulocyte membranes [11]. Cold agglutinin-induced leukopenia should be recognized as a potential cause of pseudogranulocytopenia so that WBC counts can be accurately reported and unnecessary evaluation of patients for leukopenia can be avoided.

## Cryoglobulins

Cryoglobulins are typically IgM immunoglobulins that precipitate at temperatures below 37°C, producing aggregates of high molecular weight [8]. The first clue to a diagnosis of cryoglobulinemia is laboratory artifacts detected in the automated blood cell counts [12]. The precipitated cryoglobulins of various sizes may falsely be identified as leukocytes or platelets causing pseudoleukocytosis and pseudothrombocytosis. At the same time, the RBC indexes are generally unaffected. Correction of the artifacts for automated counts can be obtained by warming the blood to 37°C or by keeping the blood at 37°C from the time of collection to the time of testing. Peripheral blood smear typically shows slightly basophilic extracellular material, and leukocyte cytoplasmic inclusions are occasionally found.

## Pseudothrombocytopenia

Pseudothrombocytopenia is caused by various etiologies, including giant platelets [13], anticoagulant-induced pseudothrombocytopenia [14], platelet satellitism [15,16], and cold agglutinin-induced platelet agglutination [17]. Regarding giant platelets, due to their large size, they are excluded from electronic platelet counting causing pseudothrombocytopenia [13]. This scenario is of particular clinical importance in patients with rapid consumption of platelets in the peripheral circulation, such as in DIC, acute immune thrombocytopenic purpura, or thrombotic thrombocytopenic purpura. Effective platelet production by bone marrow in these cases will present with many large platelets in peripheral blood, many of which may not be identified by automated analyzers. An accurate platelet count can be obtained with a manual count using phase contrast microscopy.

Anticoagulant-induced pseudothrombocytopenia is an *in vitro* platelet agglutination phenomenon generally seen in specimens collected into EDTA [14]. It has been reported both in healthy subjects and in patients with various diseases (including collagen vascular disease and neoplasm) [12], and it has an overall incidence of approximately 0.1% [13,18]. Although the agglutination is most pronounced with EDTA, it may occasionally occur with other anticoagulants, such as heparin, citrate, and oxalate [14]. Because the platelet aggregates are large, the automated hematology counters do not recognize them as platelets, leading to lower platelet counts [14]. In some cases, the aggregates are large enough to be counted as leukocytes by automated instruments, causing a concomitant pseudoleukocytosis [14]. The platelet aggregation in pseudothrombocytopenia is usually temperature-sensitive [14], with maximal activity at room temperature. The EDTA-induced pseudothrombocytopenia is mediated by autoantibodies of IgG, IgM, and IgA subclasses [19] directed at an epitope on glycoprotein IIb [20]. This epitope is normally hidden in the membrane GP IIb/IIIa due to ionized calcium maintaining the heterodimeric structure of the GP IIb/IIIa complex [20]. Through its calcium chelating effect, EDTA dissociates the GP IIb/IIIa complex with GP IIb epitope exposure [20]. It has been noted that in Glanzmann's thrombasthenia, a disorder characterized by the quantitative and/or qualitative abnormality of glycoprotein IIb/IIIa, pseudothrombocytopenia does not occur [20]. Interestingly, in recent years, Abciximab (a GP IIb/IIIa antagonist) has been found to be associated with pseudothrombocytopenia [21]. If anticoagulant-induced pseudothrombocytopenia is suspected, a peripheral blood smear should be examined for platelet clumping [14].

Platelet satellitism has features similar to anticoagulant-induced pseudothrombocytopenia. In the presence of EDTA, platelets bind to leukocytes and form rosettes [15,16]. The binding is usually to neutrophils [15], but binding to other leukocytes has also been reported [16]. The automated analyzers do not identify platelets that bind to leukocytes, resulting in pseudothrombocytopenia. Platelet satellitism is mediated by autoantibodies of IgG type directed at GP IIb/IIIa on the platelet membrane and to an Fcγ receptor III on the neutrophil membrane [22].

Platelet agglutination due to cold agglutinins causing pseudothrombocytopenia is a rare condition. The platelet agglutination is anticoagulant-independent, usually occurs at 4°C, and is mediated by IgM autoantibodies directed against GP IIb/IIIa [17]. Because these autoantibodies have little activity at temperatures above 30°C, they are not associated with any clinical significance [1].

## Spurious Leukocytosis

The presence of microorganisms in the peripheral blood can result in spuriously high WBC counts or differentials by automated analyzers. Organisms that have been shown to be associated with this artifact include *Histoplasma capsulatum*, *Candida* sp., *Plasmodium* sp., and *Staphylococcus* sp. [23]. Spurious leukopenia due to EDTA is sometimes encountered [24,25]. Leukoagglutination has been reported as a transient phenomenon in neoplasia (especially lymphoma), infections (infectious mononucleosis, acute bacterial infection, etc.), alcoholic liver diseases, and autoimmune diseases (rheumatoid arthritis, etc.). It can also occur in the absence of any obvious underlying disease, even though an inflammatory condition is often found. Other well-known EDTA-dependent counting errors are platelet clumps and platelet-to-neutrophil satellitism. Association of neutrophil clumping and platelet satellitism has also been observed [26].

### False-Positive Osmotic Fragility Test

The osmotic fragility test is useful for diagnosis of hereditary spherocytic hemolytic anemia [27]. Spherocytes are osmotically fragile cells that rupture more easily in a hypotonic solution than do normal RBCs. Because they have a low surface area:volume ratio, they lyse at a higher solution osmolarity than do normal RBCs with discoid morphology. After incubation in a hypotonic solution, a further increase in hemolysis is typically seen in hereditary spherocytosis. Cells that have a larger surface area:volume ratio, such as target cells or hypochromic cells, are more resistant to lysing in a hypotonic solution.

Conditions associated with immunologically mediated hemolytic anemias may present with many microspherocytes in peripheral blood. Consequently, the fragility test can be positive in immunologically mediated hemolytic anemias other than hereditary spherocytosis, but the former would have a positive direct Coombs test and the latter would not.

### Errors Related to Sample Collection, Transport, and Storage

EDTA is the typical anticoagulant used in blood collection tubes. It can be in a dry format or as a solution. The amount and concentration of EDTA require that blood should be collected up to a specific mark on the tube. If too little blood is collected, dilution of the sample can become an issue with alteration of parameters. Relative excess EDTA in such cases also affects the morphology of blood cells. Transport of specimen should ensure that high temperatures are avoided. Red cell fragmentation is a feature of excess heat [28].

Prolonged storage will result in degenerative changes in WBCs. This is best illustrated in neutrophils, in which WBCs have a round pyknotic nucleus. To the casual observer, these cells may appear as nucleated red cells. Abnormal lobulation of the lymphocyte nuclei is another established phenomenon with prolonged storage of blood [29]. These cells may be considered as atypical lymphocytes, with an incorrect implication of an underlying lymphoproliferative disorder.

Table 19.1 summarizes sources of laboratory errors in hematology testings.

## COAGULATION TESTING

Patients with coagulation disorders may either bleed or form thromboses. Hemostasis involves

**TABLE 19.1** Sources of Laboratory Errors in Hematology

| |
|---|
| **Falsely high hemoglobin** |
| Turbid sample (hyperlipidemia, parenteral nutrition, hypergammaglobulinemia, cryoglobulinemia, marked leukocytosis) |
| Smokers (high caboxyhemoglobin) |
| **Falsely low hemoglobin** |
| Rare |
| **Falsely high RBC count** |
| Large platelets |
| Red cell fragments |
| **Falsely low RBC count** |
| Cold agglutinin |
| **Falsely high MCV** |
| Cold agglutinin |
| Hyperosmolar state (uncontrolled diabetes mellitus) |
| **Falsely low hemoglobin** |
| Large platelets |
| Hypoosmolar state |
| **Falsely high WBC count** |
| Nucleated red cells |
| Nonlysis of red cells (due to target cells in hemoglobinopathy) |
| Giant platelets or platelet clumps (due to EDTA) |
| Cryoglobulins |
| Microorganisms |
| **Falsely low WBC count** |
| Leukoagglutination (due to EDTA) |
| Cold agglutinin |
| **Falsely increased lymphocyte count** |
| Nucleated red cells |
| Nonlysis of red cells (due to target cells in hemoglobinopathy) |
| Giant platelets or platelet clumps (due to EDTA) |
| Malarial parasites |
| Dysplastic neutrophils (hypolobated neutrophils) |
| Basophilia |
| **Falsely decreased lymphocyte count** |
| Rare |
| **Falsely high neutrophil count** |
| Rare |
| **Falsely low neutrophil count** |
| Neutrophil aggregation |
| Neutrophil with hemosiderin granules (counted as eosinophils) |
| **Falsely increased eosinophil count** |
| Neutrophils with hemosiderin granules (counted as eosinophils) |
| Red cells with malarial pigments |
| **Falsely low eosinophil count** |
| Hypogranular eosinophils |

(*Continued*)

TABLE 19.1 (Continued)

| |
|---|
| Falsely increased monocyte count |
| Large reactive lymphocytes<br>Lymphoblasts<br>Lymphoma cells<br>Immature granulocytes |
| Falsely low monocyte count |
| Rare |
| Falsely high platelet count |
| Fragmented red cells (in microangiopathic hemolysis)<br>Fragmented white cells (in leukemia)<br>Microorganisms<br>Cryoglobulin |
| Falsely low platelet counts |
| Partial clotting or platelet activation<br>Giant platelets<br>Platelet clumps and platelet satellitism (due to EDTA)<br>GP IIb/IIIa antagonists<br>Platelet agglutination (due to cold agglutinin) |
| False-positive fragility test |
| Immunologically mediated hemolytic anemias |

MCV, mean corpuscular volume.

activation of the clotting factors and platelets [30]. Evaluation of platelet events may include a CBC, examination of the peripheral smear, bleeding time, and platelet aggregation test. Evaluation of the clotting factors is typically done by partial prothrombin time (PT) and activated partial thromboplastin time (aPTT) measurements. Abnormal PT or aPTT will usually lead to mixing studies to determine whether the abnormal result is due to factor deficiency or inhibitors. If there is correction of the prolonged clotting time in the mixing study, then factor assays will be performed to identify the deficient factor(s). Inhibitors include specific clotting factor inhibitors as well as lupus anticoagulants. Inhibitor screen and inhibitor assays or confirmatory tests for lupus anticoagulants will follow. Blood samples for coagulation tests are typically obtained in tubes with sodium citrate buffer. There are various sources of erroneous test results in coagulation testing, and these are addressed in the following sections.

## ERRORS IN PT AND aPTT MEASUREMENTS

PT measures the time required for a fibrin clot to form after addition of tissue thromboplastin and calcium to platelet-poor plasma collected in a citrated tube. PT measures the activity of VII, X, V, II, and fibrinogen. If aPTT is normal, then a prolonged PT is due to factor VII deficiency. PT is relatively insensitive to minor reductions in the clotting factors. aPTT is prolonged with deficiencies of XII, XI, X, IX, VIII, V, II, and fibrinogen. Just like PT, aPTT can be normal in minor deficiencies. In general, the deficient factor has to be approximately 20–40% to cause a prolonged aPTT. Most laboratories use automated methods for PT and aPTT measurements. Either optical or mechanical methods are employed to monitor clot formation. If measured with optical methods, shortened times may be seen with turbid plasma (e.g., hyperlipidemia and hyperbilirubinemia). aPTT is a test conventionally used to monitor heparin therapy. It is important to properly separate plasma from platelets as soon as possible. Platelet factor 4 can neutralize heparin, thus spuriously reducing aPTT values. Factor VIII levels are reflected in aPTT measurements. Factor VIII is an acute phase reactant. Again, if aPTT is being used to monitor heparin therapy in a patient who has an underlying cause for acute phase reactants to be elevated, aPTT values may be falsely lower than expected.

## ERRORS IN THROMBIN TIME MEASUREMENT

Thrombin time (TT) measures the time to convert fibrinogen to fibrin. Dysfibrinogenemia, elevated levels of fibrin degradation product, and paraproteins can interfere with fibrin polymerization, thus falsely prolonging TT. Amyloidosis can inhibit the conversion of fibrinogen to fibrin, also prolonging TT. In certain malignancies, heparin-like anticoagulants have been known to be the cause of prolonged TT. Next, specific selected test errors that are often seen in the coagulation laboratory are discussed in more detail.

### Incorrectly Filled Tubes

Citrate tubes for coagulation tests are designed for a 9:1 ratio of blood to citrate buffer. Both underfilling and overfilling of the citrate tube result in imbalances in this blood-to-buffer ratio and produce artificially prolonged or shortened clotting times, respectively [31]. Both underfilling and overfilling result in too little or too much blood sample for fixed amount of anticoagulant in the tube, respectively. The amount of blood that fills the citrated tubes is controlled by vacuum, which maintains the proper 9:1 ratio of blood to anticoagulant [32]. Underfilling may be caused by air bubbles in the tube, vacuum loss, or not allowing the tube to completely fill during the blood collection process.

If the tube stopper is removed, it also becomes difficult to obtain the correct amount of blood and attain the proper 9:1 ratio. If the patient's hematocrit is known in advance of blood collection to be greater than 55% (e.g., in patients with polycythemia) or below 21% (e.g., in patients with severe anemia), the amount of sodium citrate must be adjusted using the following formula:

$$C = 0.00185 \times (100\ H) \times V$$

where $C$ is the volume of 3.2% sodium citrate in milliliters, $H$ is the hematocrit in percentage, and $V$ is the volume of blood in milliliters.

Errors in clotting tests due to hematocrit changes without adjustment in the citrate volume are most significant with elevated hematocrits because even severe anemia does not significantly change PT or aPTT.

## Dilution or Contamination with Anticoagulants

Blood collection from indwelling lines or catheters could be a potential source of testing error. Sample dilution from incomplete flushing or hemolysis caused by improper catheter insertion can alter coagulation test results. Heparinized lines should be avoided if blood must be drawn from an indwelling catheter. If using a heparinized line is absolutely necessary, adequate line flushing must be achieved before blood collection. The National Committee for Clinical Laboratory Standards recommends flushing lines with 5 mL of saline [33]. At least 5 mL or six times the dead space volume of the catheter should be discarded before blood collection. The Intravenous Nursing Standards of Practice recommends that manufacturers' instructions should always be followed regarding the appropriate discard volume. These guidelines also state that blood should not be acquired from various types of indwelling cannula, venous administration sets, and indwelling cardiovascular or umbilical lines [34]. Even when the initial volume drawn is discarded before blood collection according to these guidelines, specimens drawn from a heparinized line are still easily contaminated with heparin. Consequently, blood for coagulation tests should be drawn directly from a peripheral vein, avoiding the arm in which heparin, hirudin, or argatroban is being infused for therapy [30]. Before coagulation testing, heparin may be removed or neutralized with polybrene in the coagulation laboratory; however, residual heparin may continue to cause testing interferences [35].

By enhancing antithrombin activity, heparin inhibits activated factors II (thrombin), X, IX, XI, XII, and kallikrein. In contrast, lepirudin, danaparoid, and argatroban inhibit only activated factor II [30]. These anticoagulants (heparin, lepirudin, and argatroban) prolong aPTT and interfere with coagulation tests such as factor assays and lupus anticoagulant assays. Factor assays may yield falsely low levels, whereas lupus anticoagulant may be falsely positive.

## Traumatic Phlebotomy

Traumatic phlebotomy can result in artificially shortened coagulation results such as PT and aPTT. This is due to excessive activation of coagulation factors and platelets by release of tissue thromboplastin from endothelial cells [36]. A proper free-flowing puncture technique will avoid this release of tissue thromboplastin and avoid this artifact.

## Fibrinolysis Products and Rheumatoid Factor

Fibrinolysis is mediated by plasmin, which degrades fibrin clots into D-dimers and fibrin degradation products. Plasmin also degrades intact fibrinogen, generating fibrinogen degradation products. Fibrin degradation products and fibrinogen degradation products are collectively known as fibrin/fibrinogen degradation products (FDPs) or fibrin/fibrinogen split products (FSPs). Assays for D-dimer and FDPs are semiquantitative or quantitative immunoassays.

### *Latex Agglutination*

Patient plasma is mixed with latex particles that are coated with monoclonal anti-FDP antibodies [37]. If FDP is present in the patient plasma, the latex particles agglutinate as FDP binds to the antibodies on the latex particles. These agglutinated clumps are detected visually. Various dilutions of patient plasma can be tested to provide a semiquantitative result known as FDP titer. Latex agglutination assays are also available for D-dimers. Various automated and quantitative versions of this assay are commercially available for D-dimers in which the agglutination is detected turbidimetrically by a coagulation analyzer rather than visually by a technologist [38,39].

### *Enzyme-Linked Immunosorbent Assays*

Quantitative enzyme-linked immunosorbent assays (ELISAs) are also available for FDPs and D-dimers. The traditional ELISA method is accurate but is not useful due to long analytical time. An automated, rapid ELISA for D-dimers is also available (VIDAS, bioMerieux) [40–42].

One of the most important limitations of D-dimer and FDP assays is interference by high rheumatoid factor levels. This may cause false-positive results with almost all available assays. The most useful clue to detect this interference is evaluation of the DIC panel.

If all values in this panel (PT, aPTT, TT, and fibrinogen) are normal except for FDP or D-dimer, the presence of rheumatoid factor is most likely the cause.

## PLATELET AGGREGATION TESTING WITH LIPEMIC, HEMOLYZED, OR THROMBOCYTOPENIC SAMPLES

Platelet aggregation measures the ability of platelets to adhere to one another and form the hemostatic plug, which is the key component of primary hemostasis [30]. It can be performed using either platelet-rich plasma or whole blood. Substances such as collagen, ristocetin, arachidonic acid, adenosine 5′-diphosphate, epinephrine, and thrombin can stimulate platelets and hence induce aggregation. Response to these aggregating agents (known as agonists) provides a diagnostic pattern for different disorders of platelet function. Measurement of aggregation response is typically based on changes in the optical density of the sample.

Platelet aggregation is affected by a number of confounding variables. Lipemic and hemolyzed samples complicate aggregation measurements because they obscure spectral changes due to platelet aggregation. Thrombocytopenia also makes platelet aggregation evaluations difficult to interpret because a low platelet count by itself may yield an abnormal aggregation pattern.

## CHALLENGES IN ANTICOAGULANTS AND LUPUS ANTICOAGULANT TESTS

The International Society on Thrombosis and Haemostasis Scientific Subcommittee on Lupus Anticoagulant recommended two sensitive screening tests for lupus anticoagulants that assess different components of the coagulation pathway: clotting time-based assays, such as the dilute Russell viper venom time (DRVVT), and aPTT-based assays, such as kaolin clotting time and dilute prothrombin time (tissue thromboplastin inhibition test) [43]. Lupus anticoagulants prolong various phospholipid-dependent clotting times in the laboratory because they bind to phospholipid and thereby interfere with the ability of phospholipid to serve its essential co-factor function in the coagulation cascade. Lupus anticoagulant screening assays usually have a low concentration of phospholipid to enhance sensitivity. Any abnormal (prolonged) screening result typically requires a 1:1 mixing study in which the patient plasma is mixed with one equal volume of normal plasma to demonstrate that the clotting time remains prolonged upon mixing. Confirmatory assays are performed if the screening assay remains abnormal after the 1:1 mixing. Confirmatory assays typically demonstrate that upon addition of excess phospholipid, the clotting time shortens toward normal. The platelet neutralization procedure is a confirmatory assay in which the source of the excess phospholipid is freeze–thawed platelets. The hexagonal phospholipid neutralization procedure is also based on the same principle—that is, the clotting time becomes corrected after addition of phospholipid in hexagonal phase. Note that aPTT may or may not be prolonged, depending on the amount of phospholipid in the reagent.

In many lupus anticoagulant assays, heparin (including subcutaneous low-dose heparin) may cause false-positive lupus anticoagulant results. By enhancing antithrombin activity, heparin inhibits activated factors II (thrombin), X, IX, XI, XII, and kallikrein. Subsequently, clotting times such as PT and aPTT are prolonged and interfere with lupus anticoagulant assays. Lepirudin, danaparoid, and argatroban inhibit activated factor II and can also prolong clotting times. Before coagulation testing, heparin may be removed or neutralized with polybrene in the coagulation laboratory; however, residual heparin may continue to cause testing interferences [44]. Results for lupus anticoagulant assays can be interpreted correctly in patients on Coumadin. Table 19.2 summarizes important laboratory coagulation errors due to various entities.

## CASE STUDIES

Two case studies highlight the issues discussed in this chapter.

### Case Study 1

A cardiac surgeon was following up on his patient during the first week of surgery. A CBC was done to assess current hematologic parameters. The hemoglobin level was acceptable at 12.5 g/dL. However, the RBC count was low at 2.9 million/mm$^3$ of blood. The values for MCV and MCH were also high. The surgeon was naturally concerned with the low RBC count and found the values for hemoglobin and RBC count discrepant. He called the pathologist to discuss the findings. The pathologist reviewed the smear and found red cell agglutination. Red cell agglutination can be seen in cold hemagglutinin disease. Because hemoglobin levels are measured after lysing red cells, whether the red cells are agglutinated or not does not matter. However, with red cell agglutination, the total RBC count would be reduced. The MCV and MCH values would be falsely high. The high MCH value

TABLE 19.2 Sources of Laboratory Errors in Coagulation

| |
|---|
| **Falsely prolonged clotting times** |
| Underfilling of citrate tube |
| Polycythemia |
| Sample from indwelling catheters (dilution or contamination with anticoagulant) |
| **Falsely shortened clotting times** |
| Overfilling of citrate tube |
| Traumatic phlebotomy |
| Turbid plasma (e.g., hyperlipidemia, hyperbilirubinemia) in optical instrument |
| **Falsely shortened aPTT in patients on heparin** |
| Delay in separation of plasma from platelets |
| Elevated factor VIII (acute phase reactant) |
| **Falsely prolonged TT** |
| Dysfibrinogenemia |
| Elevated levels of FDPs and paraproteins |
| Amyloidosis |
| Heparin-like anticoagulants (in malignancy) |
| **Falsely high FDPs and D-dimer** |
| Rheumatoid factor |
| **Falsely abnormal platelet function** |
| Lipidemia |
| Hemolysis |
| Thrombocytopenia |
| **Falsely low factor levels** |
| Heparin |
| Lepirudin |
| Danaparoid |
| Argatroban |
| **False-positive results of lupus anticoagulant tests** |
| Heparin |
| Lepirudin |
| Danaparoid |
| Argatroban |

aPTT, activated partial thromboplastin time; FDPs, fibrin/fibrinogen degradation products; TT, thrombin time.

should have been an indication for the lab technologist to preview the smear and to warm the blood prior to a repeat CBC on the hematology analyzer.

## Case Study 2

A 52-year-old female who has long-standing rheumatoid arthritis is under the care of an oncologist due to a recently diagnosed soft tissue sarcoma. The oncologist is concerned about chronic DIC and decides to evaluate her. Her CBC results show thrombocytopenia. Her PT and aPTT results are within normal limits. However, her D-dimer values are elevated. The oncologist calls the pathologist to discuss the findings in this case. The pathologist reviews her peripheral smear and observes platelet clumping. The pathologist explains that this phenomenon may be seen especially with samples collected in EDTA tubes. Recollection in heparin or sodium citrate tubes should result in an accurate and higher platelet count. Rheumatoid factor is an example of a false-positive D-dimer test. Ultimately, it was proven that this patient does not have DIC.

# CONCLUSIONS

In this chapter, common tests performed in the hematology and coagulation section of the laboratory were discussed. Sources of errors can potentially include all steps in testing—collection of samples, transportation, storage, and methodology used—as well as intercurrent issues of the patients. It is imperative to follow procedures and protocols for all concerned to attempt to obtain meaningful, accurate values. Laboratory technologists and pathologists need to be aware of situations in which erroneous results may be obtained. Correlation with clinical information provided or from the medical records is required in certain situations. If aware of issues related to possible erroneous results, clinicians will also contribute to providing appropriate interpretation of laboratory results. In essence, it is a team effort of laboratory personnel and clinicians to provide an accurate interpretation of laboratory tests for better clinical decisions and patient management.

## References

[1] Vajpayee N, et al. Basic examination of blood and bone marrow. In: McPherson R, Pincus M, editors. Henry's clinical diagnosis and management by laboratory methods. Philadelphia: Saunders; 2007.
[2] International Committee for standardization in Haematology. Recommendations for reference method for haemoglobinometry in human blood (ICSH Standard EP6/2:1977) and specification for international haemoglobinocyanide reference preparation (ICSH Standard EP6/3:1977). J Clin Pathol 1978;31:139–43.
[3] Nosanchuk JS, Roark MF, Wanser C. Anemia masked by triglyceridemia. Am J Clin Pathol 1977;62:838–9.
[4] Nicholls PD. The erroneous hemoglobin–hyperlipidemia relationship. J Clin Pathol 1977;30:638–40.
[5] Straucher JA, Altson W, Anderson J, Gustafson Z, Fadjardo LF. Inaccuracy in automated measurement of hematocrit and corpuscular indices in the presence of severe hyperglycemia. Blood 1981;57:1065–7.
[6] Hoffmann J. EDTA induced pseudo-neutropenia resolved with kanamycin. Clin Lab Haematol 23:193–6.
[7] Arnold JA, Jowzi Z, Bain BJ. Images in haematology: *candida glabrata* in a blood film. Br J Haematol 1999;104:1.

[8] Hoffman R. Hematology. Basic principles and practice. 3rd ed. New York: Churchill Livingstone; 2000. p. 622–623
[9] Bessman JD, Banks D. Spurious macrocytosis, a common clue to erythrocyte cold agglutinins. Am J Clin Pathol 1980;74:797–800.
[10] Lawrence C, Zozicky O. Spurious red-cell values with the Coulter Counter. N Engl J Med 1983;13(309):925–6.
[11] Robbins SH, Conly MA, Oettinger J. Cold-induced granulocyte agglutination: a cause of pseudoleukopenia. Arch Pathol Lab Med 1991;115:155–7.
[12] Fohlen-Walter A, Jacob C, Lecompte T, Lesesve JF. Laboratory identification of cryoglobulinemia from automated blood cell counts, fresh blood samples, and blood films. Am J Clin Pathol 2002;117:606–14.
[13] Garcia Suarez J, Merino JL, Rodriguez M, Velasco A, Moreno MC. Pseudothrombocytopenia: incidence, causes and methods of detection. Sangre (Barc) 1991;36:197–200.
[14] Schrezenmeier H, Muller H, Gunsilius E, Heimpel H, Seifried E. Anticoagulant-induced pseudothrombocytopenia and pseudoleucocytosis. Thromb Haemost 1995;73:506–13.
[15] Shahab N, Evans ML. Platelet satellitism. N Engl J Med 1998;338:591.
[16] Cohen AM, Lewinski UH, Klein B, Djaldetti M. Satellitism of platelets to monocytes. Acta Haematol 1980;64:61–4.
[17] Schimmer A, Mody M, Sager M, Garvey MB, Hogarth M, Freedman J. Platelet cold agglutinins: a flow cytometric analysis. Transfus Sci 1998;19:217–24.
[18] Bartels PC, Schoorl M, Lombarts AJ. Screening for EDTA-dependent deviations in platelet counts and abnormalities in platelet distribution histograms in pseudothrombocytopenia. Scand J Clin Lab Invest 1997;57:629–36.
[19] Bizzaro N. EDTA-dependent pseudothrombocytopenia: a clinical and epidemiological study of 112 cases, with 10-year follow-up. Am J Hematol 1995;50:103–9.
[20] Van Vliet HH, Kappers-Klunne MC, Abels J. Pseudothrombocytopenia: a cold autoantibody against platelet glycoprotein GP IIb. Br J Haematol 1986;62:501–11.
[21] Stiegler H, Fischer Y, Steiner S, Strauer BE, Reinauer H. Sudden onset of EDTA-dependent pseudothrombocytopenia after therapy with the glycoprotein IIb/IIIa antagonist c7E3 Fab. Ann Hematol 2000;79:161–4.
[22] Bizzaro N, Goldschmeding R, von dem Borne AE. Platelet satellitism is Fc gamma RIII (CD16) receptor-mediated. Am J Clin Pathol 1995;103:740–4.
[23] Marshall BA, Theil KS, Brandt JT. Abnormalities of leukocyte histograms resulting from microorganisms. Am J Clin Pathol 1990;93:526–32.
[24] Lesesve JF, Haristoy X, Lecompte T. EDTA-dependent leukoagglutination. Clin Lab Haem 2002;24:67–9.
[25] Hillyer CD, Knopf AN, Berkman EM. EDTA-dependent leukoagglutination. Am J Clin Pathol 1990;94:458–61.
[26] Deol I, Hernandez AM, Pierre RV. Ethylenediaminetetraacetic acid-associated leukoagglutination. Am J Clin Pathol 1995; 103:338–40.
[27] Palek J, Jarolin P. Hereditary spherocytosis. In: Williams WJ, Beutler E, Erslev AJ, Lichtman. MA, editors. In hematology. 4th ed. New York: McGraw-Hill; 1990. p. 558–69.
[28] Bain BJ, Diamond L. Pseudopyropoikilocytoisis: a striking artefact. J Clin Pathol 1996;49:772–3.
[29] Bain BJ., Blood cells: a practical guide, 4th ed. vol. 63. Wiley-Blackwell; Hoboken, NJ.
[30] Elizabeth M, Van Cott MD, Michael Laposata Ph.D. MD. Coagulation. In: Jacobs DS, et al., editors. The laboratory test handbook. 5th ed. Cleveland, OH: Lexi-Comp; 2001. p. 327–58.
[31] Fritsma GA, Quales LA. Top 10 problems in Coag. Adv Med Lab Prof 1997;9(24):8–13.
[32] Ens GE, et al. Specimen collection and pre-analytical variables. Coagulation handbook. Hemostase Resource Inc; 1998. p. 6–7
[33] National Committee for Clinical Laboratory Standards: Collection, Transport and Processing of Blood Specimens for Coagulation Testing and Performance of Coagulation Assays. Approved Guideline 3rd ed. December 1998, Document H21-A3, 18, No. 20, p. 2–3.
[34] Intravenous Nurses Society. Revised intravenous nursing standards of practice [Standard 33] J Infus Nurs 1998;21:51–2.
[35] Jenson R, Fritzma GA. Pre-analytical variables in the coagulation laboratory. Adv Admin Lab 2000;9(7):90–4.
[36] J. Kay Levens, BS, MT(ASCP): how reliable are your coagulation results? Adv Med Lab Prof 2001;22:12.
[37] Mirshahi M, Soria J, Soria C, et al. A latex immunoassay of fibrin/fibrinogen degradation products in plasma using a monoclonal antibody. Thromb Res 1986;44(6):715–28.
[38] Escoffre-Barbe M, Oger E, Leroyer C, et al. Evaluation of a new rapid D-dimer assay for clinically suspected deep venous thrombosis (Liatest D-dimer). Am J Clin Pathol 1998;109 (6):748–53.
[39] Bates SM, Grand'Maison A, Johnston M, et al. A latex D-dimer reliably excludes venous thromboembolism. Thromb Haemost 1999;82(Suppl):258.
[40] van der Graaf F, van den Borne H, van der Kolk M, et al. Exclusion of deep venous thrombosis with D-dimer testing: comparison of 13 D-dimer methods in 99 outpatients suspected of deep venous thrombosis using venography as reference standard. Thromb Haemost 2000;83(2):191–8.
[41] Perrier A, Desmarais S, Miron MJ, et al. Noninvasive diagnosis of venous thromboembolism in outpatients. Lancet 1999;353 (9148):190–5.
[42] Pittet JL, de Moerloose P, Reber G, et al. VIDAS D-dimer: fast quantitative ELISA for measuring D-dimer in plasma. Clin Chem 1996;42(3):410–15.
[43] Brandt JT, Triplett DA, Alving B, et al. Criteria for the diagnosis of lupus anticoagulants: an update. Thromb Haemost 1995; 74(4):1185–90.
[44] Jenson R, Fritzma GA. Pre-analytical variables in the coagulation laboratory. Adv. Admin. Lab 2000;9(7):90–4.

CHAPTER

# 20

# Challenges in Clinical Microbiology Testing

*Laura Chandler*

Philadelphia VA Medical Center and Perelman School of Medicine at the University of Pennsylvania, Philadelphia, Pennsylvania

## INTRODUCTION

Clinical microbiology is a discipline that encompasses a broad range of testing methodologies, and it is complex in terms of organisms and methods used to isolate and identify them. Although significant improvements in testing methodologies have been made, clinical microbiology remains heavily reliant on culture-based methods and phenotypic methods for identification of culture organisms. The wide variety of pathogens and testing methods that are available makes microbiological testing challenging, and thus error detection and correction are important components of quality microbiology laboratory testing. Errors may occur at all stages of testing (pre-analytical, analytical, and post-analytical), and an error in one stage of testing is likely to overlap with or lead to errors in other stages (e.g., incorrect specimen collection can lead to culture, identification, and reporting of organisms that are not involved in the disease process and to incorrect or unnecessary antimicrobial therapy as a result). In the clinical microbiology laboratory, as in every other discipline, the frequency of analytical errors has been reduced considerably with the implementation of quality control and quality assurance programs. Despite the improvements in microbiological testing, microorganisms remain a constant challenge, and errors do occasionally occur. This chapter discusses some of the common interferences in the clinical microbiology laboratory.

## ISSUES WITH PRE-ANALYTICAL ERRORS

Pre-analytical parameters are an important component of microbiology testing. High-quality results from the clinical microbiology laboratory are directly related to specimen quality. Pre-analytical steps in clinical microbiology testing include selection and proper ordering of the appropriate test method (appropriate test usage), selection and collection of high-quality specimens that are representative of the disease process, and timely transport of the specimen to the laboratory [1]. Problems with these pre-analytical steps in the testing process can result in laboratory errors. Pre-analytical errors include those related to specimen selection, specimen collection and transport, and storage, as well as test ordering. Often, errors that occur downstream in the testing process can be traced back to inappropriate test selection or problems with specimens.

### Errors Related to Specimens

Many of the errors that occur in clinical microbiology are related to specimen quality. Selection, collection, and transport of good-quality and appropriate specimens are critical components of obtaining accurate, meaningful results in clinical microbiology. Selection of an appropriate specimen is critical to ensure isolation and identification of the pathogenic agent [2]. The following sections discuss general categories in which errors in specimen selection, collection, transport, and storage occur that can give misleading or erroneous microbiology results.

#### *Specimens Collected from the Wrong Anatomic Site*

Specimens must reflect the disease process; specimens that are collected from the wrong site can yield misleading results. A common example is wound specimens that are collected by swabbing the surface of the lesion. The pathogen causing the infection is unlikely to be recovered from these samples. Rather, a

*Accurate Results in the Clinical Laboratory.*
DOI: http://dx.doi.org/10.1016/B978-0-12-415783-5.00020-7

sample collected from deep within the wound is preferable. These samples are more likely to reveal the pathogen involved in the infectious process, and they are less likely to be contaminated with commensal organisms from the skin.

### *Specimens are Contaminated with Endogenous Flora*

Specimens that must be collected through sites that contain endogenous flora (e.g., sputum and urine) must be collected using methods that avoid or minimize contamination. Growth of normal flora may inhibit growth and identification of the pathogen, resulting in a false-negative culture. Cultures that are overgrown with contaminating flora can be difficult to interpret, delaying the reporting of culture results. The presence of contaminating organisms may also result in workup (identification and susceptibility testing) of organisms that are not involved in the disease process (false positive).

### *Specimens Were Collected after Administration of Antimicrobial Agents*

Many pathogens are highly susceptible to antimicrobial therapy and are rapidly killed following antimicrobial administration. For example, in patients with meningococcal meningitis, cerebrospinal fluid (CSF) may be sterilized within 30 min of administration of antimicrobials. In cases in which a bacterial pathogen is suspected but culture results are negative, the possibility of antimicrobial administration prior to specimen collection should be considered. Failure to isolate and identify the pathogen can lead to incorrect patient management.

### *Suboptimal Volume of Sample was Submitted for Culture*

A specimen may be subjected to a concentration procedure prior to analysis due to low bacterial or viral burden (e.g., CSF) in the specimen. A common cause of false-negative culture results, especially in the mycobacteriology laboratory, is insufficient volume. Sterile body fluids such as pleural fluid, peritoneal fluids, or CSF are usually concentrated by the laboratory as part of the initial processing procedure. An insufficient volume of specimen may lead to false-negative culture results. Most laboratories have minimal recommended and required volumes; adherence to these guidelines can ensure optimal sensitivity of the test. Correct volume of specimen is especially important in the clinical mycobacteriology laboratory. Respiratory specimens being cultured for isolation of *Mycobacterium tuberculosis* are concentrated to optimize recovery of the organism.

### *An Inappropriate Transport Device Has Been Used*

The wide variety of tests and methods used in microbiology means that many different collection and transport devices are used. Transport devices are intended to stabilize the specimen, protect the organisms from degradation, maintain an anaerobic environment (for anaerobic culture requests), and prevent overgrowth of contaminating flora. The microbiology laboratory should have specific transport devices that are intended for specific groups of organisms (e.g., anaerobic transports) or specific tests.

### *Prolonged Transport Time*

If specimens are collected but not transported to the laboratory for processing quickly, fastidious organisms and anaerobes can quickly be lost. For specimens that may be delayed in transport to the laboratory, special collection devices with transport media should be used. Cultures for anaerobes (e.g., abscess material, tissues, and biopsies) are especially subject to loss of viability unless the material is transported in appropriate anaerobic transport media. Anaerobes are rapidly killed upon exposure to oxygen; cultures submitted to the laboratory for anaerobes should be processed without a delay.

### *Storage of Specimen at Inappropriate Temperature*

If the microbiology laboratory does not process specimens on every shift, specimens that arrive at the laboratory may need to be stored temporarily; most specimens can be stored at 4°C to maintain organism integrity and suppress the growth of contaminants. However, certain organisms are very sensitive to cold temperatures, and cultures for these pathogens should be performed immediately after specimens are collected. Examples of organisms that do not survive well during prolonged transport or in a refrigerated storage are *Neisseria meningitidis*, *Neisseria gonorrhoeae*, *Streptococcus pneumoniae*, and *Haemophilus influenzae*. Specimens should be transported to the laboratory without delay and not refrigerated.

## Errors Related to Microbiology Test Ordering

The diversity of microorganisms and the wide variety of test methods available can be a challenge for laboratory customers. With rapidly changing technologies, microbiology laboratories are regularly updating and improving their testing profile. Customers are not always aware of the methods that are used by their laboratory. Good communication to laboratory customers is necessary to ensure that physicians are aware of the recommendations for testing and that they know what is or is not included in a certain test. Inappropriate tests are ordered when physicians are not aware of what a culture- or polymerase chain reaction (PCR)-based test includes. Failure to order the correct test can result in failure to isolate or detect a pathogen. Some of the more common occurrences are described here.

#### *Routine Culture Does Not Include Organism of Interest*

Stool cultures for enteric pathogens are one of the most commonly ordered cultures in the clinical microbiology. Routine stool cultures may not include special methods for isolation of uncommon pathogens such as *Yersinia enterocolitica* or *Vibrio* spp. To prevent this type of common ordering error, physicians should consult the laboratory user guide to determine what is included in the test they have ordered.

#### *Special Methods are Needed for Isolation of Organism*

Inappropriate testing may also occur when special methods are needed to identify certain pathogens. For example, some organisms cannot be isolated in routine cultures. Routine lower respiratory cultures generally do not include methods for isolation and identification of fastidious respiratory pathogens such as *Legionella* and *Bordetella*. Special collection and transport methods as well as culture methods or PCR testing should be performed for these organisms. The laboratory must make physicians aware of the diagnostic strategy used in the facility for unusual pathogens to prevent test ordering and interpretation errors. The laboratory should have guidelines available for diagnosis of infections due to these difficult-to-culture organisms. For many of these uncultivable organisms, molecular testing or serological testing may be the most appropriate method to diagnose them.

#### *Multiple Methods are Available but the Wrong Method is Ordered*

With the advent of more sensitive and specific testing methodologies, there may be several options available for testing. Two important examples are herpes simplex virus (HSV) encephalitis and cryptococcal meningitis. Less sensitive methods to diagnose these illnesses are still available (e.g., "viral culture" or "fungal culture"). For both of these organisms, highly sensitive, rapid methods are available, and the old methods are obsolete for diagnosis. Health care providers should ensure that the appropriate test is ordered. For HSV encephalitis, the recommended test is a molecular assay such as PCR rather than culture. For cryptococcal meningitis, a cryptococcal antigen test is recommended rather than CSF fungal culture or the obsolete India ink test. Good communication with the laboratory is critical, and health care providers should consult the laboratory's guidelines to ensure that the correct test is ordered [3].

## Errors in Blood Culture Collections

Pre-analytical errors associated with drawing specimens for blood cultures can have significant effects on the results of the culture. Several commonly encountered errors may occur with blood cultures submitted to the microbiology laboratory, including submission of a single set of blood cultures; inadequate volume of blood; drawing blood cultures through a catheter; and inadequate preparation of the draw site, resulting in contamination. These errors have significant impact on the laboratory as well as on patient care.

#### *Single Blood Culture Sets*

Submitting a single set (e.g., one aerobic and one anaerobic bottle) or a single bottle for the diagnosis of bacteremia has several disadvantages that can lead to errors in culture workup and interpretation of results. Most important, the sensitivity of the blood culture is reduced. Many studies have shown that optimal sensitivity is achieved with two or three sets of blood cultures [4,5]; current clinical practice guidelines for diagnosis of sepsis recommend that two blood culture sets be obtained within 24 hr of each other. When solitary bottles or sets are submitted, the likelihood of isolating a pathogen is reduced because the total volume of blood that is cultured is less than the optimal volume, effectively reducing the sensitivity of the culture. There are recommended benchmarks for solitary blood culture rates; an ideal rate is less than 5% [6]. Laboratories can monitor solitary blood culture rates as part of their quality management program and improve their solitary collection rate by implementing changes such as ensuring a blood culture order includes recommendations for two sets and re-educating phlebotomy or other staff who collect blood cultures. In a College of American Pathologists Q-Probes study performed in 2001, the lowest rate of solitary blood cultures was seen in institutions that had implemented several practices: having phlebotomists draw cultures, instituting a policy requiring two sets per order, and monitoring of the solitary collection rate [6].

Single bottles or single sets also present a challenge in interpretation when organisms that may represent contaminants are isolated (e.g., coagulase-negative staphylococci and streptococci). Interpretation of blood cultures that grow organisms that are potential contaminants can be difficult when the culture consists of a single set [7]. With certain organisms, it can be difficult to determine whether the organism represents a pathogen or a contaminant [8].

#### *Inadequate Volume of Blood Collected*

Commercially available blood culture bottles require a certain defined blood volume to be drawn into the bottles. For adults, the recommendation is 8–10 mL of blood per bottle. The correct volume of blood is important for two reasons: (1) Optimal sensitivity is achieved when the correct volume of blood is cultured, and (2) anticoagulants such as sodium polyanetholsulfonate that are included in the culture media are diluted to the level at which they will not inhibit the growth of fastidious microorganisms. For pediatric patients, blood volumes are generally calculated based on the weight of the patient.

#### *Drawing Blood Cultures through a Catheter*

Recommendations for blood cultures are to draw through a venipuncture. Drawing blood cultures through a catheter can result in several confounding problems, including culture and workup of colonizing organisms and dilution of blood by fluids in the catheter line. Drawing blood cultures through a vascular catheter should be discouraged [8–10]. Discarding an initial aliquot of blood when using a catheter to obtain the blood specimen does not reduce the rate of contamination [11].

#### *Inadequate Preparation of the Draw Site*

Preparation of the venipuncture site to inactivate skin organisms is necessary to prevent growth of contaminants. The isolation and identification of blood culture contaminants is a significant problem for clinical microbiology laboratories and for patient care providers. Workup of contaminating organisms adds extra, unnecessary work for the laboratory and, importantly, may provide misleading results to clinicians, resulting in unnecessary antimicrobial therapy for patients. Careful adherence to the laboratory's procedure for preparation of the draw site for blood cultures can reduce the growth of microbial contaminants [9,10]. A policy for workup of positive blood cultures based on the organism identification and number of bottles may help prevent excess work and misleading reports from the laboratory.

#### *Delayed Entry of Blood Culture Bottles into Automated Instruments*

Blood cultures should be placed into automated instruments immediately after collection (ideally, within 1 hr). In reality, delays in bottle entry often occur, especially when the microbiology laboratory is located off-site, such as in a central laboratory or at a referral laboratory. In these cases, delays of up to 24 hr are not uncommon. The storage conditions and the length of time for blood cultures can result in reduced recovery of organisms [12].

### Pre-Analytical Error: Microbiology Cultures Not Ordered on Surgical Specimens Submitted to the Pathology Laboratory

A common occurrence is failure to order microbiological testing potential for an infectious process when specimens are surgically obtained for submission to pathology. Often, these specimens are submitted in formalin, therefore obviating the possibility of culturing any live organisms.

### Suggestions for Reduction in Pre-Analytical Errors in the Clinical Microbiology Laboratory

Good communication between the laboratory and the physicians using the laboratory is the best way to prevent pre-analytical errors. The laboratory should routinely check and update its specimen collection and test ordering guidelines. A program of electronic updates to test menus is ideal. The laboratory supervisors and directors should also be available for consultation and guidance on test selection, ordering, and interpretation.

## ANALYTICAL ERRORS IN THE MICROBIOLOGY LABORATORY

With programs of quality control, quality assurance, and quality improvement in place, analytical errors in the clinical microbiology have been greatly reduced during the past 20 years. Clinical microbiology testing has also improved because of significant progress in testing methodologies, technical improvements, and availability of tests with better sensitivity and specificity. The frequency of analytical errors in the clinical laboratory is now quite low. Total quality management programs now in use in most clinical laboratories are designed to prevent and minimize analytical errors. By implementing and integrating training, competency assessment, proficiency testing, quality control, and quality assurance in the clinical microbiology laboratory, many potential analytical errors are prevented or detected and corrected before reporting, reducing the number of errors that impact patient results. With the availability of standards and guidelines for most of the testing performed by clinical microbiology laboratories, analytic errors have been reduced to a very low level. When errors do occur during the analytic phase of microbiology testing, they tend to occur as a result of biological variability exhibited by microorganisms. Automated systems for identification and susceptibility testing are also subject to occasional errors, especially with testing of unusual organisms, fastidious organisms, or organisms with unusual resistance to antimicrobial agents. Some of the more commonly seen analytical errors that can occur in the clinical microbiology laboratory are discussed next.

## Gram Stains Errors

The Gram stain is one of the most important methods used in the clinical microbiology laboratory. It is used for initial assessment of specimens prior to culture and is an important step in the workup and identification of pathogens isolated from clinical specimens. Gram stains guide workup pathways and identification of organisms; thus, they are extremely important to the accurate processing and workup of specimens. Gram stain errors can occur for several reasons: Organisms may not exhibit an expected Gram stain morphology, or technical errors performing the Gram stain may occur:

*Technical errors in preparation of the smear and performing the Gram stain*: Proper preparation of the Gram smear must be performed to ensure that the smear is not too thick or too thin. Thin smears can result in reduction of sensitivity, especially when organisms are rare in the specimen, and are prone to over-decolorization, causing Gram-positive organisms to appear Gram-negative. Thick smears are prone to under-decolorization, can reduce specificity, cause Gram-negative organisms to appear Gram-positive, and may retain artifacts during the staining, making reading difficult.

*Inherent variability in organism Gram characteristics*: Certain organisms do not always exhibit expected Gram stain characteristics; occasionally, Gram-positive organisms will appear Gram-negative, and Gram-negative organisms retain crystal violet during the destaining step and appear Gram-positive (Table 20.1). Two genera of Gram-positive organisms that are often isolated in blood cultures, *Clostridium* spp. and *Bacillus* spp., may appear Gram-negative in smears prepared from culture-positive bottles. *Acinetobacter*, a Gram-negative coccobacillus, may resist destaining and appear Gram-positive in a Gram smear. Medical technologists should be aware of this possibility and use other features of the organism (size, morphology, and other biochemical characteristics) to assist in reporting. Alternative methods of determining the Gram characteristic (e.g., Gram-Sure reagent) may be used to test organisms isolated in pure culture.

### *Analytical Errors in Interpretation of Gram Stains Performed on Positive Blood Cultures*

There are several types of errors that can potentially occur when Gram stains are performed on bottles that signal positive in automated systems. The two most common sources of analytical errors are incorrect reporting of the Gram characteristic (Gram positives reported as Gram negatives and Gram negatives reported as Gram positives). These were primarily due

**TABLE 20.1** Organisms That May Exhibit Unusual Gram Stain Characteristics and Suggestions for Identifying Aberrant Results

| Organism | Usual Gram Morphology | Possible Aberrant Result | Clues to Possible Aberrant Result |
|---|---|---|---|
| *Bacillus* spp. | Gram-positive bacilli | Gram negative | Use colony morphology with Gram stain result |
| | | | *Bacillus* spp. are often larger than most gram-negative bacilli |
| *Clostridium* spp. | Gram-positive bacilli | Gram negative | Clostridia are anaerobes |
| | | | Organism size is very large |
| | | | Often in short chains |
| *Acinetobacter* spp. | Gram-negative coccobacilli or bacilli | Gram positive | Coccobacillary shape is unusual for gram-positive organisms such as streptococci or staphylococci |
| *Campylobacter* | Faintly staining Gram negative; small coccobacilli with characteristic shape | No organisms seen | Use fuchsin counterstain |
| | | | Perform an acridine orange stain |
| *Legionella, Bordetella* | Faintly staining Gram negative | No organisms seen | Use fuchsin counterstain |
| *Brucella* spp. | Small Gram-negative cocci or coccobacilli | Gram positive | Use colony characteristics and growth characteristics to verify |

to the staining characteristics of the organisms as described previously [13]:

*Error in not recognizing more than one organism present in a positive smear*: The incidence of bacteremia with more than one organism is fairly low. In a study published in 2003, the rate of polymicrobial sepsis was 4.7% [14]. There is potential for failure to recognize the presence of a second morphotype when Gram stains of positive blood cultures with multiple morphotypes present are read. Positive blood cultures with mixed morphologies (e.g., Gram-positive cocci and Gram-negative bacilli) are occasionally misread as single morphologies present. Occasionally, errors can occur when two morphologies occur and one organism is predominant. It may be difficult to distinguish Gram-positive cocci in clusters, pairs, or short chains when these organisms are mixed. Occasionally, errors occur when Gram-negative bacilli or coccobacilli are overlooked in smears that also have Gram-positive organisms.

*Error in not recognizing the presence of organisms in blood cultures that signal positive but no organisms are seen on the Gram stain*: Blood cultures that are incubated in continuously monitored instruments may signal positive but appear negative when the culture medium is stained. There are two scenarios in which this may occur: false-positive signaling by the instrument (no organisms are present but the instrument sensor detects $CO_2$, giving a positive signal) and the presence of organisms that are difficult to visualize in a Gram stain. The laboratory should have methods in place to determine which of these is occurring when it has signal positive bottles that are Gram stain negative.

Certain organisms do not Gram stain well, including *Legionella*, *Campylobacter*, *Bartonella*, and *Brucella*. Alternate staining procedures should be used, depending on whether the Gram stain is being performed on direct clinical specimens or on isolated colonies growing on solid media. Carbol fuchsin stains can be used when organisms are not staining well with the Gram stain. Gram stain errors can occur with fastidious or difficult-to-stain organisms growing in blood culture bottles. These bottles may signal positive, but when the smear is examined by a Gram stain, they are negative. Occasionally, blood cultures that are incubated in a continuously monitoring system will signal positive and the Gram stain shows no visible organisms. In this case, an alternate staining method such as acridine orange should be used to determine if organisms are present. Anaerobes may not stain properly when exposed to oxygen [15].

*CASE REPORT: GRAM STAINS NEGATIVE, CULTURE POSITIVE BLOOD BOTTLES:* Four blood culture bottles signaled positive in the automated blood culture instrument. Gram staining of the culture medium was performed; no organisms were noted on the Gram smear. The blood culture medium was subcultured to solid media, and the bottles were returned to the instrument. The following day, bacterial colonies were evident on the subcultured media. The organism was identified as *Campylobacter jejuni* by sequencing.

Negative Gram stains of blood culture media that signal positive in automated instruments occasionally occur. In this case, *C. jejuni* was growing in the blood cultures but was not recognized in the Gram stain when the bottles signaled positive; recognition that an organism was present was not achieved until the subcultures were growing. Staining of the blood culture medium using an alternative method such as acridine orange can reveal the presence of organisms, allowing for a report to be generated.

### *Analytical Error: False-Positive Reports of Growth in Liquid Broth Cultures—Reporting the Presence of Organisms in Culture-Negative Broth Cultures*

Liquid culture media such as thioglycollate broth is commonly used in addition to solid media as a primary isolation media to allow for recovery of anaerobes and organisms that may be present in low numbers in a specimen. The liquid cultures are usually examined visually, and if turbidity is present, a smear is made for Gram staining. A common occurrence is to observe organisms present in the Gram stain (e.g., Gram-positive bacilli, Gram-positive cocci, or Gram-negative bacilli) and to report these. On subculture, the organisms do not grow. These false-positive reports of organisms present a challenge to the laboratory. Liquid nutrient media such as thioglycollate broth may be sterilized by methods other than filtration; organisms present in the original material are nonviable, but they may still be present in the liquid media and may appear in the Gram stain. Liquid nutrient media should not be used for initial processing of specimens (e.g., tissues) because the primary Gram stain report may be a false positive. Tissues should be ground in sterile saline, and then a portion of the ground specimen should be transferred to the liquid thioglycollate medium. Also, a Gram stain report from the broth media culture should not be made; subculture should be performed, and if organisms are recovered, they can be reported.

Liquid blood culture media can also have nonviable organisms present; if the bottles signal in an automated instrument, organisms may be present on the Gram stain, but they are nonviable. These organisms represent artifacts, but they are very problematic and can be confusing.

### *Methods to Reduce Gram Stain Errors*

Gram stain errors can be minimized by ensuring ongoing quality assurance programs, including that routine competency assessment programs are in place in the laboratory. A good way to provide ongoing education is for the laboratory to include challenging organisms in regular plate rounds, allowing everyone to review the Gram stains after final identification of the organism. For anaerobic bacteria, preparation of the smear in anaerobic conditions and prompt fixing of the smear can help ensure that organisms will display the proper Gram staining characteristic [15].

## Misidentification of Organisms

Incorrect identification of microorganisms remains one of the more common errors that occur in the clinical microbiology laboratory. Phenotypic identification using biochemical testing is still the most common method by which microorganisms are identified, but there are limitations to the use of biochemical identification of bacteria. Fastidious bacteria and anaerobic bacteria may not grow well in these systems and often are not correctly identified. Automated systems perform well for most nonfastidious Gram-negative organisms and the most commonly isolated Gram-positive cocci such as staphylococci and enterococci. Automated systems have limitations for certain groups of organisms (e.g., Gram-positive bacilli) and thus alternative methods must be available for laboratories to use when they encounter organisms that are not accurately identified in an automated system. There are several different reasons for misidentification of organisms, and as with many other types of errors in the laboratory, organism misidentification is often a result of more than one error in the identification process.

### *Incorrect Algorithm or Identification Pathway*

Identification of organisms that have been isolated in culture is performed using many tests and parameters in stepwise algorithms. Different groups of organisms may require different testing methodologies, especially for fastidious organisms, for organism identification.

### *Commercial/Automated Identification System Errors*

Commercial identification systems have limitations for several groups of organisms. The error rates for identification of nonfastidious Gram-negative bacteria, staphylococci, and enterococci are quite low. Fastidious Gram-negative organisms are often misidentified by automated systems. Organisms may not grow adequately in the panels used in these identification systems, and databases may not contain all organisms. Examples of organisms that can be misidentified in commercial and automated systems include *Brucella* spp. and *Francisella tularensis*. *Francisella* is frequently misidentified as *Haemophilus influenzae* or *Actinobacillus* spp. on automated systems. *Brucella* spp. have been misidentified as *Ochrobactrum anthropi*, *Oligella ureolytica*, or *Psychrobacter* [16–18]. Laboratories should be aware of the potential for misidentification of these organisms, and suspected *Francisella*, *Brucella*, and similar organisms should not be tested on an automated system. Alternative methods such as DNA sequencing or mass spectrometry should be considered. This approach can not only prevent misidentification but also will ensure that the laboratory is safely handling highly pathogenic organisms that can be easily aerosolized. Commercial identification systems may also have limited databases, and strains of organisms with rare phenotypes may be misidentified.

*CASE REPORT:* **BORDETELLA HOLMESII** *MISDENTIFIED AS* **ACINETOBACTER LWOFFI***:* In this report, four cases of bacteremia in asplenic children were described [19]. Blood cultures from all four children were positive; and Gram-negative bacilli were recovered from the cultures. The organisms were identified as *Acinetobacter lwoffi* (99% probability) using a Vitek2 Gram-negative card. Key reactions that were used to distinguish between *Acinetobacter* and *Bordetella* were catalase activity and growth on MacConkey agar. These isolates were catalase negative and did not grow on MacConkey—reactions that are not consistent with *Acinetobacter*. DNA sequencing of the 16 S rRNA gene was used to identify these four isolates as *Bordetella holmesii*. Investigation showed that *B. holmesii* was not in the database for the Vitek2 instrument.

### *Unexpected Positive or Negative Reactions in Biochemical Methods Used to Identify Organisms*

Most organisms have rare strains that do not exhibit expected biochemical reactions; when these identification characteristics are used in algorithms, an incorrect identification can result. Examples of this phenomenon are *Staphylococcus aureus* that are negative in the rapid latex agglutination test [20] and metabolically inactive strains of *Escherichia coli* that are misidentified as *Shigella* [21].

### *Misidentification of Rapidly Growing Mycobacteria*

The rapidly growing mycobacteria commonly isolated from specimens such as blood, sputum, or tissues (e.g., *Mycobacterium abscessus*, *Mycobacterium chelonae*, and *Mycobacterium fortuitum*) may be misidentified as diphtheroids. These species may grow in routine media such as liquid broth blood culture media or on solid agars used for routine bacteriology. These organisms can grow rapidly enough that they are recovered in routine culture. They may not be recognized as acid-fast organisms because they stain quite well with a Gram stain, and they may not always exhibit branching or beaded morphologies associated with rapid growers. If one of these organisms is recovered on routine culture, and the Gram stain shows Gram-positive bacilli, the organism may be incorrectly reported as diphtheroids, with no further workup.

### *Misidentification of* Listeria monocytogenes

*Listeria monocytogenes* has occasionally been misidentified as *Streptococcus agalactiae* (group B *Streptococcus*). *Listeria* can exhibit coccobacillary morphology in Gram stains. Both of these organisms are β-hemolytic on blood agar plates, and colony morphology is similar. Most identification algorithms call for a catalase test to be used on hemolytic colonies; occasionally, *Listeria* will not exhibit a strong catalase reaction. False-positive agglutination of *Listeria* in grouping reagents used to identify the β-hemolytic streptococci can rarely occur. Accurate identification of *Listeria* and group B *Streptococcus* is very important because both of these organisms can cause neonatal meningitis, and antimicrobial therapy differs significantly.

### *Misidentification of* Burkholderia cepacia

*Burkholderia cepacia* is an important organism when identified in patients with cystic fibrosis, and a report of *B. cepacia* is highly significant. Accurate identification of *B. cepacia*, *Stenotrophomonas maltophilia*, and the related glucose-nonfermenting organisms is especially important in this setting. Laboratories should be aware of the possibility of misidentification of organisms, especially if they are using automated phenotypic identification systems. *Burkholderia pseudomallei*, rarely identified in the United States, has also been misidentified as *B. cepacia* [22]. Isolates with a presumptive identification of *B. cepacia* should be confirmed by an alternate method before release of the identification to the clinician.

### *Prevention of Misidentification Errors*

Understanding the limitations of the various methods of organism identification is the best way to prevent misidentifications from occurring. Well-designed algorithms for workup and identification of organisms can ensure that the appropriate method for identification is selected. A laboratory should have several methods available to identify microorganisms. Newer methods such as DNA sequencing and mass spectrometry are much more accurate than biochemical methods, and they are currently available to laboratories. Not only can use of the newer methods result in more accurate identification but also results are often available much more quickly than with conventional methods. Careful review of results and correlating instrument reports with other data on the culture (source of isolate, growth on selective media, and manual biochemical test results) should be performed for every isolate.

## Failure to Isolate a Pathogen: Negative Culture Results

False-negative cultures occur for many reasons. Organisms may be rare in clinical specimens, and if the volume of material is low, culture results may be negative. This is often the case with joint aspirates, where the volume of material submitted for culture is usually extremely small. False-negative cultures may also occur if a fastidious or difficult-to-culture organism is causing the infection. Examples include *Legionella*, *Brucella*, and *Kingella*. For specimens such as sterile body fluids, which may contain rare or difficult-to-isolate pathogens, the laboratory may recommend using a more sensitive method of culture such as lysis centrifugation or placing the specimen in a blood culture bottle.

The timing of specimen collection has significant influence on the culture results, especially when antimicrobial agents are going to be used. This is critical for specimens such as CSF for diagnosis of bacterial meningitis. The CSF may be sterilized within 30 min of administration of antimicrobial agents, resulting in false-negative cultures. Obtaining a specimen prior to administration of antimicrobial agents is critical to ensure the isolation and identification of the pathogen. Whenever possible, the CSF should be obtained before antimicrobials are given.

## Antimicrobial Susceptibility Testing Errors

Antimicrobial susceptibility testing (AST) is one of the most important functions performed by the clinical microbiology laboratory. With improvements in testing methods and quality control and quality assurance programs, AST errors are infrequent and are usually detected and corrected before results are released to the clinician. Although laboratory methods are designed to accurately detect most resistance mechanisms, microorganisms are constantly evolving and in the setting of antimicrobial usage, resistance mechanisms emerge. Resistance is complex, and organisms have many methods by which they can acquire or express resistance to antimicrobials. Organisms may acquire new resistance determinants by sharing plasmids or other genetic elements with resistant bacteria, or they may express resistance mechanisms inducible in the presence of an antimicrobial agent.

The laboratory must routinely update its policies and procedures to stay current with susceptibility methods, known resistance patterns, and reporting strategies. Errors in AST can have significant impact on patient care. Although there are many potential sources of error in AST, most errors can be prevented by the use of well-standardized, well-controlled policies and procedures by the laboratory. By using well-established, standardized methods, and with good-quality control programs, error rates in AST can be held to a minimum. The laboratory should also ensure that the repertoire of agents for testing and reporting are matched to the hospital formulary.

Potential sources of error in AST occur in several areas: method differences, technical errors with performing and interpreting the assays, reporting errors, and inherent problems with certain organism–drug combinations. Each of these can be minimized by careful adherence to guidelines and recommendations, laboratory procedures, and the quality control program. There are well-established standards available through the Clinical and Laboratory Standards Institute (CLSI) that laboratories should implement. Emerging resistance is always a consideration in susceptibility testing. Finally, subtle resistance mechanisms or inducible resistance mechanisms that are not manifested until therapy is implemented should always be a consideration when the susceptibility testing results indicate an organism is susceptible but the therapy fails.

### Method Differences

The main methods for susceptibility testing are broth dilution, agar dilution, and disk diffusion. In addition, several systems for automated AST are available. It is important to note that each method may have limitations for certain organism–drug combinations and that there may be minor differences between the results generated by the automated systems. For example, certain organism–drug combinations cannot be reliably tested on automated systems. Laboratories should ensure that they have implemented the recommended changes to their software if using automated instruments.

### New Resistance Mechanisms or Emerging Resistance

AST systems may not accurately detect heteroresistance (subpopulations of bacteria with resistance mechanisms present). When resistance determinants are present but not being expressed by the organism at a high level, the isolate may appear susceptible in *in vitro* testing, but use of antimicrobials may allow for expression of the resistance gene. This phenomenon has been noted for several groups of organisms. An example is the emergence of carbapenem resistance in Gram-negative bacteria such as *Klebsiella* [23]. Automated systems may fail to detect carbapenem resistance [24–26]. Laboratories should ensure that screening and confirmatory testing are in place to detect potential carbapenem resistance. The emergence of resistance mechanisms in bacteria can result in testing and reporting errors. Unusual results should be reviewed and investigated; occasionally, aberrant results can be traced back to technical errors. If no technical errors have occurred, organisms with unusual resistance patterns should be tested with a different method. Unusual susceptibility testing results should be verified prior to release; laboratories should have documented protocols in place for verification of unusual results. Often, investigation reveals that a technical issue led to the unusual or unexpected result.

### Inducible Resistance Mechanism Present, Not Detected

Inducible resistance mechanisms can be difficult to identify in the laboratory. Methods are available for detection of inducible resistance to clindamycin in staphylococci; sensitivity and specificity of these tests are adequate, and very few errors occur. Inducible resistance to β-lactam agents can be more difficult to detect, especially in Gram-negative organisms such as *Acinetobacter*.

*CASE REPORT: FATAL* ACINETOBACTER BAUMANNII *INFECTION WITH DISCORDANT CARBAPENEM SUSCEPTIBILITY* Initial susceptibility testing of an *Acinetobacter baumannii* strain isolated from blood and respiratory cultures indicated the organism was susceptible to imipenem. The patient was treated with meropenem but did not respond clinically; the patient died. Subsequent testing of the bloodstream isolate using a disk diffusion test revealed that the organism was susceptible to imipenem but resistant to meropenem [27]. This case illustrates the potential for resistance mechanisms to be missed when only one drug is tested. Testing of only one agent in a class is very common when using panels on automated systems because of the limitations of space on the panels. Laboratories must be aware of the potential for failure of the automated susceptibility testing system to detect carbapenem resistance in Gram-negative bacteria.

## False-Positive *Mycobacterium tuberculosis* Culture Results

False-positive culture reports of *M. tuberculosis* remain a perplexing problem for laboratories that perform acid-fast bacteria testing. Specimens that contain *M. tuberculosis* at high levels have extraordinary potential to contaminate the work environment during the handling of these specimens for staining and culture. Many of the processing steps (pouring, centrifuging, pipetting, etc.) can generate aerosols or result in splashing, with the result being environmental contamination. Subsequent handling can transmit the contaminating organism to culture plates or liquid media, which can then grow the organism.

Although laboratories are required to have policies and procedures in place to prevent contamination, it occasionally occurs. There are several methods by which a false-positive report of an *M. tuberculosis* isolate can be identified. If the clinical picture does not match the culture result, the result should be questioned. A careful review of the laboratory records can assist in determining if a true-positive sample was handled at the same time. Single positive respiratory cultures may also be questioned. If the patient does have active respiratory tuberculosis, the organism should be growing in more than one culture. False-positive *M. tuberculosis* culture results may also be identified by molecular typing of the isolate. Laboratories should consider sending their *M. tuberculosis* isolates to be typed, and the data can be reviewed for matches. Public health authorities should assist with investigation of isolates that have matched genotype patterns.

Quality assurance programs should be in place in all laboratories that offer acid-fast bacteria testing [28]. The laboratory should collect and analyze data on smear positivity rates and positive culture rates, and it should keep statistics on how many *M. tuberculosis* isolates it routinely makes each year. These data will help in the investigation of any potential problems in the laboratory.

## POST-ANALYTICAL OR REPORTING ERRORS

Reporting and interpretation of test results are important components in laboratory testing. Reporting errors may occur when results of cultures or susceptibility testing are entered manually into the laboratory information system rather than automatically uploaded by instrument software. Susceptibility testing results are especially prone to transcription errors because there is typically a large amount of data associated with the results for each organism. Manual entry of data is tedious and time-consuming. Transcriptional errors can occur when susceptibility results are entered manually into the culture report. Transcriptional errors can be prevented by interfacing the test system (if an automated system is used) and using the automated software to check for errors. If an automated system is not in use, results should be reviewed for transcriptional errors prior to release.

### Reporting Inappropriate Organism–Drug Combinations

CLSI guidelines should be followed by the laboratory to prevent reporting of inappropriate organism–drug combinations. Automated susceptibility instrument software and an interface to the laboratory information system is the best method for preventing an error in reporting susceptibility test results. When panels of antimicrobial agents are used, reports of susceptibility testing should be tailored according to the organism being tested; the laboratory should release only the appropriate agents. For some organism–drug combinations, results should not be reported because of a lack of correlation between *in vitro* test results and response of the organism *in vivo*.

## ERROR PREVENTION, DETECTION, AND MONITORING

Detection and monitoring of errors in the microbiology laboratory should occur through a comprehensive program of routine monitoring, result review, and summary reports. The initial review of test results for each test method or culture type by a laboratory supervisor is the most important step during which analytical errors are detected. Detection of errors during the review step allows for immediate correction prior to release of results to physicians. Guidance for designing a comprehensive program for quality assurance in the microbiology laboratory is outlined in the *Cumitech 41: Detection and Prevention of Clinical Microbiology Laboratory-Associated Errors* [29].

Pre-analytical errors can be prevented or limited by providing laboratory customers with specimen selection and collection guidelines. An excellent method for assisting with specimen collection and test ordering is to provide periodic in-service education to customers. Written guidelines for specimen collection, transport, and storage should be easily available to physicians and staff collecting samples. Illustrated guidelines are especially helpful. For example, laboratories can post illustrations of various swab collection devices, which often differ according to test method or specimen type (e.g., swabs with transport media for routine throat culture, anaerobic jars containing pre-reduced media for tissues, and flocked swabs for respiratory viral specimens).

The laboratory's quality assurance program should be designed to allow for detection of problems. For example, routine monitoring of culture contamination rates (e.g., blood culture contamination rates and urine culture contamination rates) enables laboratories to monitor data over time. Significant increases in contamination rates can be used as an indicator that collection guidelines are not being followed, and in-service training to providers may be implemented. Quality assurance programs for acid-fast bacilli testing services should include monitoring of smear positivity rates, culture positivity rates, number of *M. tuberculosis* isolates, and turnaround times for reporting of *M. tuberculosis* [28]. Post-analytical errors may be prevented by a system of timely reporting, communication of unusual laboratory testing results or unusual organisms, etc.

Good communication between physicians and the laboratory is an important and necessary component of obtaining quality laboratory results. This is especially true for clinical microbiology. The laboratory must provide information to physicians about testing methodologies, specimen collection, timing, and transport concern. Laboratory plate rounds can be very helpful for discussion and interpretation of culture results and for sharing ideas for alternative testing that may be needed.

## CONCLUSIONS

Preventing errors in the microbiology laboratory is crucial from both a patient management and a patient safety perspective because clinicians depend on accurate identification of organisms for proper treatment. In this chapter, various sources of errors—pre-analytical, analytical, and post-analytical—were addressed that should help professionals to prevent errors in microbiology testings.

## References

[1] Barenfanger J. Quality in, quality out: rejection criteria and guidelines for commonly (mis)used tests. Clin Microbiol Newsl 2000;22(9):65–72.

[2] Barenfanger J. Quality assurances: decreasing clinically irrelevant testing from clinical microbiology laboratories, Part I. Clin Microbiol Newsl 2006;28(3):17–24.

[3] Barenfanger J. Improving the clinical utility of microbiology data: an update. Clin Microbiol Newsl 2003;25(1):1–8.

[4] Einstein MP, Reller LB, Murphy JR, Lichtenstein KA. The clinical significance of positive blood cultures: a comprehensive analysis of 500 episodes of bacteremia and fungemia in adults: I. Laboratory and epidemiologic observations. Rev Infect Dis 1983;5:35–53.

[5] Washington II JA, Ilstrup DM. Blood cultures: issues and controversies. Rev Infect Dis 1986;8(5):792–802.

[6] Novis DA, Dale JC, Schifman RB, Ruby SG, Walsh MK. Solitary blood cultures: a college of American pathologists Q-probes study of 132,778 blood culture sets in 333 small hospitals. Arch Pathol Lab Med 2001;125(10):1290–4.

[7] Weinstein MP. Blood culture contamination: persisting problems and partial progress. J Clin Microbiol 2003;41(6):2275–8.

[8] Hall KK, Lyman JA. Updated review of blood culture contamination. Clin Microbiol Rev 2006;19(4):788–802.

[9] Richter SS. Strategies for minimizing the impact of blood culture contaminants. Clin Micro Newslett 2002;24(7):49–53.

[10] Richter SS, Beekmann SE, Croco JL, Diekema DJ, Koontz FP, Pfaller MA, et al. Minimizing the workup of blood culture contaminants: implementation and evaluation of a laboratory-based algorithm. J Clin Microbiol 2002;40(7):2437–44.

[11] Dwivedi S, Bhalla R, Hoover DR, Weinstein MP. Discarding the initial aliquot of blood does not reduce contamination rates in intravenous-catheter-drawn blood cultures. J Clin Microbiol 2009;47(9):2950–1.

[12] Sautter RL, Bills AR, Lang DL, Ruschell G, Heiter BJ, Bourbeau PP. Effects of delayed-entry conditions on the recovery and detection of microorganisms from BacT/ALERT and BACTEC blood culture bottles. J Clin Microbiol 2006;44(4):1245–9.

[13] Rand KH, Tillan M. Errors in interpretation of Gram stains from positive blood cultures. Am J Clin Pathol 2006;126(5):686–90.

[14] Martin GS, Mannino DM, Eaton S, Moss M. The epidemiology of sepsis in the United States from 1979 through 2000. N Engl J Med 2003;348(16):1546–54.

[15] Johnson MJ, Thatcher E, Cox ME. Techniques for controlling variability in Gram staining of obligate anaerobes. J Clin Microbiol 1995;33(3):755–8.

[16] Batchelor BI, Brindle RJ, Gilks GF, Selkon JB. Biochemical mis-identification of *Brucella melitensis* and subsequent laboratory-acquired infections. J Hosp Infection 1992;22:159–62.

[17] Elsaghir AA, James EA. Misidentification of *Brucella melitensis* as *Ochrobactrum anthropi* by API 20NE. J Med Microbiol 2003;52 (Pt 5):441–2.

[18] Horvat RT, El Atrouni W, Hammoud K, Hawkinson D, Cowden S. Ribosomal RNA sequence analysis of *Brucella* infection misidentified as *Ochrobactrum anthropi* infection. J Clin Microbiol.;49(3):1165–8.

[19] Panagopoulos MI, Saint Jean M, Brun D, Guiso N, Bekal S, Ovetchkine P, et al. *Bordetella holmesii* bacteremia in asplenic children: report of four cases initially misidentified as *Acinetobacter lwoffii*. J Clin Microbiol.;48(10):3762–4.

[20] Szabados F, Woloszyn J, Kaase M, Gatermann SG. False-negative test results in the Slidex Staph Plus (bioMerieux) agglutination test are mainly caused by spa-type t001 and t001-related strains. Eur J Clin Microbiol Infect Dis.;30(2):201–8.

[21] Gupta S, Aruna C, Muralidharan S. Misidentification of a commensal inactive *Escherichia coli* as *Shigella sonnei* by an automated system in a critically ill patient. Clin Lab.57(9-10): 767–9.

[22] Weissert C, Dollenmaier G, Rafeiner P, Riehm J, Schultze D. *Burkholderia pseudomallei* misidentified by automated system. Emerg Infect Dis 2009;15(11):1799–801.

[23] Gordon NC, Wareham DW. Failure of the MicroScan WalkAway system to detect heteroresistance to carbapenems in a patient with *Enterobacter aerogenes* bacteremia. J Clin Microbiol 2009;47(9):3024–5.

[24] Tenover FC, Kalsi RK, Williams PP, Carey RB, Stocker S, Lonsway D, et al. Carbapenem resistance in *Klebsiella pneumoniae* not detected by automated susceptibility testing. Emerg Infect Dis 2006;12(8):1209–13.

[25] Markelz AE, Mende K, Murray CK, Yu X, Zera WC, Hospenthal DR, et al. Carbapenem susceptibility testing errors using three automated systems, disk diffusion, Etest, and broth microdilution and carbapenem resistance genes in isolates of *Acinetobacter baumannii-calcoaceticus* complex. Antimicrob Agents Chemother.; 55(10):4707–11.

[26] Tato M, Morosini M, Garcia L, Alberti S, Coque MT, Canton R. Carbapenem Heteroresistance in VIM-1-producing *Klebsiella pneumoniae* isolates belonging to the same clone: consequences for routine susceptibility testing. J Clin Microbiol.; 48(11):4089–93.

[27] Lesho E, Wortmann G, Moran K, Craft D. Fatal *Acinetobacter baumannii* infection with discordant carbapenem susceptibility. Clin Infect Dis 2005;41(5):758–9.

[28] Woods GL, Ridderhof JC. Quality assurance in the mycobacteriology laboratory: quality control, quality improvement, and proficiency testing. Clin Lab Med 1996;16(3):657–75.

[29] Amsterdam D, Barenfanger J, Campos J, Cornish N, Daly JA, Della-Latta P, et al. Cumitech 41: Detection and Prevention of Clinical Microbiology Laboratory-Associated Errors. Washington, D.C.: ASM Press; 2004.

CHAPTER

# 21

# Sources of Errors in Molecular Testing

*Laura Chandler*

Philadelphia VA Medical Center and Perelman School of Medicine at the University of Pennsylvania, Philadelphia, Pennsylvania

## INTRODUCTION

Molecular diagnostics is a rapidly growing discipline in laboratory medicine, with new methods and new applications continually becoming available and improvements being made. Molecular techniques are routinely used in the diagnosis and management of many infectious diseases and are increasingly being applied to other areas such as cancer detection. A wide variety of nucleic acid detection methods are available, including amplified and non-amplified techniques. Many assays are commercially available and approved by the U.S. Food and Drug Administration (FDA), allowing laboratories to implement them relatively easily. Improvements to molecular assays include simplified formats (extraction, amplification, and detection in a single device) and automated extraction platforms that allow large numbers of samples to be processed at one time. Qualitative as well as quantitative molecular techniques are commercially available, and some laboratories develop their own home-brew assays. However, several fundamental features of molecular testing require laboratories to be alert for problems with testing that may give rise to erroneous results. The more common errors that can occur with molecular testing are discussed in this chapter. As with other sections of the laboratory, a strong program of quality control and quality assurance is necessary for detection of problems, monitoring of errors, and implementing methods to improve quality if systematic errors are occurring in the molecular laboratory. Although analytical errors can occur with molecular testing, good-quality control programs mean that errors are kept to a minimum. This chapter outlines the most common interferences in molecular diagnostics testing and describes methods to prevent and deal with these problems.

## PRINCIPLES OF COMMONLY USED METHODS FOR MOLECULAR DIAGNOSTICS

Molecular diagnostics is a discipline that relies on the detection and characterization of nucleic acids (DNA or RNA) in clinical specimens. In the clinical laboratory, molecular methods were originally applied to infectious disease diagnostics for the detection and quantitation of microbial pathogens. Expansion of our knowledge about the human genome, and corresponding understanding of diseases [1], has resulted in the routine use of molecular testing in cancer diagnostics and the evolution of the field of pharmacogenomics, or personalized medicine [2–5]. A major increase in the techniques and testing methods that can be used as tools for molecular diagnostics has also contributed to the expansion of molecular diagnostics into many different disciplines in diagnostic medicine. The expansion of the field of molecular diagnostics has been due in part to technological improvements that have given rise to an impressive number of molecular testing methods, from simple hybridization to highly multiplexed polymerase chain reaction (PCR) and microarray testing. However, molecular methods can be quite complex, and an understanding of the scientific principles of molecular tests is needed to ensure that problems with testing can be readily identified and corrected. The principles of the most commonly used methods for molecular diagnostics are discussed here.

### Non-Amplified Methods Used for Molecular Diagnostics

Various techniques are classified in this category, including nucleic acid probe hybridization. Hybridization

*Accurate Results in the Clinical Laboratory.*
DOI: http://dx.doi.org/10.1016/B978-0-12-415783-5.00021-9

methods rely on the use of labeled DNA or RNA as probes to detect nucleic acids in specimens. Nucleic acid hybridization is based on complementary base pairing between a labeled nucleic acid probe and the target nucleic acid. Direct probe hybridization is a relatively simple method but requires the presence of the target in relatively high levels, so its use as a diagnostic tool is somewhat limited. Hybridization is used when the target nucleic acid is present in high enough concentrations that can be detected reliably. When sensitivity is not a limiting factor, nucleic acid probe hybridization methods are excellent to use because they are relatively easy to perform, results can be obtained rapidly, and they are not susceptible to contamination as are amplification methods. Hybridization techniques offer flexible formats (e.g., hybridization can be performed on solid substrates, in liquid solutions, or even *in situ* in tissues or cells).

Applications of direct probe hybridization in diagnostic microbiology include identification of bacteria in blood culture bottles using fluorescence *in situ* hybridization and identification of cultured microorganisms such as fungi and mycobacteria after transcription-mediated amplification (TMA). In diagnostic molecular pathology, *in situ* hybridization is used to detect gene duplications and gene mutations (e.g., *in situ* hybridization for HER-2 amplification).

## Amplification Methods Used for Molecular Diagnostics

### Polymerase Chain Reaction

PCR revolutionized biological and medical research, as well as diagnostic medicine, and the majority of molecular diagnostic assays are now based on some form of PCR. PCR is based on the use of thermally stable DNA polymerase enzymes to copy a sequence of DNA by extending primers bound to a single-stranded DNA target. Repetitive cycles using heat to melt the double-stranded DNA, followed by primer binding and extension of the primer, result in the synthesis of many copies of the target DNA [6,7], which can then be detected with probes. In end-point PCR, the newly synthesized DNA is detected by a method such as probe hybridization or gel electrophoresis after the end of the reaction. In real-time PCR, the newly synthesized DNA is detected via DNA binding dyes or labeled probes as it is being synthesized. Real-time PCR is highly advantageous to diagnostic medicine for the following reasons: Data are collected during the run so that results are obtained rapidly, and the potential for contamination is reduced or eliminated because reaction tubes are not opened after the PCR is complete.

### Strand Displacement Amplification

Strand displacement amplification (SDA) is a DNA amplification method that is performed without temperature cycling (isothermal). It relies on the use of a restriction enzyme to nick the DNA target, which is then copied by DNA polymerase. As the polymerase copies the target, the template strand is displaced [8]. SDA is used in the BD ProbeTec ET assay for the detection of *Neisseria gonorrhoeae* and *Chlamydia trachomatis*.

### Transcription-Mediated Amplification Methods

Transcription-mediated amplification methods include nucleic acid sequence-based amplification and TMA [9,10]. These two methods are similar and rely on the use of RNA transcription for amplification. Primers that contain sequences recognized by RNA polymerases are used to hybridize to the target sequence and then extended by the RNA polymerase. These methods exhibit good sensitivity, similar to PCR. Transcription-based methods are commercially available for infectious disease assays.

## DNA Microarrays

DNA microarrays are increasingly being used in molecular diagnostics. Microarrays are microscopic groups of DNA spots attached to a solid substrate. The most common materials used as substrates are glass, plastic, and silicon [11]. Microarrays are commonly referred to as gene chips or DNA chips. Microarrays are designed to contain a very large number of oligonucleotides on a single chip, allowing for collection of a large data set in one experiment. The principle of microarray chip testing is to apply target DNA or RNA to the chip, allow hybridization of complementary sequences, and then detect hybrids. Liquid detection formats (microbeads) analogous to flow cytometry are also available. Many different formats of labeled detection are available, including fluorescence, mass spectrometry, and electrochemical [12].

Originally developed for research, microarrays have been successfully adapted to most of the disciplines within laboratory medicine when molecular testing requires analysis of a large data set with applications including microbiology [13–15], cancer [16], and genetic disease [17,18]. Microarrays are very beneficial when a large number of probes are needed. In the clinical microbiology laboratory, several microarray tests are commercially available for genotyping of human papillomavirus and hepatitis C virus, identification of bacteria with 16 S rRNA probes, strain typing, and detection of antimicrobial resistance genes. In cancer diagnostics, gene expression microarrays are used for tissue of origin testing, copy number changes, and hematological malignancies [19–21]. Microarrays are particularly useful in genetic testing when testing for complex mutations such as cystic fibrosis. Microarrays have also been used for predicting outcomes and response to therapy for some cancers [22,23].

### Nucleotide Sequencing

Nucleotide sequencing (determining the exact composition and order of bases in a fragment of DNA) is routinely used for several applications in the clinical diagnostic laboratory. Chain terminating, or Sanger sequencing, relies on the use of dideoxynucleotides during amplification reactions. When a dideoxynucleotide is incorporated into the growing strand of DNA, the chain is terminated. This results in a mixture of terminated fragments, representing the entire length of the segment of DNA that is being analyzed. The fragments are electrophoresed and detected by their fluorescent label. The most convenient format for diagnostic laboratories is to use dideoxynucleotides that are each labeled with a different dye; the resulting fragments can be electrophoresed in a single capillary or lane on a gel.

Newer methods of nucleotide sequencing, sometimes referred to as next-generation sequencing, are coming into routine use in diagnostic laboratories [24–26]. Various methods are currently available, including pyrosequencing technology (based on the detection of the pyrophosphate released during synthesis of DNA) [27]; ion semiconductor ("ion torrent"), based on the release of $H^+$ during synthesis [28]; and image analysis of sequentially added fluorescent nucleotides [29]. Next-generation sequence technologies allow for generation of extremely large amounts of sequence data and are sometimes referred to as massively parallel sequencing because of the ability to generate massive amounts of data relatively easily [26].

## PRINCIPLES OF NUCLEIC ACID ISOLATION

Most molecular diagnostic testing methodologies require isolated nucleic acids for optimal performance. Although some methods can be performed directly on cells or tissues without an extraction step, most testing procedures will include a nucleic acid extraction and purification procedure prior to testing. Nucleic acids must be released from cells (eukaryotic cells and bacterial cells) and virus particles prior to molecular detection. Extraction and purification methods are used to ensure that high-quality, intact (not degraded) DNA or RNA is available for amplification or probe detection methods. Purification of nucleic acids after cell lysis removes cellular debris and amplification inhibitors, inactivates or destroys nucleases (DNases and RNases) that can degrade nucleic acids, and leaves the nucleic acid in solution that is optimal for any downstream applications. Purification can also be used to concentrate the nucleic acids so that rare targets such as bacteria or viruses can be detected.

Several different extraction methods are available. Crude preparations, in which cells may be lysed with detergents, may be used for some molecular procedures but are often inadequate for amplification procedures because they may contain inhibitors. For most nucleic acid detection procedures, extraction followed by purification of the DNA or RNA following cell lysis is needed. There are two main methods by which nucleic acids can be extracted and purified: organic extractions (e.g., phenol–chloroform extractions) and silica binding methods.

For phenol–chloroform extractions, cells are initially lysed with a chaotropic salt (guanidinium thiocyanate) and detergent and then phenol–chloroform is added to extract proteins, followed by precipitation of nucleic acids from the aqueous phase with alcohol [30,31]. Phenol–chloroform extractions are somewhat difficult for clinical laboratories because they are time-consuming and labor-intensive, and they have not been adapted to automated systems.

The silica binding method developed by Boom *et al.* [32] relies on the binding of nucleic acids to silica in the presence of salt. The silica-bound nucleic acid is separated from cellular debris and then washed and eluted into a low-salt solution. Silica binding methods result in highly pure nucleic acid that is largely free from contaminating proteins including nucleases and inhibitory substances. This method is versatile in that silica can be bound to many different substrates, such as membranes, filter beads, or magnetic particles, and many different products have been developed based on this technology. The Boom technology has also been successfully adapted to automated robotic systems, allowing for laboratories to process very large numbers of specimens. Automated systems are available from several manufacturers and have the advantage of workflow efficiency as well as reducing variability that can occur in manual extraction methods.

## PRE-ANALYTICAL CONSIDERATIONS IN MOLECULAR TESTING

### Specimens

Specimen collection, transport, storage, and processing are important components of molecular diagnostics testing; these parameters have direct effects on the outcome of testing. Successful results in molecular testing are highly dependent on the pre-analytical steps: If pre-analytical steps (specimen collection, transport, storage, and processing) are not performed correctly, errors in testing are likely to occur. Specimen collection, transport, storage, and processing guidelines are intended to ensure that the target nucleic acid is stabilized while also ensuring that interfering substances are not introduced into the specimen. If not performed correctly, pre-analytical steps can result in loss of target nucleic acids, introduction of substances that can inhibit downstream applications, or overfixing of cells and tissues. A guideline from the Clinical and Laboratory Standards Institute for specimen handling is available that outlines standardized methods for pre-analytical processes [33].

The type of specimen, volume of material, type of container or blood tube, and storage may vary, depending on the downstream application. Laboratories should implement standardized protocols designed to prevent pre-analytical errors. Collection devices such as blood collection tubes can contain substances that affect assay performance [34]. Some specimens (e.g., tissues, stool specimens, and cytology specimens in fixatives) may be extremely complex. These specimens may contain substances that can inhibit molecular assays, and pre-processing treatments and nucleic acid extraction can also introduce interfering substances. Because virtually any type of specimen can be submitted for molecular testing, appropriate guidelines and recommendations for specimen collection, transport, and storage should be followed [33], and this variable should be investigated for the potential to cause interference in molecular assays.

## Specimen Collection

Specimen collection guidelines should be optimized to ensure that the correct specimen is collected based on the testing to be performed and the target nucleic acid. For genetic testing, whole blood may be used; genomic DNA will be extracted from white blood cells. For molecular pathology studies (e.g., tumor samples), tissues that are likely to contain the cells representative of the tumor must be submitted. Infectious disease assays are quite variable in terms of the types of specimens that may be submitted. For routine monitoring (e.g., viral load testing for HIV or hepatitis C virus (HCV)), specimen collection guidelines are well characterized; these assays are performed on plasma or serum [34]. Laboratories should ideally adopt a single type of specimen because there are minor differences in viral loads between the two types of specimens. Viral load monitoring is done at standardized (defined) intervals, and variability due to specimen type is undesirable and can lead to misinterpretation of results. For this reason, laboratories should adopt a single method (e.g., plasma from EDTA tubes or serum) and not introduce another specimen collection method.

For infectious disease assays, many specimen types can be submitted depending on the clinical syndrome and agent being tested for. Some specimens (e.g., stool, urine, sputum, and endocervical swabs) are potential sources of inhibitory substances. Choice of specimen collection, transport, and storage should be based on the type of target (DNA or RNA). RNA targets are usually very unstable and subject to degradation. These specimens should be transported immediately to the lab, held at 4°C, and processed as soon as possible to avoid target degradation. Failure to process specimens in a timely manner is a common reason for false-negative results in infectious disease assays.

Some specimens may be collected directly into fixatives or preservatives (e.g., cytology specimens for HPV testing and urine for BK virus testing). Labs must develop their own guidelines to ensure that the fixative will be compatible with the downstream application. Fixatives are useful for preserving specimen integrity, including stabilizing nucleic acids, but overfixation may occur. Downstream processing must be optimized to ensure that amplifiable nucleic acids can be extracted and purified from specimens that are transported in fixatives or preservatives.

Sterile body fluids such as cerebrospinal fluid (CSF), pleural fluid, or synovial fluids may be submitted to the lab in clean containers or tubes, without preservatives. Nucleic acid extractions are performed directly on these specimens for downstream testing.

## Specimen Transport and Storage

Transport and storage conditions and times are variables that can result in problems in molecular testing because of differences in the stability of the target nucleic acid (e.g., viral vs. cellular and RNA vs. DNA). Procedures should recommend that time in transit be minimized and storage at appropriate temperature, usually 4°C. Freeze–thaw cycles can reduce viral titers and should be avoided when viral assays are used. In general, one freeze–thaw may be performed without significant reduction in titer.

Storage of specimens in preservative can be problematic, especially long-term storage. Preservatives such as formalin, glutaraldehyde, and alcohols may overfix tissues or cells, causing problems with nucleic acid extraction. Downstream applications must be optimized for fixed materials. Nucleic acid extraction from fixed materials can be problematic. Special lysis procedures and enzymatic digestion to release nucleic acids from cross-linked proteins may have to be performed. Labs should optimize their extraction protocol and check each sample (e.g., with ultraviolet (UV) spectrophotometry). Specimens submitted for molecular testing must be stored in controlled conditions so that nucleic acids are not degraded during storage. Storage conditions may depend on the target nucleic acid (e.g., RNA vs. DNA). To prevent degradation of target nucleic acid, tissues or fluids may be transferred into a stabilizing reagent (e.g., RNAlater or Trizol) that can then be used for nucleic acid extraction at a later time. These reagents are designed to lyse cells, inactivate proteins such as nucleases, and stabilize the nucleic acid during storage.

### Specimen Collection and Transport Devices

Anticoagulated whole blood or plasma samples are the most commonly used specimens for procedures performed on blood, such as genetic testing and viral load assays. For assays performed on blood, the tube type is important, and in general, the anticoagulants EDTA and acid citrate dextrose (ACD) should be used because they are most compatible with molecular methods. Heparin tubes should not be used for amplification methods because heparin is a potent inhibitor of polymerase enzymes. Blood specimens that are anticoagulated with heparin are not recommended for molecular testing because of the potential for inhibition. EDTA and ACD are the recommended anticoagulants for transport tubes for blood.

Other specimens, such as tissues, urine, body fluids, and stool, may be transported in clean containers to the laboratory. Specimens in fixatives such as alcohols or formalin should be extracted as quickly as possible after receipt by the laboratory. It can be more difficult to obtain good-quality DNA from fixed tissues.

### Extraction and Purification of Nucleic Acids

For most molecular diagnostic methods, nucleic acids must be isolated in pure form for downstream testing. Isolation of nucleic acids serves several purposes. It ensures that target nucleic acids are readily available to enzymes or probes, and it removes interfering substances (inhibitors) that can cause problems in testing. Extraction and purification are also used to concentrate nucleic acids to optimize assay sensitivity. Nucleic acid preparation methods are a critical component of molecular testing because for most assays, high-quality DNA or RNA is critical to assay performance; the target integrity must be maintained (especially important for RNA assays), and inhibitory substances must be eliminated. The quality of purified DNA and RNA, as well as the concentration and purity, affect results of hybridization, amplification, and detection methods. The target nucleic acid must be intact, in a sufficient concentration to be detected by the assay, and be free of inhibitory substances.

Nucleic acid extraction and purification procedures can be a source of errors in diagnostic testing. If intact nucleic acids are not isolated, false-negative results can occur. In addition, the extraction and purification procedures can be a source of interfering substances if reagents such as phenol or alcohols are carried over into the final eluted material containing the target nucleic acid.

For many procedures, assessment of the amount and quality of isolated nucleic acids should be performed prior to testing. Several methods are available; UV spectrophotometry is an excellent method to simultaneously assess the amount and purity of nucleic acids.

## INTERFERENCES IN MOLECULAR DIAGNOSTICS: FALSE-NEGATIVE RESULTS IN MOLECULAR ASSAYS

False-negative results occur when the analyte is present in the specimen but the molecular assay result is negative. In quantitative molecular assays, aberrant results may include completely negative results or inaccurate quantitation. For example, the assay may report viral loads below the threshold of detection when there is virus present in the specimen. There are several conditions in which false-negative results can be obtained in molecular diagnostics. Interferences that result in false negatives or inaccurate quantitation can occur in pre-analytical or analytical phases of testing.

### Inhibition in Amplification Assays

Inhibition of amplification (complete or partial failure to generate amplified DNA or RNA because of inhibition of enzymes used in amplification reactions) can be a cause of negative results or inaccurate quantitation [35]. The presence of inhibitors can cause false-negative results in qualitative assays or can reduce amplification efficiency, causing inaccurate results in quantitative assays. Inhibition of amplification reactions is a well-recognized phenomenon in the diagnostic laboratory, but the reasons for inhibition are not always well understood [36]. Many substances from a variety of sources can inhibit PCR reactions. Inhibitors can be present in the clinical specimen (e.g., hemoglobin and lactoferrin from blood specimens) or be carried over from the transport device (e.g., heparin from a blood collection tube) or from the post-collection procedures such as nucleic acid extraction [35].

Some inhibitors are well characterized and the mechanisms by which they function are understood. However, for some specimens and assays, inhibition may occur but the inhibitory substances are not defined and/or the mechanism of inhibition is not understood. Inhibitory substances may bind directly to the target nucleic acids, blocking enzyme binding, or they may bind directly to amplification enzymes, preventing enzymes from binding the target. Although inhibition is a well-recognized phenomenon, it is not always possible to determine exactly what the problem is when inhibition occurs. Inhibition can affect different amplification reactions differently and to different levels [35,36]. Inhibitors are especially problematic in quantitative assays because they may cause aberrant results (inaccurate quantitation) or they may inhibit amplification altogether. Commercially available quantitative PCR assays use software algorithms that normalize the signal from the internal calibrators present in each reaction vessel. Patient test results are based on normalized values; if partial or complete inhibition has occurred, it is detected by the system and interpreted correctly. Laboratories that develop in-house quantitative PCR should implement procedures using normalization to ensure accurate results in their quantitative PCR.

### *Mechanisms by which Inhibitors Function*

There are several mechanisms by which inhibitors function, usually targeting the enzymes involved in amplification or interfering with binding of the enzyme to the target nucleic acid:

1. Blocking active site of enzymes: Some substances block active sites of enzymes used in amplification reactions. Heme compounds from blood function to inhibit Taq polymerase in this manner.
2. Inactivating enzymes: Inhibitors can also act by denaturing or degrading the enzymes used in amplification (e.g., if proteinases are present in the sample).
3. Binding target nucleic acids, preventing the target from being accessible to primers or enzymes.
4. Binding divalent cations: Inhibition can occur when divalent cations ($Mg^{2+}$) necessary as co-factors for polymerase enzymes are bound by EDTA.

### *Commonly Encountered Inhibitors and their Sources*

Inhibitors are present in clinical specimens, collection and transport devices, and can be introduced during processing or nucleic acid extractions (Table 21.1). Amplification tests that are performed directly on specimens without a separate nucleic acid purification step are especially prone to inhibition.

**TABLE 21.1** Common PCR Inhibitors

| General Source | Inhibitory Substance |
|---|---|
| Specimens | Blood—hemoglobin, IgG, lactoferrin<br>Blood—DNA binding proteins<br>Muscles—myoglobin<br>Sputum[a]<br>Stool[a]—bile salts, complex polysaccharides<br>Tissues—melanin<br>Urine[a]—urea |
| Specimen collection tubes | EDTA<br>Heparin |
| Nucleic acid extraction | Detergents (Sarkosyl, SDS)<br>Proteinases<br>Protein disrupting agents<br>Guanidinium salts<br>Phenol<br>Alcohols |

[a]*For many complex clinical specimens, inhibitors have not been well-characterized.*

## INHIBITORS IN SPECIMENS

Many clinical specimens, such as urine, feces, blood, and sputum, are complex and contain inhibitory substances. Not all inhibitors have been characterized. Specimens collected in preservatives or fixed in alcohols or formaldehyde solutions are especially subject to inhibition during PCR, even when nucleic acids have been extracted and purified prior to use in the assay. Some inhibitors have been well-characterized. Hemoglobin is a potent inhibitor of PCR amplification and a well-characterized inhibitor of the Taq polymerase [37]. Complex specimens such as stool, sputum, and urine usually contain inhibitors; assays for microbes from complex specimens are optimally performed on extracted nucleic acids to reduce the chances of co-purification of inhibitors.

## INHIBITORS IN TRANSPORT DEVICES

Specimen collection devices such as blood collection tubes may contain substances that function as amplification inhibitors, including heparin and EDTA. EDTA can potentially inhibit PCR by binding divalent cations that are used as co-factors by amplification enzymes [35]. Heparin is easily co-purified with DNA and a strong inhibitor of PCR at levels as low as 0.032 U/mL in the PCR reaction [38]. If these substances are carried through the nucleic acid extraction procedure, inhibition of amplification reactions may occur.

## INHIBITORS IN EXTRACTION REAGENTS

Nucleic acid extraction procedures can be sources of potent amplification inhibitors. Detergents used to lyse cells (e.g., Sarkosyl and sodium dodecyl sulfate (SDS)), protein disrupting agents (guanidine), enzymes used to digest cellular proteins (proteinases), organic extraction compounds (phenol and chloroform), and alcohols may be inadvertently carried over from the extraction procedure into the purified nucleic acid samples. If present in high enough concentrations, these substances can degrade polymerases or physically block polymerase from copying target nucleic acids [35].

### *Co-Purification of Inhibitors*

Extraction and purification of nucleic acids prior to use in amplification procedures can usually eliminate inhibitors, unless they co-purify with the nucleic acid. Extraction methods differ in their ability to remove inhibitors from specimens. Different extraction and purification methods may need to be tried if a laboratory is developing its own assays. For example, melanin present in pigmented cells from skin and certain tumors is a potent inhibitor of PCR and usually co-purifies with standard DNA or RNA extraction procedures [39]. Addition of high concentrations of proteins such as bovine serum albumin can reverse this inhibition [40]. The nucleic acid extraction and purification procedures may be a source of inhibitors if residual materials from the extraction remain in the final sample.

### *Monitoring for Amplification Inhibition*

The best way to determine if inhibitors are present in a specimen is to include an internal control in every reaction. Monitoring for inhibition is a requirement for laboratories that are performing diagnostic assays using molecular methods. Commercially available, FDA-approved diagnostic assays include internal amplification controls to monitor for inhibitors. Several different types of amplification controls can be used, and both internal and external control materials should be included in amplification assays. Internal controls are used to monitor the entire process from extraction to assay interpretation [41,42].

For infectious disease assays, internal controls can be designed so that the same primer set is used with a different probe sequence. For real-time PCR systems, the tests should be designed to contain an internal amplification control in every reaction. The amplification plot can be used to determine if inhibition has occurred. In systems in which the amplification can be monitored using software, amplification curves can be examined to determine if inhibition is occurring. Inhibition plots can be prepared and monitored. Laboratories that are developing their own assays must design a method for monitoring inhibition. Guidelines for choosing internal control materials are available [41,42]. If the result of the internal control is negative, the report can be modified to indicate that the specimen was inhibitory rather than negative. For genetic assays, an unrelated target sequence can be amplified and detected along with the gene of interest. For example, an assay for a housekeeping gene such as β-actin can be performed simultaneously with the target sequence.

Monitoring of the rate of inhibitors can be performed as part of the laboratory's quality assurance program. Unusually high rates of inhibition can be an indication that a problem exists somewhere in the system. For genetic assays, the extracted DNA is usually quantified so the amount of DNA that is put into the reaction can be standardized. If UV spectroscopy is used, the purity of the extracted nucleic acids can be verified by checking the A260/A280 ratio. Low ratios (for DNA, a ratio of 1.8 is ideal) indicate contamination with protein or other substances. If the ratio is poor, the DNA can be re-extracted and the test re-run.

### *Strategies to Prevent Inhibition*

Laboratories that are developing their own nucleic acid amplification methods should optimize all phases of testing (specimen selection, transport devices, and nucleic acid extraction/purification) to reduce the presence of inhibitors in the samples being analyzed. Amplification assays should be developed with inclusion of substances that can facilitate amplification, such as bovine serum albumin. If the laboratory is monitoring for inhibitors using internal controls and detects an increased rate of inhibition, the assay can be re-optimized. Commercially available PCR assays now contain internal controls; inhibition of the internal control can be monitored and if significant inhibition occurs, the test may be reported as being inhibitory rather than negative. Using additives such as bovine serum albumin can alleviate or reverse the effects of inhibitors in PCR. For home-brew assays, laboratories should optimize the assay prior to routine use of any additives.

One of the best methods to prevent inhibition is to ensure that highly pure nucleic acids are used on the assay. Nucleic acid extraction/purification methods that are based on the Boom technology of silica binding [32] result in good-quality nucleic acids. The silica binding methods are widely available, in many formats, and usually result in highly pure nucleic acids. Laboratories may wish to avoid the use of assays that do not incorporate a nucleic acid extraction and purification step, especially for specimens that are known to contain inhibitors such as blood or urine.

For assays performed on plasma, guidelines for specimen collection should be followed carefully.

Drawing the correct amount of blood into the tube will ensure that the EDTA is diluted to a level that will not inhibit downstream PCR testing after extraction of nucleic acids.

Inhibition of PCR can often be alleviated by the use of substances that can facilitate amplification. Laboratories that develop their own test methods can experiment with incorporating substances into the assay during the development phase. Substances that have been successfully used to relieve inhibition include bovine serum albumin [43] and DMSO [35].

## Poor Quality of Target Nucleic Acid: Target Degradation

The quality of the target nucleic acid has a significant effect on the performance of the molecular assay. Good-quality, intact DNA or RNA is needed for optimal detection. Amplification methods generally require purified nucleic acids that have been extracted through a nucleic acid extraction and purification procedure. Several commercially available assays do not incorporate a nucleic acid extraction step prior to use in the assay; these assays are susceptible to inhibition. Optimally, the nucleic acid extraction method will give high-quality, intact nucleic acids that are free from copurified material or from residual reagents carried over from the purification process.

### *Fixed Tissues or Cells*

Fixed tissues or cells present a challenge for obtaining good-quality target nucleic acids. Depending on the condition of the tissues, and the fixative that has been used, extraction protocols may require some form of pretreatment step (e.g., to dissolve paraffin if tissues are embedded or to remove other chemicals such as preservatives or fixing agents). Pretreatment protocols can utilize heat, organic chemicals such as xylene, or enzyme digestion. Pretreatments can reduce the amount or quality of target nucleic acid and can also result in inhibition of amplification if reagents are coextracted with the target material. Laboratories should include positive and negative controls when possible, which are handled in the same manner as clinical specimens.

### *Degradation of the DNA or RNA Target*

Nucleic acids can become degraded during transport or storage. RNA is especially prone to degradation, and special procedures must be implemented in the laboratory when working with RNA. Degradation of target nucleic acids can result in false-negative results or in inaccurate quantification in quantitative assays [44]. Nucleases may be present in the specimens, and nucleic acid may be degraded by nucleases. Nucleases may be removed by using a nucleic acid extraction procedure that degrades proteins. RNA is especially prone to degradation by RNases, enzymes that are ubiquitous in the environment. Target degradation can be prevented or minimized by maintaining the samples in the correct environmental conditions prior to use in the assay. All samples should be handled using procedures that are optimized to prevent degradation of nucleic acids. If degradation of the target is suspected, the quality of the extracted nucleic acid should be checked with UV spectroscopy or some other method.

## Sequence Mismatch between Primer and Target DNA: Role of Genetic Variation

Nucleotide sequence variation in the target DNA or RNA sequence can result in false-negative test results, for example, if there are mismatches between the sequences of the primers and the target DNA or if a gene deletion occurs. Microorganisms, especially viruses, are prone to this phenomenon because they can evolve rapidly. Mutations in the primer binding area can prevent amplification altogether or reduce efficiency of amplification.

### *Sequence Variation in Primer Binding or Probe Regions*

A mismatch between the primers or the probe sequence can result in false-negative assay results or can give inaccurate quantitative results in viral load assays. Viruses are especially prone to this phenomenon because they evolve rapidly. Sequence variation should always be a consideration when designing assays, especially home-brew assays. HIV and HCV viral load assays are subject to this phenomenon. Both of these viruses exist as viral populations with significant genetic variation. Although quantitative assays for these viruses are usually targeted to conserved areas of the genome, there is enough genetic variation in the population such that occasional strains of these viruses may be missed altogether or underquantitated [45]. False-negative molecular assays for bacterial pathogens can also occur because of sequence variation. PCR-based assays for *N. gonorrhoeae* may be negative when strains of the organism with genetic variation in the primer binding region are present in the specimen [46].

*CASE REPORT: FALSE-NEGATIVE RESULT OF PCR TESTING FOR* **NEISSERIA MENINGITIDIS** A 2-year-old boy presented to the hospital with meningitis; empiric antimicrobial therapy had been started prior to obtaining CSF and blood specimens for culture.

A Gram stain performed on the CSF revealed Gram-negative diplococci; culture was negative. Real-time PCR testing for *N. meningitidis* using primers for the ctrA gene was negative. Because the Gram stain was consistent with *N. meningitidis*, a second PCR assay using 16 S rDNA primers was performed with positive results for *N. meningitidis*. During the time the testing was being performed, blood cultures grew *N. meningitidis*. The child recovered. Follow-up testing of the isolate from the blood culture showed sequence variation in the ctrA gene, with polymorphisms present on one of the primer binding sequences, and a single nucleotide substitution in the probe binding area [47].

For laboratories performing in-house developed tests, primer and probe sequences should be verified. This can be done by searching publicly available databases.

## Analyte below Limit of Detection of Assay

One of the most common reasons for false-negative results, especially in qualitative infectious disease assays, is the presence of the analyte at very low levels—below the analytical sensitivity of the assay. Although nucleic acid amplification methods can theoretically detect a single molecule of target, in reality this is rarely achieved. The analytical sensitivity of molecular amplification methods should be considered when the assay result is negative.

### *Pre-Analytical Considerations that can Influence Nucleic Acid Amplification Assay Results*

The timing of specimen collection is important when using molecular assays for diagnosis of infectious diseases. The specimens must be collected during the acute phase of illness; often, the pathogen may be undetectable by the time the patient presents for medical care. Diagnostic methods for infections due to arthropod-borne viruses such as West Nile virus are especially subject to this type of error. Viremias (virus in the blood) are transient, and with rare exceptions, testing of blood for virus by PCR is likely to yield negative results. A similar phenomenon occurs for CSF. In most cases, the recommended diagnostic strategy for arthropod-borne viruses such as West Nile will include serological testing in addition to PCR to avoid misdiagnosis due to false-negative PCR testing.

The timing of specimen collection is critical for many assays, especially viruses, in which the analyte may be present transiently. Collecting a specimen during the acute phase of illness is highly recommended for respiratory viruses, arthropod-borne viruses, and many others. For respiratory viral assays, the optimal specimen type is usually a nasopharyngeal swab or aspirate, and throat swabs are usually discouraged by the laboratory.

For some assays, concentration of the specimen prior to testing in PCR assays is a useful method for increasing the sensitivity of the assay. For example, when testing CSF for infectious pathogens such as herpes simplex virus, concentration of the analyte via nucleic acid extraction may help increase the chances of pathogen detection.

## Technical Problems with Assay

### *Pre-Analytical Considerations*

Specimen collection, processing, and storage conditions are critical steps in molecular testing, and appropriate conditions must be used for successful results. For viral load assays and for laboratory-developed assays, parameters must be worked out systematically. These usually include optimizing the concentration of nucleic acid to include in the assay, optimizing concentrations of cations that amplification enzymes need, and optimizing reaction temperatures and ramp speeds. Occasionally, amplification assays exhibit failure or reduced performance even when parameters are optimized and the assays have been functioning well in the laboratory. In these cases, thermal cycler function should be checked. Even slight variations in annealing and extension temperatures can have detrimental effects on reactions.

# INTERFERENCES IN MOLECULAR TESTING: FALSE-POSITIVE TEST RESULTS IN MOLECULAR ASSAYS

Molecular assays have excellent performance characteristics and are usually highly sensitive and specific. Occasionally, however, false-positive test results can occur. Laboratories should be aware of several important reasons why false-positive results occur. Laboratories should be routinely monitoring for false-positive results by several methods so that if unexpected results occur, they can be identified and corrected in a timely manner, before larger problems occur. Monitoring for false-positive results should include verification of primer and probe sequences, monitoring and verifying assay conditions, and the use of negative controls in their molecular assays. Some of the more common causes of false-positive test results are discussed next, with suggestions for strategies to prevent and control them.

## Contamination

In general, molecular diagnostic methods, especially infectious disease assays, require very high analytical sensitivity. Because the amount of microbial DNA or RNA present in a clinical specimen may be very low, an amplification method is used to optimize sensitivity. Although there are some nucleic acid detection tests based on direct hybridization, most nucleic acid detection methods now rely on some form of nucleic acid amplification, such as PCR. Theoretically, nucleic acid amplification methods have the capability to detect a single molecule of analyte, and many assays do approach this level of sensitivity, with some assays capable of detecting as few as 5–10 molecules of target analyte. However, the extremely high analytical sensitivity of amplification methods also increases the risk of false-positive test results due to the presence of contaminating DNA in the reaction, where even a single molecule of DNA can serve as a template for amplification in a later reaction. Contamination of samples is a major consideration for laboratories performing amplification reactions, and procedures must be in place to prevent, monitor, and control contamination. A false-positive PCR result can have serious implications; there are numerous reports of false-positive results in the literature, many of which have significant consequences [48]. The molecular diagnostics laboratory must ensure that methods are in place to prevent, detect, monitor, and correct for contamination events.

### *Sources of Contamination*

Contaminating DNA may be introduced into reactions from a number of different sources:

*Amplicon DNA*: An important source of false-positive results in nucleic acid amplification testing is amplified DNA from a previous amplification reaction. If reaction tubes are opened after amplification (e.g., for gel analysis, microarray testing, and sequencing), amplified DNA can be aerosolized, contaminating the environment, and it can be carried directly into reagents and consumables. This contaminating DNA can then be carried directly into new reactions the next time assays are set up, causing false-positive results.

*Specimens*: Nucleic acids from specimens may also contaminate the environment during specimen handling, for example, when pipetting specimens into tubes for nucleic acid extraction procedures. Cross-contamination of samples containing the analyte prior to or during sample processing can lead to false-positive results. When multiple samples are processed in batches, there is potential for aerosolization, splashing, etc.

*Positive control material*: Incorrect handling of positive controls (e.g., plasmid DNA) may also result in contamination of patient samples, leading to false-positive results. The second potential source of contamination is amplified material from previous reactions. If tubes are opened for post-amplification analysis (e.g., gel electrophoresis and microarray testing), amplicons can contaminate the environment and be carried over into the next reactions, causing false-positive results. Contaminating nucleic acids can be present in the laboratory environment (on clothing, benchtop, waste containers, etc.). DNA contamination may occur in reagents, enzyme mixes, master mixes, etc.

### *Methods to Prevent and Control Contamination*

Laboratory practices to prevent contamination in the molecular diagnostics laboratory include both physical design and layout of the laboratory, workflow practices, and policies and procedures designed to prevent contamination. Many references are available for designing and implementing contamination control programs in the molecular diagnostic laboratory [49]:

1. Laboratory design: Optimally, a molecular diagnostic laboratory should be designed so that separate work areas for different parts of a procedure are used. Although separate rooms are optimal (e.g., specimen processing room, PCR master mix room, amplification room, and post-PCR DNA handling room), separate work areas within a large space can function to prevent contamination. Limiting traffic in the molecular laboratory can help prevent contamination.
2. Work practices and workflow: Workflow practices should be implemented to ensure a one-way workflow from clean to amplified area. Specimens are extracted in one area and then amplified in a different area. PCR products or other amplified materials are not carried into the specimen handling or pre-amplification areas. Work practices that should be implemented include use of dedicated pipettors, use of dedicated packages of consumables such as tips and tubes, and use of dedicated lab coats. Laboratory policies and procedures should include specific methods for contamination prevention. Cleaning the work areas should be performed prior to and after every procedure. Cleaning agents specifically designed to destroy nucleic acids should be used.
3. DNA degrading enzymes such as uracil DNA-glycosylase (UNG) can be used. Uracil is incorporated in the PCR master mix; during amplification reactions, uracil is incorporated into the amplified DNA. UNG is added to the new PCR reaction tubes; if any DNA is carried over from the amplification into the new PCR reaction tubes, the UNG will degrade the contaminating DNA prior to the start of the new PCR reaction. This method is

utilized in some commercially available PCR diagnostic assays and is quite beneficial in preventing carryover contamination.

4. Converting end-point assays to real-time assays that do not require manipulation after the amplification is an excellent method for preventing problems with environmental contamination. Much of the risk for contaminating the environment with target DNA comes from opening amplification tubes after reactions are complete. Splashing, spilling, and aerosol formation can all be avoided when tubes for real-time assays are discarded unopened.

### *Methods for Detection and Monitoring of Contamination*

Laboratories can include in their quality assurance programs methods to ensure that any contamination issues are promptly detected and corrected:

*No-template or negative controls*: A no-template reaction should be included in every batch of amplified reactions. Typically, the laboratory will set up a no-template control using water or buffer without DNA. The control reaction should be handled exactly the way specimens are handled; that is, it should be processed through the extraction and purification and then assayed. Positive results in a negative control reaction indicate possible contamination. The laboratory should have a procedure in place to follow up and investigate a positive result in water or other negative controls.

*Environmental testing for contamination*: For laboratories that are at risk of contamination (e.g., when high-volume testing with open reaction conditions) or that have experienced contamination, routine environmental monitoring studies (wipe testing) can be implemented. There are no standardized procedures for environmental testing; the laboratory must establish its own policies and procedures for doing this type of monitoring. Examples include using swab testing of the environment and then processing the swab through the extraction and testing procedure, handling it as if it were a specimen. Any positive results should trigger a review of policies and procedures and implementation of cleaning and decontamination. Environmental contamination can be difficult to control if the environment has become contaminated with amplified DNA.

*Monitoring assay positivity rates*: Monitoring assay positivity rates can be a useful method for determining if a problem is occurring with an assay. Monitoring positivity rates is quite useful for tests such as *N. gonorrhoeae* or *Chlamydia trachomatis* PCR. An unusual increase in the number of positive tests may indicate contamination has occurred. Monitoring can be done as part of the laboratory's quality assurance program.

## Primer or Probe Cross-Reactivity Resulting in Nonspecific Amplification or Hybridization: Assay Specificity

False-positive PCR results in molecular testing can occur when primers or probes exhibit cross-reactivity and nonspecifically bind to sequences that are present in the specimen or when specific sequences are present in unrelated organisms that are not pathogens.

### *Mispriming*

Hybridization of primers to nontarget sequences (mispriming) can occur under several conditions. Primers may bind to identical sequences present in nucleic acids that are not the specific targets of the assay. For laboratory-developed tests, primer sequences should be checked by searching against a database such as GenBank to verify that they will not cross-react to other organisms. Mispriming can also occur when assay conditions are not optimized (annealing and extension temperatures, composition of amplification mix, ion concentration, etc.).

### *Cross-Reactivity in Primer or Probe Binding Regions*

Nucleotide sequence homology in primer or probe binding regions can give false-positive results. Hybridization probes used to detect amplified product may exhibit cross-reactivity to closely related species. For amplification assays that use labeled probes for detection of product, probe sequences should be checked for potential cross-reactivity to related species or genes, including pseudogenes if DNA is targeted. False-positive results can have significant negative consequences. For assays that use hybridization probes (e.g., ribosomal RNA probes), if nonspecific hybridization is occurring, assay parameters should be checked. Incorrect temperatures during the hybridization of the labeled probe are a common reason for nonspecific hybridization.

*CASE REPORT: FALSE-POSITIVE REPORT OF* **TROPHERYMA WHIPPLEI** *IN CSF AND INTESTINAL BIOPSY SAMPLES DUE TO HOMOLOGY IN PRIMER BINDING REGIONS* A 13-year-old male presented to the emergency room with a 2-week history of central nervous system symptoms. Multiple tests were performed; a PCR-based assay for *Tropheryma whipplei* performed on a CSF specimen was positive. The primers used in the

PCR were targeted to a *T. whipplei*-specific portion of the 16 S rRNA gene; product detection was by melting curve analysis and agarose gel electrophoresis. Clinically, the diagnosis of Whipple's disease was not consistent with the results of the PCR, and follow-up investigation was pursued. PCR testing with a different primer set was performed and was negative. The amplicons from the original PCR were sequenced. Sequencing results revealed that the amplified material was a human gene; the original PCR test was a false positive due to primer cross-reactivity [50]. Ultimately, the child was diagnosed with chronic lymphocytic meningitis of unknown origin.

### *Examples of Known Problems with Cross-Reactivity in Molecular Assays*

#### AMPLIFICATION OF NONPATHOGENIC *NEISSERIA* SPECIES IN PCR: FALSE-POSITIVE RESULTS FOR *NEISSERIA GONORRHOEAE*

Several different assays using molecular methods are available for detection of *N. gonorrhoeae* (GC) in urine or urogenital sites. Commercially available tests include PCR, SDA, and TMA. The molecular targets for probes and primers used in these assays varies [51]; these genes or sequences may be present in the closely related, nonpathogenic species such as *N. cinerea* and *N. lactamica*. These nonpathogenic *Neisseria* species are common colonizers of the oropharyngeal cavity, rectal, and urogenital sites. False-positive results (due to the presence of identical DNA sequences in nonpathogenic *Neisseria* species) can occur with some of these assays [52]. There are several recommendations for avoiding the possibility of a false-positive report of GC. Testing of specimens collected from extragenital sites such as throat or rectal specimens should be performed with culture rather than using molecular amplification assays. Laboratories should consider performing some form of confirmatory testing for specimens that test positive with molecular assays for GC and *Chlamydia trachomatis* (CT). Ideally, a different method or different molecular target would be used. Repeat testing of positive specimens with the same method is often used for practical reasons and can be cost-effective. Laboratories performing molecular testing for GC and CT should monitor their positivity rates as part of their quality management program for molecular diagnostics. A significant increase in the positivity rate can trigger an investigation into the potential for contamination or an investigation of the performance of the assay. Screening guidelines can also help; ensuring that testing is performed for only those patients for whom screening is recommended will help to ensure that the positive predictive value of the test is adequate.

#### FALSE-POSITIVE PCR RESULTS FOR METICILLIN-RESISTANT *STAPHYLOCOCCUS AUREUS*

Surveillance testing for meticillin-resistant *Staphylococcus aureus* (MRSA) is routinely performed in many health care settings. Culture-based methods have been shown to be slightly less sensitive than PCR assays, and many laboratories have begun using PCR-based assays for detection of MRSA in screening programs. Several commercially available assays are in use, most of which are based on the detection of the staphylococcal cassette chromosome mec (SCCmec) element. Resistance to meticillin is due to an altered penicillin binding protein, PBP2a, which is encoded by the mecA gene. The mecA gene is located on a mobile genetic element, called the SCCmec, which was originally acquired from a non-*S. aureus* species. The SCCmec is integrated into an *S. aureus*-specific gene called orfX. Using primers directed to the orfX/mecA junction region allows for simultaneous identification of *S. aureus* and mecA. Certain strains of *S. aureus* contain deletions in a portion of the mecA cassette; functionally, these isolates do not express mecA and are thus meticillin susceptible in phenotypic assays. These strains are phenotypically not meticillin resistant, but they give positive results in PCR assays that are designed to target the orfX/mecA region. The incidence of false-positive MRSA reports varies and is most likely regional. Laboratories that perform both culture- and PCR-based methods for MRSA should investigate the possibility of mecA mutants. Laboratories may wish to monitor their positive PCR surveillance screens, monitor the rate of discordance between the PCR assay results and phenotypic results, and follow up on discordant results to determine if such strains are present in their population.

## SOURCES OF ERRORS IN DNA MICROARRAY TESTING

Microarrays are attractive tools for molecular diagnostic testing because of the ability to test for a large number of targets in a single experiment. Microarrays have enormous potential for microbiology, cancer, and pharmacogenetics testing. DNA microarrays are ideal for testing for the presence of multiple species of organisms, genotyping viruses (e.g., HCV and HPV), testing for multiple single nucleotide polymorphisms, or performing gene expression studies. Consequently, microarrays have received much attention in the molecular diagnostic laboratory and are currently undergoing significant expansion in terms of different technologies available and targets.

When using microarrays for diagnostic testing, common sources of error should be considered. Problems can occur with probe sequences, probe labels, and detection methods:

*Autofluorescence*: For fluorescently labeled probe assays, autofluorescence can interfere with correct data collection from fluorescently labeled probes on microarray substrates. Background signal may come from nonspecifically bound probes or from autofluorescence (intrinsic signal from the substrate).

*Nonspecific hybridization of probes:* Nonspecific hybridization may occur if stringency of hybridization is not optimized for all of the probes used on the microarray.

*Contamination*: Instruments that automate microarray testing are subject to contamination during liquid handling steps. Pipettors, trays, and microarray substrates can become contaminated with DNA from previous or current experiments if splashing or aerosolization occurs within the instrument.

# QUALITY MANAGEMENT

A comprehensive program for quality management specifically designed for the molecular diagnostics laboratory should be in place to ensure the accuracy of testing, accurate performance of test, tracking of trends, and rapid identification and correction of problems. Establishment of a quality management program is the principal method by which common interferences in the molecular diagnostic laboratory can be controlled, monitored, identified, and corrected. The program should be comprehensive, including methods for validation of assays, ongoing quality control, and monitoring for problems. Assay validation is especially important because the performance characteristics of the assay will be verified through this process [53] and any potential problems with the assay should be identified prior to clinical use. Ongoing quality management should include a strong program of quality control to identify problems in analytical phases of testing and monitoring of results.

## Quality Control

Quality control is the portion of the quality management program devoted to detecting problems at the testing level. In molecular laboratories, quality control must include a method to ensure correct performance of nucleic acid isolation steps, amplification and probe controls, and detection steps. Quality control must also be included to monitor for inhibition if amplification assays are used and to check for contamination.

### *Quality Control for Qualitative Assays*

For qualitative assays, external control material should include both positive and negative controls. For qualitative assays with a single molecular target (e.g., infectious disease assays), positive and negative controls should be included in each run of the assay. Positive and negative controls ensure that probes and primers are working correctly, and they allow monitoring for contamination events. External controls should monitor the entire process (nucleic acid extraction and purification, amplification, and detection).

For more complex molecular testing (e.g., multiplexed PCR, microarrays, and nucleotide sequencing results), selection of quality control materials can be a challenge. External control materials may not be readily available for all of the markers being tested for and reported. Performing an external control for every analyte in the assay is also not practical. An option for quality control materials for these assays is to rotate different materials with each run, covering all the targets over a defined number of runs or kits. Statistical strategies and software for dealing with multidimensional data QC are being developed [54,55].

Sources of materials for molecular quality control are somewhat limited. Ideally, quality control material should be well-characterized and optimized for the assays that are performed in the laboratory. For genomic assays, obtaining DNA with well-characterized mutations can be difficult.

### *Quality Control for Quantitative Assays*

Quantitative assays should include positive controls at several levels in the reportable range of the assay—for example, a low positive, midrange positive, and a high positive control. Results of the positive control samples should be monitored using statistical software to search for trends and to ensure that problems are detected.

### *Quality Control for Thermal Cyclers*

Thermal cyclers can be a source of error in PCR assays if they are not performing correctly. Thermal cyclers should be monitored annually with temperature checks. Incorrect cycling temperatures can be a source of variation in amplification assays. Calibration of cyclers that exhibit incorrect temperatures should be performed by the instrument manufacturer.

## Result Reporting

Reporting of the results from the molecular diagnostics laboratory can be a source of error. Results that are generated from complex molecular assays can be difficult to report in a simplified laboratory report format.

Laboratory information systems differ in their ability to handle many of the interpretive or narrative results that may be written into a result of molecular testing.

Many of the tests, especially in molecular pathology, generate large amounts of data (e.g., multiplexed PCRs, DNA microarrays, and gene rearrangement assays). In addition to the large amount of data that must be reported, these tests often require interpretive comments and linking of information between different sections of the laboratory. This can be a challenge for both the laboratory and the customers reading the reports.

For genetic assays, it can be helpful to include literature citations or links to websites or databases that contain more detail on the mutations and the current nomenclature. Guidelines for appropriate reporting of genetic testing are available [56].

Concerning reporting of viral load assays, errors in interpreting the results of viral load assays can be prevented with reports that state the assay method used, the reportable range, and how the assay correlates to an international standard (e.g., for HCV and HIV).

## CONCLUSIONS

There are many sources of errors in molecular testing, and this chapter addressed ways to avoid such errors in molecular testing. It is important to pay attention to each specific step of molecular testing, including sample collection and pre-analytical issues, in order to avoid errors in molecular diagnostics laboratories.

## References

[1] Stankiewicz P, Lupski JR. Structural variation in the human genome and its role in disease. Annu Rev Med 2010;61:437–55.
[2] Wang L, McLeod HL, Weinshilboum RM. Genomics and drug response. N Engl J Med 2011;364(12):1144–53.
[3] Weber WW. Pharmacogenetics: from description to prediction. Clin Lab Med 2008;28(4):499–511.
[4] Al-Ghoul M, Valdes Jr R. Fundamentals of pharmacology and applications in pharmacogenetics. Clin Lab Med 2008;28(4): 485–97.
[5] Weinshilboum R. Inheritance and drug response. N Engl J Med 2003;348(6):529–37.
[6] Mullis KB. The unusual origin of the polymerase chain reaction. Sci Am 1990;262:56–65.
[7] Saiki RK, Gelfand DH, Stoffel S, Scharf SJ, Higuchi R, Horn G, et al. Primer-directed enzymatic amplification of DNA with a thermostable DNA polymerase. Science 1988;239:487–91.
[8] Little MC, Andrews J, Moore R, Bustos S, Jones L, Embres C, et al. Strand displacement amplification and homogeneous real-time detection incorporated in a second-generation DNA probe system, BDProbeTecET. Clin Chem 1999;45(6):777–84.
[9] Romano JW, Williams KG, Shurtliff RN. NASBA technology: isothermal RNA amplification in qualitative and quantitative diagnostics. Immunol Invest 1997;26(1/2):15–28.
[10] Hill CS. Molecular diagnostic testing for infectious diseases using TMA technology. Expert Rev Mol Diagn 2001;1(4): 445–55.
[11] Hadd AG, Brown JT, Andruss BF, Ye F, WalkerPeach CR. Adoption of array technologies into the clinical laboratory. Expert Rev Mol Diagn 2005;5(3):409–20.
[12] Heller MJ. DNA microarray technology: devices, systems, and applications. Annu Rev Biomed Eng 2002;4:129–53.
[13] Miller MB, Tang YW. Basic concepts of microarrays and potential applications in clinical microbiology. Clin Microbiol Rev 2009;22(4):611–33.
[14] Bodrossy L, Sessitsch A. Oligonucleotide microarrays in microbial diagnostics. Curr Opin Microbiol 2004;7(3):245–54.
[15] Lucchini S, et al. Microarrays for microbiologists. Microbiology 2001;147:1403–14.
[16] Geyer FC, Lopez-Garcia MA, Lambros MB, Reis-Filho JS. Genetic characterization of breast cancer and implications for clinical management. J Cell Mol Med 2009;13(10):4090–103.
[17] Chu T, Burke B, Bunce K, Surti U, Allen Hogge W, Peters DG. A microarray-based approach for the identification of epigenetic biomarkers for the noninvasive diagnosis of fetal disease. Prenat Diagn 2009;29(11):1020–30.
[18] Salvado C, Cram D. Microarray technology for mutation analysis of low-template DNA samples. Methods Mol Med 2007;132: 153–73.
[19] Dumur CI, Lyons-Weiler M, Sciulli C, Garrett CT, Schrijver I, Holley TK, et al. Interlaboratory performance of a microarray-based gene expression test to determine tissue of origin in poorly differentiated and undifferentiated cancers. J Mol Diagn 2008;10(1):67–77.
[20] Mrozek K, Radmacher MD, Bloomfield CD, Marcucci G. Molecular signatures in acute myeloid leukemia. Curr Opin Hematol 2009;16(2):64–9.
[21] Stancel GA, Coffey D, Alvarez K, Halks-Miller M, Lal A, Mody D, et al. Identification of tissue of origin in body fluid specimens using a gene expression microarray assay. Cancer Cytopathol 2011;120(1):62–70.
[22] Kapur P. Tailoring treatment of rectal adenocarcinoma: immunohistochemistry for predictive biomarkers. Anticancer Drugs 2011;22(4):362–70.
[23] Vitucci M, Hayes DN, Miller CR. Gene expression profiling of gliomas: merging genomic and histopathological classification for personalised therapy. Br J Cancer 2010;104(4):545–53.
[24] Voelkerding KV, Dames SA, Durtschi JD. Next-generation sequencing: from basic research to diagnostics. Clin Chem 2009;55(4):641–58.
[25] Voelkerding KV, Dames S, Durtschi JD. Next generation sequencing for clinical diagnostics-principles and application to targeted resequencing for hypertrophic cardiomyopathy: a paper from the 2009 William Beaumont hospital symposium on molecular pathology. J Mol Diagn 2010;12(5):539–51.
[26] Metzker ML. Sequencing technologies: the next generation. Nat Rev Genet 2010;11(1):31–46.
[27] Ronaghi M, Uhlen M, Nyren P. A sequencing method based on real-time pyrophosphate. Science 1998;281(5375):363–5.
[28] Rothberg JM, Hinz W, Rearick TM, Schultz J, Mileski W, Davey M, et al. An integrated semiconductor device enabling non-optical genome sequencing. Nature 2011;475(7356):348–52.
[29] Bentley DR, Balasubramanian S, Swerdlow HP, Smith GP, Milton J, Brown CG, et al. Accurate whole human genome sequencing using reversible terminator chemistry. Nature 2008;456(7218):53–9.
[30] Chomczynski P, Sacchi N. Single-step method of RNA isolation by acid guanidinium thiocyanate-phenol-chloroform extraction. Anal Biochem 1987;162(1):156–9.

[31] Chomczynski P, Sacchi N. The single-step method of RNA isolation by acid guanidinium thiocyanate-phenol-chloroform extraction: twenty-something years on. Nat Protoc 2006;1(2): 581–5.

[32] Boom R, Sol CJ, Salimans MM, Jansen CL, Wertheim-van Dillen PM, van der Noordaa J. Rapid and simple method for purification of nucleic acids. J Clin Microbiol 1990;28(3):495–503.

[33] Clinical and Laboratory Standards Institute. Collection, transport, preparation, and storage of specimens for molecular methods: approved guideline. Wayne, PA: Clinical and Laboratory Standards Institute; 2005.

[34] Lew J, Reichelderfer P, Fowler M, Bremer J, Carrol R, Cassol S, et al. Determinations of levels of human immunodeficiency virus type 1 RNA in plasma: reassessment of parameters affecting assay outcome. TUBE meeting workshop attendees. Technology utilization for HIV-1 blood evaluation and standardization in pediatrics. J Clin Microbiol 1998;36(6):1471–9.

[35] Wilson IG. Inhibition and facilitation of nucleic acid amplification. Appl Environ Microbiol 1997;63(10):3741–51.

[36] Huggett JF, Novak T, Garson JA, Green C, Morris-Jones SD, Miller RF, et al. Differential susceptibility of PCR reactions to inhibitors: an important and unrecognised phenomenon. BMC Res Notes 2008;1:70.

[37] Akane A, Matsubara K, Nakamura H, Takahashi S, Kimura K. Identification of the heme compound copurified with deoxyribonucleic acid (DNA) from bloodstains, a major inhibitor of polymerase chain reaction (PCR) amplification. J Forensic Sci 1994;39(2):362–72.

[38] Yokota M, Tatsumi N, Nathalang O, Yamada T, Tsuda I. Effects of heparin on polymerase chain reaction for blood white cells. J Clin Lab Anal 1999;13(3):133–40.

[39] Eckhart L, Bach J, Ban J, Tschachler E. Melanin binds reversibly to thermostable DNA polymerase and inhibits its activity. Biochem Biophys Res Commun 2000;271(3):726–30.

[40] Giambernardi TA, Rodeck U, Klebe RJ. Bovine serum albumin reverses inhibition of RT-PCR by melanin. Biotechniques 1998;25(4):564–6.

[41] Hoorfar J. Making internal amplification control mandatory for diagnostic PCR. J Clin Microbiol 2003;41(12):5835.

[42] Hoorfar J, Malorny B, Abdulmawjood A, Cook N, Wagner M, Fach P. Practical considerations in design of internal amplification controls for diagnostic PCR assays. J Clin Microbiol 2004;42(5): 1863–8.

[43] Kreader CA. Relief of amplification inhibition in PCR with bovine serum albumin or T4 gene 32 protein. Appl Environ Microbiol 1996;62(3):1102–6.

[44] Ginocchio CC, Wang XP, Kaplan MH, Mulligan G, Witt D, Romano JW, et al. Effects of specimen collection, processing, and storage conditions on stability of human immunodeficiency virus type 1 RNA levels in plasma. J Clin Microbiol 1997;35 (11):2886–93.

[45] Jenny-Avital ER, Beatrice ST. Erroneously low or undetectable plasma human immunodeficiency virus type 1 (HIV-1) ribonucleic acid load, determined by polymerase chain reaction, in West African and American patients with non-B subtype HIV-1 infection. Clin Infect Dis 2001;32(8):1227–30.

[46] Whiley DM, Tapsall JW, Sloots TP. Nucleic acid amplification testing for *Neisseria gonorrhoeae*: an ongoing challenge. J Mol Diagn 2006;8(1):3–15.

[47] Jaton K, Ninet B, Bille J, Greub G. False-negative PCR result due to gene polymorphism: the example of *Neisseria meningitidis*. J Clin Microbiol 2010;48(12):4590–1.

[48] Borst A, Box AT, Fluit AC. False-positive results and contamination in nucleic acid amplification assays: suggestions for a prevent and destroy strategy. Eur J Clin Microbiol Infect Dis 2004;23(4):289–99.

[49] Lo YM, Chan KC. Setting up a polymerase chain reaction laboratory. Methods Mol Biol 2006;336:11–18.

[50] Goyo D, Camacho A, Gomez C, de Las Heras RS, Otero JR, Chaves F. False-positive PCR detection of *Tropheryma whipplei* in cerebrospinal fluid and biopsy samples from a child with chronic lymphocytic meningitis. J Clin Microbiol 2009;47(11):3783–4.

[51] Palmer HM, et al. Evaluation of the specificities of five DNA amplification methods for the detection of *Neisseria gonorrhoeae*. J Clin Microbiol 2003;41(2):835–7.

[52] Katz AR, Effler PV, Ohye RG, Brouillet B, Lee MV, Whiticar PM. False-positive gonorrhea test results with a nucleic acid amplification test: the impact of low prevalence on positive predictive value. Clin Infect Dis 2004;38(6):814–9.

[53] Jennings L, Van Deerlin VM, Gulley ML. Recommended principles and practices for validating clinical molecular pathology tests. Arch Pathol Lab Med 2009;133(5):743–55.

[54] Kricka LJ, Master SR. Validation and quality control of protein microarray-based analytical methods. Mol Biotechnol 2008;38(1): 19–31.

[55] Kauffmann A, Gentleman R, Huber W. arrayQualityMetrics—a bioconductor package for quality assessment of microarray data. Bioinformatics 2009;25(3):415–16.

[56] Chen B, Gagnon M, Shahangian S, Anderson NL, Howerton DA, Boone JD. Good laboratory practices for molecular genetic testing for heritable diseases and conditions. MMWR Recomm Rep 2009;58(RR-6):1–37 [quiz CE-1–4]

CHAPTER

# 22

# Problems in Pharmacogenomics Testing


*Dina N. Greene*, Cecily Vaughn†, Elaine Lyon†‡*

*Northern California Kaiser Permanente Regional Laboratories, The Permanente Medical Group, Berkeley, California

†ARUP Institute for Clinical and Experimental Pathology, Salt Lake City, Utah

‡University of Utah, Salt Lake City, Utah


## INTRODUCTION

Pharmacogenetics is the key for personalized medicine. The goal of pharmacogenetic testing is to ensure that the right drug is given to the right patient at the right dose. Pharmacogenetic targets are often enzymes involved in drug metabolism; inherited mutations or variants in genes coding for these proteins can alter their function. Inherited variants also play a role in treatments for infectious diseases. In addition, pharmacogenetics applies to drugs developed for cancer therapies targeting molecular mechanisms of tumor biology; however, mutations within oncological pathways can render these therapies ineffective.

Molecular technologies are at the forefront of pharmacogenetic testing. From detecting single known mutations to mutation panels and sequencing and to quantitative polymerase chain reaction (PCR), molecular techniques can detect and quantify relevant targets. This chapter provides brief descriptions of commonly used methods. Current applications of pharmacogenetic tests are described, along with the challenges of detecting the genetic alterations and interpreting the results. Inherited and somatic variations are addressed.

## METHOD DESCRIPTIONS

Various techniques are applied in pharmacogenomics testing.

### Targeted Single Mutation Detection

Multiple methods are available for detection of single nucleotide base changes. A traditional method, still in use today, is the use of restriction enzymes that cut DNA at specific palindromic sequences. Mutations may create or destroy a restriction site. In this method, PCR amplifies a specific region of interest, and products are treated with the restriction enzyme. The resulting differences in the PCR product sizes are resolved by electrophoresis.

Allele-specific PCR is a mutation detection technique based on the principle that DNA extension is efficient when the 3′ nucleotide of the primer matches the template, but extension is poor or nonexistent when the terminal base is mismatched. Allele-specific PCR is also known as amplification refractory mutation system or allele-specific primer extension (ASPE). It is generally coupled to other technologies such as real-time (or quantitative) PCR (qPCR) or arrays. Incorporating qPCR into the allele-specific analysis bolsters the technique by reducing contamination and decreasing turnaround time. Importantly, qPCR reduces the number of false positives due to background or nonspecific primer binding. A threshold difference is allowed between the quantity of housekeeping gene amplified and the quantity of the selected allele. Thus, by monitoring the amplification, a clear population of mutant alleles can be distinguished from wild type or background. Allelic ratios of the fluorescence emitted by the wild-type and mutant reactions can determine homozygosity (of either the wild type or the mutant) or heterozygosity.

Probe chemistries have been used with many variations for targeted mutation detection. Two methods commonly employed in clinical laboratories are hybridization probes (LightCycler technology) and hydrolysis probes (TaqMan technology). In LightCycler technology, probes are designed to be a perfect match to either the mutation or the wild-type sequence. They are


*Accurate Results in the Clinical Laboratory.*
DOI: http://dx.doi.org/10.1016/B978-0-12-415783-5.00022-0

typically 15–20 bases long, with the mutation located toward the center of the probe. Probes have been described with Förster (or fluorescence) resonance energy transfer as the detection system [1,2]. In this approach, two probes—one over the mutation of interest (the reporter probe) and one adjacent to the first (the anchor probe)—are each fluorescently labeled, one with an acceptor fluorophore and the other with a donor fluorophore. When the two probes are hybridized (at lower temperatures), the energy of the donor is transferred nonradiatively to the acceptor. The acceptor then emits at a longer wavelength, which is detected by the instrument. Single-labeled probes as well as unlabeled probes have also been described [3].

To detect mutations using LightCycler technology, PCR is performed, followed by a "melt" analysis. In a melt analysis, the temperature is slowly increased, with continual monitoring of fluorescence. As the temperature rises, the probes disassociate from their template, or melt, at a characteristic temperature, which is seen as a loss of fluorescence. A mismatch between the probe and the DNA amplicon will lower the temperature at which the probe melts from the DNA amplicon. A derivative of the melting curve easily visualizes melting "peaks," resolving the perfectly matched allele and the mismatched allele (Figure 22.1). Other variants within the region of the probes may be detected by a shift in the expected melting temperature. However, if melting temperatures between the targeted mutation and another variant are similar, the result may be a false positive for the targeted mutation [4].

In hydrolysis (TaqMan) chemistry, probes are dually labeled with a fluorophore and quencher molecule. When the probe is hybridized to the product, the fluorescence is quenched. As PCR primers are

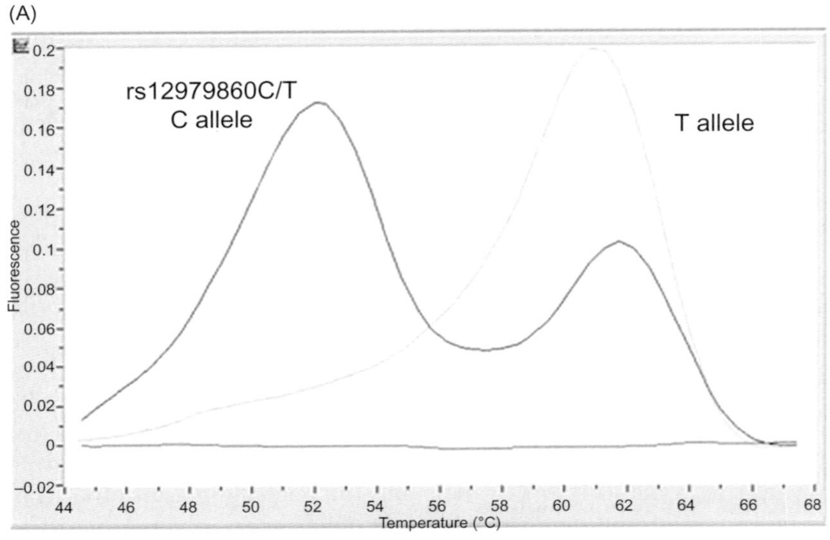


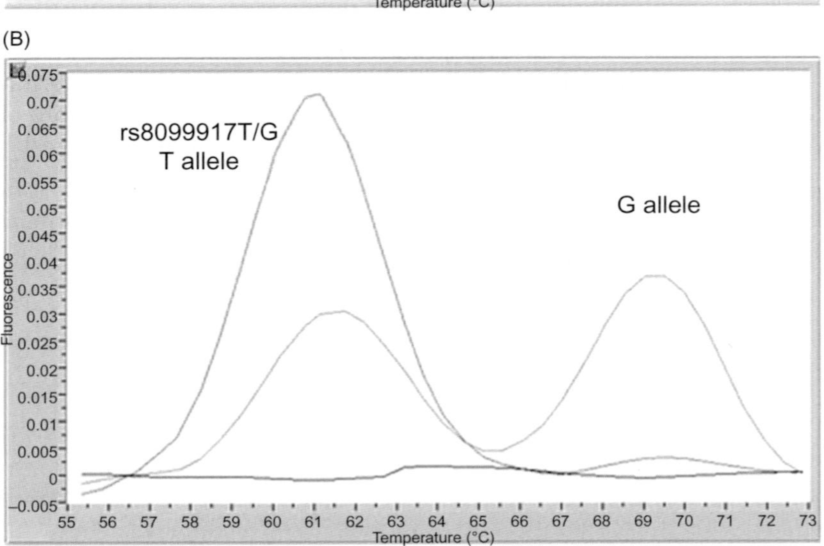


**FIGURE 22.1 Hybridization probes for IL28B-related variants.** (A) Melting curve analysis of rs12979860 with the probe being a perfect match to the T allele. A sample homozygous for the T allele exhibits a single peak at the higher melting temperature. A heterozygous sample peaks corresponding to the C and T alleles. (B) Melting curve of rs8099917 with the probe being a perfect match to the G allele. A heterozygous sample exhibits two peaks, corresponding to the T and G alleles, whereas a sample homozygous for the T allele exhibits only the peak at the lower melting temperature.

extended, replicating the DNA template, the polymerase displaces and hydrolyzes the probe. As the probe is hydrolyzed, the fluorophore is released from the effect of the quencher, resulting in a fluorescent signal. In TaqMan systems, the wild-type and mutant allele are amplified separately by allele-specific PCR as described previously. The ratio of the fluorescent signal of each amplification is used to determine zygosity. Other variants within the region of the probe may not be detected but may interfere with amplification of one allele.

## Mutation Panels

When testing for more than three or four mutations simultaneously, methods capable of highly multiplexed reactions are used. Several commercial platforms are available for pharmacogenetic tests, namely Roche's AmpliChip, Luminex's Tag-It, and AutoGenomics' Infiniti. Each of these platforms is briefly discussed.

The AmpliChip (Roche Molecular Systems, Branchburg, NJ) is an oligonucleotide microarray hybridization method, with probes synthesized on a glass substrate. The microarray uses several hundred wild-type and mutant probe sets, with redundancy for each variant detected. After PCR, amplicons are fragmented and denatured to generate small single-stranded DNA fragments. The fragmented DNA is labeled with biotin at the 3′ termini and hybridized to the oligonucleotides on the microarray. After staining with a streptavidin-conjugated fluorescent dye (phycoerythrin), the microarray is scanned by a laser that excites the fluorescent label bound to the hybridized target DNA fragments. The emitted light is detected and is proportional to bound target DNA at each probe site. The fluorescence ratio of the wild-type to mutant probe determines the genotype.

Tag-It (Luminex, Austin, TX) is a bead array with oligonucleotides bound to microspheres. As with the AmpliChip, it begins with PCR, followed by ASPE. Biotin-dCTP is included in the subsequent amplification reaction and will only be incorporated into the product if the 3′ end of the primer perfectly matches the allele. The 5′ end of the primer has a tag sequence that hybridizes to the microspheres. The microspheres are labeled with fluorophores and tagged with a universal anti-tag sequence complementary to the allele-specific primer tags. The bead–ASPE hybridization products are incubated with an R-phycoerythrin–streptavidin conjugate, and fluorescence intensities are measured. Again, the allelic ratio between the wild type and variant provides the genotype [5].

The Infiniti platform (AutoGenomics, Carlsbad, CA) is a film microarray. Similar to the other platforms, DNA is first PCR amplified for the specific regions of interest and the products are cleaned with exonuclease I and shrimp alkaline phosphatase (EXO/SAP). Allele-specific primer extension follows, and the fluorescently labeled products are hybridized to the film array. Labeled products bind specifically to their complementary microarray locations. Fluorescence levels are detected by a microscope and are converted into genotype data using allelic ratios.

## Sequencing

Unlike mutation-specific analyses, DNA sequencing allows for the detection of any variations within the amplified region of the gene. Classically, dideoxy (Sanger) sequencing has been the most popular sequencing method. Pyrosequencing has also been adopted for use in the clinical lab.

Sanger sequencing relies on the formation of the phosphodiester bond during DNA extension. Through DNA replication, the terminal 3′-hydroxyl group of the DNA polymer reacts with the α-phosphate group on the incoming nucleotide. Pyrophosphate is released, and a covalent phosphodiester bond is formed. If the 3′-hydroxyl group is absent, the adjacent complementary nucleotide can bind to the polymerase, but a phosphodiester bond cannot be formed. Dideoxynucleotide triphosphates (ddNTPs) do not contain 2′- or 3′-hydroxyl groups on the ribose ring of the nucleotide. Thus, these modified bases can be incorporated into the DNA polymer but will terminate further polymerization.

Amplified DNA is used as a template for a unidirectional reaction; a single primer is added to each sequencing reaction. Fluorescently labeled ddNTPs (ddTTP, ddATP, ddGTP, and ddCTP) are each conjugated to a different fluorophore; therefore, each nucleotide provides a unique signal. At every base extended during the sequencing reaction, a minority of the polymers will incorporate the complementary ddNTP rather than dNTP. Incorporation of ddNTP will terminate further polymerization, leading to an assembly of products that can be ordered from shortest to longest.

The products of the sequencing reaction are separated by capillary electrophoresis wherein smaller DNA molecules migrate faster than the larger molecules. A fluorescence detector is utilized to decipher the terminal nucleotide of the fragment. Single nucleotide resolution is achieved and a chromatogram is generated that assigns each nucleotide a specific color, facilitating interpretation of the sequencing result.

Sanger sequencing is the current gold standard for mutation detection because its accuracy surpasses that of any other molecular technique. Sanger sequencing will recognize both rare and common mutations, as well as small insertions, deletions, and duplications, although large deletions and duplications will not be detected. The technique is amenable to high-throughput production, and data analysis and interpretation are relatively straightforward. However, Sanger sequencing tends to be more time-consuming compared to mutation-specific detection. In addition, for applications in which the population of target DNA is in the minority, such as during tumor characterization, the lack of analytical sensitivity can result in false-negative typing.

Pyrosequencing, or real-time DNA sequencing, uses detection of pyrophosphate release [6]. Like Sanger sequencing, it exploits the natural mechanism of DNA polymerization to infer the DNA sequence. However, in contrast to Sanger sequencing, detection is accomplished as each nucleotide is incorporated rather than by separation of a final set of DNA products. When the DNA phosphodiester bond is formed, pyrophosphate (PPi) is released. Pyrosequencing utilizes this PPi as a substrate for a series of enzymatic reactions that ultimately produce light.

As with Sanger sequencing, targeted pyrosequencing is preceded by PCR. The amplification reaction includes a biotinylated primer, which facilitates subsequent isolation of the amplicon strand to be sequenced. Pyrosequencing proceeds by sequential addition of each nucleotide to a sequencing primer. If the nucleotide added is complementary to the next base in the DNA sequence, it is incorporated into the polymer and PPi is released. Sulfurylase catalyzes the formation of ATP from PPi and adenosine phosphosulfate. Luciferase then catalyzes the reaction between ATP, $O_2$, and luciferin, ultimately releasing light. The light is collected by a CCD camera and recorded as peaks into a graph known as a pyrogram (Figure 22.2). The amount of light generated is directly proportional to the number of the specific nucleotide incorporated. For example, if two GTPs are incorporated in tandem, they will produce twice as much light compared to the incorporation of a single GTP. Pyrosequencing can be applied to a range of pharmacogenomics applications, including detection of single nucleotide polymorphisms (SNPs),

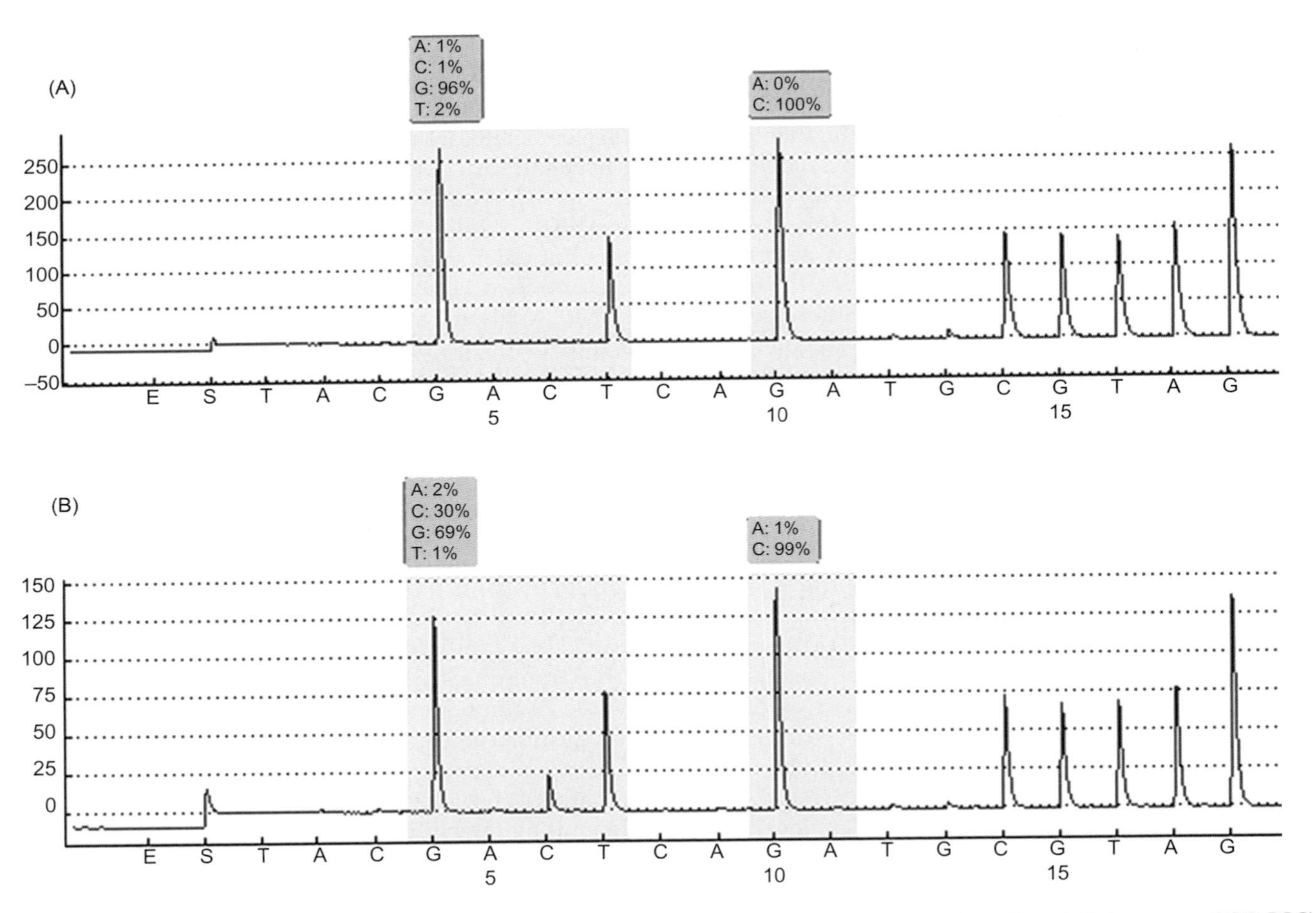


**FIGURE 22.2** Pyrosequencing results for *KRAS* codons 12 and 13 in (A) a tumor wild type for codons 12 and 13 (sequence: GGT GGC) and (B) a tumor harboring a c.35 G>C, p.Gly12Ala mutation, as evidenced by a reduced "G" peak at dispensation 4 and a novel "C" peak at dispensation 6.

copy number, triallelic polymorphisms, tandem repeats, and insertions/deletions [7]. Like dideoxy sequencing, it can detect all mutations within a targeted region, but sequencing lengths are much shorter, typically less than 100 bases, and often limited to only a few codons. Novel variants outside the targeted region will not be detected with this methodology. However, the analytical sensitivity of pyrosequencing, which is typically approximately 5%, outperforms that of dideoxy sequencing, which is typically approximately 25% [8]. This is particularly useful for oncology applications, in which heterogenetic cell mixtures are common. Because all tumor tissue will have some portion of normal cells that likely do not harbor the somatic mutation, increased analytical sensitivity can be essential. The added analytical sensitivity provided by pyrosequencing is the fundamental reason why it is utilized in the clinical lab. In addition, targeted pyrosequencing is cost-effective and less time-intensive compared to other sequencing methodologies. Pyrosequencing is easily automated, accurate, and depletes the need for labeled nucleotides and a separation matrix.

# APPLICATIONS OF PHARMACOGENETICS TESTINGS

Applications of pharmacogenetics may involve the detection of somatic mutations acquired through mutagenesis or germline mutations inherited parentally. Somatic mutations are important in oncology, whereas germline mutations are often relevant for drug metabolism. Applications for each with the challenges in testing are described here. Table 22.1 summarizes the applications, the commonly tested variants, and common platforms.

## Tumor/Somatic Mutation Detection

One of the hallmarks of cancer cells is the insubordinate activity of signaling pathways [9]. In almost all malignant tumors, a tumor suppressor will be inactivated and/or an oncogene will be upregulated. The mechanism by which these somatic mutations allow tumor cells to activate pro-growth signaling can affect the tumor's response to therapeutic agents. Examples of such a relationship are *KRAS* mutations in epidermal growth factor receptor (EGFR)-directed therapy for colorectal cancer, *BCR-ABL* translocations for the effectiveness of tyrosine kinase inhibitors against chronic myelogenous leukemia (CML), and *KIT* mutations in gastrointestinal stromal tumors treated with imatinib.

There are benefits for recognizing these mechanism-specific mutations. First, it benefits the patient by administering certain chemotherapeutic agents, which often have deleterious side effects, only when there is a greater possibility of the drug being effective. Second, it provides pharmaceutical companies a

**TABLE 22.1** Genes Important in Pharmacogenomics Testing

| Gene | Indication | Commonly used Methods | Commonly Tested Variants | Comments |
|---|---|---|---|---|
| *KRAS* | Response to EGFR therapies | Targeted sequencing; pyrosequencing; allele-specific PCR | Exons 12, 13, 61 | |
| *BCR-ABL1* | Response to imatinib | qRT-PCR; sequencing | Quantification of bcr-abl translocation, mutations in the kinase domain | Quantitative PCR used for residual disease monitoring; sequencing used to predict resistance to therapy |
| *KIT* | Imatinib or tyrosine kinase inhibitors | Single variant detection; sequencing | D816V, targeted exons 8, 9, 11, 13, 17 | Exons tested depend on cancer type |
| *HLA-B**5701 | Abacavir sensitivity | Single variant detection | Presence or absence of HLA-B*5701 or HCP5 variant | |
| *IL28B* | HCV response | Single variant detection | rs12979860C/T, rs8099917T/G | Laboratories may test one or the other or both variants |
| *CYP2D6* | Multiple drugs | Targeted mutations | *2, *3, *4, *5, *6, *7, *8, *9, *10, *11, *12, 14, *15, *17, *29, *35, *41 | Many rare variants will not be detected |
| *CYP2C19* | Clopidogrel sensitivity | Targeted mutations | *2,*3,*17 | |
| *CYP2C9/VKCOR1* | Warfarin sensitivity | Targeted mutations | CYP2C9 *2, *3 VKCOR1 c-1639 G>A | For sensitivity testing only; will not predict warfarin resistance |

foundation to select patients for clinical trials and also for targeting future drug discovery projects.

## *KRAS*

EGFR is a transmembrane tyrosine kinase receptor that functions to activate signaling pathways involved in survival and proliferation. Activation of EGFR occurs when a ligand binds to the extracellular domain, triggering dimerization and consequential autophosphorylation of the cytoplasmic domain. EGFR activation stimulates both the mitogen-activated protein kinase (MAPK) and the phosphoinositol-3 kinase (PI3K) pathways, ultimately triggering the expression of genes involved in growth and metastasis. Multiple metastatic cancers—namely colorectal, lung, and pancreatic cancers—are known to have aberrant expression of proteins involved in the EGFR pathway, namely due to gene duplications and point mutations. The EGFR serves as the gatekeeper for the MAPK and PI3K pathway activation, and therefore some cancers will harbor mutations specific to the EGFR and not to downstream pathway components. Because the ligand binding site of the receptor is located on the extracellular membrane, EGFR has been an attractive therapeutic target using monoclonal antibodies. In particular, cetuximab and panitumumab, which function pharmaceutically to block the receptor, are frequently used in the treatment of colorectal cancer patients. However, only a subset (8–23%) of patients with metastatic colorectal cancer respond to these drugs. The primary factor governing the response or resistance to EGFR-specific drugs is mutations in proteins downstream of EGFR activation that result in constitutive activity, keeping the pathway turned on without initial signaling through the EGFR receptor [10]. Mutations in *KRAS* that lead to such constitutive activity have shown the highest correlation to anti-EGFR drug resistance. *KRAS* genotype analysis of tumor tissue can therefore help predict if a patient will respond to EGFR inhibitors. Specifically, whereas a wild-type *KRAS* suggests the patient may respond, a mutation in *KRAS* indicates drug resistance. It is important to note that the response rate to anti-EGFR therapy in patients with wild-type *KRAS* is less than 50%, suggesting that there are other factors involved.

*KRAS* mutations that are known to convert the gene into an active oncogene are localized to hot spots in codons 12, 13, and 61. Multiple molecular assays have been developed to detect mutations in these regions [8]. The most commonly utilized methods are Sanger sequencing, pyrosequencing, and allele-specific PCR. Less commonly utilized are melt-curve analysis, StripAssay, and multiplex assays. These techniques differ in their diagnostic and analytical sensitivity.

For detection of somatic mutations, interpretation of Sanger sequencing becomes more difficult, and often the analytical sensitivity is inadequate. This is a function of the starting population of cells, which is a mixture of tumor cells and nontumor or normal cells. Often, the signal-to-noise ratio is not high enough to distinguish if a mutation is, in fact, present. The analytical sensitivity of Sanger sequencing is approximately 20%, which can lead to false-negative interpretations in heterogeneous cell populations. Thus, Sanger sequencing is less attractive for utilization in detection of somatic mutations.

To increase analytical sensitivity in detecting somatic mutations, other methods are often employed. For example, pyrosequencing has a sensitivity of approximately 5%. Monitoring the incorporation of each sequential nucleotide heightens the ability to detect small amounts of DNA polymerization. Consequently, the lower limit of analytical sensitivity is enhanced. In addition, this allows an estimate of the percentage of cells in the population analyzed that harbor the detected mutation. Multiple studies have shown that even in tissues with low tumor cell content, pyrosequencing provides sufficient analytical sensitivity and specificity to assess *KRAS* mutation status in metastatic colorectal cancer tissue. In contrast to Sanger sequencing, most *KRAS* pyrosequencing assays will only analyze the codons of interest plus a few flanking nucleotides. Although mutations not previously described may be missed, the clinical significance of *KRAS* mutations other than those in codons 12, 13, and 61 is unclear. An example of a *KRAS* pyrogram is shown in Figure 22.2.

Like pyrosequencing, allele-specific PCR is highly sensitive and can detect as few as 10 copies of mutant alleles in proportions as low as 1% of the total DNA. Therefore, compared with Sanger sequencing, the lower limit of detection is at least 10-fold more sensitive. In addition to an increased analytical sensitivity, allele-specific PCR has the advantage of eliminating the post-PCR processes required by sequencing, thus eliminating a major contamination source and reducing turnaround time. The primary disadvantage of allele-specific PCR is that mutations outside the targeted nucleotide will not be detected. The allele-specific PCR methods to detect *KRAS* mutations have focused solely on the seven most common mutations in codons 12 and 13, which account for at least 90% of all *KRAS* mutations. Approximately 10% of mutations may be overlooked, particularly mutations in codon 61, which can represent up to 8% of *KRAS* mutations. Although the clinical implications of mutations in codon 61 are not fully understood, at least one study has associated such mutations with anti-EGFR therapy resistance. In general, mutation detection between allele-specific PCR and Sanger sequencing is highly concordant if the tumor cell percentage in the sample is sufficient. As such, the method of choice is typically a function of the proportion of tumor cells in the sample.

The typical allele-specific PCR assay utilizes conventional probes such as TaqMan or molecular beacon to detect amplification. Scorpion primers are designer oligonucleotides that function as both the primer and the probe [11]. When utilized for *KRAS* detection, Scorpion probes provide an additional 10-fold decrease in the lower limit of detection (~0.1% or one target allele) compared to conventional probes. Although the sensitivity exhibited by Scorpion primers may initially seem attractive, the clinical significance of such a low limit of detection remains controversial. Furthermore, the Scorpion assay is only available through a single commercial vendor and the cost is significantly higher compared to that of the reagents required for conventional allele-specific RT-PCR or DNA sequencing.

### *BCR-ABL1*

CML is a myeloproliferative disorder that results in the proliferation of mature granulocytes and their precursors. The signature characteristic of CML is a unique chromosomal abnormality resulting from a reciprocal translocation between the long arms of chromosome 9 and 22, known as the Philadelphia (Ph1) chromosome [12]. The Ph1 chromosome is present in nearly 100% of CML cases because it is the causative molecular abnormality of CML. The Ph1 chromosome fuses the 5′ end of the breakpoint cluster region (*BCR*) on chromosome 22 with the *ABL1* kinase protooncogene on chromosome 9, resulting in the loss of the native 5′ end of *ABL1*, which regulates kinase activity. The gene fusion results in an actively transcribed and translated protein called BCR-ABL1. BCR-ABL1 retains the kinase activity of ABL1, using phosphorylation events to trigger responses in multiple anti-apoptotic and pro-growth signaling pathways. However, the fusion protein loses its ability to be regulated, allowing for constitutive activity of ABL1, which ultimately leads to the pathogenicity.

Detection of the Ph1 chromosome is essential not only for diagnosis of CML but also because targeted therapy with imatinib, a tyrosine kinase inhibitor with specificity toward BCR-ABL1, has proven to be a highly effective first-line therapy. The methods utilized to detect the Ph1 chromosome are conventional cytogenetics, fluorescence *in situ* hybridization (FISH), and RT-PCR. Cytogenetics uses a karyotype to visually scrutinize chromosomes and detect chromosomal abnormalities. Although cytogenetics is the gold standard for Ph1 detection, it requires a bone marrow biopsy and the manual evaluation of metaphase cells, making it a cumbersome and time-consuming technique with a sensitivity of approximately 1:20 cells. FISH, a probe-based detection method, has a slightly higher sensitivity, detecting from 1:100 to 1:500 Ph1-positive bone marrow cells, but it is prone to false-positive results. In contrast, quantitative (q)RT-PCR, which is based on extracting mRNA from whole blood and detecting the *BCR-ABL1* transcript, is relatively efficient, sensitive, and specific. However, qRT-PCR will not provide any additional information about bone marrow morphology or other chromosomal abnormalities.

The primary goals of CML treatment are to eliminate the Ph1 chromosome and *BCR-ABL1* gene expression. If treatment with imatinib is successful, both will be achieved, but they will still need to be monitored periodically. If treatment with imatinib is not successful, or if relapse is detected, therapy must be altered. Conventional cytogenetics is not sensitive enough to be used in disease monitoring. qRT-PCR extends the limit of detection for cells expressing the Ph1 chromosome down to 1:100,000–1:1,000,000 and thus is four or five orders of magnitude more sensitive than cytogenetics [13]. In addition, qRT-PCR eliminates the need for the continued collection of bone marrow.

Using qRT-PCR rather than RT-PCR allows for a quantitative measurement to detect response and progression. Studies have shown that the concentration of *BCR-ABL1* transcript predicts the most important clinical outcomes: disease progression, drug resistance, and survival [13,14]. qRT-PCR is a powerful and efficient method for monitoring CML, but a few cautionary notes are necessary. Importantly, RNA is an inherently difficult macromolecule to extract and analyze because it is degraded by RNase, which is an exceptionally stable and abundant enzyme. In addition, RNA yields can vary from person to person. The latter is overcome by standardizing the amounts of BCR-ABL1 transcripts in a sample to the amount of an endogenous expressed housekeeping gene. This is necessary for accurate *BCR-ABL1* quantification, but it can also be an additional source of bias in an assay. Depending on which housekeeping gene is measured, the relative expression, and hence quantitative assessment, of *BCR-ABL1* can vary.

A critical component to all tumor marker assays is precision. The assays must exhibit minimal fluctuations in order for consecutive results to translate into therapeutic meaning. The current limit of analytical precision for most *BCR-ABL1* qRT-PCR assays is approximately 0.5 logs. Therefore, if a patient has a rise in *BCR-ABL1* transcript that is less than or equal to 0.5 logs, it may simply be a result of analytical precision. Studies have shown that remaining within 0.5 log fluctuations of *BCR-ABL1* is a good prognostic indicator, and conversely a poor prognostic indicator if the transcript increases by more than 0.5 logs [15]. However, improving the precision of the assays may allow for a more fine-tuned assessment of patient status.

A universal problem with tumor markers is the lack of standardization and the large interlaboratory variation observed. Currently, this also holds true for the *BCR-ABL1* transcript, in which up to eightfold differences in concentration are observed between laboratories evaluating identical specimens [16]. The variability is a result of procedural and/or vendor differences in RNA extraction, instrumentation, reagents, and housekeeping genes. Extensive efforts are being brought forth to create an international scale for *BCR-ABL1* to harmonize quantitative BCR-ABL1 measurements across tests and laboratories. This will enhance patient care by facilitating inter-laboratory studies and allowing for patient portability [17].

A rise in *BCR-ABL1* transcript greater than 0.5 logs suggests that treatment is unsuccessful. Multiple mutations in the kinase domain of BCR-ABL1 have been shown to infer drug resistance to imatinib [13]. Identifying the specific mutations present can guide therapy because certain mutations are known to respond to alternative tyrosine kinase inhibitors, whereas other mutations are known to be resistant to all currently available kinase inhibitors. Samples suspected of harboring such mutations are subjected to nested PCR with the first round of PCR amplifying across the translocation, followed by a second round that amplifies the ABL1 region of interest from the fusion transcript. Amplicons are sequenced to identify mutations.

## *KIT*

KIT is a tyrosine kinase receptor involved in the development of melanocytes, interstitial cells of Cajal, hemopoietic progenitor cells, mast cells, and primordial germ cells [18]. Like all tyrosine kinase receptors, KIT is activated via ligand-specific binding, with subsequent receptor dimerization, autophosphorylation, and downstream signaling leading to gene activation/repression and a physiological response. Although KIT follows the general convention of receptor signaling, KIT is relatively complex when implicated in cancer and scrutinized for its pharmacogenomics associations. This complexity is multifaceted and a result of KIT association with multiple tumor types, mutation-specific responses, and the appearance of secondary resistance. Furthermore, an alternative tyrosine kinase receptor, PDGFR$\alpha$, is implicated in some of the same cancers.

Because KIT has several roles in physiological homeostasis, it is not surprising that *KIT* mutations are associated with multiple tumor types [18]. *KIT* mutations are at the pathological root of many gastrointestinal stromal tumors (GISTs), melanomas, seminomas, mastocytosis, and acute myeloid leukemia (AML) cases. The primary goal of detecting specific *KIT* alterations in these cancers is to provide prognostic and/or potential therapeutic information. In general, cancers with wild-type *KIT* alleles will not be responsive to imatinib or other tyrosine kinase inhibitors. However, the presence of a *KIT* mutation does not necessarily confer responsiveness. Multiple *KIT* mutations observed both within and between the cancer types further enhance the complexity. Pathogenic *KIT* alleles have been shown to encompass the whole spectrum of genetic variation; deletions, point mutations, internal tandem duplications, insertions, and complex mutations have all been reported. The type of cancer to be tested, and thus the expected mutation(s), must be considered when designing the assay. In GISTs and melanomas, exon 11 is the most common for mutations, followed by exons 9, 13, and 17. The exon that harbors the mutation also influences drug dosing and prognosis. In GISTs with wild-type *KIT*, one may also want to evaluate the *PDGFRA* genotype [19]. D842V is the most common GIST-associated *PDGFRA* mutation and is analogous to the D816V KIT mutation described later. Cancers harboring this *PDGFRA* mutation are completely resistant to all of the currently available tyrosine kinase inhibitors. In seminomas and mastocytosis, identifying the *KIT* point mutation D816V is critical [20]. D816V allows for constitutive kinase activity leading to cell proliferation and inhibition of apoptosis and infers resistance to tyrosine kinase inhibitors. For AML patients, mutations in *KIT* confer poor prognosis [21]. These mutations occur in exons 8 and 17. Most exon 8 mutations are insertions/deletions, whereas the exon 17 mutations are a mixture of point mutations and insertions.

Secondary mutations have been best defined in GISTs, in which resistance to kinase inhibitors was observed in more than 85% of patients who were initially responsive to the drugs [19]. This resistance often develops in the original lesion, suggesting a clonal selection. Specifically, such mutations were present in a subset of cells or a single clone within the tumor when treatment with the kinase inhibitor was initiated. By selectively killing the tumor cells that harbor only the primary mutation, the minor clones are given the opportunity to expand. Screening for such minor clones before the initiation of therapy may allow the physician to predict the best first- and second-line therapies by understanding to what drugs the tumor will eventually become resistant.

Techniques utilized for detection of *KIT* mutations will vary depending on the type of cancer and whether primary or secondary mutations are implicated. One confounding issue in KIT testing is that laboratories need to design testing panels that facilitate appropriate physician ordering and bolster subsequent result interpretation. Tiered testing might be an excellent cost- and time-efficient method to define the KIT status in a tumor population. In these cancer-specific cases, the most common mutation would be evaluated first. If that mutation is absent, assessment of the alleles should continue for the second most common mutation. Alternatively, if that mutation is present, the need for further testing is obsolete (unless secondary resistance is suspected). These algorithms should also direct testing to *PDGFRA* when implicated.

## Host Factors Influencing Response to Infectious Disease

Various factors influence treatment of a patient suffering from an infectious disease.

### *HLA 5701*

Inherited factors can affect how humans respond to treatment of infectious diseases. One example is the Human leukocyte antigen (HLA)-B locus *5701, which has been associated with abacavir sensitivity in treating HIV infections [22]. Symptoms related to abacavir sensitivity include fever, rash, fatigue, and respiratory and gastrointestinal problems. Testing for the locus *HLA-B**5701 will identify those who are HLA-B*5701 positive and likely to develop a severe reaction to abacavir and in whom abacavir treatment should be avoided [23]. The *HLA-B**5701 sequence is nearly identical to those of other HLA types, such as *HLA-B**5702 or *HLA-B**5706. High-resolution HLA typing or HLA sequencing can be performed to definitively genotype this region. However, these tests are expensive and labor-intensive. To find a solution that is simpler and less expensive, carefully designed primers to amplify only the *HLA*-B*5701 allele can be used to detect the presence or absence of the allele. This type of a test may not be able to distinguish between one copy (heterozygous) or two copies (homozygous). Care should be taken to not mistakenly call a closely related allele because there may also be slight, although inefficient, amplification of related HLA alleles. The *HLA-B**5701 allele is in linkage disequilibrium with the single-nucleotide variant rs2395029 in the *HCP5* gene [24]; therefore, this can be used as a surrogate and tested by a variety of methods suitable for single variant detection. The HCP5 variant has close to 100% negative and positive predictive value when used to screen for abacavir sensitivity. The exception is a rare allele that is positive for the variant but negative for *HLA-B**5701. This result leads to a false positive, resulting in the avoidance of treatment that may be safe. In addition, deletions that cover the HCP gene region have been reported [25]. If an undetected deletion is present, the result may be a false negative or a false positive, depending on which allele is deleted.

### *IL28B*-Associated Variants

A second example of a host factor influencing responses to treatments is *IL28B*-associated variants and hepatitis C (HCV) infection. The standard treatment for HCV is a combination of PEG-interferon (PEG-IFN) and ribavirin. The combination of these therapies eliminates the HCV RNA in 70–90% of individuals with HCV type 2 but less than half of individuals with HCV type 1. Treatment success is measured by sustained virological response, defined as the absence of detectable HCV RNA in the serum 24 weeks after the end of treatment.

It has long been known that there are ethnic differences in HCV clearance, suggesting that host genetic factors influence treatment. Genome-wide association studies have identified two single-nucleotide variants that are independently associated with response to PEG-IFN and ribavirin in Caucasians [26,27]. These two variants, rs12979860C/T and rs8099917T/G, are located upstream of the *IL28B* gene, which encodes for λ or type III interferons (IFN-λ). These variants can be used to predict response to PEG-IFN/ribavirin therapy, although with few other options, this treatment would not be denied. The favorable alleles are the C allele for rs12979860 and the T allele for rs8099917. One or two copies of risk alleles (T allele for rs12979860 and G allele for rs8099917) predict higher risk of chronic hepatitis C and increased risk for treatment failure. The effect is strongest in HCV type I infections.

Methods suitable for single variant or multiplexed variants will detect the favorable/unfavorable genotypes. Figure 22.1 shows an example of both variants detected by hybridization probes. However, individuals' genotypes may be considered favorable by one variant but unfavorable or at-risk by the other variant. Further studies are needed to determine if one or the other variant is a better predictor.

## Polymorphism of Metabolic Enzymes

Polymorphisms of certain isoforms of the cytochrome P450 mixed function family of enzymes play key roles in metabolism of certain drugs that are substrates for such isoforms.

### *CYP2D6*

CYP2D6 is an enzyme from the cytochrome P450 superfamily involved in the metabolism of approximately 25% of commonly prescribed drugs, including tamoxifen, β-receptor blockers, analgesics, anticonvulsants, antidepressants and antipsychotics, antihypertensives, and norepinephrine reuptake inhibitors.

Variants in this enzyme can produce nonfunctional enzymes or enzymes with reduced-function. These variants are referred to as "*" alleles, representing a combination of nucleotide variants. The *1 allele is considered the normal, functional allele. Genotyping can identify variant alleles, although many are rare and therefore may not be included in a clinically available assay. The combination of alleles results in four phenotypic categories: ultrafast metabolizer, extensive (normal) metabolizer, intermediate metabolizer, and poor metabolizers. Depending on the drug, CYP2D6 may inactivate an active drug or metabolize a prodrug to its active form. If CYP2D6 activity is reduced, an active drug may build up toxic compounds; alternatively, a prodrug may result in ineffective treatment. Molecular assays detect common variants in the *CYP2D6* gene that affect protein function.

Several commercially available and U.S. Food and Drug Administration (FDA)-cleared tests for *CYP2D6* are available and were discussed previously. These include the AmpliChip CYP450 test, which detects 20 *CYP2D6* alleles (including the *5 deletion) and 7 *CYP2D6* duplications [28]; the Luminex Tag-It assay, which interrogates 17 allelic variants including the *5 deletion and duplication [5]; and the AutoGenomics CYP2D6I assay [29], which detects 15 allelic variants including the *5 deletion and duplication alleles.

All the assays described are genotyping assays; they do not confirm if the variants detected are on the same chromosome (*in cis*) or on different chromosomes (*in trans*). Assumptions are therefore made to assign the variants based on reported combinations of mutations on a single chromosome. For example, the nonfunctional *4 allele contains both the c.1846G>A mutation, which affects splicing and results in a frameshift, and the c.100C>T mutation *in cis*. In contrast, the reduced functional allele, *10, is defined by the c.100C>T mutation alone. Although c.1856G>A can exist on a chromosome without c.100C>T (named *4M), this allele is rare. Thus, if the assay utilized detects both c.1856G>A and c.100C>T mutations, it is presumed that these mutations are present *in cis*. Heterozygosity for each of these mutations predicts a *4 heterozygous allele (one copy of a nonfunctional allele). If allelic combinations are not taken into consideration, this result could be misinterpreted as a *4/*10 compound heterozygous (one copy of a nonfunctional allele and one copy of a decreased-function allele). A true *4/*10 combination shows homozygosity for c.100C>T and heterozygosity for c.1846G>A.

Gene duplications add another level of complexity to *CYP2D6* genotype interpretations. Although most molecular assays can detect the presence of duplication, not all can assess which allele is duplicated (i.e., if the duplication arose from the wild-type or mutant allele). In addition, the number of copies is not determined. The AmpliChip assay was designed to reduce the diagnostic complexity associated with such genotypes by detecting seven duplications also known to contain nonfunctional or reduced-function variants (e.g., c.1846G>A and/or c.100C>T). Knowing which allele is duplicated is important in predicting the phenotype. An example is a patient with the *CYP2D6**1/*4 genotype and gene duplication. If the *1 is duplicated (*1XN/*4), the patient has at least two functional copies of the gene, predicting an extensive metabolizer phenotype. However, if *4 is the duplicated allele (*1/*4XN), the duplicated allele would still be nonfunctional and the patient would be predicted to have only one functional allele, consistent with an intermediate-to-extensive metabolizer.

Laboratories may report the * allele designation detected or the nucleotide positions of base changes. If none of the targeted mutations are detected, the wild-type or *1 allele is presumed. However, unless sequencing of *CYP2D6* is performed to detect all variants, rare alleles may be missed by targeted mutation analysis. Alternatively, reports should state that "none of the targeted mutations were detected."

### Hemostasis (CYP2C19, CYP2C9, and VKORC1)

Hemostasis is the ability of the body to properly control the spatial and temporal behavior of the clotting cascade. In certain pathologies, the risk of thrombotic events is high, and therefore pharmaceuticals are prescribed to reduce the risk of clotting. Dosing of such drugs is critical because there is a delicate balance between the risk of clotting and the risk of bleeding. Two of the most commonly prescribed oral anticoagulants are clopidogrel and warfarin, and the efficacy of both is known to be affected by genetic variations.

For more than 60 years, warfarin has been front-line therapy for the treatment or prevention of thromboembolisms in patients at risk for clotting. Despite its widespread use, adverse events resulting from interindividual variability are common. Mechanistically, warfarin executes its function by inhibiting the activation of vitamin K, required for the $\gamma$-carboxylation of many of the factors involved in the coagulation cascade. In the absence of $\gamma$-carboxylation, these factors are unable to cleave their substrates, and as such, the coagulation cascade is partially arrested. Two distinct genes are implicated in the pharmacological response to warfarin: *CYP2C9* and *VKORC1* [30]. The FDA has mandated that the warfarin package insert contain a table that integrates *CYP2C9* and *VKORC*1 genotype with drug dose.

CYP2C9, a member of the cytochrome P450 family, is responsible for the rate-limiting step to break down warfarin into its inactive metabolite. *CYP2C9* is a highly polymorphic gene, and some of these SNPs reduce the catalytic efficiency. *CYP2C9**2 and *3 are well-defined alleles that cause a decreased warfarin clearance rate. Cumulatively, they are present in approximately 20% of Caucasians. Patients harboring either of these alleles are predicted to require a lower loading and maintenance warfarin dose, and they are also at greater risk for bleeding complications as a result of warfarin administration.

Vitamin K is a co-factor for the enzymatic addition of $\gamma$-carboxylate functional groups onto various proteins, including many of the principal components of the coagulation cascade. The result of this reaction is the oxidation of vitamin K. To function as a co-factor for subsequent $\gamma$-carboxylation reactions, vitamin K must be converted back into its reduced form. VKORC1 is the enzyme responsible for such vitamin K recycling, and it is the mechanistic target of warfarin; warfarin binds to VKORC1 and prevents VKORC1 from reducing vitamin K. Polymorphisms within the *VKORC1* warfarin binding site can alter the binding affinity of the enzyme for the drug. The most common of these, −1639G>A, is associated with warfarin sensitivity.

Combined, *CYP2C9* and *VKORC1* have been shown to be predictive of warfarin dosage requirements for approximately 50% of patients. Multiple FDA-cleared platforms for detection of the common *CYP2C9* and *VKORC1* variant alleles are available, and all function with extremely high (>99%) sensitivity and specificity [31,32]. The fundamental issues regarding CYP2C9 and VKORC1 testing are bureaucratic in nature [33]. Controversy exists regarding the clinical utility, and hence reimbursement poses a major issue. Despite the endorsement for genotype testing by the FDA, American Association for Clinical Chemistry, National Academy of Clinical Biochemistry, College of American Pathologists, and others, the Centers for Medicare and Medicaid Services is reluctant to provide any reimbursement for *CYP2C9* or *VKORC1* testing until a large-scale, prospective, randomized study has been completed.

Further studies are also needed to design the most effective scheme for incorporating *CYP2C9/VKOR1* genotype testing into the treatment plan [34]. Algorithms have been proposed, but it is unclear whether it is imperative for the genotyping to be completed with expedited turnaround time or if it is sufficient to begin a patient on a dose appropriate for his or her height/weight/age and adjust the dose as the genotype results become available. In either case, the international normalized ratio should be used in tandem with the genotype results to ensure that the proper therapeutic target is achieved.

The use of platelet antagonists, such as aspirin and clopidogrel, has been shown to improve outcomes in patients with acute coronary syndrome and percutaneous coronary interventions [35]. Clopidogrel functions by irreversibly binding to the P2Y12 platelet receptor, preventing ADP activation of platelets. Approximately 25% of patients do not respond to clopidogrel; a large part of the variability is attributed to the *CYP2C19* gene. Clopidogrel is a prodrug, requiring a two-step enzymatic reaction to be converted into the active metabolite. CYP2C19 is a catalyst for both of these reactions, and like many members of the *CYP* family, it is a highly polymorphic gene. Although multiple polymorphisms have been identified in *CYP2C19*, only five have been shown to alter the kinetic properties of the enzyme. *CYP2C19*2, *3, *4,* and **5* are associated with loss of function and a decreased metabolic rate of clopidogrel; *CYP2C19*17* is associated with a gain of function and an increased metabolic rate of clopidogrel. Although robust and accurate FDA-cleared and laboratory-developed tests can identify these mutations, the main challenge in testing for *CYP2C19* variants lies in the clinical relevance of these alleles, which is currently highly controversial [36,37].

In 2010, the FDA mandated a box label warning be placed in the clopidogrel package insert to caution patients that there is diminished effectiveness of the drug in patients who are poor metabolizers, but available genetic testing for the *CYP2C19* gene can identify such persons. The label suggests that if genetic testing reveals that the patient is a poor metabolizer, alternative treatment or treatment strategies should be considered. Immediately following the FDA endorsement, multiple national cardiovascular associations contested the decision, stressing that there was not sufficient evidence to support these claims. The controversy over the utility of *CYP2C19* testing remains [38]. More than 30 clinical trials and at least a dozen meta-analyses reveal conflicting data, bolstering the debate. In summary, clopidogrel is a widely prescribed drug that can reduce adverse thrombotic events in at-risk patients. There is an association between *CYP2C19* genotype and clopidogrel response: Poor metabolizers will have a decreased serum concentration of the active metabolite and will have platelets with higher *ex vivo* reactivity to ADP than predicted of a patient on clopidogrel. However, the association of genotype with the risk of cardiovascular events remains incredibly controversial.

## CASE STUDIES

The following scenarios illustrate a few of the challenges in pharmacogenetic testing. They are typical types of cases received in the clinical laboratory.

### Case Study 1 Involving CYP2D6

A male in his 30s is nonresponsive to psychiatric medication, and a sample is sent for *CYP2D6* testing.

His genotype showed a *17 decreased function allele and the *2a functional allele. The assay also detected a duplication but could not determine which allele was duplicated. Without knowing the duplicated allele, a phenotype could not be accurately predicted. If the *2a allele was duplicated, he would have at least two copies of a functional allele and one allele with decreased function. The predicted phenotype would be an extensive or rapid metabolizer. If the *17 allele was duplicated, he would have a duplicated decreased function allele and one normal function allele. The predicted phenotype would be an extensive metabolizer. A family study was performed and demonstrated that the *2a allele and the duplication were inherited from his mother, whereas the *17 allele was inherited from his father. With this information, he was predicted to have at least an extensive metabolizer phenotype and is likely a rapid metabolizer. A rapid metabolizer phenotype could explain his nonresponsiveness because the medication may be quickly eliminated, before it is able to be effective.

### Case Study 2 Involving IL28B-Associated Variants

An individual has been diagnosed with an HCV infection, and a sample is sent for genotyping the *IL28B*-associated variants to predict the risk of treatment failure with PEG-IFN and ribavirin. Genotyping showed that she has two favorable alleles (T/T) for rs8099917 but one risk allele and one favorable allele (C/T) for rs12979860. One risk allele predicts treatment failure independently for each locus. In combining the two loci, she has only one risk allele and three favorable alleles. Because it is not known if the two variants have the same predictive power, the predicted risk for treatment failure is indeterminate. Further clinical studies are necessary to determine her risk of treatment failure.

## CONCLUSIONS

Pharmacogenetic testing is available for a number of inherited and somatic applications. However, challenges remain in the technologies to accurately detect the variations, as well as in their interpretation. The main challenge is to show clinical utility—that patient outcomes are improved by pharmacogenetic testing. Evidence is easier to obtain for the oncology applications, as is seen in the number of tests developed and used clinically. Showing improved outcomes is difficult for metabolic enzymes because of pathway redundancy with other enzymes involved and rarity of variants to show differential effects between heterozygotes and homozygotes. As studies continue to show validity and utility, pharmacogenetic testing will be at the forefront of personalized medicine.

## References

[1] Lyon E. Mutation detection using fluorescent hybridization probes and melting curve analysis. Expert Rev Mol Diagn 2001;1(1):92–101.

[2] Lyon E, Wittwer CT. LightCycler technology in molecular diagnostics. J Mol Diagn 2009;11(2):93–101.

[3] Crockett AO, Wittwer CT. Fluorescein-labeled oligonucleotides for real-time PCR: using the inherent quenching of deoxyguanosine nucleotides. Anal Biochem 2001;290(1):89–97.

[4] Greene DN, Procter M, Grenache DG, Lyon E, Bornhorst JA, Mao R. Misclassification of an apparent alpha 1-antitrypsin "Z" deficiency variant by melting analysis. Clin Chim Acta 2011;412 (15-16):1454–6.

[5] Melis R, Lyon E, McMillin GA. Determination of CYP2D6, CYP2C9 and CYP2C19 genotypes with Tag-It mutation detection assays. Expert Rev Mol Diagn 2006;6(6):811–20.

[6] Nyren P, Pettersson B, Uhlen M. Solid phase DNA minisequencing by an enzymatic luminometric inorganic pyrophosphate detection assay. Anal Biochem 1993;208(1):171–5.

[7] Langaee T, Ronaghi M. Genetic variation analyses by pyrosequencing. Mutat Res 2005;573(1-2):96–102.

[8] Tsiatis AC, Norris-Kirby A, Rich RG, et al. Comparison of Sanger sequencing, pyrosequencing, and melting curve analysis for the detection of KRAS mutations: diagnostic and clinical implications. J Mol Diagn 2010;12(4):425–32.

[9] Hanahan D, Weinberg RA. The hallmarks of cancer. Cell 2000;100(1):57–70.

[10] Monzon FA, Ogino S, Hammond ME, Halling KC, Bloom KJ, Nikiforova MN. The role of KRAS mutation testing in the management of patients with metastatic colorectal cancer. Arch Pathol Lab Med 2009;133(10):1600–6.

[11] Thelwell N, Millington S, Solinas A, Booth J, Brown T. Mode of action and application of Scorpion primers to mutation detection. Nucleic Acids Res 2000;28(19):3752–61.

[12] Cortes J, Quintas-Cardama A, Kantarjian HM. Monitoring molecular response in chronic myeloid leukemia. Cancer 2011;117(6):1113–22.

[13] Press RD. Major molecular response in CML patients treated with tyrosine kinase inhibitors: the paradigm for monitoring targeted cancer therapy. Oncologist 2010;15(7):744–9.

[14] Press RD, Willis SG, Laudadio J, Mauro MJ, Deininger MW. Determining the rise in BCR-ABL RNA that optimally predicts a kinase domain mutation in patients with chronic myeloid leukemia on imatinib. Blood 2009;114(13): 2598–605.

[15] Press RD, Galderisi C, Yang R, et al. A half-log increase in BCR-ABL RNA predicts a higher risk of relapse in patients with chronic myeloid leukemia with an imatinib-induced complete cytogenetic response. Clin Cancer Res 2007;13(20):6136–43.

[16] Branford S, Fletcher L, Cross NC, et al. Desirable performance characteristics for BCR-ABL measurement on an international reporting scale to allow consistent interpretation of individual patient response and comparison of response rates between clinical trials. Blood 2008;112(8):3330–8.

[17] White HE, Matejtschuk P, Rigsby P, et al. Establishment of the first world health organization international genetic reference panel for quantitation of BCR-ABL mRNA. Blood 2010;116(22): e111–17.

[18] Antonescu CR. The GIST paradigm: lessons for other kinase-driven cancers. J Pathol 2011;223(2):251–61.

[19] Corless CL, Barnett CM, Heinrich MC. Gastrointestinal stromal tumours: origin and molecular oncology. Nat Rev Cancer 2011;11(12):865–78.

[20] Kemmer K, Corless CL, Fletcher JA, et al. KIT mutations are common in testicular seminomas. Am J Pathol 2004;164(1): 305–13.

[21] Wakita S, Yamaguchi H, Miyake K, et al. Importance of c-kit mutation detection method sensitivity in prognostic analyses of t(8;21)(q22;q22) acute myeloid leukemia. Leukemia 2011;25(9): 1423–32.

[22] Mallal S, Nolan D, Witt C, et al. Association between presence of HLA-B*5701, HLA-DR7, and HLA-DQ3 and hypersensitivity to HIV-1 reverse-transcriptase inhibitor abacavir. Lancet 2002;359(9308):727–32.

[23] Mallal S, Phillips E, Carosi G, et al. HLA-B*5701 screening for hypersensitivity to abacavir. N Engl J Med 2008;358 (6):568–79.

[24] Colombo S, Rauch A, Rotger M, et al. The HCP5 single-nucleotide polymorphism: a simple screening tool for prediction of hypersensitivity reaction to abacavir. J Infect Dis 2008;198 (6):864–7.

[25] Melis R, Fauron C, McMillin G, et al. Simultaneous genotyping of rs12979860 and rs8099917 variants near the IL28B locus associated with HCV clearance and treatment response. J Mol Diagn 2011;13(4):446–51.

[26] Ge D, Fellay J, Thompson AJ, et al. Genetic variation in IL28B predicts hepatitis C treatment-induced viral clearance. Nature 2009;461(7262):399–401.

[27] Suppiah V, Moldovan M, Ahlenstiel G, et al. IL28B is associated with response to chronic hepatitis C interferon-alpha and ribavirin therapy. Nat Genet 2009;41(10):1100–4.

[28] Rebsamen MC, Desmeules J, Daali Y, et al. The AmpliChip CYP450 test: cytochrome P450 2D6 genotype assessment and phenotype prediction. Pharmacogenomics J 2009;9(1):34–41.

[29] Savino M, Seripa D, Gallo AP, et al. Effectiveness of a high-throughput genetic analysis in the identification of responders/non-responders to CYP2D6-metabolized drugs. Clin Lab 2011;57(11-12):887–93.

[30] Eriksson N, Wadelius M. Prediction of warfarin dose: why, when and how? Pharmacogenomics 2012;13(4):429–40.

[31] King CR, Porche-Sorbet RM, Gage BF, et al. Performance of commercial platforms for rapid genotyping of polymorphisms affecting warfarin dose. Am J Clin Pathol 2008;129(6):876–83.

[32] Lyon E, McMillin G, Melis R. Pharmacogenetic testing for warfarin sensitivity. Clin Lab Med 2008;28(4):525–37.

[33] Stack G. Education committee of the academy of clinical laboratory physicians and scientists. Pathology consultation on warfarin pharmacogenetic testing. Am J Clin Pathol 2011;135 (1):13–19.

[34] Anderson JL, Horne BD, Stevens SM, et al. A randomized and clinical effectiveness trial comparing two pharmacogenetic algorithms and standard care for individualizing warfarin dosing (CoumaGen-II). Circulation 2012;125(16):1997–2005.

[35] Ahmad T, Voora D, Becker RC. The pharmacogenetics of antiplatelet agents: towards personalized therapy? Nat Rev Cardiol 2011;8(10):560–71.

[36] Pare G, Eikelboom JW, Sibbing D, Bernlochner I, Kastrati A. Testing should not be done in all patients treated with clopidogrel who are undergoing percutaneous coronary intervention. Circ Cardiovasc Interv 2011;4(5):514–21.

[37] Sibbing D, Bernlochner I, Kastrati A, Pare G, Eikelboom JW. Current evidence for genetic testing in clopidogrel-treated patients undergoing coronary stenting. Circ Cardiovasc Interv 2011;4(5):505–13.

[38] Holmes MV, Perel P, Shah T, Hingorani AD, Casas JP. CYP2C19 genotype, clopidogrel metabolism, platelet function, and cardiovascular events: a systematic review and meta-analysis. JAMA 2011;306(24):2704–14.

# Index

*Note*: Page numbers followed by "*f*" and "*t*" refer to figures and tables, respectively.

## B



## C

## T